REVEL enables students to read and interact with course material on the devices they use, **anywhere** and **anytime**. Responsive design allows students to access REVEL on their tablet devices, with content displayed clearly in both portrait and landscape view.

Highlighting, **note taking**, and a **glossary** personalize the learning experience. Educators can add **notes** for students, too, including reminders or study tips

REVEL's variety of **writing** activities and assignments develop and assess concept **mastery** and **critical thinking**.

Superior assignability and tracking

REVEL's assignability and tracking tools help educators make sure students are completing their reading and understanding core concepts.

REVEL allows educators to indicate precisely which readings must be completed on which dates. This clear, detailed schedule helps students stay on task and keeps them motivated throughout the course.

REVEL lets educators monitor class assignment completion and individual student achievement. It offers actionable information that helps educators intersect with their students in meaningful ways, such as points earned on quizzes and time on task.

Cognition

Seventh Edition

Gabriel Radvansky
University of Notre Dame

Mark Ashcraft
University of Nevada Las Vegas

 330 Hudson Street, NY, NY 10013

Portfolio Manager: Bimbabati Sen
Portfolio Manager Assistant: Anna Austin
Product Marketer: Jessica Quazza
Content Developer: Aphrodite Knoop
Content Development Manager: Gabrielle White
Art/Designer: iEnergizer/Aptara®, Ltd.
Digital Studio Course Producer: Elissa Senra-Sargent
Full-Service Project Manager: iEnergizer/Aptara®, Ltd.
Compositor: iEnergizer/Aptara®, Ltd.
Printer/Binder: LSC Communications/Willard
Cover Printer: LSC Communications/Willard
Cover Design: Lumina Datamatics, Inc.
Cover Art: Aeyaey/Fotolia

Acknowledgements of third party content appear on page 460–465, which constitutes an extension of this copyright page.

Copyright © 2018, 2011, 2008 by Pearson Education, Inc. or its affiliates. All Rights Reserved. Printed in the United States of America. This publication is protected by copyright, and permission should be obtained from the publisher prior to any prohibited reproduction, storage in a retrieval system, or transmission in any form or by any means, electronic, mechanical, photocopying, recording, or otherwise. For information regarding permissions, request forms and the appropriate contacts within the Pearson Education Global Rights & Permissions department, please visit www.pearsoned.com/permissions/.

PEARSON, ALWAYS LEARNING, and REVEL are exclusive trademarks owned by Pearson Education, Inc. or its affiliates, in the U.S., and/or other countries.

Unless otherwise indicated herein, any third-party trademarks that may appear in this work are the property of their respective owners and any references to third-party trademarks, logos or other trade dress are for demonstrative or descriptive purposes only. Such references are not intended to imply any sponsorship, endorsement, authorization, or promotion of Pearson's products by the owners of such marks, or any relationship between the owner and Pearson Education, Inc. or its affiliates, authors, licensees or distributors.

Library of Congress Cataloging-in-Publication Data

Names: Ashcraft, Mark H., author. | Radvansky, Gabriel A., author.
Title: Cognition / Mark Ashcraft, University of Nevada Las Vegas, Gabriel Radvansky, University of Notre Dame.
Description: Seventh edition. | Hoboken, NJ : Pearson Education, [2018]
Identifiers: LCCN 2017011076 | ISBN 9780134478029
Subjects: LCSH: Memory. | Cognition.
Classification: LCC BF371 .A68 2018 | DDC 153—dc23 LC record available at https://lccn.loc.gov/2017011076

This work is solely for the use of instructors and administrators for the purpose of teaching courses and assessing student learning. Unauthorized dissemination or publication of the work in whole or in part (including selling or otherwise providing to unauthorized users access to the work or to your user credentials) will destroy the integrity of the work and is strictly prohibited.

1   17

ISBN 10:   0-13-447689-1
ISBN 13: 978-0-13-447689-6

Brief Contents

1 Cognitive Psychology	1
2 Cognitive Neuroscience	24
3 Sensation and Perception	45
4 Attention	78
5 Short-Term Working Memory	105
6 Learning and Remembering	132
7 Knowing	159
8 Memory and Forgetting	187
9 Language	212
10 Comprehension	247
11 Reasoning and Decision Making	275
12 Problem Solving	309
13 Emotion	335
14 Cognitive Development in Infants and Children	354
15 Cognitive Aging	375

Contents

Preface   x
Acknowledgments   xii

1 Cognitive Psychology   1

- 1.1 Thinking About Thinking   2
- 1.2 Memory and Cognition Defined   3
- 1.3 An Introductory History of Cognitive Psychology   5
  - 1.3.1 Anticipations of Psychology   6
  - 1.3.2 Early Psychology   6
  - 1.3.3 Behaviorism   8
  - 1.3.4 Emerging Cognition   8
- 1.4 Cognitive Psychology and Information Processing   11
- 1.5 Measuring Information Processes   12
  - 1.5.1 Interpreting Graphs   13
  - 1.5.2 Time and Accuracy Measures   13
- 1.6 The Standard Theory and Cognitive Science   15
  - 1.6.1 The Standard Theory   16
  - 1.6.2 A Process Model   17
  - 1.6.3 Revealing Assumptions   18
  - 1.6.4 Cognitive Science   20
- 1.7 Themes of Cognition   21
- **Summary: Cognitive Psychology**   **22**

2 Cognitive Neuroscience   24

- 2.1 The Brain and Cognition Together   24
  - 2.1.1 Dissociations and Double Dissociations   25
- 2.2 Basic Neurology   25
  - 2.2.1 Neuron Structure   26
  - 2.2.2 Neural Communication   26
  - 2.2.3 Neurons and Learning   29
- 2.3 Important Brain Structures and Function   30
  - 2.3.1 Subcortical Brain Structures   30
  - 2.3.2 Cortical Brain Structures   31
  - 2.3.3 Principles of Functioning   32
  - 2.3.4 Split-Brain Research and Lateralization   34
  - 2.3.5 Cortical Specialization   35
  - 2.3.6 Levels of Explanation and Embodied Cognition   37
- 2.4 Neuroimaging   37
  - 2.4.1 Structural Measures   38
  - 2.4.2 Electrical Measures   38
  - 2.4.3 Metabolic Measures   40
  - 2.4.4 Other Methods   41
- 2.5 Connectionism   42
- **Summary: Cognitive Neuroscience**   **43**

3 Sensation and Perception   45

- 3.1 Psychophysics   45
  - 3.1.1 Detection and Absolute Thresholds   45
  - 3.1.2 Discrimination   46
  - 3.1.3 Decisions About Physical and Mental Differences   46
  - 3.1.4 Signal Detection Theory   47
- 3.2 Visual Sensation and Perception   50
  - 3.2.1 Gathering Visual Information   51
  - 3.2.2 Synesthesia   53
  - 3.2.3 Visual Sensory Memory   54
  - 3.2.4 The Early Parts of a Fixation   55
  - 3.2.5 Visual Attention   55
  - 3.2.6 Trans-saccadic Memory   56
- 3.3 Pattern Recognition   57
  - 3.3.1 Gestalt Grouping Principles   57
  - 3.3.2 The Template Approach   58
  - 3.3.3 Visual Feature Detection   59
- 3.4 Top-Down Processing   61
  - 3.4.1 Conceptually Driven Pattern Recognition   61
  - 3.4.2 Connectionist Modeling   62
- 3.5 Object Recognition and Agnosia   66
  - 3.5.1 Recognition by Components   66
  - 3.5.2 Context and Embodied Perception   68
  - 3.5.3 Agnosia   68
  - 3.5.4 Implications for Cognitive Science   71
- 3.6 Auditory Sensation and Perception   71
  - 3.6.1 Auditory Sensory Memory   72
  - 3.6.2 Auditory Pattern Recognition   74
- **Summary: Sensation and Perception**   **76**

4 Attention   78

- 4.1 Multiple Meanings of Attention   78
  - 4.1.1 Attention as a Mental Process   79
  - 4.1.2 Attention as a Limited Mental Resource   80
- 4.2 Basic Input Attentional Processes   81
  - 4.2.1 Alertness and Arousal   81
  - 4.2.2 Orienting Reflex and Attention Capture   83
  - 4.2.3 Visual Search   85
  - 4.2.4 Contrasting Input and Controlled Attention   88
  - 4.2.5 Video Games as Mechanisms for Improving Attention   88
  - 4.2.6 Hemineglect   88
- 4.3 Controlled, Voluntary Attention   91
  - 4.3.1 Selective Attention and the Cocktail Party Effect   93
  - 4.3.2 Selection Models   93

4.4	Attention as a Mental Resource	97
	4.4.1 Automatic and Controlled Processing	97
	4.4.2 The Role of Practice in Automaticity	101
	4.4.3 Disadvantages of Automaticity	101
Summary: Attention		**103**

5 Short-Term Working Memory 105

5.1	A Limited-Capacity Bottleneck	105
	5.1.1 Short-Term Memory Capacity	106
	5.1.2 Forgetting From Short-Term Memory	107
5.2	Short-Term Memory Retrieval	110
	5.2.1 Serial Position Effects	110
	5.2.2 Short-Term Memory Scanning	110
5.3	Working Memory	114
	5.3.1 The Components of Working Memory	116
	5.3.2 The Central Executive	117
	5.3.3 The Phonological Loop	117
	5.3.4 The Visuo-Spatial Sketch Pad	119
	5.3.5 The Episodic Buffer	121
	5.3.6 Engle's Controlled Attention Model	121
5.4	Assessing Working Memory	123
	5.4.1 Dual Task Method	123
	5.4.2 Working Memory Span	125
	5.4.3 Improving Working Memory	126
5.5	Working Memory and Cognition	127
	5.5.1 Working Memory and Attention	127
	5.5.2 Working Memory and Long-Term Memory	127
	5.5.3 Working Memory and Reasoning	128
	5.5.4 Sometimes Small Working Memory Spans Are Better	129
	5.5.5 Working Memory Overview	129
Summary: Short-Term Working Memory		**130**

6 Learning and Remembering 132

6.1	Preliminary Issues	133
	6.1.1 Mnemonics	133
	6.1.2 The Ebbinghaus Tradition	135
	6.1.3 Memory Consolidation	135
	6.1.4 Metamemory	136
6.2	Storing Information in Episodic Memory	138
	6.2.1 Rehearsal	138
	6.2.2 Depth of Processing	140
	6.2.3 Challenges to Depth of Processing	141
6.3	Boosting Episodic Memory	143
	6.3.1 The Self-Reference Effect	143
	6.3.2 Generation, Production, and Enactment	143
	6.3.3 Organization in Storage	144
	6.3.4 Improving Memory	144
	6.3.5 Imagery	144
	6.3.6 Adaptive Memory	144
6.4	Context	146
	6.4.1 Encoding Specificity	146
	6.4.2 Source Monitoring	147
6.5	Facts and Situation Models	148
	6.5.1 The Nature of Propositions	148
	6.5.2 Situation Models	150
6.6	Autobiographical Memories	152
	6.6.1 Psychologists as Subjects	152
	6.6.2 Infantile Amnesia	153
	6.6.3 Reminiscence Bump	153
	6.6.4 Involuntary Memory	154
6.7	Memory for the Future	155
	6.7.1 Prospective Memory	155
	6.7.2 Episodic Future Thinking	156
Summary: Learning and Remembering		**156**

7 Knowing 159

7.1	Semantic Memory	160
	7.1.1 Persistence of Semantic Knowledge	161
	7.1.2 Semantic Networks	161
	7.1.3 Feature Comparison Models	162
	7.1.4 Tests of Semantic Memory Models	163
	7.1.5 Semantic Relatedness	165
7.2	Connectionism and the Brain	168
	7.2.1 Connectionism	168
	7.2.2 The Benefits of Connectionist Models	169
7.3	Semantic Priming	171
	7.3.1 Nuts and Bolts of Priming Tasks	171
	7.3.2 Empirical Demonstrations of Priming	172
	7.3.3 Automatic and Controlled Priming	173
	7.3.4 Priming Is Implicit	174
7.4	Schemata and Scripts	175
	7.4.1 Bartlett's Research	176
	7.4.2 Schemata	176
	7.4.3 Scripts	177
7.5	Concepts and Categorization	180
	7.5.1 Classic View of Categorization	181
	7.5.2 Characteristics of Human Categories	182
	7.5.3 Probabilistic Theories of Categorization	183
	7.5.4 Explanation-Based Theories	184
Summary: Knowing		**185**

8 Memory and Forgetting 187

8.1	The Seven Sins of Memory	187
8.2	Forgetting Through Decay and Interference	188
	8.2.1 Paired-Associate Learning	189
	8.2.2 Associative Interference	190
	8.2.3 Situation Models and Interference	191
	8.2.4 Overcoming Forgetting from Interference	192
	8.2.5 Retrieval Cues	192
	8.2.6 Part-Set Cuing Effect	193
8.3	False Memories, Eyewitness Memory, and "Forgotten Memories"	195
	8.3.1 False Memories	195
	8.3.2 Integration	196
	8.3.3 Leading Questions and Memory Distortions	199

8.3.4	The Misleading Information Effect	199
8.3.5	Source Misattribution and Misinformation Acceptance	200
8.3.6	Stronger Memory Distortion Effects	202
8.3.7	Repressed and Recovered Memories	202
8.3.8	The Irony of Memory	203

8.4 Amnesia and Implicit Memory   206
    8.4.1 Dissociation of Episodic and Semantic Memory   206
    8.4.2 Anterograde Amnesia   207
    8.4.3 Implicit and Explicit Memory as Revealed by Amnesia   208

Summary: Memory and Forgetting   210

9 Language   212

9.1 Linguistic Universals and Functions   212
    9.1.1 Defining Language   213
    9.1.2 Language Universals   213
    9.1.3 Animal Communication   215
    9.1.4 Levels of Analysis   218

9.2 Phonology   219
    9.2.1 Sounds in Isolation   219
    9.2.2 Combining Phonemes into Morphemes   222
    9.2.3 Speech Perception and Context   222
    9.2.4 The Effect of Context   223
    9.2.5 Top-Down and Bottom-Up Processes   224
    9.2.6 Embodiment in Speech Perception   225
    9.2.7 The Puzzle of Apparent Segments in Speech   225

9.3 Syntax   226
    9.3.1 Chomsky's Transformational Grammar   227
    9.3.2 Limitations of Transformational Grammar   228
    9.3.3 The Cognitive Role of Syntax   230
    9.3.4 Prosody   231

9.4 Lexical Factors   232
    9.4.1 Morphemes   232
    9.4.2 Lexical Representation   232
    9.4.3 Polysemy   233

9.5 Semantics   235
    9.5.1 Case Grammar   236
    9.5.2 Interaction of Syntax and Semantics   236
    9.5.3 Evidence for the Semantic Grammar Approaches   238

9.6 Brain and Language   239
    9.6.1 Language in the Intact Brain   239
    9.6.2 Aphasia   240
    9.6.3 Generalizing from Cases of Brain Damage   244

Summary: Language   244

10 Comprehension   247

10.1 Conceptual and Rule Knowledge   247
    10.1.1 Comprehension Research   247
    10.1.2 Online Comprehension Tasks   247
    10.1.3 Metacomprehension   248
    10.1.4 Comprehension as Mental Structure Building   248
    10.1.5 Levels of Comprehension   249

10.2 Reading   250
    10.2.1 Gaze Duration   251
    10.2.2 Basic Online Reading Effects   253
    10.2.3 Benefits of Online Reading   255
    10.2.4 Factors That Affect Reading   256

10.3 Reference, Situation Models, and Events   258
    10.3.1 Reference   258
    10.3.2 Situation Models   260
    10.3.3 Events   265

10.4 Conversation and Gesture   265
    10.4.1 The Structure of Conversations   266
    10.4.2 Cognitive Conversational Characteristics   266
    10.4.3 Empirical Effects in Conversation   270
    10.4.4 Metaphors and Idioms   271
    10.4.5 Gesture   272

Summary: Comprehension   273

11 Reasoning and Decision Making   275

11.1 Formal Logic and Reasoning   275
    11.1.1 Categorical Syllogisms   276
    11.1.2 Theories of Syllogistic Reasoning   278
    11.1.3 Conditional Reasoning   279
    11.1.4 Hypothesis Testing   283

11.2 Decisions   283
    11.2.1 Algorithms and Heuristics   284

11.3 Classic Heuristics, Biases, and Fallacies   286
    11.3.1 The Representativeness Heuristic   287
    11.3.2 The Availability Heuristic   289
    11.3.3 The Simulation Heuristic   290
    11.3.4 Elimination by Aspects   290
    11.3.5 The Undoing Heuristic   291

11.4 Framing and Risky Decisions   293
    11.4.1 Risk Aversion and Seeking   293
    11.4.2 Outcome Magnitude   294

11.5 Adaptive Thinking and "Fast and Frugal" Heuristics   295
    11.5.1 Some Fast and Frugal Heuristics in Detail   296
    11.5.2 The Ongoing Debate   297

11.6 Other Explanations   298
    11.6.1 Bayesian Theories   301
    11.6.2 Quantum Theory   301

11.7 Limitations in Reasoning   302
    11.7.1 Limited Domain Knowledge   302
    11.7.2 Limitations in Processing Resources   305

Summary: Reasoning and Decision Making   306

Appendix: Algorithms for Coin Tosses and Hospital Births   307

12 Problem Solving — 309

- 12.1 Studying Problem Solving — 310
- 12.2 Basics of Problem Solving — 310
  - 12.2.1 Characteristics of Problem Solving — 310
  - 12.2.2 A Vocabulary of Problem Solving — 311
- 12.3 Gestalt Psychology and Problem Solving — 315
  - 12.3.1 Early Gestalt Research — 316
  - 12.3.2 Difficulties in Problem Solving — 316
- 12.4 Insight and Analogy — 319
  - 12.4.1 Insight — 319
  - 12.4.2 Analogy — 321
  - 12.4.3 Neurocognition in Analogy and Insight — 323
- 12.5 Means–End Analysis — 325
  - 12.5.1 The Basics of Means–End Analysis — 325
  - 12.5.2 The Tower of Hanoi — 326
  - 12.5.3 General Problem Solver — 328
- 12.6 Improving Your Problem Solving — 329
  - 12.6.1 Increase Your Domain Knowledge — 329
  - 12.6.2 Automate Some Components of the Problem-Solving Solution — 330
  - 12.6.3 Follow a Systematic Plan — 330
  - 12.6.4 Draw Inferences and Develop Subgoals — 330
  - 12.6.5 Work Backward and Search for Contradictions — 331
  - 12.6.6 Search for Relations Among Problems — 331
  - 12.6.7 Find a Different Problem Representation — 332
  - 12.6.8 If All Else Fails, Try Practice — 332
- **Summary: Problem Solving — 333**

13 Emotion — 335

- 13.1 What Is Emotion? — 335
  - 13.1.1 Neurological Underpinnings — 335
- 13.2 Emotion and Perception — 337
  - 13.2.1 Emotional Guidance of Attention — 337
  - 13.2.2 Visual Search — 339
  - 13.2.3 Emotional Stroop — 340
  - 13.2.4 Emotion and Self-Control — 341
- 13.3 Emotion and Memory — 341
  - 13.3.1 Making Memory Better — 343
  - 13.3.2 Making Memory Worse — 345
- 13.4 Emotion and Language — 347
  - 13.4.1 Prosody — 347
  - 13.4.2 Words and Situations — 348
- 13.5 Emotion and Decision Making — 349
  - 13.5.1 Stress Impairs Performance — 349
  - 13.5.2 Stress Improves Performance — 352
- **Summary: Emotion — 352**

14 Cognitive Development in Infants and Children — 354

- 14.1 A Lifespan Perspective — 354
- 14.2 Neurological Changes — 355
- 14.3 Perception and Attention — 356
  - 14.3.1 Perceptual Memory — 356
  - 14.3.2 Attention Processes — 356
- 14.4 Memory Development — 359
  - 14.4.1 Memory Systems Present at Birth — 359
  - 14.4.2 Memory Systems That Improve with Age — 359
- 14.5 Language Acquisition — 362
  - 14.5.1 Stages of Language Acquisition — 363
  - 14.5.2 Competence and Performance — 364
  - 14.5.3 Learning New Words — 365
- 14.6 Learning Numbers and Arithmetic — 366
  - 14.6.1 Numerical Magnitude — 367
  - 14.6.2 Counting — 367
  - 14.6.3 Arithmetic — 368
- 14.7 Decision Making and Problem Solving — 369
  - 14.7.1 Piaget — 369
  - 14.7.2 Vygotsky — 370
  - 14.7.3 Bruner — 371
  - 14.7.4 Children as Rational Constructivists — 372
- **Summary: Cognitive Development in Infants and Children — 373**

15 Cognitive Aging — 375

- 15.1 Neurological and Cognitive Changes in Older Adults — 375
  - 15.1.1 Cognitive Aging Studies — 376
  - 15.1.2 Age-Related Neurological Changes — 377
  - 15.1.3 Neurological Preservation — 377
  - 15.1.4 Successful Cognitive Aging — 378
- 15.2 Perception and Attention in Older Adults — 378
  - 15.2.1 Attention and Aging — 379
  - 15.2.2 Attention Preservation and Improvement — 380
- 15.3 Memory in Older Adults — 380
  - 15.3.1 Short-Term Working Memory — 380
  - 15.3.2 Episodic Long-Term Memory — 381
  - 15.3.3 Prospective Memory — 381
  - 15.3.4 Semantic Long-Term Memory — 382
  - 15.3.5 Metamemory — 383
  - 15.3.6 Age-Related Stereotypes — 383
- 15.4 Language Processing Changes — 383
  - 15.4.1 Anaphoric and Syntactic Complexity — 384
  - 15.4.2 Discourse Processing — 384
  - 15.4.3 Situation Model Processing — 384
- 15.5 Reasoning, Decision Making, and Problem Solving — 386
  - 15.5.1 Reasoning in Older Adults — 386
  - 15.5.2 Decision Making in Older Adults — 386
  - 15.5.3 Problem Solving in Older Adults — 388
- 15.6 Aging and Emotion — 388
- **Summary: Cognitive Aging — 389**

Glossary — 391
References — 408
Credits — 460
Name Index — 466
Subject Index — 477

Preface

The psychology of human memory and cognition is fascinating, dealing with questions and ideas that are inherently interesting: how we think, reason, remember, and use language, to name just a few. When cognitive psychologists talk about research at conventions, they are agitated, intense, and full of energy. However, in contrast to this enthusiasm, undergraduate texts often portray the field as dull, too concerned with the minutiae of experimental method and technical jargon, and not concerned enough with the interesting issues. Without slighting the empirical foundation of the field, we have tried to capture some of the excitement of the area. All professors want their students to understand the material, of course, but we also want you to appreciate cognitive psychology as one of the most interesting and memorable topics of your student career. Several features of the text are designed to accomplish this:

- To engage your interest and understanding, examples of the main points are sprinkled throughout the text. Each of the chapters has a box that asks you to "Prove It." This feature gives you a demonstration project that can be done quickly to illustrate the points being made.

- Mastering the terminology of a new field can be difficult. To help you with the jargon, critical terms are bold-faced in the text and linked to a glossary entry.

- Each major section of a chapter ends with a brief Section Summary. This, along with the glossary terms and other learning guides, should help you check your understanding and memory as you study. Note that some people find it helpful to read the Section Summaries first as a preview of the section's content.

- We try to use a more colloquial style than is customary in the field (or in texts in general). Our students have told us that these features make the text more enjoyable to read. One said, "It's interesting—not like a textbook," which we take as a compliment. Some professors may expect a more formal, detached style, of course. We would rather have you read and remember the material than have you cope with a text selected because of a carefully pedantic style. Besides, you will have plenty of time to deal with boring texts elsewhere.

- Although "how people think" is a topic that is likely to be of basic interest to just about everyone, most of you will not end up being cognitive scientists. So, although the material is written to be useful to people going on to a career some field of cognition, the exposition is also written to give insights to applications outside of formal cognitive science, in careers that more of you are likely to pursue.

New to the Edition

Like the first six editions, this seventh edition is directed primarily toward undergraduates at the junior and senior level, who are probably taking their first basic course in memory and cognition. It has also been used successfully in introductory graduate surveys, especially when first-year students need a more thorough background in memory and cognition. There is much continuity between the sixth edition of *Cognition* and this one: The foundation areas in cognition are still covered thoroughly, as you'll see in the Contents.

But this revision has several new features that you'll want to note:

- There continue to be tremendous increases in the study of memory and cognition with the technologies and perspectives of cognitive neuroscience. This was reflected in prior editions, and this emphasis continues to grow in the seventh edition. The chapter devoted to issues of neuropsychology has been expanded.

- The presentation of the material has been updated to better suit the REVEL platform, and make the learning of the material smoother and better.

- Two new modules have been added to cover issues of cognitive development that were previously allocated to a single module. One of these chapters is on the developmental cognition of infants and children, and the other is on issues of cognitive aging. These are capstone chapters that recapitulate the topics in the text and can be used—or not—as desired by individual instructors wanting to give different flavors or emphases in their course.

- The text has been thoroughly updated, adding and expanding on important topics and developments that are central to the field across a range of topics. As always, there has also been some careful pruning of topics and streamlining of presentation to make room for the new material. Specific example changes include:

Chapter 1: Updating consideration of issues of the history of cognitive psychology; inclusion of issues of replicability in psychological research; an updating of the themes of cognitive psychology to include the future-oriented nature of much of thought.

Chapter 2: Continued development of issues and methods of cognitive neuroscience.

Chapter 3: In-depth coverage of issues related to psychophysics; expanded explanation of signal detection theory; consideration of misreading effects.

Chapter 4: Added discussion of local versus global processing; inclusion of discussion of the default mode network; inclusion of a discussion of whether video game playing can improve attention.

Chapter 5: Expanded coverage of Engle's attentional control model; coverage of issues of working memory enhancement attempts.

Chapter 6: Coverage of the story mnemonic is now included; expanded discussion of the process of memory consolidation; discussion of the self-reference effect; inclusion of a discussion of episodic future thinking.

Chapter 7: Expanded discussion of the persistence of knowledge in memory; reorganization of topics on memory in Chapters 6–8.

Chapter 8: Explicit focus on the issues of forgetting; coverage of misinformation acceptance.

Chapter 9: Expanded coverage of the distinctly human nature of language; inclusion of a discussion of prosody.

Chapter 10: Inclusion of the issue of grammatical aspect and cognition; broader consideration of event cognition, including a discussion of how comics are used to study cognition; coverage of the comprehension of idioms and metaphors.

Chapter 11: Expansion of the coverage or heuristics and errors in reasoning; inclusion of coverage of the elimination by aspects heuristic; coverage of issues of decision framing and risky decisions, such as risk aversion for gains and risk seeking for loses; added coverage of decision making as being Bayesian or being governed by principles of quantum theory.

Chapter 12: Reorganization of problem solving issues to provide a better grounding for students as they progress through the chapter.

Chapter 13: Expanded coverage of issues of emotion and memory consolidation; more in-depth discussion of how choking under pressure can occur.

- As in the first six editions, we have tried to strike a balance between basic, core material and cutting-edge topics. As cognitive psychology continues to evolve, it is important to maintain some continuity with older topics and evidence. Students need to understand how we got here, and instructors cannot be expected to start from scratch each time they teach the course. We've preserved the overall outline and organization of the text, while updating the sections to reflect newer material.

We hope that the balance between classic research and current topics, the style we have adopted, and the standard organization we have used will make the text easy to teach from and easy for students to read and remember. More important, we hope you will find our portrayal of the field of cognitive psychology useful. As always, we are delighted to receive the comments and suggestions of those who use this text, instructors and students alike. You can contact G.A. Radvansky by writing in care of the Department of Psychology, University of Notre Dame, Notre Dame, IN 46556, or e-mail him at gradvans@nd.edu. You can contact Mark Ashcraft by writing in care of the Psychology Department, University of Nevada Las Vegas, 4505 S. Maryland Pkwy, Box 455030, Las Vegas, NV 89154-5030, or e-mail him at mark.ashcraft@unlv.edu.

REVEL™

Educational technology designed for the way today's students read, think, and learn.

When students are engaged deeply, they learn more effectively and perform better in their courses. The simple fact inspired the creation of REVEL: an immersive learning experience designed for the way today's students read, think, and learn. Built in collaboration with educators and students nationwide, REVEL is the newest, fully digital way to deliver respected Pearson content.

REVEL enlivens course content with media interactives and assessments—integrated directly within the author's narrative—that provide opportunities for students to read about and practice course material in tandem. This immersive educational technology boosts student engagement, which leads to better understanding of concepts and improved performance throughout the course.

Learn more about REVEL—http://www.pearsonhighered.com/revel

Available Instructor Resources

The following resources are available for instructors. These can be downloaded at http://pearsonhighered.com/irc. Login required.

- **PowerPoint**—provides a core template of the content covered throughout the text. Can easily be added to to customize for your classroom.
- **Instructor's Manual**—includes in-class discussion questions and research assignments for each chapter.
- **Test Bank**—includes additional questions beyond the REVEL in multiple choice and open-ended—short and essay response—formats.
- **MyTest**—an electronic format of the Test Bank to customize in-class tests or quizzes. Visit: http://www.pearsonhighered.com/mytest.

Acknowledgments

The list of students, colleagues, and publishing professionals who have helped shape the project continues to grow.

For editorial support and assistance, we thank Bimbabati Sen, Sponsoring Editor and Aphrodite Knoop, Development Editor.

Professional colleagues who have assisted across the years include R. Reed Hunt, John Jonides, Michael Masson, James S. Nairne, Marjorie Reed, Gregory B. Simpson, Richard Griggs, Richard Jackson Harris, Donald Homa, Paul Whitney, Tom Carr, Frances Friedrich, Dave Geary, Mike McCloskey, Morton Gernsbacher, Art Graesser, Keith Holyoak, George Kellas, Mark Marschark, Randy Engle, Fred Smith, Pamela Ansburg, Jeremy Miller, J. L. Nicol, Jennie Euler, Bob Slevc, and Joe Magliano.

In addition to our undergraduate classes, who have tested many of the ideas and demonstrations in the text, we'd like to thank a few special students who have helped in a variety of ways, from reading and critiquing to duplicating and checking references: Mike Faust, David Fleck, Elizabeth Kirk, David Copeland, Don Seyler, Tom Wagner, Paul Korzenko, and Jeremy Krause. We're very grateful to all.

Gabriel Radvansky
University of Notre Dame

Mark Ashcraft
University of Nevada Las Vegas

Chapter 1
Cognitive Psychology

 ## Learning Objectives

1.1: Analyze the mental processes behind our thoughts

1.2: Differentiate memory and cognition

1.3: Summarize the history of cognitive psychology

1.4: Interpret how planning guides behaviors involved in problem solving

1.5: Compare human information processing to the operations of a computer program

1.6: Explain the mental processes that take place while doing a task

1.7: Describe the themes of cognition

This course is about human memory and cognition; more specifically, the scientific study of it. For the moment, consider memory and cognition to be the mental events and knowledge we use when we recognize an object, remember a name, have an idea, understand a sentence, or solve a problem. In this course, we consider a broad range of subjects, from basic perception to complex decision making, and from seemingly simple mental acts such as recognizing a letter of the alphabet to very complicated acts such as having a conversation. We ask questions such as:

- "How do we read for meaning?"
- "How do we memorize facts?"
- "What does it mean to forget something?"
- "How do we know that we don't know something?"

The unifying theme behind all this is one of the most fascinating and important questions of all time:

How do people think?

We are interested in a scientific approach to memory and thought. This is **cognitive psychology**. One of the central features of modern cognitive psychology is its allegiance to objective, empirical methods of investigation. We are experimentalists, and you will read about this approach in this module. Although we present many studies, we also try to make connections with your everyday experiences and how they are relevant to the discussion of pertinent issues.

Within the boundaries of science, cognitive psychology is asking a wide range of fascinating questions. There has been an explosion of interest in cognition both in and outside psychology proper. Questions that were on the back burner for too long are now active areas of research. For example: "How do we read?" "How do we use language?" The pent-up interest in these questions, unleashed during the **cognitive revolution** of the late 1950s and early 1960s, has yielded tremendous progress. Furthermore, we now acknowledge, seek, and sometimes participate in the important contributions of disciplines such as linguistics, computer science, anthropology, and the neurosciences. This interdisciplinary approach is called **cognitive science**, the scientific study of thought, language, and the brain. In other words, this is the scientific study of the mind.

This course aims to share what has been discovered about human memory and cognition and the insights those discoveries provide about human thought. Human memory—your memory, with its collection of mental processes—is the most highly sophisticated, flexible, and efficient computer available. How does it work? As amazing as electronic computers are, their abilities are primitive compared to what you do routinely in even a single minute of thinking. We have a basic need to understand ourselves, including how our mind works.

This course also aims to describe how cognitive psychology has made these discoveries. You will appreciate this information more if you also understand how one conducts research and acquires knowledge. Few of you will become cognitive scientists, but presumably most of you are majoring in psychology or a related field. Because the cognitive approach influences many areas of psychology, your understanding of cognitive psychology will enhance your mastery of psychology as a whole. Indeed, cognitive psychology is the core and "the most prominent school of thought" in psychology (Robins, Gosling, & Craik, 1999).

Finally, this course will also illustrate the pervasiveness of cognitive psychology and its impact on fields outside psychology. Cognitive science is a multidisciplinary field. This fusion and cross-pollination of ideas stems from the conviction that researchers in linguistics, artificial intelligence, the neurosciences, economics, and even anthropology can contribute important ideas to psychology and vice versa. Psychology has a long tradition of influencing educational practice, and it is important that it continue to do so. Even fields as diverse as medicine, law, and business use findings from cognitive psychology. For example, a cognitive psychologist named Daniel Kahneman won the Nobel Prize in Economic Sciences in 2002 for his work on decision making. But it should not surprise you that cognitive psychology is relevant to so many other fields. After all, what human activity doesn't involve thought?

1.1: Thinking About Thinking

OBJECTIVE: Analyze the mental processes behind our thoughts

What is going on when we are thinking? What are the cognitive processes that shape our thoughts? The science of cognitive psychology attempts to study not only what we are thinking but also why and how we are thinking it. Memory, perception, emotions, beliefs, reasoning, imagination, and how we acquire knowledge all factor into cognitive processes.

Let's begin to develop a feel for cognitive psychology by considering three examples. For all three, you should read and answer the question, but more important, try to be as aware as possible of the thoughts that cross your mind as you consider the question.

> **The first question is easy: How many hands did Aristotle have?**

Here we are not particularly interested in the correct answer: two. We are more interested in the thoughts you had as you considered the question. Most students report a train of thought something like this: "Dumb question. Of course he had two hands. Wait a minute, why would a professor ask such an obvious question? Maybe Aristotle

Table 1.1 Summary of the Intuitive Cognitive Analysis
An informal analysis will uncover some of the thoughts you had. These are tracked below. Bear in mind that Table 1.1 illustrates the intuitive analysis and is not a full description of these processes.

Processes	Topic
Sensory and perceptual	
Focus eyes on print	Visual perception, sensory memory
Encode and recognize printed material	Pattern recognition, reading
Memory and retrieval	
Look up and identify words in memory	Memory retrieval
Retrieve word meanings	Semantic retrieval
Comprehension	
Combine word meanings to yield sentence meaning	Semantic retrieval, comprehension
Evaluate sentence meaning, consider alternative meanings	Comprehension
Judgment and decision	
Retrieve answer to question	Semantic retrieval
Determine reasonableness of question	Comprehension, conversation
Judge speaker's intent and knowledge	Decision making and reasoning
Computational (Question 2)	
Retrieve factual knowledge	Semantic retrieval
Retrieve knowledge of how to divide and execute procedure	Procedural knowledge

had only one hand. Nah, I would have heard of it if he had had only one hand—he must have had two."

First, perceptual processes were used for the written words of the question to focus your eyes on the printed line, then move your focus across the line bit by bit, registering the material into a memory system. Smoothly and rapidly, other processes took the material into memory to identify the letters and words. Of course, few college readers consciously attend to the nuts and bolts of perceiving and identifying words unless the vocabulary is unfamiliar or the print is bad. Yet your lack of awareness does not mean that these processes did not happen. Ask any first-grade teacher about the difficulties children have identifying letters and putting them together into words.

We have encountered two important lessons already. First, mental processes such as reading can occur with little conscious awareness, especially if they are highly practiced. Second, even though these processes can operate very quickly, they are complex. Their complexity makes it even more amazing how efficient, rapid, and seemingly automatic they are.

As you identified the words in the question, you were looking up their meanings and fitting them together to understand the question. Surely, you were not consciously aware of looking up the meaning of *hands* in a mental dictionary. But just as surely, you did find that entry, along with your general knowledge about the human body.

Now we are getting to the meat of the process. With little effort, we retrieve information from memory that *Aristotle* refers to a human being, a historical figure from the past. Many people know little about Aristotle beyond the fact that he was a Greek philosopher. Yet this seems to be enough, combined with what we know about people in general, to determine that he was probably just like everyone else: He had two hands.

At a final (for now) stage, people report thoughts about the reasonableness of the question. In general, people do not ask obvious questions, at least not of other adults. If they do, it is often for another reason—a trick question, maybe, or sarcasm. So, students report that for a time they decided that maybe the question was not so obvious after all. In other words, they returned to memory to see whether there was some special knowledge about Aristotle that pertains to his hands. The next step is truly fascinating. Most students claim to think to themselves, "No, I would have known about it if he had had only one hand," and decide that it was an obvious question after all. This lack-of-knowledge reasoning is fascinating because so much everyday reasoning is done without benefit of complete knowledge. In an interesting variation, if students are asked, "How many hands did Beethoven have?" their knowledge of Beethoven's musical fame typically leads to the following inference: "Because he was a musician, he played the piano, and he could not possibly have been successful at it with only one hand. Therefore, he must have had two." An occasional student goes even further with, "Two, but he did go deaf before he died."

Now that's interesting! Someone found a connection between the disability implied by the question "How many hands?" and a related idea in memory, Beethoven's deafness. Such an answer shows how people can also consider implications, inferences, and other unstated connections as they reason: The thinking process can consider a great deal of knowledge, and this illustrates the role of prior knowledge in reasoning, where richer knowledge about Beethoven can lead to an inference.

One other thing to note from this example is that there are different cognitive processes that are all operating at the same time or similar times—perception, attention, memory, language comprehension, and so forth. These processes are also providing input and influencing one another. In essence, cognition is a complex and interactive thing, and it is going to take a lot of time and effort to tease it all apart and understand how it works.

The second question: What is 723 divided by 6?

This question uses your knowledge of arithmetic. Just as with the first question, many of your mental processes happened more or less automatically: identifying the digits, accessing knowledge of arithmetic procedures, and so on. Yet you may be aware of the steps in doing long division: Divide 6 into 7, subtract 6 from 7 to get the first remainder, bring down the 2, then divide 12 by 6, and so on. These steps are mentioned at the bottom of Table 1.1, "Computational," which includes your knowledge of how to do long division. Cognitive psychology is also interested in your mental processing of arithmetic problems and knowledge you acquired in school, not just the kind of reasoning you used in the Aristotle question.

The third question: Does a robin have wings?

Most adults have little to say about their train of thought when answering this question. Many people insist, "I just knew the answer was yes." The informal analysis for Question 1 showed how much of cognition occurs below awareness. The assertion that "I just knew it" is not useful, however certain you are that no other thoughts occurred. You had to read the words, find their meanings in memory, check the relevant facts, and make your decision as in the previous examples. Each of these steps is a mental act, the very substance of cognitive psychology. Furthermore, each step takes some amount of time to complete.

Question 3 takes adults about one second to answer. However, the question "Does a robin have feet?" takes a little longer, around 1.2 to 1.3 seconds. Even small time differences can give us a wealth of information about cognition and memory. The difference in Question 3 is that most of the mental processes do not require much conscious activity; the question seems to be processed automatically. Because such automatic processes are so pervasive, we are particularly interested in understanding them.

WRITING PROMPT

Studying Human Cognition

How can we approach the study of human cognition and thought in a way that does not bias us with our preconceptions?

▶ The response entered here will appear in the performance dashboard and can be viewed by your instructor.

Submit

1.2: Memory and Cognition Defined

OBJECTIVE: Differentiate memory and cognition

To better understand the topic of this title, we need to be more explicit about the terms we use. Just what do we

Understanding the Terms Memory and Cognition

Now that you have an idea of the topics under cognitive psychology, we need more formal definitions of the terms *memory* and *cognition*.

Memory | **Cognition**

The term *cognition* is a much richer term. In Ulrich Neisser's landmark book *Cognitive Psychology* (1967), he stated that cognition "refers to all the processes by which the sensory input is transformed, reduced, elaborated, stored, recovered, and used . . . [including] such terms as sensation, perception, imagery, retention, recall, problem solving, and thinking" (p. 4).

For the present, we use the following definition: *Cognition* is the collection of mental processes and activities used in perceiving, remembering, thinking, and understanding, as well as the act of using those processes.

Cognitive psychology is largely, though not exclusively, interested in what might be everyday, ordinary mental processes. These processes are entirely commonplace—not simple, by any means, but certainly routine. Our definition should not include only "normal" mental activities, however. Although cognitive psychology generally does not deal with psychologically "abnormal" states, such as schizophrenia, such "non-normal" processes, although unusual or rare, may enrich our science.

mean when we use the terms **memory** and **cognition**? In this section, we will address these two concepts.

Most cognitive research deals with the sense modalities of vision and hearing and focuses heavily on language. Some people may be concerned that the reliance on seemingly sterile experimental techniques and methods, techniques that ask simple questions, may yield overly simple-minded views about cognition. This reflects a concern that cognitive research lacks **ecological validity**, or generalizability to the real-world situations in which people think and act (e.g., Neisser, 1976). To some this criticism is sensible, and it is definitely true that the findings derived from work in cognitive psychology should, in some way, find value and applicability in the real world, even if that value may be several steps removed from the original study. A primary reason that cognitive psychologists often do not try to do studies that have an immediate and direct implication for real-world activities is the glaring fact that cognition is complex, even when using artificially simple tasks. At our current level of sophistication, we would be quickly overwhelmed if tasks were very complex or if we tried to investigate the full range of a behavior in all its detail and nuance. In this stage of investigation, it is reasonable for scientists to take an approach called **reductionism**, attempting to understand complex events by breaking them down into their components. An artificially simple situation can reveal an otherwise obscure process. Once the basic processes and components of cognition are understood, then better accounts of how they work together can be put forward. The greater goal is for scientists to eventually put the pieces back together and deal with the larger events as wholes.

WRITING PROMPT

Studying Memory and Cognition

What aspects of human experience fall under the categories of memory and cognition? Are there limits to the scientific study of these concepts?

▶ The response entered here will appear in the performance dashboard and can be viewed by your instructor.

Submit

1.3: An Introductory History of Cognitive Psychology

OBJECTIVE: Summarize the history of cognitive psychology

Let's now turn to cognitive psychology's history and development (for an excellent history of cognitive psychology, see Mandler, 2007). Figure 1.1 summarizes the main patterns of influence that produced cognitive psychology and cognitive science, along with approximate dates.

To a remarkable extent, the bulk of the scientific work on memory and cognition is quite recent, although some elements, and many experimental tasks, appeared even in the earliest years of psychology. However, interest in memory and cognition—thinking—is as old as recorded history. Aristotle, born in 384 B.C., considered the basic principles of memory and proposed a theory in his treatise *De Memoria* (*Concerning Memory*; see Hothersall, 1984). Even a casual reading of ancient works such as Homer's *Iliad* or *Odyssey* reveals that people have always wondered how the mind works and how to improve it (in Plato's *Phaedrus*, Socrates fretted that the invention of written language would weaken reliance on memory and understanding, just as modern parents worry about the Internet). Philosophers of every age have considered the nature of thought. Descartes even decided that the proof of human existence is our awareness of our own thought: *Cogito ergo sum*, "I think, therefore I am" (Descartes, 1637/1972, p. 52).

The critical events at the founding of psychology, in the mid- to late 1800s, converged most strongly on one man, Wilhelm Wundt, and on one place, Leipzig, Germany. In 1879, Wundt established the first laboratory for psychological experiments that had a lasting impact, at the University of Leipzig. Yet Wundt's was not the first psychology laboratory. For example, Ferdinand Ueberwasser founded a psychology laboratory in 1783. However, for various reasons, it did not have a lasting or widespread impact (Schwarz & Pfister, 2016). Also, several people had already been doing psychological research, but did not fully identify themselves as psychologists, but more as physiologists and the like (e.g., Weber's and Fechner's work in psychophysics, Helmholtz's studies of the speed of neural impulses, and Broca's and Wernicke's identification of linguistic brain regions). American psychologist William James even established an early laboratory in 1875,

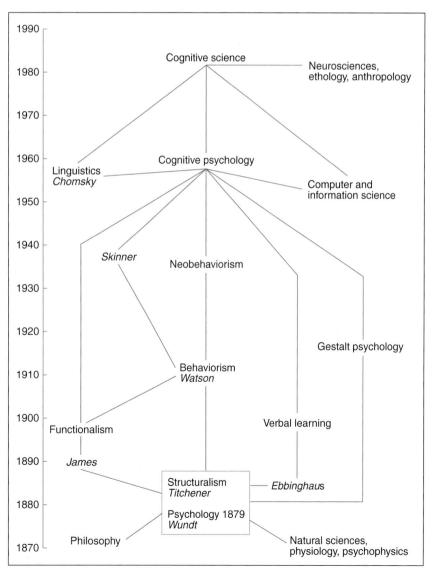

Figure 1.1 The Main Patterns of Influence That Produced Cognitive Psychology and Cognitive Science

although apparently he used it more for classroom demonstrations than for genuine experiments. Still, the consensus is that 1879 is the beginning of the discipline of psychology, separate from philosophy and physiology.

1.3.1: Anticipations of Psychology

Aristotle, for two reasons, is one of the first historical figures to advocate an empirically based, natural science approach. First, although he was certainly not the only great thinker to insist on observation as the basis for all science, he was the first to express this—a position known as **empiricism**. Second, Aristotle's inquiry into the nature of thought led him to a reasonably objective explanation of how learning and memory take place. The basic principles of association he identified have figured prominently in many psychological theories. Equally important was Aristotle's insistence that the mind is a "blank slate" at birth, a *tabula rasa*, or clean sheet of paper (Watson, 1968). The idea is that experience, rather than inborn factors, "writes" a record onto the blank sheet.

There have been many fits and starts in the study of memory over time since Aristotle. For example, St. Augustine, in Chapter 10 of his *Confessions*, presented a quite modern account of memory. Most other anticipations of psychology date from the Renaissance and later periods and are largely developments in scientific methods and approaches. By the mid-1800s, more observational or empirical methods were adopted. By the time psychology appeared, the general procedures of scientific inquiry were well developed. Given the progress in scientific fields such as physics, biology, and medicine by the mid-1800s, it is not surprising that the early psychologists thought the time was ripe for a science of the mind.

1.3.2: Early Psychology

Four early psychologists are of particular interest for cognitive psychology. These early psychologists from the late 19th and early 20th centuries worked to develop scientific methods for studying thought and behavior, which had not been explicitly or emphatically done before.

WILHELM WUNDT To a large extent, the early psychologists were students of Wilhelm Wundt (1832–1920) (Benjamin, Durkin, Link, Vestal, & Acord, 1992). Beginning in 1875, Wundt directed more than 200 doctoral theses on psychological topics (Leahey, 2000). Wundt continually updated his book *Principles of Physiological Psychology*, reporting new results from his laboratory. He also founded the first psychology journal, *Philosophical Studies* (neither of these titles matches its modern connotations). Unfortunately, Wundt's later interests went largely unrecognized until recently (Leahey, 2000). His work on language, child

Wilhelm Wundt

psychology, and other applied topics foreshadowed some modern insights but was rejected or ignored at the time.

In terms of psychology, Wundt believed that the study of psychology was "of conscious processes and immediate experience"—what today we consider areas of sensation, perception, and attention. To study these, in addition to extensive use of response time measures, Wundt used the method of *Selbst-Beobachtung*. Translated literally as "self-observation," this generally is known as **introspection**, a method in which one looks carefully inward, reporting on inner sensations and experiences. Wundt intended this to be a careful, reliable, and scientific method in which the observers (who were also the participants) needed a great deal of training to report only the elements of experience that were immediate and conscious. Reports in which memory intruded were to be excluded.

EDWARD TITCHENER For American psychology in Wundt's tradition, the most important figure was Edward Titchener, an Englishman who came to Cornell University in 1892. Working with Wundt convinced Titchener that psychology's progress depended critically on introspection. Topics such as mental illness and educational and social psychology (including Wundt's broader interests) were "impure" because they could not be studied this way. Titchener insisted on rigorous training for his introspectors, who had to avoid "the stimulus error" of describing the physical stimulus rather than the mental experience of it. Moreover, Titchener made himself the final authority on whether introspection reports were correct or not. By these means, Titchener attempted to study the structure of the conscious mind: the sensations, images, and feelings that, for Titchener, were the very elements of the mind's structure. He called this **structuralism**, an early movement or school of psychological thought (see Figure 1.1). Such a system was destined for difficulties. For example, it is unscientific for one person, Titchener, to be the ultimate authority to validate observations. As other researchers used introspective methods, differences and contradictory results began to crop up, producing disputes that hastened the decline of Titchener's once-powerful structuralism.

HERMANN VON EBBINGHAUS In contrast to Titchener's structuralism, there was the theoretically modest but eventually more influential work of Hermann von Ebbinghaus. Ebbinghaus was a contemporary of Wundt's in Germany, although he never studied with Wundt in person. Ebbinghaus's achievements in studying memory and forgetting are all the more impressive because he worked outside the establishment of the time. Historical accounts suggest that Ebbinghaus read Wundt's book, decided that a study of the mind by objective methods was possible, and set about the task of figuring out how to do it.

Hermann von Ebbinghaus

Lacking a formal laboratory and in an academic position with an absence of sufficiently like-minded colleagues, Ebbinghaus had to rely on his own resources, even to the extent that he alone served as a subject in his research. Ebbinghaus's aim was to study memory in a "pure" form. To do this, he needed materials that had no preexisting associations, so he constructed lists of *nonsense syllables*, consonant–vowel–consonant (CVC) trigrams that, by definition, had no meaning. Ebbinghaus would learn a list (e.g., of 16 items) to a criterion of mastery (e.g., two perfect recitations), then set the list aside. Later, he would relearn the same list, noting how many fewer trials he needed to relearn it. His measure of learning was the "savings score," the number (or proportion) of trials that had been saved in memory between the first and second sessions. His savings measure of memory is based on the idea that if information is stored in memory, even in a form that is not strongly consciously available, it can still ease the relearning of that material. This is in contrast to modern memory researchers who place a greater emphasis on methods such as recall and recognition, and miss some of the potential advantages of the savings method. Using his savings method, Ebbinghaus was able to study retention and forgetting of memories as a function of time, degree of learning or overlearning, and even the effect of nonsense versus meaningful material (he compared forgetting curves for nonsense syllables and meaningful poetry).

Ebbinghaus's work, described in his 1885 book, gained wide acclaim as a model of scientific inquiry into memory. For instance, Titchener praised Ebbinghaus's work as the most significant progress since Aristotle (cited in Hall, 1971). It is difficult to point to another psychologist of his day whose contributions or methods continue to be used. The field of verbal learning owes a great deal to Ebbinghaus. The Ebbinghaus tradition, depicted in Figure 1.1, is one of the strongest influences on cognitive psychology.

WILLIAM JAMES American philosopher and psychologist William James, a contemporary of Wundt, Titchener, and Ebbinghaus's, provided at Harvard an alternative to Titchener's rigid system. His approach, influenced by the writings of Darwin, was **functionalism**, in which the functions of consciousness, rather than its structure, were of interest. Thus, James asked questions such as "How does the mind function?" and "How does it adapt to new circumstances?"

James's informal analyses led to some useful observations. For example, he suggested that memory consists of two parts: an immediately available memory that we are currently aware of and a larger memory that is the

William James

repository for past experience. The idea of memory being divided into parts, based on different functions, is popular today. Indeed, the first serious models of human cognition included the two kinds of memory James discussed in 1890.

Probably because of his personal distaste for experimentation and his broad interests, James did not do much actual research. However, his far-reaching ideas were more influential than any of Titchener's work, as evidenced by his classic 1890 book *Principles of Psychology*. James's influence on the psychology of memory and cognition was delayed, however, for it was John B. Watson, in 1913, who solidified a new direction in American psychology away from both the structuralist and functionalist approaches. This new direction was behaviorism.

1.3.3: Behaviorism

Not all of American psychology from 1910 through the 1950s was behaviorist. The fields of clinical, educational, and social psychology, to name a few, continued in their own development in parallel to behaviorism. Furthermore, there were changes within behaviorism that smoothed the transition to cognitive psychology. This was a kind of neobehaviorism with some unobservable, mediating variables. Nonetheless, it was still a behaviorist environment.

Most people who take introductory psychology know of John B. Watson, the early behaviorist who stated in his 1913 "manifesto" that observable, quantifiable behavior was the proper topic of psychology, not the fuzzy and unscientific concepts of thought, mind, and consciousness. Attempts to understand the "un-observables" of the mind were inherently unscientific, in his view, and he pointed to the unresolved debates in structuralism as evidence. Thus, psychology was redefined as the scientific study of observable behavior, the program of **behaviorism**. There was no room for mental processes because they were not observable behaviors.

Why did such a radical redefinition of psychology's interests have such broad appeal? Part of this was a result of the work that Pavlov and others were doing on conditioning and learning. Here was a scientific approach that was going somewhere compared to the endless debates in structuralism. Furthermore, the measurement and quantification of behaviorism mirrored successful sciences such as physics. Modeling psychology on the methods of these sciences might help it become more scientific (Leahey, 2000, calls this mentality "physics envy"). One of behaviorism's greatest legacies is the emphasis on methodological rigor and observables, traditions that continue to be in force to this day.

During the behaviorist era, there were a few psychologists who pursued cognitive topics—Bartlett of Great Britain, for example—but most American experimental psychology focused on observable, learned behaviors, especially in animals (but see Dewsbury, 2000, for a history of research on animal cognition during the behaviorist era). Even the strongly cognitive approach of Tolman—whose article "Cognitive Maps in Rats and Men" (1948), a molar (as opposed to molecular) approach to behaviorism, is still worth reading—included much of the behaviorist tradition:

- Concern with the learning of new behaviors
- Animal studies
- Interpretation based closely on observable stimuli

Gestalt psychology, which immigrated to the United States in the 1930s (Mandler & Mandler, 1969), always maintained an interest in human perception, thought, and problem solving but never captured the imaginations of many American experimentalists.

Thus, the behaviorist view dominated American experimental psychology until the 1940s, when B. F. Skinner emerged as a vocal, even extreme, advocate. In keeping with Watson's earlier sentiments, Skinner also argued that mental events such as thinking have no place in the science of psychology—not that they are not real, but that they are unobservable and hence unnecessary to a scientific explanation of behavior.

1.3.4: Emerging Cognition

It is often difficult to determine precisely when historical change takes place. Still, many psychologists favor the idea that a cognitive revolution occurred in the mid- to late 1950s, with a relatively abrupt change in research activities, interests, scientific beliefs, and a definitive break from behaviorism (Baars, 1986). Because of the nature and scope of these changes, some see the current approach as a revolution that rejected behaviorism and replaced it with cognitive psychology. However, some historians claim that this was not a true scientific revolution but merely "rapid, evolutionary change" (see Leahey, 1992). In either case, the years from 1945 through 1960 were a period of rapid reform in experimental psychology. The challenges to neobehaviorism came both from within its own ranks and from outside, prodding psychologists to move in a new direction.

WORLD WAR II Lachman, Lachman, and Butterfield (1979) made a point about the growing dissatisfaction among the neobehaviorists. They noted that many academic psychologists were involved with the U.S. war effort during World War II. Psychologists accustomed to studying animal learning in the laboratory were "put to work on the practical problems of making war . . . trying to understand problems of perception, judgment, thinking, and decision making" (p. 56). Many of these problems arose because of soldiers' difficulties with sophisticated technical devices: skilled pilots who crashed their aircraft, radar and sonar operators who failed to detect or misidentified enemy blips, and so on.

Tasks, such as the vigilance needed for air traffic control, require cognitive processes at a fundamental level.

Lachman et al. (1979) were very direct in their description of this situation:

> Where could psychologists turn for concepts and methods to help them solve such problems? Certainly not to the academic laboratories of the day. The behavior of animals in mazes and Skinner boxes shed little light on the performance of airplane pilots and sonar operators. The kind of learning studied with nonsense syllables contributed little to psychologists trying to teach people how to operate complex machines accurately. In fact, learning was not the central problem during the war. Most problems arose after the tasks had already been learned, when normally skillful performance broke down. The focus was on performance rather than learning; and this left academic psychologists poorly prepared. (pp. 56–57)

As Bruner, Goodnow, and Austin (1956) put it, the "impeccable peripheralism" of stimulus–response (S–R) behaviorism became painfully obvious in the face of such practical concerns.

To deal with practical concerns, wartime psychologists were forced to think about human behavior very differently from how they had been up until that point. The concepts of attention and vigilance, for instance, were important to understand sonar operators' performance. Experiments on the practical and theoretical aspects of vigilance began (see especially Broadbent, 1958). Decision making was a necessary part of this performance too, and from this came such developments as signal detection theory. These wartime psychologists rubbed shoulders with professionals from different fields—those in communications engineering, for instance—from whom they gained new outlooks and perspectives on human behavior. Thus, these psychologists returned to their laboratories after the war determined to broaden their own research interests and those of psychology as well.

VERBAL LEARNING **Verbal learning** was the branch of experimental psychology that dealt with humans as they learned verbal material composed of letters, nonsense syllables, or words. The groundbreaking research by Ebbinghaus started the verbal learning tradition, which derives its name from the behaviorist context in which it found itself. Thus, verbal learning was defined as the use of verbal materials in various learning paradigms. Throughout the 1920s and 1930s there was a large body of verbal learning research, with well-established methods and procedures. Tasks such as serial learning, paired-associate learning, and, to an extent, free recall were the accepted methods.

Proponents of verbal learning were similar to the behaviorists. For example, they agreed on the need to use objective methods. There also was widespread acceptance of the central role of learning, conceived as a process of forming new associations, much like the learning of new associations by a rat in a Skinner box. From this perspective, a theoretical framework was built that used a number of concepts that are accepted today. For example, a great deal of verbal learning was oriented around accounts of interference among related but competing newly learned items.

The more moderate view in verbal learning circles made it easy for people to accept cognitive psychology in the 1950s and 1960s: There were many indications that an adequate psychology of learning and memory needed more than just observable behaviors. For instance, the presence of meaningfulness in "nonsense" syllables had been acknowledged early on: Glaze (1928) titled his paper "The Association Value of Nonsense Syllables" (and apparently did so with a straight face). At first, such irksome associations were controlled for in experiments to avoid contamination of the results. Later, it became apparent that the memory processes that yielded those associations were more interesting.

In this tradition, Bousfield (1953; Bousfield & Sedgewick, 1944) reported that, with free recall, words that were associated with one another (e.g., *car* and *truck*) tended to cluster together, even though they were arranged randomly in a study list. There were clear implications that existing memory associations led to the reorganization. Such evidence of processes occurring between the stimulus and the response—in other words, mental processes—led proponents of verbal learning to propose a variety of mental operations such as rehearsal, organization, storage, and retrieval.

The verbal learning tradition led to the derivation and refinement of laboratory tasks for learning and memory. Its advocates borrowed from Ebbinghaus's example of careful attention to rigorous methodology to develop tasks that measured the outcomes of mental processes in valid and useful ways. Some of these tasks were more closely associated with behaviorism, such as the paired-associate learning task that lent itself to tests of S–R associations in direct ways.

Nonetheless, verbal learning gave cognitive psychology an objective, reliable way to study mental processes—research that was built on later (e.g., Stroop, 1935)—and a set of inferred processes such as storage and retrieval to investigate. The influence of verbal learning on cognitive psychology, as shown in Figure 1.1, was almost entirely positive.

LINGUISTICS The changes in verbal learning were a gradual shifting of interests and interpretations that blended almost seamlessly into cognitive psychology. In contrast, 1959 saw the publication of an explicit, defiant challenge to behaviorism. Watson's 1913 article was a behaviorist manifesto, crystallizing the view against introspective methods. To an equal degree, Noam Chomsky's 1959 article was a cognitive manifesto, a rejection of a purely behaviorist explanation of the most human of all behaviors: language.

Noam Chomsky

In 1957, B. F. Skinner published a book titled *Verbal Behavior*, a treatment of human language from the radical behaviorist standpoint of reinforcement, stimulus–response associations, extinction, and so on. His central point was that the psychology of learning—that is, the conditioning of new behavior by means of reinforcement—provided a useful and scientific account of human language. In oversimplified terms, Skinner's basic idea was that human language, "verbal behavior," followed the same laws that had been discovered in the animal learning laboratory: A reinforced response increased in frequency, a nonreinforced response should extinguish, a response conditioned to a stimulus should be emitted to the same stimulus in the future, and so on. In principle, then, it is possible to explain human language, a learned behavior, by the same mechanisms given knowledge of the current reinforcement contingencies and past reinforcement history of the individual.

Noam Chomsky, a linguist at the Massachusetts Institute of Technology, reviewed Skinner's book in the journal *Language* in 1959. The first sentence of his review noted that many linguists and philosophers of language "expressed the hope that their studies might ultimately be embedded in a framework provided by behaviorist psychology" and therefore were interested in what Skinner had to say. Chomsky alluded to Skinner's optimism that the problem of verbal behavior would yield to behavioral analysis because the principles discovered in the animal laboratory "are now fairly well understood . . . [and] can be extended to human behavior without serious modification" (Skinner, 1957, cited in Chomsky, 1959, p. 26).

But by the third page of his review, Chomsky stated that "the insights that have been achieved in the laboratories of the reinforcement theorist, though quite genuine, can be applied to complex human behavior only in the most gross and superficial way. . . . The magnitude of the failure of [Skinner's] attempt to account for verbal behavior serves as a kind of measure of the importance of the factors *omitted* from consideration" (p. 28, emphasis added). The fighting words continued. Chomsky asserted that if the terms *stimulus, response, reinforcement*, and so on are used in their technical, animal laboratory sense, then "the book covers almost no aspect of linguistic behavior" (p. 31) of interest. To Chomsky, Skinner's account used the technical terms in a nontechnical, metaphorical way, which "creates the illusion of a rigorous scientific theory [but] is no more scientific than the traditional approaches to this subject matter, and rarely as clear and careful" (pp. 30–31).

To illustrate his criticism, Chomsky noted the operational definitions that Skinner provided in the animal laboratory, such as for the term *reinforcement*. But, unlike the distinct and observable pellet of food in the Skinner box, Skinner claimed that the person exhibiting the behavior could administer reinforcement for his or her verbal behavior; that is, self-reinforcement. In some cases, Skinner continued, reinforcement could be delayed for indefinite periods or never be delivered at all, as in the case of a writer who anticipates that her work may gain her fame for centuries to come. When an explicit and immediate reinforcer in the laboratory, along with its effect on behavior, is generalized to include non-explicit and non-immediate (and even nonexistent) reinforcers in the real world, it seems that Skinner had brought along the vocabulary of scientific explanation but left the substance behind. As Chomsky bluntly put it, "A mere terminological revision, in which a term borrowed from the laboratory is used with the full vagueness of the ordinary vocabulary, is of no conceivable interest" (p. 38).

Chomsky's own view of language emphasized its novelty and the internal rules for its use. Language *was* an important behavior—and a learned one at that—for psychology to understand. An approach that offered no help in understanding this was useless.

To a significant number of people, Chomsky's arguments summarized the dissatisfactions with behaviorism that had become so apparent. The irrelevance of behaviorism to the study of language and, by extension, any significant human behavior, was now painfully obvious. In combination with the other developments—the wartime fling with mental processes, the expansion of the catalog of such processes by verbal learning, and the disarray within behaviorism itself—it was clear that the new direction for psychology would take hold.

> **WRITING PROMPT**
>
> **Development of Cognitive Psychology as Field of Study**
>
> How did the development of cognitive psychology emerge out of different fields of study, and what is the value of understanding how the mind works by taking a broad perspective?
>
> ▶ The response entered here will appear in the performance dashboard and can be viewed by your instructor.
>
> Submit

1.4: Cognitive Psychology and Information Processing

OBJECTIVE: Interpret how planning guides behaviors involved in problem solving

If we had to pick a date to mark the beginning of cognitive psychology, we might pick 1960. This is not to say that significant developments were not present before this date, for they were. This is also not to say that most experimental psychologists who studied humans became cognitive psychologists that year, for they did not. As with any major change, it takes a while for the new approach to catch on and for people to decide that the new direction is worth following. However, several significant events clustered around 1960 that were significant departures from what came before. Just as 1879 is considered the formal beginning of psychology, and 1913 the beginning of behaviorism, so 1960 approximates the beginning of cognitive psychology.[1]

In his 1959 review, Chomsky made a forceful argument against a purely behaviorist position. He argued that the truly interesting part of language was exactly what Skinner had omitted: mental processes and cognition. Language users follow rules when they generate language, rules that are stored in memory and operated on by mental processes. In Chomsky's view, it was exactly there, in the organism, that the key to understanding language would be found.

Researchers in verbal learning and other fields were making the same claim. As noted, Bousfield (1953) found that people cluster or group words together based on the associations among them. Memory and a tendency to reorganize clearly were involved. Where were these associations? Where was this memory? And where was this tendency to reorganize? They were in the person, in memory and mental processes.

During the 1950s, certain reports on attention, first from British researchers such as Colin Cherry and Donald Broadbent, pertained to the wartime concerns of attention and vigilance. Again, mental processes were being isolated and investigated. No one could deny their existence any longer, even though they were unseen mental processes. A classic paper, Sperling's monograph on visual sensory memory, appeared in 1960. (MacLeod [1991] noted an increase around 1960 of citations to the rediscovered Stroop [1935] task.)

Another startling development of this period was the invention of the modern digital computer. At some point in the 1950s, certain psychologists realized the relevance of computing to psychology. In some interesting and possibly useful ways, computers behave like people (not surprising, according to Norman, 1986, p. 534, because "the architecture of the modern digital computer . . . was heavily influenced by people's naive view of how the mind operated"). They take in information, do something with it internally, and then produce some observable product. The product gives clues to what went on internally. The operations done by the computer were not unknowable because they were internal and unobservable. They were under the control of the computer program, the instructions given to the machine to tell it what to do.

The realization that human mental activity might be understood by analogy to this machine was a breakthrough. The computer was an existence proof for the idea that unobservable processes could be studied and understood. Especially important was the idea of symbols and their internal manipulation. A computer is a symbol-manipulating machine. The human mind might also be a symbol-manipulating system, an idea attributed to Allen Newell and Herb Simon. According to Lachman et al. (1979), their conference in 1958 had a tremendous impact on those who attended. Newell and Simon presented an explicit analogy between information processing in the computer and that in humans. This important work was the basis for the Nobel Prize awarded to Simon in 1978 (see Leahey, 2003, for a full account of Simon's contributions).

Among the indirect results of this conference was the 1960 publication of a book by Miller, Galanter, and Pribram

---

[1] Gardner (1985, p. 28) stated, "There has been nearly unanimous agreement among the surviving principals that cognitive science was officially recognized around 1956. The psychologist George A. Miller . . . has even fixed the date, 11 September 1956." Miller recalled a conference from September 10 to 12, 1956, at MIT, attended by leading researchers in communication and psychology. On the second day of the conference, there were papers by Newell and Simon on the "Logic Theory Machine," by Chomsky on his theory of grammar and linguistic transformations, and by Miller himself on the capacity limitations of short-term memory. Others whom Gardner cited suggest that, at a minimum, the five-year period of 1955 to 1960 was the critical time during which cognitive psychology emerged as a distinct and new approach. By analogy to psychology's selection of 1879 as the starting date for the whole discipline, 1960 is special in Gardner's analysis: In that year, Jerome Bruner and George Miller founded the Harvard Center for Cognitive Studies at Harvard.

called *Plans and the Structure of Behavior*. The book suggested that human problem solving could be understood as a kind of planning in which mental strategies or plans guide behavior toward its goal. The mentalistic plans, goals, and strategies in the book were not just unobservable, hypothetical ideas. Instead, they were ideas that in principle could be specified in a program running on a lawful, physical device: the computer.

WRITING PROMPT

Comparing Cognition to Digital Computer

In what ways can cognition be usefully thought of as being like a digital computer? In what ways might this be inappropriate?

The response entered here will appear in the performance dashboard and can be viewed by your instructor.

Submit

1.5: Measuring Information Processes

OBJECTIVE: Compare human information processing to the operations of a computer program

The aim of cognitive psychology is to reverse engineer the brain in much the same way that engineers reverse engineer devices that they cannot get into. Putting it simply, we want to know:

- What happens in there?
- What happens in the mind—or in the brain, if you prefer—when we perceive, remember, reason, and solve problems?
- How can we peer into the mind to get a glimpse of the cognition that operates so invisibly?
- What methods can we use to obtain some scientific evidence on mental processes?

Guiding Analogies

With the development of cognitive psychology, the seemingly unrelated fields of communications engineering and computer science supplied psychology with some intriguing ideas and useful analogies that were central to developing the cognitive science.

Channel Capacity

To highlight one, psychologists found the concept of **channel capacity** from communications engineering useful (a similar, more popular term would be *bandwidth*). In the design of telephone systems, for instance, one of the built-in limitations is that any channel—any physical device that transmits messages or information—has a limited capacity. In simple terms, one wire can carry just so many messages at a time, and it loses information if capacity is exceeded. Naturally, engineers tried to design equipment and techniques to get around these limitations to increase overall capacity.

Psychologists noticed that, in several ways, humans are limited-capacity channels, too. There is a limit on how many things you can do, or think about, at a time. This insight lent a fresh perspective to human experimental psychology. It makes sense to ask questions such as "How many sources of information can people pay attention to at a time?" "What information is lost if we overload the system?" "Where is the limitation, and can we overcome it?"

The Computer Analogy

1.5.1: Interpreting Graphs

If you are good at interpreting data in graphs, go ahead and just study the figures. Some students struggle with graphs, not understanding what is being shown. Because you will encounter many graphs in this text, you need to understand what you are looking at and what it means.

Take a moment to go through these graphs to see how they present data and to what you should pay attention.

Figure 1.2 is a graph of response time data, the time it takes to respond to an item. We abbreviate response time as RT, and it is usually measured in milliseconds (ms), thousandths of a second (because thought occurs so fast).

In the figure, the label on the *y*-axis is "Vocal RT"; these people were making vocal responses (speaking), and we measured the time between the onset of a multiplication problem and the vocal response. The numbers on the *y*-axis show you the range of RTs that were observed. The dependent variable is always the measure of performance we collected—here it is vocal RT—and it always goes on the *y*-axis.

The *x*-axis label in the left panel is "Multiplication problems," and we have plotted two problems: 2 × 3 and 6 × 9. It is customary to show a more general variable than this on the *x*-axis, as shown in the right panel. There you see a point for a whole set of small multiplication problems, from 2 × 3 up to 4 × 5; a set of medium-size problems, such as 2 × 7 and 8 × 3; and a set of large problems, such as 6 × 8 and 9 × 7. So the *x*-axis label in the right panel is "Size of problem." Notice that the *y*-axis is now in whole seconds, to save some space.

Now the data. The points in the graph are often a mean or average of the dependent variable, RT in this case. Both panels show two curves or lines each, one for college students and one for fourth-grade students (Campbell & Graham, 1985), for multiplication problems. Notice that the curves for fourth-graders are higher. Looking at the *y*-axis in the left panel, the average fourth-grader took 1,940 ms to answer "6" to the problem 2 × 3, compared to 737 ms for the average college student. In the right panel, the average fourth-grader took about 2,400 ms to respond to small problems, 4,100 ms to medium, and 4,550 ms to large. Compare this much greater increase in RT as the problems get larger with the pattern for college students: There was still an increase, but only from 730 ms to about 900 ms.

Why did Campbell and Graham find that fourth-graders were slower? No doubt this is because college students have had more practice in doing multiplication problems than fourth-graders. In other words, college students know multiplication better, have the facts stored more strongly in memory, and so can retrieve them more rapidly. It is a sensible cognitive effect that the strength of information in memory influences the speed of retrieval. And it is easily grasped by looking at and understanding the graphed results.

1.5.2: Time and Accuracy Measures

How we peer into the mind to study cognition depends on using acceptable measurement tools to assess otherwise unseen, unobservable mental events. Other than Wundt's method of introspection, what can we use? There are many measures, but two of the most prominent behavioral measures are

1. The *time* it takes to do some task
2. The *accuracy* of that performance

Because these measures are so pervasive, it is important to discuss them at the outset.

Figure 1.2 Vocal Response Times (RTs) to Multiplication Problems
SOURCE: Data from Campbell and Graham (1985).

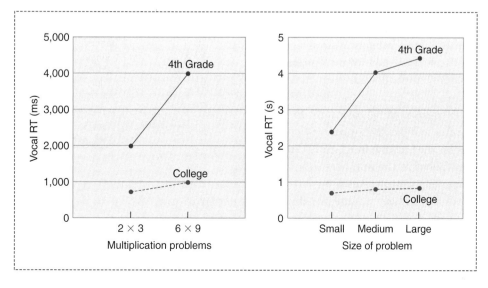

RESPONSE TIME Many research programs in cognitive psychology rely heavily on **response time (RT)**, a measure of the time elapsed between some stimulus and the person's response to the stimulus (RT is typically measured in milliseconds, abbreviated *ms*; a millisecond is one thousandth of a second).

Why is this so important, especially when the actual time differences can seem so small, say, on the order of 40 to 50 ms?

It has been long known that RT measures can reveal individual differences among people. In 1868, the Dutch physiologist Donders (1868/1969) observed that RT is more informative and can be used to study the "speed of mental processes." A moment's reflection reveals why cognitive psychology uses response times: *Mental events take time*. That is important—the mental processes and events we want to understand occur in real time and can be studied by measuring how long they take. Thus, we can "peer into the head" by looking at how long it takes to complete certain mental processes.

Here is an example of this kind of reasoning from measuring RTs. Research in mathematical cognition studies how we remember and use mathematical knowledge. Consider two simple arithmetic problems, such as $2 \times 3 = ?$ and $6 \times 9 = ?$ The left panel of Figure 1.2 shows the time it took some fourth-graders and some college adults to solve these problems (Campbell & Graham, 1985).

There are two important effects: an obvious age difference in which children were slower than adults, and an effect related to the problems, longer RT for 6×9 than for 2×3. The right panel of the figure shows comparable functions for a range of multiplication problems, from small ones such as 2×3 to medium (e.g., 7×3) and large (e.g., 6×9) problems. For both age groups, the curves increase as the size of the problems increases, commonly known as the problem size effect (e.g., Stazyk, Ashcraft, & Hamann, 1982).

Think of the basic assumption again: Mental processes take time. The implication is that longer time is evidence that some process or subprocess took longer in one case than in the other. What could account for that? Most adults agree that 6×9 is harder than 2×3, but that by itself is not very useful; of course a harder decision will take longer to make. But why would 6×9 be harder? After all, we learned our multiplication facts in grade school. Haven't we had sufficient experience since then to equalize all the basic facts, to make them pretty much the same in difficulty? Apparently not.

So what accounts for the increase in RT? It is unlikely that it takes longer to perceive the numbers in a larger problem—and also unlikely that it takes longer to start reporting the answer once you have it. One possibility is that smaller problems have a memory advantage, perhaps something to do with knowing them better. This might date back to grade school, such as the fact that problems with smaller numbers (from 2 to 4) occur more frequently in grade-school textbooks (Ashcraft & Christy, 1995; Clapp, 1924). Another possibility is that smaller problems are easier to figure out or compute in a variety of ways. Aside from simply remembering that 2×3 is 6, you could also count up by 2s or 3s easily and rapidly. But counting up by 6s or 9s would take longer and be more error prone (LeFevre et al., 1996).

The point here is not to explain exactly why solving one kind of problem takes longer than another (see Ashcraft, 1995, or Geary, 1994). Instead, the point is to show how we can use time-based measures to address interesting questions about mental processing (for an exposition of how to use more than just the mean response time to explore cognition, see Balota & Yap, 2011).

ACCURACY In addition to RT measures, we are often interested in **accuracy**, broadly defined. An early use of accuracy as a measure of cognition was the seminal work by Ebbinghaus, published in 1885. Ebbinghaus compared correct recall of information in a second learning session with recall of the same material during original learning as a way of measuring how much material had been saved in memory.

Figure 1.3 is a classic serial position graph, showing the percentage of items correctly recalled. The *x*-axis indicates each item's original position in the list.

In this experiment (Glanzer & Cunitz, 1966), the list items were shown one at a time, and people had to wait 0, 10, or 30 seconds before they could recall the items. Making it even more difficult, the retention interval was filled with counting backward by 3s. Here, it is clear that memory was influenced by an item's position in the list—recall was better for early items than for those in the middle. Also notice the

Figure 1.3 Hypothetical Serial position curves, showing the decrease in accuracy at the end of the list when 0, 10, or 30 seconds of backward counting intervenes between study and recall.

SOURCE: Based on Glanzer & Cunitz, 1966.

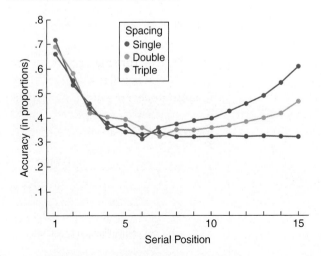

big effect that delaying recall with backward counting had at late positions. So, we cannot conclude that early list items *always* had an advantage over late list items—look how accurately the very last items were recalled when there was no delay. Instead, the overall bowed shape of the graph tells us something complex and diagnostic about memory: Recalling the items from the end of the list may depend on a different kind of memory than recalling the early words, and maybe that memory that can be disrupted by activity-filled delays.

More modern variations on simple list-learning tasks look not only at proportion correct on a list but also at incorrect responses, such as any recalled words that were not on the studied list (called intrusions). Did the person remember a related word such as *apple* rather than the word that was studied, *pear*? Was an item recalled because it resembles the target in some other way, such as remembering *G* instead of *D* from a string of letters? This approach is similar to the Piagetian tradition of examining children's errors in reasoning, such as failure to conserve quantity or number, to examine their cognitive processes.

In more complex situations, accuracy takes on richer connotations. For instance, if we ask people to read and paraphrase a paragraph, we do not score the paraphrase according to verbatim criteria. Instead, we score it based on its meaning, on how well it preserves the ideas and relationships of the original. Preserving the gist is something memory does well (e.g., Neisser, 1981). However, remembering exact, verbatim wording is something we seldom do well, perhaps because this level of detail is often not needed.

REPLICATION Although the various measures for studying the mind that are available to cognitive psychologists are all well and good, it is important to remember that any one study is often insufficient to provide the amount of evidence needed to adequately support a given model or theory. The results from any one study, considered in isolation, can often be interpreted in multiple ways and be consistent with multiple theories. To provide solid support for a particular theory might require multiple experiments to rule out various alternative explanations and triangulate on a theory that best captures the truth about how cognition works.

Foremost in this search for additional evidence is the need for findings—particularly surprising and important findings—to be replicated. Psychological research has taken some heat recently for a large number of published findings that are difficult, if not impossible, to replicate (Open Science Collaboration, 2015). A finding is only scientifically useful and meaningful if it can be replicated. The simplest type of replication would be for the researchers who first found a result to try to replicate it themselves. Better yet is if the replication can be done by other researchers at other institutions. This would promote confidence that the result is not due to some, typically unintentional, implicit bias or anomaly in the first lab that might be producing the result. Replications are even more convincing if the basic pattern is found even when various aspects of the original study are changed, such as the specific materials, the modality of the presentation, the precise instructions used, and so on. The persistence of a finding under a wide variety of conditions and materials would be the hallmark of a robust finding. Robust findings can regularly be replicated.

Another way of dealing with the issue of replicability in psychology is the increasing reliance on measures of effect size. Traditional inferential statistics in psychology, such as *t*-tests, analysis of variance, correlation, regression, and the like, have been interpreted using some measure of statistical significance, often abbreviated as *p*. By convention, many effects are considered statistically significant when $p < .05$ (that is, there is an estimated less than 5% probability that an observed effect in a study is due to chance). The *p*-value tells the researcher only whether the effect is statistically significant or not. In comparison, a measure of effect size (e.g., Cohen, 1988) provides the researcher with an index of how extensive the effect is. For example, a study with a large number of observations may find a statistically significant difference between two conditions, although the effect size itself may be very small. This may lead researchers to question the utility and theoretical importance of the findings of their studies. Typically, it is more likely to easily replicate study findings of a difference between different conditions when the *p*-values are quite small and the effect sizes are quite large.

WRITING PROMPT

Empirical Approach to Study of Cognition

Why is it important to take an empirical approach to studying cognition, and why is it important to use different methods and approaches for this study?

The response entered here will appear in the performance dashboard and can be viewed by your instructor.

Submit

1.6: The Standard Theory and Cognitive Science

OBJECTIVE: Explain the mental processes that take place while doing a task

Here we present a standard theory of human cognition along with major outlines that are widely accepted. Although its details are inaccurate, this theory is generally accurate enough to provide a useful heuristic or guide to thinking about human memory and cognition.

Because many researchers make wide use of it, it is generally known as the standard model or **modal model of memory**.

1.6.1: The Standard Theory

The basic structure of the standard model is useful because it is relatively simple, direct, and is easy to think about and use when discussing ideas about cognition.

The basic system includes three components:

1. **Sensory memory**
2. **Short-term memory**
3. **Long-term memory**

At the input end, environmental stimuli enter the system, with each sense modality having its own sensory register or memory. Some of this information is selected and forwarded to short-term memory, a temporary working memory system with several **control processes** at its disposal. The short-term store can transmit information to and retrieve information from long-term memory. It is also the component responsible for response output, for communicating with the outside world. If consciousness is anywhere in the system, it is here.

Let's use the multiplication example described earlier to trace the flow of processing through Figure 1.2.

You read "2 × 3 = ?" and encode the visual stimulus into a visual sensory register. **Encoding** is the act of taking in information and converting it to a usable mental form. Because you are paying attention, the encoded item is passed to short-term memory (STM). This STM is a working memory system where the information you are aware

Figure 1.4 Information Flow Through the Memory System in the Atkinson and Shiffrin (1968, 1971) Model, the Original Standard Theory in the Information-Processing Approach

Figure 1.4 illustrates the standard model of memory (Atkinson & Shiffrin, 1968, 1971), often called the modal model. It is one of the first models to receive widespread acceptance.

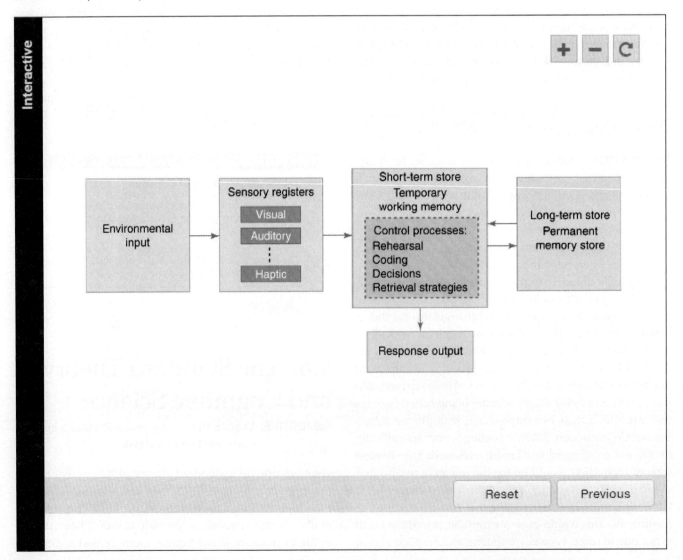

of is held and manipulated. For this example, the system may search long-term memory (LTM) for the answer. A control process in working memory initiates this search, while others maintain the problem until processing is completed. After the memory search, LTM "sends" the answer, 6, to STM, where the final response is prepared and output, say by speech.

Each step in this sequence consumes some amount of time. By comparing these times, we start to get an idea of the underlying mental processing. As you saw in Figure 1.2, a problem such as 2 × 3 takes about 700ms to answer, compared to more than 1,000 ms for 6 × 9 (values are taken from Campbell & Graham, 1985). The additional 300ms might be due to long-term memory retrieval, say because of differences in how easily the problems can be located in LTM.

1.6.2: A Process Model

Although the modal model provides a useful summary, we often need something more focused to explain our results. A common technique is to conceptualize performance in terms of a **process model**, a small-scale model that delineates the mental steps involved in a task and makes testable predictions. Formally, a process model is a hypothesis about the specific mental processes that take place when a particular task is performed.

A PROCESS MODEL FOR LEXICAL DECISION A task that is often used in research in cognitive psychology to explore process models is the **lexical decision task**, a timed task in which people decide whether letter strings are or are not English words (see Meyer, Schvaneveldt, & Ruddy, 1975). In this task people are shown a series of letter strings. The task is to decide on each trial whether they form a word. So, the letter string might be a word, such as MOTOR, or a nonword, such as MANTY. People are asked to respond rapidly but accurately, and response time is the main dependent measure.

Logically, what sequence of events must happen in this task?

In the process model shown in Figure 1.5, the first stage involves *encoding*, taking in the visually presented letter string and transferring it to working memory. Working memory polls long-term memory to assess whether the letter string is stored there. Some kind of *search* through long-term memory takes place. The outcome of the search is returned to working memory and forms the basis for a decision: either "yes, it's a word," or "no, it's not." If the decision is yes, then one set of motor responses is prepared and executed, say, pressing the button on the left; the alternative set of responses is prepared and executed for pressing the other button.

LEXICAL DECISION AND WORD FREQUENCY Say that our results revealed a relationship between RT and the frequency of the words (word frequency is almost always an influence on RT in lexical decision).

We might test words at low, medium, and high levels of frequency in the language:

Figure 1.5 **A.** A general process model, adapted from Sternberg (1969). **B.** A list of the memory components and processes that operate during separate stages of the process model. **C.** A process analysis of the lexical decision task, where RT to each letter string is the sum of the durations of the separate stages. Note that for the three-word trials, the only systematic difference arises from the search stage; encoding, decision, and response times should be the same for all three-word trials, according to the logic of process models and the assumptions of sequential and independent stages of processing.

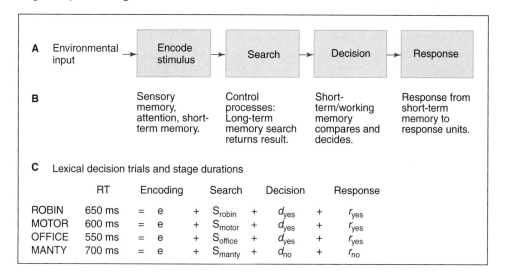

- ROBIN occurs infrequently, about twice per million words.
- MOTOR is of moderate frequency, occurring 56 times per million.
- OFFICE is of high frequency, occurring 255 times per million (Kucera & Francis, 1967; the most frequent printed word in English is THE, occurring 69,971 times per million).

It takes longer to judge words of lower frequency than higher-frequency words (Allen & Madden, 1990; Whaley, 1978). This is the **word frequency effect**. Other variables also affect response times, but word frequency is enough for our example.

For the sake of argument, say that average responses to low-frequency words, such as ROBIN, took 650 ms; those to medium-frequency words, such as MOTOR, took 600 ms; and those to high-frequency words, such as OFFICE, took 550 ms.

What does the process model in Figure 1.5 tell us about such a result?

Logically, we would not expect that word encoding would be influenced by frequency, with high-frequency words being easier to see. So, we assume that encoding is unaffected by word frequency and is relatively constant.

Likewise, all three cases will net a successful search. So, we would not expect time differences in the decision stage because the decision is the same (yes). And finally, "yes" responses should all take about the same amount of time for any word. Thus, the encoding, decision, and response stage times are constants, regardless of word frequency.

The only stage left is the search stage. On reflection, this seems likely to be influenced by word frequency. For instance, it could easily be that words used more frequently are stored more strongly in memory, or stored repeatedly (e.g., Logan, 1988); either possibility could yield shorter search times. Thus, we can tentatively conclude that word frequency has an effect on the search stage. Any factor that affects long-term memory search should influence this stage and should produce a time or accuracy difference. Using the numbers from earlier, the search process would take an extra 50 ms for each change from high to medium to low word frequency.

1.6.3: Revealing Assumptions

Several assumptions are made when doing a process analysis. It is important to understand those sorts of assumptions so you can better appreciate how theories and models of cognition are derived. We will use the forgoing example as a guide.

PARALLEL PROCESSING As research has been done, evidence has accumulated that casts doubt on the assumptions of serial, nonoverlapping stages of processing. Instead, some evidence exists that multiple mental processes can operate *simultaneously*—which is termed **parallel processing**. One example involves typing. Salthouse (1984) did a study of how skilled typists type and how performance changes with age. His data argued for a four-process model. The input stage encoded the to-be-typed material, a parsing stage broke large reading units (words) into separate characters, a translation stage transformed the characters into finger movements, and an execution stage triggered the keystrokes. Significantly, his evidence indicated that these stages operate in parallel: While one letter is typed, another is translated into a finger movement, and the input stage is encoding upcoming letters, even as many as eight characters in advance of the one being typed. Moreover, older adults counteracted the tendency toward slower finger movements by increasing their "look ahead" span at the upcoming letters (see Townsend & Wenger, 2004, for a thorough discussion of serial versus parallel processing).

In moving away from the simpler computer analogy, the cognitive science approach embraced the idea that we need to understand cognition with some reference to the brain. An important lesson we have learned from neuroscience is that the brain shows countless ways in which different cognitive components and processes operate simultaneously, in parallel. Furthermore, there is now ample neurological evidence that different regions of the brain are more specialized for different processing tasks, such as encoding, responding, memory retrieval, and controlling the stream of thought (Anderson, Qin, Jung, & Carter, 2007).

CONTEXT EFFECTS A second difficulty with the early assumptions of sequential stages and nonoverlapping processes arose when context effects were taken into account. A simple example of this is the speedup in deciding—that you are faster to decide MOTOR is a word if you have seen MOTOR recently. A more compelling demonstration comes from work on lexical ambiguity—the fact that many words have more than one meaning. As an example, Simpson (1981) had people do a modified lexical decision task, judging letter strings such as DUKE or MONEY (or MANTY or ZOOPLE) after they had read a context sentence. When the letter string and sentence were related—for instance, "The vampire was disguised as a handsome count," followed by DUKE—the lexical decision on DUKE was faster than normal. The reason involved **priming**, the idea that concepts in memory become activated and hence easier to process. In this case, because the context sentence primed the royalty sense of the word *count*, the response time to DUKE was speeded up.

Process Analysis Assumptions

> **Interactive**
>
> **Sequential Stages of Processing**
>
> The first is the assumption of **sequential stages of processing**. It was assumed that there is a sequence of stages or processes, such as those depicted in Figure 1.5, that occur on every trial, a set of stages that completely accounts for mental processing. More important, the order of the stages was treated as fixed on the grounds that each stage provides a result that is used for the next one. More to the point, this assumption implies that one and only one stage can be done at a time, which may not be the case in reality. The influence of the computer analogy is very clear here. Computers have achieved high speeds of operation, but use **serial processing**: They do operations one by one, in a sequential order. And yet there is no a priori reason to expect that human cognition has this quality in all situations. It may well be that some operations are done in parallel, rather than sequentially.
>
> **Independent and Nonoverlapping**

This was an issue for earlier cognitive models because there was no mechanism to account for priming. Look again at Figure 1.5. Is there any component that allows a context sentence to influence the speed of the processes? No, you need a meaning-based component to keep track of recently activated meanings that would speed up the search process when meanings matched but not when they were unrelated.

Let's look more deeply at the influence of context on cognition. Information that is active in long-term memory, for example, can easily have an effect right now on *sensory memory*, the input stage for external stimuli. Here is a simple example:

> As you read a sentence or paragraph, you begin to develop a feel for its meaning. Often you understand well enough that you can then skim through the rest of the material, possibly reading so rapidly that lower-level processes such as proofreading and noticing typograpical errors may not function as accurately as they usually do. Did you see the mistake?

What? Mistake? If you fell for it, you failed to notice the missing *h* in the word *typographical*, possibly because you were skimming but probably because the word *typographical* was expected, predictable based on meaning. You may have even "seen" the missing *h* in a sense. Why? Because of your understanding of the passage, its meaningfulness to you may have been strong enough that the missing *h* was supplied by your long-term memory.

We call such influences **top-down** or **conceptually driven processing** when existing context or knowledge influences earlier or simpler forms of mental processes. It is one of the recurring themes in cognition. For another example, adapted from Reed (1992), read the following sentence:

> FINISHED FILES ARE THE RESULT OF YEARS OF SCIENTIFIC STUDY COMBINED WITH THE EXPERIENCE OF MANY YEARS.

Now, read it a second time, counting the number of times the letter *F* occurs.

If you counted fewer than six, try again—and again, if necessary. Why is this difficult? Because you know that function words such as OF carry very little meaning, your perceptual input processes are prompted to pay attention only to the content words. Ignoring function words, and consequently failing to see the letter *F* in a word such as OF, is a clear-cut example of conceptually driven processing (for an explanation of the "missing letter effect," see Greenberg, Healy, Koriat, & Kreiner, 2004).

OTHER ISSUES Another issue with these assumptions involves other, often slower mental processes of interest to cognitive psychology. Some process models are aimed at accuracy-based investigations—percentage correct or the nature of one's errors in recall. In a similar vein, many cognitive processes are slower and more complex. Studies of decision making and problem solving often involve processing that takes much longer than most RT tasks; for example, some cryptarithmetic problems (e.g., substitute digits for letters in the problem SEND + MORE) can take 15 to 20 minutes! A more meaningful measure of these mental processes involves a verbal report or **verbal protocol** procedure, in which people verbalize their thoughts as they solve the problems. This type of measure in cognitive research is less widely used than time and accuracy measures, but is important nonetheless (see Ericsson & Simon, 1980, for the usefulness of verbal protocols).

1.6.4: Cognitive Science

Cognitive psychology is now firmly embedded in a larger multidisciplinary effort focused on the study of mind. This broader perspective is cognitive science. As noted earlier, *cognitive science* draws from, and influences, a variety of disciplines such as computer science, linguistics, and neuroscience, and even such far-flung fields as law and anthropology (e.g., Spellman & Busey, 2010). It is a true study of the mind, in a broad sense, as illustrated in Figure 1.6.

In general, cognitive science is the study of thought, using available scientific techniques and including all relevant scientific disciplines for exploring and investigating cognition. One of the strongest contributions to this expanded body of evidence has been the consideration of the neurological bases and processes that underlie thought.

Figure 1.6 Cognitive Psychology and the Various Fields of Psychology and Other Disciplines to Which It Is Related

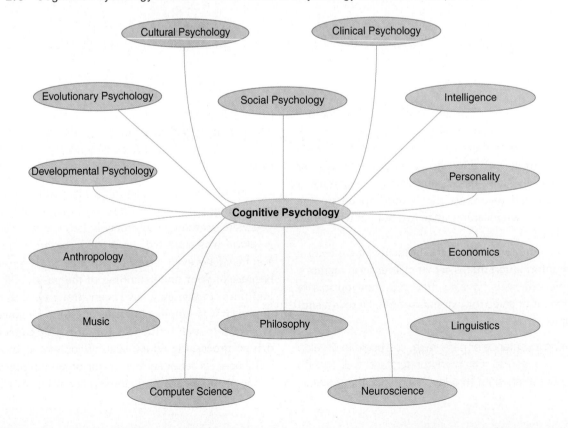

> **WRITING PROMPT**

Component Steps of Cognitive Process

What is the utility of breaking a cognitive process down into its component steps?

▶ The response entered here will appear in the performance dashboard and can be viewed by your instructor.

Submit

1.7: Themes of Cognition

OBJECTIVE: Describe the themes of cognition

Across the various topics in cognition, a number of themes appear repeatedly. You will not necessarily find sections in this text labeled with them. Instead, they crop up across several areas of cognitive science, in different contexts. If you can read a section and identify and discuss the themes that pertain to it, then you probably have a good understanding of the material.

> **WRITING PROMPT**

Comparing Themes of Cognition

A number of themes of cognition were listed that cut across the modules in this title. Which of these has the greatest application to problems that might be encountered in the real world, and why?

▶ The response entered here will appear in the performance dashboard and can be viewed by your instructor.

Submit

Seven Themes of Cognition

Here is a list and brief description of seven important themes that occur throughout the text.

1. Data-driven versus conceptually driven processing

Some processes rely heavily on information from the environment (data-driven or bottom-up processing). Others rely heavily on our existing knowledge (conceptually driven or top-down processing). Conceptually driven processing can be so powerful that we often make errors, from mistakes in perception up through mistakes in reasoning. But could we function without it? As you will see, much of cognition involves a combination of both data-driven and conceptually driven processes, although their balance may vary depending on the circumstances.

2. Representation

3. Implicit versus explicit memory

4. Metacognition

5. Brain

6. Embodiment

7. Future Orientation

Summary: Cognitive Psychology

1.1 Thinking About Thinking

- Cognitive psychology is the scientific study of human mental processes. This includes perceiving, remembering, using language, reasoning, and solving problems.
- Intuitive analysis of examples such as "How many hands did Aristotle have?" and "Does a robin have wings?" indicates that many mental processes occur automatically (very rapidly and below the level of conscious awareness).

1.2 Memory and Cognition Defined

- Memory is composed of the mental processes of acquiring and retaining information for later use (encoding), the mental retention system (storage), and then using that information (retrieval).
- Cognition is the collection of mental processes and activities used in perceiving, remembering, thinking, and understanding, as well as the act of using those processes.

1.3 An Introductory History of Cognitive Psychology

- The modern history of cognitive psychology began in 1879 with Wundt and the beginnings of experimental psychology as a science.
- The behaviorist movement rejected the use of introspection and substituted the study of observable behavior.
- Modern cognitive psychology, which dates from approximately 1960, rejected much of the behaviorist position but accepted its methodological rigor. Many diverse viewpoints, assumptions, and methods converged to help form cognitive psychology. This was at least a rapid, evolutionary change in interests, if not a true scientific revolution.

1.4 Cognitive Psychology and Information Processing

- Cognitive psychology began as a separate field around 1960 with the decline of behaviorism; the developments in linguistic and verbal learning; and important papers by researchers on attention, visual processing, and memory.
- The advent of the modern digital computer played a key role in the development of cognitive psychology and served as the basic metaphor for how the mind processes knowledge. From this, models of the mind as a means of processing information were developed.

1.5 Measuring Information Processes

- Although channel capacity was an early, useful analogy in studying information processing, a more influential analogy was later drawn between humans and computers: that human mental processing might be analogous to the sequence of steps and operations in a computer program. Computers still provide an important tool for theorizing about cognitive processes.
- Measuring information processes, the mental processes of cognition, has relied heavily on time and accuracy measures. Differences in response time (RT) can yield interpretations about the speed or difficulty of mental processes, leading to inferences about cognitive processes and events. Accuracy of performance, whether it measures correct recall of a list or accurate paraphrasing of text, also offers evidence about underlying mental processes. The findings of individual studies can be interesting and informative. However, in order for a finding to be truly valuable, it is best if it has been replicated to show that it is stable and robust. Good science is rooted in being able to accurately predict outcomes.

1.6 The Standard Theory and Cognitive Science

- The modal model of memory suggests that mental processing can be understood as a sequence of independent processing stages, such as the sensory, short-term, and long-term memory stages.
- Process models are appropriate for fairly simple, rapid tasks that are measured by response times, such as the lexical decision task.
- There is substantial evidence to suggest that cognition involves parallel processing and is influenced by context; for example, research on skilled typing shows a high degree of parallel processing. Also, slower, more complex mental processes, such as those in the study of decision making and problem solving, may be studied using verbal protocols.
- Cognitive psychology is better understood as residing within the context of a broader cognitive science. This approach describes cognition as the coordinated, often

parallel operation of mental processes within a multi-component system. The approach is deliberately multidisciplinary, accepting evidence from all the sciences interested in cognition.

1.7 Themes of Cognition

- A number of themes running throughout the study of cognition cut across many of the subdomains. These include data-driven versus conceptually driven processes, representation, implicit versus explicit memory, metacognition, the brain, embodiment, and a future orientation to cognition.

SHARED WRITING

Studying Cognitive Psychology

Cognition, memory, and thought are central to who we are as human beings. Given that, please discuss why it has taken so long to study this in a principled, scientific way, and how might the knowledge gained from the developing fields of cognitive science influence emerging technologies and everyday life.

A minimum number of characters is required to post and earn points. After posting, your response can be viewed by your class and instructor, and you can participate in the class discussion.

Post   0 characters | 140 minimum

Chapter 2
Cognitive Neuroscience

 Learning Objectives

2.1: Explain the neurological connection of cognition

2.2: Describe the functionality of neurons

2.3: Identify the neural anatomy of the human brain

2.4: Describe the four ways of understanding the human brain

2.5: Summarize connectionism as a computer-based method in cognitive science

An important aspect of modern *cognitive science* is an understanding of neural processes and how they relate to thought. In this module, we provide a basic overview of the fundamental principles of neural processing and various brain structures that are important for a wide range of cognitive processes. We begin with an exposition of the characteristics of individual neurons, how they communicate information, and the influence of some basic classes of neurotransmitters. After this, we look at various structures of the brain, outline the major regions of the cortex, and cover some of the major methods of assessing how neural processing corresponds to cognitive processing. Finally, we discuss a computational approach, called neural net or connectionist modeling, which tries to capture the basic principles of neural processing in computer-based mathematical models of thought.

2.1: The Brain and Cognition Together

OBJECTIVE: Explain the neurological connection of cognition

We start with a stunning story of cognitive disruption to motivate our interest in **cognitive neuroscience**. Tulving (1989) described a patient he called K. C., a young man who had sustained brain damage in a motorcycle accident. Some nine years after the accident, he still showed pervasive disruption of long-term memory. The fascinating thing about his memory impairment was that it was selective: K. C. remained competent with language, his intelligence was normal, and he was able to converse on a number of topics. But K. C. explained that he could not remember any experience from his own past—in the sense of bringing back to conscious awareness, a single thing that he had ever done or experienced in the past. For example, even though he could remember how to play chess, he could not remember ever having played it before. He knew his family had a vacation house on a lake, but he did not have any recollection of having been there. K. C.'s brain damage seemed to destroy his ability to access what we will call **episodic memory**, his own autobiographical knowledge, while leaving his general knowledge system—his **semantic memory**—intact. This pattern is called a **dissociation**, a disruption in one component of mental functioning but no impairment of another. Can these two forms of long-term memory, episodic and semantic memory, be the same, given K. C.'s dissociation between the two? Probably not. So how must the cognition be organized neurologically for disruptions such as these to take place?

Although the evidence of brain damage is important, we also need other kinds of evidence—for example, information about the neurochemical and neurobiological activities that support normal learning and thought, or the changes in the brain that accompany aging. Therefore, we are interested in contributions from all the various neurosciences—neurochemistry, neurobiology, neuroanatomy, and so on—as they relate to human cognition.

Although it is important to understand cognitive handicaps and the value of rehabilitation and retraining for patients with brain damage, our interest in cognitive science extends to study of normal cognition from the standpoint of the human brain. We need to learn about normal cognition through any means available. Thus, more and more investigators are examining the behavioral and cognitive effects of brain damage (e.g., McCloskey, 1992), using those observations to develop and refine theories of normal cognition (Martin, 2000). As you will see throughout this course, the great misfortune of brain damage sometimes leads to a clearer understanding of normal processes.

2.1.1: Dissociations and Double Dissociations

The concept of dissociation—the opposite of association—is important, so we should spend a little more time on it.

DISSOCIATIONS AND THE LACK OF DISSOCIATION
Consider two mental processes that "go together" in some cognitive task, called process A and process B (see Figure 2.1). By looking at these processes as they may be disrupted in brain damage, we can determine how separable the processes are.

Figure 2.1 Process A and Process B

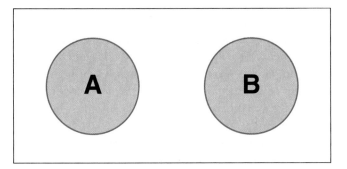

Complete separation is a **double dissociation**. Evidence of a double dissociation requires at least two patients, with "opposite" or reciprocal deficits. For example, Patient X has a brain lesion that has disrupted process A. His performance on tasks that use process B is intact, not disrupted at all.

Patient Y has a lesion that has damaged process B, but tasks that use process A are normal, not disrupted by the damage.

Think of a double dissociation as illustrated in this simple diagram. If these circles were to depict actual brain regions, such as those used in language processes, then damage to either one of them could easily leave the other one unaffected.

In a simple dissociation, process A could be damaged while process B remains intact, yet no other known patient has the reciprocal pattern. For example, *semantic retrieval* (retrieving the meaning of a concept) could be intact while *lexical retrieval* (finding the name for the concept) could be disrupted: this is called **anomia**. In this situation, lexical retrieval is dissociated from semantic retrieval, but it is probably impossible to observe the opposite pattern. How can you name a concept if you cannot retrieve the concept in the first place?

In a full or complete **association** (lack of dissociation), disruption of one of the processes always accompanies disruption in the other process. This pattern implies that process A and process B rely on the same region or brain mechanism, such as recognizing objects and recognizing pictures of those objects.

Likewise, cognitive science is now putting to good use the new high-tech brain imaging capabilities we have adopted from medicine. We can now use brain images based on positron-emission tomography (PET) scans and magnetic resonance imaging (MRI) to localize regions of activity during different kinds of cognitive processing, testing not just the brains of brain-damaged individuals but normal, intact ones as well (see Poldrack & Wagner, 2004; Sarter, Berntson, & Cacioppo, 1996).

Just as the cognitive influence on neuroscience has been a rejuvenating one, the cross-fertilization for cognition has been crucial and will become more so with time. What has become clear is that cognition and neuroscience are highly interlinked. This is particularly evident when you consider how embodied views of cognition critically depend on neurological accounts of perceptual and motor process and how they influence even seemingly abstract tasks, such as using semantic memory or solving problems (Marshall, 2009). We focus only on a few of the many highlights in the interrelation of cognition and neuroscience.

WRITING PROMPT

Value of Double Dissociations

What is the value of double dissociations while studying the neuropsychology of cognition? Explain the type of errors that could be made without them.

▶ The response entered here will appear in the performance dashboard and can be viewed by your instructor.

Submit

2.2: Basic Neurology

OBJECTIVE: Describe the functionality of neurons

At birth, the human brain weighs approximately 400 g (about 14 oz). It grows to an average of 1,450 g in adults, slightly more than 3 lb. and is roughly the size of a ripe grapefruit.

The basic building block of the brain (indeed of the entire nervous system) is the **neuron**. The neuron is a specialized cell that receives and transmits a neural impulse.

How many neurons are there in the brain? Available estimates vary. Williams and Herrup (1988) reported that there are about 100 billion neurons in the brain, although humans will lose about 10% of this mass over the course of a lifetime. To put that figure in perspective, consider that the Milky Way galaxy has about 100 billion stars. Moreover, there is a very high degree of interconnectivity in the brain among these cells, with these neurons making about 100 trillion connections. Talk about a complex mental machine! In addition to neurons, there are many other types of cells in the brain, including glial cells and connective and circulatory tissue.

2.2.1: Neuron Structure

Neurons are the components that form nerve tracts throughout the body and in all brain structures. In this section, we examine neuron structure in more detail.

At one end of the neuron, many small branchlike fingers called **dendrites** gather a neural impulse into the neuron itself. In somewhat more familiar terms, the dendrites are the *input* structures of the neuron, taking in the message that is being passed along in a particular neural tract.

The central portion of each neuron is the cell body, or **soma**, where the biological activity of the cell is regulated. Extending from the cell body is a longish extension or tube, the **axon**, which ends in another set of branchlike structures called *axon terminals* or sometimes *terminal arborizations*—the latter term derives from the treelike form of these structures. The axon terminals are the *output* end of the neuron, the place where the neural impulse ends within the neuron itself. Obviously, this is the location where an influence on the next neuron in the pathway must take place.

Some neurons also have a fatty coating on the axon. This coating is called the **myelin sheath** and serves as an insulator for the axon. The myelin sheath is typically not continuous, but has gaps along the way. These gaps are called the **nodes of Ranvier**.

An important job of the myelin sheath is to speed neural communication. It can do this by having a nerve impulse jump from one of these nodes to the next, rather than traveling down the entire length of the axon. This functionally shortens the distance the signal travels, thereby speeding neural communication. Note that people with multiple sclerosis are suffering from damage to the myelin that would normally be present on their neurons.

Not all neurons are myelinated. Often, myelin is found in neurons that need to transmit signals a relatively long distance, such as out in the periphery of the nervous system. Most of the cortical neurons from which cognition arises are not myelinated. You can easily see differences between bundles of myelinated and unmyelinated neurons even without the aid of a microscope. Neurons that have myelin sheaths appear to be white, because the myelin is made of fat. However, neurons that do not have a myelin sheath appear to be gray. So, when people talk about "white matter" and "gray matter" in a person's brain, they are talking about bundles of myelinated and unmyelinated neurons, respectively. Also, neurons themselves do not produce the myelin. Instead, myelin is produced by the glial cells in the nervous system that support the neurons.

Receptor cells react to the physical stimulus and trigger a pattern of firing down *sensory neurons*. These neuron tracts pass the message along into the spinal cord. For a simple reflex, the message loops quickly through the spinal cord and goes back out to the arm muscles through a tract of motor neurons that terminate at *effector cells*, which connect directly to the muscle fibers and cause the muscles to pull your arm away.

As the reflex triggers the quick return of a message out to the muscles, it simultaneously routes a message up the spinal cord and into the brain. Thus, the second route involves only the central nervous system—the spinal cord and brain. There is only one kind of neuron in the central nervous system, called an *interneuron* or *association neuron*.

Because we are focusing on the brain here, we are interested only in the interneurons of the central nervous system. For simplicity, we will just refer to them here as neurons.

2.2.2: Neural Communication

As noted earlier, the primary job of neurons is to communicate information. This is done by two basic components—one that is primarily electrical and occurs within the

Figure 2.2 The Various Structures of the Neuron

Figure 2.2 illustrates an idealized or prototypical neuron. The details of structure vary, but each neuron within the nervous system has the same general features.

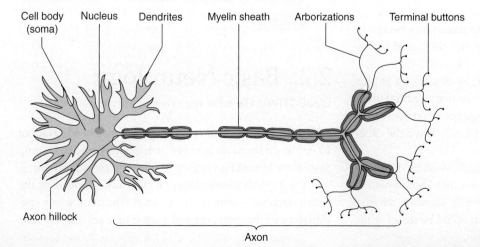

Figure 2.3 An Illustration of a Sensory–Motor Reflex Arc

Figure 2.3 is a diagram of the elements of the nervous system that are activated during a simple reflex, such as jerking your arm away when you accidentally touch a hot stove.

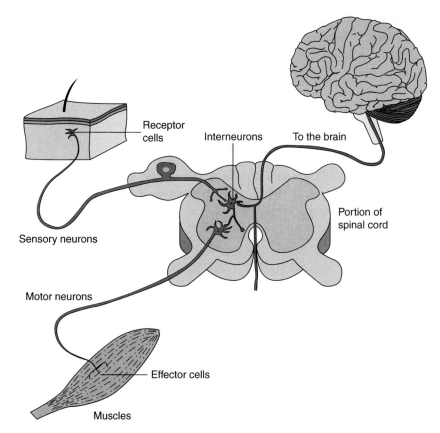

interior of the neuron, and another that is primarily chemical and occurs between neurons. The first type involves the action potential and is discussed in this section. The second involves the synapse and neurotransmitters and is discussed in the following sections.

ELECTRICAL COMPONENT Again, the electrical part of neural communication is called the **action potential**, which is a change in the electrical charge of the neuron (relative to the charge outside of it). We are all familiar with the electrical component of neural transmission. This is where the tingling sensation comes from if you happen to touch a live wire or touch your tongue on both poles at the end of a 9-volt battery. When a neuron is not being stimulated, it has a resting electrical charge of −70 mV (millivolts). When a neuron is stimulated, there is a depolarization of its electrical potential, and the neuron is said to "fire" (see Figure 2.4).

At this point there is a change in the charge of the neuron to +40 mV. Note that the entire neuron does not change its electrical charge. Instead, the action potential is **propagated** from the dendrites, through the soma, and down the axon over time, almost like a wave of electrical charge. After the neuron has fired, there is a recovery period in which the neuron resets itself and is ready to fire again. It should be noted that action potentials follow the **all-or-none principle**. That is, there is either an action potential, which is always the same anywhere in the nervous system, or there is no action potential. It is *not* the case that some action potentials are larger or smaller than others.

An action potential does not occupy the entire body of the neuron at once. Instead, a wave of electrical charge flows down the axon, with sodium ions entering the interior of the cell as the ion gates on the cell membrane open in sequence. These sodium ions have a positive electrical charge, thereby causing the action potential to be electrically positive. Behind this wave of sodium ions, there is a second wave of potassium ions being forced out of the cell. Potassium is also positively charged. However, because it is being pushed out of the neuron, it renders the base charge of the neuron negative again.

CHEMICAL COMPONENT There may be relatively few or many axon terminals emanating from a single neuron. In either case, these terminals in the brain are adjacent to dendrites from many other neurons. Thus, the impulse of the action potential within a neuron terminates at the axon terminals and is taken up by the dendrites of the next neurons in the pathway. These are the neurons whose dendrites are adjacent to the axon terminals. The region where the axon terminals of one neuron and the dendrites of another come together is the **synapse**. For the most part, the neurons do not actually touch one another (some regions of the brain contradict this rule). They are essentially islands unto themselves.

Figure 2.4 Action Potential

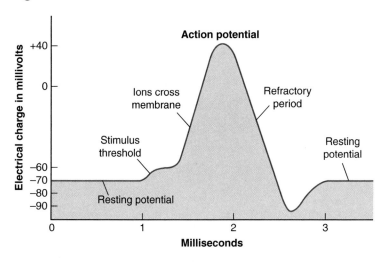

The synapses in the nervous system are extremely small physical gaps or clefts between the neurons. Each synapse is 100 to 200 angstroms wide, with an angstrom being one ten-thousandth of a millimeter. Note that the word *synapse* is also used as a verb: A neuron is said to synapse on another, meaning that it passes its message on to that other neuron.

A general law of the nervous system, especially in the brain, is that any single neuron synapses on a large number of other neurons. The evidence for this *divergence* is that a typical neuron synapses on anywhere from 100 to as many as 15,000 other neurons. Likewise, many different neurons can synapse on a single destination neuron, a principle known as *convergence*. So, as you can see, there is a great deal of complexity in how neurons connect to one another. This complexity is important in regulating the flow of the vast different types of knowledge that our brains need to handle at any one time.

By some counts, more than 100 different neurotransmitters have been identified in the human brain. Many seem to have rather ordinary functions, such as maintaining the physical integrity of the living organism. Others, especially **acetylcholine** and possibly **norepinephrine**, seem to have major influences on cognitive processes such as learning and memory (Drachman, 1978; Sitaran, Weingartner, Caine, & Gillin, 1978; Squire, 1987). For example, acetylcholine can enhance memory (Mishkin & Appenzellar, 1987). Acetylcholine may do this by enhancing the strength of synaptic potentials during long-term potentiation (see later discussion). Also, decreased levels of acetylcholine have been found in the brains of people with Alzheimer's disease, with very low levels of acetylcholine associated with more severe dementia (Samuel, 1999). It is tempting to suggest this as part of the explanation for the learning and memory deficits observed among such

---

Chemical Substances and Their Effects

For the bulk of the nervous system, the bridge across the synaptic cleft involves chemical activity within the synaptic cleft itself. Here we examine the two key substances: neurotransmitters and neuromodulators.

Neurotransmitters

A neuron releases a chemical transmitter substance, or simply a **neurotransmitter**, from small buttons or sacs in the **axon terminals**. This chemical fits into specific receptor sites on the dendrites of the next neuron and thereby causes some effect on that next neuron. The relationship between those chemicals and the receptor sites is like a key-and-lock system in which different chemicals bond only with certain receptor sites. Two general effects of these chemicals are possible: excitation and inhibition. That is, the effect can be excitatory, in which the neurotransmitters encourage the next neuron to fire, or it can be inhibitory, in which the neurotransmitters encourage the postsynaptic neuron to not fire.

So, why do we have both excitatory and inhibitory neurotransmitters? One reason is that one of the ways that the nervous system codes information is by the pattern of activation across a wide set of neurons, with some neurons needing to fire and others not. As a rough analogy, computers code information as a pattern of 1s (on) and 0s (off), although the nervous system does this differently and with much greater complexity. You do not want all your neurons firing at once. This would be like having all the bits in your computer set to 1; it does not convey any meaningful information. The closest your brain gets to having all the neurons fire at once is in a seizure.

Neuromodulators

patients, although it could be a side effect of the disease instead of a cause (e.g., Banich, 2004; Riley, 1989). In either case, the result suggests that acetylcholine plays an essential role in normal learning and memory processes.

Another important neurotransmitter for human cognition is **glutamate**, which is an excitatory neurotransmitter. This is important for creating or strengthening the connections between neurons. This is how we learn, by changing how neurons are connected to one another, and glutamate is an important neurotransmitter for doing this. A crucial neurotransmitter that is derived from glutamate is *gamma-aminobutyric acid*. However, rather than remembering such a long name, most neuroscientists simply refer to this neurotransmitter as **GABA**. GABA is an inhibitory neurotransmitter that works by weakening the connections between neurons and, in some sense, it is the opposite of glutamate.

2.2.3: Neurons and Learning

So, what exactly do these neurotransmitters change when learning occurs? Although it is not completely understood, one of the basic processes involved is something called **long-term potentiation (LTP)** (Bliss & Collingridge, 1993; Bliss & Lomo, 1973; Gustafsson & Wigstrom, 1988).

LONG-TERM POTENTIATION LTP is the process by which connections between neurons are strengthened. This strengthening essentially changes the ease with which two connected neurons will fire. More specifically, how strongly the presynaptic neuron encourages the postsynaptic neuron to fire is a function of LTP.

Let's look at the processes that bring about LTP during learning in a bit more detail, following work done by Pittenger and Kandel (2003). The process described here is illustrated in Figure 2.5.

When a neuron is stimulated, it can release glutamate into the synapse. This is taken up by the next neuron. Some of these glutamate neurons are taken up by what are known as *AMPA receptors*. These are fast-acting receptor sites that will result in a depolarization of the neuron at that moment and may cause it to fire. What is more interesting here is that some of the glutamate will also be taken up by what are known as *NMDA receptors*. The consequence of these is a series of chemical and genetic reactions that ultimately results in the creation of new AMPA receptors. What these new AMPA receptors do is to make it more likely that the neuron will take up glutamate that is released in the future, making it more likely that the neuron will fire. In other words, a simple way of thinking about this is that the AMPA receptors cause the neuron to fire, but the NMDA receptors lead to a strengthening of the connection between the neurons. This idea of changing the strength of the connection between neurons will be important when we discuss the development of connectionist models of cognition later in the module.

Often LTP can last for days or weeks.

CONSOLIDATION We need to be able to remember things for longer periods of time. This is accomplished by a process known as **consolidation** (Abraham, 2006), which takes place over long spans of time lasting days, weeks, and years. Consolidation makes memories more and more permanent over time. Interestingly, one of the processes that occur when we sleep can aid this consolidation process. People have been shown to remember more information if they had slept during the delay between study and test than if they had not (Hu, Stylos-Allan, & Walker, 2006), particularly if they experienced slow-wave (stages 3 and 4) sleep, as compared to REM (dreaming) sleep (Rasch & Born, 2008).

At a small scale, LTP captures change in individual neural connections. At a larger scale, consolidation captures changes in assemblies of neurons over long periods of time. At an even larger scale, changes in brain connections from experience lead to different ways of processing information over a person's lifetime, even to the point where different cultures—and hence different collections and complexes of experience—lead to different ways of neurologically processing information (Park & Huang, 2010).

Although LTP, a form of consolidation, is an important process involved in the storage of new information in the brain, it is not the only process that is operating.

NEUROGENESIS An example of another neural process that contributes to new learning is **neurogenesis**, the

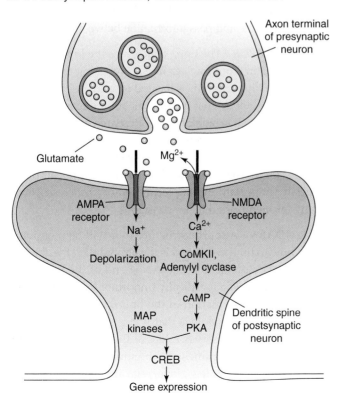

Figure 2.5 Short- and Long-Term Effects of Glutamate on a Postsynaptic Neuron, Which Can Result in LTP

creation of new neurons, particularly in the hippocampus. Even though the heavy lifting portion of creating new neurons in your brain is now over, and most of the neurons you will ever have are already made, there does continue to be some neurogenesis in the brain, and this can influence new learning. For example, some research has found that most of the new neurons that are created in the hippocampus each day end up not making it—they die. However, the more learning that is done during the day, the more likely these new neurons will be recruited in encoding that new information, and the more likely they will be to survive (Shors, 2014). In other words, the more you learn each day, the more neurons you can keep.

Significant research is being done on various psycho-biochemical properties of the neural system, such as the direct influence of various chemical agents on neurotransmitters and the resulting behavioral changes. As an example, Abraham (2006) and Thompson (1986) described progress in identifying neural changes believed to underlie memory storage and retrieval. Just as various psychoactive drugs affect the functioning of the nervous system in a physical sense, research is now identifying the effects of drugs and other treatments on the functioning of the nervous system in a psychological or cognitive sense (e.g., the effect of alcohol intoxication; Nelson, McSpadden, Fromme, & Marlatt, 1986).

WRITING PROMPT

Learning and Memory at a Neural Level

What is the fundamental nature of learning and memory at a neural level? How might our behaviors enhance or impede these neural processes?

The response entered here will appear in the performance dashboard and can be viewed by your instructor.

Submit

2.3: Important Brain Structures and Function

OBJECTIVE: Identify the neural anatomy of the human brain

Ignoring many levels of intermediate neural functioning and complexity, we now take a tremendous leap from the level of single neurons to the level of the entire brain, the awesomely complex "biological computer." To account for all human behavior, including bodily functions that occur involuntarily (e.g., digestion), entails an extensive discussion of the central and peripheral nervous systems. But to explore cognitive neuroscience, we can limit ourselves to just the central nervous system—the brain and spinal cord. In fact, our discussion even omits much of the central nervous system, save for the **neocortex** (or **cerebral cortex**), which sits at the top of the human brain, and a few other nearby structures.

2.3.1: Subcortical Brain Structures

The physically lower brain structures are collectively called the old brain or brain stem.

The subcortical brain structures include the thalamus, corpus callosum, hippocampus, and amygdala—all of which play key roles in cognition, illustrated in the following video in Figure 2.6.

Many of these subcortical (meaning "below the cortex") structures are important in understanding cognition. Let's consider four of them here.

THALAMUS First, deep inside the brain is the **thalamus**, meaning "inner room" or "chamber." It is often called the gateway to the cortex because almost all messages entering the cortex come through the thalamus (a portion of the olfactory sense of smell is one of the very few exceptions). In other words, the thalamus is the major relay station from the sensory systems of the body into the neocortex. The ability to coordinate information from different parts of the brain, such as remembering what was said and who said it, involves effective functioning in the thalamus.

CORPUS CALLOSUM Second, just above the thalamus is a broad band of nerve fibers called the **corpus callosum**. As described later, the corpus callosum ("callous body") is the primary bridge across which messages pass between the left and right halves—the hemispheres—of the cortex. This is the broadband connection of the brain that allows the two halves to communicate and pass information between each other.

HIPPOCAMPUS The third structure is the **hippocampus**, from the Latin word for "sea horse," referring to its curved shape. The hippocampus is located immediately interior to the temporal lobes; that is, underneath the temporal lobes but in the same horizontal plane. Research on the effects of hippocampal damage is described later in the course, including one of the best-known case histories in cognitive neuroscience, that of patient H. M. The hippocampus is crucially important in long-term memory processes, especially for memories that are conscious.

AMYGDALA A fourth structure that will be mentioned from time to time in this text is the **amygdala**. This almond-shaped structure is adjacent to one end of the hippocampus. The amygdala is critically important in the processing of emotional information in the brain. One of the unique aspects of the amygdala is that it gets nearly direct inputs from the olfactory nerves (Herz & Engen, 1996), which is why how something smells can be strongly associated with emotional responses (e.g., love, relaxation, or disgust). The amygdala is also strongly connected to the hippocampus,

Figure 2.6 Lower Brain Structures

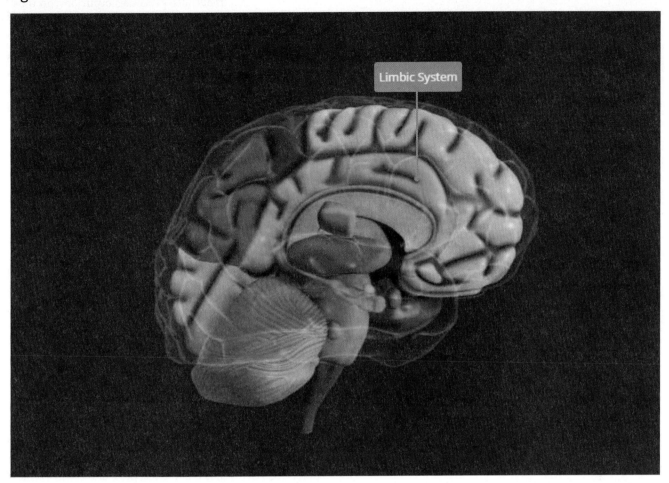

which is why emotional experience can sometimes be remembered very well.

2.3.2: Cortical Brain Structures

Now, let's move up a level in the brain. Figure 2.7 shows the neocortex, or cerebral cortex, the top layer of the brain, responsible for higher-level mental processes.

The cortex is a wrinkled, convoluted structure that nearly surrounds the old brain. The two halves or hemispheres cover about 2,500 cm^2 and are from about 1.5 to 3 mm thick. The wrinkling comes about by trying to get such a large surface area in a small volume. It compares to trying to get a piece of paper into a cup. To get the paper in, you wrinkle it up. The neocortex is the most recent structure to have evolved in the human brain (*neo* means "new") and is much larger in humans than in other animals. Compare the average weight of it in humans, 1,450 g, with that in the great apes, 400 g. In addition, because the neocortex is primarily responsible for higher mental processes such as language and thought, it is not surprising that it is so large relative to the rest of the brain—about three fourths of the neurons in the human brain are in the neocortex.

The side, or lateral, view (*lateral* simply means "to the side") in Figure 2.7 reveals the four general regions, or **lobes**, of the neocortex.

Clockwise from the front, these are the **frontal lobe**, **parietal lobe**, **occipital lobe**, and **temporal lobe**, named after the skull bones on top of them (e.g., the temporal lobes lie beneath your temples). Note that these lobes are not separate from one another in the brain. Instead, each hemisphere of the neocortex is a single sheet of neural tissue. The lobes are formed by the larger folds and convolutions of the cortex, with the names used as convenient reference terms for the regions. As an example, the central fissure, or fissure of Rolando, shown in the figure is merely one of the deeper folds in the brain, serving as a convenient landmark between the frontal and parietal lobes.

Note that the lobes of the cortex are generally specialized for processing different kinds of information. For example:

- The occipital lobe is used for visual processing.
- The temporal lobe is used for auditory, linguistic, and memory processing.

Figure 2.7 The Four Lobes of the Neocortex

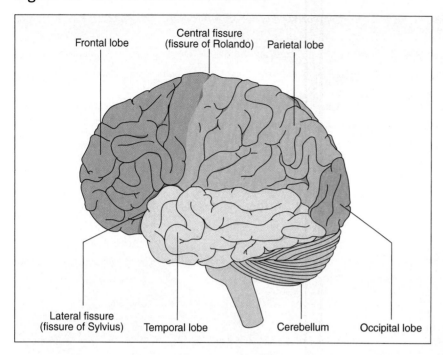

- The parietal lobe is used for spatial and sensory processing.
- The frontal lobe is used for cognitive control.

Throughout this course, we may refer to mental functions critically involving different lobes of the cortex. However, we know that a more fine-grained consideration of the localization of different brain activities than at the level of just the four lobes is needed. Rather than have you memorize many names and terms, whenever we refer to a part of the cortex as being involved in cognition, we will also give you some numbers. These numbers refer to what are called **Brodmann's areas**. There are maps of the outside and inside of the cortex laid out with the different numbered areas to help orient you to different locations in the brain. So, for example, if you were to read that attention can involve activity in the anterior cingulate cortex (B.A. 24), you would know that you could click on that B.A. number to reveal the Brodmann map with that area highlighted to know just where this is.

As different topics are covered in this course, various smaller areas of the cortex will be identified as being critically involved for doing specific tasks. For example, the fusiform face area (B.A. 37) is important for the processing of human faces, and damage in this area can lead to a deficit known as prosopagnosia (an inability to identify faces). Similarly, the parahippocampal place area (also B.A. 37) is important when processing spatial locations, and damage there can lead to an inability to process location information (Epstein & Kanwisher, 1998). Thus, different parts of the brain are critically involved in different types of cognitive processes. This does not mean that those processes occur only in those parts of the brain, but that those parts of the brain are important and critically involved in that type of thinking. An analogy would be with a car. The gas pedal is critically involved in the speed of the car, but by itself it does not make the car go (you need an engine, wheels, and many other components for that).

2.3.3: Principles of Functioning

Two important principles of functioning in the neocortex are described here. This is necessary background knowledge for understanding the effects of brain function on cognitive processes—that is, how the mind relates to the brain. These principles involve the ideas of contralaterality and hemispheric specialization.

CONTRALATERALITY When viewed from the top, the neocortex is divided into two mirror-image halves—the left and right **cerebral hemispheres**. This follows a general law of anatomy that, except for internal organs such as the heart, the body is basically bilaterally symmetrical. What is somewhat surprising, however, is that the receptive and control centers for one side of the body are in the opposite hemisphere of the brain. This is **contralaterality** (*contra* means "against" or "opposite"). In other words, for evolutionary reasons that will probably remain obscure, the right hemisphere of the brain receives its input from the left side of the body and controls the left side. Likewise, the left hemisphere receives input from and controls output to the right side of the body. As an example, people who have a stroke in the left hemisphere will

Different parts of the brain are involved depending on the kind of activity a person is engaged in.

often have some paralysis in the right half of the body. There are a few exceptions, such as the olfactory nerves, in which there are ipsilateral (same side) connections.

HEMISPHERIC SPECIALIZATION The second issue concerning lateralization involves different specializations within the two cerebral hemispheres, that is **hemispheric specialization**. Despite their mirror-image appearance, the two hemispheres do not completely mirror one another's functions and emphases. Instead, each hemisphere tends to specialize in different abilities and tends to process different kinds of information. This is the full principle of **cerebral lateralization** and specialization: Different functions or actions within the brain tend to rely more heavily on one hemisphere or the other or tend to be performed differently in the two hemispheres. This is not to say that a process or function can happen only in one hemisphere. It merely says that there is often a tendency, sometimes strong, for one or the other hemisphere to be especially dominant in different processes or functions.

Many people have heard of these "left brain versus right brain" issues, often from the popular press. Such treatments are notorious for exaggerating and oversimplifying laterality and specialization. For instance, in these descriptions, the left hemisphere ends up with the rational, logical, and symbolic abilities—the boring ones—whereas the right hemisphere gets the holistic, creative, and intuitive processes—the sexy ones! Corballis (1989, p. 501) noted that the right hemisphere achieves "a certain cult status" in some treatments.

But even ignoring that oversimplification, it is far too easy to misunderstand the principles of lateralization and specialization; too easy to say "process X happens in *this* hemisphere, process Y in *that* one." Even the simplest act of cognition, say naming a picture, involves multiple components, distributed widely across both hemispheres, and

Dominant Functions of the Left and Right Hemispheres

The most obvious evidence of lateralization in humans is the overwhelming incidence of right-handedness across all cultures and apparently throughout the known history of human evolution (Corballis, 1989).

Left Hemisphere

The most obvious evidence of lateralization in humans is the overwhelming incidence of right-handedness across all cultures and apparently throughout the known history of human evolution (Corballis, 1989). Accompanying this tendency toward right-handedness is a particularly strong left-hemispheric specialization in humans for language. So, for most people, language ability is especially lateralized in the left hemisphere. Countless studies have demonstrated this general tendency (see Provins, 1997, for a review of the handedness–speech relationship). The left lateralization of language in the brain also applies to many left-handers, so handedness and brain lateralization are not completely synced.

Right Hemisphere

complex coordination of the components. Disruption of any one of those could disrupt picture naming. Thus, several different patients, each with dramatically different localized brain damage, could show an inability to name a picture, each for a different reason relating to different lateralized processes.

Nonetheless, there is a striking division of labor in the neocortex, in which the left cerebral hemisphere is specialized for language. This is almost always true; it characterizes up to 85% or 90% of the population. However, the percentages are this high only if you are a right-handed male with no family history of left-handedness and if you write with your hand in a normal rather than inverted position (Friedman & Polson, 1981). If you are female, if you are left-handed, if you write with an inverted hand position, and so on, then the "left hemisphere language rule" is not quite as strong. In such groups, the majority of people have the customary pattern, but the percentages are not as high. For example, Bryden's (1982) review indicated that 79% of women have language lateralized to the left hemisphere. Thus, directing language input to the left cerebral hemisphere is optimal and efficient for many people, but not for all. (For simplicity, however, we rely on the convenient fiction that language is processed in the left hemisphere for most people; see Banich, 2004, for useful discussions.)

2.3.4: Split-Brain Research and Lateralization

Despite the exaggerated claims you often read, there has been a good deal of careful work on the topic of lateralization and specialization of different regions in the two hemispheres. Among the best known is the research on **split-brain** patients.

Research on Split-Brain Patients

Before about 1960, evidence of hemispheric specialization was rather indirect. Neurologists and researchers simply noted the location and kind of head injury that was sustained and the kind of behavioral or cognitive deficit that was observed after the injury. Sperry (1964; Gazzaniga & Sperry, 1967), however, put the facts of anatomy together with a surgical procedure for severe epilepsy.

> In the surgical operation, the corpus callosum is completely severed to restrict the epileptic seizure to just one of the cerebral hemispheres. For patients who had had this radical surgery, a remarkably informative test could be administered, one that could reveal the different abilities and actions of the two hemispheres. That is, from the standpoint of brain functioning, when a patient's corpus callosum is surgically cut, the two hemispheres cannot communicate internally with each other—information in one hemisphere cannot cross over to the other. Sperry's technique was to test such people by directing sensory information to one side or the other of the body (e.g., by placing a pencil in the left or right hand of such a patient or presenting a visual stimulus to the left or right visual field), then observing their behavior.

Although we have discussed some localization of function in this module, implying that different hemispheres, lobes, or parts of the brain are largely responsible for different types of cognition, it should be kept in mind that there is some individual variability in this. Some people are different. For example, some people have the distribution of processes associated with the left and right hemisphere reversed (and it is not just being left-handed that does this). Also, there can be influences of culture on cognitive neurological processing (Ambday & Bharucha, 2009). For example, Tang et al. (2006) found that whereas English speakers tended to use more language areas when processing Arabic numerals, Chinese speakers tended to use more visual/spatial areas of the brain, suggesting that different cultures can lead people to use different parts of the brain to process the same kinds of information.

2.3.5: Cortical Specialization

Not only are the different hemispheres of the cortex differentially able to do different things, but also the different parts of the cortex are tuned for different kinds of processes. As with lateralization of brain function, the fact that a part of the brain is implicated in a certain kind of mental process, such as remembering, does not mean that that part of the brain does only that process, nor does it mean that other parts of the brain are not involved in that activity. The localization or specialization of function in the cortex means that that part of the brain is critically important for a certain kind of cognition. There will be numerous examples of the localization of function throughout this course for different types of cognition. To give you a better feel for this, we will cover three types of cortical specialization and the implications they have for human thought.

SENSORY AND MOTOR CORTICES First, consider a kind of cortical specialization in which a section of cortex is specialized for a certain type of processing. As you know, your brain is connected to the rest of your body. When you sense things touching your body or move part of your body, this involves specific parts of your cortex. These are called the sensory and motor cortices, respectively, and are shown below in Figure 2.8.

The **sensory cortex** is a band of cortex at the front of the parietal lobes, just behind the central sulcus (B.A.s 1, 2, and 3), and is responsible for processing sensory information from throughout the body. In comparison, the **motor cortex** is a band of cortex at the back of the frontal

Figure 2.8 The Locations of the Sensory and Motor Cortices

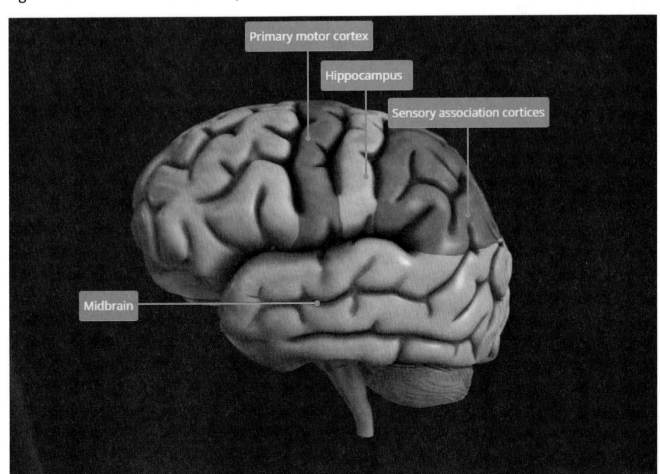

lobe, just in front of the central sulcus (B.A. 4), and is responsible for controlling all of your voluntary muscle movements. But it is not the case that the whole of the sensory or motor cortices is responsible for any and all sensations and voluntary movements you make. Instead, there are different places along these cortices that are specialized for different parts of the body, such as the fingers, leg, or mouth. So, not only is there a localization of sensory and motor processes in the brain, but there is even a localization of different body parts within each of those parts of the cortex.

MIRROR NEURONS A second kind of specialization in the cortex does not involve a region of the brain, but instead involves certain kinds of neurons that seem to be used to do a specific kind of mental processing. Here, this specialization is with neurons found in association with processing in or near the motor cortex. These neurons are active when a person is performing an action or, more interestingly, watching another person do something (see Glenberg, 2011). These were originally discovered in work involving monkeys (Di Pellegrino, Fadiga, Fogassi, Galese, & Rizzolatti, 1992). Because it was initially thought that these neurons are involved when people mentally imitate or mirror another person's actions, they were called **mirror neurons**. These neurons appear in several places in the cortex, such as the inferior frontal cortex (B.A. 44 and 45). Although the exact nature and role of mirror neurons is not yet completely understood (see the July 2011 issue of *Perspectives on Psychological Science* for a lively debate on the issue), they are used in some way to plan and execute movements and to understand what people are doing.

DORSAL AND VENTRAL PATHWAYS A third type of specialization of function in the cortex does not involve certain regions of the cortex, or certain types of cells, but a chain or sequence of cells and areas that are all working toward a processing goal. A good example of this involves the visual pathways in the cortex and the processing of different types of visual information. Many initial, early visual processes occur at the back of the brain in the occipital lobe, a region called V1 (for "visual 1") (B.A. 17). The visual pathways that emerge from the early visual processing done in this and subsequent areas diverge into two pathways, depending on the type of visual information being processed (Mishkin, Ungerleider, & Macko, 1983; Ungerleider & Haxby, 1994), as shown below in Figure 2.9.

Figure 2.9 The Dorsal and Ventral Pathways

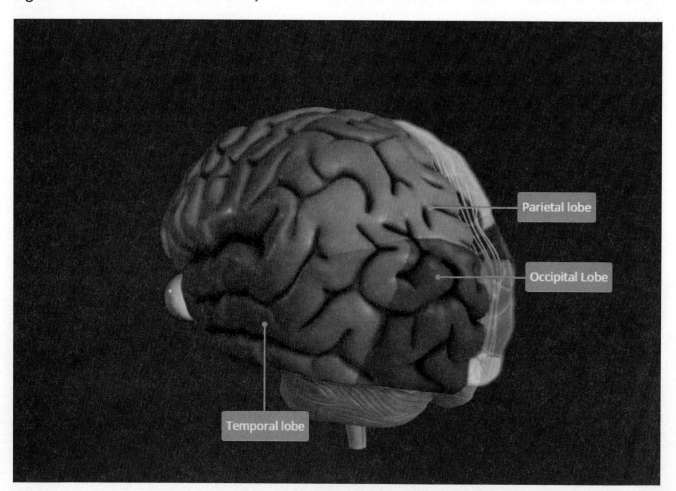

One of these pathways is the **dorsal pathway**, which moves along the top of the cortex across the parietal lobe. The dorsal pathway is primarily involved in determining *where* things are in space. As such, it is also called the "where" system in vision. The other of these is the **ventral pathway**, which moves along the bottom of the cortex into the temporal lobe. The ventral pathway is primarily involved in determining *what* things are. As such, it is also called the "what" system in visual. Although it seems odd when you first come across this fact, the visual system does process where a thing is and what it is differently, although the system does bring this information together later on.

2.3.6: Levels of Explanation and Embodied Cognition

Although understanding how neurons communicate with each other is important for a deeper comprehension of human cognition, it is not necessarily the case that all our understanding of thought can be reduced to understanding how neurons do their job. Instead, assemblies of neurons working together can produce cognitive processes that are **emergent properties**: They are present when neurons work together but are absent in the individual neurons (Minsky, 1986). A more everyday example of an emergent property is that of containment. Although no piece of wood has the property of containment, if you arrange several pieces into a box shape, then the property of containment emerges. Similarly, some aspects of cognition are emergent properties of collections of neurons working together.

Because of this, cognitive science will often use levels of explanation to capture how a process is operating. Thus, we can sometimes understand how problem solving may be progressing without knowing what the individual neurons are doing (let alone the individual molecules, protons, quarks, and such). That said, cognitive scientists understand that there are important bridges across these different levels, and that understanding what is happening at one level can help explain and give an understanding of what is happening at another level. For example, lacking appropriate levels of a certain neurotransmitter may have dramatic consequences on a person's ability to think clearly (see Marshall, 2009, for a clear exposition of these ideas).

One example of understanding how the brain processes information can inform our understanding of how cognition works comes from research on embodied or grounded cognition. As a reminder, one of the basic tenets or ideas about embodied cognition is that the way we think about things reflects how our bodies work and interact with objects in the world. Research in cognitive neuroscience has observed that much of cognition is oriented around allowing us to use our bodies effectively in the environments we find ourselves in—after all, the body is an action-oriented system made to do things. Consistent with an embodied cognition perspective, it has been reported that certain parts of the motor cortex (B.A. 4, as noted above) become activated when people are motionless but are processing words that refer to actions. Moreover, different parts of the motor cortex become active for different kinds of verbs. For example, a verb such as *kick* would be more likely to activate the part of the cortex responsible for controlling the feet, whereas a verb such as *punch* would be more likely to activate the part responsible for the hand (e.g., Kemmerer, Castillo, Talavage, Patterson, & Wiley, 2008).

WRITING PROMPT

Brain Structure and Flow of Thought

How does the organization of the brain and its various substructures influence the flow of thought?

  The response entered here will appear in the performance dashboard and can be viewed by your instructor.

Submit

2.4: Neuroimaging

OBJECTIVE: Describe the four ways of understanding the human brain

There are several methods used to look at the brain and how it processes information. The various forms of neuroimaging are important for cognitive psychology because the data derived from such work can be used to understand which brain systems are involved in cognitive process and to test cognitive theories of how mental processes unfold (Coltheart, 2013; Mather, Cacioppo, & Kanwisher, 2013). Some sort of understanding of the cognitive processes and representations involved in a process is necessary to even begin to understand any neuroimaging data (Frank & Badre, 2015).

We cannot know what different parts of the brain do unless we know what people are thinking about. In this part of the module, we look at a number methods that help with this. In later modules, many of these methods are used to help understand how thinking works in one context or another.

We divide our discussion into four basic areas:

1. Structural measures that look at the structure of the brain itself
2. Electrical measures that capitalize on the electrical properties of the action potential when neurons talk to one another

3. Processing measures that assess changes in metabolic processes, such as blood flow, because of different types of thinking
4. Other neuroimaging measures

2.4.1: Structural Measures

One of the first things we may be interested in is the structure of the brain itself. This is of particular concern if there is reason to believe that there has been some sort of brain damage or disease, or if there is some reason to suspect that someone may have an unusually small or large brain structure.

2.4.2: Electrical Measures

Although it is sometimes important to know about the structure of the brain, in terms of memory, cognition, and thinking, it is often vital to know about *how* the brain is processing information.

USING ELECTRICAL ASPECTS TO MEASURE NEURAL ACTIVITY One way of getting at this issue is to look at the amount of activity in different parts of the cortex and how this changes depending on whether one is doing a particular task or not. This can be done by capitalizing on the change in electrical charge that occurs during an action potential.

The most basic way this is done is by inserting a microelectrode into the brain to record the activity of a single cell. This is called **single cell recording**, which looks at how the firing rate of an individual cell changes, by seeing whether it fires more or less often as a function of the given task. Because it is so invasive, this procedure is typically used with animals. However, it is being increasingly used on people who have electrodes implanted in their brains as

---

Techniques to View the Brain Structure (Figure 2.10)

The simplest way to assess the brain structure is to open the skull and look at the brain. However, this is a rather dramatic line of action and there are other less invasive methods.

Computerized Axial Tomography (CT) Scans

One way to assess brain structure less intrusively is by using **computerized axial tomography** or **CT scans**. With this technique, people place their heads in a big mechanical doughnut that takes a series of X-ray images. These images are then assembled by a computer to allow various "slices" of the brain to be studied as X-ray images. This fairly common technique allows for a quick assessment of general brain structure.

Magnetic Resonance Imaging (MRI) Scans

part of a clinical treatment. This method was used to map visual cells in the occipital lobe.

There are less invasive ways to measure neural activity. This can be done by measuring brain wave patterns using **electroencephalogram (EEG)** recordings. In this technique, electrodes are placed on the person's scalp, and the device records the patterns of brain waves. Typically, researchers have focused on **event-related potentials (ERPs)**, the momentary changes in electrical activity of the brain when a stimulus is presented to a person (e.g., Donchin, 1981; Rugg & Coles, 1995). As an example, read the following sentence (adapted from Banich, 2004, based on Kutas & Hillyard, 1980): "Running out the door, Patty grabbed her jacket, her baseball glove, her cap, a softball, and a skyscraper." Of course, you noticed that *skyscraper* does not fit the context of the sentence; it is called a semantic anomaly. What is fascinating is that the ERP recording of your brain wave activities would show a marked change about 400 ms after you read *skyscraper*, an electrically negative wave called N4 or N400. The N4 would be present, though smaller, if the last word in the sentence had been *lamp*, and at baseline if the last word had been *bat*.

Osterhout and Holcomb (1992) found this N4 effect for semantic anomalies and a similar effect when the grammar or syntax of the sentence violated normal language rules. They showed their participants control sentences along with sentences that contained grammatical or syntactic anomalies:

(Control) John told the man to leave.

(Anomalous) John hoped the man to leave.

With this manipulation of syntactic anomaly, Osterhout and Holcomb found a pronounced P6 or P600 effect, an electrically positive change in activity, roughly 600 ms after reading the word *to* (the word that signals the syntactic violation). Figure 2.11 shows their result.

Clearly, our language comprehension system reacts differently—and very quickly—when we encounter unusual or incorrect sentences.

ERP technology is especially good at telling when mental mechanisms operate in an online task, but assessments of where this process occurs is much trickier and not very precise. By carefully controlling surrounding conditions and measuring the elapsed time since a stimulus was presented, we can begin to see how the electrical activity of the brain changes moment by moment when the person is processing a stimulus.

Another way of capitalizing on the electrical aspect of neural processing is to use a method known as **transcranial magnetic stimulation (TMS)**. With this technique, the apparatus is positioned near a person's head with a part of the brain targeted. Then, when the machine is turned on, it produces a magnetic field that stimulates, and possibly disrupts, the electrical activity in a part of a person's brain. Think of this as inducing a mild seizure in a person's brain. Essentially, this process may have the

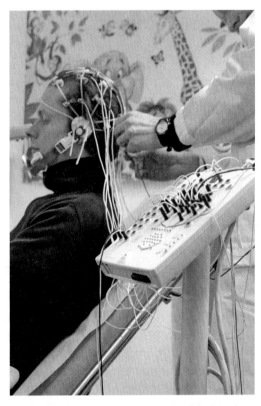

A person having his EEGs recorded for an ERP study.

Figure 2.11 Mean ERPs to syntactically acceptable sentences (solid curve) and syntactically anomalous sentences (dotted curve). The P600 component, shown as a dip in the dotted curve, shows the effect of detecting the syntactic anomaly.

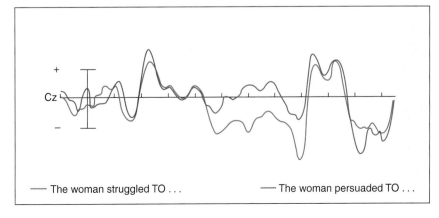

effect of giving a person a temporary lesion by temporarily knocking out the ability of part of the brain to effectively do a task, and assessing the effects on cognition when that part of the brain is no longer operating. As one example of this, Grafman et al. (1994) used TMS to disrupt the ability of people to recall lists of words when the TMS stimulation was applied over temporal-lobe and frontal-lobe areas of the brain, which are involved in memory retrieval.

2.4.3: Metabolic Measures

In addition to electrical activity, other things that change in the brain when we think are metabolic processes, such as blood flow. For example, when a part of the brain is working hard, the neurons there need more energy, such as what comes from oxygen, to keep working. This change in demands of different parts of the brain can be reflected in the amount of blood flowing through the capillaries to that part of the brain.

So, this shows that we can look at brain activity and make reliable connections to how people think.

An advantage to these techniques, from the perspective of cognitive science, is that they show the brain in action rather than just the physical structures. Such scans are called *functional* because they show the brain as it is functioning while performing some mental task. A second advantage is that they can be applied with (apparently) minimal risk to normal people.

The set of color pictures from a PET scan study assessing the processing of different types of information (i.e., language and/or music) relied on a similar procedure, computer-assisted detection of blood flow patterns in a patient injected with an irradiated substance that binds to oxygen in the blood. However, imaging techniques that measure blood

---

Using Metabolic Measures to Understand the Brain Functioning (Figure 2.12)

What the metabolic measures can do is tap into this information to help figure out which parts of the brain are more active under various circumstances.

Interactive

Positron Emission Tomography (PET) Scans

One means of doing this is with a method called **positron emission tomography (PET) scans.** With PET scans, a person is injected with a radioactive isotope that is taken up by the bloodstream. This isotope has a very short half-life, so the impact of the radiation on the body is minimal. After the isotope is injected, the person is placed into a scanner and asked to do a kind of task. What is found is that certain parts of the brain will work harder than they normally do. After an analysis is done, that part of the brain will "light up" in the PET scan image, suggesting that that part of the cortex is more involved in that type of thinking.

Functional MRI (fMRI) Scans

flow have a time lag drawback. The increase in metabolic activity can lag anywhere from several hundred milliseconds to several seconds after the cognitive activity.

Neuroimaging data can be used not only to verify and expand theories of cognition but also to help solve more applied and clinical problems. Take the example of dyslexia. Some work has shown that people with dyslexia show different patterns of neural processing as revealed by various neuroimaging technologies (Shaywitz, Mody, & Shaywitz, 2006). For example, some people have a genetically based form of dyslexia in which there are problems in the occipital-temporal region (where the occipital and temporal lobes meet) of the left hemisphere, a region that is strongly associated with language processing. Specifically, there appears to be a disruption of processing in this region of the brain that makes it difficult to accurately assemble sets of letters into words. In comparison, another group of dyslexics show very adequate processing in the left occipital-temporal region. Their dyslexia, instead, seems environmentally based, related to poorer educational support and other factors, and shows a stronger connection to the right hemisphere prefrontal region, which is associated with memory retrieval. Essentially, rather than mentally sounding words out, they treat words more as wholes, attempting to retrieve word meanings directly. Thus, less frequent and new words become much more difficult to process, and dyslexics of this type show persistent difficulties with reading of a very different sort.

2.4.4: Other Methods

Apart from looking at the structure of the brain, or the electrical or metabolic activity that a person exhibits when doing a certain kind of thinking, other methods are available to cognitive neuroscientists to relate mental function to the underlying brain structures.

Methods Used to Relate Mental Function to Underlying Brain Structures

In this section, we consider a few of these, including lesions, direct stimulation, and special populations.

Deliberate Lesioning of the Brain

An example of the investigation technique used by Sperry, discussed earlier in the section on lateralization and split-brain patients, involved deliberate lesioning of the brain. Although such techniques are limited in their usefulness, they can sometimes be revealing of the secrets of cognitive processing. **Lesions** are used with only two kinds of subjects—laboratory animals and patients with medical conditions requiring brain surgery. However, a long-standing tradition reports case studies of people or groups of people who by disease or accident have experienced damage or lesions to the brain. Much of the evidence described throughout this course comes from victims of strokes, diseases, aneurysms, head injuries, and other accidental circumstances. In all cases, the site and extent of the brain lesion are important guides to the kind of disruption in behavior that is observed and vice versa (a clear description of the lesion method is found in Damasio & Damasio, 1997).

Direct Stimulation

Special Populations

> **WRITING PROMPT**
>
> **Value of Neuroimaging Methods**
>
> What valuable insights do neuroimaging methods provide cognitive science? Describe the limitations of such methods.
>
> ▶ The response entered here will appear in the performance dashboard and can be viewed by your instructor.
>
> [Submit]

2.5: Connectionism

OBJECTIVE: Summarize connectionism as a computer-based method in cognitive science

To better understand how the nervous system codes and processes information, it is helpful to have scientific models that capture these sorts of processes. To this end, we conclude the module with a brief presentation on **connectionism**, an important computer-based method in cognitive science.

Connectionist models are also called **neural net models** or **parallel distributed processing (PDP) models**; for our purposes the three terms are treated as synonymous. They refer to a computer-based technique for modeling complex systems that is inspired by the structure of the nervous system. A fundamental principle in connectionist models is that the simple nodes or units that make up the system are interconnected. Knowledge, all the way from the simplest to the most complex, is represented in these models as simple interconnected units. The connections between units can either be excitatory or inhibitory. In other words, the connections can have positive or negative weights. The basic units receive positive and negative activation from other units; depending on these patterns, they in turn transmit activation to yet other units. Furthermore, the interconnectedness of these basic units usually is described as "massive" because there is no particular restriction on the number of interconnections any unit can have. In principle, any bit of knowledge or information can be connected or related to an almost limitless number of other units.

Figure 2.13 illustrates an early connectionist model by McClelland and Rumelhart (1981), a model that dealt with word recognition.

The bottom row of nodes or units represents simple features, simple patterns such as a horizontal line and a vertical line, each connected to letters at the next higher level, which in turn are connected to words at the top level. For simplicity, look at the feature on the far left, the horizontal line. The connection directly up from that to the capital letter *A* would be a positive, excitatory connection because the letter *A* has a horizontal line. The connection from this feature up to the letter *N*, however, would be a negative, inhibitory one: If the feature detection system detects a horizontal line, this works against recognition of the letter *N*. In the same fashion, the capital *A* would have a positive connection up to the word ABLE because *A* is in the first position there, but it would have a negative, inhibitory connection to TRAP because TRAP does not begin with the letter *A*.

Referring to such models by the term *parallel distributed processing* highlights a different facet of the brain and the computer system. Mental processes operate in a thoroughly parallel fashion and are widely distributed across the brain. Likewise, processing in a PDP model is thoroughly parallel and distributed across multiple levels of knowledge. As an example, even as the feature detectors at the bottom of the model are being matched to an incoming stimulus, word units at the top of the model may already be activated. Thus, activation from higher levels may

Figure 2.13 An Illustration of Part of McClelland and Rumelhart's (1981) PDP Model of Feature, Letter, and Word Recognition

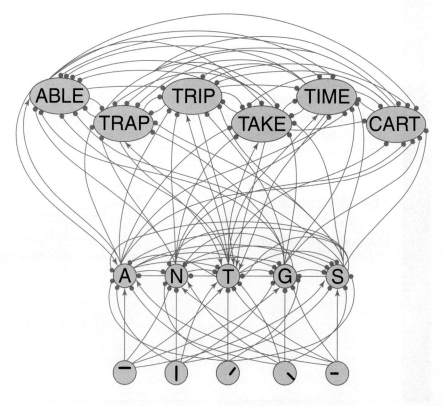

influence processing at lower levels, even as the lower levels affect activation at higher levels.

Consider how the system would recognize the word TRAP in the sentence "After the bear attacked the visiting tourists, hunters went into the forest to set a trap." Even as the feature and letter detector units would be working on the *T*, then the *R*, and so forth, word units at the top would have already received activation based on the meaning and context of the sentence. With bears attacking tourists and hunters going into the forest, the word TRAP is highly predictable from the context, but CART would not be (just as *bat* was predictable but *skyscraper* was not in the example given earlier). Given this context, TRAP would be more easily recognized, perhaps because the features within the letters would have been activated by the context already. In this fashion, the comprehension and word recognition systems would be operating in parallel with the feature and letter detection systems, each system making continuing, simultaneous contributions to the other systems and to overall mental processing.

The similarity of the **connectionist** scheme to the functioning of the brain is obvious and vitally important—it is widely believed that connectionist models operate on the same, or at least very similar, basic principles as the brain (McClelland, Rumelhart, & Hinton, 1986). In other words, the connectionist framework may give us an excellent way of modeling and simulating cognitive processes (for a more recent example, see Monaghan & Pollmann, 2003).

WRITING PROMPT

Applications of Connectionist Models

What application might there be for connectionist models outside of psychology?

> The response entered here will appear in the performance dashboard and can be viewed by your instructor.

Submit

Summary: Cognitive Neuroscience

2.1 The Brain and Cognition Together

- By understanding the structure and function of the nervous system, we can gain better insight into the structure and function of cognition. Patients, such as K. C., have brain damage that reflects differences in various components of cognition.
- The best evidence for a brain–cognition linkage is if there is a double dissociation with one function being affected, but not a second; one with one type of damage, and the reverse with another type of brain damage.

2.2 Basic Neurology

- Neurons are the basic building blocks of the nervous system. Understanding how these elements work can provide a better understanding of the machinery that underlies cognition.
- Neurons communicate information via an electrical component and a chemical component. The electrical component is the action potential by which, obeying the all-or-none principle, a signal is eventually sent down the axon to communicate with other neurons.
- The chemical component involves neurotransmitters released at the synapse. There is a wide variety of neurotransmitters, each involved in a different type of neural process. Some of these neurotransmitters are excitatory, stimulating the postsynaptic neuron, and some are inhibitory, suppressing the postsynaptic neuron.
- Information is stored in the nervous system via a pattern of interconnections among neurons and how strong these interconnections are. Early on in memory formation, adjustments to this pattern are affected by processes such as long-term potentiation, but later memory solidification is affected by a process of consolidation. The storage of new memories may also be affected by the processes involved in neurogenesis.

2.3 Important Brain Structures and Function

- Many subcortical structures make a meaningful contribution to cognition, including the thalamus, the corpus callosum, the hippocampus, and the amygdala.
- The cerebral cortex is the largest brain structure in humans and is the seat of many cognitive functions. The cortex is divided into four lobes and two hemispheres, with some localization of function in these various brain regions, such as the occipital lobe being primarily involved in visual processing. Damage to certain areas of the brain can cause specific cognitive deficits, such as prosopagnosia.
- There are contralateral connections of the brain with the body, as well as some lateralization of cognitive function in the two hemispheres. This separation of different mental tasks by the two halves of the cortex is

- most clearly illustrated in split-brain patients who had the corpus callosum cut, thereby separating the two hemispheres.
- Although there are different levels of analysis and explanation, a good understanding of neurophysiology can be very helpful in comprehending human cognition. A good example of this is how embodied cognition processes may involve neural processes used in actual physical action.

2.4 Neuroimaging

- There are a number of neurological measures for assessing how brain structure and function relate to cognitive processes. These provide insight into how the mind and the brain interact. It is important to note that these measures provide insight into how the nervous system is functioning, but they do not let us know just what a person is thinking.
- Structural measures, such as CT and MRI scans, allow evaluation of the structure of living brains. These measures are useful in identifying brain abnormalities. The consequences of such abnormalities, such as from a stroke, tumor, or other trauma, may affect thinking in principled ways.
- Electrical measures, such as single cell recording, EEGs, ERPs, and TMS, take advantage of the electrical component of neural transmission. Neurological measures based on these kinds of signals have the potential to provide moment-to-moment information about the intensity of neural processes.
- Metabolic imaging methods take advantage of changes in blood flow to assess how hard different regions of the brain are working. The great advantage of this kind of imaging technology is that it allows us to more precisely pinpoint when different regions in a person's brain are active for different types of mental activity.
- There are numerous other ways of assessing neurological function and how it relates to cognition. These include looking at people who have sustained a lesion to the brain, direct stimulation of the cortex, case studies, and testing special populations, among many others that are not discussed here. Each of these different approaches has the potential to reveal new and interesting aspects of human cognition.

2.5 Connectionism

- The notion that human cognition is analogous to processing in a computer system has largely been abandoned at the detailed level, especially because of evidence of widespread parallel processing in humans. Connectionist (neural net, PDP) models can simulate such parallel processes and therefore may be excellent ways of modeling human cognitive processes. These models use the nervous system, and how it represents and processes information, as the inspiration for how they are structured and developed.

SHARED WRITING

Cognitive Neuroscience and Cognitive Science

Understanding some basic neurophysiology can be helpful in understanding cognition. In your own words, described the concept of emergent properties, link this to how cognition emerges out of the nervous system, and use an analogy of another emergent property that you may find in everyday life (e.g., how containment emerges out of putting several flat surfaces together to form a box).

> A minimum number of characters is required to post and earn points. After posting, your response can be viewed by your class and instructor, and you can participate in the class discussion.

Post   0 characters | 140 minimum

Chapter 3
Sensation and Perception

 Learning Objectives

3.1: Explain the process of relating mental experience to physical reality

3.2: Distinguish the processes of visual sensation and perception

3.3: Evaluate the process by which we identify patterns and objects

3.4: Explore ideas about how what we know influences what we see

3.5: Summarize the science of object recognition and agnosia

3.6: Describe the science of hearing

It is truly a wonder we can see or hear anything at all given the implausible, even backward, structure of the eye and the indirect, unlikely structure of the ear. Yet, we can see the flame of a single candle on a dark night from a distance of 20 miles. We can hear a watch ticking 20 feet away in a large, quiet room. In addition (some of you please note!), we can smell a single drop of perfume diffused into the entire volume of a six-room apartment (Galanter, 1962). Yet the complexity of the mental processing of perception exceeds these sensitivities. Because we "understand" what we have seen so quickly, with seemingly little effort, "we can be deceived into thinking that vision should therefore be fairly simple to perform" (Hildreth & Ullman, 1989). This module (and this title) repeatedly emphasizes that a rapid and unconscious process is neither necessarily simple nor simple to investigate. If anything, just the opposite is probably true.

This module presents a basic study of perception in vision and hearing. We focus on the visual and auditory sensory registers because they are our most prominent intersections with the world (some animals make more use of other senses, such as the sense of smell in dogs). This module covers several theories, including an elaboration of the connectionist model. In a later section, we consider brain-related disruptions in perception. We see what perceptual deficits tell us about the normal processes we take for granted.

3.1: Psychophysics

OBJECTIVE: Explain the process of relating mental experience to physical reality

How do you know what exists out in the world and what does not? In other words, how does mental experience compare with physical reality? In what ways do our sensations and perceptions capture the world as it is, and in what ways do they fall short? The relating of mental experience with physical reality is a specialized sub-domain of research on sensation and perception known as **psychophysics**.

In this section, we cover two issues of interest to psychophysicists: the detection and the discrimination of sensory experiences. Studies of detection tell us what information a person can sense or perceive, and what is outside of one's awareness. In comparison, studies of discrimination tell us how much the world needs to change before a person notices that it is different.

3.1.1: Detection and Absolute Thresholds

One basic question that psychophysicists ask is "How much physical energy (i.e., light, heat, sound, etc.) is needed for a person to detect that something is present in the environment?" The minimal amount of energy needed to detect a

Table 3.1 Some Sensory Thresholds

To give you an idea of how sensitive different sensory modalities are, Table 3.1 presents various sensory thresholds.

Some Sensory Thresholds	
Vision	Detection of a candle flame 20 miles away on a dark, clear night
Hearing	Detection of the ticking of a watch 20 feet away
Touch	Detection of the wing of a fly falling on one's cheek from a height of 1 cm
Smell	Detection of one drop of perfume diffused into the volume of a six-room apartment
Taste	Detection of 1 teaspoon of sugar into 2 gallons of water.

stimulus—whether it is visual, auditory, tactile, or any other type—is referred to as a *sensory threshold*. Conventionally, a sensory threshold is defined as the amount of physical energy needed for a person to detect the presence of a stimulus 50% of the time over many trials. For example, imagine that you are in a completely dark environment. How bright would a light have to be for you to notice it had been turned on? The brightness at which you can perceive the light half of the time is said to be at your threshold. Stimuli that are detected more than half the time are said to be *supraliminal*, and stimuli that are so weak that they are detected less than half the time are said to be *subliminal*.

3.1.2: Discrimination

Another basic question that psychophysicists ask is "How much does the world need to change before a person notices that it is different?" The smallest amount of physical change that a person can detect is the ren **just noticeable difference**, or **JND**.

3.1.3: Decisions About Physical and Mental Differences

Although psychophysics primarily refers to the comparison of sensory magnitudes, it is possible to extend its

Capturing the JND by Psychophysicists
Psychophysicists have made several attempts to capture the JND.

Weber Fraction

The earliest attempt was the *Weber fraction*. The formula for this is $k = \Delta I / I$, where k is the constant, and I is the intensity of a stimulus. Essentially, this formula states that the ability to notice a change in the intensity of a sensation is a proportion of what the sensation was before.

Fechner's Law

Steven's Power Law

basic principles to making any cognitive comparisons, including those between two mental symbols. One basic finding is that the greater the distance between two items, the easier it is to make a discrimination. This distance effect also holds for both physical and symbolic differences.

Briefly, the *symbolic distance effect* is that we judge differences between symbols more rapidly when they differ considerably on some symbolic dimension, such as value.

Consider the stimuli in panels A and B of Figure 3.1.

Which dot is higher? Despite its simple nature, it takes some amount of time to make this decision. The time it takes to decide which dot is higher depends on the separation of the dots; the greater the separation, the faster the response. This is the symbolic distance effect: Two stimuli can be discriminated more quickly when they differ more (Moyer & Bayer, 1976).

Now consider the bottom two illustrations. For panel C in Figure 3.1, which balloon is higher? For panel D, which yo-yo is lower?

It is probably not obvious to you at a conscious level, but when the question is "Which balloon is higher?" people's responses depend not only on the discriminability of the two heights but also on the semantic dimension needed (Banks, Clark, & Lucy, 1975). In other words, semantic knowledge that balloons are held at the bottom by strings, float up in the air, and are therefore oriented in terms of highness is a significant influence on the decision times; describing the illustrations as balloons led people to treat the pictures symbolically rather than as merely physical stimuli. When the same pictured display was accompanied by the question "Which balloon is lower?" responses were much slower. And, as you would expect, the situation was reversed when people judged items such as those in panel D, the yo-yos. "Which yo-yo is lower?" yielded faster responses than "Which yo-yo is higher?" because knowledge about yo-yos is that they hang down from their strings.

The name for this is the *semantic congruity effect* (Banks, 1977; Banks et al., 1976) in which a person's decision is faster when the dimension being judged matches or is congruent with the implied dimension. In other words, the implied dimension in the balloon illustration is height because balloons float up. When asked to judge "how high" some "high" object is, the judgment is faster because "height" is congruent with "high." Likewise, "lowness" is implied in the yo-yo display, so judging which of two "low" things is lower is also a congruent decision. Figure 3.2 displays the general form of the symbolic distance effect and the semantic congruity effect (Banks, 1977).

3.1.4: Signal Detection Theory

A fundamental method of cognitive psychology developed through research in psychophysics is *signal detection theory*, originally developed in the field of communication theory (Peterson, Birdsall, & Fox, 1954) and then applied in psychology (Tanner & Swets, 1954).

Figure 3.1 Comparison of Different Heights

Stimuli used by Banks, Fujii, and Kayra-Stuart (1976) in a study of physical and symbolic comparisons. In the top two panels, people were asked which dot is higher or lower. In the bottom two panels, people were asked which balloon is higher/lower and which yo-yo is higher/lower.

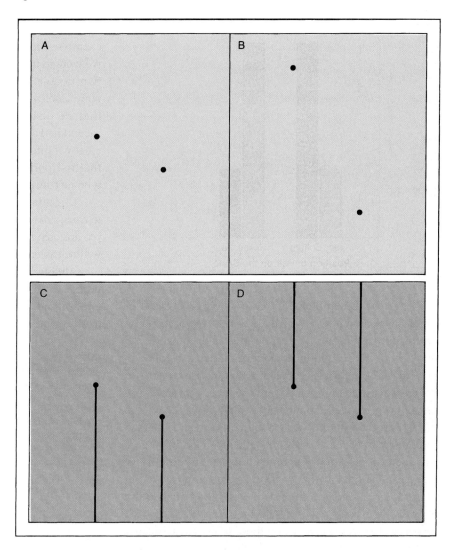

Figure 3.2 Idealized data (A) the symbolic distance effect; (B) the semantic congruity effect

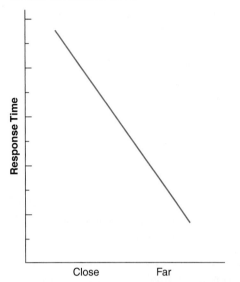

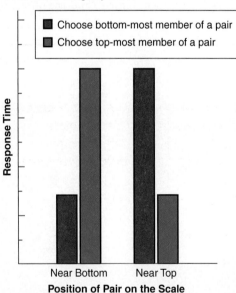

As can be seen in Table 3.2, these four types of responses can be laid out in a two-by-two grid, with type of response on one dimension, and type of stimulus on the other. If you think about it for a minute, to understand what is going on in terms of how people are making their decisions, you do not need to use all four pieces of information. That is because half of the grid is redundant. For example, if you know that on 50 trials the stimulus was actually present, and a person responded "yes" on 43 trials (hits), then you know for sure that there were 7 "no" responses (misses) on those trials. Similarly, if you know that on 50 trials the stimulus was not present, and a person responded "yes" on 12 trials (false alarms), then you know for sure that there were 38 "no" responses (correct rejections) on those trials. Thus, because

Table 3.2 The Relationship Between Different Types of Stimuli and Responses in Signal Detection Theory

When people engage in a psychophysical task, they are often making yes–no judgments, such as assessing whether a stimulus is present or not, whether a stimulus has changed or not, and so on. In these situations, there are four possible types of responses, as shown in Table 3.2.

	ITEM TYPE	
RESPONSE TYPE	True	False
Yes	Hit	False Alarm
No	Miss	Correct Rejection

The first is if a person responds "yes" in cases when that is appropriate (e.g., the stimulus is actually present). This is called a *hit*. The second is if a person responds "yes" when that is inappropriate (e.g., the stimulus is not there). This is called a *false alarm*. The third is if a person responds "no" when that is inappropriate (e.g., the stimulus is actually present). This is called a *miss*. Finally, the fourth is if a person responds "no" when that is appropriate (e.g., the stimulus is not there). This is called a *correct rejection*.

using the hit and false alarm rates would make using the miss and correct rejection rates redundant, signal detection calculations typically use only the hit and false alarm rates.

The intuitive way to analyze the data from a study might be to look simply at the hit rate, such as the number of times a person correctly identified the presence of a stimulus. However, this can be problematic. As an extreme example, suppose two people both had hit rates of 86%. If we were to use only hit rates, we might conclude that these two people were performing similarly. However, if we were also to include the false alarm rates, we might find that the first person had a false alarm rate of 5% and the second had a false alarm rate of 82%. By using both of these pieces of information, we can more correctly conclude that the first person was far more accurate than the second, who just seemed to have a bias toward saying "yes."

Using both the hit and false alarm rates, it is possible to calculate two *independent* indexes of how accurately people are using the hit and false alarm rates. This exposition will be easier to understand if you consult Figure 3.3.

Although psychophysicists doing sensation and perception research primarily developed these measures, or independent indexes, in psychology, their use has been extended to other domains of cognition, such as research on memory.

WRITING PROMPT

Usefulness of Signal Detection

What aspects of how people report memories to another (such as an experimenter) make the use of signal detection useful? How does this approach help us get at whether something is actually remembered or not, as opposed to a guess?

 The response entered here will appear in the performance dashboard and can be viewed by your instructor.

Submit

Figure 3.3 Signal and Noise Distributions

Here, the distribution on the right stands for the signal items, and the one on the left stands for the noise distribution. The difference between the two (d') represents the ease of distinguishing the signal from the noise, whereas the bias criterion (β) represents the threshold of signal strength at which a person is willing to indicate that a signal is present.

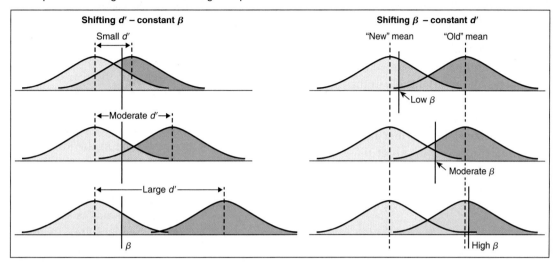

Signal Detection Measures

Here, we will look at measure of discrimination and measure of bias as two signal detection measures. These measures provide cognitive psychologists with clearer assessments of how accurate a person may be on a task.

Measure of Discrimination

The first signal detection measure that can be calculated is a measure of *discrimination*, typically d'. This measure provides information about how easy it is to discriminate between the two curves. For the second pair of curves, they are easier to discriminate, and d' would be greater than for the top set of curves. Conversely, for the third pair of curves, they are harder to discriminate and d' would be smaller.

Measure of Bias

3.2: Visual Sensation and Perception

OBJECTIVE: Distinguish the processes of visual sensation and perception

The human eye is a complex structure with several elements working together to allow vision to happen. Each of its structures plays an important role in the transduction of the proximal stimulus (the light waves entering the eye and being focused on the **retina**) into the neural impulses that result in **sensation**. Farther along the visual pathway, perceptual processes interpret these sensory signals.

Figure 3.4 illustrates the basic sensory equipment of human vision. Light waves enter the eye, are focused and inverted by the lens, and are projected onto the retina. The retina has three layers of neurons: rods and cones, bipolar cells, and ganglion cells (see Figure 3.4B).

The rods and cones are neurons stimulated by light, beginning the process of vision. Patterns of neural firing from these cells pass on to a second layer, the bipolar cells, which then collect the messages and move them along to a third layer, the ganglion cells. The axons of the ganglion cells converge at the rear of the eye, forming the bundle of fibers that makes up the optic nerve. The optic nerve signal exits the eye and continues through various structures, eventually projecting to the visual cortex of the occipital lobe in the lower rear portion of the brain.

A brief explanation is in order about how the eyes transmit information to the brain. The contralaterality principle is not as simple as "left eye to right hemisphere." Instead, each eye transmits to both hemispheres. Most important, each half of the retina gathers information from the contralateral visual field. As shown in Figure 3.5, where you are looking is your fixation point.

The left half of the retina in each eye receives images from the right visual field (the house), and the right half of each retina receives images from the left visual field (the tree). Thus, stimuli in the right visual field—the solid lines in the figure—project to the left half of the retina in both eyes, and this is then transmitted to the left hemisphere. Similarly, stimuli in the left visual field—the dotted lines—project to the right half of both retinas and are then sent to the right hemisphere.

At each step—from rods and cones to bipolar cells, then from bipolar cells to ganglion cells—there is some loss of information. Because there is a great deal of compression of information in the early stages of vision, the message that finally reaches the cortex is an already processed and summarized record of the original stimulus (Haber & Hershenson, 1973). There are about 120 million rods and about 7 million cones on each retina. Most of the cones are in a small area known as the **fovea**, which provides our most accurate, precise vision. Some of the cones in the fovea have "private" bipolar cells for relaying impulses: One cone synapses with one bipolar cell. In contrast, in peripheral vision, tens to hundreds of rods converge on a single

Figure 3.4 Basic Sensory Equipment of Human Vision

A. The structure of the human eye, fovea, optic nerve, and other structures. B. The retina, rods and cones, and ganglion cells. Based on Hothersall (1985).

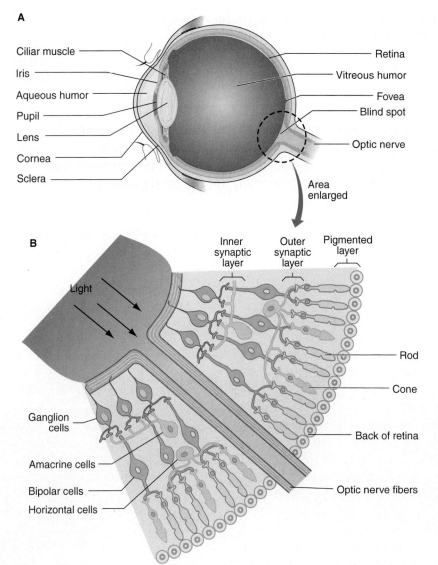

Figure 3.5 Contralaterality Principle

Binocular pathways of information flow from the eyes into the visual cortex of the brain. The patterns of stimulus-to-brain pathways demonstrate the contralaterality of the visual system.

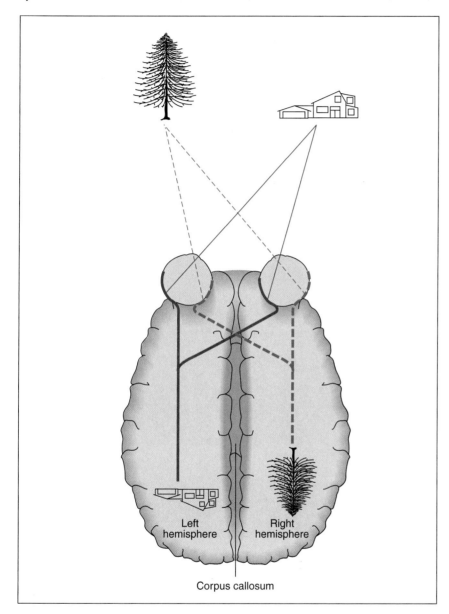

bipolar cell. Such convergence results in a loss of information because a bipolar cell cannot "know" which of its many rods triggered it. Finally, about 1 million ganglion cells combine to form the optic nerve. So, essentially, vision is compressed from 120 million bits of information down to 1 million. Despite this compression, human vision is still amazingly sensitive and acute. Like all good summaries, the visual system preserves the most useful information—the edges, the contours, and any kind of change—and omits the less useful, steady-state information.

Although much of the coverage here is on the visual processing of individual objects, we must keep in mind that sensory and perceptual processes even influence cognition based on general visual properties or impressions of a scene as a whole. For example, some evidence suggests that people experience and perceive rooms with lighter colored ceilings and/or walls (but not floors) as being higher than rooms with darker ceilings or walls (Oberfeld, Hecht, & Gamer, 2010). Thus, sensation and perception are not just about bits and pieces of a scene; ultimately, the vision system integrates all the information into a greater whole that influences cognition on a broad scale.

With regard to early visual processing in the retina and optic nerve, we are talking about *sensation*, the reception of stimulation from the environment and the encoding of it into the nervous system. Our primary interest of course is in what we do with this information. This is **perception**, the process of interpreting and understanding sensory information. As such, we need to explore the stages of visual perception. We begin with how the eye gathers information from the world, and then turn to the memory that retains that information: visual sensory memory.

3.2.1: Gathering Visual Information

To start out, let us eliminate an apparently common misunderstanding about vision. Winer, Cottrell, Gregg, Fournier, and Bica (2002) asked college students, "How does vision work?" using several variations of the task. In every variation, they found a substantial percentage of college students exhibited *extramission*, the belief that that vision involves some kind of ray or wave going out from the eyes to the object being perceived (think of the rays emanating from Superman's eyes). For instance, 69% of the students drew outward-pointing arrows on diagrams. Another 33% gave extramission responses when asked about looking at a shining lightbulb, when the correct answer should be obvious (the bulb emits light, which goes to the eyes). We correct this common misunderstanding here—vision is not the result of anything coming out from our eyes toward what we are looking at. Instead, it is triggered when the reflection of light from an object hits our eyes.

Saccades and Fixations

Here are the facts. The eyes sweep from one point to another in fast movements called **saccades** (French for "jerk," pronounced "suh-KAHD"). Pauses called **fixations** interrupt these movements.

> **Interactive**
>
> A saccade is not only quite rapid but also quite variable, taking anywhere from 25 ms to 175 ms (Rayner, 1998). Furthermore, it takes up to 200 ms just to plan and start the movement (Haber & Hershenson, 1973), and this planning involves area MT (B.A. 19) along the dorsal stream of visual processing. So, assume something in the range of 250 ms to 300 ms for an entire fixation–saccade cycle. At that rate, there are three or four visual cycles per second. Each cycle registers a separate visual scene. During the saccade, there is suppression of the normal visual processes, even those that do not involve the current scene. In essence, some types of thinking stop when we move our eyes. For example, if given a task of rotating mental images, people cease rotating the images during a saccade (Irwin & Brockmole, 2004; see Irwin, 2014, for a similar disruption in cognition because of eye blinks).

A subtler misbelief is that we take in visual information in a continuous fashion whenever our eyes are open. After all, our visual experience is of a connected, coherent visual scene that we scan and examine at will. However, this is largely an illusion. If you watch people's eyes as they read, you will see that their eyes do not sweep smoothly across the page. Instead, they jerk, bit by bit, with pauses between successive movements.

Several studies have shown that we sometimes fail to see an object we are looking at directly because our attention is directed elsewhere (Mack, 2003); this **inattention blindness** is due, in some sense, to our lack of attention to an object. In a dramatic demonstration of this, Haines (1991) tested experienced pilots in flight simulators. A few of them proceeded to land the simulator, paying attention to the gauges and dials on the instrument panel yet failing to notice that another airplane was blocking the runway (see Mack & Rock, 1998).

> **Prove It**
>
> Yogi Berra supposedly once said something to the effect that "you can observe a lot just by watching." Very little of the evidence you are reading about in this module can be observed easily without specialized apparatus. For instance, you need to present visual stimuli in a highly controlled, precise fashion. However, you can make some important and revealing observations just by watching someone's eyes (see Figure 3.6).

Figure 3.6 Tracking Eye Movements

Get close to a friend and watch as he or she reads a passage of text silently and as he or she looks at a photo or drawing, maybe something like the photo in Figure 3.6.

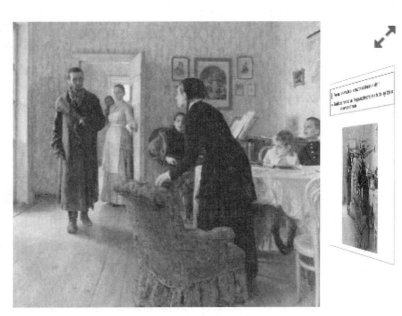

3.2.2: Synesthesia

Most people experience a fairly stable mapping of sensory experiences. For example, seeing a letter elicits a perception of that letter, and that is all. That said, some people have a condition known as *synesthesia*. A person with synesthesia will have inappropriate and involuntary sensory experiences in addition to typical sensory experience (see Grossenbacher & Lovelace, 2001; Hochel & Milán, 2008; Hubbard & Ramachandran, 2005; Rich & Mattingly, 2002, for reviews). For example, a person may experience seeing different colors when hearing different people's voices, or may have certain taste experiences when feeling different textures. For some people with synesthesia (called projectors), their additional sensory qualia are experienced as if they were out there in the world; whereas for others (called associators), their additional experiences are in their mind, but not necessarily out there in the world (Dixon, Smilek, & Merikle, 2004). Perhaps the most common form of synesthesia is when a person reports experiencing colors (photisms) when reading words or numbers. This is called grapheme–color synesthesia.

Most people with synesthesia (called synesthetes) either are born this way or develop the condition early on. If you are not a synesthete, you might think of this as a bewildering and confusing way to experience the world. However, synesthetes do not experience such distraction or confusion. Keep in mind that their additional sensory experiences of the world have always been there! Many synesthetes do not even realize that they have synesthesia until high school or college when they figure out that most other people are *not* having those additional experiences. An analogy for what synesthesia may be like is to imagine that you are person with color vision in a world in which most people are color blind. You would be receiving additional sensory and perceptual information that most others do not. Although you would not find it at all confusing, and might even find it helpful, your accounts of your experiences would baffle others.

How does the condition of synesthesia come about? There are two primary explanations for this phenomenon. First, there is the idea that synesthesia is caused by a lower ability to suppress inappropriate feedback loops in perceptual processing (Grossenbacher & Lovelace, 2001). That is, sensory signals that would normally be dampened are allowed to spread, resulting in additional sensory experiences under certain circumstances. The second explanation is that there is an incomplete pruning of extra cortical connections during development (Maurer, 1997). The normal pruning of inappropriate and unnecessary neural connections as infants mature does not happen in some synesthetes.

In addition to the differences in sensory and perceptual experiences, synesthesia may influence other aspects of cognition. For example, some work with grapheme–color synesthetes has shown that memory for materials, such as lists of words, that elicit the synesthetic experience is superior to that of normal controls, either through subjective reports or through experimental verification (Luria, 1968; Mills, Innis, Westendorf, Owsianiecki, & McDonald, 2006; Radvansky, Gibson, & McNerney, 2011; Smilek, Dixon, Cudahy, & Merikle, 2002; Ward, 2008; Yaro & Ward, 2007). Thus, the heightened sensory-perceptual experiences of synesthesia can actually improve some kinds of memory.

3.2.3: Visual Sensory Memory

We turn our attention now to **visual sensory memory**. Because this memory system has such a short duration, we have few useful intuitions about its operation. Although our primary concern here is with the normal operation of visual sensory memory, unusual circumstances can give us some clues. We begin with such a circumstance.

Everyone has seen a flash of lightning. Think about that for a moment, and guess what the duration of a bolt of lightning is. Most people guess that the flash of light lasts a little more than a half second or so, or maybe closer to a whole second. If your estimate was in this neighborhood, then it is reasonable—but wrong. The bolt of lightning is actually three or four separate bolts. Each one lasts about 1 millisecond (ms), and there is a separation of about 50 ms between bolts. Thus the entire lightning strike lasts no more than about 2/10 of a second, or 200 ms, and is composed of several individual flashes (Trigg & Lerner, 1981).

Our perception of lightning is a mental event that reflects visual persistence.

Given that an estimate of a half second to a second was so off, what was reasonable about it? It was the perception of a flash of light that extended in time. This phenomenon is **visual persistence**, the apparent persistence of a visual stimulus beyond its physical duration. The neural activity on the retina that the lightning flash causes does not outlast the flash itself. The eye does not continue to send "lightning" messages after the flash is over. Your perception of the lightning is a mental event that reflects visual persistence in that you perceive a lighted scene that then quickly fades away. Because any persistence of information beyond its physical duration defines the term *memory*, the processes of visual perception (as opposed to sensation) must begin with memory, a temporary visual buffer that holds visual information for brief periods. This type of memory is *visual sensory memory* or **iconic memory** (Neisser, 1967).

AMOUNT AND DURATION OF STORAGE Sperling and his coworkers (Averbach & Sperling, 1961; Sperling, 1960) reported the characteristics and processes of visual sensory memory in their classic research.

See Figure 3.7 for a schematic diagram of a typical trial.

INTERFERENCE A related series of experiments explored the loss of information from iconic memory. The original research suggested that forgetting was a passive process like fading or **decay**; that is, the mere passage of time degraded the icon. However, in normal vision we look around continuously, shifting our gaze from one thing to another. What happens to iconic memory when one visual scene immediately follows another? The answer, in short,

Figure 3.7 Schematic of Sperling's Classic Study of Iconic Memory

Sperling presented a visual stimulus for a controlled period, usually in milliseconds, to study "the information available in brief visual presentations," the title of his paper in 1960.

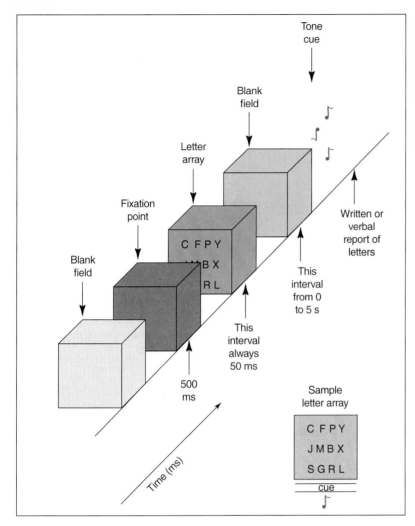

is **interference**, forgetting caused by the effects of intervening stimulation or mental processing.

3.2.4: The Early Parts of a Fixation

The evidence reported by Sperling, Averbach and Coriell, and many others, led cognitive psychology to propose that iconic memory existed and that it was an early phase in visual perception (Neisser, 1967). However, later evidence shows results that are even more fascinating. A study by Rayner, Inhoff, Morrison, Slowiaczek, and Bertera (1981) examined performance during reading, using an eye tracker for precise timing measurements. After people had fixated a word for a mere 50 ms, the word was replaced with an irrelevant stimulus, which remained in view for another 175 ms to fill up the rest of the fixation time. Surprisingly, this did not affect reading at all—people often did not notice that the word had even changed!

This is very important, so stop and think about it for a moment. Despite what it feels like, as you read these words on this page, you are not viewing them continuously. You are seeing them in brief bursts, extracting information quickly and then devoting mental energy to processing them further, unaware of your "downtime" during the fixation and of your blindness during the following saccade.

Several investigators have also collected evidence on what might be called "dynamic icons," that is, iconic images that contain movement (see Finke & Freyd, 1985; Irwin, 1991, 1992; Loftus & Hanna, 1989). Treisman, Russell, and Green (1975) presented a brief (100 ms) display of six moving dots and asked people to report the direction of movement. Partial report performance was superior to whole report performance; it is more accurate. It starts out higher and declines over time. In short, the moving images of the dots were decaying just as the static letter grid had in Sperling's procedures. Visual perception, therefore, is not a process of flipping through successive snapshots, with three or four snapshots per second. Instead, it is a process of focusing on the visually attended elements of successive fixations, where each fixation encodes a dynamic segment of the visual environment.

In fact, integration across brief intervals of time can occur even without eye movements. Loftus and Irwin (1998) showed that temporal integration—perceiving two separate events as if they had occurred at the same time—transpires seamlessly when visual events occur within about 20 ms of each other. This happens without any conscious awareness that the two separate events have occurred. Events separated by 40 ms or more, or separate events that last for 40 ms or more, tend not to be integrated as completely, however. With these longer durations, people can more easily detect that two separate events happened rather than just one.

3.2.5: Visual Attention

How do all these different results make sense: the wholesale input of visual stimulation; the persistence, decay, interference, and integration of information; and the concept of visual attention? The entire sequence of encoding visual information—selecting part of it for further processing, planning subsequent eye movements, and so on—is very active and rapid. The visual continuity we experience—our feeling that we see continuously—is due to the

A Study by Averbach and Coriell on Interference

One study on interference was by Averbach and Coriell (1961; reprinted in Coltheart, 1973).

> **Study**
>
> Averbach and Coriell presented a display of two rows of letters, eight letters per row, for 50 ms. A blank white postexposure field, varying in duration, followed the display and then was followed by a partial report cue. Unlike Sperling, Averbach and Coriell used a visual cue, either a vertical bar marker or a circle marker. The bar marker was just above (or below) the position of the to-be-reported letter, and the circle marker surrounded the position where the to-be-reported letter had just been.

constant updating of visual sensory memory and to our focus on attended information. As we attend to a visual stimulus, we are examining the readout from iconic memory. In the meantime, a new visual scene is being registered in iconic memory. Our mental processes then pick up the thread of the newly registered scene, providing a smooth transition from one display to the next.

Focal attention was Neisser's (1967) term for this mental process of visual attention, such as the mental redirection of attention when the partial report cue is presented. It seems that focal attention, or simply visual attention, helps bridge between successive scenes in visual sensory memory. This prevents us from sensing the blank space of time occupied by the eye's saccades by directing focal attention to elements of the icon. Although we sense a great deal of visual information, what we perceive is the part of a visual scene selected for focal, visual attention. To exaggerate a bit, what you perceive right now is not the printed screen in front of you, but the processed and attended portions that were registered in sensory memory, your iconic trace, as processed by visual attention.

3.2.6: Trans-saccadic Memory

To build up an understanding of the visual world, we need to move our eyes, head, and body, gathering visual information across each successive fixation (see Higgins & Rayner, 2015, for a review). First, it is important to point out that vision is not continuously processing input from the eyes. Instead, during a saccadic eye movement, the processing of visual information is suppressed, which is why you do not see a blur from your eyes moving. This is saccadic suppression.

Next, we turn to the question of how we put the information from all of these fixations together. This is done using iconic memory in a way known as **trans-saccadic memory** (e.g., Irwin, 1996), the memory that is used across a series of eye movements. How does iconic memory track information to figure out how to put together information from different fixations? It does not use retinal coordinates (where the images fall on the eyes) or spatial coordinates (where things are in space) to do this (Irwin, Yantis, & Jonides, 1983). Instead, trans-saccadic memory works by using object files (Kahneman, Triesman, & Gibbs, 1992), which are iconic representations of individual objects used to track what is going on in the world. Evidence for their use comes from studies that ask people to detect changes in objects after a saccadic eye movement (Henderson & Anes, 1994)—for example, detecting whether a letter changed to a plus sign in a display. In general, people's responses are fairly accurate in noticing changes in objects they focused on. This does not occur for all objects in a visual scene, but only those to which people attend.

To accomplish trans-saccadic memory and integration effectively, our brain assumes that everything that we are not attending to is more or less stable, which is why we may miss those changes. Moreover, consistent with the theme that some of cognition is future oriented, some evidence suggests that trans-saccadic memory is predictive. That is, memory is, to some degree, predicting that the world is likely to be stable as a default assumption (Higgins & Rayner, 2015). What is noticed and processed further are those circumstances that violate this prediction of stability; hence, the world is different from what it had been before, which may result in visual attention capture.

WRITING PROMPT
Perception as an Active Process

How can perception be thought of as an active process? Are there any examples that you can think of to back up your ideas?

▶ The response entered here will appear in the performance dashboard and can be viewed by your instructor.

Submit

3.3: Pattern Recognition
OBJECTIVE: Evaluate the process by which we identify patterns and objects

We turn now to one of the most intriguing and debated topics in visual perception, the identification of patterns and objects. As you will see, pattern recognition does not occur instantly, although it does happen automatically and spontaneously. Perceptual pattern recognition is, in many ways, a problem-solving process, with much of the mental work occurring subconsciously and very rapidly. Essentially, during perception a person needs to identify the nature of the distal objects in the world based on the proximal images reaching the retina. Often these images are compromised in some way, such as being occluded by another object or being against a complex visual background. Vision parses the visual image in a number of ways to extract information about the objects that are actually present, and it follows a number of perceptual principles in doing so.

3.3.1: Gestalt Grouping Principles
Perhaps the best known and established of these perceptual principles are the **Gestalt** grouping principles laid out by the Gestalt psychologists in the early to mid-20th century. Although the roots of Gestalt psychology are in the 20th century, it still has important implications today (for a companion paper review, see Wagemans, Elder, et al., 2012; Wagemans, Feldman, et al., 2012). These principles identify those characteristics of perception in which ambiguities in a stimulus are resolved to help determine which objects are present. They are also aimed more at processing information about the whole of an object, rather than simply, and only, building up a mental representation from more basic elements. In fact, obtaining a Gestalt for a perceptual whole of an object may actually disrupt the perception of the parts of an object (Poljac, de-Wit, & Wagemans, 2012). fMRI neuroimaging work has supported the Gestalt-based processing of objects (Kubilius, Wagemans, & Op de Beeck, 2011). This neuroimaging has shown that some parts of the brain, namely the lateral occipital lobe (B.A. 19) and the posterior fusiform gyrus (B.A. 37), are involved in processing whole objects, apart from the occipital brain areas involved in processing individual elements (see the section on feature detection later).

FIGURE-GROUND PRINCIPLE One of these Gestalt principles is the **figure-ground** principle. When viewing an image, part of the image is treated as the figure or foreground (the object identified), which is segregated from the visual information upon which it is set (the background).

Classic examples illustrating difficulties in determining figure-ground are reversible figures, such as the one shown in Figure 3.8 in which a person shifts back and forth between what is the foregrounded object and what is the background. At one moment it might be two faces, whereas at the next it might be a vase.

CLOSURE AND OTHER PRINCIPLES The aim of some Gestalt grouping principles is to provide a more complete percept from an incoming image that may be fragmentary

Figure 3.8 Figure-Ground

An illustration of the Gestalt figure-ground perceptual grouping principle.

Figure 3.9 Gestalt Grouping Principles

Illustration of the Gestalt perceptual grouping principles of proximity (A), similarity (B), closure (C), and good continuation (D).

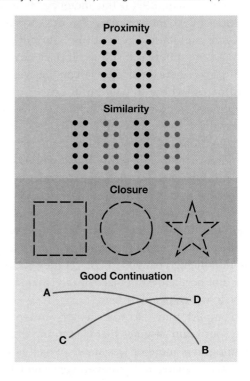

or incomplete. They follow the principle of **closure**, in which a person "closes up" an image that has gaps or parts missing, perhaps because they are being occluded (blocked) by some other object.

An example of closure can be seen in Figure 3.9C.

Although there is just a collection of dashed lines, they can be joined using the Gestalt principle of closure to give the impression of a circle and a square.

In the Gestalt principle of **proximity**, elements that are near to one another tend to be grouped together. This is shown in Figure 3.9A. Because of this principle, you see a flock of geese in flight as forming a classic V. Similarly, you may perceive groups of dots moving together as a person walking (Johansson, 1973).

Another Gestalt principle is **similarity**, in which elements that are visually similar in some way, such as having a similar color or texture, tend to be grouped together. This is shown in Figure 3.9B.

Similarity is seeing the individual dots on a television or computer screen as being part of the same object if they have a similar color or visual texture. Certain Gestalt grouping principles take into account some form of trajectory. In some cases, the trajectory is the edge of an object. The principle of **good continuation** assumes that when an edge is interrupted, people assume that it continues along in a regular fashion. In the example shown in Figure 3.9D people tend to organize this as lines from A to B and from C to D, not A to D, B to C, and so on.

Finally, with **common fate**, entities that move together are grouped together. For example, when an animal moves in the forest, it is easier to spot than if it remains motionless. The movement allows perception to group those points together because they are moving together.

Although we use Gestalt grouping principles to some degree, we need to go deeper to understand how to recognize a visual stimulus as a familiar pattern. How does cognition manage to input a visual stimulus such as *G* or a tree and end up recognizing it as familiar and meaningful? How do we recognize patterns of handwriting, or different printed fonts, despite incredible variability? The following sections present some ideas about how this occurs, by looking at the case of written language.

3.3.2: The Template Approach

As Neisser (1967) pointed out, pattern recognition would be simplified, though still thorny, if all the equivalent patterns we saw were identical. That is, it would be easier to explain pattern recognition if there were one and only one way for the capital letter *G* to appear. However, the visual environment is not like this. An enormous variety of visual patterns, in countless combinations of orientation and size, can all be categorized as the capital letter *G*, and likewise for all other letters, figures, shapes, and so on.

The use of **templates**, stored models of all categorizable patterns, may help with pattern recognition. When the computer at your bank reads your checking account number, it is using a template matching process, making physical identity matches between the numbers on your check and its stored templates for the digits 0 through 9. When the computer recognizes a pattern, it has matched it to one of its stored templates (think also of bar codes).

The template approach has simplicity and economy on its side. However, beyond this simplicity and economy, the template approach has little else to recommend it, and it has some serious flaws. We have already covered the primary reason, the enormous variability in the patterns that we can recognize. Also, just think of how long it would take you to learn the infinite number of possible patterns (for all of the objects in the world, the different orientations they can be in, the various distances they can be from you, etc.) and then search through those patterns in memory. Would you have time left for anything else?

3.3.3: Visual Feature Detection

An improvement over the template approach is **feature analysis** or **feature detection**. A feature is a very simple visual element that can appear in combination with other features across a variety of stimulus patterns. A good example of a visual feature might be a straight horizontal line, which appears in capital letters such as *A*, *G*, *H*, and *L*; others would be vertical or diagonal lines, curves, and so on. In general, feature theories claim that we recognize patterns by first identifying their features. Rather than matching an entire template-like pattern for capital *G*, we identify the elemental features that are present in the *G*. When we detect "circle opening right" and "straight horizontal" segments, the features match with those stored in memory for capital *G*.

Several investigators have proposed theories of feature-based pattern recognition. We will discuss one such model, Pandemonium, in detail. Understanding Pandemonium will also help you understand the reasons behind interactive, connectionist approaches.

PANDEMONIUM Selfridge (1959), an early advocate of feature detection, described a model of pattern recognition called **Pandemonium**; an illustration of the model is shown in Figure 3.10. This model has many similarities to the connectionist models that will be discussed later.

In Selfridge's imaginative description, the mechanisms involved are little mental demons that shout aloud as they attempt to identify patterns. As the figure shows, a set of data or image demons encodes the pattern. Next, the computational demons begin to act. They are the feature analyzers in Selfridge's model; each one has a single simple feature it is trying to match in the stimulus pattern. For instance, one demon might be trying to match a simple horizontal line, another a vertical line, another a curve opening to the right, and so on. When a computational demon matches a stimulus feature, it begins to shout excitedly.

At the next level up, listening to all this shouting is a set of cognitive demons. These demons represent the different letters of the alphabet, one for each letter. Each one is listening for a particular combination of demons: For instance, the G demon is listening for the "open curve" and the "horizontal bar" feature analyzers or demons to shout. Any evidence from the computational demons that suggests a match with the stimulus also causes a cognitive demon to begin shouting: Based on the feature analysis evidence, it thinks that it is the matching pattern. Several of the cognitive demons will be shouting at once because several letters usually share some features (e.g., C and G). Thus, the one that shouts the loudest is the one

Figure 3.10 Selfridge's Pandemonium Model

Data or image demons encode the visual pattern. Computational demons try to match the simple features present in the pattern. Cognitive demons represent the combination of features that are present in different letters of the alphabet; each tries to match the several computational demons that match the stimulus input. Finally, the decision demon identifies the pattern by selecting the loudest cognitive demon, the one whose features most nearly match the encoded pattern. Adapted from Selfridge (1959).

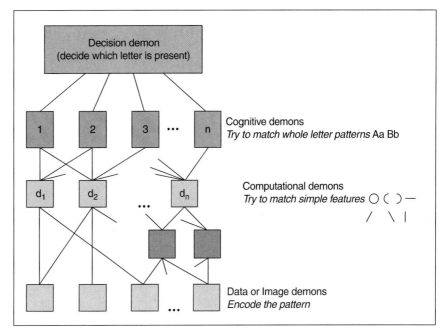

whose pattern most nearly matches the stimulus. The decision demon hears the loudest cognitive demon and has the final say in recognizing and categorizing the pattern.

THREE IMPORTANT IDEAS The utility of our thinking about perception as a process that captures some of the stages of cognition highlights a number of important points. Three of these ideas are feature detection, parallel processing, and problem solving.

Thinking about how we interact with the world further emphasizes the inadequacy of template approaches, and the advantage of feature-based approaches, for visual feature detection. Specifically, humans are often in motion—walking, running, driving, and so on. Moreover, the things we are looking at also are often moving. Thus, our views of objects can be constantly shifting. The area of the brain that leads us to see objects as constant and stable across such movement and different viewing angles is area MT (B.A. 19), part of the dorsal stream of visual processing, at the conjunction of the occipital, parietal, and temporal lobes. This area is also involved in the perception of apparent movement (Larsen & Bundesen, 2009), as well as real movement, when cognition must make inferences about what goes on in between. For example, watching a movie or television involves a series of still pictures presented to you in rapid fashion. The pictures change faster than your iconic memory can decay, replacing one another. Your brain fills in any jumps in position, producing the illusion of motion. This mental perception of illusory motion is **beta movement** (Wertheimer, 1912). A related perceptual illusion occurs when you see lights moving or flowing around on a movie marquee or chasing Christmas lights. This is the **phi phenomenon** (Wertheimer, 1912).

Selfridge's Model for Pattern Recognition

Aside from the vividness of the model's description of scores of shouting demons producing a noisy Pandemonium, Selfridge's model incorporates several important ideas about pattern recognition.

Feature Detection

First, it is a feature detection model. The demons detect and report elementary, simple features. There are several related lines of evidence for feature detection in visual pattern recognition (e.g., Pritchard, 1961). Especially convincing are neurophysiological studies showing that special visual cortex cells exist for simple visual features. These are found in area V1 (B.A. 17) of the occipital lobe (the "V" stands for "visual"). The most widely known evidence of this kind comes from the pioneering research of Hubel and Wiesel (1962), for which they won a Nobel Prize. Using electrode implant procedures, these researchers found neurons in cats' brains that respond only to vertical lines, others that respond only to diagonals, and so on. On the assumption that the human brain is not radically different from a cat's brain (at least for vision), this suggests that feature detection may even have a physiological status in the nervous system. Furthermore, it means that psychological theories must be compatible with this neuro-logical evidence.

Parallel Processing

Problem Solving

Essentially, when iconic memory receives visual images in relatively close proximity in space and time, it will infer virtual movement.[1] Beta movement occurs when making inferences from one picture to the next, as in a movie, but phi movement involves illusory tracking of an object in space.

WRITING PROMPT

Perception Likened to the Pandemonium Model

Why would perception use a visual system that might be something like the Pandemonium model? What are the benefits of this approach? Discuss how we receive information from the world, and about how the world actually is.

▶ The response entered here will appear in the performance dashboard and can be viewed by your instructor.

Submit

3.4: Top-Down Processing

OBJECTIVE: Explore ideas about how what we know influences what we see

The typical way that most people think about vision is that there is stuff out there in the world and we see what it is. This is the bottom-up, data-driven view of visual processing. The information in the world, the data, drives what we perceive as being out there. Although this is true, it is only half the story. Because the information in the world that reaches us may be fragmentary or incomplete, cognition needs to fill in what is missing. The topic of this section is the filling in of that information.

3.4.1: Conceptually Driven Pattern Recognition

Even Selfridge knew that Pandemonium was missing an important ingredient. Basically, Pandemonium is a completely **bottom-up, data-driven processing** system in which processing is driven by the stimulus pattern, the incoming data. The patterns to be recognized came in to the image demons at the bottom and then were processed at higher and higher levels until the top-level demon finally recognized the pattern. Yet, Selfridge presented examples like those shown in Figure 3.11, showing that **context** influences pattern recognition.

How adequate is the bottom-up approach as a sole explanation of visual pattern recognition? Did you "see" the words *went* and *event* despite the fact that they are written identically in the two sentences? So what was the missing ingredient in Selfridge's model?

The missing ingredient was context and a person's expectations. Such effects are called top-down or conceptually driven effects, in which context and higher-level knowledge influence lower-level processes (remember "typograpical"?). In Figure 3.11 your knowledge of English words and your understanding of the meaning of sentences lead you to perceive what is written as either *went* or *event*, depending on the context.

Note that top-down, conceptually driven processing can be influenced not only by things such as knowledge of a language or the context of a sentence. Even one's culture can influence perception to some degree. For example, evidence suggests that when Westerners view a scene, they are more likely to focus perceptual processes on a foreground object and its visual properties. Eastern Asians, by comparison, are more likely to focus on the context and the background of the scene, taking into account the relations

---

[1] The phi phenomenon was important in the establishment of automated railroad crossings. Years ago, important railroad crossings had a railroad employee who would swing a lantern to warn vehicles that they were approaching a crossing. Then automated crossings were designed with two lights that would blink at the appropriate rate to produce the phi phenomenon and produce the perception of a swinging lamp. However, when the first automated crossings were built, the timing was off–it just looked like two lights alternating on and off. Nevertheless, the railroad companies stuck with that timing.

Figure 3.11 Context and Pattern Recognition

As an illustration of the power of top-down processing, consider these two sentences. If you look carefully, the way the word *went* is written in the first sentence is identical to the way the word *event* is written in the second sentence. Although they are perceptually identical, your expectations based on the context bias you to see the beginning part as either a *w* or an *e* and a *v*.

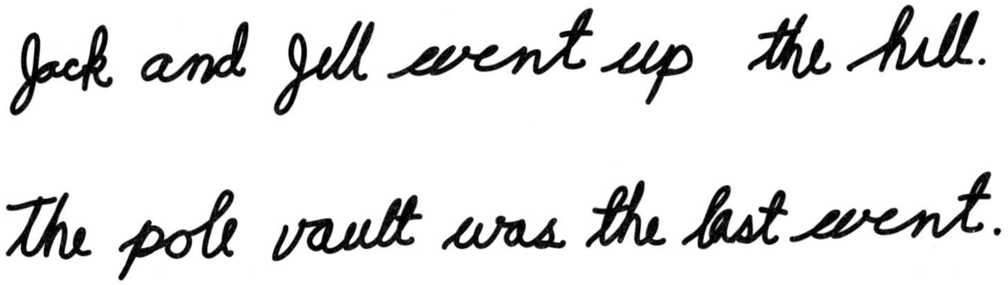

Jack and Jill event up the hill.

The pole vault was the last event.

Evidences for Conceptually Driven Processing (Figure 3.12)

Let us examine evidence that not only supports the feature theory approach but also makes the case for conceptually driven processing.

Visual Search

In Neisser's (1964) classic work on visual search, people saw pages of characters, 50 lines of printed letters, with four to six letters per line. One task was to scan the page as rapidly as possible to find the single occurrence of a prespecified letter (in another task, people had to find the line without a certain letter). As an illustration, do the visual searches presented in Figure 3.12, on the next slide, timing yourself as you find the targets.

among the various elements (e.g., Nisbett & Masuda, 2003). Evidence of different levels of neural activity in perceptual processing regions of the brain supports this conclusion (Gutchess, Welsh, Boduroglu, & Park, 2006). Contextual effects of top-down processing are not limited to linguistic materials. For example, Gartus, Klemer, and Leder (2015) reported that artistic judgments of modern art were higher when the art was embedded in a museum context than in a street context (although judgments of graffiti were similar in both contexts).

Conceptually driven and data-driven processes work in combination in most situations involving pattern recognition. An excellent way to model this, to explore how this combination works, is within the connectionist model. Think of this model as Pandemonium Plus, a bottom-up model like Selfridge's with an added top-down processing effect.

3.4.2: Connectionist Modeling

Connectionist modeling is a computational approach often used in cognitive science. Connectionist models involve a massive number of mathematical computations. Essentially, each unit in a connectionist layer connects with many or all of the units in the next layer. The impact of each experience on each of these connections needs to be computed. Even if the number of units in a layer is fairly small, the number of separate computations in a single run of the model is staggeringly large because of very large number of connections among the units.

To flesh out the word recognition model (McClelland & Rumelhart, 1981; Rumelhart & McClelland, 1986), we will use a model that recognizes four-letter words, such as *tree*.

The connectionist approach to modeling and understanding cognition differs in a number of ways from more

Basic Statement of Parallel Distributed Processing Principles

We start with certain basics of connectionist modeling, including some of the vocabulary.

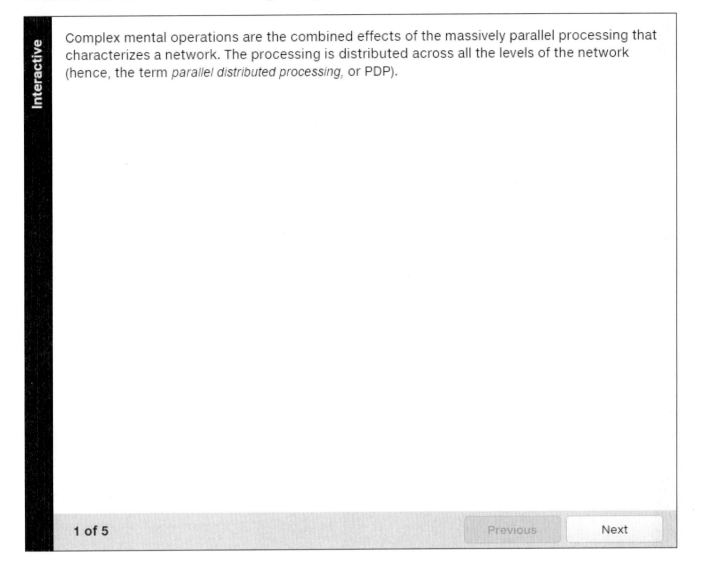

Complex mental operations are the combined effects of the massively parallel processing that characterizes a network. The processing is distributed across all the levels of the network (hence, the term *parallel distributed processing,* or PDP).

traditional and psychological approaches. Connectionism comes more from a computer science approach to issues of the mind and thinking. Because of this, certain sets of terms may be unfamiliar, but you will need to learn them to understand this approach to perceptual processes.

TOP-DOWN AND BOTTOM-UP INFLUENCES IN A CONNECTIONIST MODEL It is important to distinguish between top-down and bottom-up effects in the connectionist model. In this section, we will look at the connectionist modeling of four-letter words to better understand top-down and bottom-up influences.

For simplicity, the figure shows only a handful of four-letter words. Note, however, that three of the word-level units are consistent with the letter detection performed on *T* in the first position; that is, three of the words begin with a *T*. Now think about the fuller model that identifies four-letter words. Each of the four input unit segments performs as described earlier, forwarding both activation and inhibition to the **hidden units**, which in turn forward activation and inhibition to the **output units**. At the end, one of the several output units will have received enough activation to exceed its threshold. When this happens, that unit responds by answering the question "What is this word?"

There is one more complexity needed to get top-down effects into the model. Reflect for a moment on how likely the spelling pattern *TZ* is at the beginning of English words. Not very likely, is it? On the other hand, *TA, TE, TI,* and other consonant–vowel pairs are likely, as are a few consonant–consonant pairs such as *TH* and *TR.* The network displays these likelihoods; to distinguish them visually from the other connections in the figure, they are shown with curve-shaped connections. The overall effect

Table 3.3 A Primer of Connectionist Terminology
Below is a list of basic terms and some explanations to help you.

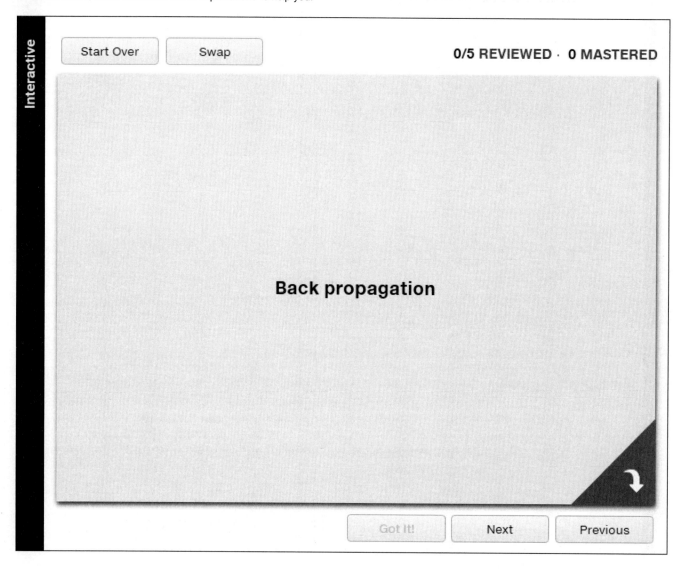

of these letter-to-letter weights is that the activations in the system can make up for missing features at the perceptual level.

Figure 3.14 shows the final levels of activation for three possible words, given the partially obscured stimulus pattern at the bottom.

This shows an important feature of connectionist models: Enough knowledge is represented in the system, in the weights for letter-to-letter sequences, that the model identifies the word *fork* even when the last letter could also be an *R*.

This is important because it illustrates the general theme of top-down, conceptually driven processing. If you saw the partially obscured pattern in Figure 3.14, you would identify the word as *fork*, based on your knowledge that *forr* is not an English word. Your knowledge of English assists your perception. This is what happens in the connectionist model; higher-level knowledge, coded as weighted connections in the network, participates in the lower-level task of identifying letters.

Note that the weights of the connections change with experience. Every experience that you have changes you because of the changes in the connections among the neurons, even if in very subtle ways. For example, a study by Hussain, Sekuler, and Bennett (2011) presented people with long series of faces or patterns. The task was to select which of eight options they had seen before. As people progressed through the task, their performance improved—they gained expertise in this kind of perceptual processing. When they had started, their accuracy was around 20% to 40% correct, but after two days of practice, their accuracy increased to around 60% to 80%. Importantly, this change was not temporary. When the researchers asked the same people to do the task over a year later, their performance

Connectionist Modeling of Four-Letter Words (Figure 3.13A–Figure 3.13C)

Let us build the simple model for recognizing four-letter words using connectionist modeling. piece by piece. Here there are three levels of units.

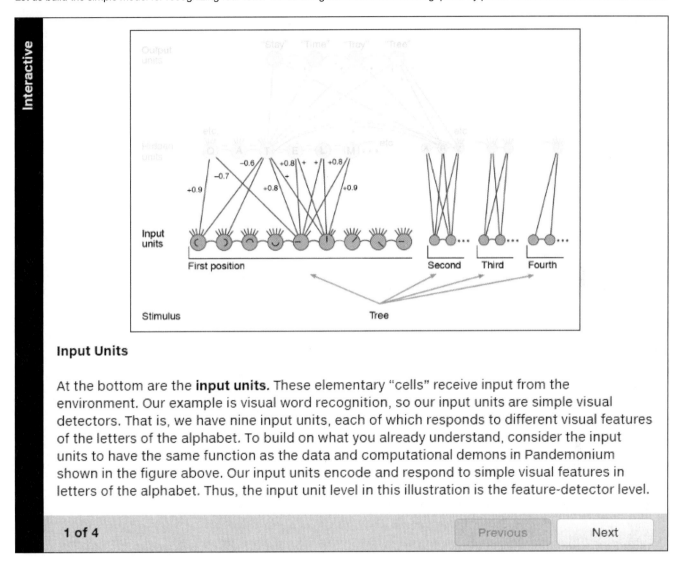

Input Units

At the bottom are the **input units.** These elementary "cells" receive input from the environment. Our example is visual word recognition, so our input units are simple visual detectors. That is, we have nine input units, each of which responds to different visual features of the letters of the alphabet. To build on what you already understand, consider the input units to have the same function as the data and computational demons in Pandemonium shown in the figure above. Our input units encode and respond to simple visual features in letters of the alphabet. Thus, the input unit level in this illustration is the feature-detector level.

1 of 4

Figure 3.14 Connectionist Model of Word Recognition and the Resulting Activations

A possible display that might be presented in the connectionist model of word recognition and the resulting activations of selected letter and word units. The letter units are for the letters indicated in the fourth position of a four-letter display. Based on Rumelhart & McClelland (1986).

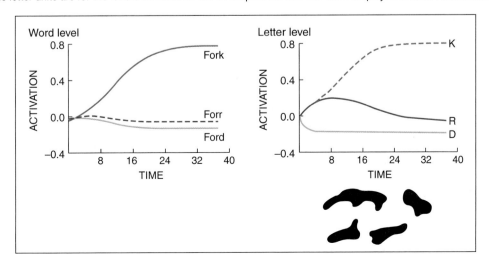

only dropped 10% to 20% for items they had seen before, and a bit more for new items. The point is that they did not have to start back at the beginning. The experience of being in this study changed the connections between their neurons that persisted for over a year (and probably longer).

Such connectionist models satisfy a difficulty you read about earlier: the need for top-down processing. In Figure 3.13, the top-down effect is prominent in the curved connections, which represent mutual excitation and inhibition. As you will read at several points in this text, connectionist accounts of a whole range of processes can provide new insights into ways of modeling and understanding human cognition. Indeed, connectionist models are finding applications in a stunningly large number of fields (e.g., see Corder, 2004, on a neural net application to landing a crippled airliner).

WRITING PROMPT

Neural Nets Versus Hard-Coding Solutions

What are the advantages of neural nets over hard-coding solutions to problems, both perceptual and otherwise? Briefly outline some of these advantages.

▶ The response entered here will appear in the performance dashboard and can be viewed by your instructor.

Submit

3.5: Object Recognition and Agnosia

OBJECTIVE: Summarize the science of object recognition and agnosia

How does this approach to identifying letters and words extend to other objects, such as recognizing a tree, a briefcase, a human face, or a knife hidden in a carry-on bag going through airport security (McCarley, Kramer, Wickens, Vidoni, & Boot, 2004; Smith, Redford, Washburn, & Taglialatela, 2005)? Some of the most significant work on the topic of object recognition involves a process very similar to the feature detection ideas you have been studying.

3.5.1: Recognition by Components

The idea in Biederman's (1987, 1990) recognition by components (RBC) theory is that we recognize objects by breaking them down into their parts, and then look up this combination in memory to see which object matches it. Here, pattern recognition has a small number of basic "primitives," simple three-dimensional geometric forms like those shown in Figure 3.15.

These forms, called **geons**, are a combined form of geometric ions (remember ions from chemistry?). Recognizing a briefcase, for example, involves analyzing the object into its two geons, the rectangular box (geon 2 in the figure) and the curved cylinder (geon 5). By itself, the rectangular box geon would match the memory for brick or box. But when that component and the curved part on top are detected, the combination matches the memory for briefcase or suitcase.

Biederman (1987) argued that mental representations of three-dimensional objects are composed of geons, much as written language is composed of letters, combined and recombined in different ways. Thus, when we recognize objects, we break them down (*parse* is the technical term) into their components and note where the components join together. We match this pattern to information stored in memory to yield recognition. Two aspects of these patterns are particularly important. First, we find the edges of objects. This enables us to determine which edges maintain the same relationships to one another regardless of viewing orientation. However you look at a brick, the two long edges that are visible remain parallel to one another.

Second, we scan regions of the pattern where the lines intersect (vertices), usually places that form deep concave angles. Look at the deep concave angles on the briefcase in Figure 3.15 where the curved component joins the rectangle. Examining the edges and the areas of intersection

Figure 3.15 Primitives in Pattern Recognition

Geons (components) and the objects they make.

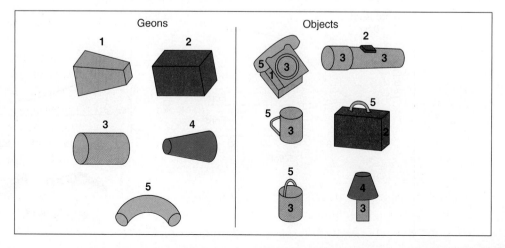

enables us to determine which components are present in the pattern: rectangular solid joined on the upper surface by a curved segment. Then we compare this description with stored descriptions in memory, something like "briefcase: rectangular solid joined on the upper surface by a curved segment." When we find a match between the identified components and the stored representation, we recognize the pattern.

EVIDENCE FOR RBC In his investigations of the RBC model, Biederman (1987, 1990) discovered several facts about object recognition. For one, the emphasis on the importance of vertices turns out to be critical.

Figure 3.16 shows several drawings for which people either cannot recover from the deletions or take longer before recognizing the objects. Look at these carefully and try to figure out what the objects are. It is difficult because of the deletion of the vertices.

Figure 3.16 Non-recoverable Objects

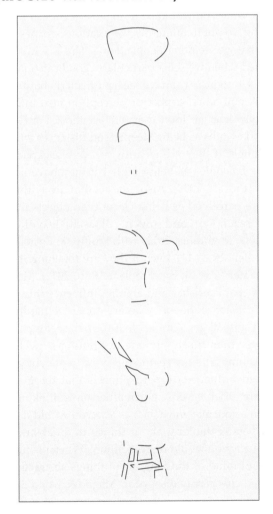

Now look at Figure 3.17.

The front of the flash cards shows drawings for which people either cannot recover from the deletions or take longer before recognizing the objects. Look at the images carefully and try to figure out what the objects are. It is difficult because of the deletion of the vertices.

On the other side of the flash cards, you see the recoverable versions of the drawings that were degraded to the same degree as the images on the front. However, parts of the continuous edges were deleted, leaving the vertices still visible. It is relatively easy to identify the original objects in these images (you can identify them, can't you?).

In Biederman's data (Biederman & Blickle, 1985), people never made more than 30% errors in identifying recoverable patterns, even when 65% of the continuous line contours were deleted and the pattern was shown for only 100 ms. But when the same percentage of the junctions or intersections were deleted as in Figure 3.16, people made errors in the 100-ms condition almost 55% of the time.

SHORTCOMINGS OF RBC AND EMBODIED PERCEPTION As useful as it is, RBC is incomplete. First, one difficulty is its ties to bottom-up processing, and object recognition is strongly influenced by context and prior knowledge (e.g., Biederman, Glass, & Stacy, 1973; Palmer, 1975). For example, Tanaka and Curran (2001) tested people with some expertise, "bird experts" and "dog experts" who had more than 20 years of experience in local bird and dog organizations. These people showed neurological evidence of enhanced early recognition in their areas of expertise, compared to how they recognized objects outside of those areas (e.g., plants).

There is also evidence that retrieval of an object's identity (at least in terms of the category it belongs to) occurs as fast as identifying that there is even something there—that is, the presentation of a stimulus (Grill-Spector & Kanwisher, 2005; but see Bowers & Jones, 2008). Indeed, Dell'acqua and Job (1998) claimed that object recognition is automatic, given that judgments of a perceptual feature (is this picture elongated horizontally or vertically?) were strongly affected by the top-down knowledge of the identity of the object in that picture.

Second, the model suggests that perceiving components is the first major step in object recognition, thus claiming that to perceive the whole it is necessary to first identify the components. Certain data, however, show that people can perceive the overall shape of an object as rapidly and as accurately as the components (e.g., Cave & Kosslyn, 1989). And, again, there are even data from Dell'acqua and Job (1998) indicating that the whole object is recognized automatically using stored knowledge about it, without necessarily identifying components or features. All of these contradict the features-first aspect of the RBC model.

Finally, neuropsychological evidence shows that object recognition is a joint effort between two mental processes and two different regions of the brain, one for features and components—"bits and pieces," as it were—and another for overall shape and global patterns—the Gestalt or overall form.

Figure 3.17 Recoverable Objects

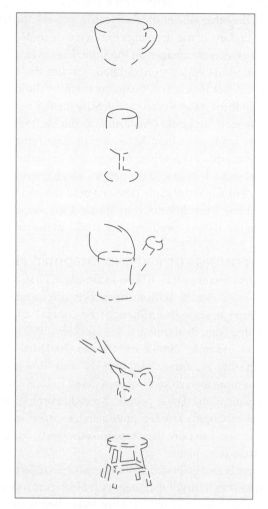

Most of this neuropsychological evidence comes from studying people who, because of some kind of brain damage, have lost the fundamental ability we have been discussing here: the ability to look at something and rapidly recognize what it is. We will consider the challenges of these people in detail after taking a more in-depth look at the impact of context and embodied perception on perceptual processes.

3.5.2: Context and Embodied Perception

One idea that has affected a broad range of thinking about cognition is the influence of embodied processes—that is, how the structure of our bodies and how we interact with the world influence how we think. At first, it might seem that this would not be relevant to perception. After all, perception is about taking in information from the environment, not about acting on it. However, even if a person does not actually do anything, there are clear embodied influences on perception (Tipper, 2010). Ambrosini, Sinigaglia, and Costantini (2012) noted a striking example of this. They found that when they monitored the eye movements of people as they watched videos of human actions (e.g., reaching for an object), the viewers were less efficient in their eye gaze behavior if their hands were tied behind their backs than if their hands were simply resting on the tabletop. So, even though the people did not have to do anything with their hands, just watching the video with their hands tied behind their backs made it harder to perceptually process their watching of others doing things with their hands. Most of the embodied effects on perception have to do with the motor programs (patterns of muscle movements) that people would engage in if they were actually interacting with what they were seeing.

There are a number of neural systems for programming motor activity, including the motor cortex and mirror neurons. Work on perception has shown that activation of these areas occurs under a variety of circumstances when people are looking at objects, including when they are perceiving and predicting the actions of others (Wilson & Knoblich, 2005). So, when you are watching another person, you arrive at a perceptual understanding in part because you are mentally simulating how you would perform that activity and, perhaps, what you would be trying to accomplish by doing that thing. To put it differently, we vicariously experience or enact what we see other people doing. This is why we may feel elated when we see a particularly good catch in a football game or wince when we see a particularly hard tackle.

This embodied aspect of perception also influences how we look at our immediate environment. For example, people view hills as being steeper and distances walked as being longer if they are wearing a heavy backpack; they perceive balconies as being higher if they have a fear of heights (Proffitt, 2006). Also, objects that are within their grasp are perceived as being closer than objects that they cannot reach out and touch (Bloesch, Davoli, Roth, Brockmole, & Abrams, 2012; Witt, Proffitt, & Epstein, 2005; Yang & Beilock, 2011), even if they are touching it with a laser pointer (Davoli, Brockmole, & Witt, 2012). Moreover, people's characteristics and abilities influence this perception. For instance, people who are doing well at hitting a ball in a softball game report that the ball seems perceptually larger than those who are not hitting so well (Witt, 2011). An important caveat to all of this work showing an embodied influence on perception is that these sorts of effects are observed only when a person is thinking about things in a concrete way (such as *how* you would do something), versus thinking about things in an abstract way (such as *why* you would do something). When people are in a frame of mind of thinking about things abstractly, these embodied effects may disappear (Maglio & Trope, 2012).

3.5.3: Agnosia

You have been reading about perception, studying the use of mental mechanisms such as feature detection and top-down processing to recognize objects around us. But we have not questioned that it happens or thought that there

A Study by Tucker and Ellis on Context and Embodied Perception (Figure 3.18)

There is also behavioral evidence that people activate mental representations of the muscle activities they would engage in if they were to interact with an object or the environment.

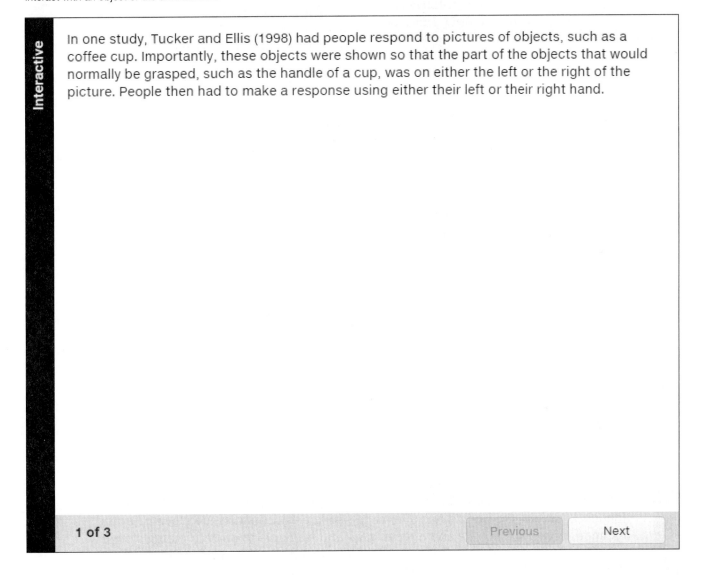

might be problems in actually recognizing a simple, ordinary object. It can be amazing to learn that a person may lose the ability to glance at something and immediately know what it is. Say there is an object, a cup, on the kitchen counter or a briefcase on the floor. We encode the stimulus, the set of features or geons, into the visual system. It is then an automatic, seemingly instantaneous step from encoding to identification: You see the thing, and you immediately know that it is a cup or a briefcase—right?

Wrong. You will now read about a deficit, caused by brain damage, whereby people can no longer do the seemingly instantaneous mental steps of pattern recognition. There are certainly types of brain damage that can disrupt the recognition of printed language, letters, and words, which we will discuss in a later module. But for now, we look at a different kind of disruption of recognition, when the recognition of objects—real-world things—is disrupted. This is **agnosia**, defined as a failure or deficit in recognizing objects. This happens either because the person cannot synthesize the pattern of features into a whole or because the person cannot then connect the whole pattern to meaning (from the prefix *a*, meaning "not," and the Greek root *gnostic*, meaning "to know"). Often this is associated with damage to the left occipital and/or temporal lobes (B.A. 37).

When this disruption affects a person's recognition of faces, sometimes while leaving object recognition intact, it is **prosopagnosia**, a disruption of face recognition. This often is a result of damage to the fusiform gyrus (also B.A. 37), also along the ventral stream. The fact that there are separate conditions involving objects and faces is important because it shows that perception is complex, and that different brain areas emphasize different qualities of information. Perception is not a one-size-fits-all system, but rather a number of specialty systems that typically work in seamless harmony.

Bear in mind that when we talk about agnosia and agnosics (patients with agnosia), we are not talking about people whose basic sensory systems are damaged. In testing an agnosic patient, the person can see and detect visual stimuli; this is not blindness. Instead, it is a cognitive mental loss. The agnosic can input the basic visual stimulus, but then cannot do anything with that encoded information.

A famous case of agnosia—and prosopagnosia, too—is related in *The Man Who Mistook His Wife for a Hat* (Sacks, 1970), about an elderly music professor (called Dr. P.) who had lost his ability to recognize objects and faces. At the end of a session with his doctor, he reached over and grasped his wife's head as if reaching to pick up his hat. In another meeting with the doctor, he was able to describe the components or elementary features of an object yet was unable to identify the object he was looking at:

> "About six inches in length," he commented. "A convoluted red form with a linear green attachment."
>
> "Yes," I said encouragingly, "and what do you think it is, Dr. P.?"
>
> "Not easy to say." ...
>
> "Smell it," I suggested.
>
> "Beautiful!" he exclaimed. "An early rose. What a heavenly smell!" (Sacks, 1970, pp. 13–14)

Dr. P. mistook the grandfather clock in the hall for a person and started to greet it with an outstretched hand. Although he could describe the parts of an object (there were five "out-pouchings," and so forth), he could not identify a glove that the doctor held in front of him. Dr. P. had serious and pervasive visual agnosia, a profound loss in the ability to visually recognize things.

Subtypes of Visual Agnosia

Likewise, there are subtypes of visual agnosia, each with a somewhat different type of deficit, each involving different regions of the brain.

Apperceptive Agnosia

One form of visual agnosia is **apperceptive agnosia**, a disruption in perceiving patterns. That is, although there is no disruption in the ability to process rudimentary visual features, say color or brightness, "the ability to coalesce this basic visual information into a percept, an entity, or a whole is lost" (Banich, 2004, p. 195). The region usually associated with apperceptive agnosia is in the right hemisphere, in the parietal lobe. If the agnosia is severe, the person has almost no ability to discriminate between objects—for instance, between a square and a rectangle—and is unable to copy or match simple shapes. In less severe cases, there can still be difficulties with patterns like those in below figure, the patterns that you probably had little or no difficulty identifying (Warrington & James, 1988). People with apperceptive agnosia somehow cannot fill in the missing contours to perceive the whole form or pattern.

Associative Agnosia

Although agnosia is not limited to vision—there can be auditory agnosias, for example—an agnosia is modality specific. That is, a person with visual agnosia has disrupted recognition of objects presented visually but no disruption of hearing, touch, or other sensory systems (Dr. P. recognized the rose by smelling it).

3.5.4: Implications for Cognitive Science

What do these neurological disruptions mean for our understanding of normal perception? How does evidence like this advance our understanding of cognition?

Start with the deficit of apperceptive agnosia, where the *a* prefix to *perceptive* denotes some kind of perceptual failure. Here we have a serious disruption in an early stage of perception. It is a disorder of feature detection, a malfunction in the process of extracting features from visual stimuli. Biederman's (1990) geons, for instance, are not being identified or at least not processed much beyond noticing small segments or junction points. Furthermore, it may be important that **apperceptive agnosia** seems to result from damage in the right hemisphere, in the parietal region; growing evidence exists that the right hemisphere is more involved in global processing to include forming global patterns, and that the left hemisphere plays more of a role in local processing (i.e., processing small components and features). If this is so, then it seems reasonable to talk about a disrupted mechanism for forming a Gestalt from the features, where this disrupted mechanism would correspond to the symptoms of apperceptive agnosia.

Associative agnosia is a deeper dysfunction: Although the Gestalt or pattern has been formed, it seems to have lost the pathway to the meaning and name of the object. The damaged regions in associative agnosia are lower, more toward the temporal lobe, and in both hemispheres. This pathway, from the vision centers in the occipital lobe forward and down toward the temporal lobe, is the "what" pathway, which is activated when you look at something to decide what it is. The temporal lobes are particularly associated with language and word meaning. The ability to connect a perceived pattern to its meaning and name is the impairment in associative agnosia.

In conclusion, the varieties of agnosia tell us at least three important things about the perception:

1. Detecting the features in a visual stimulus is a separate (and later) process from sensation. The basic features—whether horizontal lines in a capital *A*, geons, or something else—must be extracted from the sensory signal.
2. Detecting visual features is critical in constructing a perceived pattern, a percept. If the person cannot extract the features, then he or she cannot "get" the Gestalt, cannot form an overall pattern or percept.
3. There is a separate step for hooking up the pattern with its meaning and name in memory. This is different from knowing the meaning and name of an object in verbal form. Indeed, given that P. T. only later realized what his pantomime meant, the visual association path can be isolated from all of the other ways of knowing about objects and patterns.

In short, simple, "immediate" recognition of objects—the cup, the briefcase—is neither simple nor immediate. The disruptions of agnosia, whether caused by difficulties in feature detection or in associating patterns with meaning, provide additional evidence of the complexity of perception.

> **WRITING PROMPT**
> **The Nature of Perception**
> What do the different kinds of processing (features and shapes) suggest about the nature of perception? What would perception be like if people had access to only some of this information?
>
> The response entered here will appear in the performance dashboard and can be viewed by your instructor.
>
> Submit

3.6: Auditory Sensation and Perception

OBJECTIVE: Describe the science of hearing

Auditory stimuli consist of sound waves moving the air. Human hearing responds to these stimuli using an awkward combination of components, a Rube Goldberg–type mechanism that translates the sound waves into a neural message. (Google Rube Goldberg if you do not know about the contraptions he drew.) First, the sound waves funnel into the ear, causing the tympanic membrane, or eardrum, to vibrate. This in turn causes the bones of the middle ear to move, which then sets in motion the fluid in the ear's inner cavity. The moving fluid then moves the tiny hair cells along the basilar membrane, generating the neural message, which is sent along the auditory nerve into the cerebral cortex (e.g., Forgus & Melamed, 1976). Thus, from the unpromising elements of funnels, moving bones and fluid, and the like (Figure 3.19) comes our sense of hearing, or **audition**.

Interestingly, both ears project auditory information to both cerebral hemispheres, although the majority of the input obeys the principle of contralaterality.

The sensitivity of our hearing defines our auditory world. A pure tone, such as that generated by a tuning fork, is a traveling sound wave with a regular frequency, a smooth pattern of up-and-down cycles per unit of time.

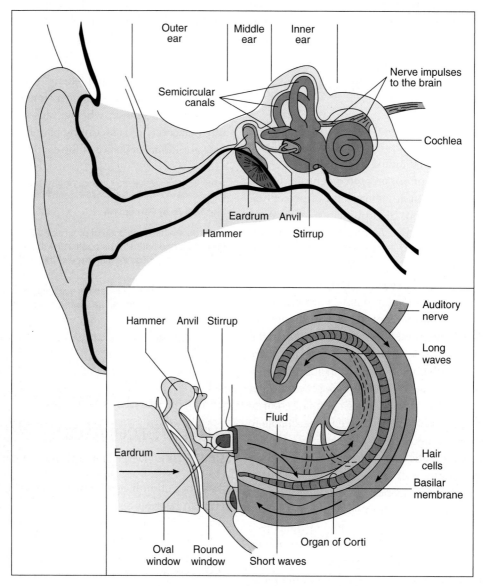

Figure 3.19 The Human Ear
Gross structure of the human ear and a close-up of the middle and inner ear structures. From Price (1987).

Generally, humans are sensitive to patterns as low as 20 cycles per second (cps) and as high as 20,000 cps—although the upper limits decline with age, which is why older adults cannot hear very high pitches. Most of the sound patterns we are interested in, such as those in spoken speech or music, are very complex, combining dozens of different frequencies that vary widely in intensity or loudness. In terms of the sound wave patterns, these different frequencies are superimposed and can be summarized in a spectrum.

In one sense, human hearing is not that impressive: Dogs, for instance, are sensitive to higher frequencies than humans are. Yet in a quite different sense, our hearing is very complex. For instance, we can accurately discriminate between highly similar sounds even from birth: Newborns notice the slight difference between the sounds *pah* and *bah* (Eimas, 1975). Also, we routinely convert the continuous stream of sounds in speech into a meaningful message with little or no effort, at a rate of about two or three words per second. How does audition work? How does it coordinate with our knowledge of language to work so rapidly?

3.6.1: Auditory Sensory Memory

Auditory sensory memory is also called **echoic memory** (Neisser, 1967). Both terms refer to a brief memory system that receives auditory stimuli and preserves them for some amount of time. Neisser's argument for the existence of echoic memory is still airtight:

> Perhaps the most fundamental fact about hearing is that sound is an intrinsically temporal event. Auditory information is always spread out in time; no single millisecond

The Primary Auditory Cortex

The primary auditory cortex, a region in the superior (upper) medial (midway back) temporal lobe, also extends somewhat farther rearward in the brain, into the parietal lobe. Auditory input to the brain is sent primarily to this auditory cortex, although at least four other nearby zones and several secondary areas are also affected.

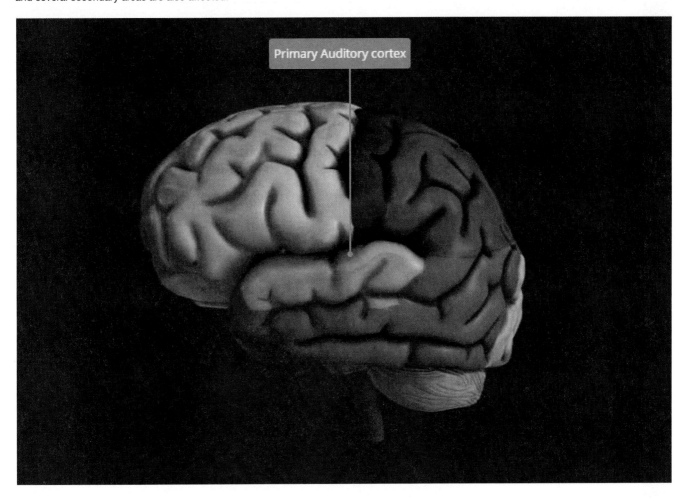

contains enough information to be very useful. If information were discarded as soon as it arrived, hearing would be all but impossible. Therefore, we must assume that some "buffer," some medium for temporary storage, is available in the auditory cognitive system (pp. 199–200).

The function of echoic memory is to encode sensory stimulation into memory and hold it just long enough for the rest of the mental system to gain access to it.

AMOUNT AND DURATION OF STORAGE What is the duration of echoic memory? Darwin, Turvey, and Crowder (1972; see also Moray, Bates, & Barnett, 1965) answered this question using an auditory analogue of Sperling's work.

They used a task that presents auditory stimuli briefly, in different locations, so that we can cue selected parts for partial report.

PERSISTENCE AND INTERFERENCE OF AUDITORY INFORMATION Without the process of redirected attention, the echoic memory degrades with the passage of time, similar to the loss in iconic memory. However, remember that there is also another kind of forgetting in iconic memory due to interference by subsequent stimuli. Is there any evidence of this in auditory sensory memory? In a word, yes, although a straightforward parallel with iconic persistence and interference may be misleading. We consider the original evidence, and then discuss the controversy over the current status and understanding of auditory sensory memory.

INTERFERENCE OF AUDITORY INFORMATION Having established the persistence of auditory traces, Crowder (1972) went on to investigate auditory interference. After people had heard the items in the list, people in the three suffix groups (auditory, verbal, tone) then heard an additional auditory stimulus: the word *zero* or a simple tone. The groups were told that this final item was merely a cue to begin recalling the list. In reality, the intent of the auditory suffix was to interfere with the lingering auditory trace for the last items in the list. As predicted, the verbal suffix group showed a higher error rate on the last items.

The "Three-Eared Man" Procedure (Figure 3.20)

Darwin et al. (1972) used what they called the **three-eared man** procedure, in which three different spoken messages come from three distinct locations.

> **Procedure**
>
> People heard recorded letters and digits through stereo headphones, with one message played to the left ear, one message to the right ear, and the final message to both ears. The message played into both ears seemed to be localized in the middle of the head, at the "third ear." Each of the messages had three stimuli, say, T 7 C on the left ear, 4 B 9 on the right ear, and so on. Each sequence lasted 1 second on the recording, and all three sequences were played simultaneously. Thus, in the space of 1 second, three different sequences of letter and digit combinations were played, for a total of nine separate stimuli.

However, the tone suffix group had very few errors. The auditory suffix had degraded the auditory trace for the last digits in the list when the suffix was similar to the list. Figure 3.21 summarizes this program of research.

This **suffix effect** is inferior recall of the end of the list in the presence of an additional meaningful nonlist auditory stimulus. In general, the more a suffix is like the information on the list, the greater the effect (Ayers, Jonides, Reitman, Egan, & Howard, 1979). It is also important what the person thinks the suffix is. In a study by Neath, Surprenant, and Crowder (1993), people heard a *baa* sound at the end of a list of words. If they were told that a person made the sound, there was a larger suffix effect than if they were told a sheep made it.

In summary, auditory sensory memory is generally similar to visual sensory memory. Both register sensory information and hold it for a brief period of time: 250 ms to 500 ms in vision but 2,000 ms to 4,000 ms in audition. This duration for auditory sensory memory, however, varies with the complexity of stored information. Generally, more information is encoded than can be reported. The items held in both sensory systems are prone to loss over short periods of time. Finally, by redirecting attention during the retention interval, information can be sent to short-term memory, preventing it from being lost. Just as in vision, our auditory world usually is one of continuous stimulation, not bursts of sound followed by empty intervals.

3.6.2: Auditory Pattern Recognition

We postpone an in-depth discussion of auditory pattern recognition until later after our discussion of attention and language.

Auditory Persistence and Interference (Figure 3.21)

The best-known evidence on auditory persistence was done by Crowder and Morton (1969; also Crowder, 1970, 1972).

> **Interactive**
>
> In their work, a list of nine digits was presented visually, at the fairly rapid rate of two items per second. In the Silent Vocalization condition, people saw the nine numbers and read them silently as they appeared. In the Active Vocalization condition, people not only saw the list but also named the digits aloud. In the Passive Vocalization condition, people heard an accompanying recording that named the digits for them.
>
> 1 of 3   Previous   Next

Prove It

The influence of top-down processing on perception is very powerful, and we typically do not notice it. In cognitive psychology we can manipulate whether people use top-down knowledge by giving some people—but not others—knowledge they can later use. However, there are cases in our everyday lives when the influence of knowledge is clear, such as when we listen to songs.

Whereas the lyrics of many songs are clear from the first time you hear them, there are also songs that take some time to figure out just what the singer is singing. Everyone has had the experience of listening to a song for the first time. However, after you read the lyrics in the liner notes, it can be amazing how clear the words seem when you hear the song the next time. You can demonstrate this with some of your friends by playing songs to each other that you know are difficult to get the first time around, and then listening a second time after reading the lyrics. What happens here is that your prior knowledge (top-down processing) can now help organize the incoming auditory perceptual stream, allowing you to correctly parse the lyrics.

We do not always have a lyric sheet to help us figure out what a singer is singing and so are left to figure the words out with other knowledge. Although we usually get this right, we can get it wrong, sometimes with hilarious results. There is a website called "kissthisguy," which is an archive of misheard lyrics (sometimes called "mondegreens") that chronicles the failures of top-down processing on auditory perception.

Recognizing Auditory Patterns (Table 3.4)

For now, here is the basic information about recognizing auditory patterns.

> **Interactive**
>
> **Templates**
>
> Attempts to understand how we recognize sounds, especially language, have paralleled the work on visual pattern recognition. However, the attempts at explaining auditory pattern recognition by using stored templates in memory were not successful. For example, in language, different people produce different language sounds. Think of how different the same spoken sentence would sound, in terms of pitch, coming from a man versus a woman, or from an adult versus a child. Furthermore, even the same sounds produced by the same speaker can vary widely from time to time. Think of how different you would sound if you were speaking a sentence in your normal voice, versus shouting, whispering, or if you had a cold. Moreover, the "same" sound varies from word to word, even when spoken by the same speaker. In psycholinguistics, this is the **problem of invariance**. The problem is that the sounds of speech are not invariant from one time to the next. Instead, any particular sound in a word changes depending on the sounds that precede and follow it. As another example, consider melodies in music. If people were to identify a melody using a template, then if the melody were transposed up or down in pitch, and all of the notes changed, people should not be able to recognize it as the same melody. But, they clearly do.
>
> **Feature Detection**
>
> **Conceptually Driven Processing**

WRITING PROMPT

Knowledge and Perception

How much of your perception do you think is driven by what you already know about the world, and the context in which your experiences are embedded? Explain briefly.

The response entered here will appear in the performance dashboard and can be viewed by your instructor.

Submit

Summary: Sensation and Perception

3.1 Psychophysics

- The mental world depends to a great extent on physical reality. However, the two do not always directly correspond. The realm of psychophysics tries to assess the relationship between our mental experience and the realities of the world we inhabit.
- The most basic process that psychophysics investigates is the sensory threshold, or the minimal amount of something that we can detect. That is, how can we tell whether something is there or not? This is operationalized as the ability to detect the presence of a low energy stimulus half the time over a large number of trials. Anything above this threshold is said to be supraliminal, and anything below it is said to be subliminal.
- Psychophysics is also concerned with the ability of a person to detect when something has been changed

from what it was before. This is operationalized as the just noticeable difference, or JND. In other words, how much does the world have to change before a person notices that it is now no longer the same as it was? Several formulae have been devised to capture this ability.

- The principles of sensory processing that have been developed in the arena of psychophysics can be and have been extended to other domains of cognition, such as making decisions about physical and symbolic, or mental, differences. This also includes the use of signal detection measures, which account for measures of both discrimination and response bias.

3.2 Visual Sensation and Perception

- The eye sweeps across the visual field in short, jerky movements called saccades, taking in information during brief fixations. The information encoded in these fixations is stored in visual sensory memory for no more than about 250 ms. This iconic image, which may include movement, is lost rapidly or can be interfered with by subsequent visual stimulation. More information is stored in visual sensory memory than can be reported immediately. Information that is reported has been selected by focal attention.

- We do not continuously extract information from the visual scene around us, but instead extract most of the information we need within the first 50 ms of fixation. Thus, visual sensory memory is a fast-acting and rapidly adapting system, ideally suited for processing information in real time in a continuously dynamic world.

- To build up a complete mental representation of the world, we use trans-saccadic memory. This integration tracks the various entities in the world using object files of what they are doing and how they might be changing. However, this also requires that a person be actively attending to those objects.

3.3 Pattern Recognition

- Recognizing visual patterns follows principles that have been known for quite some time. The most familiar of these are the Gestalt grouping principles, including figure-ground segregation, closure, proximity, similarity, and good continuation.

- Recognition of visual patterns is not a process of matching stored templates to a visual stimulus. Feature detection is a much more convincing account. The features detected are elementary patterns that can be combined to form more complex visual stimuli.

3.4 Top-Down Processing

- A feature detection account of pattern recognition, such as Pandemonium, must be augmented by conceptually driven processes to account for the effects of context in visual recognition. Current models of this sort include the connectionist approach.

3.5 Object Recognition and Agnosia

- The recognition by components (RBC) theory claims that we recognize objects by extracting or detecting three-dimensional components, geons, from visual stimuli, then accessing memory to determine what real-world objects contain these components. The most informative parts of objects tend to be where the components join together; people have difficulty recognizing objects when these intersections are degraded.

- Studies of patients with visual agnosia demonstrate the complexity of perception. Patients with apperceptive agnosia sometimes are unable to detect even elementary features and therefore have difficulty in perceiving a whole pattern or Gestalt. Those with associative agnosia can perceive the whole but still cannot associate the pattern with stored knowledge.

3.6 Auditory Sensation and Perception

- Generally, the last items in a list presented auditorily are recalled better than items presented visually (the modality effect); furthermore, a suffix added to the end of the list degrades performance on the last list items, demonstrating erasure from echoic memory.

- Theories of auditory pattern recognition resemble those of vision; that is, they involve feature detection plus a substantial role for top-down processing.

SHARED WRITING

Iconic and Echoic Memories

Iconic and echoic memories are important for everyday perception of the world we live in. Illustrate the importance of these very brief memory systems by describing what the world would be like for a person who did not possess such systems.

A minimum number of characters is required to post and earn points. After posting, your response can be viewed by your class and instructor, and you can participate in the class discussion.

Post   0 characters | 140 minimum

Chapter 4
Attention

Learning Objectives

4.1: Identify the different meanings of attention

4.2: Explain the processes of input attention

4.3: Contrast controlled attention with voluntary attention

4.4: Compare automatic processing with conscious processing

Attention—one of cognitive psychology's most important topics and one of our oldest puzzles. What does it mean to pay attention to something? To direct your attention to something? To be unable to pay attention because of boredom, lack of interest, or fatigue? What sorts of things grab or capture our attention? How much control do we have over our attention? (Cognitive science uses "attend to," meaning "pay attention," even though some dictionaries cite this is an archaic usage.) We have to work at paying attention to some things (most topics in a faculty meeting, for example). But for other topics, it seems effortless: A good spy novel can rivet your attention.

4.1: Multiple Meanings of Attention

OBJECTIVE: Identify the different meanings of attention

Attention is one of the thorniest topics in cognitive psychology, possibly because the term means so many different things. This wide reach of the term *attention* has even led some to speculate that the word is so broad as to be almost meaningless; that attention is not a "thing" in itself; and that it would be better to talk of attentional effects instead of attention as a process in and of itself (Anderson, 2011).

Here, we do not take such an extreme position, but use the term *attention* to describe a wide range of phenomena, from the basic idea of arousal and alertness all the way up to consciousness and awareness. Some attention processes are extremely rapid, so we are aware only of their outcomes, whereas other processes are slow enough that we seem to be aware of them—and able to control them—throughout.

In some cases, attention is reflexive. Even when we deliberately concentrate on something, that concentration can be disrupted and redirected by an unexpected, attention-grabbing event, such as the sudden loud noise in the otherwise quiet library. In other cases, we are frustrated that our deliberate attempts to focus on some task are so easily disrupted by another train of thought. For example, we try very hard to pay attention to a lecture, only to find ourselves daydreaming about last weekend's party.

For organizational purposes, this module is structured around the six meanings of attention to impose some coherence on the field and prevent you from getting lost in the trees. The final type of attention in the table is nearly synonymous with short-term or working memory, so it is not discussed until later. Although other organizational schemes are possible, this approach should help you develop an understanding of attention and see how some topics flow into others. The list will also help avoid some confusion that arises when the term *attention* is used for processes or mechanisms more precisely described by another term, such as *arousal*.

For example:

- *Alertness* is associated with the neurotransmitter norepinephrine, and with activity in the brain stem, the right frontal lobe, and portions of the parietal cortex.
- *Orienting* is associated with the neurotransmitter acetylcholine, and with activity in the tectum (in the midbrain), the superior parietal lobe (B.A. 7), and the temporal parietal junction (B.A. 19).
- Finally, *executive attention*, which encompasses spotlight, selective, and resource attention, is associated

with the neurotransmitter dopamine, and with activity in the anterior cingulate (B.A. 24), the prefrontal cortex (B.A. 10), and the basal ganglia.

At every turn in considering attention, we confront four interrelated ideas:

1. We are constantly presented with more information than we can attend to.
2. There are serious limits in how much we can attend to at once.
3. We can respond to some information and perform some tasks with little, if any, attention.
4. With sufficient practice and knowledge, some tasks become less and less demanding of our attention.

Let's start by giving two general metaphors for attention, both of which apply throughout the list in Table 4.1.

Table 4.1 Six Meanings of Attention

Here the six different connotations of the term *attention* are shown as they are categorized within either input attention or controlled attention.

Input Attention	Controlled Attention
Alertness or arousal	Selective attention
Orienting reflex or response	Mental resources and conscious processing
Spotlight attention and search	Supervisory attentional system

4.1.1: Attention as a Mental Process

Attention can be thought of as the mental process of concentrating effort on a stimulus or a mental event. By this, we mean that attention is an activity that occurs within cognition.

Concentrating Effort on Data-Driven Versus Conceptually Driven Processes (Figure 4.1)

Initial identification of the pattern relies almost exclusively on data-driven processing, whereas later identification relies heavily on conceptually driven processing.

When it refers to an external stimulus, attention is the mental mechanism by which we actively process information in the sensory registers pertaining to that entity.
Focus your attention on the picture above and look at it for a few seconds. What do you see? The next slide shows a potential, most common response.

This activity focuses a mental resource—effort—either on an external stimulus or an internal thought. When it refers to an external stimulus, attention is the mental mechanism by which we actively process information in the sensory registers pertaining to that entity. When we examine a picture like that in Figure 4.1, we focus our mental energies on an external stimulus, the splotches and patches of black and white in the photograph.

4.1.2: Attention as a Limited Mental Resource

Now consider attention as a mental resource, a kind of mental fuel. In this sense, attention is the limited mental energy or resource that powers cognition. It is a mental commodity, the stuff that gets used when we pay attention. According to this metaphor, attention is the all-important mental resource needed to run cognition.

A fundamentally important idea here is that of limitations: Attention is limited, finite. We usually state this by talking about the limited capacity of attention. Countless experiments, to say nothing of everyday experiences, reveal the limits of our attention (the capacity to attend to stimuli, to remember events that just happened, to remember things we are supposed to do). In short, there is a limit to how many different things we can attend to and do at once.

It does not take long to think of daily situations that reveal these limitations. You can easily drive down an uncrowded highway in daylight while carrying on a conversation. You can easily listen to the news on the radio under normal driving conditions. However, in the middle of a heavy rainstorm, you cannot talk to the person sitting in the passenger seat; in rush hour traffic, you cannot (and should not try to) do business on the cell phone. Under such demanding circumstances, the radio and the conversations are annoyances or irritating—and dangerous—distractions. You must turn down the volume or turn off the phone.

ATTENTION TO PLACES AND OBJECTS So, what do we pay attention to? Do we attend to locations in space, such as what is happening on the road in front of us instead of in the cow pasture to the side? Or do we attend to objects, such as what that other car is doing as it moves down the road?

Well, the answer is that we do a bit of both. Although these seem to be different kinds of attention, they interact intimately (Leonard, Balestreri, & Luck, 2015). Some attention is clearly oriented toward specific locations in space. This will be clear later when we talk about attention as a spotlight. However, other kinds of attention are clearly object-based.

Not only can attention be directed to either location or objects, but also the breadth of attention to either of these can vary under different circumstances. One attends to an area of space that can be either large or small. For example, we can either attend to a large amount of the video game display, looking for anything moving that might be of importance, or focus on just one part of it to solve a puzzle. Similarly, when attending to objects, we can attend to either the object whole or its parts: the forest or the trees. This is often referred to as the **global–local distinction** (Förster, 2012; Navon, 1977), and an example of it is shown in Figure 4.2.

NOT ATTENDING At this point, let's talk about when we are not concentrating our attention. What is our mind doing at that time? Surely, it is not shut off.

Instead, a portion of the brain is operating, processing information in a different way than when attention is being concentrated. This is not a single structure in the brain. Like the network of brain areas that make up the attention network, the portions of the brain that are functioning when a person is not doing anything in particular is a network of allied brain structures called the **default mode network**, or **DMN** (Buckner, Andrews-Hanna, & Schacter, 2008).

Activity in the DMN is negatively correlated with activity in the attention network (Andrews-Hanna, 2012). That is, when one of these brain networks is active, the other is not, and vice versa. The DMN is often active when people are daydreaming, their minds are wandering, or they are thinking about past events in their lives (spontaneous autobiographical remembering). The DMN is also active when people are doing routine activities that do not require intense concentration, such as watching a television show or film (Hasson, Furman, Clark, Dudai, & Davachi, 2008; Lerner, Honey, Silbert, & Hasson, 2011; Regev, Honey, Simony, & Hasson, 2013). This may be the part of the brain we use in our everyday comprehension of stories and events.

WRITING PROMPT

What Is the Mind Doing When You Are Inattentive?

Attention appears as many different things when looked at from different perspectives. Is there any one thing called attention or is it something that emerges out of many different cognitive processes?

 The response entered here will appear in the performance dashboard and can be viewed by your instructor.

Submit

Figure 4.2 Global or Local Processing
Look at the image below. What do you see?
SOURCE: Gabriel Radvansky, 2016.

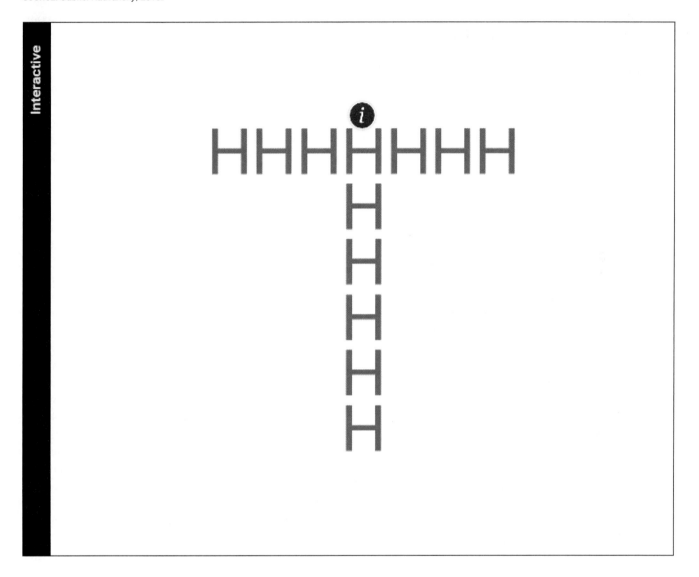

4.2: Basic Input Attentional Processes

OBJECTIVE: Explain the processes of input attention

Returning to the main topic of attention, we will start with a section on the more basic types of attention listed in Table 4.1, those occurring early in the stream of processing. These processes seem either reflexive or **automatic**, are low-level in terms of informational content, and occur rapidly. They are especially involved in getting sensory information into the cognitive system, so they can generally be called forms of **input attention**.

4.2.1: Alertness and Arousal

It almost seems axiomatic to say that part of what we mean by attention involves the basic capacity to respond to the environment. This most basic sense refers to **alertness and arousal** as necessary states of the nervous system: The nervous system must be awake, responsive, and able to interact with the environment. We cannot attend to something while we are unconscious. Certain things can impinge on us and rouse us to a conscious state (e.g., alarm clocks, smoke detectors, or other loud noises).

Despite the importance of consciousness, there also needs to be some element of alertness. We need to monitor the environment for new, interesting, and important events. Sometimes this can be difficult, especially when this alertness must be strung out over a long period during which nothing much happens. The maintenance of attention for infrequent events over long periods of time is known as **vigilance** or **sustained attention**. The study of vigilance began on British radar operators during World War II (Mackworth, 1948). However, vigilance is important in many other domains, including air traffic control, sonar detection, and nuclear power plant operations (Warm, 1984). Even quality inspections in a factory involve some degree of vigilance as workers constantly monitor for important but relatively infrequent flaws in products (Wiener, 1984). The maintenance of attention during vigilance is neurologically complicated, involving a network of brain structures, primarily localized in the right hemisphere of the brain (Langner & Eickhoff, 2013). The need for this sustained coordination of many brain regions may explain why sustained attention can be difficult at times.

Several fundamental vigilance phenomena have been observed over the years (see See, Howe, Warm, & Dember, 1995, for a review). For instance, there is a decline in performance as time on the task wears on, showing that people have difficulty maintaining attention on a single task over long periods of time. This decline takes place after about 20 to 35 minutes. Interestingly, the problems that occur with a decline in vigilance do not appear to involve people failing to notice the signal in the task they are doing. Instead, people have difficulty making the decision to respond that they have detected something, a shift in response bias (such as being more or less willing to say that they have seen something). Vigilance is also affected by the neurological and physiological state of a person, such being too hot or cold, the level of arousal, or drug use (Warm, 1984).

Finally, several aspects of the task can influence how effective people are, such as how long the signal is (longer is better), how often there is a signal (more frequent is better), and how busy the background is (less busy is better) (Warm & Jerison, 1984). Performance on vigilance tasks can be improved by giving people rest breaks or the opportunity to do some other task for a brief period before returning to the primary vigilance task (Helton & Russell, 2015). There is also some evidence that meditation training can improve certain aspects of vigilance performance (MacLean et al., 2010).

EXPLICIT AND IMPLICIT PROCESSING Although nobody disputes that arousal and alertness are a necessary precondition for most cognitive processes, this may overemphasize a kind of thinking known as **explicit processing**. Explicit processes are those involving conscious awareness that a task is being actively done, and usually conscious awareness of the outcome. The opposite is known as **implicit processing**, processing with no necessary involvement of conscious awareness (Schacter, 1989, 1996). As you will see, the distinction between implicit and explicit is often in terms of memory performance, especially long-term memory. When you are asked to learn a list of words and then name them back, that is a more explicit cognitive task: You are consciously aware of being tested and aware that you are remembering words you just studied. By contrast, you can also demonstrate memory for information without awareness, which is a more implicit cognitive task. For example, you can reread a text more rapidly than you read it the first time, even if you have no recollection of ever having read it before (Masson, 1984).

Evidence shows that some mental processing can be done with only minimal attention. Much of this is discussed later.

For now, consider a study by Bonebakker et al. (1996; see Andrade, 1995, for a review of learning under anesthesia) in which they played recorded lists of words to surgery patients. One list was played just before and another during surgery, then a patient's memory was tested up to 24 hours later. Despite the fact that all the patients were given general anesthesia and were unconscious during the surgery itself, they showed memory for words they had heard. However, they remembered only 6% to 9% more words compared to a control condition of new words. They certainly did not learn any complex ideas. So, you do need to pay attention to learn well. It is just that certain small amounts of learning can sometimes occur unconsciously.

A powerful part of the study was that performance was based on an implicit memory task, the **word stem completion task**. Patients were given word stems and told to complete them with the first word they thought of. To ensure the task was measuring implicit memory, patients were further asked to exclude any words they explicitly remembered hearing, such as those they remembered hearing before receiving the anesthesia. For example, say that they heard *BOARD* before surgery and *LIGHT* during surgery. When tested 24 hours after surgery, the patients completed the word stems (e.g., *LI_ _ _*) with words they had heard during surgery (*LIGHT*) more frequently than they did with presurgery words (*BO_ _ _*) or with control words that had never been presented. In other words, they remembered hearing *BOARD* and excluded it on the word stem task. Because they did not explicitly remember *LIGHT*, they finished *LI_ _ _* with *GHT*, presumably because their memory of *LIGHT* was implicit. The results demonstrated that the patients had implicit memory of the words they had heard while under the anesthesia.

Implicit Memory Versus Explicit Memory

Here is a more in-depth version of the task to aid your understanding of the procedure.

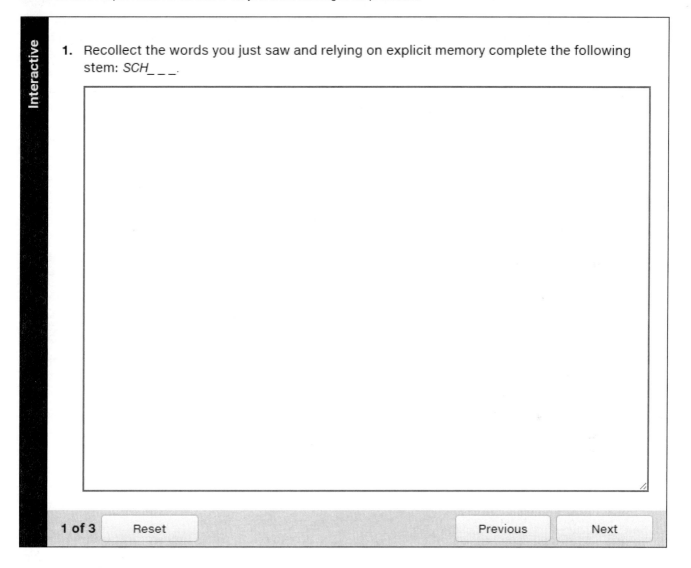

4.2.2: Orienting Reflex and Attention Capture

Now consider another kind of attention, the kind caused by a reflexive response in the nervous system.

In a quiet room, an unexpected noise grabs your attention away from what you were doing and may involve a reflexive turning of your head toward the source of the sound. In vision, you move your eyes and head toward the unexpected stimulus, the flash of light or sudden movement in your peripheral vision. This is the **orienting reflex**, the reflexive redirection of attention toward the unexpected stimulus. This response is found at all levels of the animal kingdom and is present very early in life.

Although a host of physiological changes accompany the orienting reflex, including changes in heart rate and respiration (Bridgeman, 1988), we focus on its more mental aspects. The cognitive manifestation of all of this is a redirection of attention toward something, even if the eyes and body do not actually move toward the source. We refer to this process as **attention capture**, which is the spontaneous redirection of attention to stimuli in the world based on physical characteristics.

The orienting reflex is a location-finding response of the nervous system. An unexpected stimulus, a noise or a flash of light, triggers the reflex so that you can locate the stimulus—find where it is in space. This allows you to protect yourself against danger, in the reflexive, survival sense. After all, what if the unexpected movement is from a rock thrown at you or some other threat (e.g., Öhman, Flykt, & Esteves, 2001)? Note that this system also allows you to monitor for more positive survival stimuli, such as noticing a baby's face (Brosch, Sander, Pourtois, & Scherer, 2008). In general, people are more likely to have their attention captured by something important to them in some way (Anderson & Yantis, 2013).

However, people may miss things that are important to them, such as pedestrians or cyclists in a street scene, if several other things capture their attention, such as other cars and trucks. This is especially true if they are near the more vulnerable road users (Sanocki, Islam, Doyon, & Lee, 2015). Thus, our attention can be helpful and guide us in many situations, but it is not perfect and may let us down in critical circumstances.

The "where" pathway projects from the visual cortex to upper (superior) rearward (dorsal) regions of the parietal lobe (and the "what" pathway is also called the ventral pathway).

SOCIAL CUES Attention not only is directed by objects and entities in the environment but also is directed by social cues. Perhaps the main cue is noticing where other people are looking (Birmingham, Bischof, & Kingstone, 2008; Kingstone, Smilek, Ristic, Friesen, & Eastwood, 2003). It has even been suggested that our face and eyes have evolved in such a way to communicate this sort of attention-directing information (Emery, 2000). Even our language can influence how we direct attention. For example, a study by Estes, Verges, and Barsalou (2008) showed that attention can be directed based on the meanings of words activated in long-term memory. In this study, people saw a cue word in the middle of the screen, which was soon followed by either an X or an O at either the top or the bottom of the screen. The task was to indicate which of these two letters was seen by pressing one of two buttons. Researchers found that people were faster to respond to a letter probe if the meaning of the cue word signified a direction consistent with the location of the letter. So, if the cue word was *hat*, people would respond to the letters faster if they were on the top of the screen rather than the bottom. Similarly, if the word was *boot*, the opposite was true (see Figure 4.3).

MODULATING ATTENTION CAPTURE We also orient toward things when something unexpected occurs: the unexpected sound in the quiet library, sudden and startling movement (Abrams & Christ, 2003; Franconeri & Simons, 2003), the abrupt onset of a new object (Davoli, Suszko, & Abrams, 2007; Yantis & Jonides, 1984), a change

Figure 4.3 The *HAT* and *BOOT* Figure

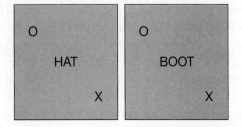

in the color of an object (Lu & Zhou, 2005), an animate object moving (Pratt, Radulescu, Guo, & Abrams, 2010), the change in pitch in a professor's voice during a lecture, or maybe the word *different* in italics in a textbook paragraph. Notice that what seems to capture attention is the occurrence of something *unexpected*, not just something *new* (Vachon, Hughes, & Jones, 2012). Thus, not all visual changes capture attention equally. Moreover, a visual offset, when something suddenly disappears from a scene, is much less likely to capture attention than a visual onset, something suddenly appearing (e.g., Cole & Kuhn, 2010).

A clearer picture of the image in Figure 4.1, in case you had trouble identifying the object.

Orienting focuses us so we can devote deliberate attention to the stimulus if warranted; Cowan (1995) called these *voluntary attentive processes*. In this sense, orienting is a future-oriented preparatory response, one that prepares the system for further voluntary processing. In **visual attention**, fMRI (functional magnetic resonance imaging) neurological scanning has shown that the attention capture process itself seems to involve retinotopic (specific places on the retina in your eye) portions of the occipital lobe, the part of the brain dedicated to vision. This contrasts more controlled aspects of attention that involve portions of the dorsal parietal and frontal cortex, farther down the stream of neural brain activity (Serences et al., 2005; Yantis, 2008).

However, if the stimulus that triggers an orienting reflex occurs repeatedly, it is no longer novel or different; it has become part of the normal, unchanging background. The process of **habituation** begins to take over, a gradual reduction of the orienting response back to baseline. For example, if the unexpected noise in the quiet library is the ventilation fan coming on, you first notice it but then grow accustomed to it. You have oriented to it, and then that response habituates to the point that you will probably orient again when the fan stops running. When the constant noise stops, that is a change that triggers the orienting response.

4.2.3: Visual Search

The last sense of attention to be considered among the input attentional processes is a kind of visual attention. It is related to perceptual space—the spatial arrangement of stimuli in your visual field and the way you search that space. It is different from the orienting response in that there is no necessary movement of the eyes or head, although there is a strong correlation with eye movements (researchers often exploit this relationship to have a general idea of where attention is directed, using eye-tracking devices). Instead, there is a mental shift of attentional focus, as if a spotlight beam were focused on a region of visual space, enabling you to pick up information in that space more easily (think of a "Superman beam").

Numerous studies that have been done on this kind of visual attention include work that has found regions of the brain that seem to be involved in focused, visual attention.

Further analysis suggested that the cost of people having directed their attention to the wrong place resulted from a three-part process:

1. Disengaging attention from its current focus
2. Moving the attentional spotlight to the target's true location
3. Engaging attention at that new location

Posner et al. (1980) concluded from this and other related experiments that the attentional focus being switched was a cognitive phenomenon. It was not tied to eye movements but to an internal, mental mechanism. They suggested that attention is like a spotlight that highlights the objects and events it shines on. Thus, **spotlight attention** is the mental attention-focusing mechanism that prepares you to encode stimulus information. Furthermore, Posner et al. (1980) suggested that this shift in attention is

---

Posner's Spatial Cuing Task (Figure 4.4–Figure 4.5)

Consider Figure 4.4, which depicts three kinds of displays in Posner's spatial cuing task (Posner, Nissen, & Ogden, 1978; Posner, Snyder, & Davidson, 1980).

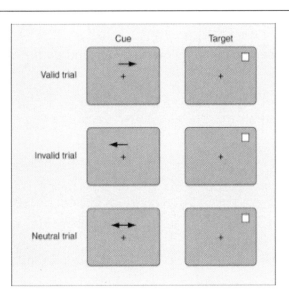

Figure 4.4: In Posner's spatial cuing task, the person fixates on the plus sign in the center of the screen, then sees an arrow pointing left or right or a two-headed arrow. For the targets shown in the figure, with a target appearing on the right, the right-pointing arrow is a valid cue, the left-pointing arrow an invalid cue, and the two-headed arrow a neutral cue. In this experiment, one-headed arrow cues were valid on 80% of the trials.

People in this task are first asked to fixate the centered plus sign on the display, are then shown a directional cue, and finally see a simple target (the thing they are supposed to respond to). For Posner, the directional cue was an arrow. However, attention can be directed in many other different ways, such as by eye gaze or language (e.g., "left"), although these have more minor influences on attention (Gibson & Sztybel, 2014). The task had people press a button when they detected the target.

essentially the same as the redirection of attention in the orienting reflex, with one big difference: It is voluntary. Therefore, it can happen before a stimulus occurs and can be triggered by cognitive factors such as expectations.

SPOTLIGHT METAPHOR IN VISUAL ATTENTION As Cave and Bichot (1999) pointed out, countless studies of visual attention, many of them inspired by Posner's work, have adopted the spotlight metaphor. Much of that work has explored the characteristics and limits of visual attention, attempting to evaluate the usefulness of the metaphor. The evidence suggests that the mental spotlight does not sweep, enhancing the intermediate locations along the way. Instead, it jumps (much as the saccade does). Conversely, some evidence supports the similarity between a real spotlight and spotlight attention. For example, the size of the spotlight beam can be altered, depending on circumstances (see Cave & Bichot, 1999, for an extensive review).

VISUAL SEARCH In the typical result (Treisman & Gelade, 1980, Experiment 1), people could search rapidly for an item identified by the presence of a unique feature. It made little or no difference whether they searched through a small or a large display. For instance, people were able to search through as few as 5 items or as many as 30 in about the same amount of time, approximately 500 ms. The target object just seemed to pop out of the display. Thus, this is called a **pop-out effect**. Because there was no increase in RT across the display sizes, Treisman and Gelade concluded that visual search for a dimension such as shape or

---

Visual Search: Demonstration and Analysis (Figure 4.6)

A series of studies by Treisman and her associates (Treisman, 1982, 1988, 1991; Treisman & Gelade, 1980) examined **visual search**. Typically, people were told to search a visual display for either of two simple features (e.g., a letter *S* or a blue letter) or a conjunction of two features (e.g., a green *T*). The search for a simple feature was called a *feature search*: People responded "yes" when they detected the presence of either of the specified features, either a letter S or a blue letter. In the *conjunction search* condition, they had to search for the combination of two features: *T* and the color green.

The Task
Look at the figures presented in the following slides and do the quick demonstrations.

color occurs in parallel across the entire region of visual attention. Such a search must be largely automatic and must represent very early visual processing. In the results shown in Figure 4.7, this is the flat, low function of the graph. (See Finlayson & Grove, 2015, for an extension of visual search research into three dimensions, whereby things like distance from the person becomes a factor).

But when people had to do a conjunction search, such as a green *T*, they took more time, up to 2,400 ms, as more and more distractors filled the display (distractors for both conditions were brown *T*s and green *X*s). Such a conjunction search seems a more serial, one-by-one process and a far more conscious, deliberate act. This is the steeply increasing function in Figure 4.7.

INHIBITION OF RETURN Because a visual search can be complex, people need a way to track what they have already checked and what they have not. A big problem would occur if a person kept checking the same items repeatedly without checking others. To help people from returning to inappropriate locations, there is a special attention process. This is called **inhibition of return** (Klein, 2000; Posner & Cohen, 1984), in which recently checked locations are mentally marked by attention as places that the search would not return to. This process is guided by the operations of the superior colliculus (which is part of the midbrain structure known as the tectum) and the parietal lobe (Klein, 2000; Vivas, Humphreys, & Fuentes, 2006). This is consistent with the idea that inhibition of return is an important visual process (involving the superior colliculus) as well as knowledge of where things are in space, the "where" neural pathway (involving the parietal lobe). When people are not searching for something, but are simply scanning or memorizing a picture, the opposite is shown, with people being more likely to return to a previously fixated location—a *facilitation of return* (Dodd, Van der Stigchel, & Hollingworth, 2009). Thus, attention operates in different ways depending on what a person's goals are at the time.

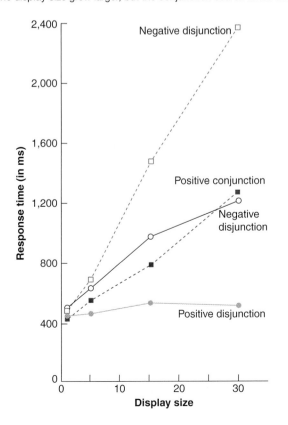

Figure 4.7 Search Times When Targets Were of a Specified Color or Shape

The dashed lines are for the disjunction search conditions (e.g., search for either a capital *T* or a boldfaced letter). The solid lines show search times for the conjunction condition (e.g., search for a boldfaced *T*). The important result is that disjunction search times did not increase as the display size grew larger, but the conjunction search times did.

> **Prove It**
>
> **The Visual Search Task**
> Here we have a simple task to do with a friend and a stopwatch (perhaps as an app on your phone) that illustrates the difference between feature and conjunction searches in visual attention. Using two different colors of marker pens, make up a few sheets of paper, or some 4 × 6 index cards, on which you draw letters in two distinct colors such as red and green. For simplicity's sake, let's restrict ourselves to *X*s and *T*s. Have your participant search for a *green T*.
>
> For the *feature* search trials, you will have several red *X*s and *T*s all over the paper. On the "yes" sheets, you will put a green *T* in one spot on the paper; on the "no" sheets, you will just have red *X*s and *T*s. Make a separate sheet, one for "yes" trials and one for "no" trials, with 4 *X*s and *T*s, then do the same for 6 *X*s and *T*s, then 8, 10, and 12 (do not forget to go back and put the green *T* in for your "yes" trials). For the *conjunction* search trials, in addition to the green *T* for the "yes" sheets, you will put green *X*s and red *T*s on the sheets. As before, you will also have "no" sheets that only have green *X*s and red *T*s.
>
> Tell your participant that the task is to find the green *T* as fast as possible. When she finds the green *T*, have her raise her right hand. However, if she does not think the green *T* is there, have her raise her left hand. Time the participant (the second hand/display on your watch or phone is fine) each time. The standard result is that the feature search items should show a constant search rate regardless of the number of distractors. This is because the target letter should pop out under these circumstances. In comparison, for the conjunction search items, there should be an average increase in response time as the number of distractors increases.

In some sense, the locations are inhibited from or kept out of the search pattern. These items were highly activated in cognition because they were attended to, and inhibition of return turns down this activation level. So, you only continue to search through those locations that are likely to still have the item you are looking for. A consequence of this inhibition of return process is that people are slower to respond to events (such as a change in brightness) in locations that have recently been searched, and inhibited, relative to other locations. It is like searching for a friend at the airport when many people are arriving from many different flights. You search visually through the faces as they come out of security, but you do not keep scanning those same faces repeatedly. Inhibition of return keeps you from returning to those faces already scanned, having your visual search move on to other faces and ideally allowing you to find your arriving friend faster.

Although this description of visual search heavily emphasizes the influence of visual features, and the bottom-up processes that can use such features, most of our real-world everyday searching is typically not this difficult. For example, when you walk into a new kitchen and are looking for the sink, you do not just scan the room for all of the objects, searching for the one with the most sink-like features. Instead, you also draw on your prior knowledge of how things are typically organized in space (Huang & Grossberg, 2010). The sink is more likely to be on an outside wall, perhaps by a window, than somewhere else. Thus, knowledge about the object exerts a top-down influence on the tracking system involved in the visual search.

Also, attention can be influenced by embodied characteristics. For example, people find it easier to spot a target during visual search when their hands are placed near the display as compared to when they are farther away (Davoli, Brockmole, & Goujon, 2012). The idea is that hand location corresponds to locations in space that can be easily manipulated, so we are more likely to direct our attention to those locations. Directing attention in this way facilitates even tasks that do not require manipulating the environment, such as visual search. Furthermore, it is important not only *where* your hands are but also *what* they are holding. Biggs, Brockmole, and Witt (2013) found that if people are holding a weapon, such as gun, they are more likely to direct attention to people than objects. Thus, a person's action capabilities can influence where he or she directs attention.

4.2.4: Contrasting Input and Controlled Attention

Treisman's two conditions provide clear evidence of both a very quick, automatic attentional process—essentially the capture of attention due to "pop-out"—and a much slower, more deliberate attention, the type used for the conjunction search. In line with Johnston, McCann, and Remington's (1995) suggestion, we use the term *input attention* for the fast, automatic process of attention. The slower one, in Johnston et al.'s terms, is **controlled attention**.

The geometric shapes in the interactive figure below refer to different regions of the brain that are involved in attention.

4.2.5: Video Games as Mechanisms for Improving Attention

Much of what we have learned from research in cognitive psychology tells about the human mind's limits and capabilities. It would be ideal to use this scientific knowledge to find ways to improve how we think. Some attempts at improving attention have focused on the use of action video games (such as first-person shooter games). These are of interest because they require a person to use attention in atypical ways, yet place demands on *how* to use that attention. Moreover, some people find them engaging and fun. So, can playing certain kinds of video games actually improve cognition?

Some work on this topic shows improved attention processing in people who regularly play video games than in people who do not (e.g., Chisholm & Kingstone, 2015). Other studies suggest that people who do not typically play such video games can exhibit improved attention processing after they spend some time playing such games. For example, people who play action video games may have a larger attentional spotlight (Feng, Spence, & Pratt, 2007) or make their eye movements (and possibly move attention) faster (Heimler, Pavani, Donk, & van Zoest, 2014). Outside of video game playing, evidence exists that training in the visual arts can influence visual attention, such as whether focus is on local or global level of processing (Chamberlain & Wagemans, 2015).

With that said, there is also evidence that playing such video games does not have a strong and lasting influence on attention (e.g., Gobet et al., 2015). Part of the argument against studies that have found differences between players and nonplayers is their methodological shortcomings, such as using a limited number of cognitive tasks to compare players and nonplayers (Boot, Blakely, & Simons, 2011; Latham, Patston, & Tippett, 2013).

4.2.6: Hemineglect

In many cases, cognitive science has gained insight into a process when there has been some disruption to the system. The study of attention is no exception. For example, under the influence of alcohol, the operation of attention is compromised. Specifically, people attend to a narrow range of information after consuming alcohol (Harvey, Kneller, & Campbell, 2013).

Here is a quotation from Banich (1997) about Bill, who has an interesting neurological condition related to attention:

> As he did every morning after waking, Bill went into the bathroom to begin his morning ritual. After squeezing toothpaste onto his toothbrush, he looked into the mirror and began to brush his teeth. Although he brushed the teeth on the right side of his mouth quite vigorously, for the most part he ignored those on the left side. . . . He shaved all the

Treisman's Two Conditions of Attention

Note that research has suggested that these two forms of attention operate with some degree of independence (Berger, Henik, & Rafal, 2005).

Spotlight Attention

Consider the early, rapid stages of feature detection as relying on spotlight attention (Posner & Cohen, 1984). The spotlight is directed toward a visual display and enhances the detection of objects and events within it (Kanwisher & Driver, 1992). It provides the encoding route into the visual system. This attentional focus mechanism provides early, extremely rapid feature detection for the ensuing process of pattern recognition. It is especially visual. For instance, it has been called *posterior attention* because the earliest stages of visual perception occur in the posterior region of the brain, in the occipital lobe, as illustrated in Figure 4.8, as well as involving the superior colliculus, a midbrain structure (Berger et al., 2005).

The *spotlight attention* we are talking about (we presume there is an equivalent mechanism for other senses) appears to be rapid, automatic, and perceptual. It is distinguished from the slower, controlled or conscious *attention* process that matches the more ordinary connotation of the term attention. The "regular" kind is the conscious attention that we have loosely equated with awareness. Based on some neurophysiological evidence, we might even call this frontal or anterior attention because activity in the frontal regions of the brain seems to accompany elements of conscious awareness, such as awareness of the meaning of a word (Posner et al., 1992).

Controlled Attention

stubble from the right side of his face impeccably but did a spotty job on the left side. . . . [After eating at a diner,] when Bill asked for the check, the waitress placed it on the left side of the table. After a few minutes, he waved the waitress over and complained, saying "I asked for my tab 5 minutes ago. What is taking so long?" (Banich, 1997, p. 235)

Bill suffers from **hemineglect**, a syndrome that leads to behavior such as brushing only the teeth on his right, washing only his right arm, and shaving only the right side of his face. To many people, this phenomenon is almost too bizarre to believe, maybe because the processes of mental attention have always been so closely tied to perception and voluntary movement and so automatic that we think they are indivisible parts of the same process. Look at yourself in a mirror, then look at the left side of your face. No problem: You merely move your eyes, shift your direction of gaze, and look at it. If I ask you to stare straight ahead and then attend to something in your left field of vision, say the letter X on a computer screen, your normal response is to shift your eyes toward the left and focus on the target. You simply look at the X and pay attention to it. You can even shift your mental attention to the left without moving your eyes.

The syndrome known as *hemineglect* (or hemi-inattention) is a disruption in the ability to refocus your attention to one side of your face or the other, say to the X on the left of the computer screen. It is a disruption or decreased ability to attend to something in the (often) left field of vision. *Hemi* means "half," and *neglect* mean "to ignore" or "to fail to perceive." Thus, hemineglect is a disorder of attention in which one half of the perceptual world is neglected to some degree and cannot be attended to as completely or accurately as normal. Some form of hemineglect is often observed in stroke victims, even if it

Figure 4.8 Lateral and Medial View of the Left and Right Hemispheres of the Brain

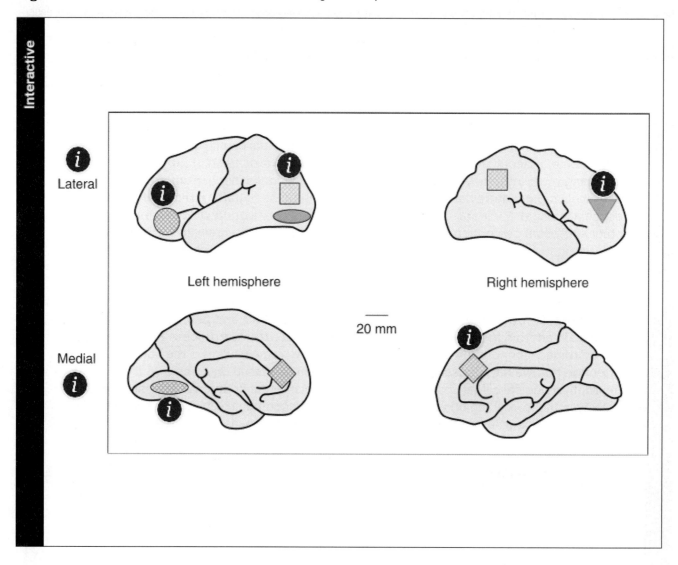

is in a more limited and temporary form. Very often, the neglect is of the left visual field, for stimuli to the left of the current fixation, the current focus of attention. And because of the principle of contralaterality, it is not surprising that the brain damage leading to hemineglect is often in the right hemisphere, in particular, certain regions of the right parietal lobe (see Intriligator & Cavanagh, 2001, for evidence that localizes selective attention in the parietal lobe).

Here are the facts (see Banich, 1997, or Rafal, 1997, for complete treatments): A patient with hemineglect cannot voluntarily direct attention to half of the perceptual world, whether the to-be-perceived stimulus is visual, auditory, or any other type of sensation. In some cases, the neglect is nearly total, as if half of the perceptual world has simply vanished, is simply not there in any normal sense of the word. In other cases, the neglect is partial, so for such people it is more accurate to say that they are less able to redirect their attention than are normal people. Either way,

there is a disruption in the ability to control attention. Note that this is not a case of sensory damage like blindness or deafness. The patient with hemineglect receives input from both sides of the body and can make voluntary muscle movements on both sides. And in careful testing situations, such patients can also respond to stimuli in the neglected field. But somehow, the deliberate devotion of controlled attention to one side is deficient.

WRITING PROMPT

Disruptions in Attention

Think of situations in which the operation of attention is compromised. What can be done to improve the effectiveness of how you use your attention?

▶ The response entered here will appear in the performance dashboard and can be viewed by your instructor.

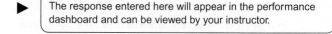

Disruptions in Patients With Hemineglect (Figure 4.9)

Bisiach and Luzzatti (1978) present a compelling description of hemineglect. The afflicted individuals were from Milan, Italy, which they were quite familiar with before their brain damage. This study focused on the main piazza in town, a broad, open square with buildings and shops along the sides and a large cathedral at one end. These patients were asked to imagine themselves standing at one end of the piazza, facing the cathedral, and to describe what they could see. They uniformly described only the buildings and shops on the right side of the piazza. When asked to imagine themselves standing on the steps of the cathedral, facing back the opposite way, they once again described what was on their right side. From this second view, they described exactly what they had omitted from their earlier descriptions. Likewise, they then omitted what they had described earlier.

> **Interactive**
>
> Critically important here is the observation that these reports, based on memory, were exactly the kind of reports patients with hemineglect give when actually viewing a scene. If these patients had been taken to the piazza, they probably would have seen and described it the same way as they did from memory. (For a similar account, see "Eyes Right!" in Sacks, 1970; the patient there eats the right half of everything on her dinner plate, then complains about not getting enough food.) Figure 4.9 shows some drawings made by a patient with hemineglect. Click or tap "Next" to view the figure.

4.3: Controlled, Voluntary Attention

OBJECTIVE: Contrast controlled attention with voluntary attention

We turn now to several senses of the term *attention* that point to the controlled, voluntary nature of attention. *Controlled attention*, in contrast to what you have just been studying, refers to a deliberate, voluntary allocation of mental effort or concentration. You decide to pay attention to this stimulus and ignore others. Paying attention this way involves effort.

Cognitive psychology has always been intrigued by the fact that at any moment, scores of different sensory messages are impinging on us. We cannot attend to all of them (we would be overwhelmed instantly), nor can we afford for our attention to be captured by one, then another, then another of the multiple sensory inputs (we would lose all coherence, all continuity). Therefore, it makes sense to ask questions about **selective attention**, the ability to attend to one source of information while ignoring other ongoing stimuli around us. How do we do this? How do we screen out the surrounding noises to focus on just one? How can we listen covertly to the person on our right, who is gossiping about someone we know, while overtly pretending to listen to a conversational partner on our left? (And how did we notice that the person on our right was gossiping?) Somewhat the converse of selective attention is the topic of divided attention: How do we divide or share our attentional capacity across more than one source of information at a time, and how much information are we picking up from the several sources?

An example of how these questions involve real-world problems is the issue of whether we really can talk (or text message) on a cell phone and drive at the same time, dividing our attention between two demanding tasks (Kunar, Carter, Cohen, & Horowitz, 2008; Spence & Read, 2003; Strayer & Johnston, 2001). Driving itself can already be taxing on attention with the need to monitor multiple vehicles and road conditions at one time (Lochner & Trick, 2014). Can attention be divided between driving and a cell phone conversation? In short, the general answer is, "No, it really can't." Talking on the cell phone can lead to *inattention blindness*, in which people fail to attend to or process information about traffic, even if they are looking directly at it. This is equally true for both handheld and hands-free cell phone conversations, but not for listening to the radio or music (Strayer & Drews, 2007). This is because we are actively involved in cell phone conversations, but not in what is going on over the radio.

Controlled, Voluntary Attention

A person's ability to control his or her attention is a function of external sources that are competing for cognitive processing.

Dangerously divided attention.

In these studies, people drive a simulator while having their eye movements monitored (so the experimenter knows what they are looking at and for how long). Under these circumstances, people who are having phone conversations are less likely to recognize road signs or other important traffic events, even when they look directly at them. This is even revealed in electroencephalographic (EEG) recordings of drivers' brains, not just what they consciously report. When you are actively involved and interacting in a conversation, your limited-capacity attention is drawn away from your immediate environment. You have less attention to devote to driving; consequently, your driving suffers and becomes more dangerous. Moreover, people are often unaware of how impaired their driving becomes when they are talking on a cell phone (Sanbonmatsu, Strayer, Biondi, Behrends, & Moore, 2016). In an interesting twist, it has even been found that people find listening to cell phone conversations more distracting to attention than listening to conversations in which they hear both sides (Galván, Vessal, & Golley, 2013). At a more general level, the question is, when do we start reaching the limits of our attentional capacity?

4.3.1: Selective Attention and the Cocktail Party Effect

When there are many stimuli or events around you, you may try to focus on just one. The ones you are trying to ignore are distractions that must be excluded. The mental process of eliminating those distractions is called **filtering** or **selecting**. Some aspect of attention seems to filter out unwanted, extraneous sources of information so we can select the one source we want to attend to.

The process of selective attention seems straightforward in vision: You move your eyes, thereby selecting what you attend to. However, attention is separate from eye movements: You can shift your attention even without eye movements. But in hearing, attention has no outward, behavioral component analogous to eye movements, so cognitive psychology has always realized that selective attention in hearing is thoroughly cognitive. This accounts for the heavy investment in filter theories of auditory perception. If we cannot avoid hearing something, we then must select among the stimuli by some mental process, filtering out the unimportant and attending to the important.

DUAL TASK OR DUAL MESSAGE METHODS A general characteristic of many attention experiments involves the procedure of *overload*. We can overload a sensory system by presenting more information than it can handle at once and then test accuracy for some part of the information. This has usually involved a **dual task method**. Two tasks are presented such that one task captures attention as completely as possible. Because attentional resources are consumed by the primary task, few, if any, resources are left over for attention to the other tasks. By varying the characteristics or content of the messages, we can make the listener's job easier or harder. For instance, paying attention to a message spoken in one ear while trying to ignore the other ear's message is more difficult when both messages are spoken by the same person.

Going a step further, when we examine performance to the attended task, we can ask about the accuracy with which a message is perceived and about the degree of interference caused by a second message. We can also look at accuracy for information that was not in the primary message, the unattended message in the other ear. If there is any evidence of remembering the unattended message, or even some of its features, we can discuss how unattended information is processed and registered in memory.

SHADOWING EXPERIMENTS Some of the earliest cognitive research on auditory selective attention was done by E. Colin Cherry (1953; Cherry & Taylor, 1954). Cherry was interested in speech recognition and attention. Cherry characterized his research procedures, and the question he was asking, as the **cocktail party effect** (although you can think of it as a dorm party problem): How do we pay attention to and recognize what one person is saying when we are surrounded by other spoken messages?

4.3.2: Selection Models

It appears that a physical difference between the messages permits people to distinguish between them and eases the job of selectively attending to the target task (Johnston & Heinz, 1978). Investigators routinely call this **early selection**. This refers to some of the earliest phases of perception, an acoustic analysis based on physical features of the message. The evidence is that people can select a message based on sensory information, such as loudness, location of the sound source, or pitch (Egan, Carterette, & Thwing, 1954; Spieth, Curtis, & Webster, 1954; Wood & Cowan, 1995a). These attentional processes are controlled by specific brain regions, such as the frontal lobes, that have unique EEG signals (e.g., Cavanagh & Frank, 2014).

EARLY SELECTION THEORY This evidence, indicating that people could somehow tune their attention to one message over the other, prompted Donald Broadbent

Cherry's Shadowing Task

To simulate this real-world situation in the laboratory, Cherry (see also Broadbent, 1952) devised the workhorse task of auditory perception research, the **shadowing task**.

> **Interactive**
>
> **The Task**
>
> In this task, Cherry recorded different types of spoken messages, then played them to a person who was wearing headphones. The task was to "shadow" the message coming into the right ear, repeating the message out loud as soon as it was heard. In the research domain of auditory attention, this is known as *stream segregation* and *streaming and auditory scene analysis* (Bregman, 1990). In most of the experiments, people were also told to ignore the other message, the one coming into the left ear. (It makes no difference which ear is shadowed or ignored. For simplicity, assume that the right ear gets the to-be-shadowed attended message and the left ear gets the unattended message.)

(1958) to propose an early selection theory of attention. In this view, attention acts as a selective filter, as shown in Figure 4.10.

LATE SELECTION THEORY Treisman (1960, 1964) did a series of studies to explore this slippage more closely. She used the standard shadowing task but varied the nature of the unattended message across a subtler range of differences. She first replicated Cherry's findings that selective attention was easy when physical differences existed. Then she turned to the situation in which physical differences were absent—both messages were recorded by the same speaker. Because the same pitch, intonation, stress, and so on were in both messages, early selection should not be possible. Yet she found that people could shadow accurately. The basis for the selection was *message content*, what the message was about rather than what it sounded like. In this situation, the grammatical and semantic features are the basis for selection (*semantic* refers to meaning). Because attentional selection occurs after all the initial processing of the message is done, this is called **late selection**. It is later in the stream of processing than early selection based on sensory features, yet before the moment of having to respond aloud with the shadowed speech.

INHIBITION AND NEGATIVE PRIMING Most of the discussion of attention to this point has focused on what

Broadbent's Early Selection Theory of Attention (Figure 4.10)

Attention can be made selective by focusing in on physical features of a stimulus, as is the case with early filter theories of attention.

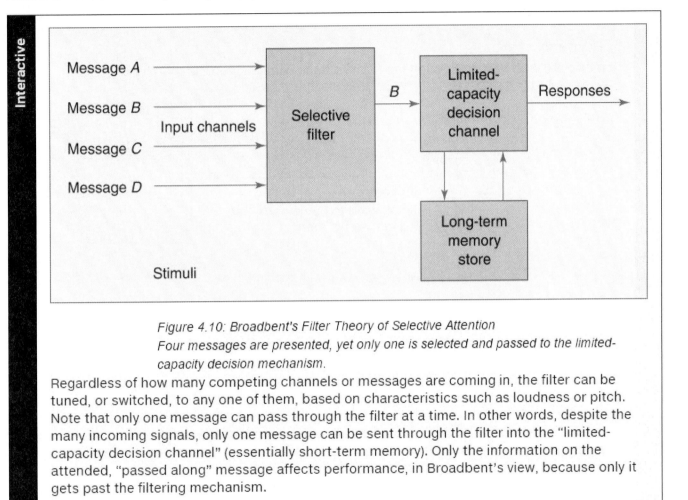

Figure 4.10: Broadbent's Filter Theory of Selective Attention
Four messages are presented, yet only one is selected and passed to the limited-capacity decision mechanism.

Regardless of how many competing channels or messages are coming in, the filter can be tuned, or switched, to any one of them, based on characteristics such as loudness or pitch. Note that only one message can pass through the filter at a time. In other words, despite the many incoming signals, only one message can be sent through the filter into the "limited-capacity decision channel" (essentially short-term memory). Only the information on the attended, "passed along" message affects performance, in Broadbent's view, because only it gets past the filtering mechanism.

gets attended to and activated in cognition, as well as some filter to keep irrelevant information from entering the stream of processing. At this point, we would like to discuss a proposed cognitive mechanism that goes beyond the idea of a filter keeping out irrelevant information and only allowing selected information to be processed further. This is the cognitive attention mechanism of **inhibition**. Inhibition is thought to actively suppress mental representations of salient but irrelevant information so the information's activation level is reduced, perhaps below the resting baseline level. You already encountered an example of this when we discussed inhibition of return. Here, we look further at how inhibition may be operating to help people select relevant information and filter out irrelevant information.

For inhibition to operate, there needs to be a salient source of interfering and irrelevant information—the irrelevant information needs to be strong and wrong. Under such circumstances, inhibition will be involved. A study by Tipper (1985) provided a classic demonstration of this.

In this study, people were presented with a series of pairs of line drawings of objects, with one object presented in green and the other in red. The task was to name the red object as quickly as possible. The important condition here involved target trials on which the red object had appeared in green on the previous trial (called the *prime* trial). It was observed that people were slower to respond to the target trials (red object) when the trials had been preceded by the to-be-ignored distractor primes

Treisman's Late Selection Theory of Attention (Figure 4.11)

To show the power of late selection, Treisman did a study now considered a classic (1960).

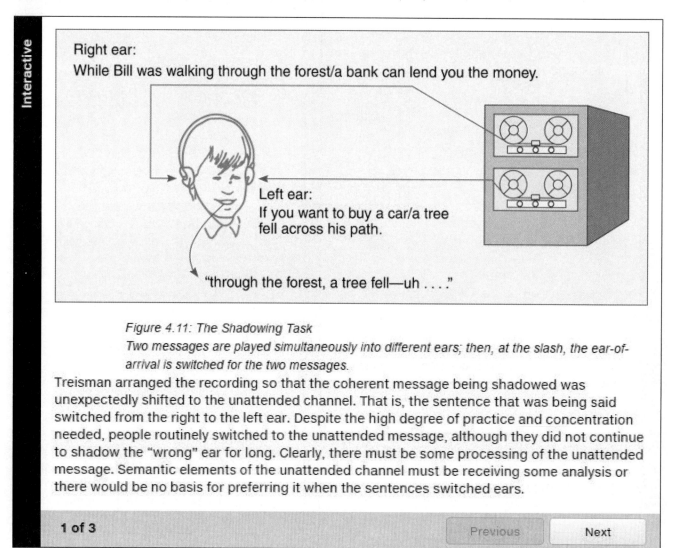

Figure 4.11: The Shadowing Task
Two messages are played simultaneously into different ears; then, at the slash, the ear-of-arrival is switched for the two messages.

Treisman arranged the recording so that the coherent message being shadowed was unexpectedly shifted to the unattended channel. That is, the sentence that was being said switched from the right to the left ear. Despite the high degree of practice and concentration needed, people routinely switched to the unattended message, although they did not continue to shadow the "wrong" ear for long. Clearly, there must be some processing of the unattended message. Semantic elements of the unattended channel must be receiving some analysis or there would be no basis for preferring it when the sentences switched ears.

(same object in green) compared to control trials where the ignored object on the prime trial was some other object. This response time slowdown is called **negative priming** (Neill, 1977; Tipper, 1985; for a review, see Frings, Schneider, & Fox, 2015).

This is because when people are looking at the display, in addition to their processing of the red object, there is also some activation and processing of the green object because people are looking directly at it. The mental representation of the object becomes activated. However, because the identity of this object is irrelevant to the task—it was green, so it did not need to be named—attention actively inhibited and suppressed the object's representation. Then, when the person needed this information on the next trial (because it was now the red object), it took longer to activate and use because it had been inhibited. So the inhibition process slowed down the person's response time. However, see Mayr and Buchner, 2006, and Neill, Valdes, and Terry, 1995, for alternative accounts of negative priming that do not involve an active inhibitory mechanism.

The idea that inhibition is an important part of attention has been extended to many other areas of psychology, particularly those dealing with individual differences. For example, in developmental psychology, attentional inhibition is thought to develop slowly with age (Diamond & Gilbert, 1989), meaning that it is difficult, especially for young children, to maintain focus. In contrast, in older adults there is increased difficulty suppressing irrelevant information (Hasher & Zacks, 1988). Inhibitory problems are also thought be present in schizophrenia (Beech, Powell, McWilliams, & Claridge, 1989), with schizophrenics having trouble keeping unwanted thoughts out of consciousness. In less extreme cases, people who are depressed also have trouble inhibiting irrelevant information (MacQueen, Tipper, Young, Joffe, & Levitt, 2000), leading to trouble focusing on the task at hand.

WRITING PROMPT

Role of Inhibition in Attention

Can you think of a situation in which you selected relevant information and filtered out irrelevant information? Did this process of inhibition help you streamline your thought process?

> The response entered here will appear in the performance dashboard and can be viewed by your instructor.

Submit

4.4: Attention as a Mental Resource

OBJECTIVE: Compare automatic processing with conscious processing

An important and far-reaching meaning of the term *attention*—this one may be closer to our everyday meaning—treats attention as mental effort, as a mental resource that fuels cognitive activity. If we selectively attend to one particular message, we are deliberately focusing mental energy on that message, concentrating on it to the exclusion of other messages. This sense involves the idea that attention is a limited resource, that there is only so much mental fuel to be devoted here or there at any one time. Kahneman (1973) also suggested that capacity might be elastic, in that increasing the task load might also increase a person's arousal, thus making additional resources available. Approaches that emphasize this meaning of attention are called resource theories.

A corollary to this idea of limited capacity is that attention, loosely speaking, is the same as consciousness or awareness. After all, if you can be consciously aware of only one thing at a time, doesn't that illustrate the limited capacity of attention? Even on a smaller scale, when we process very simple stimuli, there is evidence of this limit to attention. If you are asked to respond to a stimulus and then immediately to a second one, your second response is delayed a bit. This is the **psychological refractory period** or **attentional blink**, which is a brief slowdown in mental processing due to having processed another very recent event (e.g., Barnard, Scott, Taylor, May, & Knightley, 2004; Pashler & Johnson, 1998). The implication is that allocating attention to a first stimulus momentarily deprives you of the attention needed for a second stimulus. However, this blink can be overcome if the target item that occurs during the period of the blink is particularly important, such as the future thinking involved in predicting what to do on the next trial (Livesey, Harris, & Harris, 2009). The blink may also be overcome if people need to make eye movements (saccades) between the two stimuli. This possibly occurs because the eye movement creates an event boundary so that the two stimuli are processed as part of different events. In such a situation, the blink may not occur (Kamienkowski, Navajas, & Sigman, 2012).

In addition, the strain on attention can be reduced when people view nature scenes (which only subtly demand attention) as compared to urban scenes (which have many elements that grab our attention), thereby freeing up attention for other tasks (Berman, Jonides, & Kaplan, 2008; Joye, Pals, Steg, & Evans, 2013).

A related idea is that this kind of attention is deliberate, willful, intended—*controlled attention*. *You* decide to pay attention to a signal, or you decide *not* to attend to it. *You* decide to pay attention to the lecture instead of your memory of last night's date, and when you realize your attention has wandered, *you* redirect it to the lecture, determined *not* to daydream about last night until class is over.

An interesting insight that William James had about attention is the idea that we may do more than one thing at a time if the other processes are habitual. Yet when processes are less automatic, then attention must oscillate among them if they are done simultaneously, with no consequent gain of time. The key point is the idea of automatic processes—that some mental events can happen without draining the pool of attentional resources. Putting it simply, the germ of James's idea, **automaticity**, has become central to cognitive psychology's views on attention, pattern recognition, and a host of other topics. Cognitive science has devoted a huge effort to recasting James's ideas about automaticity and attention into more formal, quantifiable concepts.

4.4.1: Automatic and Controlled Processing

In place of the former approach, the limited-capacity attentional mechanism and the need for filtering in selective attention, the current view is that a variety of cognitive processes can be done automatically, with little or no conscious involvement necessary. Two such theories of automaticity have been proposed, one by Posner and Snyder (1975) and one by Shiffrin and Schneider (1977; Schneider & Shiffrin, 1977). These theories differ in some of their details but are similar in their overall message (see also Logan & Etherton, 1994; for discussions that oppose the idea of mental resources, see Navon, 1984, and Pashler, 1994).

Automatic and Controlled Processes of Attention Vary Along a Continuum

What differentiates automatic from controlled processing? One way of assessing this is the ease with which a process can be interrupted or stopped. If it continues even when a person wishes to be doing something else, it is more automatic.

Automatic Processing

Posner and Snyder described four diagnostic criteria for an automatic process. First, an automatic process occurs without intention. In other words, you cannot prevent it from happening, and once it does start, you cannot stop it. A compelling example of this is the Stroop effect (named after the task described in Stroop, 1935). Words such as RED GREEN BLUE YELLOW were presented visually, written in mismatching colors of ink (e.g., RED printed in green ink). When people have to name the ink color, they must ignore the printed words themselves. This leads to tremendous interference, a slowing of the ink color naming, caused by the mismatching information and the contradictory impulses to name the word and the ink color (this is an extremely easy demonstration to do). This is another case in which inhibition is operating during attention, as the highly salient, but irrelevant, semantic meaning of the word must be suppressed. Note that this requires that a person can automatically read. People who are illiterate would not show a Stroop effect. That said, it is also the case that poor readers can show larger Stroop effects than good readers (Protopapas, Archonti, & Skaloumbakas, 2007), perhaps because, under certain circumstances, better readers have greater executive control over their attentional resources. In Posner and Snyder's terms, accessing the meaning of the written symbol RED is automatic: It requires no intention; it happens regardless of whether you want it to. In the research that demonstrates automatic access to word meaning, the term we use is *priming*. A word activates or primes its meaning in memory and, thus, primes or activates meanings closely associated with it. This priming makes related meanings easier to access: Because of priming, they are boosted up, or given an extra advantage or head start (just as well water is pumped more easily when you prime the pump). See also Dunbar and MacLeod, 1984, and MacLeod, 1991, for an explanation of Stroop interference based on priming.

Prove It

The Stroop Task

An almost fail-safe demonstration of automaticity involves the **Stroop task**. With several different colors of marker pens, write a dozen or so color names on a sheet of paper, making sure to use a *different* color of ink than the word signifies (e.g., write *red* in green ink). Alternatively, create a deck of 3 × 5 cards, with one word per card. Make a control list of noncolor words (e.g., *hammer, card,* and *wall*), again in colored inks. (And try it yourself right now—name the color of the ink for the words in the Stroop color word list and control list.)

Explain to your participant that the task is to name the *ink color* as rapidly as possible. Time the person (the second hand/display on your watch or app on your phone is more than sufficient). Or, track naming errors on each kind of list. The standard result is that the color word list will require substantially longer for ink color naming than the control list. Other useful control lists are simple blotches of color, to check on the speed of naming the colors, and pseudowords (*manty, zoople,* and the like) written in different ink colors.

Per several studies (e.g., Besner & Stolz, 1999; Manwell, Roberts, & Besner, 2004; Vecera, Behrmann, & McGoldrick, 2000), you should be able to eliminate the Stroop effect by getting people to focus on just *part* of the word or to say the first letter position (this might be easier if you used the 3 × 5 card method) or by printing only one letter in color. This work suggests that reading the whole word is a kind of "default" setting for visual attention, which might be changed depending on the task and instructions, and that our selective attention mechanism can select either whole objects (words) or their parts (letters) as the focus.

Diagnostic Criteria for Automatic and Conscious Processing

Let's contrast the diagnostic criteria for automaticity with those for conscious or controlled processing.

> **Interactive**
>
> _____ The process occurs without intention, without a conscious decision.
>
> _____ The mental process is not open to conscious awareness or introspection.
>
> _____ The process occurs only with intention, with a deliberate decision.
>
> _____ The process consumes few, if any, conscious resources; that is, it consumes little, if any, conscious attention.
>
> _____ The process is open to awareness and introspection.
>
> _____ The process uses conscious resources; that is, it drains the pool of conscious attentional capacity.
>
> _____ (Informal) The process operates very rapidly, usually within 1 second.
>
> _____ (Informal) The process is slow, taking more than 1 or 2 seconds for completion.
>
> Start Over   Check Answers

INTEGRATION WITH CONCEPTUALLY DRIVEN PROCESSES We can go one step further, integrating this explanation into the idea of conceptually driven processing. Think back to the shadowing research you read about.

Attending to one of two incoming messages and shadowing that message aloud demands controlled attention. Such a process is under direct control. The person is aware of doing the process, and it consumes most of a person's available mental resources.

Presumably, no other conscious process can be done simultaneously with the shadowing task without affecting performance in one or the other task (or both). When the messages are acoustically similar, then people must use differences of content to keep them separate. But by tracking the meaning of a passage, the person's conceptually driven processes come into play in an obvious way. Just as people "restore" the missing sound in "the *eel was on the axle" (Warren & Warren, 1970), the person in the shadowing task "supplies" information about the message from long-term memory. Once you have begun to understand the content of the shadowed message, then your conceptually driven processes assist you by narrowing down the possible alternatives and suggesting what might come next.

Saying that conceptually driven processes "suggest what might come next" is an informal way of referring to priming. You shadow, "While Bill was walking through the forest." Your semantic analysis primes related information and thereby suggests the likely content of the next clause in the sentence. It is likely to be about trees, and it is unlikely to be about banks and cars. At this instant, your "forest" knowledge is primed or activated in memory. It is ready (indeed, almost eager) to be perceived because it is so likely to be contained in the rest of the sentence. Then *tree* occurs on the unattended channel. Because we access the meanings of words in an automatic fashion, the extra boost given to *tree* by the priming process pushes it over into the conscious attention mechanism. Suddenly, you are saying "a tree fell across" rather than sticking with the right-ear message. Automatic priming of long-term memory has exerted a top-down influence on the earliest of your cognitive processes, auditory pattern recognition and attention.

THE ROLE OF PRACTICE AND MEMORY If accessing word meaning is automatic, then you might wonder about some of the shadowing research described earlier in which people failed to detect a word presented 35 times, the reversed speech, and so on. If word meaning access is automatic, why didn't these people recognize the words on the unattended channel? A plausible explanation is practice. It seems likely that the inability to be influenced by the unattended message was caused by a lack of practice on the shadowing task.

The role of practice in automaticity.

Theories on the Role of Practice and Memory

With greater degrees of practice, even a seemingly complex and attention-consuming task becomes easy, or less demanding of attention's full resources.

Logan and Klapp's Theory

Logan and Klapp (1991; see also Zbrodoff & Logan, 1986) suggested that the effect of practice is to store the relevant information in memory—that the necessary precondition for automatic processing is memory. Once a process or procedure has become automatic, devoting explicit attention to it can even lead to worse performance (e.g., Beilock & Carr, 2001; Logan & Crump, 2009).

Perfect, Andrade, & Eagan's Theory

Shiffrin and Achneider's Theory

4.4.2: The Role of Practice in Automaticity

Attention, in its usual everyday sense, is equivalent to conscious mental capacity. We can devote attention to only one demanding task at a time or to two somewhat less demanding tasks simultaneously, so long as they do not exceed the total capacity available. This devotion of resources means that few, if any, additional resources are available for other demanding tasks.

Alternatively, if a second task is performed largely at the automatic level, then it can occur simultaneously with the first because it does not draw from the conscious resource pool (or, to change the metaphor, the automatic process has achieved a high level of skill; see Hirst & Kalmar, 1987). The more automatically a task can be done, the more resources are available for other tasks.

The route to automaticity is practice and memory. With repetition and overlearning comes the ability to perform automatically what formerly needed conscious processing. A particularly dramatic illustration of the power of practice was done by Spelke, Hirst, and Neisser (1976). With extensive practice, two people could read stories at normal rates and with high comprehension, while they simultaneously copied words at dictation or even categorized the dictated words according to meaning. Significantly, once practice has yielded automatic performance, it seems especially difficult to undo the practice, to overcome what has now become an automatic and, in a sense, autonomous process (Zbrodoff & Logan, 1986).

4.4.3: Disadvantages of Automaticity

We have been talking as if automaticity is always a positive, desirable characteristic; as if anything that reduces the drain on the limited available mental capacity is a good thing. This is not entirely true.

Sometimes we *should* be consciously aware of information or processes that have become too routine and automatic. Barshi and Healy (1993) provided an excellent example, using a proofreading procedure that mimics how we use checklists. People in their study scanned pages of simple multiplication problems. Five mistakes such as "7 × 8 = 63" were embedded in the pages of problems. People saw the same sets of 10 problems over and over. But in the fixed order condition, the problems were in the same order each time. In the varied order condition, the problems were in a different order each time. Those tested in the fixed order condition missed more of the embedded mistakes than those in the varied order condition; an average of 23% missed in fixed order, but only 9% missed in varied order. Figure 4.12 shows the result across the five embedded errors.

The demands on attention and memory in flying a jet airplane are enormous. The pilot must simultaneously pay conscious attention to multiple sources of information while relying on highly practiced, automatic processes and overlearned actions to respond to others.

Figure 4.12 Results of Barshi and Healy's Experiment

The results below show the percentage of participants detecting the five embedded errors in proofreading multiplication problems. Problems were presented in fixed or varied order.

SOURCE: Data from Barshi & Healy (1993).

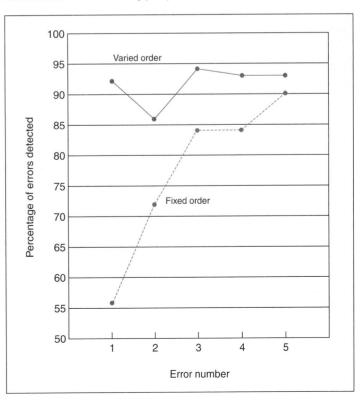

Some Disadvantages of Automaticity

There are several situations in which achieving automaticity can lead to difficulties (Reason, 1990).

> **Action Slips**
>
> You may experience **action slips**, which are unintended, often automatic, actions that are inappropriate for the current situation (Norman, 1981).
>
> Action slips can occur for many reasons, each involving a lapse of attention. The action seems foolish, and you would not have done it if you had been paying more attention. In some cases, the environment has been altered from the way it would normally be, such as pressing a button to open a door that is already (unusually) propped open. Sometimes action slips are brought about by a change that requires people to relearn something they have become accustomed to doing another way. For example, your new car may have some of its controls in a different location from where they were on your older one, so you must overcome the habit of reaching to the left dashboard to turn on the lights (this is why some controls, e.g., accelerator and brake pedals, do not change position). Often action slips occur when people have started something but are distracted (Botvinick & Bylsma, 2006). They then lapse into a more automatic pattern of behavior of doing the wrong thing, or forgetting a needed step, such as turning on the coffee machine.
>
> **Doing Things at Inappropriate Times**

Performance did improve in the fixed order condition, as more and more of the mistakes were encountered. But the first multiplication error was detected only 55% of the time, compared with the 90% detection rate for the varied order group.

The fixed order of problems encouraged automatic proofreading, which disrupted accuracy in detecting errors. In fact, it took either an earlier error that was detected or a specific alerting signal (Experiment 3) to overcome the effects of routine, automatic proofreading.

The implications of this kind of result should be clear. To ensure safety, pilots are required to go through checklist procedures for take-off, landing, and so forth. Yet because the items on the checklist are in a fixed order, repeated use of the list probably leads to a degree of automaticity and a tendency to miss errors. This is exactly what happened in March 1983: A plane landed in Casper, Wyoming, without its landing gear down, even though the flight crew had gone through its standard checklist procedure and had "verified" that the wheels were down. In Barshi and Healy's words, this incident "reminded the crew and the rest of the aviation community that the countless repetition of the same procedure can lead to a dangerous automatization" (1993, p. 496). It is interesting to wonder which is worse: too much automatization of procedures, as exemplified by the Barshi and Healy study, or too much attention paid to the procedures.

MIND WANDERING Perhaps one of the most obvious and ubiquitous examples of not being able to use our attention the way that we want to is when our minds drift from the task we are supposed to be focusing on to some other, irrelevant idea. **Mind wandering** is the situation in which a person's attention and thoughts wander from the

Why the Mind Wanders

Have you ever daydreamed about a significant other when sitting at a traffic light, or gotten to the bottom of a page and realized you had no idea what you just read?

In these cases, we have decoupled our attention from the environment to focus more exclusively on our own internal thoughts, often without an awareness that our mind is wandering until we catch ourselves (Smallwood, McSpadden, & Schooler, 2007, 2008). This idea is now supported by ERP (*event-related potential*) recordings (Barron, Riby, Greer, & Smallwood, 2011). During reading, we can now detect when a person's mind is wandering because his or her pupils become more dilated (Franklin, Broadway, Mrazek, Smallwood, & Schooler, 2013), eye movements become more erratic, eyeblinks increase, and eye movements are less tied to characteristics that should influence reading patterns (such as word frequency) (Reichle, Reineberg, & Schooler, 2010; Smilek, Carriere, & Cheyne, 2010).

current task to some other, inappropriate line of thought (for a review, see Randall, Oswald, & Beier, 2014).

WRITING PROMPT

Why Does the Mind Wander?

Narrate a situation when your attention and thoughts wandered from the current task to some other inappropriate line of thought. Did it occur when you were concentrating on something or when you were performing a task that did not take all your attention? Why do you think your mind wandered?

> The response entered here will appear in the performance dashboard and can be viewed by your instructor.

Summary: Attention

4.1 Multiple Meanings of Attention

- Attention is a pervasive and complex topic, with meanings and connotations ranging from alertness and arousal up through the notions of automatic and conscious processing. Attention can be thought of as a mental process or mechanism or as a limited mental resource.

- Controlled directed attention depends on specific neural networks to function well, and disruptions of these can lead to neurological disorders. However, it should be noted that when people are not attending in this way, the flow of the stream of thought is governed by a collection of neurological structures known as the default mode network.

4.2 Basic Input Attentional Processes

- Three basic senses of the term *attention* refer to alertness and arousal, the orienting reflex, and the spotlight attention. These correspond to input attention, a fast process involved in encoding environmental stimuli into the mental system. Interestingly, in vision, the mental spotlight attention can be shifted without any movement of the eyes, confirming the mental rather than perceptual nature of attention.
- Modern life has resulted in an increase in video game play, often involving action games that require players to redirect their attention in atypical ways. According to some studies, video gaming influences the cognitive process of attention, allowing it to be used more effectively by giving people a larger attentional spotlight. With that said, there is also evidence suggesting that such effects may be limited in scope or even nonexistent.
- A disorder known as hemineglect shows how attention can be affected by brain damage, thus informing us about normal attention. In hemineglect, a patient is unable to direct attention voluntarily to one side of space, so he or she neglects stimuli presented on that side. The evidence suggests that this arises from an inability to disengage attention from a stimulus on the non-neglected side, hence disrupting the process of shifting attention to the opposite side.

4.3 Controlled, Voluntary Attention

- Controlled or conscious attention is slower and more voluntary. Selective attention, the ability to focus on one incoming message while ignoring other incoming stimuli, is a complex ability, one investigated since the beginnings of modern cognitive science. The evidence shows that we can select one message, and reject others, based on physical characteristics or on more semantic characteristics. The later the process of selection acts, the more demanding it is of the limited capacity of the attention mechanism.
- We are able to use the attention mechanism of inhibition to keep information that would otherwise be highly active, but is irrelevant, out of the current stream of processing. Keeping inappropriate information out helps us focus on whatever it is that we want to be processing.

4.4 Attention as a Mental Resource

- When attention is viewed as a limited mental resource, issues of task complexity become concerned with how automatic or controlled different mental processes are. Automatic processes are rapid, are not dependent on intent, are unavailable to conscious awareness, and place few, if any, demands on limited attentional resources. Conscious or controlled processes are the opposite: rather slow, requiring intention, open to conscious awareness, and heavily demanding of attentional resources.
- Mental processes become more automatic as a function of practice and overlearning. One disadvantage of automaticity is that it is difficult to reverse the effects of practice in an automated task. Automaticity can also lead to errors of inattention, including action slips. When our attention is not fully engaged, our minds can wander off topic. Mind wandering is more likely to occur when there is mental capacity left over and available. Moreover, when our minds wander, the things that we allow our attention to drift to are typically things that we have enduring concerns about, such as things we are anxious or excited about.

SHARED WRITING

Putting Stress on Attention

Attention is important for effective thinking throughout the day. Describe a daily activity that puts stress on attention, how it is taxing attention (based on theories of attention), and how the strain on attention is affecting your performance in the world.

▶ A minimum number of characters is required to post and earn points. After posting, your response can be viewed by your class and instructor, and you can participate in the class discussion.

Post   0 characters | 140 minimum

Chapter 5
Short-Term Working Memory

Learning Objectives

5.1: Explain how short-term memory can be a bottleneck in the memory traffic of the brain

5.2: Summarize the process through which short-term memory is recalled

5.3: Explain how working memory functions

5.4: Identify two ways to assess working memory

5.5: Evaluate the centrality of working memory in the overall cognitive functions

Primary memory, elementary memory, immediate memory, short-term memory (STM), short-term store (STS), temporary memory, supervisory attention system (SAS), working memory (WM)—all these terms refer to the same general type of memory where the present moment is held in consciousness. It is the seat of conscious attention. This is where comprehension occurs: short-term working memory. In this module, we will look at what it is, what it does, and how it does it.

A short-term working memory is the memory of which we are conscious. Many of our intuitions and introspections about it match what has been discovered empirically. However, some mental processes of short-term working memory are not open to consciousness: They are automatic. These processes yield no useful introspections. Indeed, people often feel that they do not exist, which is why short-term working memory is only *roughly* the same as consciousness. We are aware of its contents, but not necessarily the *processes* that occur in it.

Research on short-term working memory came hard on the heels of selective attention studies of the mid-1950s.[1] George Miller's (1956) classic article is an excellent example of this upsurge in interest. A common observation, that we can remember only a small number of isolated items presented rapidly, began to take on greater significance as psychology groped toward a new approach to human memory. Miller's insightful remarks were followed shortly by the Brown (1958) and Peterson and Peterson (1959) reports. Amazingly, simple three-letter stimuli, such as *MHA*, *GPR*, or *KCD*, were forgotten almost completely within 15 seconds if a person's attention was diverted by a distractor task of counting backward by 3s. Such reports were convincing evidence that the limited capacity of memory was finally being pinned down and given an appropriate name: short-term memory.

5.1: A Limited-Capacity Bottleneck

OBJECTIVE: Explain how short-term memory can be a bottleneck in the memory traffic of the brain

If you hear a string of 10 digits, read at a rapid rate, and are asked to reproduce those digits in order, generally you cannot recall more than about 7. The same result is found with unrelated words. This is roughly the amount you can say aloud in about 2 seconds (Baddeley, Thomson, & Buchanan, 1975) or the amount you can recall in 4 to 6 seconds (Dosher & Ma, 1998; see also Cowan et al., 1998). This limit has been recognized for so long, it was included in the earliest

---

[1] There was also some research on short-term memory before the behaviorist period. For instance, Mary W. Calkins, the first woman to serve as president of the American Psychological Association, conducted work in the 1890s and reported several important effects that were then "discovered" in the 1950s and 1960s. See Madigan and O'Hara (1992) for an account of the "truly remarkable legacy" (p. 174) of this pioneering woman.

The Difference Between Short-Term Memory and Working Memory

As we proceed, we shift from the term *short-term memory* to *working memory*. Why the two terms? Basically, they have different connotations.

Short-Term Memory

Short-term memory conveys a simpler idea. It is the label we use to focus on the input and storage of new information. For example, when a rapidly presented series of digits is tested for immediate recall, we generally refer to short-term memory. Likewise, when we focus on the role of rehearsal, we are examining the short-term memory maintenance of new information. Short-term memory is observed whenever short retention is tested—no more than 15 or 20 seconds. It is also observed when little, if any, transfer of new information to long-term memory is involved.

Working Memory

---

intelligence tests (e.g., Binet's 1905 test; see Sattler, 1982). Young children and people of subnormal intelligence generally have a shorter span of apprehension, or memory span. In the field of intelligence testing, it is almost unthinkable to devise a test *without* a memory-span component. Note that this is a general aspect of short-term memory, not something special about spoken words or digits. For example, a similar finding is observed with letters in American Sign Language (ASL) (Wilson & Emmorey, 2006; Wilson & Fox, 2007), which clearly is more visual/motor relative to spoken English.

5.1.1: Short-Term Memory Capacity

For our purposes, the importance of this limitation is that it reveals something fundamental about human memory. Our immediate memory cannot encode a vast quantity of new information and hold it accurately. It has a severe limit. Miller stated this limit aptly in the title of his 1956 paper: "The Magical Number Seven, Plus or Minus Two: Some Limits on Our Capacity for Processing Information" (for a historical consideration of the paper, see Cowan, 2015). However, more recent work suggests that the situation is worse than this— that people can maintain only 4 plus or minus 1 **units** of information (Cowan, 2010), and that 7 plus or minus 2 is actually a result of some chunking. We process large amounts of information in the sensory memories, and we can hold vast quantities of knowledge in permanent long-term memory. Yet, short-term memory is the narrow end of a funnel, the four-lane bridge with only one open tollgate. It is the bottleneck in our information-processing system.

OVERCOMING THE BOTTLENECK And so this limitation remains unless what we are trying to remember is made richer and more complex by grouping it in some way, as in the 3–3–4 grouping of a telephone number or

Tollbooths force a bottleneck in a highway's traffic flow.

the 3–2–4 grouping of a Social Security number. In Miller's terms, a richer, more complex item is called a **chunk** of information. By chunking items together into groups, we can overcome this limitation.

The following is a simple example of the power of chunking:

> **BYGROUPINGITEMSINTOUNITS WEREMEMBERBETTER**

No one can easily remember 40 letters correctly if they are treated as 40 separate, unrelated items. But by chunking the letters into groups, we can retain more information. You can more easily remember the eight words because they are familiar ones that combine grammatically to form a coherent thought. You can remember a Social Security number more easily by grouping the digits into the 3–2–4 pattern. And you can remember a telephone number more easily if you group the last four digits into two two-digit numbers (of course the point generalizes beyond U.S. Social Security and phone numbers).

The term for this process of grouping items together, then remembering the newly formed groups, is **recoding**. By recoding, people hear not the isolated dots and dashes of Morse code but whole letters, words, and so on. The principle behind recoding is straightforward: Recoding reduces the number of units held in short-term memory by increasing the richness, the information content, of each unit. Try recoding the longest digit list in the "Prove It" lists, later in the module, into two-digit numbers (28, 43, and so on). This illustrates the mental effort needed for recoding. Brooks and Watkins (1990) suggested that there is already a subgrouping effect in the memory span, with the first half enjoying an advantage over the second.

Two conditions are important for recoding:

1. We can recode if there is sufficient time or resources to use a recoding scheme.
2. We can recode if the scheme is well learned, as Morse code becomes with practice.

In a dramatic demonstration of this, one person in a study by Chase and Ericsson (1982), over the period of a few months, could recall 82 digits in order by applying a highly practiced recoding scheme he had invented for himself. But what about situations in which an automatic recoding scheme is not available? What is the fate of items in short-term memory? Can we merely hold the usual small number of items?

5.1.2: Forgetting From Short-Term Memory

In addition to having a limited capacity, information in short-term memory, as the name states, is only around for a brief time.

In the experiments of Brown (1958) and Peterson and Peterson (1959), a simple three-letter trigram (e.g., *MHA*) was presented to people, followed by a three-digit number (e.g., *728*). People were told to attend to the letters, then to begin counting backward by 3s from the number they were given. The counting was done aloud, in rhythm with a metronome clicking twice per second. The essential ingredient here is the distractor task of backward counting. This requires a great deal of attention (if you doubt this, try it yourself, making sure to count twice per second). Furthermore, it prevents rehearsal of the three letters because rehearsal uses the same cognitive mechanism as the backward counting. At the end of a variable period of time, the people reported the trigram. The results were so unexpected, and the number of researchers eager to replicate them so large, that the task acquired a name it is still known by: the **Brown–Peterson task**.

The surprising result was that memory of the 3-letter trigram was only slightly better than 50% after 3 seconds of counting; accuracy dwindled to about 5% after 18 seconds (Figure 5.1).

The letters were forgotten so quickly even though short-term memory was not overloaded—a 50% loss after only 3 seconds (assuming perfect recall with a zero-second delay). On the face of it, this seems evidence of a simple decay function: With an increasing period of time, less and less information remains in short-term memory.

Later research, especially by Waugh and Norman (1965), questioned some of the assumptions made. Waugh and Norman thought that the distractor task itself might be a source of *interference*. If the numbers spoken during backward counting interfered with the short-term memory trace, then longer counting intervals would have created

Figure 5.1 Relative accuracy of recall in the Brown–Peterson task across a delay interval from 0 to 18 seconds. People had to perform backward counting by 3s during the interval.

SOURCE: Peterson & Peterson (1959).

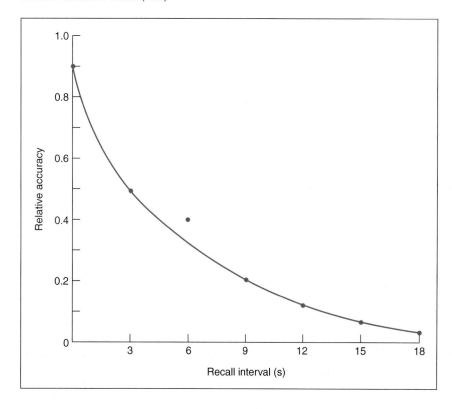

Figure 5.2 Relative accuracy in the Waugh and Norman (1965) probe digit experiment as a function of the number of interfering items spoken between the target item and the cue to recall; rate of presentation was either 1 or 4 digits per second.

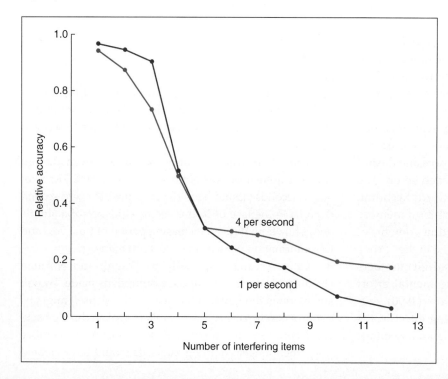

more interference. Waugh and Norman's reanalysis of several studies confirmed their suspicion. Especially convincing were the results of their own probe digit task. People heard a list of 16 digits, read at a rate of either 1 or 4 per second. The final item in each list was a repeat of an earlier item, and it was the probe or cue to recall the digit that had *followed* the probe in the original list. For instance, if the sequence *7 4 6 9* had been presented, then the probe *4* would have cued recall of the *6*.

The important part of their study was the time it took to present the 16 digits. This took 16 seconds for one group (a long time) but only 4 seconds (a short time) for the other group. If forgetting were caused by decay (a time-based process), then the groups should have differed markedly in their recall because so much more time had elapsed in the 16 seconds group. Yet, as Figure 5.2 shows, the two groups differed little.

This suggests that forgetting was influenced by the number of intervening items, not simply the passage of time. In other words, forgetting in short-term memory was caused by interference, not decay (for cross-species evidence of interference, see Wright & Roediger, 2003). Although it has been decades since the original work, the issue of decay versus interference explanations for forgetting in short-term memory continues to be of interest to cognitive psychologists (e.g., Unsworth, Heitz, & Parks, 2008). Some interesting recent work suggests that whereas the bulk of the forgetting is due to interference, there is still a small amount that can be attributed to a decay process as well (Altmann & Gray, 2002; Altmann & Schunn, 2012; Berman, Jonides, & Lewis, 2009).

PROACTIVE AND RETROACTIVE INTERFERENCE (PI AND RI) Shortly after the Peterson and Peterson report, Keppel and Underwood (1962) challenged the decay explanation for forgetting in short-term memory. They found that people forgot at a dramatic rate only after several trials. On the first trial, memory for the trigram was almost perfect.

Keppel and Underwood's explanation was that as you experience more and more trials in the Brown–Peterson task, recalling the trigram becomes more difficult because the *previous* trials generate interference. This is called **proactive interference (PI)**, in which older material interferes forward in time with your recollection of the current stimulus. This is the opposite of **retroactive interference (RI)**, in which newer material interferes backward in time with your recollection of older items.

The loss of information in the Brown–Peterson task was caused by proactive interference.

RELEASE FROM PI An important adaptation of the interference task was done by Wickens (1972; Wickens, Born, & Allen, 1963). He gave people three Brown–Peterson trials, using three words or numbers rather than trigrams. On the first trial, accuracy was near 90%, but it fell to about 40% on trial 3. At this point Wickens changed to a different kind of item for trial 4. People who had heard three words per trial were given three numbers, and vice versa. The results were dramatic. When the nature of the items was changed, performance on trial 4 returned to the 90% level of accuracy.

Wickens also included a control group who got the same kind of stimulus on trial 4 as they had gotten on the first three trials, to make sure performance continued to fall, which it did. Figure 5.3 shows this result.

The interference interpretation is clear. Performance deteriorates because of the buildup of proactive interference. If the to-be-remembered information changes, then you are released from the interference. Thus, **release from proactive interference**, or **release from PI**, occurs when the decline in performance caused by proactive interference is reversed because of a switch in the to-be-remembered stimuli. Release from PI also occurs when the change is semantic, or meaning-based, as when the lists switched from one semantic category to another (see Figure 5.4).

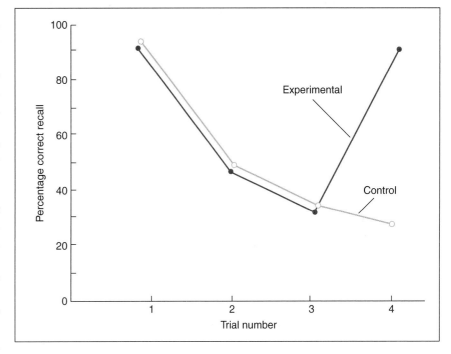

Figure 5.3 Recall Accuracy in a Release From PI Experiment by Wickens, Born, and Allen (1963)

Triads of letters are presented on the first three trials, and proactive interference begins to depress recall accuracy. On trial 4, the control group gets another triad of letters; the experimental group gets a triad of digits and shows an increase in accuracy, known as release from PI.

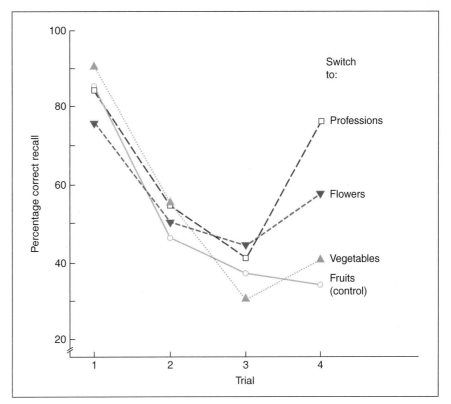

Figure 5.4 Recall Accuracy in a Release From PI experiment by Wickens and Morisano (Reported in Wickens, 1972)

All participants received word triads from the *fruits* category on trial 4. On trials 1 through 3, different groups received triads from the categories *fruits* (control condition), *vegetables*, *flowers*, and *professions*.

However, note that the more related the items on the fourth list were to the original category, the less release from PI was experienced. Thus, short-term memory, to some degree, uses semantic information.

> **WRITING PROMPT**
>
> **Capacity and Forgetting in Short-Term Memory**
>
> How do the limits on capacity and the process of forgetting in short-term memory influence what we are able to think about in the moment? What steps can people take in their day-to-day lives to make things better?
>
> ▶ The response entered here will appear in the performance dashboard and can be viewed by your instructor.
>
> Submit

5.2: Short-Term Memory Retrieval

OBJECTIVE: Summarize the process through which short-term memory is recalled

Here, we consider retrieval from short-term memory, which refers to the act of bringing knowledge to the foreground of thinking and perhaps reporting it. Our focus is on two aspects of retrieval: the serial position curve and studies of the retrieval process itself.

5.2.1: Serial Position Effects

A **serial position curve** is a graph of item-by-item accuracy on a recall task. Serial position simply refers to the original position an item had in a study list.

Before considering serial position curves per se, we will cover the two tasks used to test people: free recall and serial recall. In **free recall**, people are free to recall the list items in any order, whereas in **serial recall**, people recall the list items in their original order of presentation. Not surprisingly, serial recall is more difficult. Recalling items in order requires people to rehearse them as they are shown, trying to hold on to not only the information itself but also its position in the list. The more items there are, the harder the task becomes. By comparison, with free recall, people can recall the items in any order.

The early list positions are called the primacy portion of a serial position curve. Primacy here has its usual connotation of "first": It is the first part of the list that was studied. **Primacy effect**, then, refers to the accuracy of recall for the early list positions. A strong primacy effect means good, accurate recall of the early list items, usually because of rehearsal. The final portion of the serial position curve is the recency portion. **Recency effect** refers to the level of correct recall for the final items of the originally presented list. *High recency* means "high accuracy," and *low recency* means that this portion of the list was hardly recallable at all.

Figure 5.5 shows several serial position curves.

As Figure 5.5A shows, a strong recency effect is found across a range of list lengths; Murdock (1962) presented 20-, 30-, and 40-item lists at a rate of one item per second. Note that there is a slight primacy effect for each list length, but that the middle part had low recall accuracy. Apparently, the first few items were rehearsed enough to transfer them to long-term memory, but not enough time was available for rehearsing items in the middle of the list. For all lists, though, the strong recency effect can be attributed to recall from short-term memory.

The way to eliminate the recency effect should be no surprise. Glanzer and Cunitz (1966) showed people 15-item lists. For some people, after a list they needed to do an attention-consuming counting task for either 10 or 30 seconds before recalling the items. In contrast to people who gave immediate recall (0 seconds delay), the people who had to do a counting task showed very low recency (Figure 5.5B). However, the primacy portion of the list was unaffected. In other words, the early list items were more permanently stored in long-term memory to endure 30 seconds of counting. The most recent items in short-term memory were susceptible to interference.

Other manipulations, summarized by Glanzer (1972), showed how the two parts of the serial position curve are influenced by different factors. Note that providing more time per item during study (spacing of 3 versus 6 versus 9 seconds, in the figure) had almost no influence on recency but did alter the primacy effect (Figure 5.5C; from Glanzer & Cunitz, 1966). Additional time for rehearsal allowed people to store the early items more strongly in long-term memory. Moreover, additional time did not help the immediate recall of the most recent items. These items were held in short-term memory and recalled before interference could take place.

5.2.2: Short-Term Memory Scanning

We turn now to a different question: How do we access or retrieve the information from short-term memory? To answer this question, we turn to another memory task: recognition.

Essentially, a **recognition task** is one in which people are presented with items and are asked to indicate whether

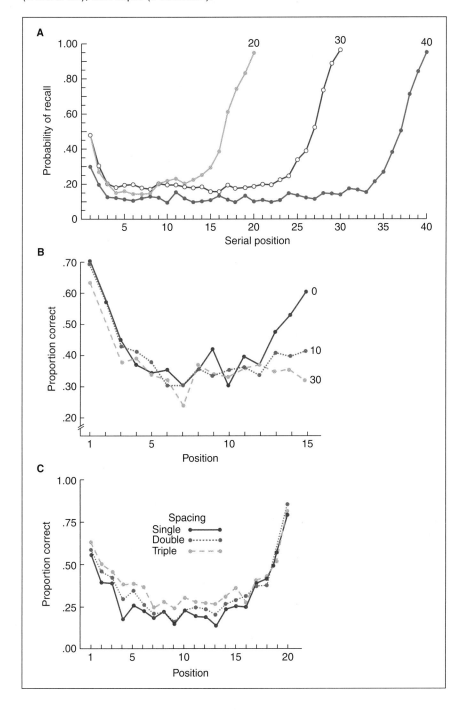

Figure 5.5 (**A**) Serial position curves showing recall accuracy across the original positions in the learned list. Rate of presentation was one item per second. (**B**) Serial position curves showing the decrease in recency when 10 or 30 seconds of backward counting is interpolated between study and recall. (**C**) Three different rates of presentation: single (3 seconds), double (6 seconds), and triple (9 seconds).

the items were part of what had been studied before. People would select "yes" to indicate "yes, I recognize that as being studied." Similarly, they would select "no" to indicate "no, I didn't study that item." Making these decisions requires people to access stored knowledge, then compare the items to that knowledge. The important angle is that we can time people as they make their "yes/no" recognition decisions, and infer the underlying mental processes used on the basis of how long they took. Saul Sternberg used this procedure in addressing the question of how we access information in short-term memory.

Sternberg (1966, 1969, 1975) began by noting that the use of *response time (RT)* to infer mental processes had a venerable history, dating back at least to Donders in the 1800s. Donders proposed a subtractive factors method for determining the time for simple mental events. For example, if your primary task involves processes A, B, and C, you devise a comparison task that has only processes A and C in it. After giving both tasks, you subtract the A + C time from the A + B + C time. The difference should be a measure of the duration of process B because it is the process that was subtracted from the primary task.

Sternberg pointed out a major difficulty with Donders's subtractive method. It is virtually impossible to make sure that the comparison task, the A + C task, contains identical A and C processes as in the primary task. There is always the possibility that the A and C components were altered when you removed process B. If so, then subtracting one from the other cannot be justified. Sternberg's solution was to arrange it so that the critical process would have to *repeat* some number of times during a single trial. Across an entire study there would be many trials in which process B had occurred only once, many in which it had occurred two times, three times, and so forth. He then examined the RTs for these conditions, and inferred the nature of process B by determining how much time was *added* to response times for each repetition of process B. This is referred to as the additive factors method.

THE STERNBERG TASK Sternberg devised a short-term memory scanning task, now simply called a **Sternberg task**. People were given a short list of letters, one at a time,

> **Prove It**
>
> ### Tests of Short-Term Working Memory
> Several tests of short-term working memory can be given with little difficulty, to confirm the various effects that you are reading about in this module.
>
> ---
>
> Simple and Working Memory Spans
> Here are some suggestions.
>
> **Simple Memory Span**
> Make several lists, being sure that the items do not form unintended patterns. Use digits, letters, or unrelated words. Read the items at a constant and rapid rate (no slower than one item per second) and have the participant name them back in order. Your main dependent variable will be the number or percentage correct.
> **Sample Lists**
> Digits
> - 8 7 0 3 1 4
> - 7 1 5 0 5 4 3 6
> - 2 8 4 3 6 1 2 9 7 5
>
> Words
> - leaf gift car fish rock
> - paper seat tire horse film beach forest brush
> - bag key book wire box wheel banana floor bar pad block radio boy

at the rate of one per second, called the memory set. People then saw a single letter, the probe item, and responded "yes" or "no" depending on whether the probe was in the memory set. For example, if you stored the set $l\ r\ d\ c$ in short-term memory and then saw the letter d, you would respond "yes." However, if the probe were m, you would respond "no."

Memory sets were from one to six letters or digits long, within the span of short-term memory, and were changed on every trial. Probes also changed on every

Table 5.1 Sample Sternberg Task

In a typical experiment, people saw several hundred trials, each consisting of these two parts, memory set then probe, as shown in the below table.

Trial	Memory Set Items	Probe Items	Correct Response
1	R	R	Yes
2	LG	L	Yes
3	SN	N	Yes
4	BKVJ	M	No
5	LSCY	C	Yes

trial and were selected so the correct response was "yes" on half the trials and "no" on the other half. This is illustrated by trials 3 and 4 in Table 5.1. Take a moment to try several of these trials, covering the probe until you have the memory set in short-term memory, then covering the memory set and uncovering the probe, then making your "yes/no" judgment. For a better demonstration, have someone read the memory sets and probe to you aloud.

Figure 5.6 illustrates the *process model* that Sternberg (1969) proposed, simply a flowchart of the four separate mental processes that occurred during the timed portion of every trial. At the point marked "Timer starts running here," the person encodes the probe. Then, the search or scan through short-term memory begins and the mentally encoded probe is compared with items in memory to see whether there was a match. A simple "yes" or "no" decision could then be made by pressing one of two buttons.

In Sternberg's task, it was the search process of the contents of short-term memory that was of interest. Notice—this is critical—that it was this process that was repeated a different number of times, depending on how many items were in the memory set. Thus, by manipulating memory set size, Sternberg influenced the number of cycles through the search process. And by examining the slope of the RT results, he could determine how much time was needed for each cycle.

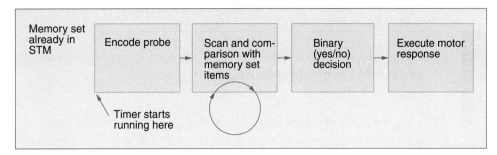

Figure 5.6 The Four-Stage Process Model for Short-Term Memory Scanning

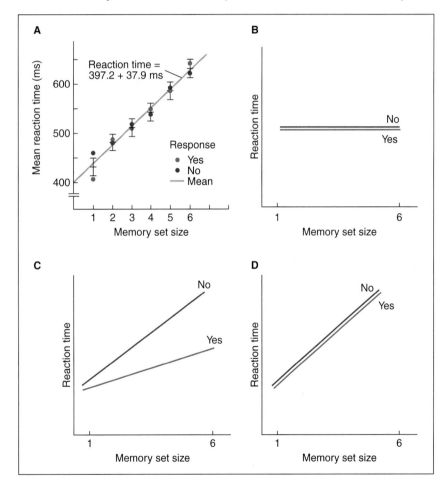

Figure 5.7 Reaction time in the short-term memory scanning task, for "yes" (shaded circles) and "no" (unshaded circles) responses. Reaction time increases linearly at a rate of 37.9 ms per additional item in the memory set.

STERNBERG'S RESULTS Figure 5.7 shows Sternberg's (1969) results. There was a linear increase in RT as the memory set got larger, and this increase was nearly the same for both "yes" and "no" trials. The equation at the top of the figure shows that the y-intercept of this RT function was 397.2 ms. Hypothetically, if there had been zero items in short-term memory, the y-intercept would be the combined time for the encoding, decision, and response stages (refer back to Figure 5.6). More important, the slope of the equation was 37.9 ms; for each additional item in the memory set, the mental scanning process took 37.9 ms. In other words, the search through short-term memory is approximately 38 ms per item—very fast.

What kind of mental search would produce these results?

Three Possibilities of Mental Search

Sternberg considered three possibilities: a serial self-terminating search, a parallel search, and a serial exhaustive search.

Interactive

1. Serial Self-Terminating Search

The most intuitively appealing was a *serial self-terminating* search in which the positions in short-term memory are scanned one by one, and the scan stops when a match is found; this is how you search for a lost object, say your car keys. On the average, the slope of the RT trials for "yes" responses should be smaller than the slope for "no" responses. On the "no" trials, all positions have to be searched before you can decide that the probe was not in the set. But on "yes" trials, people would encounter matches at all positions in the memory set, sometimes early, sometimes late, with equal frequencies at all positions. The slopes of the positive curves ("yes, I found the target") were always smaller than those of the negative curves ("no, the target was not in the display"). However reasonable such a search appears, Sternberg's data did not match the prediction—he found the same slope for both kinds of trials.

2. Parallel Search

3. Serial Exhaustive Search

LIMITATIONS TO STERNBERG'S CONCLUSIONS
Across the years, it has been suggested that increasing RTs could be the product of a parallel search in which each additional item slows down the rate of scanning for all items. This is much like a battery, which can run several motors at once, but each runs more slowly when more motors are connected (see Baddeley, 1976, for a review). Others have objected to the assumption that the several stages or processes are sequential and that one must be completed before the next one begins.

For instance, McClelland (1979) proposed that the mental stages might overlap partially in cascade fashion. Still, Sternberg's work pushed the field toward more useful ways of studying cognition. Most research based on RT tasks (e.g., visual search tasks and many long-term memory tasks) owes credit, even if only indirectly, to Sternberg's groundbreaking and insightful work.

WRITING PROMPT

Information Retrieval From Short-Term Memory

How much insight do most people have in terms of how they retrieve information from short-term memory? Explain.

▶ The response entered here will appear in the performance dashboard and can be viewed by your instructor.

Submit

5.3: Working Memory

OBJECTIVE: Explain how working memory functions

Working memory can be viewed as an augmentation of short-term memory. By the mid-1970s, all sorts of functions were being attributed to short-term memory in tasks of

Functioning of Working Memory

When assessing working memory, we have to also consider distractions and burdens on memory that affect problem solving.

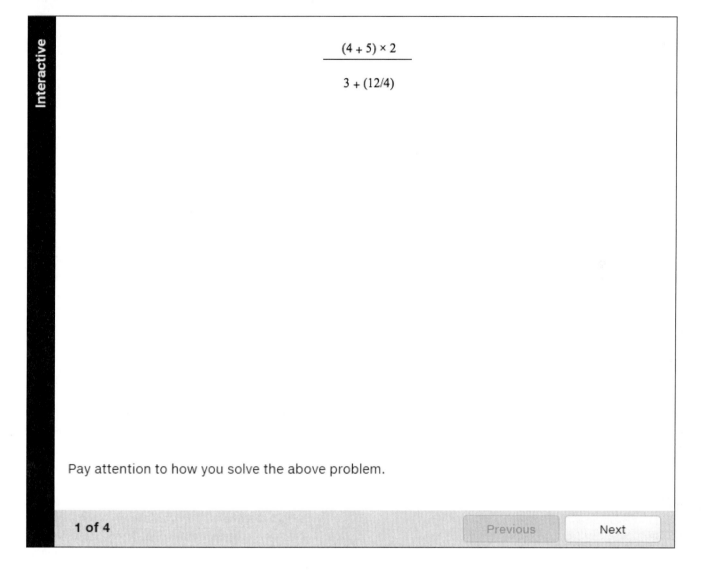

problem solving, comprehension, reasoning, and the like. Yet, as Baddeley pointed out, remarkably little work had demonstrated those functions in STM (Baddeley, 1976; Baddeley & Hitch, 1974; Baddeley & Lieberman, 1980).

Going beyond intuitive examples, Baddeley and Hitch (1974) documented their position by describing a dramatic case study, originally reported by Warrington and Shallice (1969; also Shallice & Warrington, 1970; Warrington & Weiskrantz, 1970) of a patient

> who by all normal standards, has a grossly defective STS. He has a digit span of *only two items*, and shows grossly impaired performance on the Peterson short-term forgetting task. If STS does indeed function as a central working memory, then one would expect this patient to exhibit grossly defective learning, memory, and comprehension. No such evidence of general impairment is found either in this case or in subsequent cases of a similar type.

> (Baddeley & Hitch, 1974, pp. 48–49, emphasis added; also Baddeley & Wilson, 1988; Vallar & Baddeley, 1984)

In a similar vein, McCarthy and Warrington (1984) reported on a patient who had a memory span of only one word, but could nonetheless report back six- and seven-word sentences with about 85% accuracy. Despite both types of lists relying on short-term memory, performance on one type was seriously affected by the brain damage, and the other only minimally.

Baddeley and Hitch reasoned that the problem lies with a simple maintenance theory of STM. They suggested that STM is but one component of a larger, more elaborate system, **working memory**. Because *Baddeley's working memory model* is the most well-known model, we will give extensive coverage to it here. After this, we will describe Engle's model of working memory to provide an alternative idea about the role and processing of working memory.

5.3.1: The Components of Working Memory

A description of Baddeley's working memory model provides a useful context for the studies described later (see Baddeley, 2000a; Baddeley & Hitch, 1974; Salame & Baddeley, 1982). Baddeley's working memory model has four major components.

Baddeley's model, shown in Figure 5.8, has four major components. The main part is the **central executive** (or sometimes **executive control**), assisted by two auxiliary systems: the **phonological loop** and the **visuo-spatial sketch pad**. Both auxiliary systems had specific sets of responsibilities, assisting the central executive by doing some of the lower-level processing. Thus, in the arithmetic problem mentioned earlier, the central executive would be responsible for retrieving values from memory (4 + 5, 9 × 2) and applying the rules of arithmetic. A subsystem, the phonological loop, would then hold the intermediate value 18 in a rehearsal-like buffer until it was needed again. A third auxiliary system, the **episodic buffer** (Baddeley, 2000a), is used to integrate information already in working memory with information retrieved from long-term memory. It is where different types of information are bound together to form a complete memory, such as storing together the sound of someone's voice with an image of his or her face.

The idea of working memory being divided into components is supported by neurological evidence. Smith and Jonides (1999; also Smith, 2000) reviewed several brain imaging studies to identify regions of heightened activity in various working memory tasks. In general, the thinking is that those brain regions involved in perception are also recruited by working memory for the storage of information, regions toward the posterior (back) of the brain, and that the rehearsal and processing of information is controlled by those aspects of the brain involved in motor control and attention (Jonides, Lacey, & Nee, 2005). For the Sternberg task, the scanning evidence showed strong activations in a left hemisphere parietal region and three frontal sites, Broca's area (B.A. 44), and the left supplementary motor area (SMA) and premotor area (B.A. 6).

Broca's area is important in the production of language, so finding increased activity here was not surprising. Alternatively, tasks that emphasize executive control, such as switching from one task to another, tend to show strong activity in the dorsolateral prefrontal cortex (DLPFC) (B.A. 46). (For an argument that task switching does not involve executive control, see Logan, 2003.) This area is central to understanding executive attention (Kane & Engle, 2002). The neurological basis for executive control is also supported by work showing that executive functions in cognition may have a significant genetic basis (Friedman et al., 2008).

Other studies have shown specific brain regions involved in visuo-spatial working memory. In one (Jonides et al., 1993), people saw a pattern of three random dots; the dots were then removed for 3 seconds, and a circle outline appeared; the task was to decide whether the circle surrounded a position where one of the dots had appeared earlier. In a control condition, the dots remained visible while the circle was shown, thus eliminating the need to remember the locations. Positron emission tomography (PET) scans revealed that three right hemisphere regions showed heightened activity and so were involved in spatial working memory. They were a portion of the occipital cortex, a posterior parietal lobe region, and the premotor and DLPFC region of the frontal lobe (see also Courtney, Petit, Maisog, Ungerleider, & Haxby, 1998). In related work, when the task required spatial information for responding, it was the premotor region that was more active; when the task required object rather than spatial location information, the DLPFC was more active (Jonides et al., 1993; see also Miyake et al., 2000, for a review of various executive functions attributed to working memory).[2]

Figure 5.8 An overview of Baddeley's working memory model, including the central executive, the phonological loop, the visuo-spatial sketch pad, and the episodic buffer.

SOURCE: Based on Baddeley (2000a).

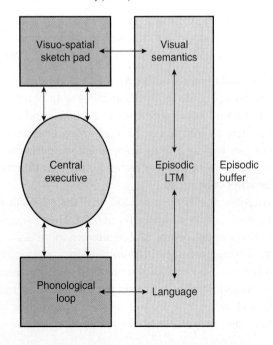

---

[2] The procedure of subtracting patterns in the control condition from those obtained in the experimental condition is straightforward conceptually, although the computations are mind boggling. But notice that conceptually it rests on the same type of logic that Donders used (and Sternberg rejected): finding a control task that contains all of the experimental tasks' components except the one of interest. It will be surprising if this method does not come under attack again, in its newer application to brain imaging.

5.3.2: The Central Executive

The *central executive* is the heart of working memory. It is like a large corporation in which the chief executive controls the tasks of planning, initiating activities, and making decisions. Likewise, in working memory, the central executive controls the planning of future actions, initiating retrieval and decision processes as necessary, and integrating information coming into the system. Some views of the central executive assume that it is composed of multiple elements that are responsible for different aspects of the control of processing (Vandierendonck, 2016).

Let's take an example to illustrate some of this. To continue with the arithmetic example, the central executive triggers the retrieval of facts such as "4 + 5 = 9" and invokes the problem-solving rules such as "how to multiply and divide." Furthermore, the central executive also "realizes" that the intermediate value 18 must be held momentarily while further processing occurs. Accordingly, it activates the phonological loop, sending it the value 18 to rehearse for a few moments until that value is needed again.

Each of the subsystems has its own pool of attentional resources, but the pools are limited. Give any of the subsystems an undemanding task, and it can proceed without disrupting activities occurring elsewhere in working memory. However, if a subsystem is given a particularly difficult task, then either it falters or it must drain additional resources from the central executive. To some degree, working memory resources are shared across processing domains, such as verbal or visual (Vergauwe, Barrouillet, & Camos, 2010), which is why closing your eyes can help you think by reducing distractions from the external environment (Vredeveldt, Hitch, & Baddeley, 2011).

The central executive has its own pool of resources that can be depleted if overtaxed. For example, people who do something that places a strain on the central executive, such as ignoring distracting information as it scrolls across the bottom of a television screen or exaggerating their emotional expressions, have greater difficulty with central executive processing immediately thereafter (Schmeichel, 2007). Moreover, damage to portions of the frontal lobes, such as the medial frontal lobes (B.A. 32) and the polar regions (B.A. 10), can disrupt working memory's executive function (Banich, 2009). This produces *dysexecutive syndrome*, in which patients continue to pursue goals that are no longer relevant, and experience heightened distractibility when they do not have clear goals (Gevins, Smith, McEvoy, & Yu, 1997).

5.3.3: The Phonological Loop

The *phonological loop* is the speech- and sound-related component responsible for rehearsal of verbal information and phonological processing. This component recycles information for immediate recall, including articulating the information in auditory rehearsal. For a debate on the articulatory versus phonological basis of this subsystem, see Baddeley (2000b); Jones, Macken, and Nicholls (2004); and Mueller, Seymour, Kieras, and Meyer (2003).

There are two components of the phonological loop, the phonological store and the articulatory loop. The **phonological store** is essentially a passive store component of the phonological loop. This is the part that holds on to verbal information. However, information in the phonological store is forgotten unless it is actively rehearsed and refreshed. Thus, rehearsal is the role of the **articulatory loop**, the part of the phonological loop involved in the active refreshing of information in the phonological store. One way of thinking about these two components of the phonological loop is that the phonological store is like your inner ear—you can hear yourself talk to yourself, or imagine hearing music. Similarly, the articulatory loop is like your inner voice, when you mentally say things to yourself.

EFFECTS OF THE PHONOLOGICAL LOOP Researchers have found several effects that provide insight into how the phonological loop works. We cover two of them here: the articulatory suppression and phonological similarity effects.

1. The **articulatory suppression effect** is the finding that people have poorer memory for a set of words if they are asked to say something while trying to remember the words (Murray, 1967). This effect is not complicated; it occurs even when you say something simple, like repeating the word *the* over and over again. What happens here is that the act of speaking consumes resources in the articulatory loop. As a result, words in the phonological store cannot be refreshed and are lost. A related phenomenon is the irrelevant speech effect (Colle & Welsh, 1976). It is hard to keep information in the phonological loop when there is irrelevant speech in the environment. This irrelevant speech intrudes on the phonological loop, consuming resources and causing you to forget verbal information. Note that it is not necessary to even actually say anything. As another illustration of the embodiment of cognition, phonological loop processing can be disrupted if people are using the same muscles as they do when they are speaking, such as by chewing gum (Kozlov, Hughes, & Jones, 2012). This is also why it is so difficult to read (and then remember what you read) when you are in a room with other people talking. (So, try to study somewhere quiet.)

2. The **phonological similarity effect** is the finding that memory is poorer when people need to remember a set of words that are phonologically similar, compared to a set of words that are phonologically dissimilar (Baddeley, 1966; Conrad & Hull, 1964). For example, it is

harder to remember the set *boat, bowl, bone,* and *bore,* compared to the set *stick, pear, friend,* and *cake*. This is because words that sound similar can become confused in the phonological store. One thing that happens is that, because the words sound similar, it is hard to keep track of what was rehearsed and what was not. As a consequence, some words may not get rehearsed and are forgotten (Li, Schweickert, & Gandour, 2000). In addition, as bits and pieces of words become forgotten or lost, people need to reconstruct them. As a result, people are more likely to make a mistake by misremembering a word that sounded like it should have been in the set, but was not—for example, recalling the word *bold* in the first set of words. In general, when people misremember words in working memory, those are the words that tend to sound similar, rather than having a similar meaning. This suggests that this aspect of working memory relies primarily on phonological rather than semantic information.

Although we have spent a great deal of time covering the spoken/heard language aspects of the phonological

Prove It

Articulatory Suppression
One of the mainstays of research on the phonological loop is the articulatory suppression task.

Articulatory Suppression Task
The basic idea behind the articulatory suppression task is that repeated talking consumes the resources of the articulatory loop, making it difficult to maintain other information.

> For this task, people are asked to repeat words aloud over and over again while trying to remember another set of verbal/linguistic information. On the face of it, the articulatory suppression task sounds very easy; that it should not be too difficult. However, actually doing it is a humbling experience that shows how limited our working memory capacity is, and how poor our ability is to do more than one thing at a time when the same part of working memory needs to be used.

loop, evidence exists that there are broader aspects of this part of working memory. For example, it has been found that memory for musical pitches shows similar characteristics to language, in that working memory has a limited capacity for what can be remembered, and people become confused by similar pitches, much like the phonological similarity effect (Williamson, Baddeley, & Hitch, 2010). Also, in a very clever study, Shand (1982) tested people who were congenitally deaf and skilled at American Sign Language (ASL). They were given five-item lists for serial recall, presented as either written English words or ASL signs. One list contained English words that were phonologically similar (*shoe, through, new*) though not similar in terms of the ASL signs. Another list contained words that were cherologically similar in ASL—that is, similar in the hand movements necessary for forming the sign (e.g., wrist rotation in the vicinity of the signer's face), although they did not rhyme in English. Recall memory showed confusions based on the cherological relatedness. In other words, the deaf people were recoding the written words into an ASL-based code and holding *that* in working memory. Their errors naturally reflected the physical movements of that code rather than verbal or auditory features of the words.

5.3.4: The Visuo-Spatial Sketch Pad

The *visuo-spatial sketch pad* is a system for visual and spatial information, maintaining that kind of information in a short-duration buffer. If you must generate and hold a visual image for further processing, the visuo-spatial sketch pad is at work. In general, support exists for the idea that the visuo-spatial sketch pad shares some of the same neural processes when manipulating mental images as are used during active perception (Broggin, Savazzi, & Marzi, 2012; Kosslyn et al., 1993).

The operation of the visuo-spatial sketch pad can be illustrated by a study by Brooks (1968). People were asked to hold a visual image in working memory (a large block capital *F*), then to scan that image clockwise, beginning at the lower left corner. In one condition, people said "yes" aloud if the corner they reached while scanning was at the extreme top or bottom of the figure and "no" otherwise. This was the "image plus verbal" condition. The other condition was an "image plus visual" search condition: While people scanned the mental image, they also had to search through a printed page, locating the column that listed the "yes" or "no" decisions in the correct order. Thus, two different secondary tasks were combined with the primary task of image scanning. All the tasks used the visuo-spatial sketch pad of working memory. The result was that making verbal responses—saying "yes" or "no"—was easy and yielded few errors. However, visual scanning of printed columns was more difficult and yielded substantial errors. This is because scanning the response columns forced the visuo-spatial sketch pad to divide its resources between two tasks, and performance suffered.

EFFECTS OF THE VISUO-SPATIAL SKETCH PAD A number of effects have been observed that illustrate basic qualities of the visuo-spatial sketch pad. Two of the overarching principles of this aspect of working memory are the influence of embodied cognition and the focus on the future. As you will see, processing in the visuo-spatial sketch pad acts as if people were actively interacting with objects in the world.

The most dramatic evidence for the visuo-spatial sketch pad comes from work on **mental rotation** (Cooper & Shepard, 1973; Shepard & Metzler, 1971). Mental rotation involves people mentally turning, spinning, or rotating objects in the visuo-spatial sketch pad of working memory. In one study, people were shown drawings of pairs of three-dimensional objects and they had to judge whether they were the same shape. The critical factor was the degree to which the second drawing was "rotated" from the orientation of the first. To make accurate judgments, people had to mentally transform one of the objects, mentally rotating it into the same orientation as the other so they could judge it "same" or "different." Figure 5.9 displays several such pairs of drawings and the basic findings of the study.

The overall result of the articulatory suppression task was that people took longer to make their judgments as the angular rotation increased. In other words, a figure that needed to be rotated 120 degrees took longer to judge than one needing only 60 degrees of rotation, much as what would be found if a person were to manually turn the objects. In fact, performance can be enhanced if people are given tactile feedback (by holding an object in their hands) when the object is the same shape and moves in the same way (Wraga, Swaby, & Flynn, 2008), consistent with an embodied interpretation. Also consistent with an embodied influence, people find it easier to mentally rotate pictures of easily manipulated objects compared to ones that are hard to physically manipulate (Flusberg & Boroditsky, 2011).

In the Cooper and Shepard (1973) report, people were shown the first figure and were told how much rotation to expect in the second figure. This advance information on the degree of rotation permitted people to do the mental rotation ahead of time. Interestingly, the mental processes seem much the same if you ask people to retrieve an image from long-term memory, then hold it in working memory while performing mental rotation on that image. Researchers have found regular time-based effects of rotation, and activation in the visual (parietal) lobes, when people are asked to retrieve an image from long-term memory and rotate it mentally in working memory (Just, Carpenter, Maguire, Diwadkar, & McMains, 2001).

Figure 5.9 Three Pairs of Drawings are Shown

1. For each, rotate the second drawing and decide whether it is the same figure as the first drawing. The A pair differs by an 80-degree rotation in the picture plane, and the B pair differs by 80 degrees in depth; the patterns in C do not match. **2.** The RTs to judge "same" are shown as a function of the degrees of rotation necessary to bring the second pattern into the same orientation as the first. Response time is a linear function of rotation.

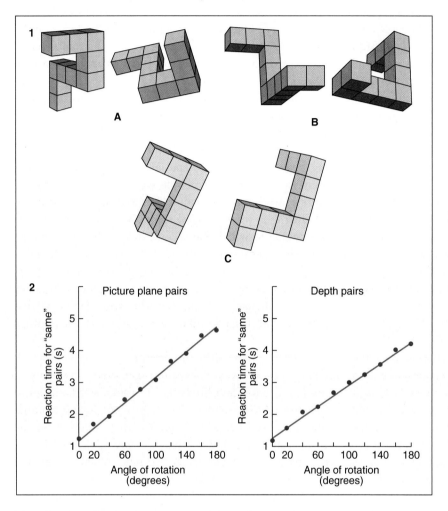

world knowledge of what is likely to be there. This is then stored in long-term memory. So, when you think back to a show you have seen on television or at the movies, you tend to remember the events as if you were actually there, with no edge to the world. You do not typically remember the image as it appeared on the screen (or even remember sitting there watching the show).

The last of the visuo-spatial phenomena considered here (and there are many more) is **representational momentum**, which is the phenomenon of misremembering the movement of an object farther along its path of travel than where it actually was when it was last seen (Hubbard, 1995, 2005). In a typical representational momentum study, people see an object moving along a computer screen. At some point the object disappears. The task is to indicate the point on the screen where the object was last seen. The typical results show a bias to misremember the object as being farther along its path of travel than it actually was (Freyd & Finke, 1984; Hubbard, 1990). It is as if visuo-spatial working memory is simulating the movement as if it were happening in the world, predicting where that object will be next in the near future. This prediction then enters into the decision process, and people place the object farther along its path.

Representational momentum can be influenced by other embodied aspects of the situation as well. For example, there is also a bias to remember objects as being farther down than they actually were, as if they were being drawn down by gravity (Hubbard, 1990). There is also evidence that visuo-spatial working memory takes into account friction (Hubbard, 1990), centripetal, and impetus forces (Hubbard, 1996), even if physics has shown these ideas to be wrong, as in the case of impetus. Finally, if an object is moving in an oscillating motion, back and forth like a pendulum, and it disappears just before it is about to swing back, people will misremember it as having started its backswing (Verfaille & Y'dewalle, 1991).

Overall, it should be clear that there is a lot of active cognition in the visuo-spatial sketch pad. This part of working memory is doing a lot of work, even if you are not consciously aware of much of it. Moreover, this work is oriented toward capturing physical aspects of the world (accurately or not) to help predict what objects will do next so you can

Another illustration of properties of the visuo-spatial sketch pad is **boundary extension**, in which people tend to misremember more of a scene than they actually viewed, as if the boundaries of an image were extended farther out (Intraub & Richardson, 1989). (For a review, see Hubbard, Hutchison, & Courtney 2010.) In boundary extension studies, people see a series of pictures. Later, memory is tested for what was seen in the pictures. This can be done by having people either draw what they remember or identify the image they saw earlier. What is typically found is that people tend to misremember having viewed the picture from farther back than was the case. That is, people misremember information from beyond the bounds of the actual picture (for example, as can be seen in Figure 5.10, the finding that people will remember more of an image than was seen previously, beyond the boundaries of the actual image). Here, visuo-spatial working memory adds knowledge of what is beyond the picture boundary, based on previous

Figure 5.10 A Pair of Pictures

If people saw the one on the left, they were more likely to select the one on the right as having been seen before.

SOURCE: Photo by Gabriel Radvansky, 2016.

better interact with them, such as intercepting or avoiding them, with a minimum of conscious cognitive mental effort.

5.3.5: The Episodic Buffer

As mentioned earlier, the *episodic buffer* is the portion of working memory whereby information from different modalities and sources is bound together to form new episodic memories (Baddeley, 2000a). This is the part of working memory where the all-important chunking process occurs, but it also includes perceptual processes, such as the integration of color with shape in visual memory (Allen, Baddeley, & Hitch, 2006).

5.3.6: Engle's Controlled Attention Model

Although Baddeley's multicomponent model is a dominant theory of working memory in research in cognitive psychology, other ideas exist about what working memory is and what it does. One such alternative perspective is **Engle's controlled attention model** (Kane & Engle, 2002).

Unlike Baddeley's theory, Engle's model does not segregate working memory into different bins and components that do different things. Instead, for this view, working memory is not a separate memory system, per se. It is merely that knowledge in long-term memory currently active and being thought about. Earlier in this course, we pointed out that attention can be thought of as being like a spotlight. Similarly, for this view we can think of working memory as being those ideas and concepts that this mental spotlight is shining on in terms of our own knowledge.

What is particularly important to consider for Engle's view of working memory is how effective working memory is at attending to and processing information. How much control a person can exercise over his or her own train of thought reflects the amount of control a person has over his or her own attention processes. Working memory

Working of the Episodic Buffer

One study that may clarify the workings of the episodic buffer was done by Copeland and Radvansky (2001). Another study, by Jefferies, Lambdon Ralph, and Baddeley (2004), illustrates the capacity needed for integrating information in the episodic buffer.

Study by Copeland and Radvansky (2001)

In this study, people were given a working memory span test (these sorts of tests are described in detail later). People read a series of sentences and had to remember the last word of the sentences in a given set. What was manipulated was the phonetic similarity of the words in a set. Sometimes, the words were phonologically similar, and other times not. The phonological similarity effect described earlier in the phonological loop section predicts that working memory performance would be worse for the phonologically similar items. However, because the words were presented at the end of meaningful sentences, rather than alone, people could use their semantic understanding of the sentences and bind this with their memory for the words. The result was that, under these circumstances, memory for the phonologically similar words was better than for the dissimilar words, much as you would find with poetry or song lyrics. Thus, the episodic buffer's use of the meanings of the words trumped the normal loss of information that happens in the phonological loop.

Study by Jefferies, Lambdon Ralph, and Baddeley (2004)

involves the information that is currently activated in working memory, and the information that is irrelevant and kept out of or suppressed from working memory. Part of the problem people have with working memory processing is not that they do not have enough information active in working memory, but that they have too much of the wrong information in the current stream of thought.

In Engle's model of working memory, two aspects of attention control are particularly important:

1. The scope of attention (i.e., how much information can be attended to at once)
2. The control of attention (i.e., how good people are at directing what is and is not attended to in working memory; Shipstead, Harrison, & Engle, 2015)

This control of attention can be improved by giving yourself less to think about to allow yourself to focus more on information that is relevant at that moment.

For example: If you are thinking about something difficult, you may be able to improve working memory processing by closing your eyes (Vredevelt et al., 2011). This relieves working memory of the need to process information coming in visually, allowing you to focus more of your attention on your internal thoughts. This control of attention involves processing of the frontal lobes of the brain (Kane & Engle, 2002).

Part of the control of attention in working memory is the suppression or inhibition of irrelevant information. This may be information that needs to be kept from entering working memory and taking up valuable working memory capacity.

For example: If you are reading about a boy fishing on a bank, you want to keep the money-bank meaning of *bank* from entering working memory, and to allow the riverbank meaning in. There is some evidence that people who do more poorly on working memory tests have difficulty with this kind of inhibitory processing (Gernsbacher & Faust, 1991).

Alternatively, people may also need to inhibit information that was recently being used in working memory, but is no longer relevant.

For example: You may have been thinking about how you need to unlock the door to get into your place. However, once you do that, you need to purge this information from working memory. There is good evidence that information that was recently used is actively suppressed in working memory (Johnson et al., 2013).

WRITING PROMPT

Working Memory Versus Attention

How is the idea of the cognitive process of working memory different from the idea of the cognitive process of attention?

▶ The response entered here will appear in the performance dashboard and can be viewed by your instructor.

Submit

5.4: Assessing Working Memory

OBJECTIVE: Identify two ways to assess working memory

In general, there are two ways to assess working memory: the dual task method and measures of working memory span.

1. In the dual task method, performance is examined by having a person do a secondary task, one that consumes working memory resources, at the same time as some primary task. It is often used to see how disruptive the secondary task is.
2. By comparison, for working memory span tests, we get a measure of a person's working memory capacity. By comparing the span scores across a range of people and abilities to performance on other tasks, we see what relationships emerge.

Let's go over each of these methods in turn.

5.4.1: Dual Task Method

For the *dual task method*, one of the tasks done by a person is identified as the primary task we are most interested in. The other is a secondary task that is done simultaneously with the first. Both tasks must rely to some significant degree on working memory. In general, we assess how well the two tasks can be done together and whether there is any competition or interference between them. Any two tasks that are done simultaneously may show complete independence, complete dependence, or some intermediate level of dependency. If neither task influences the other, then we infer that they rely on separate mental mechanisms or resources. If one task always disrupts the other, then they presumably use the same mental resources.

One-man band

Finally, if the two tasks interfere with each other in some circumstances but not others, then there is evidence for a partial sharing of mental resources. Usually such interference is found when the difficulty of the tasks reaches some critical point at which the combination of the two becomes too demanding. Researchers manipulate the difficulty of the two tasks just as you would adjust the volume controls on a stereo, changing the left and right knobs independently until the combination hits some ideal setting. In research, we vary the difficulty of each task separately—we crank up the "difficulty knobs" on the two tasks, so to speak—and observe the critical point at which performance starts to suffer.

An important aspect of working memory that the dual task method highlights is that information processed in one component may not interfere with processing in another. For example, active thinking that uses central executive resources will be relatively unaffected by processing that consumes resources in one of the subsystems.

A DUAL TASK EXAMPLE In one experiment (Baddeley & Hitch, 1974, Experiment 3), people were asked to do a reasoning task. They were shown an item such as AB and were timed as they read and responded "yes" or "no" to sentences about it. A simple sentence here would be "A precedes B," an active affirmative sentence. An equivalent meaning is expressed by "B is preceded by A," but it is more difficult to verify because of the passive construction. There were also negative sentences, such as "B does not precede A," and "A is not preceded by B" (as well as false sentences; e.g., "B precedes A"). The sentence difficulty was a way of manipulating the extent of the need for the central executive. While doing the reasoning task, people also had to do one of three secondary tasks:

- Articulatory suppression
- Repeating the numbers 1 through 6
- Repeating a random sequence of digits (the sequence was changed on every trial)

Note how the amount of articulation in the three tasks was about the same (a speaking rate of four to five words per second was enforced), but the demands on the central executive steadily increased. There was also a control condition in which there was no concurrent articulation.

Figure 5.11 shows the reasoning times for these four conditions. The control condition showed that even when reasoning was done alone, it took more time to respond to the difficult sentences. Adding articulatory suppression or repeated counting added more time to reasoning but did not change the pattern of times to any great degree; the curves for *the the the* and *one two three . . .* in the figure have roughly the same slope as the control group. This is because these tasks do not strongly consume working memory resources. However, the random digit condition yielded a different pattern. As the sentences grew more difficult, the added burden of reciting a random sequence of digits took its toll. In fact, for the hardest sentences in the reasoning task, correct judgments took nearly 6 seconds in the random digit condition, compared with only 3 seconds in the control condition. When the secondary task is difficult, the articulatory loop must drain or borrow some of the central executive's resources, thereby slowing down or sacrificing accuracy.

This dual task interference has been shown in a variety of tasks, including showing that dividing attention during driving, such as talking on a cell phone, disrupts the ability to make important judgments, such as when to brake. In general, dual task processing leads to much slower braking (Levy, Pashler, & Boer, 2006). In other words, when you tax working memory resources, it compromises the ability of the central executive to effectively process information.

ANOTHER DUAL TASK EXAMPLE Similar research has also been reported comparing the visuo-spatial sketch pad and the phonological loop.

In the visual memory span task, people saw a grid of squares on the computer screen, with a random half filled in. After a moment, the grid disappeared and was followed by an altered grid in which one of the previously filled squares was now empty. People had to point to the square that was changed, using their visuo-spatial memory of the earlier pattern. In contrast, the letter memory span task, the other primary task, should use the phonological loop. For the secondary tasks, Logie et al. used a mental addition task thought to be irrelevant to the visuo-spatial processing and an

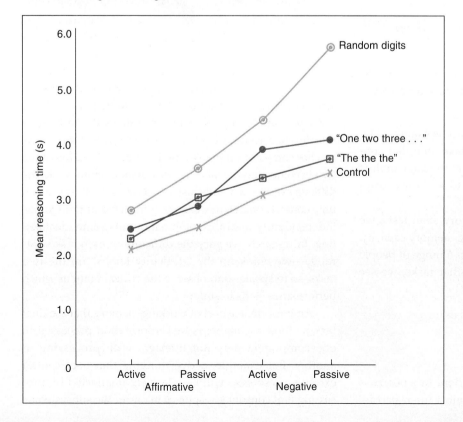

Figure 5.11 Average reasoning time is shown as a function of two variables: the grammatical form of the reasoning problem and the type of articulatory suppression task that was performed simultaneously with reasoning. In the random digits condition, a randomly ordered set of six digits had to be repeated out loud during reasoning; in the other two suppression tasks, either "the the the" or "one two three four five six" had to be repeated out loud during reasoning.

Figure 5.12 Results from Logie, Zucco, and Baddeley's (1990) Experiment on the Visuo-Spatial Sketch Pad

Two secondary tasks, adding and imaging, were combined with two primary tasks, a visual span and a letter span. The results are shown in terms of the percentage drop in performance measured from baseline; the larger the drop, the more disruption there was from the secondary task.

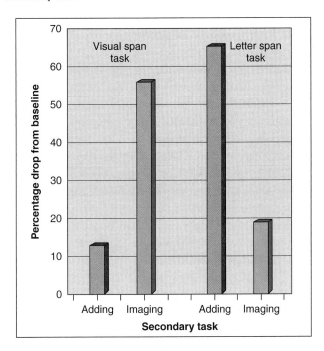

imaging task thought to be irrelevant to the phonological loop. The results are shown in Figure 5.12.

First, look at the left half of the graph, which reports the results of the visual span (grid pattern) task. Each person did the span task alone, to determine baseline, then along with the secondary tasks. The graph shows the percentage *drop* in dual task performance as compared to baseline. For instance, visual span performance dropped about 15% when the addition task was paired with it. So, dual task performance was at 85% of the single-task baseline. In other words, mental addition disrupted visual memory only modestly. But when the secondary task was visual imagery, as shown by the second bar in the graph, visual memory span dropped about 55%. This is a large interference effect, suggesting that the visuo-spatial sketch pad was stretched beyond its limits.

The right half of the figure shows performance on the letter span task. Here, the outcome was reversed in that mental addition was very disruptive to the letter span task, leading to a 65% decline. On the other hand, the imaging task depressed letter span scores only a modest 20%. Thus, only minor declines were observed when the secondary task used a different part of working memory. But substantial declines occurred when the two tasks used the same pool of resources (see Baddeley & Lieberman, 1980, for some of the original research on the visuo-spatial sketch pad). Other work suggests that the impact dual tasks such as these are having is on the encoding aspect of a task, rather than the retention of information in working memory per se (Cowan & Morey, 2007).

5.4.2: Working Memory Span

A different way to study working memory is an individual differences approach. As in any area of psychology, when we speak of individual differences, we are talking about characteristics of individuals—anything from height to intelligence—that differ from one person to the next and can be measured and related to other factors.

Individual differences in working memory are related to various cognitive processes. In this research, people are first given a test to assess their **working memory spans**. They are then given standard cognitive tasks. Consider a program of research by Engle and his coworkers (Engle, 2001; Rosen & Engle, 1997). See Engle (2002) for an excellent introduction.

First, people are given a working memory span task: The task requires simultaneous mental processing and storage of information in working memory. For example, a person might see arithmetic statements along with a word, one at a time (from Turner & Engle, 1989):

$(6 \times 2) - 2 = 10?$ SPOT
$(5 \times 3) - 2 = 12?$ TRAIL
$(6 \times 2) - 2 = 10?$ BAND

People first read the problem aloud, indicated whether the answer was correct, and then said the capitalized word, followed by the second problem and word, then the third. At that point, people tried to recall the three capitalized words, showing that they had stored them in working memory. Scores on this span task are based on the number of capitalized words recalled. Thus, someone who recalled *SPOT TRAIL BAND* and answered the arithmetic questions correctly would have a memory span of 3. In another version of the working memory span task, we use sentences instead of arithmetic problems. But both involve *processing* and *storage*: processing the problem or sentence for meaning, and storing the word for recall.

Many investigators have used span tasks to measure working memory capacity. The original work that used this method (Daneman & Carpenter, 1980) examined reading comprehension as a function of span. There were significant correlations between span scores and performance on the comprehension tasks. One of the most striking correlations was between span and verbal Scholastic Aptitude Test (SAT) scores; it was .59. However, simple span scores seldom correlated significantly with SATs. (Simple span tasks, such as remembering a string of digits, test only the storage of items, whereas working memory span tasks involve both storage and processing.) This strong correlation means that there is an important underlying relationship between one's working memory span and the verbal processing measured by the SAT.

The strongest correlation in Daneman and Carpenter's work was a .90 correlation between memory span and performance on a pronoun reference test. Here, people read sentences one by one and at some point confronted a pronoun that referred back to a previous noun. In the hardest condition, the noun had occurred up to six or seven sentences earlier. The results are shown in Figure 5.13.

Here, people with the highest working memory span of 5 scored 100% correct on the pronoun test, even in the "seven sentences ago" condition. People with the lowest spans (of 2) got 0% correct. Thus, people with higher working memory spans were able to keep more relevant information active in working memory as they comprehended the sentences.

Research since then has extended these findings. Basically, if a task relies on a need to control attention, scores on the task correlate strongly with working memory span.

Figure 5.13 The percentage of correct responses to the pronoun reference task when the antecedent noun occurred a small, medium, or large number of sentences before the pronoun, as a function of participants' working memory (reading) span.

SOURCE: Daneman & Carpenter (1980).

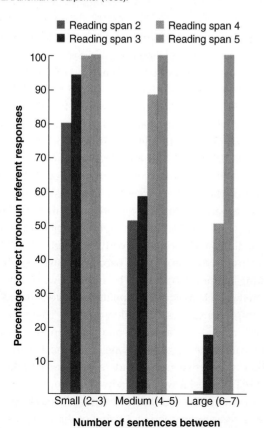

In fact, Engle (e.g., 2002) argued that working memory capacity is executive attention and offers the equation "WM = STM + controlled attention"—working memory is the combination of traditional short-term memory plus our controlled attention mechanism (Kane & Engle, 2003; see also Daneman & Merikle, 1996; Engle, 2002; Miyake & Shah, 1999; Unsworth & Spillers, 2010). This attentional control of the flow of thought involves both the maintenance of information in the short term as well as the ability to access needed information in long-term memory (Unsworth & Engle, 2007).

There is also some evidence that working memory abilities can change with practice. For example, women tend to perform less well than men on visuo-spatial tasks. However, with some training, such as 10 hours of experience playing action-based video games, the performance of females can reach the levels of males (Feng, Spence, & Pratt, 2007). Such experience can also boost performance on visuo-spatially based scientific thinking, such as understanding plate tectonics (Sanchez, 2012). It has also been suggested that musical training can boost verbal intelligence (Moreno et al., 2011). Finally, there is some evidence that certain kinds of Buddhist meditation can also improve mental imagery abilities (Kozhevnikov, Louchakova, Josipovic, & Motes, 2009). Overall, this work, along with the research by Chase and Ericsson (1982) described earlier (as well as others, e.g., Verhaeghen, Cerella, & Basak, 2004), suggests that people can develop strategies to more efficiently and effectively use their working memories over and above any base level of capacity they may have.

That said, memory span scores do not provide insight into all aspects of cognitive abilities. For example, a study by Copeland and Radvansky (2004b) gave people a variety of memory span tasks and assessed performance at more complex levels of comprehension, such as remembering event descriptions, drawing inferences about causes and effects, and detecting inconsistencies in a text. The results showed little evidence of a relation between working memory span and performance. Thus, although memory span highlights important cognitive abilities, it is not the complete story. There is individual variation that can be attributed to other factors as well.

5.4.3: Improving Working Memory

A line of work that has been gathering a great deal of attention and exploration recently is the idea that it might be possible to have people undergo working memory training and improve their overall working memory ability. This idea is similar to that of improving attention processing by having people play action video games. The issue of improving working memory ability is of such great interest because the effectiveness of working memory is strongly related to improved cognitive performance on a wide range of other mental processes, as outlined in the next section of the module. Most salient is the finding that working memory performance is often strongly correlated with measures of fluid intelligence. So, the thinking goes, if you can boost a person's working memory abilities, you should also boost his or her intelligence. And who wouldn't want that?

This question has gained so much attention and interest because certain studies appear to show that working memory training can at least modestly improve working memory performance, often within weeks, as well as performance on intelligence tests (Au et al., 2015). That said, not all the news is so rosy. Some researchers have suggested that though working memory training regimens certainly improve a person's ability to take working memory tests, it is not clear that such working memory task improvement carries over to improvements in other kinds of cognition or to general intelligence (e.g., Chooi & Thompson, 2012; Harrison et al., 2013; Waris, Soveri, & Laine, 2015). There have also been some suggestions that studies showing that working memory training improves cognition may suffer from methodological problems (Bogg & Lasecki, 2015). This does not detract from other research showing that certain kinds of activities are related to superior intelligence, such as musical instrument training (Benz, Sellaro, Hommel, & Colzato, 2015; Gordon, Fehd, & McCandliss, 2015).

WRITING PROMPT

Assessing Working Memory

What are the advantages and disadvantages of the different methods of assessing working memory? Do you think that there are ways (as of yet known or unknown) to improve a person's ability to use working memory?

The response entered here will appear in the performance dashboard and can be viewed by your instructor.

Submit

5.5: Working Memory and Cognition

OBJECTIVE: Evaluate the centrality of working memory in the overall cognitive functions

As noted earlier, working memory does not exist or operate independently of other aspects of cognition. It is the vital center of focus of a great deal of activity. In the next few sections, we discuss ways that working memory influences processing in a variety of domains, including attention, long-term memory, and reasoning.

5.5.1: Working Memory and Attention

Conway, Cowan, and Bunting (2001) examined working memory span and its relation to the classic cocktail party effect of hearing one's own name while paying attention to some other message. About 65% of the people with low memory spans detected their name in a dichotic listening task, versus only 20% of people with high spans. The idea was that high-span people were selectively attending to the shadowed message more effectively than the low-span people—so they were not as likely to detect their names on the un-shadowed message. In contrast, the low-span people had difficulty blocking out or inhibiting attention to the distracting information in the unattended message—so they were more likely to hear their own names.

In a similar demonstration, Kane and Engle (2003) used the classic *Stroop task*—naming the ink color the word is printed in, when the ink color mismatches. There was a strong Stroop effect, of course—approximately a 100-ms slowdown on mismatching items. More to the point, there was no difference in the Stroop effect for high- and low-span groups when the words were always in a mismatching color (GREEN printed in red ink) or when half of the words were presented that way. Everyone remembered the task goal—ignore the word—in these conditions. But when only 20% of the words were in mismatching colors, low-span people made nearly twice as many errors as high-span people. Because mismatching trials were rarer, the low-span people seemed less able to maintain the task goal in working memory. High-span individuals had less difficulty maintaining that goal.

In a more everyday example, Sanchez and Wiley (2006) tested people with different memory spans, giving them texts to read that included illustrations. These illustrations were often irrelevant to the main points of the text, such as having a picture of snow in a passage about ice ages—the snow is related to the topic, but does not provide or support any new information. As such, performance is better if working memory capacity were to focus on the relevant details in the text. What was found was that people with lower memory spans were more likely to be "seduced" by the irrelevant details in the pictures. They had more difficulty controlling the contents of their current stream of thought and were more apt to be led astray by attractive, but unhelpful, sources of knowledge that served as a distraction.

5.5.2: Working Memory and Long-Term Memory

Long-term memory function can also depend on working memory. Rosen and Engle (1997), for instance, had high- and low-span people do a verbal fluency task: generate members of the animal category as rapidly as possible for up to 15 minutes. High-span people outperformed their low-span counterparts, a difference noticeable even 1 minute into the task. Intriguingly, in a second experiment, both groups were tested in the fluency task alone and in a dual-task setting. While naming animals, people had to simultaneously monitor the digits that showed up, one by one, on the computer monitor and press a key whenever three odd digits appeared in the sequence. This task reduced performance on the animal naming task, but only for the high-span people, as shown in Figure 5.14.

Low-span people showed no decrease in performance.

Rosen and Engle suggested that the normal, automatic long-term memory search for animal names was equivalent in both groups. But high-span people were able to augment this with a conscious, controlled strategic search; in other words, along with regular retrieval, the high-span people could deliberately ferret out additional hard-to-find animal names using a controlled attentional process. This additional "ferreting" process used working memory. Consequently, the added digit monitoring task used up the working memory resources that had been devoted to the strategic search. This made the high-span group perform more like the low-span group when they had to perform the dual task.

Other studies have also shown the importance of working memory. For instance, Kane and Engle (2000) found that low-span people experience more proactive interference (PI) in the Brown–Peterson task than do high-span people. High-span people used their controlled attentional processes to combat PI, so they showed an increase in PI when they had to do a simultaneous secondary task. (See Bunting, Conway, & Heitz, 2004; Cantor & Engle, 1993; and Radvansky & Copeland, 2006a; for an exploration of the role of working memory span in managing associative interference during retrieval.) Generally, low-span people appear to search a wider range of knowledge, making them more prone to having irrelevant information intrude on retrieval (Unsworth, 2007). In Hambrick and Engle (2002; see also Hambrick & Meinz, 2011), high-span people had better performance than low-span people on a long-term memory retrieval task, even when both groups were equated for the rather specialized domain knowledge being tested. (In the experiment, people listened to simulated radio broadcasts of baseball games.)

5.5.3: Working Memory and Reasoning

The idea that working memory involves controlled attention can also be tied to general issues of cognitive and behavioral control, such as those needed in problem solving. People with lower memory span scores may be less effective at controlling their thought processes. One example of this is a study by Moore, Clark, and Kane (2008) that looked at working memory span and choices on moral reasoning problems. So, suppose there is a runaway trolley car. If you let it go, it will kill four unaware people a bit down the track. Or, you could push a very large person next to you in front of the trolley; it will kill him but will derail the trolley and save the other four people. So, how morally acceptable is each of these choices? Moore et al. found that moral reasoning of this type was mediated by a person's working memory capacity, with high working memory capacity people making choices on a more consistent (i.e., principled) basis.

The influence of memory span is also seen on more traditional sorts of mental reasoning, such as solving logic problems like categorical syllogisms. In a study by Copeland and Radvansky (2004a; see also Markovits & Doyon, 2004), people with various working memory spans were asked to solve a set of logic problems. There were two primary findings. First, people with greater memory spans solved more syllogisms than did people with smaller memory spans. Second, working memory span was also related to the strategies people used to reason. People with smaller memory spans used simpler strategies. It may be that having more working memory capacity allows one to keep more information active in memory, allowing a

Figure 5.14 The cumulative number of animal names generated by participants of high (open points) or low (filled points) working memory span. Dashed lines indicate performance when participants performed the secondary task of monitoring a stream of digits while generating animal names.

SOURCE: Rosen & Engle (1997).

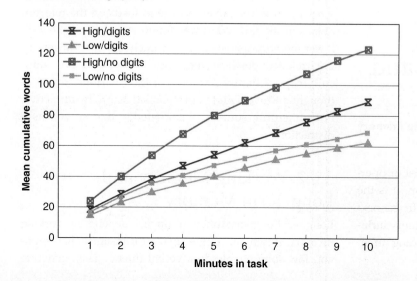

person to explore different alternatives when trying to reason and draw conclusions.

5.5.4: Sometimes Small Working Memory Spans Are Better

Intuitively, it would seem that people with more working memory capacity, or those who engage working memory resources more effectively, are more likely to succeed, and this is generally true. However, there are some interesting exceptions. One of these is illustrated in a study by Beilock and DeCaro (2007) in which high- and low-span people were given math problems to solve. Under normal conditions, high-span people tended to do better. However, people were then placed in a high-pressure situation. They were told that they were being timed, that their performance would be videotaped so math experts could evaluate them, that they would be paid for improving their performance, and so forth. In this high-pressure condition, working memory capacity was consumed with task-irrelevant anxiety-induced thoughts, and performance in both the high- and low-span groups was equivalent (and lower). Thus, when people have their working memory capacity consumed by irrelevant thoughts, they are more likely to use simpler, less effective strategies. This shift to simpler strategies equated people by causing the high-span people to solve the problems more like the low-span people, who had been using simpler strategies in the first place.

Of particular interest, in a second experiment, people were asked to do a series of word problems that required a complex series of steps (i.e., $B - A - 2 \times C$). Then under low- or high-pressure conditions, people were given a series of new problems, some of which required a simpler solution (i.e., $A - C$). Beilock and DeCaro (2007) found that the low-span people were actually more likely to use the simpler correct solution than the high-span people. The explanation was that low-span people are less likely to derive rule-based strategies for solving problems (because they have less capacity to do so) and are more likely to draw from previous similar experiences. Thus, when given problems with the simpler solutions, the low-span people were less dependent on a complex, rule-based strategy they had derived earlier, and so were more likely to use the more appropriate, simpler strategy. (For other examples of better performance by people with smaller memory spans, see Cokely, Kelley, & Gilchrist, 2006; Colflesh & Conway, 2007.)

5.5.5: Working Memory Overview

The general conclusion from all these studies is that *working memory* is a more suitable name for the attention-limited workbench system. Working memory is responsible for the active mental effort of regulating attention, for transferring information into and from long-term memory.

More important, there is a limit to the mental resources available to working memory. When extra resources are drained by the subsystems, the central executive suffers along with insufficient resources for its own work. Naturally, as processes become more automatic, fewer resources are tied down (e.g., working memory is unrelated to counting when there are only two or three things to count but is influenced for larger quantities; Tuholski, Engle, & Baylis, 2001).

Engle's view of working memory emphasizes the general nature of working memory capacity as a measure of executive attention and de-emphasizes the multicomponent working memory approach advocated by Baddeley. Part of the reason for this is the generality of the working memory effects. Working memory span predicts performance on a variety of tasks. Of particular importance, working memory span routinely correlates with measures of intelligence, especially so-called fluid intelligence (Fukuda, Vogel, Mayr, & Awh, 2010; Kane et al., 2004; Salthouse & Pink, 2008; Shelton, Elliot, Matthews, Hill, & Gouvier, 2010), and more so if metacognitive control is taken into account (McCabe, 2010). The relationship between working memory and intelligence is that people who have greater cognitive control and can manage sources of interference better, score better on intelligence tests. This is even supported by neuroimaging data in which people with higher working memory span and intelligence scores show better interference control in cortical areas tied to these processes, such as the dorsolateral prefrontal cortex (B.A. 9) and portions of the parietal cortex (B.A.s 7 and 40) (Burgess, Gray, Conway, & Braver, 2011).

Common to both the Baddeley and the Engle approaches, however, is a central set of principles. Working memory is intimately related to executive control, to the deliberate allocation of attention to a task, and to the maintenance of efficient, effective cognitive processing and behavior. However, there is a limitation in the amount of attention available at any one time. Furthermore, the ability to deliberately focus and allocate attention, and to suppress or inhibit attention to extraneous factors, is key to higher-order cognitive processing.

WRITING PROMPT

Working Memory Influence on Cognition Effectiveness

Other than those discussed in the text, what are some other ways that you think that working memory might influence the effectiveness of cognition?

The response entered here will appear in the performance dashboard and can be viewed by your instructor.

Submit

Summary: Short-Term Working Memory

5.1 A Limited-Capacity Bottleneck

- Short-term or working memory is an intermediate system between the sensory and long-term memories. Its capacity for holding information is limited, by some accounts, to only 7 +/− 2 units of information, although other accounts suggest that it may be able to hold only 4 +/ 1 chunks of information. The processes of chunking and recoding, grouping more information into a single unit, are ways of overcoming this limit or bottleneck.
- Whereas a decay explanation of forgetting from short-term memory is possible, most research implicates interference as the primary reason for forgetting. The research suggests two kinds of interference: retroactive and proactive interference.

5.2 Short-Term Memory Retrieval

- Serial position curves reveal two kinds of memory. Early positions in a list are sensitive to deliberate rehearsal that transfers information into long-term memory, called the primacy effect, whereas later positions tend to be recalled with high accuracy in the free recall task. This latter effect, called the recency effect, is due to the strategy of recalling the most recent items first. Asking people to do a distractor task before recall usually eliminates the recency effect because the distractor task prevents them from maintaining the most recent items in short-term memory.
- Sternberg's paradigm, short-term memory scanning, provided a way to investigate how we search through short-term memory. Sternberg's results indicated that this search is a serial exhaustive process occurring at a rate of about 38 ms per item to be searched.

5.3 Working Memory

- Working memory consists of a central executive and three major subsystems: the phonological loop for verbal and auditory information, the visuo-spatial sketch pad for visual and spatial information, and the episodic buffer for integrating or binding information from different parts of working memory and/or long-term memory.
- The various components of working memory are thought to operate relatively independently of one another, perhaps by using different neural substrates, although there can be some overlap for demanding tasks.
- An alternative view is that working memory is critically involved in the control of attention and thought. From this perspective, working memory is not a separate system, but the portion of long-term memory that is currently activated.
- There are capacity limits in the system.

5.4 Assessing Working Memory

- One common method for assessing working memory is to use dual task methodologies. In these tasks, people are asked to simultaneously do at least two tasks. Researchers then assess how performance on the primary task is affected by the addition of the secondary task, and the theoretical relationship between the two.
- Dual task methods can be used to study strains on individual components or on the overall capacity of working memory. For example, the subsystems may drain extra needed capacity from the central executive in situations of high working memory demands.
- An alternative research strategy is to test working memory span, then examine differences in performance as a function of span scores. This has revealed several tasks that show a relationship between span and performance. The implication is that working memory span assesses controlled attentional processes, which are significant aspects of performance.
- There have been several efforts at working memory training that aim to boost working memory performance and, hence, general intelligence. Although there has been some reported success on this score, there is also cause for concern as research continues to sort this issue out.

5.5 Working Memory and Cognition

- Working memory abilities and performance are critical to many tasks assessed by cognitive psychologists. For example, working memory capacity is strongly related to the ability to engage attention. It has also been shown to be strongly related to the efficiency with which simple facts can be retrieved from long-term memory.
- Although larger working memory capacity is associated with superior cognitive performance, there are cases in which circumstances favor smaller working memory capacity. These are typically circumstances in which it is better not to devote too much attention to a task.

- Although there is no clear view on exactly what working memory is, as evidenced by the Baddeley multicomponent model and the Engle attentional control view, there are a number of agreed-on characteristics of what working memory is able to do. These include its limited capacity, the ability to simultaneously handle certain types of noninterfering forms of information, the fact that people differ in their working memory capacities and abilities, and that these individual differences are related to performance on a variety of tasks.

SHARED WRITING

Working Memory Capacity and Performance

Working memory capacity has been linked to performance on intelligence tests. Explain why you think this might be the case. Are there circumstances under which this would not necessarily apply?

 A minimum number of characters is required to post and earn points. After posting, your response can be viewed by your class and instructor, and you can participate in the class discussion.

Post   0 characters | 140 minimum

Chapter 6
Learning and Remembering

 Learning Objectives

6.1: Analyze the factors that contribute to how well a person will remember something

6.2: Explain the process by which information is stored in episodic memory

6.3: Describe different ways that a person can improve his or her episodic memory

6.4: Explain how context plays a role in the storage and retrieval of episodic memories

6.5: Identify the influence the levels of mental representations have on memory

6.6: Characterize phenomena that has been uncovered by research on autobiographical memory

6.7: Describe how memory can be oriented toward the future

This is the first of three modules devoted to long-term memory, the storage vault for a lifetime's worth of knowledge and experience. Why three modules?

1. Long-term memory is fundamental to nearly every mental process, to almost every act of cognition. You cannot understand cognition unless you understand long-term memory.
2. Long-term memory is an enormous area of research because there is a lot to know about it.
3. People are curious about their own memories. Who has not complained at some time about forgetfulness or the unreliability of memory?

Long-term memory is divided and subdivided in various ways. Look at Figure 6.1, a taxonomy suggested by Squire (1986, 1993).

As the figure shows, a distinction can be made between declarative memory (or explicit memory) and nondeclarative memory (or implicit memory). Here, **declarative** or

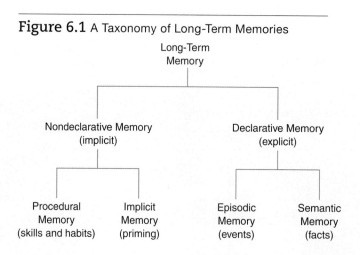

Figure 6.1 A Taxonomy of Long-Term Memories

explicit memory is long-term memory knowledge that is retrieved and reflected on consciously. In contrast, **nondeclarative** or **implicit memory** is knowledge that influences thought and behavior without any necessary involvement of consciousness. The key to this distinction is conscious awareness: Either one has it or one does not.

Here is a brief example to clarify the distinction between episodic and semantic memory. If you were asked what happened when you took your driver's license test, you would retrieve knowledge from *episodic memory*—memory of the personally experienced events. When you retrieve that information, you are conscious of it, you can talk about it, and so on. Episodic memory enables you to record your personal interactions with the world. Alternatively, if you were asked what a driver's license is, you would retrieve information from *semantic memory*, which is your general world knowledge. You retrieve the concept of a driver's license, and it is now in your conscious awareness. Whereas episodic memory is your mental slide show, semantic memory is your mental encyclopedia. Both involve knowledge that you can be consciously aware of (Tulving, 1972, 1983, 1993).

But there is more going on than what rises to the level of conscious awareness. As you read this sentence, you encounter the term *license* again. Although you are not conscious of it, you are faster at reading that term than you were the first time. The rereading speedup is called **repetition priming**. This happens at the nondeclarative, implicit level. Likewise, if we gave you some word stems an hour from now and asked you to fill in the blanks with the first word that comes to mind, you would be more likely than chance to complete *LIC___* as *LICENSE* because you had encountered it recently. This happens even if you do not consciously recall having seen the word *license* (but teasing apart conscious and unconscious influences is no simple matter; see Buchner & Wippich, 2000; Jacoby, 1991; Jacoby, Toth, & Yonelinas, 1993).

An important aspect of episodic memories is that they are integrated mental representations. Different bits and pieces of information from different parts of our conscious and unconscious mental worlds are woven together. The *episodic buffer* is the part of working memory that integrates different types of knowledge to form episodic memories. Although many of the examples given here use linguistic materials, such as word lists, episodic memories integrate various types of information including sensory, motor, spatial, language, emotional, and narrative information, as well as various other encoding and retrieval processes (Rubin, 2007). Even this is not an exhaustive list. So, as you can see, episodic memory uses a rich variety of information about a broad range of human experience.

6.1: Preliminary Issues

OBJECTIVE: Analyze the factors that contribute to how well a person will remember something

Let's start by considering four preliminary issues of episodic memory. First, we talk about a classic, ancient approach to memory, mnemonic devices. We then spend time on the first systematic research on human memory, by Ebbinghaus, published in 1885. Here is a warning: As you read the modules on long-term memory, note that just as your frustrations over your own memory problems probably are exaggerated, so is your certainty about remembering. It is a paradox; our memories are better than we often give ourselves credit for, and worse than we are often willing to believe or admit. Next, we cover the issue of the process that makes memories permanent: consolidation. This topic was mentioned earlier in the Cognitive Neuroscience module and is discussed further here because this process is vital in making long-term memories long term. Finally, we consider other issues about people's awareness of their own memory content and processes, what is known as *metamemory*.

6.1.1: Mnemonics

The term *mnemonic* (pronounced "ne-MAHN-ick") means "to help the memory"; it comes from the same Indo-European base word as *remember*, *mind*, and *think*. A **mnemonic device** is an active, strategic learning device or method. Formal mnemonics use preestablished sets of aids and require considerable practice. The strengths of mnemonics include the following:

1. The material to be remembered is practiced repeatedly.
2. The material is integrated into an existing memory framework.
3. The mnemonic provides a way to retrieve the material.

Let's consider two traditional mnemonics.

CLASSIC MNEMONICS The first historical mention of mnemonics is from the first century B.C., in Cicero's *De oratore*, a treatise on rhetoric. *Rhetoric* is the art of public speaking, which in those days meant speaking from memory.

PRINCIPLES OF MNEMONIC EFFECTIVENESS Mnemonic effectiveness involves three principles.

Another mnemonic device, the **story mnemonic**, provides an excellent illustration of these features of mnemonics. Creating a narrative or story for a set of information will improve your memory. This can be seen in Figure 6.2, from a study by Bower and Clark (1969).

Techniques of Mnemonics (Table 6.1 and Table 6.2)

The power of mnemonics is tremendous. Among other things, mnemonics enabled Greek orators to memorize and recite epics such as *The Iliad* and *The Odyssey*.

The Method of Loci

The Peg Word Mnemonic

The Peg Word Mnemonic Device

Numbered Pegs	Word to Be Learned	Image
One is a bun	Cup	Hamburger bun with smashed cup
Two is a shoe	Flag	Running shoes with flag
Three is a tree	Horse	Horse stranded in top of tree
Four is a door	Dollar	Dollar bill tacked to front door
Five is a hive	Brush	Queen bee brushing her hair
Six is sticks	Pan	Boiling a pan full of cinnamon sticks
Seven is Heaven	Clock	St. Peter checking the clock at the gates of Heaven
Eight is a gate	Pen	A picket fence gate with ballpoint pens as pickets
Nine is a vine	Paper	Honeysuckle vine with newspapers instead of blossoms
Ten is a hen	Shirt	A baked hen on the platter wearing a flannel shirt

Table 6.2

Another mnemonic is the **peg word mnemonic** (Miller, Galanter, & Pribram, 1960), in which a prememorized set of words serves as a sequence of mental "pegs" onto which the to-be-remembered material can be "hung." The peg words rely on rhymes with the numbers 1 through 10, such as "One is a bun, two is a shoe," and so on (see Table 6.2 above).

In this study, people were given 12 lists of 10 words each and asked to remember them in order. As can be clearly seen, people who formed a story from the words had much better memory of them than people who simply tried to memorize the lists as best they could. The stories people created for themselves provided structure, imagery, and effort, and served as excellent sources of **retrieval cues**. In fact, when Chao Lu set the world record in 2005 for memorizing digits of pi (67,890 of them), he created a story out of the digits to help him remember (Hu, Ericsson, Yang, & Lu, 2009). Overall, it cannot be stressed enough how the active use of retrieval cues and then practicing retrieval are important for successful remembering. Extended practice was the key in Chaffin and Imreh's (2002) study of how a pianist learned, remembered, and performed a challenging piece.

This three-step sequence may sound familiar. It is the sequence we talk about every time we consider learning and memory: the encoding of new information, its retention over time, and the retrieval of the information (Melton, 1963). Performance in any situation that involves memory depends on all three steps. A fault along any one of the three might account for poor performance. A good mnemonic device, including those you invent for yourself (e.g., Wenger & Payne, 1995), ensures success at each of the three stages. Incidentally, do not count on a magic bullet to enhance your memory. Research has found little, if any, evidence that ginkgo biloba, or any other so-called memory enhancer, has any real effect on memory (Gold, Cahill, & Wenk, 2002, 2003; Greenwald, Spangenberg, Pratkanis, & Eskenazi, 1991; McDaniel, Maier, & Einstein, 2002).

The Three Principles of Mnemonic Effectiveness

Interactive

First Principle

It provides a structure for acquiring the information. The structure may be elaborate, like a set of 40 loci, or simple, like rhyming peg words. It can even be arbitrary if the material is not extensive. (The mnemonic *HOMES* for the names of the five Great Lakes—Huron, Ontario, Michigan, Erie, and Superior—is not related to the to-be-remembered material, but it is simple.)

Second Principle

Third Principle

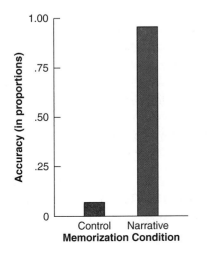

Figure 6.2 The Retention of Word Lists Illustrates the Power of the Story Mnemonic

6.1.2: The Ebbinghaus Tradition

We turn now to the first systematic research on human memory done by the German psychologist Hermann von Ebbinghaus. The Ebbinghaus tradition began with the publication of *Über das Gedächtnis* (1885). The 1964 English translation is titled *Memory: A Contribution to Experimental Psychology*. Ebbinghaus used himself as the only participant in his studies. He also had to invent his own memory task, his own stimuli, and his own procedures for testing and data analysis. Few could do as well today. In devising how to analyze his results, he even came close to inventing a within-groups *t*-test (Ebbinghaus, 1885/1913, footnote 1, p. 67).

6.1.3: Memory Consolidation

Just because you think about something and understand it does not mean that you will remember it later. Think of

Ebbinghaus's Research on Human Memory (Figure 6.3)

It is helpful to consider *why* Ebbinghaus felt compelled to invent and use nonsense syllables. His rationale was that he wanted to study the properties of memory and forgetting apart from the influence of prior knowledge. As such, words would complicate his results. If he had used words, it would be less clear whether performance reflected the simple use of memory or the influence of prior knowledge. Putting it simply, *learning* implies acquiring new information. Yet, words are not new. And a control factor he adopted, to reduce the possible intrusion of mnemonic factors, was the rapid presentation rate of 2.5 items per second.

The Relearning Task

Ebbinghaus devised the **relearning task** (or *savings task*), in which a list is originally learned, set aside for a period of time, then later relearned to the same criterion of accuracy. In most cases, this was one perfect recitation of the list, without hesitations. After relearning the list, Ebbinghaus computed a savings score as the measure of learning. The **savings score** was the reduction in the number of trials (or the time) necessary for relearning compared to original learning. Thus, if it took 10 trials to originally learn a list but only 6 trials for relearning, there was a 40% savings (4 fewer trials on relearning divided by the 10 original trials). By this method, *any* information that was left over in memory from original learning could have an influence, conscious or not (see Nelson, 1978, 1985; Schacter, 1987). Work by MacLeod (1988) indicates that relearning seems to help retrieve information that was stored in memory yet was not recallable.

---

how many things you have heard in a lecture or read in a textbook, and then had no useful memory of later when it came time to take an exam. It takes some time for memories to become relatively permanent and reliable. This is the process of *consolidation*, described earlier in this course. Consolidation does not happen all at once, but is stretched out over time and perhaps goes through two phases.

- The first is a fast *synaptic consolidation* phase in which memories may be stored for up to 2 weeks, perhaps in the hippocampus.
- The second is a slower *systems consolidation* phase in which memories are stored for up to a lifetime across the cortex.

Moreover, neurological processes that occur during sleep can aid consolidation (Fenn & Hambrick, 2013; Stickgold & Walker, 2005). Generally, the older a memory is, the more likely it is to have been consolidated, and so it is resistant to forgetting. This consolidation process begins almost immediately as you think of something and can extend for decades into the future. The reason that we do not consolidate some memories is because of factors that can lead to forgetting.

6.1.4: Metamemory

Think about the three issues of mnemonics, Ebbinghaus's work, and consolidation from a broader perspective: They

Aspects of Metamemory

Metamemory concerns your ability to assess when you have learned something, your realization that you need to remember something in the future, and even the basics of what you do and do not know. This captures the idea that memory is not primarily for recollecting about the past but for preparing a person to act in the present and the near future (e.g., Klein, Robertson, & Delton, 2010).

Interactive

Judgments of Learning

Self-Regulation

Then there is the issue of self-regulation in memory and learning. If you realize you are not doing some task particularly well, what do you do to improve? Some of the research on metacognition gives insight into difficulties people have. For example, Mazzoni and Cornoldi (1993) report that people often experience the **labor-in-vain effect**; that is, they devote more study time to difficult items and yet do not improve much at all (see also Metcalfe, 2002; Nelson, 1993). Alternatively, Thiede (1999; see also Metcalfe & Kornell, 2003) argues that when study time is used appropriately, a positive, sensible relationship between monitoring and self-regulation emerges. This is what Son and Metcalfe (2000) call the **region of proximal learning**, studying information that is just beyond one's current knowledge and saving the more difficult material for later. The problem is that people are often poor judges of what they have and have not learned and therefore make choices about what to study based on this inaccurate information (Metcalfe & Finn, 2008).

Feeling of Knowing

Tip of the Tongue (TOT) State

---

involve factors about memory, what makes remembering easier or harder. This self-awareness about memory is **metamemory**, knowledge about (*meta*) one's own memory and how it works or fails to work.

With regard to metacognitive awareness, several studies have focused on metamemory, such as people's "judgments-of-learning" and "feeling-of-knowing" estimates (Leonesio & Nelson, 1990; Nelson, 1988). Part of a person's behavior in a learning task involves self-monitoring, assessing how well one is doing and adjusting study strategies (e.g., Son, 2004). For example, these **metacognitions** guide people to know when to change their answers on multiple-choice exams (Higham & Garrard, 2005). However, metamemory can occasionally mislead us, leading to either over- or underconfidence that we have learned something (e.g., Koriat, Sheffer, & Ma'ayan, 2002).

Overall, although we do have some awareness into our memories, we are unaware of many aspects of memory. For example, many college students are unaware that they can improve their memories through various strategies such as dual coding, testing, distributed practice, and the generation effect (McCabe, 2011).

WRITING PROMPT

The Feeling of Knowing Judgment

Can you think of experiences of feeling of knowing or the tip-of-the-tongue effect? What could you have done beforehand so that that knowledge would have stayed in your memory later?

 The response entered here will appear in the performance dashboard and can be viewed by your instructor.

Submit

6.2: Storing Information in Episodic Memory

OBJECTIVE: Explain the process by which information is stored in episodic memory

How do people store information in episodic memory? And how can we measure this storage?

But what about more typical situations, when the world does not highlight material to make it more frequent or more distinctive? How do you learn and remember something new? There are three important steps to do so:

1. Rehearsal
2. Organization
3. Imagery

A summary of these will then lead us to the topic of retrieval and a discussion of forgetting.

6.2.1: Rehearsal

In Atkinson and Shiffrin's (1968) important model of human memory, information in short-term memory, was subject to **rehearsal**, a deliberate recycling or practicing of information in the short-term store. They proposed two effects of rehearsal. First, rehearsal maintains information in the short-term store. Second, the longer an item is held in short-term memory, the more likely it will be stored in long-term memory, with the strength of the item's long-term memory trace depending on the amount of rehearsal it received. In short, rehearsal transfers information into long-term memory (see also Waugh & Norman, 1965).

FREQUENCY OF REHEARSAL What evidence is there of this effect of rehearsal? Aside from Ebbinghaus, many experiments have shown that rehearsal leads to better long-term retention. For example, Hellyer (1962) used the Brown–Peterson task, with CVC trigrams and with an

---

Theories on Storing Information in Episodic Memory (Figure 6.4)
Here we discuss four theories that aim to answer these questions.

Interactive

Ebbinghaus' Theory

Ebbinghaus investigated one storage variable (repetition) and one memory task (relearning).

He found that increasing the number of repetitions led to a stronger memory, a trace of the information in memory that could be relearned faster. Thus, *frequency* has a fundamental influence on memory: Information presented more frequently is stored more strongly.

Hasher and Zacks' Theory

von Restorff's Theory

Kishiyama, Yonelinas, and Lazzara's Theory

arithmetic task between study and recall (see Figure 6.5).

REHEARSAL AND SERIAL POSITION Further evidence on the effects of rehearsal was provided in a series of studies by Rundus (1971; Rundus & Atkinson, 1970). In these experiments, Rundus had people learn 20-item lists of words, presenting them at a rate of 5 seconds per word. People were told to rehearse aloud as they studied the lists, repeating whatever words they cared to during each 5-second presentation. Figure 6.6 shows Rundus's most telling results.

In the early primacy portion of the *serial position curve*, there was a positive relationship between the frequency of rehearsal and the rate of recall. In other words, the *primacy effect* depended on rehearsal. The early items are rehearsed more frequently and so are recalled better. High recall of the late positions, the *recency effect*, was viewed as recall from short-term memory, which is why they were recalled so well despite being rehearsed so little.

Similar curves are observed in long-term memory. That is, given an event of a certain type, such as going to the movies, people are likely to remember their first and last experiences better, and not so much those in between (e.g., Sehulster, 1989). This even applies to semantic information, such as knowledge of the presidents of the United States (Roediger & Crowder, 1976). Although some people have argued that the cognitive processes involved in long-term memory differ from those involved in short-term memory (Davelaar, Goshen-Gottstein, Ashkenazi, Haarmann, & Usher, 2005), some suggestion exists that the same principles drive the serial position curves in both the short- and long-term memory (e.g., Brown, Neath, & Chater, 2007).

MASSED VERSUS DISTRIBUTED PRACTICE Rehearsing or practicing material will make it more memorable. However, not all kinds of practice are

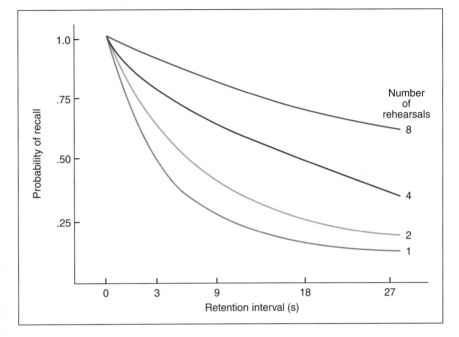

Figure 6.5 Hellyer's (1962) Recall Accuracy Results as a Function of the Number of Rehearsals Afforded the Three-Letter Nonsense Syllable and the Retention Interval

In some trials, the trigram had to be spoken aloud one, two, four, or eight times. The more frequently an item was rehearsed, the better it was retained. However, although rehearsal does improve memory, other work suggests that it is not repeated study that produces the primary memory benefit, but the repeated attempts at trying to remember (Karpicke, 2012; Karpicke & Roediger, 2007).

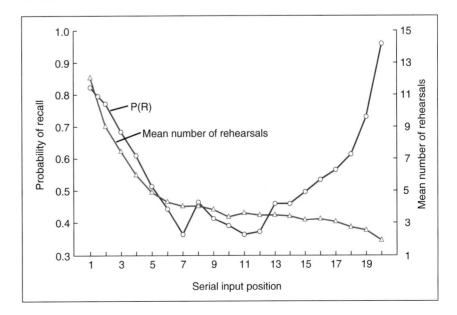

Figure 6.6 Probability of Recall

In the figure, the probability of recall, P(R), is plotted against the left axis, and the number of rehearsals afforded an item during storage is plotted against the right axis. The similar pattern of these two functions across the primacy portion of the list indicates that rehearsal is the factor responsible for primacy effects. Rundus then tabulated the number of times each word was rehearsed and compared this to the likelihood of later recalling the word.

equal. A distinction between two kinds of practice is seen in massed and distributed practice:

1. *Massed practice* is when people memorize information in one long session (remember the earlier discussion about cramming).
2. *Distributed practice* is when rehearsal is spread out across multiple, shorter occasions.

This distinction between different types of practice is important—keeping the amount of study time constant—because memory is better with distributed practice than with massed practice (Gerbier & Toppino, 2015; Glenberg & Lehmann, 1980). In other words, although you can learn if you cram, you learn better if you study a bit every day. This benefit of distributed over massed practice occurs for several reasons. The later rehearsals remind people of the prior encounters (thereby strengthening the memories) (Benjamin & Tullis, 2010), people experience the information in multiple contexts (Glenberg, 1976, 1979), and people's minds wander less when they engage in distributed practice (Metcalfe & Xu, 2016).

TESTING EFFECT It is pretty clear that when you are listening to a lecture, reading a book, or studying in some other way, you are learning new information. If you want to remember something better, it may seem intuitively that you should study as much as you can. And, for the most part, this is true.

However, it has also been found that, after some initial studying, memory may actually be better if people take a test as compared to just studying more. This is called the *testing effect* (Roediger & Karpicke, 2006). When you take a test, such as an essay exam, a fill-in-the-blank test, or a multiple-choice test, the contents of your memory are being assessed. It seems counterintuitive, but taking a test can actually improve memory more than studying. Every time you encounter information, whether you are studying it or being tested on it, it counts as a learning trial (e.g., Gates, 1917; Roediger & Karpicke, 2006).

Essentially, the testing effect is the finding that the additional experience you get from tests actually helps you remember the information better—better even than studying, especially if you take a recall test (McDaniel, Roediger, & McDermott, 2007). This testing benefit applies to recognition (multiple-choice) tests (Marsh, Roediger, Bjork, & Bjork, 2007) and to nonverbal material such as maps (Carpenter & Pashler, 2007). It also transfers to new items that were neither studied nor tested (Butler, 2010).

To make things even better, it has been found that after testing, people learn new information faster (Pastötter & Bäuml, 2014). So, an effective tool to help you study and learn the material for this or any other class is to take practice tests if they are available. If they are not available, your study group could create and take practice tests. It may sound like a lot of work, but your memory for the material will be much better than if you had just spent the same amount of time studying by yourself.

6.2.2: Depth of Processing

A more refined idea is that there are two kinds of rehearsal (Craik & Lockhart, 1972).

Craik and Lockhart (1972) proposed a theory of memory different from the stage approach of sensory, short-, and long-term memory. They embedded their two kinds of rehearsal into what they called the **levels of processing**, or **depth of processing**, framework. Essentially, information receives some amount of mental processing. Some items that get only incidental attention are processed at a shallow level—for example, hearing the sounds of the words without attending to their meaning while daydreaming during a lecture. Other items get more intentional and meaningful processing that elaborates the memory of that item—for example, by drawing relationships between already-known information and what is being processed.

Several predictions from the depth of processing framework were tested with a fair degree of initial success. For example, if information is shallowly processed, using only maintenance rehearsal, then the information should not be well remembered later. If it is only maintained, then it should not be stored at a deep, meaningful level in long-term memory. This was the kind of result that emerged. As an example, Craik and Watkins (1973) devised a monitoring task in which people heard a long list of words but only had to track the most recent word beginning with, say, a *g*. In a surprise recall test, people showed no recall differences for "G words" held a long time versus those maintained only briefly (see also Craik & Tulving, 1975).

You can think of levels of processing as being related to the idea of how much attention you pay to different things. If you pay more attention, you will remember more. This idea is illustrated in a study by Henkel (2014) in which students were asked to go to a museum and either look at objects or take pictures of them. According to the study, people remembered the objects better if they did not take pictures because the act of taking a picture detracts attention from the environment and focuses it on the act of taking a picture. Thus, the level of processing is more shallow. So, if you want to remember experiences from your life, you should spend less time taking pictures and more time in the experience.

Kinds of Rehearsal

Maintenance Rehearsal

Maintenance rehearsal is a low-level, repetitive information recycling. This is the rehearsal you use to recycle a phone number to yourself until you dial it. Once you stop rehearsing, the information is lost. In Craik and Lockhart's view, maintenance rehearsal holds information at a certain level in memory without storing it permanently. As long as information is maintained, it can be retrieved. However, when rehearsal stops, information will likely vanish.

Elaborative Rehearsal

6.2.3: Challenges to Depth of Processing

As research continued on the depth of processing framework, some difficulties emerged. One example was Baddeley's (1978) review paper, "The Trouble with Levels." A major point in this review was the problem of defining levels independently of retention scores (see Glenberg & Adams, 1978; Glenberg, Smith, & Green, 1977). Essentially, there is no method for deciding ahead of time whether a certain kind of rehearsal would lead to shallow or deep processing. Instead, we had to wait and see whether it improved recall. If it did, it must have been elaborative rehearsal. If it did not improve recall, it must have been maintenance rehearsal. The **circularity** of this reasoning was a serious problem for depth of processing ideas of memory retention.

TASK EFFECTS A second point in Baddeley's (1978) review concerned **task effects** in which difficulty arose with the levels of processing approach when different memory tasks were used. The reason was simply that very different results were obtained using one task or another.

We have known since Ebbinghaus that different memory tasks reveal different things about the variables that affect performance. Ebbinghaus used a relearning task, so even material that was hard to retrieve might still influence memory performance. In a similar vein, a substantial difference is found between performance on recall and recognition tasks. In *recognition tasks*, people are shown items that were originally studied, known as "old" or target items, as well as items that were not studied, known as "new," lure, or distractor items. They must then decide which are targets and which are distractors (multiple-choice tests are recognition tests, by the way).

Standard Memory Tasks and Terminology

Recognition accuracy usually is higher than recall accuracy. This interactive includes descriptions of these tasks. Furthermore, two different factors influence recognition: *recollection*—the actual remembering of the information—and *familiarity*—the general sense that you have experienced the information before (e.g., Curran, 2000; Yonelinas, 2002). Indeed, studies on false memory often ask people whether they actually "remember" experiencing an event or whether they just "know" that it happened.

Relearning Task
1. Original learning: Learn list items (e.g., list of unrelated words) to some accuracy criterion.
2. Delay after learning the list.
3. Learn the list a second time.

Dependent variables: The main dependent variable is the savings score: how many fewer trials are needed during relearning relative to number of trials for the original learning. If the original learning took 10 trials and relearning took 6, then relearning took 4 fewer trials. Savings score = 4/10; expressed as a percentage, savings was 40%.

Independent or control variables: Presentation rate, list item types, list length, accuracy criterion.

Recognition is easier than recall, in part because the answer is presented to a person, who then only has to make a new versus old decision. Because more information is stored in memory than can be retrieved easily, recognition shows greater sensitivity to the influence of stored information. The issue of *how much* easier recognition is than recall is difficult to resolve (see research by Craik, Govoni, Naveh-Benjamin, & Anderson, 1996, and by Hicks & Marsh, 2000).

The relevance of this to the depth of processing framework is that most of the early research that supported it used recall tasks. When recognition was used, however, maintenance rehearsal had clear effects on long-term memory. A clever set of studies by Glenberg et al. (1977) confirmed this. They used a *Brown–Peterson task*, asking people to remember a four-digit number as the (supposedly) primary task. During retention intervals that varied in duration, people had to repeat either one or three words aloud as a distractor task (don't confuse the distractor task here with distractor items, items in a recognition test that were not shown originally). Because people believed that digit recall was the important task, they devoted only minimal effort to the word repetitions, and so probably used only maintenance rehearsal. After the supposedly "main" part of the task was done, people were given a surprise recall task. The results showed the standard effect. But when they were given a surprise recognition task, the amount of time spent rehearsing *did* influence performance. Words rehearsed for 18 seconds were recognized better than those rehearsed for shorter intervals. Thus, the generalization that maintenance rehearsal does not lead to improved memory performance was disconfirmed—it did not apply when memory was tested with a recognition task.

> **WRITING PROMPT**
>
> **Recognition Accuracy vs. Recall Accuracy**
>
> What factors do you think affect recognition accuracy and recall accuracy?
>
> ▶ The response entered here will appear in the performance dashboard and can be viewed by your instructor.
>
> Submit

6.3: Boosting Episodic Memory

OBJECTIVE: Describe different ways that a person can improve his or her episodic memory

The most salient thing that many people want to do with their episodic memories is improve them so that they do not forget as much. That is, they would like to reduce the rate of forgetting that they experience. We have already gone over a few things you can do to improve your episodic memory. Specifically, you can repeatedly practice information, especially if your practice is distributed, and you can engage in deeper processing. An overarching idea in the depth of processing framework is that the more you do with information, the better it is remembered. There are numerous examples of this "hard work has its rewards" principle. In this section we look at six examples of this—six ways that show how information is processed and how the amount of effort a person puts into encoding it affects performance:

1. The self-reference effect
2. The generation effect
3. The production effect
4. The impact of enactment on memory
5. The benefits of imagery and organization
6. The consequences of taking a survival-based perspective

6.3.1: The Self-Reference Effect

The **self-reference effect** is the finding that memory is better for information that you relate to yourself in some way (e.g., Bellezza, 1992; Gillihan & Farah, 2005; Rogers, Kuiper, & Kirker, 1977; Symons & Johnson, 1997). If you think about it, you know a lot about yourself and you tend to be motivated by such information. When you relate something to yourself, say a detail you are trying to remember, you link the new knowledge to the old, yielding a more complex structure, an elaborative encoding (e.g., the locations given earlier for the method of loci—driveway, garage door, etc.—were the memorized locations one of this text's authors used for his own mnemonic device for years, based on his own house). Thus, the elaborated structure becomes more memorable.

6.3.2: Generation, Production, and Enactment

Here we discuss three ways in which episodic memory can be improved: the generation effect, the production effect, and the enactment effect.

The **generation effect** is the finding that information you generate or create yourself is better remembered compared to information you have only heard or read. This was first reported by Slamecka and Graf (1978). In their study, for the *read* condition, people simply read words printed on cards. However, for the *generate* condition, people needed to generate the word on their own. This was done by giving a word and the first letter of the word that was to be recalled, with the instruction that the to-be-generated word had to be related to the first word. For example, a person might see Long–S_____ where the word *short* needed to be generated. The results showed that people remembered words better when they had generated them as compared to when they had just read them.

In their review of work on the generation effect, Bertsch, Peta, Wiscott, and McDaniel (2007) reported that this robust finding was more likely to occur with free recall, and that the effect grew larger over longer delays. More important, the generation effect applies not only to lists of words but also to textbook material (e.g., deWinstanley & Bjork, 2004). In short, the generation effect is another example that the more effort you put into mentally processing information, the more likely you will remember it later and for a longer time.

A related finding is the *production effect* in which people actually produce information rather than simply reading or hearing it (Fawcett, 2013; MacLeod, Gopie, Hourihan, Neary, & Ozubko, 2010). For example, people given a set of materials to learn will remember those materials if they say them aloud rather than if they simply read them. The production effect occurs because actually saying something requires more effort than reading. This extra effort is deeper processing, which leads to better memory. Note that the production effect does not necessarily require that a person say something aloud. It can also be observed when people simply mouth the material, whisper it, write it, or type it out (Forrin, MacLeod, & Ozbuko, 2012).

Another way to engage in deep encoding is to take advantage of the **enactment effect**, in which there is improved memory for participant-performed tasks relative to those that are not participant-performed. In such studies, actually doing some activity is compared to just watching someone else doing it. For example, people might be told to "break the match," "point at the door," or "knock on the table," or to watch someone else do those actions. In general, people remember things better if they do them themselves (e.g., Engelkamp & Dehn, 2000; Saltz & Donnenwerth-Nolan, 1981). Essentially, the additional mental effort needed to do the task is another form of deep processing.

The value of enactment can be seen in the practical application of learning lines of dialogue. Evidence exists that even untrained nonactors (i.e., novice actors) learn dialogue, as in the script of a play, better when they rehearse the dialogue and stage movements together (Noice & Noice, 2001; see also Freeman & Ellis, 2003, and Shelton & McNamara, 2001, for other multimodality effects on learning). Physical movement, in other words, can be part of an enhanced mnemonic. Enactment improves memory by helping people better organize and structure information about their actions (Koriat & Pearlman-Avnion, 2003).

6.3.3: Organization in Storage

Another important piece of the storage puzzle involves **organization**, the structuring of information as it is stored in memory. Well-organized material can be stored and retrieved with impressive levels of accuracy. The earliest work on organization (or **clustering**) was done by Bousfield. For example, in his earliest study (Bousfield & Sedgewick, 1944), he asked people to name as many birds as they could. The result was that people tended to name the words in subgroups, such as *robin, bluejay, sparrow—chicken, duck, goose—eagle, hawk*. To study this further, Bousfield (1953) gave people a 60-item list to be learned for free recall. Unlike other work at that time, Bousfield used related words for his lists, 15 words each from the categories *animals, personal names, vegetables,* and *professions*. Although the words were shown in a random order, people tended to recall them by category; for instance, *dog, cat, cow, pea, bean, John, Bob*.

Where did this organizing come from? Obviously, people noticed that several words were drawn from the same categories. They used the strategy of grouping the items together on the basis of category (this has a nice metamemory effect as well). The consequence of this reorganization was straightforward: The way the material had been stored governed how it was recalled.

6.3.4: Improving Memory

Baddeley (1978) was one of the critics of the depth of processing framework, concluding that it was valuable only at a rough, intuitive level but not as a scientific theory. Although that may be true, it is hard to beat Craik and Lockhart's insights if you are looking for a way to improve your own memory.

SUBJECTIVE ORGANIZATION Don't misunderstand the previous section: Organization is *not* limited to sets of items with obvious, known categories. A study by Tulving (1962) showed that people can and do use **subjective organizations**—literally, organization imposed by the participant (for an update, see Kahana & Wingfield, 2000). Tulving used a multi-trial free recall task, in which the same list of words is presented repeatedly across several trials, wherein each trial had a new reordering of the words. His analysis looked at the regularities that developed in the recall orders. For example, a person might recall the words *dog, apple, lawyer, brush* together on several trials. This consistency suggested that the person had formed a cluster or chunk composed of those four items using some idiosyncratic basis. For example, a person might link the words together in a sentence or story: "The dog brought an apple to the lawyer, who brushed the dog's hair." Regardless of how they were formed, the clusters were used repeatedly during recall, serving as a kind of organized unit. Tulving called this *subjective organization*; that is, organization developed by a person for structuring and remembering information. In other words, even "unrelated" items become organized through the mental activity of a person imposing an organization.

6.3.5: Imagery

Another way you can improve memory is to use **visual imagery**, the mental picturing of a stimulus that affects later recall or recognition. We have discussed some imagery effects such as imagery-based mnemonic devices. Now we turn to visual imagery's effect on the storage of information in long-term memory and the boost it gives to material you are trying to learn. In general, visual episodic memory for pictures is very good.

Studies have shown that people can remember an astounding number of pictures after seeing them only once (Brady, Konkle, Alvarez, & Oliva, 2008). For example, Standing (1973) showed that people had over 80% memory for 10,000 pictures 2 days later.

An early contributor to understanding how imagery impacts memory was Alan Paivio.

6.3.6: Adaptive Memory

For memory to be of value to us, and be better remembered, it needs to give us something useful. It should help us survive in the world. The survival motivation is strong, and knowledge of what can either increase or decrease our

How Can Memory Be Improved? (Figure 6.7 and Table 6.3)

Think of maintenance versus elaborative rehearsal as simple recycling in short-term memory versus meaningful study and transfer into long-term memory.

> **Interactive**
>
> ### Inventing Mnemonics
> Apply this to your own learning. When you are introduced to someone, do you merely recycle that name for a few seconds, or do you think about it, use it in conversation, and try to find mnemonic connections to help you remember it? When you read, do you merely process the words at a simple level of understanding, or do you actively elaborate when you are reading, searching for connections and relationships that make the material more memorable? In other words, use the depth of processing ideas in your own metacognition. Try inventing a mnemonic (this invokes the generation effect), applying elaborative rehearsal principles, or actively doing something with the information, such as drawing a diagram (this invokes the enactment effect) to something you may need for this course. Most effective mnemonics organize and structure information in some way.

survival is important. Thus, if people bring a survival perspective to bear on what they are learning, it can improve performance. This was shown in a study by Nairne, Thompson, and Pandeirada (2007; see also Nairne & Pandeirada, 2008; Nairne, Pandeirada, & Thompson, 2008; Weinstein, Bugg, & Roediger, 2008). In this study, people were given lists of words. During the first part of the study, people rated the words for pleasantness, relevance to moving to a foreign land, personal relevance, or survival value (e.g., finding food and water or avoiding predators). Words that were rated high on survival value were more likely to be remembered later. The survival angle has such a strong impact that it can outperform the effects of other well-known memory-enhancing strategies such as imagery, self-reference, and generation (Nairne et al., 2008). Part of this benefit comes from thinking about how survival relates to us, as a kind of self-reference effect (Klein, 2012). So, if we think about how information relates to our ability to survive, endure, or otherwise be useful, this takes advantage of our fundamental motivations, which we can leverage to improve memory (Wurm, 2007; Wurm & Seaman, 2008).

WRITING PROMPT

Adaptive Memory Strategy

What memory-enhancing strategies do you adopt? How do they make learning easier for you?

 The response entered here will appear in the performance dashboard and can be viewed by your instructor.

Submit

Impact of Imagery on Memory (Table 6.4)

Paivio (1971) reviewed scores of studies that showed the generally beneficial effects of imagery on memory. These effects are beyond those caused by other variables, such as word- or sentence-based rehearsal or meaningfulness (Bower, 1970; Yuille & Paivio, 1967). One example is a paired-associate learning study by Schnorr and Atkinson (1969; see Table 6.4).

List 1 (A–B)	List 2 (C–D)	List 3 (A–B$_r$)
tall–bone	safe–fable	plan–bone
plan–leaf	bench–idea	mess–hand
nose–fight	pencil–owe	smoke–leaf
park–flea	wait–blouse	pear–kiss
grew–cook	student–duck	rabbit–fight
rabbit–few	window–cat	tall–crowd
pear–rain	house–news	nose–cook
mess–crowd	card–nest	park–few
print–kiss	color–just	grew–flea
smoke–hand	flower–jump	print–rain

List 4 (A–B')	List 5 (A–C)	List 6 (A–D)
smoke–arm	tall–bench	smoke–fable
mess–people	plan–pencil	print–idea
rabbit–several	nose–wait	mess–owe
park–ant	park–student	pear–blouse
plan–tree	grew–window	rabbit–news
tall–skeleton	rabbit–house	grew–duck
nose–battle	pear–card	park–cat
grew–chef	mess–color	nose–nest
pear–storm	print–flower	plan–just
print–lips	smoke–safe	tall–jump

Table 6.4 Lists of Paired Associates

6.4: Context

OBJECTIVE: Explain how context plays a role in the storage and retrieval of episodic memories

As noted earlier, episodic memory is for memory of personal experiences. These experiences do not occur in a vacuum. Instead, they happen in some sort of setting or *context*. Every experience you have occurs in a particular place, at a particular time, when you are thinking particular thoughts, and when you have encountered particular things. All these and more make up the context of an episodic experience. Context plays a critical role in episodic memory: It can make remembering easier or harder. In addition to the influence of context on our ability to remember, we can even look at memory for the context or source itself.

6.4.1: Encoding Specificity

In Tulving and Thompson's (1973; Unsworth, Spillers, & Brewer, 2012) view, an important influence on memory is **encoding specificity**. This phrase means that information is encoded into memory *not* as a set of isolated, individual items. Instead, each item is encoded into a richer memory representation that includes the context it was in during encoding. So, when you read *cat* in a list of words, you are likely to store not only the word *cat* but also information about the context you read it in.

In a classic study of encoding specificity, Godden and Baddeley (1975) had people learn a list of words (see Figure 6.8). Half of these people learned the list on land, and the other learned the list underwater (all of these people were scuba divers). They were then given a recall test for the list.

Figure 6.8 The Classic Encoding Specificity Result Reported by Godden and Baddeley (1975)

The figure shows better performance when the encoding context matched the retrieval context. That is, memory for things learned on land was better when tested on land as opposed to underwater, whereas things learned underwater were better remembered when the people were tested underwater as opposed to on land. The interesting finding was that memory was better when the encoding and retrieval contexts were the same, relative to when they were different.

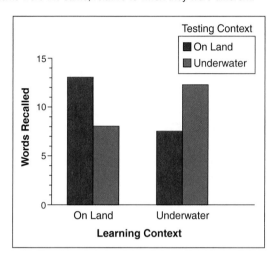

The important twist is the context in which they tried to recall the information. Half of the people recalled the items in the same context they had experienced during learning. However, the other half recalled the information in the other context.

A more everyday example of encoding specificity is the experience of going to a room in your home to do something, but when you arrive, you cannot remember why you are there. However, when you go back to where you started, you remember. So, reinstating the original context helped you remember. This is also why witnesses may return to the scene of a crime. Being there again reinstates the context, helping them remember details they might otherwise forget.

More generally, when your memory is tested, with free recall for instance, you attempt to retrieve the trace left by your original encoding. Encoding the context along with the item allows the context to serve as an excellent *retrieval cue*—a useful prompt or reminder for the information. The original context cues give you the best access to the information during retrieval, and these cues can be verbal, visual, or something else (Schab, 1990, for instance, has found that odors are effective contextual cues).

The processing of content (what it is you are to remember), context, and their binding together is more complex than it may seem initially. The brain has separate processing streams for what something is and where it is.

Another variant of the encoding specificity effect is *state-dependent learning*, which is the finding that people are more likely to remember things when their physiological state at retrieval matches that at encoding. For example, Goodwin, Powell, Bremeer, Hoine, and Stern (1969) found that people made fewer errors on a memory test when they recalled information when they were drunk (a certain physiological state) if they had learned that information inebriated, than if they tried to recall it when they were sober!

In summary, the storage of information into episodic long-term memory is affected by several factors that can lead to better memory. Moreover, the congruence between study and test contexts can be vital. Relevant rehearsal, including organizational and imaginal elements, improves performance.

6.4.2: Source Monitoring

Context is not only an important cue to access episodic memories, but it can also be used as the target of a memory search—when people seek to remember the context in which information was learned.

> **Prove It**
>
> **Cryptomnesia**
> For this demonstration you will need at least three volunteers (although you can use more). It is a variation of a study by Brown and Halliday (1991) that can result in people reporting answers that they present as their own, but which, in fact, other people gave. This is cryptomnesia. Have your three volunteers sit in a circle. Tell them that you will first read the name of a category, such as "Countries," "Flowers," "Insects," and so on (prepare a list of 10 or so category names ahead of time). Then, going around in a circle, people will need to name members of that category aloud, without using names that were said previously. It is likely that at various times people will say things that someone else has already said. Furthermore, these things should be more likely to be things that the person just before them in the circle had said as compared to the person just after him or her. To make this a bit more possible, give people a time limit of 5 seconds to say their answer. You will need to record what people say, so an audio recording might be better than writing. You should also limit the number of times you go around the circle before the task becomes very hard. Limiting the number of responses to 25 should do this.
>
> After the first part of the task is completed, write down, in random order, the responses people gave. Then ask people to pick out which ones they themselves said. Simply have them circle the ones they remember saying. This can be done an hour after the first part or a day later or a week later.
>
> If you want to get fancy, you could have several groups of volunteers and see how the amount of time that passed influences your results. What you may find is that people will pick answers that they did not actually say. Again, these are more likely to be things said by the person just before them in the circle.

What Is Source Monitoring?

Let's consider your ability to remember where information came from.

The Source of a Memory

Source monitoring (Johnson, Hashtroudi, & Lindsay, 1993) is the ability to accurately remember the source of a memory, be it something you encountered in the world or something you imagined. Failures of source monitoring can sometimes occur in which a person remembers the content of the information, but cannot accurately attribute it to a certain source. Source monitoring is a complex process that involves many parts of the brain. The hippocampus seems to be important for integrating content and source information. The prefrontal cortex and the temporal lobes are important for searching for and using source information, particularly the right dorsolateral (B.A. 10), ventrolateral (B.A. 45), and medial regions (B.A. 11) (Burgess, Scott, & Frith, 2003; Davachi, Mitchell, & Wagner, 2003; Senkfor & Van Petten, 1998). In extreme cases such as schizophrenia, people experiencing hallucinations may have trouble with source monitoring—distinguishing between what is real and what is only imagined (Ditman & Kuperberg, 2005).

An example of source monitoring might be if you have a memory of a fact about a historical figure, and it is important to know whether this is a fact that came from a history textbook or an inaccurate fictionalization that came from a novel. So, to know how reliable this fact is, you need to know where you learned it. If you misremember the source of an idea as being a textbook when it actually came from a movie, then you mistakenly think that fiction is fact—hence a source monitoring failure. This can be important in situations such as trying to determine whether something was witnessed during an accident or crime. A more everyday example of source monitoring is trying to remember which of your friends told you a secret. If you get this wrong, then you could be in big trouble with one (or more) of your friends.

WRITING PROMPT

Consequences of Source Monitoring

How often do you find yourself source monitoring? What are the consequences of that?

> The response entered here will appear in the performance dashboard and can be viewed by your instructor.

6.5: Facts and Situation Models

OBJECTIVE: Identify the influence the levels of mental representations have on memory

In many memory studies, we intentionally lure people into making mistakes. That is, we present meaningful material such as a story or a set of related sentences, then do something that invites mistakes of one sort or another. This is unlike most episodic and semantic memory studies in which people are simply asked to learn and remember material in a straightforward way.

6.5.1: The Nature of Propositions

In this section we discuss the idea that what people remember from meaningful material is the idea or gist—in other words, we do not usually remember superficial details, exact words, or exact phrasings, but we do remember the basic idea. However, psychology needs ways to represent those ideas, needs a scientific way to quantify what the vague term *meaning* means. Also, to do research on content accuracy, we need a way to score recall to see how well people remembered meaningful content. The unit that codes meaning is called a proposition. A **proposition** represents the meaning of a single, simple idea, the smallest unit of knowledge about which you can make true/false judgments.

ELABORATED PROPOSITIONS Just as some semantic theories try to capture memory in terms of a network structure, propositional theories can do the same—as networks of interconnected propositions. To illustrate, consider a sentence and its propositional representation, as presented in Figure 6.9.

REMEMBERING PROPOSITIONS Let's begin with a classic study by Sachs (1967). She was testing the idea that people tend to remember meaning rather than verbatim information. Her participants heard passages of connected text and were then tested on one critical sentence in the passage 0, 80, or 160 syllables after it had been heard.

Explore the interactive below for an example. (Why not read it now and confirm Sachs's results for yourself?)

The test was a simple recognition test among four alternatives. One alternative was a verbatim repetition, another represented a change both in surface form and in meaning, and the other two represented changes only in surface form. When recognition was tested immediately, people were very good at recognizing the exact repetition. In other words, they rejected changes in superficial structure and changes in meaning. After comprehending the next 80 syllables in the passage, however, performance was accurate only in rejecting the alternative that changed the meaning. So, after the 80-syllable delay, people showed no preference for the repetition over the paraphrases.

Figure 6.9 A propositional representation, in "node-plus-pathway" notation and in written form, of the sentence "The hippie touched the debutante in the park." Network notation after Anderson (1980) and Anderson and Bower (1973); written proposition after Kintsch (1974).

Sentence from Anderson and Bower (1973): (1) The hippie touched the debutante in the park, is represented here as a set of interrelated concepts, one for each main word in the sentence. The relationships among the words are specified by the types of pathways that connect the nodes (e.g., agent, recipient or patient, location). Sentence 1 is composed of five relationships or connections of meaning, five **semantic cases**.

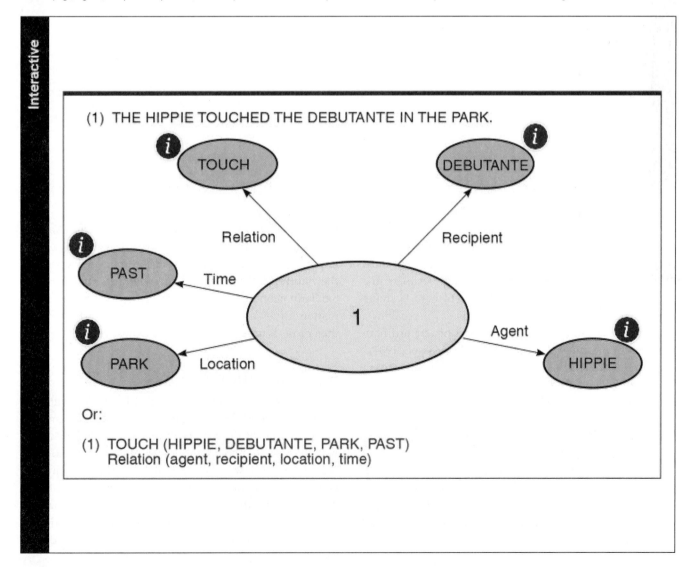

Sample Passage from Sachs (1967) Including Multiple-Choice Recognition Test for Critical Sentence

> Read the passage below at a comfortable pace but without looking back. After you have finished reading, your memory of one of the sentences in the paragraph will be tested.
>
> While most gardeners enjoy tulips, it is unlikely that they would be valued as highly today as they were in Holland during the seventeenth century. For some reason, the tulip became very popular at this time. The tulip was first brought into Europe from Turkey. In 1551, the Viennese ambassador to Turkey wrote of seeing these plants. Years later, in 1562, a cargo of tulip bulbs was sent to Antwerp. The flower then spread through Holland from there. At first, only the rich collected and traded tulips. Eventually, most of Holland was involved in the matter. Almost everyone tried to outdo their neighbors, both in the growing of tulips with rare colors and also in the high prices that were paid for them. A price of 6000 florins was paid for one bulb of the variety Semper Augustus in 1636. At this time, 6000 florins was the price of a house and grounds. Soon, everyone in Holland was working in the tulip trade, and ordinary business was being neglected. People who had been away from Holland and then returned during the craze sometimes made mistakes. A sailor is said to have mistaken a tulip bulb worth several thousand florins for an onion, and cut it up to eat with his herring.
>
> Click or tap "Next" to test your memory.

Sachs concluded that we quickly lose information about the actual, verbatim words that we hear (or read), but we do retain the meaning. We reconstruct what must have been said based on the meaning that is stored in a propositional representation. Only when there is something "special" about the verbatim words, say in recalling a joke, do we retain surface form as part of our ordinary memory for meaningful discourse (but see Masson, 1984).

Confirmation of this finding was offered by Kintsch and Bates (1977), who gave a surprise recognition test to students either 2 or 5 days after a classroom lecture. Some evidence of verbatim memory was present after 2 days, but very little persisted 5 days afterward. As expected, verbatim memory for details and extraneous comments was better than verbatim memory for general lecture statements (see also Bates, Masling, & Kintsch, 1978). However, even here, reconstructive memory seemed to play a role. Students were better at rejecting items such as jokes that had not been presented than they were at recognizing jokes and announcements that had been heard (see also Schmidt, 1994, on the effects of humor on sentence memory). For jokes, the wording can be important, so there was better surface form memory because the nature of the information constrains the wording. Similarly, people can have better memory for verbatim information if they are reading poetry, where wording, rather than prose, is important (Tillman & Dowling, 2007).

6.5.2: Situation Models

There is a great deal of evidence that basic idea units can influence memory. However, there is more to life than simple ideas. One of the themes of this text is *embodied cognition*, the idea that how we think is influenced by how we act or are otherwise involved with the world. One way that embodied cognition manifests itself is in the idea that people create models (Johnson-Laird, 1983; Zwaan & Radvansky, 1998) of the situations described, and do not just create memories of the simple propositional idea in

Levels of Representation

For van Dijk and Kintsch, there are three levels of representation when we comprehend.

Interactive

The Surface Form

The **surface form** corresponds to a verbatim mental representation of the exact words used as well as the syntax of the sentences read or heard.

The Textbase

The Situation Model

sentences. It is clear that when people move from one event to another, such as by moving from one spatial location to another, this serves as an event boundary. These distinct events are then stored separately in memory and influence how easy it is for people to retrieve information (Horner, Bisby, Wang, Bogus, & Burgess, 2016; Pettijohn, Thompson, Tamplin, Krawietz, & Radvansky, 2016). One way of thinking about different types of mental representations and their influence on memory comes from van Dijk and Kintsch's (1983) work on language comprehension.

This division of different kinds of mental representation can be seen in how well people remember information at the different levels over time. For example, in a study by Kintsch, Welsch, Schmalhofer, and Zimny (1987), people read a text and then later took a memory test. For the test, people were shown sentences and indicated whether they had read them before. Four types of memory probes were used on the recognition test:

1. Verbatim probes, which were exact sentences that had been read earlier
2. Paraphrases, which captured the same ideas as those in the text, but with a different wording
3. Inferences, which were ideas that were likely to be true, but not actually mentioned
4. "Wrongs," or incorrect probes that were thematically consistent with the passage but were incorrect if one had read and understood the passage

Kintsch et al. compared performance on these various types of memory probes to assess the strength of the representations at the various levels. For example, by comparing performance on the verbatim probes and paraphrases, one can estimate memory for the exact wording. This is because both of these probe types refer to ideas that were actually in the text, but only one uses the exact wording. So, the degree to which memory is better in the verbatim condition compared to the paraphrase condition is a measure of surface form memory.

The results of Kintsch et al.'s (1987) study are shown in Figure 6.10.

Although memory for all three levels is reasonably good immediately after reading, there are big differences later on, depending on what is being assessed. First, for the surface form, verbatim information is lost quickly from memory and reaches chance performance by the end of the 4 days. Second, memory was better for the textbase level than for the surface form. So, although people may forget the exact words they read before, they are better at remembering the ideas that were presented (cf. the work by Sachs described earlier in the module). However, even memory at this level declines over time. But for the third level, the situation model, performance starts out high and then stays high, with little evidence of forgetting over the 4-day retention interval. Thus, there is something psychologically real about looking at mental representations in this way.

A more everyday example of this would be your memory for a newspaper article you might read. Soon after reading the article, your ability to remember verbatim sentences from the article is pretty poor. Furthermore, over time, you start to forget what specific ideas were actually read in the article and what ideas are inferences you may have created when you were trying to understand it. However, you have a relatively good memory over time for the events described in the article, and this memory stays with you for a much longer period of time.

> **WRITING PROMPT**
>
> **Levels of Memory**
>
> Think of a class that you attended recently. Note your remembering from the session. What do you think you will retain for the longest time and why?
>
> ▶ The response entered here will appear in the performance dashboard and can be viewed by your instructor.
>
> Submit

6.6: Autobiographical Memories

OBJECTIVE: Characterize phenomena that has been uncovered by research on autobiographical memory

This next section addresses the study of one's lifetime collection or narrative of memories, called **autobiographical memory**. These are memories for more natural experiences and information—a self-memory system (Conway & Pleydell-Pearce, 2000). The research on autobiographical memory has revealed many phenomena that characterize it. Here, we present three of them to give you a feel for why it is important to consider autobiographical memory beyond what is known about episodic and semantic memory. More specifically, we consider infantile amnesia, the reminiscence bump, and involuntary memories.

Figure 6.10 Results of Kintsch et al.'s (1987) Study

Memory for information at the surface form, textbase, and situation model levels over the course of 4 days After Kintsch et al. (1987).

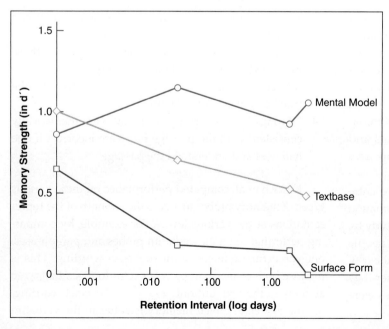

6.6.1: Psychologists as Subjects

Because autobiographical memory is memory for one's own life, this makes it an appealing topic for psychologists so they can look inward and examine their own memories for their own lives. Several modern-day Ebbinghauses have tested their own memories in carefully controlled long-term studies. One major difference from Ebbinghaus's procedure was that Linton (1975, 1978), Wagenaar (1986), and Sehulster (1989) tested their memories for naturally occurring events, not artificial laboratory stimuli. For instance, Wagenaar recorded daily events in his own life for more than 6 years, some 2,400 separate events, and tested his recall with combinations of 4 different cue types: what the event was, who was involved, and where and when it happened. Although he found that pleasant events were recalled better than unpleasant ones at shorter retention intervals, his evidence also showed that none of the events could truly be said to have been forgotten (but contrast this with the Bahrick, Hall, & Berger, 1996, evidence that bias toward pleasant things affects our memories of high school grades).

6.6.2: Infantile Amnesia

Think about the first thing you can remember. The very first thing. What is your earliest memory? Where were you? What was going on? Who were you with? How old were you? For most people, this memory is from the ages of 2 to 4 years old. But, wait a minute! What about all of the things that came before that? Why can't you remember anything earlier than that? Furthermore, if you think about it, your memory for your childhood is quite spotty, getting better only as you got older.

Infantile amnesia is the inability to remember early life events and very poor memory for your life at a very young age. One of the first people to discuss this phenomenon was Sigmund Freud (1899/1938). He thought that this was a true amnesia in which there was a catastrophic forgetting of early life events. For him, this was done to protect the ego from threatening psychosexual content. However, there has been little empirical support for this idea across the years, so we do not consider it further.

Instead, it appears more likely that infantile amnesia is not an amnesia at all. That is, it is not a massive forgetting of things that would otherwise be remembered. Instead, this reveals how memory is developing. Keep in mind that humans are born neurologically immature and quite helpless. It takes some time for the nervous system to develop to the point that we can have autobiographical memories. Clearly there is some memory, even before birth. Newborns prefer the sound of their own mother's voice, which they heard *in utero*. Moreover, implicit, procedural learning begins almost immediately as a person learns to do things such as control his or her head, arms, and legs, then more complex tasks such as sitting up, using a spoon, and dumping cereal on the floor. Children are developing semantic memories as they learn what things are and what they are called. They also have episodic memories in that they are clearly influenced by context. However, young children have a difficult time remembering specific events from their lives, especially as they move away from them.

Infantile amnesia resolves itself as one develops a sense of self and can start organizing information around this self-concept (Howe & Courage, 1993). Certain neurological correlates indicate the use of self-referential memories as opposed to more general knowledge (Magno & Allan, 2007). As we have noted earlier, information that you can relate to yourself is often remembered best. In a sense, then, the offset of infantile amnesia marks the beginning of autobiographical memory. Several things contribute to the development of the self-concept. First, this is also the time when there is a tremendous increase in the child's use of expressive language (K. Nelson, 1993). If you think about your own autobiographical memories, they are heavily influenced by language in terms of their content, their structure, and how you think about them. Second, this period of time is when the hippocampus is maturing, allowing more complex memories to be formed (Nadel & Zola-Morgan, 1984). Finally, at this age, a child has started to develop schemas and scripts that are complex enough to begin making sense of the world in a more adultlike fashion, thereby facilitating memory for individual life experiences (K. Nelson, 1993).

6.6.3: Reminiscence Bump

When you talk to people older than the typical college student, you may find that they have a bias to remember events from when they were in college (or that age, thereabouts). Why is that? Are they trying to make a social connection with you? Was that really the best time of their lives? Although we cannot answer all of these questions, we can tell you that your observation is correct and that we can provide some insight into the first question. The **reminiscence bump** is superior memory than would otherwise be expected for life events around the age of 20, between the ages of 15 and 25. This is illustrated in Figure 6.11 (Rubin, Rahhal, & Poon, 1998).

This figure shows the rate of remembering events as a function of how old the person was at the time (older adults are used here to make the reminiscence bump clearer).

Figure 6.11 Illustration of the Reminiscence Bump
To give you an idea of how sensitive different sensory modalities are, various sensory thresholds are presented in the figure below.

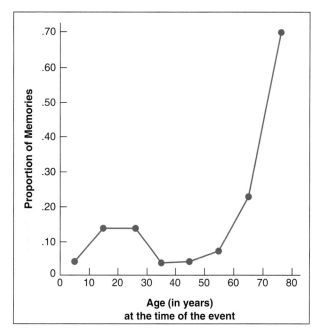

What Causes the Reminiscence Bump?

There are several ideas about what causes the reminiscence bump.

First-Time Events

One idea is that memory tends to be better for things that happen for the first time. For example, it is easier to remember your first kiss than your 27th, although you may have just as much fun during each. The early 20s area period in a person's life is when many things are happening for the first time—the first time a person lives alone, drives, votes, gets a real job, gets a speeding ticket, pays taxes, and so on. Because there are so many firsts, it is easier to remember events from this period of life. This idea is supported by the fact that people who move from one country to another, where another language is spoken, show a reminiscence bump for the time of their immigration, regardless of how old they were at the time (Schrauf & Rubin, 1998). In a similar way, recent evidence shows a relocation bump, a peak in memories surrounding the life transition of changing one's residence, occurring in middle adulthood (Enz, Pillemer, & Johnson, 2016). Presumably such moves bring on many new experiences, a number of firsts that make those times in life more memorable.

These studies often use what is known as the *Galton-Crovitz technique* (Crovitz & Shiffman, 1974; Galton, 1879), in which people are given lists of words and asked to respond with the first autobiographical memory that comes to mind. The first thing to notice is that most of the reported memories come from the recent past, and that the further away in time the event is (the older the event is, hence the younger a person was at the time), the less likely it will be recalled. However, note that around the age of 20 there is a tendency to remember more events than would be expected if this were a normal forgetting curve. So, why does this happen?

6.6.4: Involuntary Memory

Autobiographical memories not only are recalled when we actively try to remember but also happen spontaneously and involuntarily, without any clear effort to do so. These types of memories often occur (Berntsen, 2010), and they often refer to single events, rather than general periods of time (Berntsen & Hall, 2004). An example of an **involuntary memory** would be remembering a specific event that happened during a specific chemistry class when you were in high school, compared to just remembering that you took a chemistry class. These involuntary memories are often triggered by some cue—literally a retrieval cue—in the environment. For example, a person may see a highly valued object (an old toy) that then has the power to elicit a strong autobiographical memory (Jones & Martin, 2006). Emotional intensity is commonly a critical aspect of such triggered autobiographical memories (Talarico, LaBar, & Rubin, 2004).

Some of the strongest cues to spontaneously bring about autobiographical recollections are odors. Memories elicited by odors produce a strong feeling of being back in time (what it was like to experience the event long ago) (Herz & Schooler, 2002; Willander & Larsson, 2006, 2007). Moreover, odors elicit stronger emotions in the person experiencing the memory.

WRITING PROMPT

Cues for Reminiscing About Events

Which memories do you reminisce about more, first-time events or things you are expected to remember over a general period of time? Can you think of cues that may help you recollect?

> The response entered here will appear in the performance dashboard and can be viewed by your instructor.

Submit

6.7: Memory for the Future

OBJECTIVE: Describe how memory can be oriented toward the future

This final section of the module flips the way in which we typically think about memory. Most of the time when we consider memory, we are thinking about how we remember incidents that happened in the past. And certainly, that is one of the major things for which we use our memories. However, an important role of memory is not only a retrospective one that allows us to go over past information, but also a prospective one that allows us to perform well by anticipating events that have not yet happened. Consistent with one of the themes of this course, memory is just as important for the future as it is for the past. To this end, here we discuss two ways that memory for the future comes about: *prospective memory*, or remembering to do something in the future, and *episodic future thinking*, or imagining future events.

6.7.1: Prospective Memory

Prospective memory (Loftus, 1971) is the ability to remember to do something in the future.

Types of Prospective Memory

In general, there are two basic kinds of prospective memory (Einstein & McDaniel, 1990).

Interactive

Time-Based

The first is *time-based prospective memory* in which a person needs to remember to do something based on the passage of time. This can be after a certain amount of time has passed, such as needing to remember to take the pizza out of the oven in 10 minutes, remembering to take medication at 3:00, or remembering to call mom on Mother's Day. In general, time-based prospective memory is more difficult because it requires us to keep track of time, which we are generally not very good at.

Event-Based

For the improvement of daily prospective memory, a study by Trawley, Law, and Logie (2011) found that for prospective memory to do a number of tasks, it is best if people make a plan ahead of time and then stick to it. When this plan is interrupted and deviated from, people are more likely to forget something that they are supposed to do. As has been found with other types of memory, sleeping after encoding a prospective memory goal can also help performance (Scullin & McDaniel, 2010).

6.7.2: Episodic Future Thinking

In addition to remembering to do something in the future, memory is also important for imagining what the future will be like, allowing us to anticipate and plan what we will do. This is *episodic future thinking*: the mental construction of what future events might be like, of the outcome of future situations (Atance & O'Neill, 2001; Szpunar, 2010; Szpunar & Radvansky, 2016). As an example of this, when you are getting ready to ask someone you are fond of to have coffee or lunch with you, you may imagine how that situation will unfold. You can predict and prepare for how the event will unfold so you can do better when the actual event happens (Klein et. al, 2010). Episodic future thinking is actually very common. It has been estimated that this is something that we do about every 15 minutes (D'Argembeau, Renaud, & Van der Linden, 2011).

Part of the reason that this type of future thought is called episodic future thinking is because it is believed to use the same memory processes as episodic memory. This is called the *constructive episodic simulation hypothesis* (Schacter & Addis, 2007). When we try to think about a future event, we use our memories of prior, similar events to imagine what the new event might be like. For example, if we decide to fix something in the engine of our car, we may think about other times that that we have tried to fix things that are mechanical and remember any problems or troubles we had at those times. To predict how asking someone out for coffee will go, you remember how that went the last time you did it. In general, the closer the event you are trying to imagine conforms to prior events, the easier it is to construct (Szpunar & Schacter, 2013).

Episodic future thinking also exhibits some of the same patterns of behavior, although in an opposite direction, as episodic remembering. For example, when we think about the past, we have a bias to think about more recent events than older events. Similarly, when we think about the future, we have a bias to think more about future events that are closer to the current point in time than times much further ahead (Addis & Schacter, 2008; Speng & Levine, 2006).

Although episodic future thinking uses some of the same mental processes and representations as are used in retrospective episodic remembering, differences exist between the two. For instance, our imaginings of the future tend to be more positive and to focus more on events that will be important to our life stories (e.g., marriage or retirement). Also, they tend to be less vivid (Anderson, Dewhurst, & Nash, 2012; Berntsen & Bohn, 2010; Grysman, Prabhakar, Anglin, & Hudson, 2015; Rasmussen & Berntsen, 2013). Overall, it is clear that memory is important not only for recollecting what we have been through in the past but also for allowing us to do well in the future.

Summary: Learning and Remembering

6.1 Preliminary Issues

- Long-term memory is a divided between declarative memory and nondeclarative memory. Declarative memory consists of episodic and semantic memories; nondeclarative memory includes priming and procedural or motor learning. Declarative memories can be verbalized, but nondeclarative memories cannot. Conscious awareness of the memory is unnecessary for implicit memory tasks but does accompany explicit memory tasks.
- A classic method for improving memory involves mnemonics. Mnemonics, such as the method of loci, use a variety of techniques, especially visual imagery, to improve performance.
- Ebbinghaus was the first person to extensively study memory and forgetting. Working on his own, he invented methods for doing so. The relearning task revealed a sensitivity to the demands of simple recall tasks—that they measure consciously retrievable information but underestimate the amount of information learned and remembered. The classic forgetting curve he obtained, along with his results on practice effects, inspired the tradition of verbal learning and, later, cognitive psychology.
- Memory consolidation is the process that makes memories permanent. This involves two phases: a fast-acting synaptic consolidation and a slow-acting systems consolidation. The older a memory is, the most likely that it will be consolidated.
- Metamemory is one's awareness of the contents and processes of one's own memory. This is an imperfect insight. One way to assess metamemory involves

judgments of learning in which people assess whether they have learned something and whether it is in memory. Generally, these assessments are poor immediately after learning, but are better if people wait a short period to make such assessments. How well people assess whether they have learned something influences how they decide to study later. Another way to assess metamemory is examining a person's feeling of knowing whether something that is currently forgotten will be remembered later. A related phenomenon is the tip-of-the-tongue (TOT) effect wherein people cannot remember, but feel that remembering is imminent.

6.2 Storing Information in Episodic Memory

- Important variables in storage are rehearsal and organization, regardless of whether the information is verbal or perceptual. Maintenance rehearsal and elaborative rehearsal have different functions: the former for mere recycling of information, the latter for more semantically based rehearsal, which was claimed to process the information more deeply into memory. Difficulties in this depth of processing framework involve specification of the idea of depth.
- Generally, the amount of rehearsal is positively related to recall accuracy for the primacy portion of a list. Also, if people rehearse information in a distributed, rather than a massed, manner, later memory is much better. Rehearsal by taking a test can improve memory more than by simply studying alone.
- The more deeply information is processed, as with elaborative rehearsal, the better it is remembered. This elaborative rehearsal may occur not only through organization, especially by category, but also by subjectively defined chunks, to improve memory because it stores the information securely and provides a useful structure for retrieval. However, the depth of processing framework, while intuitively appealing, is hard to nail down explicitly.

6.3 Boosting Episodic Memory

- There are several ways of improving what will be remembered from episodic memory. This can be done by referring information to oneself or by taking some action with the information. This action can include the generation of the information, the production of what is heard or written by speaking or writing it, or the enactment of described actions. Memory can also be boosted by structuring and organizing the material in some way. This organization can be objective, such as categories, or can be subjectively imposed on the material by an individual. Forming mental images also improves memory by providing another memory code to allow people to access the knowledge as needed. Finally, memory improvement occurs if people relate the material to be learned to their own survival in some way, taking advantage of evolutionary pressures that have shaped memory.

6.4 Context

- Episodic memory is strongly influenced by context. One type of context is the environment one finds oneself in. According to the encoding specificity principle, memory is better when the encoding and retrieval contexts match than when they are different.
- It is also important to be able to remember the context or source in which something was learned. This can involve distinguishing between what was only imagined and what actually happened, or distinguishing between two external sources. Source monitoring issues have been observed in cases of cryptomnesia, or unconscious plagiarism.

6.5 Facts and Situation Models

- Comprehending and remembering ideas can involve constructing propositional representations in which meaningful elements are represented as nodes connected by various pathways (e.g., agent, recipient).
- We tend to remember the gist or general meaning of a passage but not the more superficial aspects such as exact wording. We routinely "recognize" a sentence as having occurred before even if the sentence is a paraphrase.
- Situation models are representations of the described situations. Whereas memories at the propositional level are forgotten rapidly, these memories persist for long periods of time.

6.6 Autobiographical Memories

- Studies of autobiographical memory, or memory in real-world settings, show the same kinds of effects as laboratory studies but sometimes more strongly. Autobiographical memory is one area in which we might see modern psychologists studying their own memories. One part of autobiographical memory is the lack of conscious memories of events from when we were infants. This is known as infantile amnesia. Another part is the superior memory people have for the time in their life around the age of 20. This is known as the reminiscence bump. Finally, autobiographical memories are not always deliberately retrieved, but are often retrieved involuntarily.

6.7 Memory for the Future

- Memory is important not only for storing and retrieving information encountered but also for allowing us to anticipate and do well in the future. Prospective memory involves remembering to do something when a future event is encountered, or a certain period of time has elapsed.
- Episodic future thinking involves using knowledge that we have in declarative memory to imagine and prepare for situations and events that we think are likely to occur.

SHARED WRITING

Episodic Memory

Given what is known about how information is learned and stored in episodic memory, what can a person expect to remember from his or her college experience right after graduation, 5 years later, 10 years later, and 30 years later? Explain.

A minimum number of characters is required to post and earn points. After posting, your response can be viewed by your class and instructor, and you can participate in the class discussion.

Post   0 characters | 140 minimum

Chapter 7
Knowing

Learning Objectives

7.1: Analyze the principle of semantic memory

7.2: Explain the connectionism approach to understanding human memory

7.3: Summarize the basic principles of semantic priming

7.4: Compare schemata and scripts

7.5: Distinguish theories of categorization of concepts as related to semantic memory

This module is about a different kind of long-term memory, namely *semantic memory* (literally "memory for meaning"), which is our permanent memory store of general world knowledge, much like an encyclopedia. Like an encyclopedia, semantic memory can be thought to contain a broad range of general, stable information on a wide range of topics. Your semantic knowledge contains information such as what Canada is, what a website is, who Tom Hanks is, what a dog is, how you order a pizza, and so on. Semantic memory is your conceptual knowledge. Some have called it *generic memory* (Hintzman, 1978), although that name might not be better in some ways (it is way too boring). Your episodic memory differs substantially from mine. Our *episodic memories* are for individual events that we experience. In comparison, our *semantic memories* are largely similar—not in exact content, depending on our cultural backgrounds, but certainly similar in terms of structure and processes (see Medin & Atran, 2004, and Nisbett, Peng, Choi, & Norenzayan, 2001, on some cultural differences). Semantic memories are knowledge you share with your community and are stable across individual events. So although you have no idea what specific things happened during *my* driver's license test, we do share similar semantic concepts of what a driver's license test is. The community of knowledge sharers may be large, such as people who have familiarity with driver's tests or people who have taken college Spanish. Members of one of those communities know what it means to take a driver's test or what the word *rojo* means. Alternatively, the community can be smaller. For example, my high school classmates and I all share an idea of what a "bink" is (a nerdy person) or who our classmates are. Just because you do not share this knowledge with me, does not mean that it is not semantic. It just means you are outside of the community of people who share that general knowledge.

The distinction between episodic and semantic memory is not simply one of convenience but reflects different kinds of mental and neurological processes. For example, Prince, Tsukiura, and Cabeza (2007) showed that episodic memory depends more heavily on some different (but related) brain regions (e.g., hippocampus and anterior prefrontal cortex) as compared to semantic memory (e.g., lateral temporal lobe and posterior prefrontal cortex). This is also evident in amnesia patients who have lost personal episodic memories but not general semantic memories.

The first use of the term *semantic memory* appears to have been in M. Ross Quillian's doctoral dissertation in 1966. Quillian set himself the task of programming a computer to understand language. Machine translation has been a long-standing goal in computer science. The overly confident predictions of the 1950s had failed to account for a subtle yet important fact: Even the simplest acts of human comprehension require vast amounts of knowledge. Thus, for computers to understand, answer questions, or paraphrase, they need this kind of knowledge base. This was Quillian's goal: to provide that extensive knowledge base, to see whether the artificial intelligence (AI) system could then "understand." The implicit point would be that humans also need this vast storehouse of knowledge—a semantic memory. This module covers the basics of semantic memory, including how concepts are stored and retrieved.

7.1: Semantic Memory

OBJECTIVE: Analyze the principle of semantic memory

A study on leading questions provides a convenient entry into the topic of semantic memory. Loftus and Palmer (1974) showed people several short traffic safety films of car accidents and then asked them to describe each accident and answer a series of questions. One question asked for an estimate of the cars' speeds (which people are notoriously poor at estimating). One group of people had to answer how fast the cars were going when they hit each other. The other groups had to answer almost the same question, except that the verb *hit* was replaced with either *smashed, collided, bumped,* or *contacted*. As you might expect, people who got the stronger verbs such, as *smashed*, gave higher estimates of speed. The wording of the question led them to a biased answer.

Why would we expect this? Why does it not surprise us that people estimated higher speeds when the question used *smashed* instead of *bumped* or *hit*? Our intuitive answer is that *smashed* implies a more severe accident than *bumped*. But consider this intuitive answer again. How did those people know that *smashed* implies a more severe accident? It is not enough merely to say that *smashed* implies more severe. We are asking a more basic question than that. We want to know what is stored in memory that tells you what *smash* and *bump* mean. How does memory represent the

---

The Enduring Nature of Semantic Knowledge (Figure 7.1 and Figure 7.2)
An apparent source of your semantic knowledge, apart from your everyday experience with the world, is the things you have learned in school. So, just how long will this knowledge last? Will it be quickly forgotten? Will it be remembered forever? What from our schooling has an impact on semantic memory?

Interactive

A study by Bahrick (1984; see also Bahrick, Bahrick, Bahrick, & Bahrick, 1993) tested people's memory for Spanish they had learned while they were in college. This assessment was made soon after graduation and at various intervals of time up to 50 years later. Bahrick had people who returned to college for their reunions take a variety of academic tests for college Spanish. (Imagine coming back for a college reunion and getting a pop quiz on material you learned while you were there and had not used since.)

fact that *smashed* implies a severe accident, that moving cars have drivers, that robins have wings, or that bananas, canaries, and daisies are all yellow? In short, what is the structure and content of semantic memory per se and how do we access that knowledge? We will spend a great deal of time in this module assessing the structure of knowledge in semantic memory. However, before we do that, let's consider the duration of semantic knowledge. Just how long does it last?

7.1.1: Persistence of Semantic Knowledge

So, just what is a platypus? Well, you can picture one in your mind. You may know it has a duck bill and webbed feet because it lives near water. The platypus lives in Australia and, although it is a mammal (of some kind), it lays eggs. You may even know that this animal is a monotreme. Even if you have not thought about a platypus in quite a while, this knowledge comes to mind quickly. This type of recall is one of the hallmarks of semantic memory: It tends to be robust and enduring. It allows you to recall knowledge without having thought about it for some time.

7.1.2: Semantic Networks

One of the earliest systematic attempts to address issues of how semantic knowledge is represented (aside from philosophical and linguistic analyses) was Quillian's (1968, 1969) work in artificial intelligence.

You should have found two characteristics of spreading activation. First, activating "ROBIN" eventually primed a node that was *also* activated by "BREATHES." This explains how information is retrieved from semantic memory. The spreads of activation intersect, and a decision is made to be sure that the retrieved link represents the idea in the

---

Quillian's Representation of Semantic Network (Figure 7.3)

The Model
His initial model of semantic memory was not a psychological model of memory but a computer program for understanding language. Shortly afterward, however, Quillian collaborated with Allan Collins, and their psychological model was the first attempt in cognitive psychology to explain the structure and processes of semantic memory.
The Collins and Quillian model of semantic memory (1969, 1970, 1972; Collins & Loftus, 1975) was based on a network metaphor of memory. At the heart of the model were fundamental assumptions about the *structure* and *retrieval processes* of semantic memory.

sentence. So, although a pathway would be found between "ANIMAL" and "RED BREAST," a decision that the idea "all animals have red breasts" would be invalid.

WRITING PROMPT
Activating Concept Nodes

As an exercise, study Figure 7.3A and mark concept nodes that are activated by a sentence like "A robin can breathe." To track the spread, write a "1" next to links and nodes that are activated by "ROBIN" and a "2" next to those activated by "BREATHES." Take this through at least two cycles. In a sense, you are "hand simulating" a memory search. What did you discover?

▶ The response entered here will appear in the performance dashboard and can be viewed by your instructor.

Submit

RELATED SEMANTIC SEARCH CONCEPTS A second characteristic of semantic search is that other concepts also become activated or primed. In our example, many other concepts were primed. There should also be a "1" next to "RED BREAST" and "BLUE EGGS" from the first cycle; a "1" next to "FLY," "FEATHERS," and "CANARY" after the second cycle; and so on. Thus, spreading activation activates related concepts. More and more distant concepts, connected by longer pathways, do not receive as much activation. Moreover, these related concepts are not activated forever because activation decays after some amount of time.

7.1.3: Feature Comparison Models
Semantic networks are not the only way semantic memory has been conceived. Another is as a collection of

---

Smith's Model of Feature Comparison (Figure 7.4)
The retrieval process in the Smith model is *feature comparison*; follow along with the sequence in Figure 7.4 as you read.

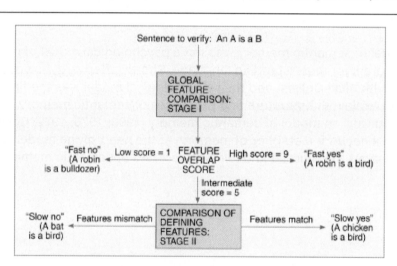

Figure 7.4 The Comparison and Decision Process in the Smith et al. Model (1974), with Sample Sentences

Stage I
Suppose you are given the sentence "A robin is a bird" and have to make a true/false judgment. For this model, you would access the concepts "ROBIN" and "BIRD" in semantic memory and then compare the two feature sets. Stage I involves a rapid, global comparison of features, to "compute" the similarity between the two concepts. For illustration, assume that these similarity scores range from 1 to 10.

For "A robin is a bird," the feature sets overlap a great deal; there are very few "ROBIN" features that are not also "BIRD" features, leading to a high overlap score (e.g., 8 or 9), so high that you confidently respond "yes" immediately. Conversely, for "A robin is a bulldozer," there is so little feature overlap (e.g., 1 or 2) that you can respond "no" immediately. These are "fast yes" and "fast no" Stage I responses.

features, as with the feature comparison model of Smith and his colleagues. In Smith's model (Rips, Shoben, & Smith, 1973; Smith, Rips, & Shoben, 1974), semantic memories were captured as **feature lists**. So, semantic memory is a collection of sets of **semantic features**: simple one-element properties of the concept. Thus, the concept "ROBIN" is a list of its features, such as *animate*, *red-breasted*, and *feathered*. Furthermore, these features vary in terms of their *definingness*, with defining features being emphasized more than less defining features. Thus, an essential feature is a **defining feature**, such as *animate* for "BIRD." Conversely, other features (e.g., "ROBIN" perches in trees) would be less so but are **characteristic features**: features that are common but not essential. Thus, characteristic features do not define "ROBIN": Robins may or may not perch in trees. But defining features are essential: If it is a robin, it has to be animate.

7.1.4: Tests of Semantic Memory Models

Some tests of semantic memory use the **sentence verification task**, in which simple sentences are presented for the yes/no decisions. Accuracy for such decisions does not tell us much; people seldom make mistakes for such simple facts. Instead, response time (RT) measures are primarily used. The following section lists the characteristics of such yes/no recognition tasks.

RECOGNITION TASKS IN SEMANTIC MEMORY The *recognition task* consisted of two basic steps: The person first learned a set of words on a list and then made yes/no decisions on a test list ("yes" if the test word was on the studied list, "no" if it was not). The important features that make a recognition test are as follows:

- People make yes/no or forced-choice decisions.
- The decisions are based on information stored in memory.
- For semantic memory research, this task includes information already in long-term memory before the beginning of the experiment.
- *Generalized recognition task (semantic)*
  - Information to be tested is already in long-term memory, such as knowledge of categories ("A robin is a bird") or words.

- People make yes/no decisions about items, presented one at a time. Unlike episodic recognition, here the "yes" response usually means the item is true. For example, in a *sentence verification task*, people say "yes" to "A robin is a bird," that is, to any sentence that is true. In a *lexical decision task*, people respond "yes" if the letter string is a word (e.g., MOTOR) and "no" if it is not (e.g., MANTY).
- *Dependent variables*: Typically, the primary dependent variable is response time (RT), although accuracy is also important. Occasionally, people are given a response deadline; that is, they are given a signal after some interval, say 300 ms, and must respond immediately after the signal. In such a task, error rate becomes the major dependent variable. Different electroencephalographic (EEG) patterns have also been used (see Kounios & Holcomb, 1992).
- *Independent variables*: An enormous range of independent variables can be tested, such as the semantic relatedness between concepts, and word length, frequency, concreteness, and the like in a lexical decision task.

Collins and Quillian (1969) tested a basic prediction of their model: Two concepts that are closer in the network should take less time to verify than two that are farther apart. Refer again to Figure 7.3 and your "hand simulation." If we assume that this part of semantic memory is accurately represented here, then some predictions can be

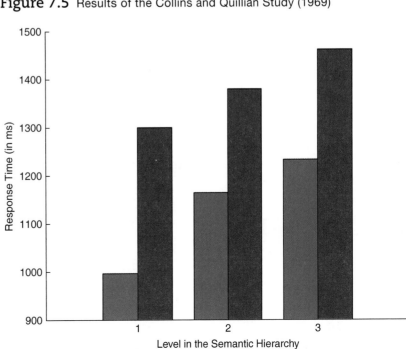

Figure 7.5 Results of the Collins and Quillian Study (1969)

made. For example, you should be quicker to verify "A robin is a bird" than "A robin is an animal," because it should take less time for activation to spread across one link rather than two. Likewise, it should take less time to verify that canaries are yellow than that they can fly or breathe, for the same reason.

As you can see, response time increased as the distance between the two concepts increased.[1] However,

---

[1] Collins and Quillian tested many more concepts than just canaries and robins, although the tradition in this area of research is to illustrate the models using these words.

Figure 7.6 An Illustration of a Portion of the Semantic Network, Taking into Account Three Empirical Effects

There is no strict cognitive economy in the hierarchy, so redundant information is stored at several different concepts. Typical members of the category are stored more closely to the category name or prototype member, and properties that are more important are stored more closely to the concept than those of lesser importance.

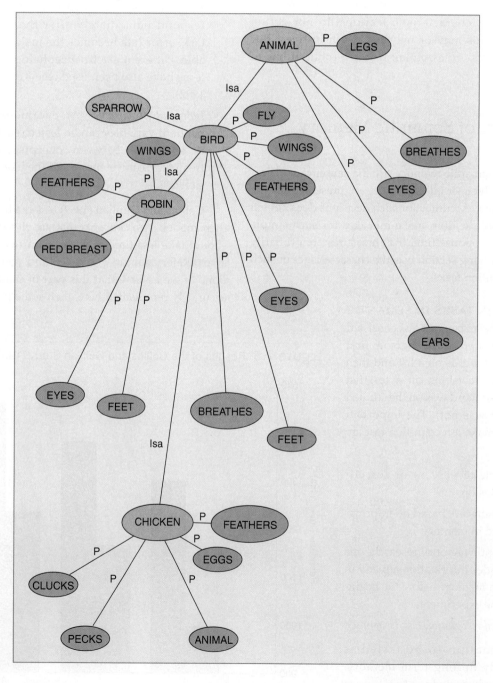

subsequent work by researchers, such as Rips et al. (1973; see also Murphy, Hampton, & Milovanovic, 2012), showed that human semantic memory is not structured as nicely as the Collins and Quillian model implies. For example, people verify the statement "A pig is an animal" faster than they verify "A pig is a mammal," although the Collins and Quillian model predicts the opposite result.

7.1.5: Semantic Relatedness

Figure 7.6 is a modified network of the BIRD category that incorporates semantic relatedness (Collins & Loftus, 1975).

For example, a strict hierarchy is incorrect. Instead, these links are of different lengths, reflecting different strengths of association, with stronger associations being verified faster than weaker ones. Moreover, for the feature of typicality (elaborated on later), more typical examples of a concept are linked by shorter pathways. Therefore, the link between *robin* and *bird* is shorter than that between *chicken* and *bird*. It is difficult in a two-dimensional figure to completely capture the associative and strength complexity of semantic memory. Thinking of semantic memory in space with three or more dimensions makes this easier to imagine, but harder to illustrate.

Regardless, the idea that performance varies directly as a function of the strength of associations is readily understood. The stronger the relation between concepts, the faster you can retrieve the idea. This is the **semantic relatedness effect**: Concepts that are more highly interrelated are retrieved faster. To demonstrate this, time yourself as you name the 12 months of the year; it takes about 5 seconds. Now time yourself as you name the 12 months again, but this time in alphabetical order. How long did that take—at least half a minute, if not longer, right? It should be obvious how this information is organized in memory—based on chronological order. Thus, retrieving the information in a way that is incongruent with its storage organization makes it difficult. The results are shown in Figure 7.7.

ADDITIONAL SUPPORT FOR SEMANTIC RELATEDNESS Kounios and Holcomb (1992; Kounios, 1996) reported further evidence of semantic relatedness. They used *event-related potential (ERP)* recordings in addition to response times. In this study, pairs of words varied in relatedness, either high (*rubies–gems*) or low (*spruces–gems*). Half of the time, a category member came before the category name (*rubies–gems*), and half the time the reverse happened (*gems–rubies*).

You can see the same increasing function as before, for example, when typicality varied from high to low in Figure 7.8.

UNEXPECTEDNESS IN SEMANTIC JUDGMENTS We can also observe changes in electrical potential as a result of cognitive processing. Figure 7.9 shows the pattern of N400s taken from one midline electrode site.

The solid curve for "Exemplar–category, related" stimuli (e.g., "All rubies are gems") continued its negative drift across time. For all three other types, however, there was a negative peak around 400 ms, especially when the subject and predicate were unrelated. In other words, the N400 is sensitive to the relatedness of the two concepts in the sentence or, more accurately, to their *un*relatedness. Kounios and Holcomb concluded that the N400 reflects retrieval in semantic memory, and coherence and integration processes in language comprehension (see also Holcomb, 1993). When two words were related, there was a substantial difference in the ERP pattern compared to when they were unrelated.

Figure 7.7 Mean N400 Amplitude Recorded From Three Midline Sites as a Function of Semantic Relatedness and Prime-Target Order

Sentences were presented with one of three quantifiers—*all*, *some*, or *no*—thus altering the meaning (e.g., "All rubies are gems" and "Some gems are rubies" are true, but "No rubies are gems" is false).

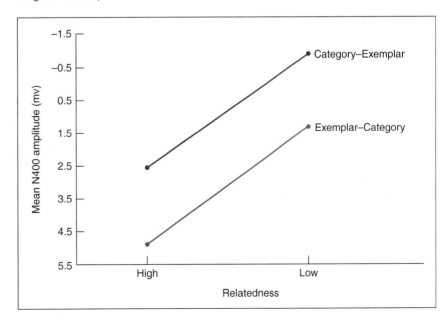

Figure 7.8 Mean Response Times to Members of Typicality Categories

The response times are much shorter than in comparable studies because the category names were given at the beginning of a block of trials and did not change within the block. Thus, each trial consisted of only the target word, and people judged whether it belonged to the given category name.

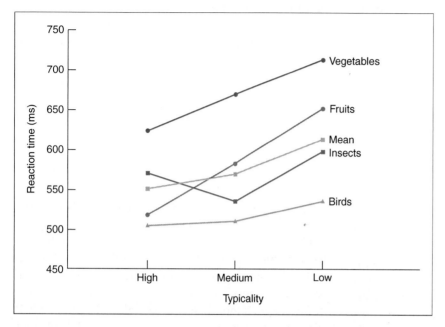

Figure 7.9 The Event-Related Potentials at One Midline Site (C_z) for Four Types of Trials

"Exemplar–category" and "Category–exemplar" refer to the order of the words. The onset of the target is shown by the vertical bar at the beginning of the time lines. In the Kounios and Holcomb study, the ERP component of interest was the N400 component, a *negative* change occurring about 400 ms after the item.

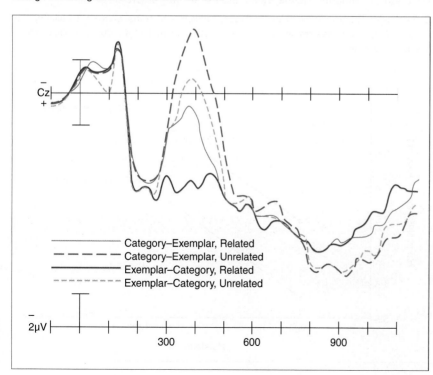

One semantic-based result addressed in Kounios (1996) makes sense given what we know about cerebral lateralization. Most research suggests that the left hemisphere is dominant for verbal and language-based tasks, whereas the right hemisphere is dominant for visual and spatial information. Kounios describes a match between ERP results and Paivio's (1990) *dualcoding hypothesis*. Recall the Paivio work showing that concrete words are learned and recalled better than abstract words because they can be represented both verbally and by a mental image (Paivio, 1971). Paivio's (1990) idea was that the word-based knowledge is located in the left hemisphere and the image-based knowledge is in the right. Thus, concrete words enjoy an advantage because they recruit processes in both hemispheres.

Kounios and Holcomb (1994) used a lexical decision task showing both concrete (e.g., *table*) and abstract (e.g., *justice*) words. Their ERPs provided support for Paivio's predictions.

EMBODIED SEMANTICS The complexity and organization of semantic memory is still being explored. Semantic memory also accounts for how we interact with the world through our senses and our actions, and so reflects embodied influences.

AMOUNT OF KNOWLEDGE The amount of knowledge we have makes a difference during memory search and retrieval—more knowledge and greater semantic relatedness go together. For example, Pexman, Holyk, and Monfils (2003) found faster reading times for concepts with more features than for those with fewer features. Also, Yates, Locker, and Simpson (2003) reported comparable effects in a lexical decision task, contrasting words with large versus "sparse" semantic neighborhoods; a large-neighborhood word like *bear* has many associates in memory, compared to a

Figure 7.10 ERPs for Concrete and Abstract Words

As shown in Figure 7.10, ERPs for concrete words were equal and high for both hemispheres. In contrast, the ERPs were markedly lower in the right than in the left hemisphere for abstract words, and both of these amplitudes were lower than those for concrete words.

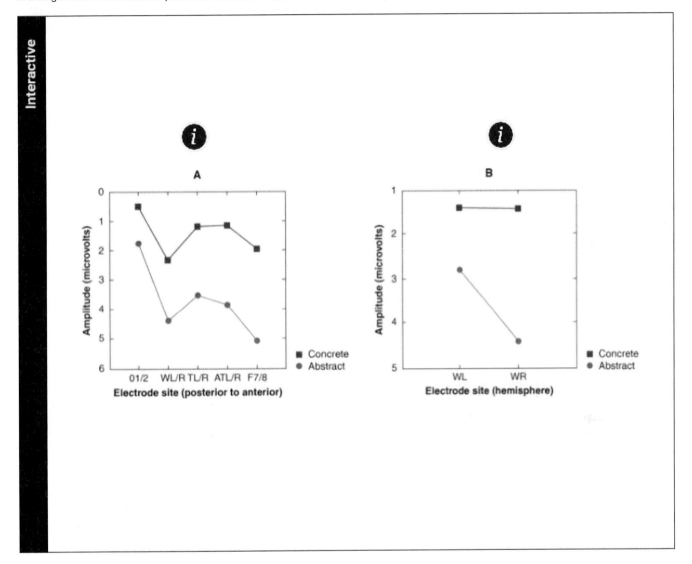

sparse-neighborhood word like *suds*, and is therefore judged more rapidly.

The influence of amount of knowledge extends to many settings and tasks. For example, Hambrick (2003) had people learn new information (about basketball) and found that the strongest influence on learning was the amount of domain knowledge they already had. In other words, the more you know about something, the easier it is to learn new things about that topic. Moreover, Westmacott and Moscovitch (2003) found enhanced performance on a variety of tasks (they used recall, recognition, fame judgment, and speeded reading) when there was some autobiographical significance to the knowledge being acquired. Presumably, the more you know about a topic, the more highly integrated and related that knowledge is in memory. This leads to more activation in memory and enhanced retrieval.

WRITING PROMPT

Semantic Relatedness

What factors do you think influence your recall, judgment, and semantics? What helps you to retrieve faster and better?

▶ The response entered here will appear in the performance dashboard and can be viewed by your instructor.

Submit

Theories Based on Embodied Semantics

As one illustration of this principle, it has been found that the **semantic distance effects** that were just discussed are found even when people retrieve knowledge of body parts, with close body parts (e.g., hand and elbow) priming each other more than distant body parts (e.g., hand and knee) (Struiksma, Noordzij, & Postma, 2011; van Elk & Blanke, 2011).

> **Interactive**
>
> ### Barsalou's Theory of Perceptual Symbols
>
> One such view is Barsalou's (1999) theory of **perceptual symbols**, which states that semantic memory is built up of sensory and motor elements derived from experience. For example, Kalénine and Bonthoux (2008) gave people sets of three pictures (e.g., either a coat, a jacket, and a stove or a coat, a hanger, and a stove), asking them to identify which of two remaining pictures went with a first. Two factors were manipulated. One was the relationship of the correct choice to the first picture. The objects were either *conceptually* related (e.g., coat–jacket) or *functionally* related (e.g., coat–hanger). The other factor was whether the objects could be interacted with and manipulated by people (e.g., a coat can be but a castle cannot). They found that people were faster to respond to conceptually related items for objects that you cannot manipulate, but were faster to respond to functionally related items for objects that you can (e.g., coat–hanger). Thus, the efficiency of using your semantic memory depends on both *how* you are using it as well as how you *interact* with the world.
>
> ### Pecher, Zeelenberg, and Barsalou's Theory
>
> ### Klatzky, Pellegrino, McCloskey, and Doherty's Theory

7.2: Connectionism and the Brain

OBJECTIVE: Explain the connectionism approach to understanding human memory

A more complex way to model human memory rather than a semantic network is to use the approach of *connectionism*. You have had small doses of this already. At the most fundamental level, *connectionist models* (also known as *PDP models* and *neural net models*) contain a massive network of interconnected nodes. Depending on the model, the nodes can represent almost any kind of information, from simple line segments and patterns in letter recognition to more complex features and characteristics in semantic memory. What makes connectionist models attractive is that, in principle, they capture the essence of the neurological processes involved in semantic memory.

7.2.1: Connectionism

Examine Figure 7.11, a connectionist network for part of the "FURNISHING" category (Martindale, 1991).

First, notice that each concept is connected to other nodes. The difference here is that each path has a number next to it. This is its weighting or strength. The weightings indicate how strongly or weakly connected two nodes are. Generally, the weighting scale goes from +1.0 to –1.0, with positive and negative numbers indicating facilitate and inhibition connections, respectively. So, for example, the

Figure 7.11 A Small Portion of a Connectionist Network

Note that the nodes at the same level exert an inhibitory influence on each other and receive different amounts of facilitation from the category name.

SOURCE: Based on Martindale (1991).

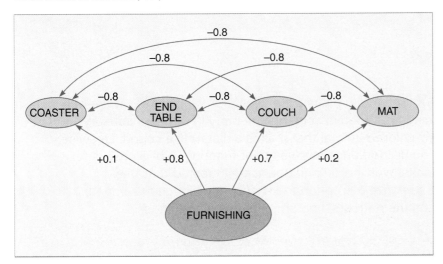

weighting between "FURNISHING" and "END TABLE" is +0.8, indicating that "END TABLE" is a central member of the category. However, "COASTER" with its +0.1 is only weakly associated with "FURNISHING." Apart from this one example, it should be noted that in the area of semantic memory, there are many connectionist models that capture the process of semantic memory. These include models that account for things like priming (Masson, 1995) and word meaning (McRae, de Sa, & Seidenberg, 1997).

The glimpse at connectionism here is very simplified. In many respects connectionist models differ from the older spreading activation models. An important distinction is that in a *connectionist* mode, a concept is defined as a pattern of activation across units in the network. *Priming* is explained by the similarity of activation patterns between a prime and a target: The "FURNISHING" concept has a similar pattern of activation to the pattern for "END TABLE," so the one serves as an effective prime for the other.

7.2.2: The Benefits of Connectionist Models

What is exciting about connectionist models is that they give us a tool for understanding the richness of cognition, a working "machine" that lets us see what happens when multiple layers of knowledge influence even the simplest acts of thought. Particularly compelling (Seidenberg, 1993) are four advantages of connectionist models. First, they are more similar to the network of neurons in the brain. The brain is a massive set of interconnected neurons, just as a connectionist network is a massive set of interconnected units. Second, the units are similar to those in the brain. In the nervous system, a neuron either fires or does not. When it does fire, it affects the neurons it synapses on. This parallels the fire/no-fire nature of connectionist units. Third, the positive and negative weights between units in connectionist models parallel excitatory and inhibitory neural synapses. Fourth, the activity of a connectionist model is massively parallel. Multiple processes are co-occurring in a model at various levels, much as there is parallel processing in the brain (McClelland, Rumelhart, & Hinton, 1986; Rumelhart, 1989).

A CONNECTIONIST MODEL OF SEMANTIC MEMORY IMPAIRMENT
A puzzling disorder in semantic memory is a **category-specific deficit**. This is a disruption in which the person loses access to one semantic category of words or concepts while not losing others. Warrington and McCarthy (1983; also Warrington & Shallice, 1984) reported the case histories of four brain-damaged patients who showed this strange dissociation. The patients they described had difficulties identifying living things but not nonliving things (including artifacts—things that people make). For instance, patient J. B. R. could identify only 6% of pictures of living things, such as *parrot* and *daffodil*, and could define only 8% of the words that named living things. But when shown pictures of nonliving things, such as a *tent* and *briefcase*, J. B. R. successfully named them 90% of the time and defined them 79% of the time.

Does this demonstration prove that impairment of patient J. B. R.'s visual semantic knowledge accounts for the dissociation? No; the model makes the correct prediction, a point in its favor. But this is not proof that the model is correct. Instead, the demonstration essentially asked whether it is possible that impairment of sensory knowledge could produce the dissociation between living and nonliving things. An answer is "yes." It is possible, because a connectionist model simulates a dissociation. In other words, the model provides a degree of assurance (probably a large degree) that the Warrington and Shallice hypothesis is reasonable and should be pursued further.

SEMANTIC MEMORY LOSS Other brain disorders affect the part of semantic memory known as **lexical memory**, the **mental lexicon** or dictionary where our word knowledge (as distinct from conceptual knowledge) is stored. One is a deficit in word finding, known as **anomia**

Studies on Semantic Memory Impairment (Figure 7.12)

How could semantic memory be splintered so that a person's categories of animate things are disrupted while categories of nonliving things are preserved? More to the point, could semantic memory be structured to reveal these two very broad categories, living and nonliving things? Why this distinction, and not some other, such as concrete versus abstract, high versus low frequency, or something else?

Warrington and Shallice's Study

Warrington and Shallice (1984) suggested a plausible explanation. Suppose that the bulk of your knowledge about living things is coded in semantic memory in terms of sensory properties: A parrot is a brightly colored animal that makes a distinctive sound. Likewise, most of what you know about nonliving things involves their function: A briefcase is for carrying around papers and books. Warrington and Shallice suggested that this dissociation could be due to a selective disruption to sensory knowledge in semantic memory. If so, that might explain the patients' impairments for living things.

Farah and McClelland's Model

(or sometimes **anomic aphasia**). At a superficial level, anomia is like a *tip-of-the-tongue (TOT) effect* in which you are unable to name the word even though you are certain you know it. However, whereas people in a TOT state generally have partial knowledge about the unrecallable word, such as the number of syllables or the first sound, anomic patients do not have such partial knowledge. Instead, there is an inability to name a semantic concept that is otherwise successfully retrieved (Ashcraft, 1993; for work on everyday forgetting—temporary inhibition—in semantic memory, see Johnson & Anderson, 2004). These people know and can use the concept—they just cannot name it. For example, Kay and Ellis's (1987) anomic patient was unable to name the word *president* but, in attempting to find the word, blurted out, "Government . . . leader . . . John Kennedy was one." Cases such as these suggest that the semantic memory of a concept is distinct from the memory of its name, such that the concept can be retrieved but the name cannot.

> **WRITING PROMPT**
>
> **Temporary Inhibition**
>
> Think of an incident in which the name of a word was on the tip of your tongue but you could not recall it. What could you remember at that time? Were you finally able to remember the word? If so, what helped in the process? If not, what do you think could have helped you recall the word better?
>
> The response entered here will appear in the performance dashboard and can be viewed by your instructor.
>
> Submit

7.3: Semantic Priming

OBJECTIVE: Summarize the basic principles of semantic priming

An important principle that has been touched on, and can be expanded on here, is the idea of the broad activation of concepts in semantic memory, namely **semantic priming**. This priming follows some basic principles:

1. The priming process takes time.
2. The activation of primed concepts is smaller the more removed concepts are from the origin.
3. The priming effect decays across time.

Priming is a fundamental consequence of retrieval from semantic memory. It is one of the most frequently tested—and discussed (e.g., McNamara, 1992; Ratcliff & McKoon, 1988)—effects in long-term memory. It is *key* to understanding semantic memory. We return to it repeatedly throughout the course. So you need to understand priming, how it affects semantic memory, and how it has been studied.

7.3.1: Nuts and Bolts of Priming Tasks

Priming is defined as the activation of concepts and their meanings. It is part of the automatic process observed in the *Stroop task* (name the color the word is printed in, not the color word itself). Priming activates all kinds of concepts that express a joint meaning of several concepts together (e.g., Mathis, 2002; Peretz, Radeau, & Arguin, 2004).

When this influence is beneficial, for instance when the target is easier or faster to process, this positive influence is called *facilitation* or a *benefit*. Often, facilitation is a faster response time compared to performance in a baseline

---

A Depiction of Two Priming Tasks (Figure 7.13A–Figure 7.13B)
Let's discuss two depictions of the priming tasks and how to set up a priming task.

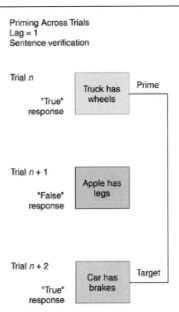

Priming Across Trials
Almost any recognition task (see figure above) can be adapted to a priming task. Here, people respond to both the **prime** and the **target**. This critical measure is the influence of the processing of the prime on the processing of the target. Care must be taken so that an equal number of related targets are true and false so people are not biased to respond "yes." In some cases, the prime and the target are separated by various intervals of time called the **stimulus onset asynchrony (SOA)**, which is the length of time between the onset of the prime and the onset of the target. If you think of the prime and target as two halves of a complete set, then the onset or beginning of the two halves occurs asynchronously, at different times. Thus, we might present a prime and 500 ms later present the target. This would be a 500-ms SOA.

Key Terminology Related to Priming

Below is a list of basic terms and some explanations to help you.

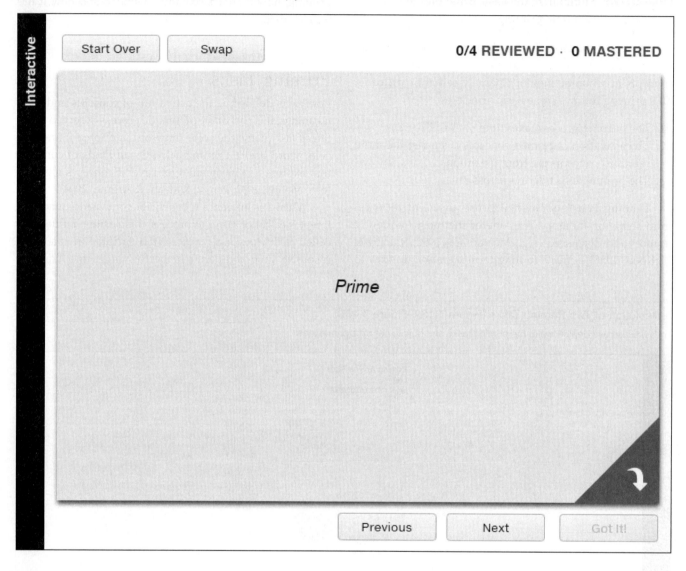

condition (when the prime and target are unrelated). Occasionally, the influence is negative, as when a prime is antagonistic to the target and so is irrelevant or misleading. When the prime impedes performance to the target, the negative influence on processing is called *inhibition* or a *cost*.

If we are interested in the influence of priming over time and/or across trials, we need to keep track of the interval between the prime and the target. This is the **lag** between the prime and target, usually indexed as the number of intervening stimuli.

7.3.2: Empirical Demonstrations of Priming

Semantic priming generally reflects a temporary heightening of related concepts. Although most studies look at this decay of the influence of priming over a few seconds (e.g., Freedman & Loftus, 1971; Loftus & Loftus, 1974), there is evidence that prior exposure can facilitate performance even 17 years later (Mitchell, 2006), although the precise cause of this very, very long-term priming is not well known. Most priming tasks in cognitive psychology involve a lexical decision task, although some other tasks are used as well. For example, priming has been reported using sentence verification (Ashcraft, 1976). A sentence such as "A sparrow has feathers" facilitates another sentence about the same category, such as "A robin can fly" (see also Anaki & Henik, 2003).

A workhorse task in cognitive science is *lexical decision task*, in which people judge whether a string of letters is a word (remember MOTOR and MANTY?). Customarily, RT

Table 7.1 Priming in the Lexical Decision Task

Type of Stimulus Pair					
Top String	Bottom String	Correct Response	Sample Stimuli	Mean RT (ms)	Mean % Errors
Word	Related word	Yes	Nurse–doctor	855	6.3
Word	Unrelated word	Yes	Bread–doctor	940	8.7
Word	Nonword	No	Book–marb	1,087	27.6
Nonword	Word	No	Valt–butter	904	7.8
Nonword	Nonword	No	Cabe–manty	884	2.6

SOURCE: From Meyer and Schvaneveldt (1971).

is the primary dependent variable. The name of the task comes from the word *lexicon*, meaning a dictionary or a list of words. So in a sense, the lexical decision task asks you whether the string of letters is an entry in your *mental lexicon*, your mental dictionary.

A huge range of topics has been investigated using lexical decision. The groundbreaking work was done by Meyer and Schvaneveldt (1971; also Meyer, Schvaneveldt, & Ruddy, 1975). They presented two letter strings at a time and told people to respond "yes" only if both were words. In addition to trials with unrelated words such as *truck–paper*, they had trials with related pairs (there were an equal number of "yes" trials and "no" trials, the latter of which involved letter strings that were not a word, such as *chair–zoople*). Meyer and Schvaneveldt found that two related words such as *bread–butter* are judged more quickly as words than two unrelated words such as *nurse–butter*. Table 7.1 shows Meyer and Schvaneveldt's results and shows the priming effect clearly. Related words were judged in 855 ms, compared to 940 ms for unrelated words.

One interesting aspect of these results is that it is not logically necessary for people to access the meanings of words. Technically, they need only "look up" the words in the mental lexicon. Yet the results repeatedly show that it is the *meaningful* connection between *bread* and *butter* that facilitates this decision, not some lexical connection. You might think because both begin with *B* that there is a lexical basis for the facilitation, but the same benefits are found for word pairs with dissimilar spellings, such as *nurse–doctor*.

7.3.3: Automatic and Controlled Priming

This facilitation in priming appears to be automatic. It happens rapidly and without intention (e.g., Smith, Bentin, & Spalek, 2001). When you see a word, you access its meaning automatically, even though you are not required to by the task. Semantic activation later may be redirected by conscious, deliberate efforts. But some priming may be more effortful and controlled. For example, in a study by Neely (1977), people saw a letter string preceded either by a neutral prime (the letter *X*) in a baseline condition, or by a category name prime that was either related or unrelated to the target. For all trials, people had to judge whether the target string, the second member of a pair, was a word (lexical decision). Table 7.2 summarizes the study and shows sample stimuli. Because the results are a bit complicated, we will go through them in stages.

Table 7.2 Conditions and Stimuli for Neely's (1977) Study

Condition	Sample Stimulus[a]
No Category Shift Expected	
No shift	BIRD–robin
Shift	BIRD–arm
Category Shift Expected From "Building" to "Body Part"; From "Body" to "Part of a Building"	
No shift	BODY–heart, BUILDING–window
Shift to expected category	BODY–door, BUILDING–leg
Shift to unexpected category	BODY–sparrow

[a]Prime is in capital letters.

First, Neely found standard semantic priming. For prime–target trials such as "BIRD–robin," there was facilitation, as shown in the left panel of Figure 7.14 (notice that any point above the dashed line at 0 ms indicates facilitation, and any point below indicates inhibition).

Because this speedup was found even at short SOAs, it suggests that normal semantic priming is automatic. Notice also that there is more facilitation as the SOA got longer. So, with more time to prepare, priming grew stronger.

Second, Neely found inhibition, a slowdown in RT when the prime was unrelated to the target. If you had seen "BIRD" as a prime, you were then slower to decide that *arm* was a word. Not surprisingly, this inhibition effect grew stronger at longer SOAs: Again, the preparation for a

Figure 7.14 Reaction Time to Lexical Decision Targets

In the left panel, people saw a prime and did not expect a shift in category; sample stimuli are "BIRD–robin" for a relevant prime and "BIRD–arm" for an irrelevant prime. In the right panel, people expected the target to come from the *building* category if they saw "BODY" as a prime and from the body part category if they saw "BUILDING" as a prime. When the shift in category occurred as expected, response time was facilitated at longer SOAs. When the expected shift did not occur, there was facilitation at the short SOA when the prime was relevant ("BODY–heart"). Inhibition occurred when the shift was completely unexpected ("BODY–sparrow").

SOURCE: From Neely (1977).

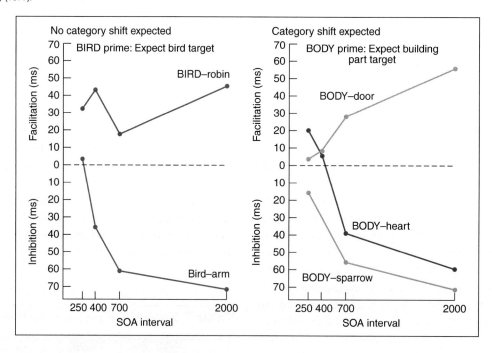

member of the *bird* category worked against you when you saw *arm* as the target.

The impressive (and complicated) part of Neely's study comes next. He told people that when they saw one particular prime, such as the category name "BODY," they should expect a target from a *different* category, such as a part of a building (e.g., *door*). Likewise, if they saw "BUILDING" as a prime, they should expect to see a body part, such as *arm*, as the target. What happened then? When the switch from one category ("BODY") to a different one (*door*) happened, there was priming only at the long SOAs, as shown in the top right panel of Figure 7.14. This makes good sense; you see "BODY," and it takes you a bit of time to remember that you should see an item from the "BUILDING" category next. When you are given that bit of time, you are ready for *door, window,* and so on. But notice that at the shortest SOA, there was no facilitation. Clearly, at short SOAs there was not enough time to prepare for the category shift.

Finally, on a small proportion of trials, a category switch was expected but did not occur—that is, when you saw "BODY" but then saw *arm* or *heart* as the target. The clever thing about this is that it should tap into normal semantic priming because *arm* and *heart* are in the "BODY" category. This was what Neely found, but only at short SOAs. With longer SOAs, people prepared for the category to switch. Then, when that *did not* happen, they were slowed, as shown in the bottom right panel of the figure.

7.3.4: Priming Is Implicit

Accessing a word's meaning is automatic (e.g., Friedrich, Henik, & Tzelgov, 1991). It occurs without conscious awareness of even having seen a word. Marcel (1980, 1983) presented primes immediately followed by a scrambled visual pattern, a visual mask. The purpose was to present the mask so soon after the prime that people were not consciously aware it at all, a form of *backward masking*.

The manipulation worked. There was no conscious awareness of the prime. Yet related primes such as "CHILD" facilitated lexical decisions about words such as *infant*. Substantial work indicates that the effect is genuine (Hirshman & Durante, 1992; McRae & Boisvert, 1998). Semantic priming can occur automatically, without conscious identification of the prime. Thus, such effects are implicit and can occur without any involvement of conscious awareness. Indeed, (implicit) priming effects can be observed in ERP patterns from cortical recordings (e.g., Heil, Rolke, & Pecchinenda, 2004), or even when the people have lost the (explicit) ability to recall information consciously (i.e., in cases of amnesia, as in Levy, Stark, & Squire, 2004). See Schacter and Badgaiyan (2001) for a readable introduction to neuroimaging results on priming.

> **WRITING PROMPT**
>
> **Semantic Priming**
>
> Do you think that semantic priming occurs automatically or are you consciously aware of the relatedness? Explain.
>
> ▶ The response entered here will appear in the performance dashboard and can be viewed by your instructor.
>
> Submit

7.4: Schemata and Scripts

OBJECTIVE: Compare schemata and scripts

Now that we have set out some basics of semantic memory, let's move on to a more specific way of using general knowledge. We often use world knowledge to understand the events and circumstances that we encounter. We start out by looking at memory for a story. The interactive below contains a story called "The War of the Ghosts." The story is important not only because of the psychological points it raises, but also for historical reasons: Bartlett (1932) used it in one of the earliest studies of remembering meaningful material. Do the demonstration in the interactive now, before reading further.

Bartlett's (1932) "The War of the Ghosts"

Interactive

1. Reproduce the story by writing it down from your memory in the space provided below.

[Reset] [Previous] [Next]

Two Retellings of Bartlett's (1932) "The War of the Ghosts"

First recall, attempted about 15 minutes after hearing the story

Two young men from Egulac went out to hunt seals. They thought they heard war-cries, and a little later they heard the noise of the paddling of canoes. One of these canoes, in which there were five natives, came forward towards them. One of the natives shouted out: "Come with us: we are going to make war on some natives up the river." The two young men answered: "We have no arrows." "There are arrows in our canoes," came the reply. One of the young men then said: "My folk will not know where I have gone"; but, turning to the other, he said: "But you could go." So the one returned whilst the other joined the natives.

The party went up the river as far as a town opposite Kalama, where they got on land. The natives of that part came down to the river to meet them. There was some severe fighting, and many on both sides were slain. Then one of the natives that had made the expedition up the river shouted: "Let us return: the Indian has fallen." Then they endeavored to persuade the young man to return, telling him that he was sick, but he did not feel as if he were. Then he thought he saw ghosts all round him.

When they returned, the young man told all his friends of what had happened. He described how many had been slain on both sides.

It was nearly dawn when the young man became very ill; and at sunrise a black substance rushed out of his mouth, and the natives said one to another: "He is dead."

Second recall, attempted about 4 months later

Given below are two retellings of the story you just attempted to retell. Compare your recalled version with those retellings. Although your version may be closer to the original, because so little time passed between reading and recalling, you should be able to see points of similarity to these retellings.

7.4.1: Bartlett's Research

Bartlett (1932), like Ebbinghaus, wanted to study memory experimentally. However, very much unlike Ebbinghaus, he wanted to study memory for *meaningful* material. So he used folktales, ordinary prose, and pictures. His typical method was to have people study material for a period of time, then recall it several times, once shortly after study and again at later intervals. By comparing successive recalls, Bartlett examined the progressive changes in what people remembered.

Using these methods, Bartlett found that memory for meaningful material is not especially reproductive; it does not strictly reproduce the original passage. Instead, Bartlett characterized this remembering as "an effort after meaning." The modern term for this is **reconstructive memory**, in which we construct a memory by combining elements from the original together with existing knowledge.

7.4.2: Schemata

Bartlett borrowed the idea of a schema to explain these adjustments and additions (although he complained about the vagueness of the term). A **schema** is a mental framework or body of knowledge about some topic. Bartlett claimed that when we encounter new material, such as the "Ghosts" story, we try to relate it to something we already know, to existing **schemata** (the plural of *schema*). If the material does not match an existing schema, then we tend to alter the

Aspects of Bartlett's Results

Two notable aspects of Bartlett's results led him to this conclusion.

> **Interactive**
>
> **Failure to Recall Information**
>
> The first aspect concerns omissions, or information people failed to recall. For the most part, people did not recall many story details, either specific names (e.g., Egulac) or events (e.g., the phrase "His face became contorted"). The level of recall for the main plot and sequence of events was not too bad, but minor events often were omitted. As a result, the retellings are shorter than the original.
>
> **Successive Recalls**

memory to make it fit. So, recall is not an exact reproduction of the original material, but a reconstruction based on elements from the original story and on existing schemata.

The use of semantic knowledge, such as schemata, to fill in our knowledge with more expected information can sometimes lead to errors. When a person encounters unusual bits of information, that person's schema may dominate cognitive processing, leading to errors that, when caught, reveal this powerful influence. Questions such as 1 and 2 show the semantic errors that can crop up:

1. How many animals of each kind did Moses take onto the ark?
2. What is the nationality of Thomas Edison, inventor of the telephone?

Read questions 1 and 2 again if you did not notice the semantic illusion (Erickson & Mattson, 1981; Reder & Kusbit, 1991). The reason we fall for the illusion, and do not immediately notice what is wrong with the sentence, should be clear. It is another illustration of the power of schemata on cognition and conceptually driven processing.

7.4.3: Scripts

One specific type of schema captures the order in which events occur. These are called **scripts**, the semantic knowledge that guides our understanding of ordered events. For example, consider the knowledge you have that guides your comprehension of even a simple story:

> Billy was excited about the invitation to his friend's birthday party. But when he went to his room and shook his piggy bank, it didn't make a sound. "Hmm," he thought to himself, "maybe I can borrow some from Mom."

SCRIPTS IN MEMORY Think for a moment about the common meaning of the word *script*: the dialogue and

Studies on the Use of Semantic Knowledge

The use of semantic knowledge, such as schemata, to fill in our knowledge with more expected information can sometimes lead to errors.

Sulin and Dooling's Research on Distortion (1974)

Subsequent research fleshed out our understanding of schemata. For example, knowledge of the theme or topic of a passage improves people's memory of the passage (Bransford & Johnson, 1972; Dooling & Lachman, 1971). On the other hand, providing a theme, say by attaching a title to a story, can also distort recall or recognition in the direction of the theme.

A demonstration of this distortion was provided by Sulin and Dooling (1974). One group of people read a paragraph about a fictitious character: "Gerald Martin's seizure of power. Gerald Martin strove to undermine the existing government to satisfy his political ambitions. Many of the people of his country supported his efforts" (p. 256). A second group read the same paragraph, but the name *Adolf Hitler* was substituted for *Gerald Martin*. After a 5-minute waiting period, people were shown a list of sentences and had to indicate whether each was from the original story or not. Preexperimental knowledge led to significant distortions in memory. People who read the Hitler paragraph rated sentences as "the same" more frequently when they matched their existing knowledge, even though the original passage did not contain such items (e.g., "Hitler was obsessed by the desire to conquer the world," p. 259). Furthermore, these thematic distortions grew stronger 1 week after reading the story.

Castel, McCabe, Roediger, & Heitman's Study (2007)

Curtis and Bharucha's Study (2009)

---

actions that are performed by actors and actresses in a play. The script for a play details what is supposed to happen on stage. Similarly, a mental script is a general knowledge about ordinary events and situations.

Prove It

Schematic Distortion

This Prove It section aims to provide an opportunity to observe how schemata can distort memory for a text and how to overcome that distortion. This demonstration is based on a study reported by Hasher and Griffin (1978). Essentially, we are looking for things that people recall from a text, but which were never actually mentioned—information that would be in a person's schema, but not what was actually read.

So, get two groups of people and have them all read the story in the simulation above. Tell each group that this story is called "The Escaped Convict." After some period of time has elapsed (at least 10 minutes), ask your groups to recall the story. Remind one group that the title of the story was "The Escaped Convict." However, with the other group, act all flustered and tell them that you were mistaken and that the actual title of the story was "The Deer Hunter."

After people are done, look at what they have recalled. One thing you should find is that people in the repeated title group should recall details about an escaped convict that were not actually in the story. This illustrates how a schema can distort memory to make it more schema consistent. In addition, people in the title switch group should have fewer intrusions of new information having to do with either the escaped

Developing Scripts

A script mentally represents what is supposed to happen in a particular circumstance. Are you going to a restaurant? Your script tells you what to expect, the order of events, who the central characters are, and what you and they are supposed to do. Are you invited to a birthday party, taking an airplane flight, or sitting in a class on human memory and cognition? Your general knowledge of what happens in these settings guides your comprehension as the events unfold, and leads to certain expectations.

> **Interactive**
>
> **The Theory Behind Scripts**
> *The Escaped Convict/The Deer Hunter*
>
> The man walked carefully through the forest. Several times he looked over his shoulder and scrutinized the woods behind him. He trod carefully trying to avoid snapping twigs and small branches that lay in his path, for he did not want to create excess noise. The chirping of the birds in the trees almost annoyed him, their loud calls serving to distract him. He did not want to confuse those sounds with the type he was listening for.
>
> The theory behind scripts is straightforward. People have a generalized memory of experienced events and this is invoked, or retrieved, when a new experience matches an old script. One function of a script is to provide a kind of shorthand for the whole event. You need not describe every element of the event but can refer to the whole by invoking the script. More important, the script provides a framework within which new experiences are understood and within which a variety of inferences can be drawn to complete your understanding (Abbot, Black, & Smith, 1985; Reiser, Black, & Abelson, 1985; Seifert, Robertson, & Black, 1985).
>
> 1 of 3   Previous   Next

convict or deer hunter theme. If so, then what you have done is make memory more accurate by discrediting the schema in semantic memory and leading people to rely more on their episodic memory of what they actually read.

WRITING PROMPT

Developing the Idea of Scripts

Develop a script around this idea:

You went to the mall. You asked the salesman for the men's department. You bought some things and left.

> The response entered here will appear in the performance dashboard and can be viewed by your instructor.
>
> Submit

PREDICTIONS If the general idea of scripts in semantic memory is correct, then we should be able to make some predictions about how they affect people's performance. Specifically, information that better conforms to a script in memory should be easier to understand, whereas deviations should provide some difficulty. Moreover, the closer information is in a mental script, the more easily it should be processed compared to information that is more distant.

Processing Difficulties (Figure 7.15)

A well-known example of processing difficulties is shown in the story in the following simulation (Bransford, 1979). Be sure to try to understand the story before you read the assumption. As this passage shows, a story may activate a script but mismatch the expected events in it, producing difficulties in comprehension and recall.

> **The Story**
>
> Jim went into the restaurant and asked to be seated in the gallery. He was told that there would be a one-half hour wait. Forty minutes later, the applause for his song indicated that he could proceed with the preparation. Twenty guests had ordered his favorite, a cheese soufflé. Jim enjoyed the customers in the main dining room. After two hours, he ordered the house specialty—roast pheasant under glass. It was incredible to enjoy such exquisite cuisine and yet still have fifteen dollars. He would surely come back soon.

EVIDENCE OF SCRIPTS Many researchers have reported evidence of people's use of scripts (Bower, Black, & Turner, 1979; Graesser, 1981; Graesser & Nakamura, 1982; Long, Golding, Graesser, & Clark, 1990; Maki, 1989).

WRITING PROMPT

Recall Task

Develop the following scripts:
1. Your cat fell sick. You take it to the vet. You take care of it.
2. You went to a new country. You didn't have the local currency. You were stranded all night.

Do you remember predictable events and actions better than unpredictable ones, or is it the other way around?

The response entered here will appear in the performance dashboard and can be viewed by your instructor.

Submit

7.5: Concepts and Categorization

OBJECTIVE: Distinguish theories of categorization of concepts as related to semantic memory

Let's look at another area of study that addresses semantic relatedness, namely the structure of concepts and categories in semantic memory. Concepts and categories are ways that semantic memory can use an abstract understanding of the world to allow us to deal with specific instances.

THEORIES OF CATEGORIZATION As you move about the world, you interact with a wide range of entities—objects, people, events, and so on. In many cases, the entities you are dealing with may be novel to you. For example, when walking down the street, you may come across a squirrel you have never encountered before. Will

Research on the Evidence of People's Use of Scripts

> **Smith and Graesser's Research**
>
> **Hannigan and Reinitz's Research**
>
> Hannigan and Reinitz (2001) studied a somewhat different script effect—cause and effect. They introduced manipulations into the scripted stories that corresponded either to a cause of an event or to an effect of some event. For example, in a slide sequence that depicted going grocery shopping, some people saw a woman taking an orange from the bottom of a pile of oranges—but they didnot see the pile of oranges rolling to the floor. In a different condition, people saw the oranges on the floor, but not the slide showing the woman taking the orange from the pile. In other words, some people saw only the cause (pull an orange from the bottom), and some saw only the effect (oranges on the floor). Their script-guided comprehension, however, made up for causes they didnot see. That is, when they saw the effect, they mistakenly judged new cause scenes (i.e., never seen before) as "old"—if you have seen the oranges on the floor, you are more likely to remember later on that you saw the woman pulling an orange from the bottom of the pile. In terms of script knowledge, and what we know to be true about cause and effect generally, remembering the woman pulling an orange from the bottom of the pile is understandable. After all, something had to cause the oranges to fall to the floor. (See Hannigan & Reinitz, 2003, for evidence that an object from one scripted scene, say a vase of flowers in a restaurant, can "migrate" to a different restaurant memory.)

it attack you? Is it food? Will it run away? Will this squirrel make a good pet? What does it eat? How does it reproduce and raise its young? Can it vote in an election? You know what it is likely to do, and how you should interact with it, by using your categorical knowledge of what a squirrel is. **Categories** are aspects of semantic memory that allow us to predict what is likely to happen in new encounters. This involves the cognitive ability to abstract away from individual instances to understand and mentally represent the invariant characteristics of the larger category (Burgoon, Henderson, & Markman, 2013). Essentially, when you use categories, you treat individual members as if they were more or less the same (all squirrels are pretty much alike, after all). This allows you to save a lot of time and mental effort. Although categorization is beneficial in most situations, it can have a few drawbacks, such as when we overextend our categories. One example is using stereotypes (which are a kind of category) to draw conclusions about people we hardly even know.

7.5.1: Classic View of Categorization

There are three general classes of theories of categorization that we cover:

1. Classic view
2. Probabilistic theories
3. Explanation-based theories

Note that each of these has strong points for how people create and use their categories, and they all also have weaknesses.

The **classic view of categorization** (e.g., Bruner, Goodnow, & Austin, 1956) takes the view that people create and use categories based on a system of rules. So if something satisfies a set of rules, then it is a member of a category, but if it does not, then it is not a member of a category. Of critical importance is that the rules identify necessary and sufficient features for something to be in a category. For example, the category BACHELOR can be defined as an unmarried adult male. These features are necessary in that if they are not present, the person is not a bachelor. A person who is a married adult male would not qualify, nor would an unmarried male child, or an unmarried adult female (although she could be a bachelorette). These features are also sufficient in that nothing more is needed to identify a bachelor. For example, people's occupations, the kind of cars they drive, how tall they are, how many legs they have—are all features that may be present, but do not contribute to identifying the category beyond the sufficient ones.

The classic view of categorization follows scientific taxonomies, such as the definition of what makes an animal a member of a species. This type of classification allows us to identify a bat as a mammal rather than as a bird, a penguin as a bird rather than as an amphibian, and a chimpanzee as an ape rather than as a monkey. It is quite clear that people can create and use categories defined by necessary and sufficient rules. When given novel stimuli, people can readily derive the features that define a category. The more important question is whether this is how semantic memory derives and uses categories. Is this a psychologically real way of describing human categorization? A great deal of evidence suggests that it is not. Before turning to other theories, let's cover some important aspects of human categorization.

7.5.2: Characteristics of Human Categories

If semantic memory does not typically use necessary and sufficient rules or features to create categories, what are human categories like?

Qualities such as typicality, family resemblance, and correlated attributes are inconsistent with the idea that categories are defined by necessary and sufficient features. In the next few sections, we look at some other theories of human categorization that are more flexible and allow for categories that have these characteristics.

Principles of Categorization in Semantic Memory

A number of aspects of how people create and use categories are inconsistent with the classic view of categorization.

Interactive

Graded Membership

Central Tendency

Some other principles of categorization are related to graded membership. The first is the idea that categories have a **central tendency**. In other words, there is some mental core or center to the category where the best members are found.

Typicality

Family Resemblance

Correlated Attributes

7.5.3: Probabilistic Theories of Categorization

Some of the theories of categorization that have been developed beyond the classic view are **probabilistic theories**.

Notice that both probabilistic approaches make the same or similar predictions—prototype theory predicts that a typical example is judged rapidly because it is highly similar to the prototype, and exemplar theory states it is because the typical example resembles so many of the stored exemplars. Because of this similarity, it is difficult to distinguish between these two views, although there have been attempts to do so (e.g., Feldman, 2003; Rehder, 2003). Chin-Parker and Ross (2004) have shown how multiple classification schemes are possible, either prototype- or exemplar-based, depending on how people are mentally oriented and process information. They oriented people to either *diagnostic* features (necessary features) or *prototypical* features (typical, but not necessary features) in a learning task. The results showed that a category could be learned in either way, but people showed sensitivity only to the kind of feature they used during learning. In other words, if they learned via diagnostic features, they did not show sensitivity to prototypes. And likewise for the other group: If they learned via prototypes, they did not show sensitivity to diagnostic features (for work on slower reasoning and decision making in this type of task, see Ross & Murphy, 1999, and Yamauchi & Markman, 2000).

So, probabilistic theories move beyond the classic view by assuming people average across their experiences, and with the need to derive hard and fast rules about what is and is not in a category. Moreover,

---

Theories of Categorization

These theories of categorization assume that categories in semantic memory take into account various probabilities and likelihoods across a person's experience.

Interactive

Prototype Theory

One way of making categories probabilistically is by using a **prototype**, which is essentially the central, core instance of a category. A prototype is an average of all your experiences with members of a category. Imagine taking many pictures of many types of dogs and then morphing all those images together. That average would be a prototype. Note that the prototype is an idealized representation that probably does not correspond to any individual member. In **prototype theory**, our mental categories are represented with reference to the prototype, with typical members stored close to the prototype and peripheral members stored farther away. Again, when you think of a *dog*, your prototype is less likely to be like a Chihuahua but instead is more likely a German shepherd, a golden retriever, or some other more "doggy dog." Similarly, for your BIRD category, the prototype is a rather ordinary, typical, nondescript bird, and the FLOWER category would be a generic flower.

Although the idea that people use prototypes is appealing, some aspects of human categories are not captured by a prototype, such as category size variability, correlated attributes, and anything else about the category that requires a person to consider the category as a whole. A prototype cannot do this because it is a mental representation of the average of the category members, not the variation among them.

Exemplar Theory

functional magnetic resonance imaging (fMRI) evidence suggests that different brain regions are involved depending on whether people are deriving categories using rules (consistent with the classic view) or based on the similarity of the items (consistent with the probabilistic theories). For example, rule-based categorization appears to involve brain regions implicated in cognitive control, such as regions of the frontal lobes. In contrast, similarity-based categorization appears to involve regions implicated in the configural processing of perceptual images (e.g., Grossman et al., 2002). Apart from the neuroimaging evidence, the prototype and exemplar probabilistic views capture the flexibility of semantic memory to understand and appropriately interact with all kinds of new instances or category members that we encounter in our experiences in the world.

7.5.4: Explanation-Based Theories

There is no question that probabilistic theories do a good job at capturing human categorization. However, there are also a few issues with which they have trouble. For one, there is a circularity problem. Specifically, how does memory know which experiences to average across to form a category without knowing what the category is ahead of time?

Another issue is the high degree of flexibility of semantic categories. For example, think of the category THINGS TO TAKE OUT OF A BURNING BUILDING. You can generate a category on the fly, so to speak, that has all of the qualities of more traditional categories, such as BIRDS, TOOLS, or FRUIT. These **ad hoc categories** are created based on situational circumstances and have characteristics of regular categories (Barsalou, 1983). Ad hoc categories have a graded structure (e.g., a baby is a

---

Categories of Explanation-Based Theories

These points have led to theories of semantic classification called **explanation-based theories**. For these views, semantic categories are essentially theories of the world we create to explain why things are the way they are.

Interactive

Structure

Explanation-based theories highlight the important aspect of categories as structures we impose on the world—structures that may or may not reflect how the world actually is. For example, *shoe* is in the same category as *brick*, but in a different category than *sock*, if your category is THINGS TO POUND A NAIL WITH IF YOU DON'T HAVE A HAMMER. This is because we have a theory about what makes something good to hammer a nail with and can then apply this to the world around us. People can use their understanding of the causal relations among category members to make inferences about the internal structure and functioning of other members of a category (Rehder & Burnett, 2005).

Embodied Cognition

Psychological Essentialism

better member of this category than a television set), have a prototype (e.g., a highly valued, irreplaceable thing that can be damaged by fire), exhibit **typicality effects** (e.g., the family dog would be highly typical, but a visiting neighbor is something that most people are unlikely to list), and so on.

So, explanation-based theories capture some of the conscious and unconscious reasoning that we go through to try to make sense of the world. We use these mini-theories for what makes a category—we do not need the hard and fast rules of the classic view and do not have to mentally average or abstract across our experiences to derive a category, as with the probabilistic views. Instead, we bring our power of reasoning to bear on the causal structure of the world—that is why things are the way they are—to understand how to organize and classify what we come in contact with. Overall, these various theories of categorization demonstrate that people have many ways to mentally characterize their world, many ways to draw on their old experiences to help them act and think in ways that increase their success and performance in new situations.

> **WRITING PROMPT**
>
> **Categorization of Experiences**
>
> How do you categorize your experiences? Do you think that the brain processes this information and helps you to better understand and appropriately interact with all kinds of new instances that you encounter in your experiences in the world? If so, give an example of such a situation. If not, what do you think contributes to categorizing the experiences?
>
> The response entered here will appear in the performance dashboard and can be viewed by your instructor.
>
> Submit

Summary: Knowing

7.1 Semantic Memory

- Semantic memory contains our long-term memory knowledge of the world. Early studies of the structure and processes of semantic memory generated two kinds of models: network approaches and feature list approaches.
- In network models, concepts are represented as nodes in a semantic network, with connecting pathways between concepts. Memory retrieval involved the process of spreading activation: Activation spreads from the originating node to all the nodes connected to it by links.
- Feature comparison models assume that semantic concepts are sets of semantic features. Verification consists of accessing the feature sets and comparing these features.
- An important effect is semantic relatedness, in that highly related concepts are more easily processed. Also, the amount of knowledge stored in memory affects performance, possibly because more knowledge leads to higher semantic relatedness.
- Neurological measures, such as the N400 component of an ERP, provide some insight into semantic processing. For example, the N400 is larger when a person encounters semantically unrelated or anomalous information.
- Recent theories of perceptual symbols assume that semantic memories are created out of our experience with the world through perception and interaction.

7.2 Connectionism and the Brain

- Connectionist approaches to memory allow one to simulate these processes in a way that is more analogous to how the brain does this.
- Connectionist approaches even furnish persuasive analyses of semantic disruptions caused by brain damage, so-called category-specific deficits.

7.3 Semantic Priming

- In a priming task, a prime is presented first, and RT to the target is measured. When the prime is related, RT to the target usually is faster, even at short time intervals (SOAs). This is evidence that semantic priming is automatic.
- Priming is an implicit memory process, although conscious processes can alter it later on.

7.4 Schemata and Scripts

- Through his early work with materials such as "The War of the Ghosts," Bartlett developed a schema theory. From this view, schemata are generalized knowledge structures, as are semantic memories, that people use as guides for common experiences. Schemata not only can help organize information to facilitate learning, but also can cause people to misremember information in a more schema-consistent fashion.
- Scripts are representations of ordered events, such as going to a restaurant or attending a birthday party.

- Script theories make predictions about comprehension and retrieval and provide a useful way to explain how people understand and interact with the real world, including how people can "remember" events because they are consistent with a stored script.

7.5 Concepts and Categorization

- The classic view of categorization is that people use necessary and sufficient rules. Although people *can* make classifications in this way, they typically do not in most situations.
- For probabilistic views, categories take into account regularities, such as typicality and correlated attributes, to produce categories with graded structures.
- For the prototype view, categories are based on an averaged representation, the prototype, against which all other members of the category are compared.
- Exemplar theory suggests that we store multiple examples or exemplars in memory, then make judgments by comparing an item to the stored exemplars.
- Explanation-based theories suggest that people are problem solving when they create categories. People develop (largely implicit) theories of why something is in a category. This is how people create ad hoc categories on the fly and exhibit psychological essentialism—the intuitive belief that members of a category share an underlying essence.

SHARED WRITING

Schema/Script Versus Mental Category

Compare and contrast the concept of a schema/script versus the concept of a mental category. How are they similar? How are they different?

▶ A minimum number of characters is required to post and earn points. After posting, your response can be viewed by your class and instructor, and you can participate in the class discussion.

Post   0 characters | 140 minimum

Chapter 8
Memory and Forgetting

 Learning Objectives

8.1: Identify the ways in which long-term memory can fail

8.2: Explain the process by which stored information is retrieved from episodic memory

8.3: Describe types of variations in memories

8.4: Differentiate between the types of amnesia

Having studied remembering from episodic and semantic memory in the previous two modules, we now look at the opposite of remembering: *forgetting*. There are a number of ways that information can be forgotten. In some cases, the forgetting occurs even though the information is in memory. The problem is that a person just cannot get at it. This is more like most of the forgetting that we experience in our lives. In other cases, the forgetting can be more extreme, where the memory no longer exists. So obviously, there can be no way to ever get at this. This extreme forgetting can happen with some cases of amnesia.

The distinction between milder and more severe forms of forgetting reflects a distinction made in cognitive psychology between memory availability and accessibility. The term **availability** refers to whether a memory trace exists somewhere in the memory system so it could be retrieved. Information that no longer exists would not be available in memory. By comparison, the term **accessibility** refers to whether a memory trace that does exist can be found, activated, and brought to mind. Forgetting may occur when a memory is available, but not accessible. An analogy that is often used to illustrate this is that of books in a library. Whether a given book is in the library refers to its availability. If the book is in the library, then it is available. The ability to find a book on the library shelves refers to its accessibility. Forgetting in which there is availability, but not accessibility, may be when the book is on a shelf in the library, but the catalog entry is missing, or the book has been put on the wrong shelf. You would not be able to recover the book unless you happened to stumble upon it.

A major difference between this module and the previous two is the greater emphasis on the *accuracy* of remembering versus the *speed* (response time, RT). In many cases of forgetting, the emphasis is on *inaccuracy*. Do people remember something or not? A stunning aspect of many of the results of remembering is inaccurate, error-prone memory such as eyewitness recollection of the details of an event that are just plain wrong. Schacter (1996) writes eloquently of the "fragile power" of memory, the paradoxical situation in which we are capable of remembering amazing quantities of information yet have a strong tendency to misremember under a variety of circumstances. We focus on the fragile part of this description here.

8.1: The Seven Sins of Memory

OBJECTIVE: Identify the ways in which long-term memory can fail

In a very approachable work, Schacter (1999) provides some specifics about the fragile nature of memory by enumerating seven sins of memory, seven ways in which long-term memory lets us down (see Table 8.1). These are ways that the normal operation of memory leads to problems.

Table 8.1 The Seven Sins of Memory

Sin	Description
1. Transience	The tendency to lose access to information across time, whether through forgetting, interference, or retrieval failure.
2. Absent-mindedness	Everyday memory failures in remembering information and intended activities, probably caused by insufficient attention or superficial, automatic processing during encoding.
3. Blocking	Temporary retrieval failure or loss of access, such as the tip-of-the-tongue effect, in either episodic or semantic memory.
4. Misattribution	Remembering a fact correctly from past experience but attributing it to an incorrect source or context.
5. Suggestibility	The tendency to incorporate information provided by others into our own recollection and memory representation.
6. Bias	The tendency for knowledge, beliefs, and feelings to distort recollection of previous experiences and to affect current and future judgments of memory.
7. Persistence	The tendency to remember facts or events, including traumatic memories, that one would rather forget; that is, failure to forget because of intrusive recollections and rumination.

8.2: Forgetting Through Decay and Interference

OBJECTIVE: Explain the process by which stored information is retrieved from episodic memory

We turn now to the cognitive mechanisms or processes that likely lead to much of the forgetting that we experience every day. Two theories of forgetting have preoccupied cognitive psychology from the very beginning: decay and interference.

DECAY It is a bit unusual for the name of a theory to imply its content as clearly as does the term **decay**. The older a memory trace is, the more likely that it has been forgotten, just as the print on an old newspaper fades into illegibility. The principle dates back to Thorndike (1914), who called it the *law of disuse*. Habits, and by extension the memories that underlie them, are strengthened when they are used repeatedly, and those that are not are weakened through disuse. Thorndike's idea was a beautiful theory, easily understood and straightforward in its predictions.

However, the decay theory of forgetting is problematic because it claims that the passage of time itself causes forgetting. McGeoch (1932) gave a scathing attack of this claim, arguing that it is the *activities* that occur during a period of time that cause forgetting, not time itself. In other words, time alone doesn't cause forgetting—it is what happens during that time that does. Although there are still some arguments for some, at least partial, influence of decay (Schacter, 1999), it is difficult to imagine a study that would provide a clean, uncontaminated demonstration of it. As time passes, there can be any number of opportunities for interference, even if by the momentary thoughts you have while your mind wanders. The time interval also gives opportunities for selective remembering and rehearsal, which would boost remembrance of old information.

INTERFERENCE Interference theory was a staple in the experimental diet of verbal learners for at least two reasons. First, the arguments against decay theory and for interference theory were convincing, on both theoretical and empirical grounds. Demonstrations such as the often-cited Jenkins and Dallenbach (1924) study made sense within an interference framework where, after identical time delays, people who had remained awake after learning recalled less than those who had slept (Figure 8.1).

The everyday activities encountered by people who stayed awake interfered with memory. Fewer interfering activities occurred for people who slept, so their memory was better.

Drosopoulos, Schulze, Fischer, and Born (2007) replicated this effect. They further concluded that the memory benefit of sleep serves to mitigate the effects of *interference*. However, this was primarily for information that was not strongly encoded (that is, not well learned) to begin with.

Figure 8.1 The Classic Jenkins and Dallenbach (1924) Result

Results from study show higher recall of nonsense syllables for two people who slept after acquisition versus remaining awake after acquisition.

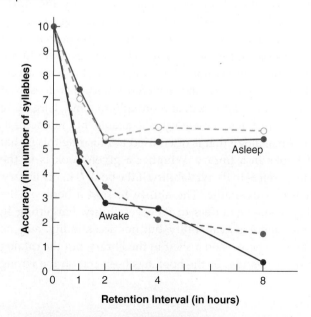

Although how much interference is experienced is linked to how related the material is, there are other reasons for long-term memory forgetting that are influenced by whether you get some sleep.

Specifically, when you create and store new long-term memories, they do not instantly appear in your brain in a well-established form. Instead, there is a period of time during which memories go through a process of *consolidation*, the more permanent establishment of memories in the neural architecture. Later we see the consequences of a dramatic disruption in consolidation that can result in amnesia. For now, it is important to note that the disruption when people are awake (and not asleep) is interference of the memory consolidation process (Wixted, 2005). New information is encoded into memory that uses the same neural parts (such as the hippocampus) that were used by the older information. This reuse of same neural networks interferes with memories for the older information, thereby disrupting consolidation and causing forgetting.

This interference is somewhat like writing messages in clay. Imagine writing a message in a bit of clay, then writing even more messages on the clay. Sometimes you will write over messages you had previously written, making the earlier ones harder to recover. This is the type of interference we are talking about. Still, sometimes a message does not get written over, and eventually the clay hardens. In this case, the message consolidates into the clay, making it harder to disrupt. Ultimately, these older memories are more robust and less prone to disruption.

A second reason for the popularity of interference studies is that these effects were easily obtained with a task already in wide use, the paired-associate learning task. This task was a natural for studying interference—and, it conformed to the behaviorist Zeitgeist or "spirit of the times" before the cognitive revolution. Unlike consolidation-disrupting interference, the interference explored using paired-associate tasks was due to *cue overload*. In these cases, the many memories related to a specific cue compete with one another during retrieval.

8.2.1: Paired-Associate Learning

A few moments studying the *paired-associate learning* task will help you understand interference theory. In other words, how does the relationship between what was learned initially and what is learned later influence how well one remembers something later?

The basic paired-associate learning task is as follows: A list of stimulus terms is paired, item by item, with a list of response terms. After learning, the stimulus terms should prompt the recall of the proper response terms.

Table 8.2 presents several paired-associate lists as a demonstration. Imagine learning List 1 to a criterion of one perfect trial (try to get a good idea of what the task is like).

Table 8.2 Lists of Paired Associates

List 1 (A–B)	List 2 (C–D)	List 3 (A–B$_r$)
tall–bone	safe–fable	plan–bone
plan–leaf	bench–idea	mess–hand
nose–fight	pencil–owe	smoke–leaf
park–flea	wait–blouse	pear–kiss
grew–cook	student–duck	rabbit–flight
rabbit–few	window–cat	tall–crowd
pear–rain	house–news	nose–cook
mess–crowd	card–nest	park–few
print–kiss	color–just	grew–flea
smoke–hand	flower–jump	print–rain

List 4 (A–B′)	List 5 (A–C)	List 6 (A–D)
smoke–arm	tall–bench	smoke–fable
mess–people	plan–pencil	print–idea
rabbit–several	nose–wait	mess–owe
park–ant	park–student	pear–blouse
plan–tree	grew–window	rabbit–news
tall–skeleton	rabbit–house	grew–duck
nose–battle	pear–card	park–cat
grew–chef	mess–color	nose–nest
pear–storm	print–flower	plan–just
print–lips	smoke–safe	tall–jump

After that, you switch to the second half of the study, which involves learning another list. The similarity of the first and second lists is critical. If you were switched to List 2, you would experience little or no interference because List 2 has terms that are dissimilar to List 1. In the lingo of interference theory, this was the *A–B, C–D* condition, where the letters *A* through *D* refer to different lists of stimulus or response terms. This condition represented a baseline condition because there is no similarity between the *A–B* and the *C–D* terms (however, you may need fewer trials on the second list because of "general transfer" effects from List 1, warm-up or learning to learn).

If you shifted to List 3, however, there would have been massive negative transfer. It would have taken you more trials to reach criterion on the second list. This is because the same stimulus and response terms were used again but in new pairings. Thus, your memory of List 1 interfered with the learning of List 3. The term for this was *A–B, A–B$_r$*, where the subscript *r* stood for "randomized" or "re-paired" items. Finally, if you switched to List 4 (the *A–B, A–B′* condition), there would have been a great deal of positive transfer; you would need fewer trials to reach criterion on the second list because List 4 (designated *B′*) is related to the earlier one (*B*). For instance, in List 1 you learned *plan–leaf*; in List 4, *plan* went with *tree*.

These are all *proactive interference* effects, showing the effects a prior task has on current learning. Table 8.3 is the general experimental design for a proactive interference (PI) study as well as for a retroactive interference (RI) study. As a reminder, *retroactive interference* occurs when a learning experience interferes with recall of an *earlier* experience; the newer memory interferes backward in time ("retro").

Table 8.3 Designs to Study Two Different Kinds of Interference

Proactive Interference (PI)				
	Learn	Learn	Test	Interference Effect
PI group	A–B	A–C	A–C	A–B list interferes with A–C; e.g., an A–B word intrudes into A–C
Control group	—	A–C	A–C	
Retroactive Interference (RI)				
	Learn	Learn	Test	Interference Effect
RI group	A–B	A–C	A–B	A–C list interferes with A–B; e.g., an A–C word intrudes into A–B
Control group	A–B	—	A–B	

Both proactive and retroactive interference have been examined thoroughly, with complex theories based on the results. Although the literature is extensive, we make no attempt to cover it in depth here (but see standard works, e.g., Postman & Underwood, 1973; Underwood, 1957; Underwood & Schultz, 1960; and Klatzky, 1980, for a very readable summary).

8.2.2: Associative Interference

Taking the idea that people can store ideas in memory and that these can be thought of as being organized in a network, we can make some predictions about how this affects performance during remembering. First, a given concept can have multiple associations with it. So, a node in a network can have multiple links to multiple concepts. Second, we can further assume that there is a limit to people's cognitive resources—how much of a network a person can search at once. We have already covered a number of these sorts of mental limits in our discussions of attention and working memory, and we are just extending that logic here. Given these two simple assumptions, we can make some predictions about memory.

In a classic study, Anderson (1974) had people memorize a list of sentences about people in locations, such as "The hippie is in the park." The important part of the study was that, across the list, he varied the number of associations with the person and location concepts. That is, a given person in the study list could be described as being in 1, 2, or 3 different locations, and each location could have 1, 2, or 3 different people in it. When we graph a network representation of this type of information, we can see various numbers of links "fanning" off of a given concept node.

Now, Anderson (1974) further assumed that the amount of activation that could spread along the links of the network was limited. As such, the more links there were fanning off of a concept node in memory, the more widely the activation was distributed, and the longer the processing of the activation along any one of those pathways took. The end result is the prediction that the more links there are off a given concept—the more links fanning off of a concept node—the slower the retrieval process. This, of course, would yield a longer response time, which, as can be seen in Figure 8.2, is exactly what happened in Anderson's (1974) study. After people had memorized the list of sentences, they were given a recognition test in which they had to indicate whether the test sentences were studied on the list or not. The **fan effect** that was found was that when more words associated with a concept, response times were longer.

The fan effect is an **associative interference** effect—the more words associated with a concept, the slower people were to retrieve any one of them. In an interesting extension of this finding, Bunting, Conway, and Heitz (2004; see also Radvansky & Copeland, 2006a) looked at the fan effect in terms of the working memory capacity of their participants. They found that people with lower working memory capacity exhibited greater interference—a larger fan effect—than people with a higher working

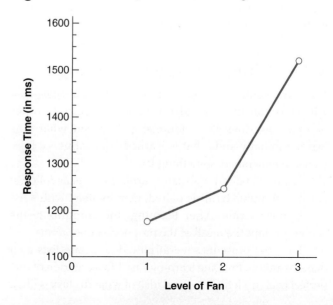

Figure 8.2 An Example of Data Showing a Fan Effect

memory capacity. That is, people with less working memory capacity were further disrupted when their capacity was divided up among many words (larger "fans"). They were working with fewer working memory resources to begin with, so a cognitive task that places a greater burden on them has a more disruptive effect because they have less extra capacity to compensate for that disruption.

8.2.3: Situation Models and Interference

Not only are situation model memories (memories for described events) remembered better over the long term, but they also have other benefits, which may contribute to this superior memory.

As you can see in Figure 8.3, Radvansky and Zacks (1991) found a fan effect when there was a single object in multiple locations, but not when there was a single location with multiple objects.

This mental organization is based on how people think about interaction with the world. In a study by Radvansky, Spieler, and Zacks (1993), students learned sentences about people in small locations that typically contain only a single person, such as witness stand, tire swing, or store dressing room. Here, a situation in which multiple people are in one of these locations is unlikely. But, because people can move from place to place, a person-based organization is plausible—and this is what is observed with a fan effect when a single location was associated with multiple people, but not when a single person was associated with multiple locations.

The Fan Effect

For the fan effect, the more things a person knows about a concept, the longer retrieval time should be. However, this basic principle can shift around somewhat when we start thinking about how the studied information might be organized into *situation models*. For example, in one study, Radvansky and Zacks (1991) had people memorize sentences about objects in locations. Perform the task below in a similar activity.

Interactive

1. In the space provided below, note down the sentences that you can remember.

Figure 8.3 Radvansky and Zacks (1991) Fan Effect

Response times to sentence memory probes as a function of the level of fan (number of associations with a sentence concept). Data are divided based on whether the shared concept is a single location with multiple objects or a single object in multiple locations.

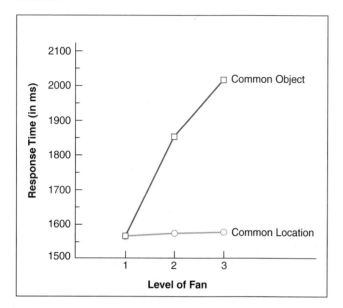

This applies not only to information that people learn from sentences but also to how we interact with the world. Specifically, if a given object is encountered in more than one location, the memory for that object will be worse. For example, in a study by Radvansky and Copeland (2006b; Pettijohn & Radvansky, 2016a, 2016b; Radvansky, Krawietz, & Tamplin, 2011; Radvansky, Tamplin, & Krawietz, 2010), people navigated a virtual environment in which they picked up objects and moved them either across a room or from one room to another. The researchers found that, compared to moving objects across a room, when people moved objects from one room to another, they had a harder time remembering. That is, walking through doorways caused forgetting. This occurred because people had memories of the object being in more than one event, more than one room. Therefore, when they needed to remember something about these objects, these different memories interfered with one another, and memory was worse. This is much like a fan effect, only with objects and the places where you interact with them.

8.2.4: Overcoming Forgetting from Interference

Beginning in the mid-1960s, a different theory came to dominate cognitive psychology's view of forgetting. Both the decay and interference theories suggest that information in long-term memory can be lost from memory. This definition of *forgetting* (loss from memory) was implicit in the mechanisms thought to account for it. *Forgetting* is now used without the idea of complete loss from memory to refer to situations in which there is difficulty remembering. In other words, we now focus on cases where the information is available in memory but may not be accessible.

For example, one line of research looks at **retrieval-induced forgetting**, the temporary forgetting of information because of having recently retrieved related information (e.g., Anderson, Bjork, & Bjork, 2000; MacLeod & Macrae, 2001). Similarly, Anderson (2003) and Storm (2011) have suggested that forgetting is an active inhibition process, designed to override mistaken retrieval of related information ("activated competitors" in Anderson's terms). Even here, the unwanted information that is causing interference is still in memory—if it weren't, there would be no need to override it. That said, the memories that are more likely to be inhibited are the ones that have recently been processed, whereas older memories are more likely to be facilitated by recently retrieved, related information (Bäuml & Samenieh, 2010).

It is possible, therefore, that there may be no *complete* forgetting from long-term memory aside from loss due to organic or physical factors such as stroke or diseases like Alzheimer's dementia. So, the information is available in memory in some form. Forgetting may be due to retrieval failure or a process of retrieval inhibition, a deliberate (though only partially successful) attempt to forget (e.g., when you try to forget an unpleasant memory or an incorrect fact; Bjork & Bjork, 2003).

8.2.5: Retrieval Cues

We have already discussed how access can be increased by reinstating the original learning context. This can be thought of as providing effective retrieval cues. So, let's look at retrieval cues more generally. Any cue that was encoded along with the learned information should increase accessibility. This is why the category cues helped the people in Tulving and Pearlstone's study recall more than they otherwise would have. Similarly, this is why recognition usually reveals higher performance than recall. In a recognition test, you merely have to pick out which of several alternatives is the correct choice. What better retrieval cue could there be than the very information you are trying to retrieve? Subsequent research has shown the power of retrieval cues in dramatic fashion. (A convincing demonstration is presented in the following sample experiment taken from Bransford & Stein, 1984; do that demonstration now, before reading further.)

Thomson and Tulving (1970) asked people to learn a list of words for later recall. Some of the words were accompanied by cue words printed in lowercase letters; people were told they need not recall the cue words but that the cues might

More About Retrieval Failure

> **An Everyday Example of Retrieval Failure**
>
> Everyone is familiar with retrieval failure. Students often claim that they knew the information but that they "blocked" on it during the exam. Sometimes this is an example of retrieval failure. A straightforward experience of this is the classic *tip-of-the-tongue (TOT) effect*. People are in the TOT state when they are momentarily unable to recall a word, often a person's name, that they know is in long-term memory. Although you may be unable to retrieve a word or name during a TOT state, you usually have access to partial information about it, such as the sound it starts with, its approximate length, and the stress or emphasis pattern in pronunciation. (See Brown & McNeill, 1966; Burke, MacKay, Worthley, & Wade, 1991; Jones, 1989; Koriat, Levy-Sadot, Edry, & de Marcus, 2003; and Meyer & Bock, 1992.) If you want to try this yourself, provide a list of questions that can be used to trigger the TOT state.
>
> However, retrieval failure, like the TOT effect, is not limited to lapses in remembering names or words. As Tulving and Pearlstone (1996) found, it is a fundamental aspect of memory.
> *"TOT is pronounced "tee-oh-tee", not like the word tot. Furthermore, it is often used as a verb: "The subject TOTed ("tee-oh-teed") seven times on the list of 20 names.*
>
> **Research on Retrieval Failure**

be helpful in learning. Some of the cue words were high associates of the list items, such as *hot–COLD*, and some were low associates, such as *wind–COLD*. During recall, people were tested for their memory of the list in one of three conditions: low-associate cues, high-associate cues, or no cues.

The results were that high associates used as retrieval cues benefited recall both when they had been presented during study and when no cue word had been given. When no cue word was given, people spontaneously retrieved the high associate during input and encoded it along with the list item. In contrast, when low associates had been given, only low associates functioned as effective retrieval cues. High-associate retrieval cues were no better than no cues at all. In other words, if you had studied *wind–COLD*, receiving *hot* as a cue for *COLD* was of no value. Thus, retrieval cues can override existing associations during recall.

Demonstrations of the effectiveness of retrieval cues are common. For instance, you hear a "golden oldie" on the radio, and it reminds you of a particular episode (a special high school dance, with particular classmates, and so forth). This even extends to general context effects. Marian and Neisser's (2000) bilingual participants remembered more experiences from the Russian-speaking period of their lives when they were interviewed in Russian, and more from the English-speaking period when interviewed in English (see also Schrauf & Rubin, 2000). Actors remember their lines better when enacting their stage movements of a performance, even 3 months later (and with intervening acting roles; Noice & Noice, 1999).

8.2.6: Part-Set Cuing Effect

There is no question that, for the most part, retrieval cues help memory. However, there are some notable exceptions. One is the **part-set cuing effect** (Slamecka, 1968) where people cued with a subset of a list have more difficulty recalling the rest of the set than if they had not been cued at

Sample Experiment by Bransford and Stein (1984)
This demonstration experiment illustrates the importance of retrieval cues. Please follow the instructions exactly.

1. In the space provided below, list and write down as many of the sentences as you can remember (you need not write "can be used" each time).

all. In other words, cuing people with part of the information impairs memory compared to doing nothing. For example, if someone asked you to name the seven dwarves (from Disney's movie *Snow White*), you would have a harder time with the last three if you were told the names of four of them than if you were simply asked to name all seven of them by yourself. One cause of the part-set cuing effect is that when people are provided with part-set cues, these items disrupt the retrieval plan that a person would normally use by imposing a different organization of the material. Also, part-set cuing involves the use of an active inhibitory mechanism (Aslan, Bäuml, & Grundgeiger, 2007), much like what would be occurring in retrieval practice. In short, when a person is given a part-set cue, this causes an implicit retrieval of those items. At that time, the related memory traces serve as competitors and are actively inhibited.

> **Prove It**
>
> ### Part-Set Cuing
> It seems surprising that giving people part of a set of information makes their performance worse compared to giving them nothing. Yet this is exactly what the part-set cuing effect says will happen. Make several lists of words. Each list should be 48 words long from 4 categories (e.g., tools, birds, countries, etc.), with 12 words from each category. Then get two groups of your friends. For both groups, read the lists of words to them, with the words in a random order (not grouped by category). Do this at a pace of about 1 word per second. Then after reading the list, have one group try to recall the entire list of 48 words. For the other group, read to them a subset of 24 words (6 from each category), and then have them try to recall the remaining 24.

When scoring the recalls for both groups, don't count the ones that were read to the second group (the part-set cue group), but only the other 24. This is because what you are trying to test is how well people do on those particular items as a function of whether they got the part-set cues or not. If everything goes well, you should find that the people to whom you read half of the list will have a harder time than the people who simply tried to recall the entire list.

WRITING PROMPT

Role of Interference in Forgetting

Given that memory is strongly influenced by interference, which causes forgetting, what can you do to help minimize its influences? What is a more applied situation where these ideas may be important to consider?

▶ The response entered here will appear in the performance dashboard and can be viewed by your instructor.

Submit

8.3: False Memories, Eyewitness Memory, and "Forgotten Memories"

OBJECTIVE: Describe types of variations in memories

Schacter (1996) spoke of the fragile nature of memory and the seven sins of memory, discussing how our memories can fail us in certain situations. Indeed, we have been discussing a variety of things that show different memory weaknesses. Is this what Schacter meant when he talked about the "sins" of memory as being the weaknesses in our memory systems? What is the weakness that Schacter pinpointed—what situations did he have in mind? A straightforward answer is that memory fails us in exactly those situations that call for absolutely accurate recall, completely correct recollection of real-world events exactly as they happened. The weakness of memory seems to be that we are often unable to distinguish between what really happened and what our existing knowledge and comprehension processes might have contributed to recollection. We will discuss two research programs that show incorrect or distorted memory, then tackle the issues raised by these results.

8.3.1: False Memories

A simple yet powerful laboratory demonstration of **false memory**, memory of something that did not happen, was reported by Roediger and McDermott (1995), based on a demonstration by Deese (1959). (See Roediger & McDermott, 2000, for an introduction to this work and the *DRM* [Deese–Roediger–McDermott] task as it is often called; see Gallo, 2010, for a review of this work.) Roediger and McDermott had people study 12-item lists made up of words such as *bed rest awake,* and *pillow,* words highly associated with the word *sleep*. The word *sleep* was never presented. Instead, it was the **critical lure** word, a word that was highly related to the other words in the list but that never actually appeared. In immediate free recall, 40% of the people recalled *sleep* from the list and later recognized it with a high degree of confidence. This is a false memory.

In a second study, people studied similar lists, recalling the list either immediately or after a distractor task (doing arithmetic for 2 minutes). Then, everyone was given a recognition task. During free recall, 55% of the people recalled the lure. The recognition results, shown in Figure 8.4, were even more dramatic. Of course, a few people "recognized" nonstudied words that were unrelated to the study list words (e.g., *thief* for the *sleep* list). More important, correct recognition for studied words increased to well above chance for study/arithmetic lists and even higher for study/recall lists. However, false recognition of the critical lure was higher than correct recognition of words actually shown on the list, showing the same pattern of increases across conditions. There was an 81% false alarm rate for critical lures when the lists had been studied and recalled. In other words, falsely

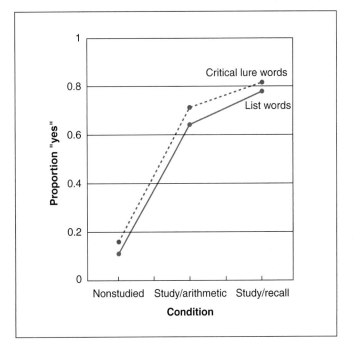

Figure 8.4 Roediger and McDermott's (1995) Results Showing the Occurrence of False Memories After Hearing Lists of Related Words

remembering the lure during recall strengthened memories of the lure word, leading to a higher false recognition rate. When questioned further, most people claimed to "remember" the critical lure word rather than merely "know" it had been on the list.

In terms of *content accuracy*—memory for the ideas—performance is good, exactly what we would expect; you see a list of words such as *bed, rest, awake,* and *pillow*, and, because the list is "about" sleep, you then recall *sleep*. But in terms of *technical accuracy*, memory for the exact experience, performance is poor because people came up with the word *sleep* based on their understanding of the list and then could not distinguish between what had really been there and what was supplied from memory. These sorts of meaning-based false memories are formed quickly, in as little as 4 seconds (Atkins & Reuter-Lorenz, 2008). It appears that the critical process is not the automatic spread of activation, such as that observed in lexical decision priming, but the use of the thematic information in the study lists during efforts to remember what had been heard earlier (Meade, Watson, Balota, & Roediger, 2007). Roediger and McDermott's (1995, p. 812) conclusion about this compelling memory illusion summarized the situation aptly:

> All remembering is constructive in nature. The illusion of remembering events that never happened can occur quite readily. Therefore, the fact that people may say they vividly remember details surrounding an event cannot, by itself, be taken as convincing evidence that the event actually occurred.

Prove It

False Memory

This is a fairly easy demonstration to perform if you have several volunteers and about 15 minutes available. It is an adaptation of the Deese/Roediger and McDermott (1995) method, shorter than the original experiment while demonstrating the same effect; see Stadler, Roediger, and McDermott (1999) for additional word lists that can be used. Prepare enough copies of your distractor task (e.g., a page full of simple arithmetic problems) to have one for each volunteer.

1. Tell your volunteers they will hear three lists and afterward will be asked to recall as many of the words as they can. The order of recall is not important.
2. Read the three lists to your volunteers at an "easy" speaking rate, about one word per 2 seconds. Pause only briefly between lists.
3. After finishing the third list, have your volunteers do 2 minutes of arithmetic, finishing as many problems as they can.
4. Ask your volunteers to write down as many words as they can remember from the three lists. Give ample time (approximately 3 minutes) so they can get as many words as possible.
5. When everyone is done, have them all turn over their sheets of paper and make recognition decisions, one by one, to the 20-word recognition test. For each word, they should say "no" if the word was not on the list and "yes" if it was on the list. When they say "yes," also have them note whether they remember the word specifically or whether they just "know" it was on the list.
6. Look especially for recall of the words *sleep, thief,* and *chair*, because these are the non-presented critical lures. On recognition, look for false alarms (saying "yes") to the critical lures in positions 5, 13, and 16.

Word Lists

1. bed, rest, awake, tired, dream, wake, snooze, blanket, doze, slumber, snore, nap, peace, yawn, drowsy
2. steal, robber, crook, burglar, money, cop, bad, rob, jail, gun, villain, crime, bank, bandit, criminal
3. table, sit, legs, seat, couch, desk, recliner, sofa, wood, cushion, swivel, stool, sitting, rocking, bench

Recognition List
1-dream, 2-fork, 3-weather, 4-bracelet, 5-chair, 6-robber, 7-stool, 8-traffic, 9-snooze, 10-couch, 11-radio, 12-jail, 13-sleep, 14-sand, 15-blanket, 16-thief, 17-bed, 18-boy, 19-skin, 20-cushion

Scoring Key

"Yes" words	1, 6, 7, 9, 10, 12, 15, 17, 20
"No" words	2, 3, 4, 8, 11, 14, 18, 19
Critical lures	5, 13, 16

8.3.2: Integration

False memories can also be created by inappropriately combining information from different sources or events, so that the combined information becomes linked or fused in memory. To illustrate this, let's look at a classic set of studies by Bransford and Franks (1971, 1972).

SAMPLE EXPERIMENT OF BRANSFORD AND FRANKS (1971)—PART 1 Read the following sentences carefully and attempt the activity that follows.

1. The girl broke the window on the porch.
2. The tree in the front yard shaded the man smoking his pipe.
3. The hill was steep.
4. The sweet jelly was on the kitchen table.
5. The tree was tall.
6. The old car climbed the hill.

7. The ants in the kitchen ate the jelly.
8. The girl who lives next door broke the window on the porch.
9. The car pulled the trailer.
10. The ants ate the sweet jelly that was on the table.
11. The girl lives next door.
12. The tree shaded the man who was smoking his pipe.
13. The sweet jelly was on the table.
14. The girl who lives next door broke the large window.
15. The man was smoking his pipe.
16. The old car climbed the steep hill.
17. The large window was on the porch.
18. The tall tree was in the front yard.
19. The car pulling the trailer climbed the steep hill.
20. The jelly was on the table.
21. The tall tree in the front yard shaded the man.
22. The car pulling the trailer climbed the hill.
23. The ants ate the jelly.
24. The window was large.

Bransford and Franks (1971) were interested in the general topic of how people acquire and remember ideas, not merely individual sentences but integrated wholes. They asked people to listen to sentences like those in Part 1 of the sample experiment you just read, one by one, and then (after a short distractor task) answer a simple question about each sentence. After going through this procedure for all 24 sentences and taking a 5-minute break, people were given another test. During this second test, people had to make yes/no recognition judgments, saying "yes" if they remembered reading the sentence in the original set and "no" otherwise. They also had to indicate, on a 10-point scale, how confident they were about their judgments: Positive ratings (from 1 to 5) meant they

---

Sample Experiment of Bransford and Franks (1971)—Part 1

This activity repeats the sentences you just read.

Interactive

1. The girl broke the _____ on the porch. (Broke what?)
2. The tree in the _____ shaded the man smoking his pipe. (Where?)
3. The _____ was steep. (What was?)
4. The sweet jelly was on the _____. (On what?)
5. The tree was _____. (Was what?)
6. The _____ climbed the hill. (What did?)
7. The ants in the _____ ate the jelly. (Where?)
8. The girl who lives _____ broke the window on the porch. (Lives where?)
9. The car _____ the trailer. (Did what?)
10. The _____ ate the sweet jelly that was on the table. (What did?)
11. The _____ lives next door. (Who does?)
12. The _____ shaded the man who was smoking his pipe. (What did?)
13. The sweet jelly was on the _____. (Where?)
14. The girl who lives next door broke the large _____. (Broke what?)
15. The _____ was smoking his pipe. (Who was?)
16. The _____ climbed the steep hill. (The what?)
17. The large window was on the _____. (Where?)
18. The tall tree was _____. (Was where?)
19. The car pulling the trailer _____ the steep hill. (Did what?)
20. The _____ was on the table. (What was?)
21. The tall tree in the front yard _____ the man. (Did what?)
22. The car _____ climbed the hill. (Which car?)
23. The ants ate the _____. (Ate what?)
24. The _____ was large. (What was?)

WORD BANK
- front yard
- table
- pulled
- in the front yard
- old car
- man
- window
- climbed
- pulling the trailer
- window
- ants
- porch
- old car
- hill
- girl
- shaded
- window
- tall
- jelly
- kitchen
- kitchen table
- jelly
- tree
- next door

Start Over   Check Answers

Table 8.4 Sample Experiment of Bransford and Franks (1971) Part 2

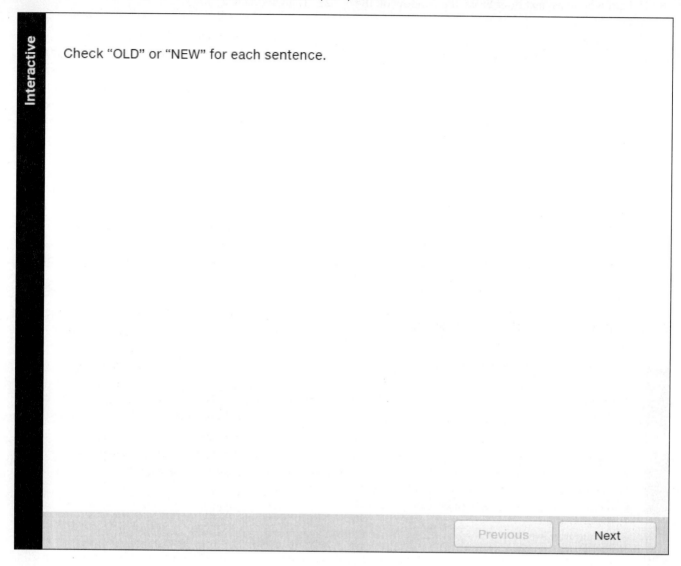

were sure they had seen the sentence; negative ratings (from –1 to –5) meant they were sure they had not. *Without looking back at the original sentences,* take a moment now to make these judgments about the sentences in Part 2 of the sample experiment below; "OLD" means "Yes, I've seen it before" and "NEW" means "No, I didn't see it before."

All 28 sentences in this recognition test are related to the original ideas in Part 1. The clever aspect of the recognition test is that only 4 of the 28 sentences had in fact appeared on the original list; the other 24 are new. As you no doubt noticed, the separate sentences were all derived from four basic idea groupings, such as "The ants in the kitchen ate the sweet jelly that was on the table." Each of the complete idea groupings consisted of four separate simple *propositions*; for example,

The ants were in the kitchen.
The ants ate the jelly.
The jelly was sweet.
The jelly was on the table.

The original set of sentences in Part 1 of the sample experiment of Bransford and Franks presented six sentences from each idea grouping:

- Two of the six were called "ones": simple, one-idea propositions such as "The jelly was on the table."
- Two of the six were "twos": two merged simple propositions as in "The ants in the kitchen ate the jelly."
- Two of the six were "threes": as in "The ants ate the sweet jelly that was on the table."

In Bransford and Franks' experiments, the final recognition test (Part 2) presented ones, twos, threes, and the overall four for each idea grouping. Their findings are shown in Figure 8.5.

Figure 8.5 Confidence Ratings for People's Judgments of New and Old Sentences

Just as your performance probably indicated, people overwhelmingly judged threes and fours as having been on the study list (just as you probably judged question 20, the 4, as old). Furthermore, they were confident in their ratings, as shown in the figure.

SOURCE: From Bransford & Franks (1971).

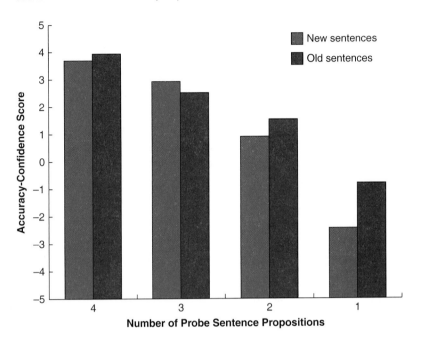

People were recognizing the sentences that expressed the overall idea most thoroughly, even when they had not seen exactly those sentences during study. Such responses are called **false alarms**, saying "OLD" when the correct response is "NEW."

Moreover, people were not confident about having seen old sentences; the only sentences they were sure about were sentences in which ideas from different groupings had been combined. Furthermore, they were fairly confident that they had not seen ones, even though they had seen several sentences that were that short. Because the shorter sentences did not express the whole idea (which would be captured by the single situation model that captured all of the ideas), people believed that they had not seen them before.

These results (Bransford & Franks, 1971, 1972) suggest that people had acquired a more general idea than any of the individual study sentences had expressed. In essence, people were reporting a *composite* memory in which related ideas were fused together, forming one memory of the whole idea. Therefore, later recognition performance was entirely reasonable. They were matching the combined ideas in the recognition sentences to their composite memory. Rather than verbatim memory, Bransford and Franks found "memory for meaning," memory based on the **integration** of related material.

8.3.3: Leading Questions and Memory Distortions

Another line of research provides a simple yet powerful demonstration of how inaccurate memories can be. This was begun by Elizabeth Loftus and her colleagues on the topic of leading questions and memory distortions (see Loftus, 2003, 2004, for highly readable introductions to this area).

8.3.4: The Misleading Information Effect

Investigators have developed several tasks to test for the effects of **misleading information** (e.g., Zaragoza, McCloskey, & Jamis, 1987). In a typical study, people see the original event in a film or set of pictures (e.g., pictures depicting a car accident, with one showing a stop sign). Later, they are exposed to additional information, such as a narrative about the accident. Some people receive only neutral information, whereas others get a bit of misinformation (the narrative mentions "the yield sign," for instance). Finally, there is a memory test, often a yes/no recognition task that asks about the critical piece of information: Was there a stop sign or a yield sign?

A common result is that some people incorrectly claim to remember the misinformation, the yield sign here. This is the **misinformation effect**. Belli (1989), for instance, found that misled people had more than 20% lower accuracy than did control groups who were not exposed to the misinformation. Furthermore, Loftus, Donders, Hoffman, and Schooler (1989) found that misled groups were faster in their incorrect judgments—picking the yield sign, for example—than in their correct decisions. This suggests a high degree of misplaced confidence on the part of the misled people. What is particularly troubling is that misinformation effects can persist even if the information is later corrected. For example, a study by Lewandowsky, Stritzka, Oberauer, and Morales (2005) looked at misinformation about the Iraq war that was reported by the press and then later retracted. However, people continued to inappropriately remember the misinformation as true. This was particularly true for Americans who had more support for the war than for Australians or Germans, who had less support. These latter groups better remembered that the misinformation was false and had been retracted.

The misleading information effect does not always result from incorrect information that was encountered

A Study by Elizabeth Loftus on Leading Questions and Memory Distortions (Figure 8.6)

Loftus started by examining the effects of *leading questions*; that is, questions that tend to suggest to a person what answer is appropriate. She wondered whether there were long-term consequences of leading questions in terms of what people remember about events they have witnessed.

> **Part 1**
>
> In one study, Loftus and Palmer (1974) showed several short traffic safety films depicting car accidents to college classes. They asked the students to describe each accident after seeing the film and then answer a series of questions about what they had seen. One of the questions asked for an estimate of the car's speed, something people are notoriously poor at. The longer-term importance of this effect gets to the heart of issues about eyewitness testimony and memory distortion. Loftus and Palmer wondered whether the question about speed altered the people's memories of the filmed scene. In other words, if participants are exposed to the implication that the cars had "smashed" together, would they remember a more severe accident than they had actually seen? This is called **memory impairment**: a genuine change or alteration in memory of an experienced event as a function of some later event.

after a previously witnessed event. Another way that misinformation can influence memory is if people encounter incorrect information that contradicts knowledge that they previously knew. For example, although most people know that the Pacific Ocean is the world's largest ocean, people can be misled by simply reading a passage that presents incorrect information. In a study by Fazio, Barber, Rajaram, Ornstein, and Marsh (2013), if people read a text that (incorrectly) stated that the Atlantic is the largest ocean, they were more likely 2 weeks later to mistakenly report that the Atlantic is the largest ocean (71% correct) compared to people who were not misled (82% correct), even if they had previously known that the Pacific is the largest. Thus, encountering new but incorrect information can also distort our knowledge.

8.3.5: Source Misattribution and Misinformation Acceptance

Several reviews and summaries (Ayers & Reder, 1998; Loftus, 1991; Loftus & Hoffman, 1989; Roediger, 1996) outline the overall message of this research. As Loftus (1991) noted, alteration of the original memory may be only one part of memory distortion. There seem to be three important memory distortion effects:

Two of the Three Important Memory Distortion Effects

> **Source Misattribution**
>
> Sometimes people come to believe that they remember something that never happened. This is *source misattribution*, the inability to distinguish whether the original event or some later event was the true source of the information. Source misattribution suggests a confusion in which we cannot clearly remember the true origin of a piece of knowledge (Zaragoza & Lane, 1994). Using the stop sign/yield sign example, source misattribution occurs when we cannot correctly distinguish whether memory of the yield sign came from the original film or from another source, maybe the mistaken narrative that was read later or maybe from prior knowledge and memory (Lindsay, Allen, Chan, & Dahl, 2004).
>
> Another example of source misattribution can be seen in studies of the false fame effect. In these studies, people read a list of nonfamous names, which increases the familiarity of those names. Later, people are more likely to judge the names as famous, essentially confusing familiarity with fame (Jacoby, Woloshyn, & Kelly, 1989). They have lost memory for the source of the feeling of familiarity (that it had been read on a list explicitly labeled as nonfamous names), so they make their decisions solely on the basis of how familiar the names were. This confusion is particularly likely when people did not remember reading the original list of names, suggesting that the effect occurred at an implicit level (Kelley & Jacoby, 1996; see also Busey, Tunnicliff, Loftus, & Loftus, 2000).
>
> **Misinformation Acceptance**

1. Source misattribution
2. Misinformation acceptance
3. Overconfidence in the accuracy of memory

IMPLANTED MEMORIES Yet another way to examine the acceptance of misinformation is by creating **implanted memories** of events that never happened.

OVERCONFIDENCE IN MEMORY Despite our feeling that we remember events accurately ("I saw it with my own eyes!"), we often misremember experiences. And, as you have just read, we can be induced to form memories for events that never happened on the basis of suggestion, evidence, or even just related information (e.g., the class photos).

As if this weren't bad enough, we often become overly confident in the accuracy of our memories (see Wells, Olson, & Charman, 2002, for an overview) and surprisingly unaware of how unreliable memory can be (a classic illustration is shown in Figure 8.8).

As you read a moment ago, Roediger and McDermott's (1995) participants not only recalled (falsely) and recognized the critical lure, the majority claimed that they genuinely remembered it and had explicit, "vivid memory" of hearing the word in the list. Aside from a basic belief in ourselves, this overconfidence seems to involve two factors:

1. **Source memory**, our memory of the exact source of information
2. **Processing fluency**, the ease with which something is processed or comes to mind: as if you thought to yourself, "I remembered 'sleep' too easily to have just imagined that it was on the list, so it must have been on the list" (see Kelley & Lindsay, 1993)

Studies on Acceptance of Misinformation and Pseudo-events (Figure 8.7)

Not everything we remember actually happened. Sometimes these memories come from incorrect information we hear from other people, yet we often believe and accept these false memories as real.

> Early use of this approach (e.g., Hyman, Husband, & Billings, 1995; Loftus & Hoffman, 1989) involved telling people childhood stories about themselves that their parents had supplied to the researchers, then questioning them about their memory for the episodes. Unknown to the person, one of the stories was a fictional, although plausible, event (called a *pseudo-event*; for instance, "when you were 6, you knocked over a punch bowl at a wedding reception"). A large number of people came to accept the bogus story as true and claimed to "remember" it. For example, none of Hyman et al. participants claimed to remember the pseudo-event when they were first told about it, but 25% "recalled" it by their third session of questioning.

8.3.6: Stronger Memory Distortion Effects

Can something as simple as this false memory effect in the laboratory explain real-world inaccuracies in memory? Probably.

It doesn't take much to realize the implications of this work: Memory is pliable. Memories of events can be altered and influenced, both by the knowledge people have when the event happens and by what they encounter afterward. People report that they remember events that did not happen. And in many cases, they become confident about their accuracy for those events. Unfortunately, at this point, because true and false memories overlap so much on so many dimensions, there is no reliable way to assess whether any given memory is true or false without other corroborating evidence (Bernstein & Loftus, 2009).

8.3.7: Repressed and Recovered Memories

There are broad, disturbing implications of these findings for when people are trying to remember real-world events (Mitchell & Zaragoza, 2001). If we can "remember" things with a high degree of confidence and conviction, even though they never happened, then how seriously should eyewitness testimony be weighed in court proceedings? Juries usually are heavily influenced by eyewitnesses. Is this justified? Should a person be convicted of a crime

Figure 8.8 Which Penny Drawing Is Accurate?

Look at the pennies below. Which one do you think is accurate?

SOURCE: From Nickerson & Adams (1979).

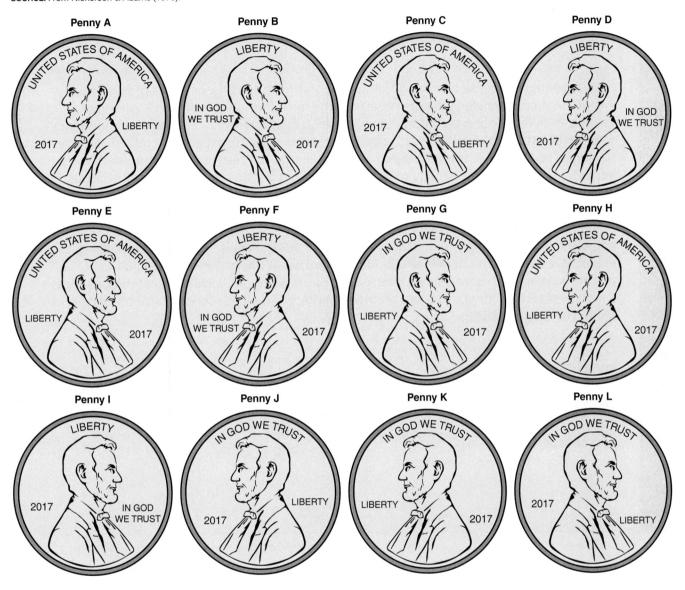

based solely on someone's memory of a criminal act? The controversy over recovered memories is an obvious and worrisome arena in which our understanding of human memory is critical.

This sort of work is difficult because there are no reliable indicators of true versus false memories, although some overall patterns distinguish them. For example, people overall show less confidence in false than true memories, tend to recall them later, and provide less detail about false than true memories (Frost, 2000; Heaps & Nash, 2001). Moreover, certain evidence indicates different physiological processing for true versus false memories. For example, it has been shown that true memories are more likely to produce distinct patterns of gamma oscillations in certain regions of the brain (Sederberg et al., 2007). (EEG recordings show regular oscillations at different frequencies, and gamma oscillations occur in the 28–100 Hz range.)

8.3.8: The Irony of Memory

Memory presents us with great irony. How can this powerful, flexible system be so fragile, so prone to errors? We complain about how poor our memories are, how forgetful we are, how hard it is to learn and remember information. We deal with the difficulties of the transience of our memories, our absent-mindedness, the occasionally embarrassing blocking we experience when trying to remember. Are these accurate assessments of our memories (e.g., Roediger & McDermott, 2000)?

Factors That Influence False Memories and Memory Distortions

Consider just a sampling of influences on false memories and memory distortions.

Interactive

Repeated Exposure to Misinformation

First, repeated exposure to information can distort memories. Repeated exposure to misinformation increases memory reports of the misinformation (Mitchell & Zaragoza, 1996) and, at least when the information was read in story format, increases the tendency to believe that the information was known before being in the experiment (Marsh, Meade, & Roediger, 2003). There has been some suggestion that the retrieval of misinformation may actually suppress the availability of accurate information, thereby increasing the impact of misinformation (MacLeod & Saunders, 2008). Moreover, repetition increases confidence in the misinformation (Roediger, Jacoby, & McDermott, 1996; Schreiber & Sergent, 1998), as do repeated attempts to remember (Henkel, 2004) or any efforts that involve the person actually generating information (Lane & Zaragoza, 2007). Finally, repeated questioning can enhance recall of some details and induce forgetting of oth-ers, even when no misinformation is present (Shaw, 1996; Shaw, Bjork, & Handal, 1995). Repeated questioning also increases confidence in one's memories, whether they are correct or not (Heaps & Nash, 2001; Shaw, 1996).

Imagination Inflation

Warnings About Misleading Information

Social Aspects

Well, they are probably exaggerated. First, as Anderson and Schooler (1991) note, when we complain about memory failures, we neglect the huge stockpile of facts and information that we expect memory to store and to provide immediate access to. We underestimate the complexities, not to mention the sheer volume, of information stored in memory. For example, some have estimated that the typical adult has at least a half-million–word vocabulary, and it is almost impossible to estimate how many people we have known across the years.

We also fall into the trap of equating remembering with recall. When we say we have forgotten something, we probably mean we are unable to recall it at that moment. However, recall is only one way of testing memory. Recognition and relearning are far more forgiving in terms of showing that information has indeed been retained in memory. Finally, we focus on the failures of retrieval, without giving credit for the countless times we remember accurately.

How much cognitive psychology will you remember in a dozen years? A study by Conway, Cohen, and Stanhope (1991) examined exactly that: students' memory of the concepts, specific facts, names, and so on from a cognitive psychology course taken up to 12 years earlier (see Figure 8.9). Recall of material dwindled quite a bit across the 12 years, from 60% to 25% for concepts, for example. But recognition for the same material dropped only a bit, from 80% to around 65% to 70%. Correct recognition for all categories of information remained significantly above chance across all 12 years. Your honest estimate—your metacognitive awareness of having information in storage—can be quite inaccurate.

A Look at Recovered Memories and Repressed Memories

In It is important to understand recovered and repressed memories and how they can mislead us.

Recovered Memories

Here is a summary of a recovered memory case. A person "recovers" a memory, possibly a horrible childhood memory of abuse. The absence of that memory for many years is said to indicate that the experience was repressed or intentionally forgotten. Although the recovery sometimes is spontaneous, it can also be an outcome of efforts by (not very good) therapists trying to bring a memory into awareness. Now that the awful memory is "recovered," the person may seek restitution, such as having the "perpetrator" brought to trial. There is often no objective way to determine whether the recovered memory is real, no sure way to determine whether the remembered event actually happened. Therefore, these cases often simply become one person's word against another's, both people claiming the truth.

There have been many court cases involving recovered memories, and several people have been convicted of crimes based on someone's recovered memory (Loftus, 1993; Loftus & Ketcham, 1991). Cognitive science has become involved in this controversy for the obvious reason: our understanding of how memory works. As the research has developed, certain aspects of the recovered memory situation have fallen under greater scrutiny.

Repressed Memories

Figure 8.9 Memory for Materials Learned in a College Course on Cognitive Psychology Over Many Years

SOURCE: From Conway et al. (1991).

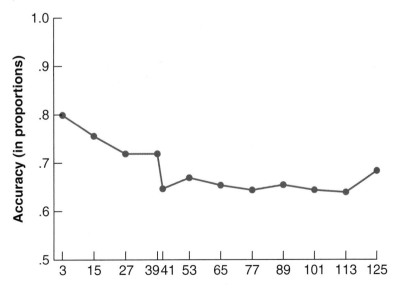

WRITING PROMPT

Reflecting on the Trustworthiness of Memory

How trustworthy is memory? Discuss how much of what we remember actually happened and how much of it we make up.

▶ The response entered here will appear in the performance dashboard and can be viewed by your instructor.

Submit

8.4: Amnesia and Implicit Memory

OBJECTIVE: Differentiate between the types of amnesia

We study dysfunctions caused by brain damage to understand cognition and its organization. Sometimes the patterns of disruptions and preserved abilities can tell us a great deal about how cognition works. This has been especially fruitful for understanding long-term memory in cases of amnesia.

Amnesia is the catastrophic loss of memories or memory abilities caused by brain damage or disease. Amnesia is one of the oldest and most thoroughly studied mental disruptions caused by brain disorders, as well as a common result of brain injury and damage. Although some amnesias are temporary, due to a blow to the head or even acute emotional or physical stress (e.g., transient global amnesia; Brown, 1998), the amnesias in which we are interested here are relatively permanent, caused by enduring changes in the brain.

8.4.1: Dissociation of Episodic and Semantic Memory

We begin with a case history. Tulving (1989) described patient K. C., who as the result of a motorcycle accident

---

Kinds of Amnesia

Many kinds of amnesias have been studied, and we have space to discuss only a few. A few bits of terminology will help you understand the material and alert you to the distinctions in memory that are particularly relevant.

Interactive

Amnesia Related to the Time of Injury

First, the loss of memory in amnesia is always considered in relation to the time of the injury. If a person suffers loss of memory for events before the brain injury, this is **retrograde amnesia**. Interestingly, retrograde amnesia commonly shows a temporal gradient—memories that are more distant in time from the injury are less impaired (e.g., Brown, 2002; Wixted, 2004). This temporal gradient is called **Ribot's law** (Ribot, 1882). Another form of amnesia is **anterograde amnesia**, disruption in acquiring new memories for events occurring after the brain injury. A person can show both forms of amnesia, although the extent of the memory loss usually is different for events before and after the damage. For example, anterograde amnesia often seems more extensive, simply because it disrupts learning from the time of the brain damage on to the present. The cases we talk about here are extreme in that the memory disruption is so extensive, which is uncharacteristic of most cases of amnesia.

Amnesia Related to Dissociations

Amnesia Related to Focal Brain Lesions

experienced serious brain injury, especially in the frontal regions. As a result of this injury, K. C. shows a seemingly complete loss of episodic memory: He is completely amnesic for his own autobiographical knowledge. K. C. has profound retrograde and anterograde amnesia. He shows great difficulty in both storing and retrieving personal experiences in long-term memory.

Although K. C.'s episodic memory no longer worked, his semantic memory did. He was adept at answering questions about his past by relying on general, semantic knowledge. When asked about his brother's funeral, he responded that the funeral was very sad, not because he remembers attending the funeral (he did not even remember that he had a brother) but because he knew that funerals are sad events.

K. C.'s memory disruption, intact semantic memory yet damaged episodic retrieval, is evidence of a dissociation between episodic and semantic memory. This suggests that episodic and semantic memories are separate systems, enough so that one can be damaged while the other stays intact. In Squire's (1987) taxonomy, K. C. has lost one of the two major components of declarative knowledge, his episodic memory.

FUNCTIONAL NEUROIMAGING EVIDENCE There are limitations on what can be learned about normal cognition from data from brain-damaged patients. Brain-damaged patients may be unique. The data from them may not be generalizable. Because we might worry about the generality of such results—K. C. could have been atypical before his accident—Tulving presented further support for his conclusions, studies of brain functioning among normal individuals (Nyberg, McIntosh, & Tulving, 1998).

When brain activity is measured in a scanner, such as is done with a PET scan (as in Figure 8.10), certain areas may become more active, relative to their baseline activity. This is often taken as an indication that those parts of the brain are more involved in the task being done at the time.

Tulving and his colleagues developed what they called the Hemispheric Encoding/Retrieval Asymmetry model, or **HERA model** (Habib, Nyberg, & Tulving, 2003; Nyberg, Cabeza, & Tulving, 1996). Data from PET studies, such as the Nyberg et al. (1998) study, show that the left frontal lobe is more likely to be involved in the retrieval of semantic memories and the encoding of episodic memories. This makes sense, because when you encounter a new event, to create a new episodic memory you need to understand the event using your semantic knowledge. By comparison, the right frontal lobe is more likely to be involved in the retrieval of episodic memories. So, based on the HERA model, different parts of the brain are involved in different types of memory processing (see also Buckner, 1996; Shallice, Fletcher, & Dolan, 1998). The HERA model is *not* making the claim that episodic memories are stored in the right hemisphere and semantic memories in the left. Instead, it is just that those brain regions are more involved in those kinds of activities. Even researchers who do not agree with the HERA model do agree that the brain processes semantic and episodic memories differently (e.g., Ranganath & Pallar, 1999; Wiggs, Weisberg, & Martin, 1999).

8.4.2: Anterograde Amnesia

The story of anterograde amnesia begins with a classic case history. A popular theoretical stance in 1950 was that memories are represented throughout the cortex, rather than concentrated in one place. Karl Lashley articulated this position in his famous 1950 paper "In Search of the Engram." Three years later, neurosurgeon William Scoville made an accidental discovery. Scoville performed radical surgery on a patient, Henry Molaison (1926–2008), more commonly known as H. M., sectioning (lesioning) H. M.'s hippocampus in both the left and right hemispheres in an attempt to gain control over his severe epilepsy. To Scoville's surprise, the outcome of this surgery was pervasive anterograde amnesia: H. M. was unable to learn and recall anything new. Although his memory of events before

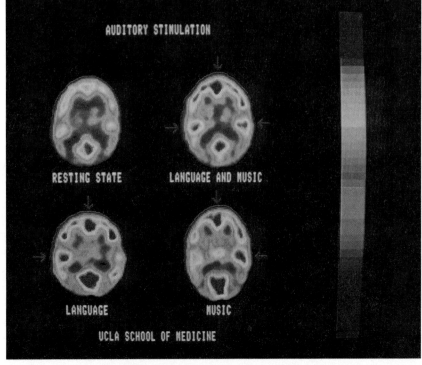

Figure 8.10 Illustration of Blood Flow as Revealed in PET Scan Images

the surgery remained intact, as did his overall IQ (118, well above average), he lost the ability to store new information in long-term memory.

Across the years, H. M. served as a participant in hundreds of tasks (e.g., Milner, Corkin, & Teuber, 1968) documenting the many facets of his anterograde amnesia. His memory of events prior to the surgery, including his childhood and school days, was quite good, with some gaps. His language comprehension was normal, and his vocabulary was above average. Yet any task that required him to retain information across a delay showed severe impairment, especially if the delay was filled with an interfering task. These impairments applied equally to nonverbal and verbal materials. For instance, after a 2-minute interference task of repeating digits, he was unable to recognize photographs of faces. He was unable to learn sequences of digits that went beyond the typical short-term memory span of seven. In a conversation reported by Cohen (in Banich, 1997), he told about some rifles he had (it was a childhood memory). This reminded him of some guns he had also had, so he told about them. Telling about the guns took long enough, however, that he forgot he had already talked about the rifles, so he launched into the rifle story again, which then reminded him of the guns—and so on until his attention was diverted to some other topic.

H. M.'S IMPLICIT MEMORY Interestingly, the evidence from the H. M. studies also suggests that H. M.'s memory was normal when it involved more unconscious *implicit memory* because he was able to learn a motor skill, mirror drawing. This task requires a person to trace between the lines of a pattern while looking at it and the pencil only in a mirror (Figure 8.11). H. M.'s performance showed a normal learning curve, with very few errors on the third day of practice.

Note, though, that on days 2 and 3, H. M. did not remember having done the task before; he had no explicit memory of ever having done it, despite his normal performance based on implicit memory.

Likewise, H. M. showed systematic learning and improvement on solving the Tower of Hanoi problem in which a stack of discs needs to be moved from one position to another following specific rules. Although he did not remember the task itself, his performance nonetheless improved across repeated days of practice. Such empirical demonstrations confirm what clinicians working with amnesia patients have known or suspected for a long time: Despite profound difficulties in what we normally think of as memory, aspects of patients' behavior do demonstrate a kind of memory—in other words, implicit memory (see Schacter, 1996). All the subtypes underneath "nondeclarative (implicit)" memory—skill learning, priming, simple classical conditioning, and nonassociative learning—represent different aspects of implicit memory; different forms and types of performance in which implicit memories can be displayed (Squire, 1993; see Gupta & Cohen, 2002, and Roediger, Marsh, & Lee, 2002, for reviews).

IMPLICATIONS FOR MEMORY What do we know about human memory as a function of H. M.'s disrupted and preserved mental capacities? How much has this person's misfortune told us about memory and cognition?

The most apparent source of H. M.'s amnesia was a disruption in the transfer of information to long-term memory. H. M.'s retrieval of information learned before surgery was intact, indicating that his long-term memory per se, including retrieval, was unaffected. Likewise, his ability to answer questions and do other simple short-term memory tasks indicates that his attention, awareness, and working memory functions also were largely intact. But he had a widespread disability in transferring new declarative information into long-term memory. This disability affected most or all of H. M.'s explicit storage of information in long-term memory (Milner et al., 1968), including real-world episodic material.

It is a mistake to conclude from this that H. M.'s memory disruption—say, the process of explicit rehearsal—takes place in the hippocampus. Instead, it seems more likely that the hippocampus is on a critical pathway for successful transfer to long-term memory. Other research on patients with similar lesions (e.g., Penfield & Milner, 1958; Zola-Morgan, Squire, & Amalral, 1986) confirms the importance of the hippocampus to this process of storing new information in long-term, explicit memory. In some sense, then, the hippocampus is a gateway into long-term memory. Thus, the hippocampus is essential for declarative or explicit memory. See Eichenbaum and Fortin (2003) for an introduction to the relationship between the hippocampus and episodic memory, and see Barnier (2002) for an extension of these effects to posthypnotic amnesia.

8.4.3: Implicit and Explicit Memory as Revealed by Amnesia

To repeat a point made earlier, the operative word in the definitions of explicit and implicit memories is *conscious*. Explicit memories, whether episodic or semantic, come to us with conscious awareness and therefore have an explicit effect on performance, an effect that could be verbalized. For example, name the third letter of the word meaning "unmarried man." The very fact that you can say *c* and name the word *bachelor* attests to the fact that this is an explicit memory. In contrast, fill in the following word stems: "ki__, swe__, fl__." Even without any involvement of conscious awareness, you may have filled these in with the words *kitchen, sweet, flag* with greater likelihood than

Figure 8.11 H. M. Studies

A. In this test, people trace between the two outlines of the star while viewing their hand in a mirror. The reversing effect of the mirror makes this a difficult task initially. Crossing a line constitutes an error. **B.** Patient H. M. shows clear improvement in the motor learning star task, an influence of implicit learning and memory.

SOURCE: After Blakemore (1977).

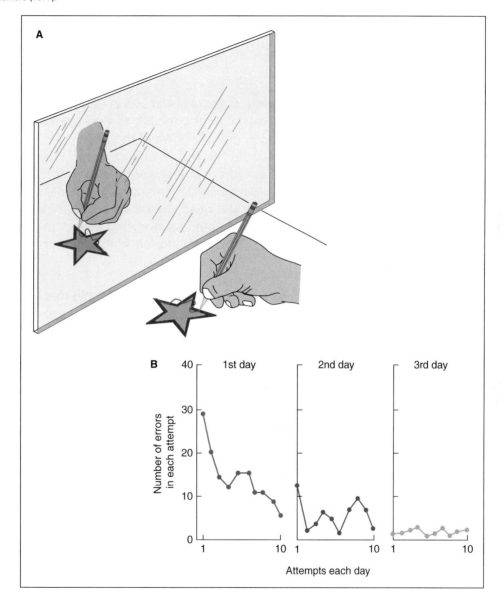

would have been expected by chance. The words *kitchen*, *sweet*, and *jelly* occurred in this module. Even reading material a second time through is done more rapidly, an effect called *repetition priming*. That is, when information is encountered, it unconsciously activates prior memories of the same thing, making it easier to process. We all demonstrate such implicit effects as repetition priming, amnesic or not (Graf & Schacter, 1987; Kolers & Roediger, 1984).

In addition, cognitive science learned an important lesson from patients such as H. M. and K. C. If we had stuck to laboratory-based experiments alone, and never paid attention to patients with amnesia, we would have failed to realize the importance of that second, less obvious kind of long-term memory, the kind not dependent on conscious recollection. We would have missed implicit memory.

WRITING PROMPT

Amnesia and Normal Forgetting

How does amnesia differ from normal forgetting, and what does amnesia tell us about the normal operation of memory?

> The response entered here will appear in the performance dashboard and can be viewed by your instructor.

Submit

Repetition Priming Tests

Repetition priming has been established in several tasks, such as word identification and lexical decisions (Morton, 1979), word and picture naming (Brown, Neblett, Jones, & Mitchell, 1991), and rereading fluency (Masson, 1984). In all these, a prior encounter with the stimulus yields faster performance on a later task, even though you may not consciously remember having seen it before (see Logan, 1990, for the connection of repetition priming).

> **Interactive**
>
> **Explicit Memory Test**
>
> In a classic demonstration of repetition priming, Jacoby and Dallas (1981) had people study a list of familiar words, answering a question about each as they went through the list. Sometimes the question asked about the physical form of the word, as in, "Does it contain the letter *L*?" Sometimes it asked about the word's sound, as in, "Does it rhyme with *train*?" And sometimes, the question asked about a semantic characteristic, as in, "Is it the center of the nervous system?" This was a direct manipulation of *depth of processing*. Asking about the physical form of the word should induce shallow processing, leading to poor memory. Asking about rhymes demands somewhat deeper processing, and asking about semantic characteristics should demand full, elaborative processing of the list words.
>
> At test, explicit memory was assessed by a yes/no recognition task ("Did this word occur in the study phase?"). Here, recognition accuracy was affected by the type of question answered during study. When a question related to the physical form, recognition was at chance, 51%. When it had asked about the sound of the word, performance improved. And when semantic processing had been elicited, recognition accuracy was high, 95%. This a test of explicit memory because people had to say "yes" or "no" based on whether they had seen the word earlier. As expected, more elaborative processing led to better explicit memory performance.
>
> **Implicit Memory Test**

Summary: Memory and Forgetting

8.1 The Seven Sins of Memory

- Forgetting may occur because the knowledge no longer exists in memory. This is forgetting due to a lack of availability. Alternatively, forgetting may occur because the knowledge is in memory, but cannot be successfully retrieved. This is forgetting due to a lack of accessibility.
- Schacter has defined seven "sins" in which memory performance is compromised. Three of these are sins of omission, namely transience, absent-mindedness, and blocking. Three more are sins of commission, namely, misattribution, suggestibility, and bias. Finally, the sin of persistence is when we wish to forget knowledge, but it persists in memory.

8.2 Forgetting Through Decay and Interference

- Although there has been some suggestion that forgetting occurs through a decay process associated with the passage of time, most normal forgetting appears to be due to a process of interference. Using paired-associate lists, memory researchers have found that this can be proactive interference, in which older memories impair the ability to access newer memories, or retroactive interference, in which newer memories impair the ability to access older memories.

- Associative interference, or a fan effect, can be observed as more things are learned about a concept. This is brought about by activation being divided among the various associations. This type of interference can be attenuated if information can be integrated in some way, as with a situation model. This is the cognitive process leading to such phenomena as the finding that walking through doorways causes forgetting.
- The loss of memories through forgetting can also be reduced if people have the appropriate retrieval cues available. These allow people to access those memories more directly than if people were left to their own devices. Note that memory retrieval cues can sometimes backfire, as with the part-set cuing effect.

8.3 False Memories, Eyewitness Memory, and "Forgotten Memories"

- Several paradigms give clear evidence of false memories, such as the Roediger and McDermott (1995) list presentation studies and eyewitness memory research by Loftus and others. "Remembering" in such situations is affected by source misattribution, the acceptance of misinformation, and bias. People tend to be overconfident about their memories, regardless of the distortions that might be involved.
- In eyewitness memory and testimony, any new information about an event is integrated with relevant existing knowledge. Thus, we are less than accurate when we attempt to retrieve such knowledge because we are often unable to discriminate between new and original information.
- Cases of "forgotten" and "recovered" memories are particularly difficult to assess because of the fragile, reconstructive nature of memories. It is a concern that therapeutic techniques used to assist in "recovering" memory of trauma are so similar to variables such as repetition and repeated questioning, variables that increase the false memory effect.

8.4 Amnesia and Implicit Memory

- Studies of people with amnesia caused by brain damage have taught us a great deal about long-term memory. Patient K. C. shows total amnesia for episodic information, although his semantic memory is unimpaired, suggesting a dissociation between episodic and semantic memories. Patients like H. M., a person with anterograde amnesia, typically are unable to acquire new explicit memories, but have intact implicit memory. The medial temporal area and especially the hippocampus are very important for the formation of new explicit memories, but different brain structures underlie implicit learning.
- Recognition memory for information acquired across an extended period is remarkably accurate across many years, whereas recall performance begins to decline within months.

SHARED WRITING

The Unreliability of Autobiography

Given what is known about human memory, in what ways can parts of a person's written autobiography be viewed as trustworthy, and in what ways might parts of it be suspect? What can be done to make what is reported more reliable?

> A minimum number of characters is required to post and earn points. After posting, your response can be viewed by your class and instructor, and you can participate in the class discussion.

Post   0 characters | 140 minimum

Chapter 9
Language

Learning Objectives

9.1: Describe the ways language is defined

9.2: Explain how phonology influences spoken language

9.3: Explain the rules governing syntax

9.4: Analyze lexical factors to understand the meaning of language

9.5: Compare semantics and syntax to understanding intended meaning

9.6: Summarize the processes the brain uses to comprehend language

Language, along with music, is one of the most common and universal features of human society. Language pervades every facet of our lives, from our most public behavior to our most private thoughts. We might imagine a society that has no interest in biology, or even one with no formal system of numbers and arithmetic. But a society without language is inconceivable. Every culture, no matter how primitive or isolated, has language. Every person, unless deprived by nature or accident, develops skill in the use of language. Human use of language is astounding because we are typically able to process approximately three words per second, drawing on a vocabulary of about 75,000 words.

In this module, we will delve into the basics of language including characteristics, functions, structure, and form.

Linguistics is the study of language and has had a profound influence on cognitive psychology. It was a major turning point when Chomsky rejected behaviorism's explanation of language. Because approaches such as Chomsky's seemed likely to yield new insights and understanding, psychology renewed its interest in language in the late 1950s and early 1960s, borrowing heavily from linguistic theory.

And yet, as psychologists began to apply and test linguistic theory, they discovered an important limitation. Language is a purposeful activity. It is there to do something: to communicate, to express thoughts and ideas, to make things happen. However, linguistics focused on language itself as a formal, almost disembodied system so that the *use* of language by humans was seen as less interesting, tangential, or even irrelevant. This view denied a fundamental interest of psychology—behavior. Thus, a branch of cognitive psychology evolved called **psycholinguistics**, which is the study of language as it is learned and used by people.

In this module, we present only a brief survey of psycholinguistics.

9.1: Linguistic Universals and Functions

OBJECTIVE: Describe the ways language is defined

One might define *language* as "the use of words and how they are spoken in various combinations so a message can be understood by other people who also speak that language." That is not a bad start. For example, one critical idea in the definition is that meaning and understanding are *attributed* to the words and their pronunciations, rather than being part of those words. As an illustration, the difference in sound between the words *car* and *cars* is the *s* sound, denoting plural in English. But this meaning is not inherent in the *s* sound, any more than the word *chalk* necessarily refers to the white stuff used on blackboards. This is an important idea: Language is based on arbitrary connections between linguistic elements, such as sounds (pronunciations) and the meanings they denote.

9.1.1: Defining Language

This definition of *language* is a bit confining and restricts language to human speech. By this rule, writing would not be language, nor would sign language for the deaf. It is true that writing is a recent development, dating back only about 5,000 years, compared to the development of articulate speech, thought to have occurred some 100,000 years ago (e.g., Corballis, 2004). It is equally true that the development of writing depends critically on a spoken language. Thus, the spoken, auditory form of a language is more basic than the written version. Is there any doubt that children would fail to acquire language if they were only exposed to books instead of speech? Nonetheless, we include written language because reading and writing are major forms of communication.

9.1.2: Language Universals

There are a large number of differences between languages. Take word order, for example. English is largely a *subject-verb-object (SVO) language* in which the sentence subject comes first, followed by the verb and then the object of the sentence. Other languages have other structures. Japanese, for instance, is an SOV language in which the verb typically comes at the end. Despite the many differences among human languages, they all share some universal properties.

Hockett (1960a, 1960b, 1966) proposed a list of 13 **linguistic universals**, features or characteristics that are common to all languages. To distinguish human language from animal communication, Hockett proposed that only human language contains all 13 features. Several of the universals he identified, such as the vocal–auditory

---

Language as a Shared Symbolic System for Communication

Let's offer a definition that is more suitable here. **Language** is a shared symbolic system for communication.

Interactive

1. Language is symbolic.

It consists of units (e.g., sounds that form words) that symbolize or stand for the referent of the word. The referent, the thing referred to by the final *s*, in words such as *cars*, for example, is the meaning "plural."

2. The symbol system is shared by all users of a language culture.

3. The system enables communication.

Hockett's Linguistic Universals

Here are 12 of Hockett's linguistic universals, along with short explanations. We limit our discussion to four of these, plus two others implied, but absent from the list.

> **Interactive**
>
> **Vocal–Auditory Channel**
>
> The channel or means of transmission for all linguistic communications is vocal–auditory. Hockett excluded written language by this universal because it is a recent invention and not found in all language cultures.
>
> **Broadcast Transmission and Directional Reception**
>
> **Rapid Fading**
>
> **Interchangeability**
>
> **Total Feedback**
>
> **Specialization**
>
> **Semanticity**
>
> **Discreteness**
>
> **Displacement**
>
> **Productivity**
>
> **Duality of Patterning**
>
> **Cultural or Traditional Transmission**

requirement, are not essential characteristics of human language, although they were likely essential to its evolution. Other features are critically important to our analysis.

SEMANTICITY An important aspect of language is **semanticity**: that it conveys meaning. For example, the sounds of human language carry meaning, whereas other sounds, such as coughing or whistling, are not part of our language because they do not convey meaning in the usual sense. We are ignoring here a situation such as a roomful of students coughing in unison at a professor's boastful remark to indicate their collective opinion. The coughing sound in this case is *paralinguistic* and functions much as rising vocal pitch does to indicate anger. Then again, it could just be a roomful of coughing students.

ARBITRARINESS **Arbitrariness** means there is no inherent connection between the units (sounds, words) used in a language and their meanings. There are a few exceptions, such as **onomatopoeias** like *buzz*, *hum*, and *zoom*. But as Pinker (1994) notes, some units we consider onomatopoetic, such as a pig's oink, are not because they are different sounds in other languages. In Japanese, for example, a pig's sound is "boo-boo." But far more common, the language symbol bears no relationship to the thing itself. The word *dog* has no inherent correspondence to the four-legged furry creature, just as the spoken symbol *silence* does not resemble its referent, true silence. Hockett's example drives the point home: "*Whale* is a small symbol for a very big thing, and *microorganism* is a big symbol for an extremely small thing."

Because there are no built-in connections between symbols and their referents, knowledge of language must involve learning and remembering the arbitrary connections. Thus, we refer to language as a shared system. We all have learned essentially the same connections, the same set of word-to-referent associations, and stored them in memory as part of our knowledge of language. By convention—by agreement with the language culture—we all know that *dog* refers to one particular kind of physical object. Obviously, we have to know which word goes with a particular referent because there is no way to look at an object and decide what its name must be.

Two important consequences of the arbitrariness of language deserve special attention, partly because they help to distinguish human language from animal communication and partly because they tell us about the human language user. These two consequences concern flexibility and the principle of naming. Hockett listed neither of these, although they are derived from his point about arbitrariness.

FLEXIBILITY OF SYMBOLS Note that arbitrariness makes language symbolic. *Desk* and *pupitre* are the English and French symbols for a particular object. Were it not for the history of our language, we might call that object a *zoople* or a *manty*. A consequence of this *symbolic* aspect of language is that the system demonstrates tremendous **flexibility**. Because the connection between symbol and meaning is arbitrary, we can change those connections and invent new ones. We routinely shift our terms for the things around us, however slowly such change takes place.

Although our names for things such as *dog* seem so strongly tied to the object, the word is actually arbitrary and can vary from language to language. For example, instead of *dog*, the same things would be called *chien*, *gǒu*, or *canis*, in French, Chinese, and Latin.

Contrast this flexibility with the characteristics of the opposite of a symbolic system, called an iconic system. In an *iconic system*, each unit has a physical resemblance to its referent, just as a map is physically similar to the terrain it depicts. In such a system there is no flexibility because changing the symbol for a referent would make the connection arbitrary.

NAMING A corollary to arbitrariness and flexibility involves **naming** (Glass & Holyoak, 1986). We assign names to all the objects in our environment, to all the feelings and emotions we experience, to all the ideas and concepts we conceive of. So wherever it is you are sitting right now as you read this text, each object in the room has a name. In an unfamiliar or unusual place (an airport control tower or a car repair shop), you may not know the name of something, but it rarely, if ever, occurs to you that the thing might not have a name.

Furthermore, we do not stop by naming just the physical objects around us. We have a vocabulary for referring to unseen characteristics, privately experienced feelings, and other intangibles and abstractions. Terms such as *perception, mental process spreading activation,* and *knowledge* have no necessary physical referent, nor do words such as *justice, cause, truth, likewise,* and *however* refer to concrete objects. Indeed, we even have words such as *abstractions* and *intangibles* that refer to the *idea* of being abstract. Going one step further, we generate or invent names for new objects, ideas, activities, and so forth. For instance, think of the new vocabulary that had to be invented and mastered to describe the various actions and operations for using the Internet and modern technology. Because we need and want to talk about new things, new ideas, and new concepts, we invent new terms. (See Kaschak & Glenberg, 2000, on how we invent new verbs from already known words; e.g., *to crutch* or *to google*.)

DISPLACEMENT One of the most powerful language tools is the ability to talk about something other than the present moment, a feature called **displacement**. By conjugating verbs to form past tense, future tense, and so on, we can communicate about objects, events, and ideas that are not present but are remembered or anticipated. When we use constructions such as "If I go to the library tomorrow, then I will be able to . . . ," we demonstrate a particularly powerful aspect of displacement: We can communicate about something that has never happened and indeed might never happen, while anticipating future consequences of that never-performed action. To illustrate the power and importance of displacement, try speaking only in the present tense for about 5 minutes. You will discover how incredibly limiting it would be if we were "stuck in the present."

PRODUCTIVITY By most accounts, the principle of **productivity** (also called **generativity**) is important because it gives language a notable characteristic—novelty. Indeed, the novelty of language, and the productivity that novelty implies, formed the basis of Chomsky's (1959) critique of Skinner's book and the foundation for Chomsky's own theory of language (1957, 1965).

9.1.3: Animal Communication

In contrast to flexible and productive human language, animal communication is neither. Animal communication

Importance of Productivity

It is an absolute article of faith in both linguistics and psycholinguistics that the key to understanding language and language behavior lies in an understanding of *novelty*, an understanding of the productive nature of language.

> **Interactive**
>
> Consider the following: Aside from trite phrases, customary greetings, and so on, hardly any of our routine language is standardized or repetitive. Instead, the bulk of what we say is novel. Our utterances are not memorized, are not repeated, but are new. This is the principle of *productivity*, that language is a productive and inherently novel activity, that we generate utterances rather than repeat them. We (the authors of your text) lecture on the principle of productivity every time we teach our memory and cognition classes, each time uttering a stream of sounds, a sequence of words and sentences, that is novel, new, literally invented on the spot—the ideas we talk about may be the same semester after semester, but the sentences are new each time. Even in somewhat stylized situations, as in telling a joke, the language is largely new. Only if the punchline requires a specific wording, do we try to remember the exact wording of a previously used sentence.

is seen in a wide range of circumstances, from insects to primates. For example, bees communicate the location of honey (a form of displacement) through a waggle dance (Dyer, 2002; Sherman & Visscher, 2002; von Frisch, 1967). Essentially, they orient themselves within the hive to the relative position of the sun, and then act out a dance that conveys how the flight will progress to get to the source of the nectar. This is even more impressive given that it is fairly dark in a beehive, and the dance is performed on a vertical surface.

Closer to humans, consider the signaling system of vervet monkeys (Marler, 1967). This consists of several distress and warning calls, alerting an entire troupe to imminent danger. These monkeys produce a guttural "rraup" sound to warn of eagles, one of the monkey's natural predators; they "chutter" to warn of snakes and "chirp" to warn of leopards. The system thus exhibits *semanticity*, an important characteristic of language. That is, each signal in the system has a different, specific referent (eagle, snake, and leopard). Furthermore, these seem to be arbitrary connections: "Rraup" doesn't resemble eagles in any physical way.

But as Glass and Holyoak (1986) note, the troupe of monkeys cannot get together and decide to change the meaning of "rraup" from *eagle* to *snake*. The arbitrary connections to meaning are completely inflexible. (This inflexibility results at least in part from genetic influence; compare this with Hockett's last universal, cultural transmission.) Furthermore, there is a vast difference between naming in human languages and in animal communication. There seem to be no words in the monkey system for other important objects and concepts in their environment, such as "tree." And as for displacement and productivity, consider the following quotation from Glass and Holyoak (1986, p. 448): "The monkey has no way of saying 'I don't see an eagle,' or 'Thank heavens that wasn't an eagle,' or 'That was some huge eagle I saw yesterday.'" Even human infants can use displacement by pointing to refer to things that are not immediately present, whereas chimpanzees cannot (Liszkowski, Schäfer, Carpenter, & Tomasello, 2009), suggesting that nonhuman animals lack even the basic cognitive abilities required by language.

Although there are no true languages among the animal communication systems, this is not to say that nothing can be learned about language from studying animals. As one illustration, work by Hopkins, Russell, and Cantalupo (2007) used magnetic resonance imaging (MRI) with chimpanzees to show that there was a lateralization of function as a consequence of tool use. Moreover, those regions of the brain that were more affected corresponded to Broca's and Wernicke's areas in humans, which correspond to critical areas of human language production and comprehension (as you will see later in the module). This suggests that our development of language may be tied, to some extent, to the development of tool use by our ancestors. Similarly, there is evidence that some great apes and nonverbal human infants can think about and signal absent objects (Bohn, Call, & Tamasello, 2015).

In short, beyond a level of arbitrariness, animal communication does not exhibit the characteristics that appear to be universally true of human language. There are no genuine languages in animals, although there may be genuine precursors to human language among various apes. In human cultures, genuine language is the rule. (For a more up-to-date discussion of animal cognition, see Bekoff, Allen, & Burghardt, 2002.) Note that there has been some suggestion that our close relatives, the Neanderthals, also possessed speech (D'Anastasio et al., 2013; Dediu & Levinson, 2013), although this view is controversial (Berwick, Hauser, & Tattersall, 2013).

Table 9.1 Miller's (1973) Five Levels of Language Analysis

Miller (1973) proposed that language is organized on five levels (see Table 9.1). In addition to the three traditional levels of phonology, syntax, and lexical or semantic knowledge, Miller suggested two higher levels as well. He called these the *conceptual knowledge* and *belief* levels. For organizational purposes, we focus primarily on the first three of the levels in this module.

Level	Explanation
1. Phonology	Analysis of the sounds of language as they are articulated and comprehended in speech
2. Syntax	Analysis of word order and grammaticality (e.g., rules for forming past tense and plurals, rules for determining word ordering in phrases and sentences)
3. Lexical or semantic	Analysis of word meaning and the integration of word meanings within phrases and sentences
4. Conceptual	Analysis of phrase and sentence meaning with reference to knowledge in semantic memory
5. Beliefs	Analysis of sentence and discourse meaning with reference to one's own beliefs and one's beliefs about a speaker's intent and motivations

Check Your Understanding

9.1.4: Levels of Analysis

The traditional view of language from linguistics is that it is the set of all acceptable, well-formed utterances. In this scheme, the set of rules used to generate the utterances is called a **grammar**. In other words, the grammar of a language is the complete set of rules that will generate all the acceptable utterances and will not generate any unacceptable, ill-formed ones. According to most linguists (e.g., Chomsky, 1965), such a grammar operates at three levels: **Phonology** of language deals with the sounds of language; **syntax** deals with word order and grammaticality; and **semantics** deals with accessing and combining the separate word meanings into a sensible, meaningful whole.

A CRITICAL DISTINCTION BETWEEN COMPETENCE AND PERFORMANCE Chomsky (1957, 1965) insisted that there is an important distinction in any investigation of language, the distinction between competence and performance. **Competence** is the internalized knowledge of language and its rules that fully fluent speakers of a language have. It is an ideal knowledge, to an extent, in that it represents a person's complete knowledge of how to generate and comprehend language. **Performance** is the actual language behavior a speaker generates, the string of sounds and words that the speaker utters.

When we produce language, we are not only revealing our internalized knowledge of language, our competence; we are also passing that knowledge through the cognitive system. So, it is not surprising that our performance sometimes reveals errors. Speakers may lose their train of thought as they proceed through a sentence, and so may be forced to stop and begin again. We pause, repeat ourselves, stall by saying "ummm," and so on. We can attribute all these **dysfluencies**, these irregularities or errors in

---

The Strong and Weaker Versions of Sapir-Whorf Hypothesis

In its strongest version, the hypothesis claims that language controls both thought and perception to a large degree, so you cannot think about ideas or concepts your language does not name. In its weaker version, the hypothesis claims that your language influences and shapes your thought, making it merely more difficult, rather than impossible, to think about ideas without having a name for them.

The Strong Sapir-Whorf Hypothesis

In a series of studies assessing the Sapir-Whorf hypothesis, Eleanor Rosch tested members of the Dani tribe in New Guinea on a perceptual and memory test (Rosch-Heider, 1972). She administered both short- and long-term memory tasks, using chips of different colors as the stimuli. She found that the Dani learned and remembered more accurately when the chips were "focal" colors rather than "non-focal" colors, as when the learning trial presented a "really red" red as opposed to a "sort-of-red" red. In other words, the central, perceptually salient, "good" red was a better aid to accuracy than the non-focal "off-red." The compelling aspect of these studies involved the language of the Dani people, which contains only two color terms, one for "dark" and one for "light." Nothing in their language expresses meanings such as "true red" or "off-red," and yet their performance was influenced by the centrality of focal versus non-focal colors. Thus, in this example, a person's language could have affected cognition (the language had very few color terms), and yet did not. Moreover, others have suggested that some reported linguistic relativity effects are not reliable (Wright, Davies, & Franklin, 2015). That said, there is evidence that language may influence the shape of perceptual color spaces (Regier, Kay, & Khetarpal, 2009). Other research shows that different ways of referring to objects—such as the distinction between count and mass nouns in English, which does not occur in Japanese and Chinese—do not influence object perception (Barner, Li, & Snedeker, 2010). Results such as these seem to disconfirm the strong Sapir-Whorf hypothesis.

otherwise fluent speech, to the language user. Lapses of memory, momentary distractions, intrusions of new thoughts, "hiccups" in the linguistic system—all of these are imperfections in the language user rather than in the user's basic competence or knowledge of the language. Chomsky, as a linguist, was not particularly interested in these performance-related aspects of language. Psychology, on the other hand, views them as rich sources of evidence for understanding language and language users.

THE SAPIR-WHORF HYPOTHESIS We tend to think of mental processes, including those related to language, as universal, as being equally true of all languages. Even slight familiarity with another language, however, reveals at least some of our beliefs to be misconceptions.

An organizing issue in studies of cultural influences on language and thought is how one's language affects one's thinking. This topic is called the **Sapir-Whorf hypothesis**, or more formally the **linguistic relativity hypothesis** by Whorf (1956). This idea comes out of work by Edward Sapir, a linguist and anthropologist, and his student Benjamin Whorf. The basic idea is that the language you know shapes the way you think about events in the world around you.

> **WRITING PROMPT**
>
> **Defining Language**
>
> What makes something language, and what is it about language that is unique to humans? How does having language influence what kind of thinking we can do?
>
> The response entered here will appear in the performance dashboard and can be viewed by your instructor.
>
> Submit

9.2: Phonology

OBJECTIVE: Explain how phonology influences spoken language

In any language interaction, the task of a speaker is to communicate an idea by translating that idea into spoken sounds. The hearer goes in the opposite direction, translating from sound to intended meaning. Essentially, a person is transferring the contents of his or her mind to another person (a lot like ESP, only in a plausible—spoken—way). Among the many sources of information available in the spoken message, the most obvious and concrete one is the sound of the language itself, the stream of speech signals that must be decoded. Other sources such as the gestures and facial expressions of the speaker are also available, but we focus on the speech sounds here. Thus our study of the grammar of language begins at this basic level of *phonology*, the sounds of language and the rule system for combining them.

9.2.1: Sounds in Isolation

To state an obvious point, different languages sound different: They are composed of different sets of sounds. The basic sounds that compose a language are called *phonemes*. If we were to conduct a survey, we would find around 200 different phonemes across all known spoken languages. However, no single language uses even half that many phonemes. English uses about 46 phonemes (experts disagree on whether some sounds are separate phonemes or blends of two phonemes; the disagreement centers on diphthong vowel sounds, as in *few*, seemingly a combination of "ee" and "oo"). Hawaiian uses only about 15 phonemes (Palermo, 1978). There is little significance to the total tally of phonemes in a language; no language is

Table 9.2 English Consonants and Vowels

Table 9.2 shows the typology of the phonemes of English based on the characteristics of their pronunciation.

English Consonants								
Manner of Articulation		**Bilabial**	**Labio-dental**	**Dental**	**Alveolar**	**Palatal**	**Velar**	**Glottal**
(oral) Stops	Voiceless	P (*put*)			t (*tuck*)		k (*cap*)	
	Voiced	b (*but*)			d (*dug*)		g (*got*)	
Nasal (stop)		m (*map*)			n (*nap*)		ŋ (*song*)	
Affricatives	Voiceless					č (*churn*)		
	Voiced					ǰ (*jump*)		
Fricatives	Voiceless		f (*fit*)	Q (*think*)	s (*sad*)	š (*fish*)		h (*had*)
	Voiced		v (*vote*)	ð (*them*)	z (*zip*)	ž (*azure*)		
Glides		w (*won*)				y (*yes*)		
Liquids					l (*lame*)	r (*rage*)		

(Continued)

Table 9.2 Continued

English Vowels			
	Front	Center	Back
High	i (b*ee*f)		u (b*oo*m)
			U (b*oo*k)
	I (b*i*t)		
Middle		I (b*i*rd)	o (b*ow*l)
	e (b*a*be)	(s*o*fa)	
	ɛ (b*e*d)		(b*ough*t)
	æ (b*a*d)	(b*u*s)	
Low			
			a (p*a*lm)

SOURCE: Based on Glucksberg & Danks (1975).

superior to another because it has more (or fewer) phonemes.

For consonants, three variables are relevant: place of articulation, manner of articulation, and voicing.

ARTICULATION *Place of articulation* is the place in the vocal tract where the disruption of airflow occurs. As shown in Figure 9.1, a bilabial consonant such as /b/ disrupts the airflow at the lips, whereas /h/ disrupts the column of air at the rear of the vocal tract, at the glottis. *Manner of articulation* is how the airflow coming up from the lungs is disrupted. If the column of air is completely stopped and then released, it is called a *stop consonant*, such as the consonant sounds in *bat* and *tub*. A *fricative consonant*, such as the /f/ in *fine*, involves only a partial blockage of airflow. Finally, **voicing** refers to whether the vocal cords begin to vibrate immediately with the obstruction of airflow (for example, the /b/ in *bat*) or whether the vibration is delayed until after the release of air (the /p/ in *pat*).

Vowels, by contrast, involve no airflow disruption. Instead, they differ on two dimensions: placement in the mouth (front, center, or back) and tongue position in the mouth (high, middle, or low).

PHONEMES Let's develop a few more conscious intuitions about phonemes.

Figure 9.1 The Vocal Tract and Places of Articulation

Places of articulation: 1, bilabial; 2, labiodental; 3, dental; 4, alveolar; 5 and 6, palatoalveolar; 7, velar; 8, uvular; 9, glottal.

SOURCE: Based on Fromkin and Rodman (1974).

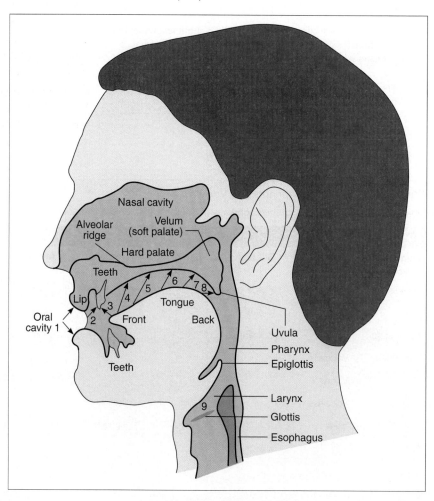

Figure 9.2 Spectrographs of Different Phonemes

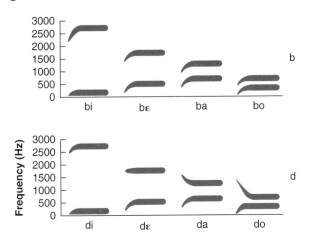

Stop for a moment and put your hand in front of your mouth. Say the word *pot* and then *spot*. Did you notice a difference between the two /p/ sounds? Most speakers produce a puff of air with the /p/ sound as they say *pot*; we puff very little (if at all) for the /p/ in *spot* if it is spoken normally. Given this, you would have to agree that these two /p/ sounds are different at a purely physical level. And yet you hear them as the same sound in these two words. Figure 9.2 shows hypothetical spectrograph patterns for two families of syllables, the /b/ family on the top and the /d/ family on the bottom. Note how remarkably different "the same" phoneme can be.

For psycholinguistics, the two /p/ sounds, despite their physical differences, are both instances of the same phoneme, the same basic sound group. That is, the fact that these two sounds are treated as if they were the same in English means that they represent one phoneme. So, let's redefine the term *phoneme* as the category or group of language sounds that are treated as the same, despite physical differences among the sounds. In other words, the English word *spot* does not change its meaning when pronounced with the /p/ sound in *pot*.

A classic illustration of phoneme boundaries is shown in Figure 9.3, from a study by Liberman, Harris, Hoffman, and Griffith (1957).

When the presented sound crossed a boundary, such as between stimulus values 3 and 5 and between 9 and 10 in Figure 9.3, identifications of the sound switched rapidly from /b/ to /d/ and then from /d/ to /g/. Variations within the boundaries did not lead to different identifications; despite the variations, all the sounds from values 5 to 8 were identified as /d/.

There are two critical ideas here. First, all of the sounds falling within a set of boundaries are perceived as the same despite physical differences among them. This is called **categorical perception**. Because English speakers discern no real difference between the hard /k/ sounds in *cool* and *keep*, they are perceived categorically, as belonging to the same category, the /k/ phoneme. Second, different phonemes are the sounds that are perceived as different by speakers of the language. The physical differences between /s/ and /z/ are important in English. Changing from one to the other gives you different words, such as *ice* and *eyes*. Thus, the /s/ and /z/ sounds in English are different phonemes.

An interesting side effect of such phonemic differences is that you can be insensitive to differences of other languages if your own language does not make that distinction. Spanish does not use the /s/ versus /z/ contrast, so native speakers of Spanish have difficulty distinguishing or pronouncing *ice* and *eyes* in English. Conversely, the hard /k/ sounds at the beginning of *cool* and *keep* are interchangeable in English; they are the same phoneme. But, this difference is phonemic in Arabic. The Arabic words for

Figure 9.3 Illustration of Phoneme Boundaries

One person's labeling data for synthesized consonants ranging from /b/ to /g/. Note that small changes in the stimulus value (e.g., from values 3 to 4) can result in a complete change in labeling, whereas larger changes (e.g., from values 4 to 8) that do not cross the phoneme boundary do not lead to a change in labeling.

SOURCE: From Liberman et al. (1957).

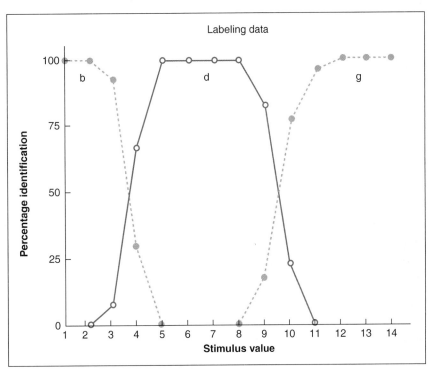

heart and *dog* differ only in their initial sounds, exactly the two different hard /k/ sounds in *cool* and *keep*.

9.2.2: Combining Phonemes into Morphemes

From a stock of about 46 phonemes, English generates all its words, however many thousands that might be. Phonemes combine into meaningful units called **morphemes**. Morphemes are not just words. Some morphemes are individual words, whereas other morphemes are not words by themselves, but must be combined with other morphemes to make words. This fact, that a small number of units can be combined so flexibly into so many words, is the linguistic universal of productivity at the level of phonology. So, from a small set of phonemes we can generate a functionally infinite number of words.

9.2.3: Speech Perception and Context

Here, we will approach the question of how people produce and perceive speech. Do we hear a word and segment it in some fashion into its separate phonemes? When we speak, do we string phonemes together, one after another, like stringing beads on a necklace?

The answer to both questions is "no." Even when the "same" sound is being pronounced, it is not physically identical to other instances of that "same" sound. The sounds *change*—they change from speaker to speaker and from one time to the next within the same speaker. Most

---

Rules of Combining Phonemes into Words

Recall also that the essential ingredient of productivity is rules. We now turn to the rules of combining phonemes into words.

Interactive

Phoneme Combinations

Let's work with a simple example of phoneme combinations. There are three phonemes in the word *bat*: the voiced stop consonant /b/, the short vowel sound /ae/, and the final voiceless /t/. Substitute the voiceless /p/ for /b/, and you get *pat*. Now rearrange the phonemes in these words, and you will discover that some of the arrangements do not yield English words, such as *abt, *tba,* and *atp.* Why? What makes *abt and *atp illegal strings in English?

Although it is tempting to say that syllables like *abt cannot be pronounced, a moment's reflection suggests this is false. After all, many such "unpronounceable" strings in English are pronounced in other languages. For example, the initial *pn-* in the French word for *pneumonia* is pronounced, whereas English makes the *p* silent. Instead, the rule is more specific. English usually does not use a "voiced–voiceless" sequence of two consonants within the same syllable. It only seldom uses any two-consonant sequence when both are in the same "manner of articulation" category. Of course, if the two consonants are in different syllables, then the rule does not apply.

Phonemic Competence and Rules

prominently, they change or vary from one word to another, depending on the preceding and following sounds.

This variability in sounds is the *problem of invariance*. This term is somewhat peculiar because the problem in speech perception is that sounds *are not* invariant; they change all the time. You saw an illustration of this in Figure 9.2 where the initial /b/ and /d/ sounds looked very different in the spectrographic patterns depending on the vowel that followed. A second illustration of the problem of invariance is in Figure 9.4, which shows the influence of each of the three phonemes in the word *bag*. To pronounce *bag*, do you simply articulate the /b/, then /ae/, then /g/? No! As the figure shows, the /ae/ sound influences both /b/ and /g/, the /g/ phoneme (dotted lines) exerts an influence well back into the /b/ sound, and so on.

The term for this is **coarticulation**: More than one sound is articulated at the same time. As you type the word *the* on a keyboard, your right index finger starts moving toward *h* before your left index finger has struck the *t*.

Figure 9.4 Coarticulation and the Problem of Invariance

Coarticulation is illustrated for the three phonemes in the word *bag*; solid diagonals indicate the influence of the /b/ phoneme and dotted diagonals, the influence of /g/.

SOURCE: From Liberman (1957).

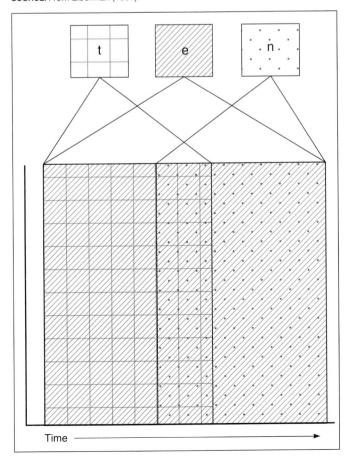

In like fashion, your vocal tract begins to move toward the /ae/ before you have articulated /b/ and toward /g/ before even finishing the /b/. This is another illustration of the problem of invariance where each phoneme changes the articulation of each other phoneme and does so depending on what the other phonemes are. The problem of invariance is made clearer by considering what we do when we whisper. Whispering changes some of the vocal characteristics of the phonemes. For example, voiced phonemes become voiceless. Yet, we typically have little trouble understanding what is being whispered to us.

In short, the sounds of language, the phonemes, vary widely as we speak them. Yet we tolerate a fair degree of variability for the sounds within a phoneme category, both when listening and decoding from sound to meaning and also when speaking, converting meaning into spoken sound. This categorical perception of phonemes in spoken language is a decision-making process that requires some cognitive control to take into account a variety of factors to make this categorization decision. Studies using functional MRI (fMRI) scans have found that the left inferior frontal sulcus (B.A. 44) is critically involved in this process (Myers, Blumstein, Walsh, & Eliassen, 2009), supporting the idea that some mental control is need to make these decisions.

9.2.4: The Effect of Context

But how do we do this? How do we tolerate the degree of variability—how do we make these decisions? The answer is context and conceptually driven processing. If we had to rely entirely on the spoken signal to figure out what was being said, then we would be processing speech in an entirely data-driven, bottom-up fashion. We would need some basis for figuring out what every sound in the word was and then retrieve it from memory. This is almost impossible, given the variability of phonemes. Instead, *context*—in this case the words, phrases, and ideas already identified—leads us to correct identification of new incoming sounds.

Pollack and Pickett (1964) cleverly demonstrated this. They recorded several spontaneous conversations, spliced out single words, then played them to people. When the words were isolated, people identified them correctly only 47% of the time. But performance improved when longer and longer segments of speech were played, because more and more supportive syntactic and semantic context was then available.

In a related study, Miller and Isard (1963) presented three kinds of sentences: fully grammatical sentences such as "Accidents kill motorists on the highways," semantically anomalous sentences such as "Accidents carry honey between the house," and ungrammatical strings such as "Around accidents country honey the shoot." They also varied the loudness of the background noise, from the difficult –5 ratio, when the noise was louder than the speech, to the easy ratio of +15, when the speech was much louder than the noise. People shadowed the strings they heard, and correct performance was the percentage of their shadowing that was accurate. As shown in Figure 9.5, accuracy improved going from the difficult to easy levels of speech-to-noise ratios. More interestingly, the improvement was especially dramatic for grammatical sentences, as if grammaticality helped counteract the background noise. For instance, at the ratio labeled 0 in the figure, 63% of the grammatical sentences were shadowed accurately, compared with only 3% of the ungrammatical strings. Indeed, even at the easiest ratio of +15, less than 60% of the ungrammatical strings could be repeated correctly.

People use their linguistic knowledge even to the point of hearing things that are not there. In a study by Richard Warren (1970), people heard sentences in which part of the sentence was removed from the recording and replaced with a cough. For example, in the sentence "The state governors met with their respective legislatures convening in the capital city," the "s" sound was replaced with a cough at the point indicated by the asterisk (*). The vast majority of people did not report that any speech sounds were missing and could not report the location of the cough when asked to do so on a printed version of the sentence later on. DeWitt and Samuel (1990) reported a similar finding with music, with people reporting hearing music notes or tones that were obscured by noise. So, when people are listening, they are actively using their knowledge to interpret what they hear.

Figure 9.5 Percentage of Strings Shadowed Correctly
SOURCE: From Miller & Isard (1963).

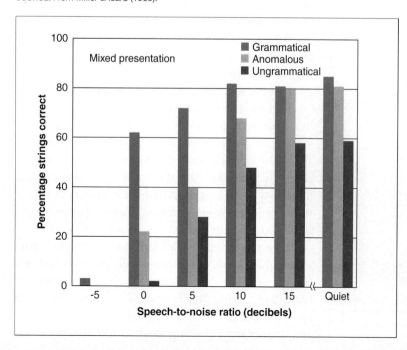

9.2.5: Top-Down and Bottom-Up Processes

More recent evidence is largely consistent with these early findings. That is, there is a combination of data-driven and conceptually driven processing in speech recognition, a position called the *integrative* or *interactive approach* (Rapp & Goldrick, 2000). At a general level, a variety of conceptually distinct language processes, from the perception of the sounds up through integration of word meanings, operate simultaneously, each having the possibility of influencing the ongoing activity of other processes. While features of the speech signal are analyzed perceptually, a listener's other linguistic knowledge is also called into play at the same time. These higher levels of knowledge and analysis operate in parallel with the phonemic analysis and help to identify the sounds and words (Dahan, 2010; Dell & Newman, 1980; Pitt & Samuel, 1995; Samuel, 2001). Moreover, to overcome the relative dearth of invariant information in the speech signal, it also appears that language perception relies heavily on characteristic knowledge of the speaker, such as whether the person speaks with a lisp (Kraljik, Samuel, & Brennan, 2008).

As a concrete example, imagine a sentence that begins "The grocery bag was. . . ." You are processing the *bag* segment of this speech signal. Having already processed the previous word to at least some level of semantic interpretation, you have developed a useful context for the sentence. To be simple about it, the grocery topic limits the number of possibilities that can be mentioned later in the sentence. Similar evidence of the role of context was reported by Marslen-Wilson and Welsh (1978) in a task that asked people to detect mispronunciations, and by Dell and Newman (1980) in a task that asked people to monitor spoken speech for the occurrence of a particular phoneme (recall also the demonstrations of context effects in Treisman's shadowing experiments, e.g., 1960, 1964).

Such results are so powerful that theories of speech recognition must account for both aspects of performance, the data driven and the conceptually driven. McClelland and Elman (1986) proposed a specific connectionist model that does exactly that. In their TRACE model, information is continually being passed among the several levels of analysis. Lexical or semantic knowledge, if activated, can alter the ongoing analysis at the perceptual level by "telling" it what words are likely to appear next; the model's predictions of what words are likely to appear are based on semantic knowledge. At the same time, phonemic information is passed to higher levels, altering the patterns of activation there (see Dell, 1986, for a spreading activation network theory of sentence production, and Tyler, Voice, & Moss, 2000, for a useful review).

9.2.6: Embodiment in Speech Perception

The perception of speech is critical for language. Intuitively this may seem like an odd place for aspects of embodied cognition to show up. However, speech perception is actually where one of the first embodied theories of cognition came from (although it was not labeled as such at the time). This is the **motor theory of speech perception** (see, e.g., Liberman, Cooper, Shankweiler, & Studdert-Kennedy, 1967; Liberman & Mattingly, 1985).

According to the motor theory of speech perception, people perceive language, at least in part, by comparing the sounds that they are hearing with how they themselves would move their own vocal apparatus to make those sounds. That is, we create embodied representations of how those sounds might be spoken to help us perceive speech. There are several lines of evidence for this idea (for an excellent review, see Galantucci, Fowler, & Turvey, 2006). According to Galantucci et al., people find it much easier to understand synthesized speech if it takes issues of coarticulation into account, rather than simply presenting a string of phonemes. Also, the parts of the cortex that are more active during speech perception overlap substantially with those involved in speech production. This is similar to the idea of mirror neurons that fire when primates observe actions of others. Finally, people find it easier to comprehend speech if they can see the person talking, which gives them more information about how the sounds are being made. It is also true that people can better understand song lyrics if they can see a person singing (Jesse & Massaro, 2010). This theory does not explain all aspects of speech perception, such as how people who could never speak can understand spoken language. However, it does illustrate how the structure of our bodies, and how we use parts of our bodies in the environment (in this case moving the air around with our vocal apparatus), influences cognition.

Related to the idea that motor programs are involved in speech perception, some evidence also exists that people activate mental motor programs just by thinking about words to themselves. Take the example of tongue twisters, which involve difficult movements of the vocal apparatus when they are spoken aloud. What is interesting is that people are likely to show evidence of articulation difficulty, such as reading times, when they are simply asked to read tongue twisters. That is, even when the people were only "saying" (inner speech) the twisters silently to themselves in their minds, the language processing system takes into account and simulates the muscle movements that would be involved if the person were actually speaking, and these simulated movements produce the normal tongue twister difficulty (Corley, Brocklehurst, & Moat, 2011; but see Oppenheim & Dell, 2010).

9.2.7: The Puzzle of Apparent Segments in Speech

As if the preceding sections were not enough to convince you of the need for conceptually driven processing, consider one final feature of the stream of spoken speech. Despite coarticulation, categorical perception, and the problem of invariance, we naively believe that words are somehow separate from each other in the spoken signal—that there is a physical pause or gap between spoken words, just as there is a blank space between printed words.

This is not true. Our intuition is entirely wrong. Analysis of the speech signal shows that there is almost no consistent relationship between pauses and the ends of words. If anything, the pauses we produce while speaking are longer within words than between words. As evidence of this, see Figure 9.6, a spectrograph recording of a

Figure 9.6 A Spectrogram Recording of a Spoken Sentence

A spectrogram from the sentence "John said that the dog snapped at him," taken from fluent spoken speech. Note that the pauses or breaks do not occur regularly at the ends of words; if anything, they occur more frequently *within* the individual words (e.g., between the /s/ and /n/ sounds, between the /p/ and /t/ sounds; compare with the end of *the* and the beginning of *dog*).

SOURCE: From Foss & Hakes (1978).

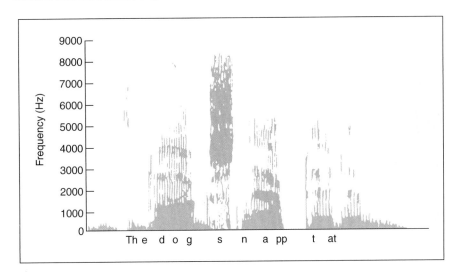

spoken sentence. Inspection of the patterns in correspondence to the words listed at the bottom illustrates the point: The pauses in the spectrograph bear no particular relationship to the ends of words. There must be other kinds of information that human cognition uses to decode spoken language.

How can our intuitions about our own language, that words are articulated as separate units, be so wrong? (Note that our intuitions about foreign languages—they sound like a continuous stream of speech—are more accurate.) How do we segment the speech stream and come to know what the words and phrases are? Part of the answer is our knowledge of words in the language and the fact that some phoneme combinations simply cannot or do not form words (Norris, McQueen, Cutler, & Butterfield, 1997). Another part of the answer to these questions is syntax, the second level of language analysis and the topic we address next.

WRITING PROMPT

Speech Sounds and Cognition

How are speech sounds used to construct language? How are phonemes the same and different across speakers, and what challenges does this produce for cognition?

 The response entered here will appear in the performance dashboard and can be viewed by your instructor.

Submit

9.3: Syntax

OBJECTIVE: Explain the rules governing syntax

At the second level of analysis we have *syntax*, the arrangement of words as elements in a sentence to show their relationship to one another. We have already studied how

---

Three Elements of Syntax

The three elements of syntax help us order words to create acceptable, well-formed sentences.

Word Order

Phrase Order

There is more to syntax than just word order, however. We also rely on the ordering of larger units such as phrases or clauses to convey meaning. Consider the following sentences:

(3) Bill told the men to deliver the piano on Monday.
(4) Bill told the men on Monday to deliver the piano.

In these examples, the positioning of the phrase "on Monday" helps us determine the intended meaning, whether the piano was to be delivered on Monday or whether Bill had told the men something on Monday. Thus, the sequence of words, phrases, and clauses contains clues to meaning, clues that speakers use to ex-press meaning and clues that listeners use to decipher meaning.

Number Agreement

sounds combine to form meaningful words. At this level of analysis, we are interested in how the words are sequenced to form meaningful strings, the study of syntax.

If you have a connotation associated with the word *syntax*, it probably is not the psycholinguistic sense of *grammar* but the "school grammar" sense instead. In school, if you said, "He ain't my friend no more," your teacher might have responded, "Watch your grammar." To an extent, this kind of school grammar is irrelevant to the psycholinguistic study of syntax. Your teacher was being *prescriptive* by teaching you what is proper or prestigious according to a set of cultural values. In another way, though, school grammar does relate to the psycholinguistic study of language; language is for expressing ideas, and anything that clarifies this expression, even arbitrary rules about *ain't* and double negatives, improves communication. (And finally, your teacher was sensitive to another level of language: People DO judge others on the quality of their speech.)

In general, we need to understand what these syntactic clues are and how they are used. We need to explore the syntactic rules that influence comprehension. We begin by looking at the underlying syntactic structure of sentences, taking a piece-by-piece approach to Chomsky's important work.

9.3.1: Chomsky's Transformational Grammar

At a general level, Chomsky intended to "describe the universal aspects of syntactic knowledge" (Whitney, 1998), that is, to capture the syntactic structures of language. He noted that language has a hierarchical phrase structure: The words do not simply occur one after the other. Instead, they come in groupings, such as "on Monday," "the men," and "deliver the piano." Furthermore, these groupings can be altered, either by moving them from one spot to another in the structure or by modifying them to express different meanings (e.g., by changing the statement into a question). These two ideas—words come in phrase structure groupings, and the groupings can be modified or transformed—correspond to the two major syntactic rule systems in Chomsky's theory.

PHRASE STRUCTURE GRAMMAR An important point in Chomsky's system is that the **phrase structure** grammar accounts for the constituents of the sentence, the word groupings and phrases that make up the whole utterance, and the relationships among those constituents.

THE INADEQUACY OF PHRASE STRUCTURE GRAMMAR ALONE Chomsky's theory relied heavily on a phrase structure approach because it captures an important aspect of language—its productivity. This kind of grammar is generative; by means of such phrase structure rules, an entire family of sentences can be generated. Furthermore, the phrase structure grammar is joined with two other components, the *lexical entries* (the words of a sentence) and the *lexical insertion rules* (the rules for putting the words into their slots). These components generated the first representation of the sentence, the **deep structure** representation. In Chomsky's view, the deep structure is an abstract syntactic representation of the sentence being constructed with only bare-bones lexical entries (words).

The deep structure is critical for two reasons:

1. It is the representation passed to the transformational fix-it rules to yield the surface structure of the sentence.
2. The deep structure is also submitted to a semantic component that "computes" meaning. This takes the deep structure and produces a semantic representation that reflects the underlying meaning.

Because of the separate treatment of the semantic component, a sentence's true meaning might not be reflected accurately in the **surface structure**. A surface structure might be **ambiguous**, or have more than one meaning. For instance, consider two classic examples of ambiguous sentences:

(5) Visiting relatives can be a nuisance.

(6) The shooting of the hunters was terrible.

A moment's reflection reveals the ambiguities. These alternative meanings are revealed when we **parse** the sentences, when we divide the sentences into phrases and groupings, much the way the phrase structure grammar does. The two meanings of sentence 5—the two deep structures—correspond to two different phrase structures. For sentence 5, the ambiguity boils down to the grammatical function of *visiting*, whether it is used as an adjective or a verb. These two grammatical functions translate into two different phrase structures (*verb + noun* versus *adjective + noun*).

Noam Chomsky, developer of the theory of transformational grammar.

Example Illustrating the Phrase Structure Grammar (Figure 9.7A–9.7D)

Let's start with the phrase structure grammar that generates the overall structure of sentences.

1. S ⟶ NP + VP
2. NP ⟶ D + N
3. VP ⟶ V + NP
4. N ⟶ superhero, criminal,… etc.
5. V ⟶ caught,… etc.
6. D ⟶ the, a,… etc.

Rewrite Rules of the Grammar

To illustrate this point, consider the following sentence:

The superhero caught a criminal.

In a phrase structure grammar, the entire sentence is symbolized by an *S*. In this grammar, the sentence *S* can be broken down into two major components, a noun phrase *(NP)* and a verb phrase *(VP)*. Thus the first line of the grammar illustrated in the above figure shows S → NP + VP, to be read, "The sentence can be rewritten as a noun phrase plus a verb phrase." In the second rule, the *NP* can be rewritten as a determiner *(D)*, an article such as *the* or *a*, plus a noun *(N)*: NP → D + N. In other words, a noun phrase can be rewritten as a determiner and a noun. In rule 3 we see the structure of a verb phrase; a *VP* is rewritten as a verb *(V)* plus an *NP*: VP → V + NP.

1 of 4

Sentence 6, however, has only one phrase structure; there is only one way to parse it: {[the shooting of the hunters] [was terrible]}. Thus sentence 6 is ambiguous at the level of surface structure. Because phrase structure rules can generate such ambiguous sentences, Chomsky felt this illustrated a limitation of the pure phrase structure approach: There must be something missing in the grammar. If it were complete, it would not generate ambiguous sentences.

A second difficulty Chomsky pointed out involves examples such as the following:

(7a) Pierre bought a fine French wine.

(7b) A fine French wine was bought by Pierre.

According to phrase structure rules, there is almost no structural similarity between these two sentences. Yet they mean nearly the same thing. The phrase structure approach does not capture people's intuitions in which active and passive paraphrases are identical at the level of meaning.

TRANSFORMATIONAL RULES Chomsky's solution to such problems was to postulate a second component to the grammar, a set of **transformational rules** that handle the many specific surface forms that can express an underlying idea. These transformational rules convert the deep structure into a surface structure, a sentence ready to be spoken.

9.3.2: Limitations of Transformational Grammar

A great deal of early psycholinguistic research was devoted to structural aspects of language. For example, many studies focused on testing the *derivational complexity hypothesis*. This hypothesis suggests that the difficulty of comprehending a sentence is directly related to the number of grammatical transformations applied. So, if a deep structure has two transformations applied to it,

Applying the Transformational Rules

By applying different transformations, we can form an active declarative sentence, a passive voice sentence, a question, a negative, a future or past tense, and so on. With still other transformations, phrases can exchange places, and words can be inserted and deleted.

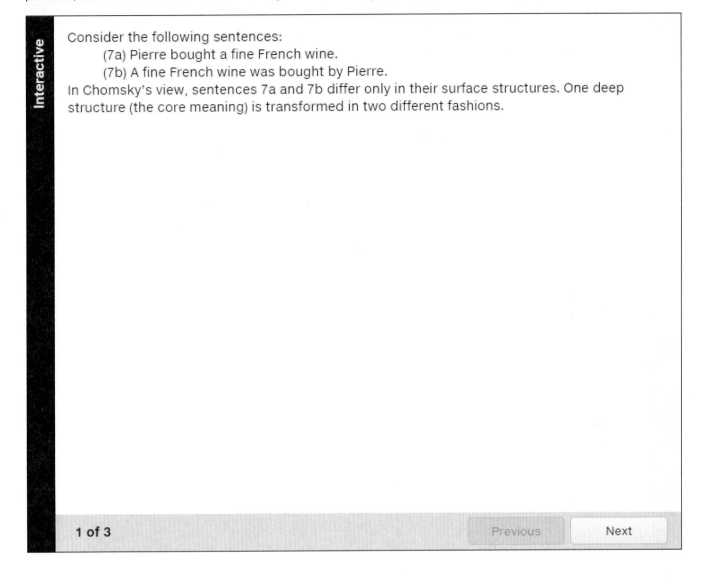

it is more difficult to comprehend than if only one transformation is applied. Some results tended to support this theory (e.g., Palermo, 1978). However, on the whole, psychology became dissatisfied with this approach. Work by Fodor and Garrett (1966) was especially instrumental in dimming the enthusiasm. They noted that much of the support for the derivational complexity hypothesis failed to control potentially important factors. For instance, a derivationally more complex sentence generally has more words in it than a simpler one (contrast sentences 9a and 9c).

Moreover, there was a *metatheoretical point of view*. To oversimplify a bit, the major components were said to be the syntactic rules for generating first a deep then a surface structure. Meaning was literally off to the side. This depicts the difficulty psychology had with linguistic theory: It seemed that meaning was secondary to syntax. It is almost as if the theory, as it was applied to language use, suggested that we first make up our minds about which phrase constituents we are going to use and only then decide what we are going to talk about. To psychologists concerned with how we use language to express meaning, this theory seemed wrong.

This oversimplification made it seem as if Chomsky encouraged linguists to avoid meaning. It was not that extreme, as Chomsky repeatedly emphasized the importance of both syntax and semantics. He pointed out that even a perfectly grammatical, syntactically acceptable sentence may be semantically anomalous. His most famous example is "Colorless green ideas sleep furiously." The sentence is grammatically acceptable—consider a sentence with completely parallel syntax, such as "Tired young children sleep soundly." But Chomsky's sentence has no meaning in any ordinary sense.

Still, psychologists felt Chomsky's work never dealt with meaning satisfactorily. Furthermore, trying to apply his theory to the actual use of language—turning his competence-based theory into a performance theory of language production and comprehension—only made it more apparent that a different approach was needed.

9.3.3: The Cognitive Role of Syntax

From a psychological perspective, what is the purpose of syntax? Why follow syntactic rules? Essentially, we use syntax to determine or find meaning. If an infinite number of sentences are possible, then the one sentence being said to us right now could be about *anything*. Syntax helps listeners extract meaning and helps speakers convey it.

Bock's (1982) article on a cognitive psychology of syntax discusses several important issues that psycholinguistics must explain. She notes that the syntactic burden falls more heavily on the speaker than the listener. When you have to produce a sentence rather than comprehend it, you must create a surface structure, a string of words and phrases to communicate your idea as well as possible. Thus, syntax is a feature of language related to the speaker's mental effort.

AUTOMATIC PROCESSING Two points Bock raises should illustrate some issues in the psycholinguistic study of syntax. First, consider the issues of automatic and conscious processes as they apply to language production. *Automatic processes* are the product of a high degree of practice or overlearning. Bock noted that several aspects of syntactic structure are consistent with the idea of automaticity. For instance, children rely heavily on regular word orders even if the native language they are learning has irregular word order. By relying repeatedly on the same syntactic frames, children can generate and use them more automatically. Similarly,

---

Planning and Production of Speech

Language does not emerge from thought fully formed and ready to go. There are a series of steps that go into the transformation from idea to spoken utterance.

> Ferreira and Swets (2002) demonstrated this tendency for language order to be influenced by memory retrieval in a clever experiment by asking people to state the answer to easy and hard addition problems, in sentence frames like "The answer is __." They found that people delayed nearly a half a second more before they started talk-ing when the problem was hard (e.g., 23 + 68) than when it was easy (e.g., 21 + 22). Clearly speech production is sensitive to the ease of memory retrieval.

adults tend to use only a few syntactic structures with regularity, suggesting that those structures can be called into service rapidly and automatically.

Interestingly, the syntax you use can be influenced by a previous sentence, quite literally *syntactic priming* (Bock, 1986; West & Stanovich, 1986). Bock's later work (Bock & Griffin, 2000) found evidence that a particular syntactic construction can prime later ones up to lag 10 (i.e., with 10 intervening sentences), and there is some evidence that it can last up to a week later (Kaschak, Kutta, & Schatschneider, 2011). Syntactic priming has even been found in written language and American Sign Language (Branigan, Pickering, & Cleland, 1999; Branigan, Pickering, Stewart, & McLean, 2000; Hall, Ferreira, & Mayberry, 2015). Interestingly, syntactic priming is not affected by anterograde amnesia, which affects declarative memory (Ferreira, Bock, Wilson, & Cohen, 2008).

PLANNING In Bock's second point, she reviewed evidence of an important interaction between syntax and meaning. In general, we tailor the syntax of our sentences to the accessibility of the lexical or semantic information being conveyed, known as the **given-new strategy** (Clark & Clark, 1977). Phrases that contain more accessible information, or given information, tend to occur earlier in sentences. This is information that is either well known or has been recently discussed in a discourse (and so is more available). By comparison, less accessible, newer concepts tend to come later, possibly to provide extra time for retrieval (but see Clifton & Frazier, 2004, for an alternative account).

9.3.4: Prosody

Syntax is an important aspect of understanding the structure of an utterance, in terms of the role that different

---

Importance of Prosody

Language is typically not delivered in a monotone, but involves a moving up and down in pitch, somewhat like a melody. In fact, prosody in language may use some of the same cognitive processes as music (Perrachione, Fedorenko, Vinke, Gibson, & Dilley, 2013).

Interactive

Conveys Meaning in Language

Prosody conveys meaning in language, although a different kind of meaning than the semantic meanings of the words themselves. For example, the sentences "Those are my shoes." and "Those are my shoes?" contain exactly the same words but are spoken with different prosodies. One conveys a statement and the other a question, often with a rise in pitch at the end of the sentence. If you listen to yourself say these two sentences, you will notice a difference in their pitch pattern, even though they contain the same words and syntactic structure. These differences in pitch information are prosody and convey different information about the intent of the speaker.

Serves to Help Speakers Place Emphasis on Concept

words play and how they relate to an underlying meaning. However, word order is not the only way to convey information about meaning and intent. Another major clue, particularly for spoken language, is **prosody**, which is the change in pitch (either higher or lower) across the phonemes and morphemes of an utterance to convey different meanings.

WRITING PROMPT

Syntax Structure and Cognition

How does the structure of syntax in language help cognition create and build mental representations of the ideas that language is conveying? Is it possible to have a language without syntax?

▶ The response entered here will appear in the performance dashboard and can be viewed by your instructor.

Submit

9.4: Lexical Factors

OBJECTIVE: Analyze lexical factors to understand the meaning of language

We now turn to lexical and semantic factors, which relate to the level of meaning in language. In particular, we refer to retrieval from the *mental lexicon*, the mental dictionary of words and their meanings. After rapid perceptual and pattern recognition processes, the encoded word provides access to the word's entry in the lexicon and to the semantic representation of the concept. The evidence you have read about throughout this course, such as results from the Stroop and the lexical decision tasks, attests to the close relationship between a word and its meaning and the seemingly automatic accessing of one from the other. Recall in the Stroop task that seeing the word *red* printed in green ink triggers an interference process with naming the ink color, clear evidence that *red* is processed to the level of meaning (MacLeod, 1992). Likewise, the lexical decision task does not require that you access the word's meaning but only that you identify a letter string as a genuine word. Nonetheless, identifying *doctor* as a word primes your decision to *nurse*.

Prove It

Speech Errors

Work by Fromkin (1971), Garrett (1975), and others (e.g., Ferreira & Humphreys, 2001; for work on error monitoring, see Hartsuiker & Kolk, 2001) has tabulated and made sense of speech errors that occur when we substitute or change sounds, syllables, words, and so on. Speech errors are not random but are quite lawful. For instance, when we make an exchange error, the exchange is between elements at the same linguistic level; initial sounds exchange places with other initial sounds, syllables with syllables, words with words (e.g., "to cake a bake"). If a prefix switches places, its new location will be in front of another word, not at the end.

Collect a sample of speech errors, say, from radio news broadcasters or your professors' lectures, then analyze them in terms of the linguistic level of the elements involved and the types of errors such as these (intended phrase in parentheses):

Sample of Speech Errors	
Shift	She decide to hits it. (decides to hit it)
Exchange	Your model renosed. (your nose remodeled)
Perseveration	He pulled a pantrum. (tantrum)
Blend	To explain clarefully. (clearly/carefully)

9.4.1: Morphemes

A *morpheme* is the smallest unit of language that has meaning. To return to an earlier example, the word *cars* is composed of two morphemes: *Car* refers to a concept and a physical object, and *-s* is a meaningful suffix, denoting "more than one of." Likewise, the word *unhappiness* is composed of three morphemes: *happy* as the base concept, the prefix *un-* meaning "not," and the suffix *-ness* meaning "state or quality of being." In general, morphemes that can stand on their own and serve as words are called *free morphemes*, such as *happy, car,* and *legal*, whereas morphemes that need to be linked onto a free morpheme are called *bound morphemes*, such as *un-, -ness,* and *-s*. Although the concept of a morpheme is important, there is some debate as to whether the meaning of more common words such as *unhappiness* may be stored directly in memory or "computed" from the three morphemes (see Carroll, 1986; Whitney, 1998).

9.4.2: Lexical Representation

Think about the word *chase* as an example of how free morphemes might be represented in the mental lexicon. The representation of *chase* must specify its meaning—indicate that it means "to run after or pursue, in hopes of catching." Like other semantic concepts, *chase* can be represented in reference to related information, like *run, pursue,* the idea of *speed*. Given this, along with what you know about events in the real world from schemas and scripts, you can easily understand a sentence like

(11) The policeman chased the burglar through the park.

Embodied Aspect to Lexical Memory

Lexical knowledge can include information that can capture embodied characteristics of cognition, with certain parts of the brain becoming more active for certain types of words.

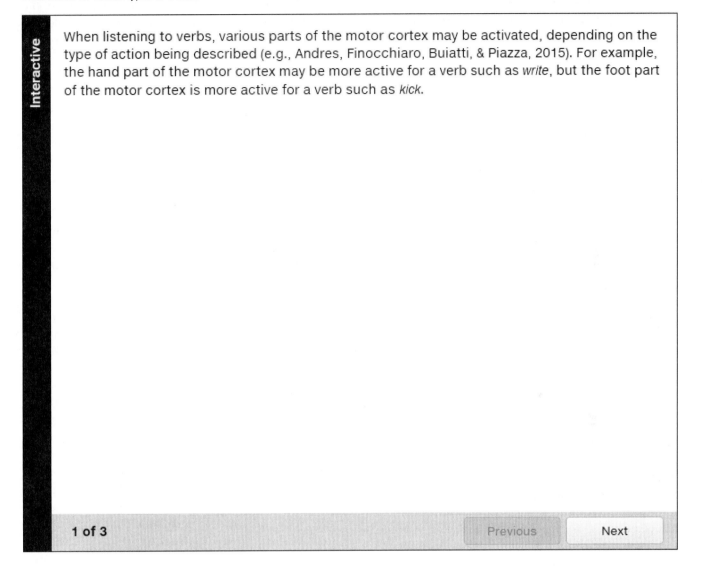

From a more psycholinguistic perspective, however, you know more about *chase* than just its basic meaning. For one thing, you know it is a verb specifying a kind of physical action. Related to that, you have a clear idea of how *chase* can be used in sentences, the kinds of things that can do the chasing, and the kinds of things that can be chased (e.g., McKoon & Macfarland, 2002). Imagine, then, that your lexical representation of *chase* also includes this knowledge; *chase* requires some animate thing to do the chasing, some other kind of thing to be chased, and a location where the chasing takes place.

9.4.3: Polysemy

Whereas our understanding of words like *chase* is very clear, it is not too long before we run into cases of **polysemy**, the fact that many words in a language may have multiple meanings. The task of the language processing system is to figure out which meaning is the intended one. Though a word may be polysemous, not all meanings are equal. Generally, there is one primary meaning that people typically would think of first when they hear the word or that would likely be listed first in a dictionary. This is the *dominant* meaning of a word. Other meanings then would be the *subordinate* meanings. So, take a simple word like *run*. The dominant meaning has something to do with using our legs to move fast. However, there are many subordinate meanings, too, such as having a run in our stockings, a movie having a run at the theater, having our nose run, to cut and run (retreat), to run our engine, to watch paint colors run, and so on. The way we distinguish which specific meaning to use from the mental lexicon would depend on the context a word is in.

POLYSEMY AND PRIMING Let's consider an example of how context can resolve polysemy to determine the intended meaning. As one example, the word *count* is ambiguous by itself. Putting the word in a sentence may not help: "We had trouble keeping track of the count." You still cannot tell the intended meaning. What is missing is context, some conceptual framework to guide the interpretation of the polysemous word. With an adequate context, you can determine which sense of the word *count* is intended in these two sentences:

My dog wasn't included in the final count.

The vampire was disguised as a handsome count.

These sentences, taken from Simpson's (1981, 1984) work on polysemy, point out the importance of context: Context can help determine the intended meaning. With neutral contexts such as the "We had trouble" sentence, word meanings are activated as a function of their dominance. The number sense of *count* is dominant, so that meaning is more activated. But a context that biases the interpretation in any way results in a stronger activation for the biased meaning. With *vampire*, you activated the meaning of *count* related to nobility and Count Dracula (see also Balota & Paul, 1996; Klein & Murphy, 2002; Piercey & Joordens, 2000; but cf. Binder, 2003; Binder & Rayner, 1998).

The resolution of lexical ambiguity with polysemous words is important for successful comprehension. If you do not get the intended meaning of a word, then you will not get the intended message. It appears that ambiguity resolution works, in part, in a two-stage process. When people encounter an ambiguous word, they activate all the meanings, at least to some degree. Then in the second stage, they deactivate the inappropriate ones based on the information from the rest of the discourse context. However, not everyone does this equally well. Work by Gernsbacher and Faust (1991) shows that good readers suppress inappropriate meanings faster. By comparison, poor readers maintain multiple meanings for a much longer period, which may contribute to their problems.

CONTEXT AND ERPS Let's consider another example of the effects of context, an offshoot of the Kounios and Holcomb work with event-related potentials (ERPs). In one study, Holcomb, Kounios, Anderson, and West (1999; see also Laszlo & Federmeier, 2009; Lee & Federmeier, 2009; Sereno, Brewer, & O'Donnell, 2003) recorded ERPs in a simple sentence comprehension task. People saw sentences one word at a time and were asked to respond after seeing the last word, with "yes" if the sentence made sense and "no" if it did not. The experimental sentences varied along two dimensions, whether the last word was concrete or abstract and whether it was congruent with the sentence meaning or anomalous (i.e., made no sense). As an example, "Armed robbery implies that the thief used a weapon" was a concrete–congruent sentence. Substituting *rose* for *weapon* made the sentence concrete but anomalous. Likewise, "Lisa argued that this had not been the case in one single instance" was an abstract–congruent sentence, and substituting *fun* for *instance* made it abstract–anomalous.

Figure 9.8 shows some of the ERP patterns obtained. In the left panel, you see the "normal" ERP patterns for the congruent, sensible sentences.

The three profiles, from top to bottom, came from the three midline electrode sites shown in the schematic drawing (frontal, central, and parietal). In the right panel are the ERP patterns when the sentences ended in an anomalous word. Notice first in the left panel that the solid and dotted functions, for concrete and abstract sentences, tracked each other very closely: Whatever neural mechanisms operated during comprehension, they generated similar ERP patterns. But now make a left-to-right comparison of the patterns, seeing the differences in the right panel when the sentences ended in a nonsensical, anomalous word (*rose* in the armed robbery sentence, for example). Here, there were marked changes in the ERP profiles. For example, at the central location, there was a steady downward trend (in the positive direction, in terms of electrical potentials) for sensible sentences but a dramatic reversal of direction for anomalous words.

In short, the neural mechanisms involved in comprehension generated dramatically different patterns when an anomalous word was encountered. The mismatch between the context, the already-processed meaning of the sentence, and the final word yielded not only an overt response (the response indicating "no, that sentence makes no sense"), but also a neural response signifying the brain-related activity that detected the anomalous ending of the sentence. (Don't get confused about directions here. The functions underneath the gridline are electrically positive, so deflection upward in these graphs is a deflection toward the negative, a deflection going in a negative direction; this is what the *N* in **N400** signifies, a "negative going" pattern.) Even at the level of neural functioning, there is a rapid response to nonsensical ideas that follow sensible context, a kind of "something's wrong here" response that the brain makes some 400 ms after the nonsensical event.

WRITING PROMPT

Lexical Factors in Memory Retrieval

How are words and other lexical elements stored and used by cognition? How are they indexed for memory retrieval? How are they combined to form new words?

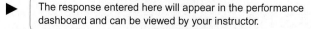

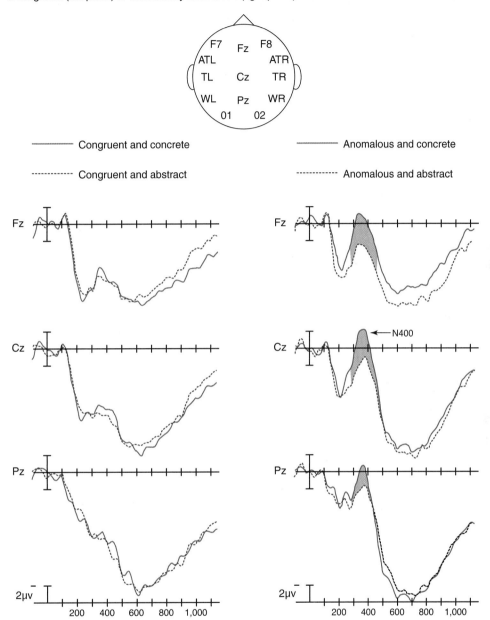

Figure 9.8 ERPs in Simple Comprehension Task
ERP profiles at three midline sites (frontal [F_z], central [C_z], parietal [P_z]) for sentences ending with a congruent (left panel) or contextually anomalous (right panel) word.

9.5: Semantics

OBJECTIVE: Compare semantics and syntax to understanding intended meaning

In this section, we consider issues of how the language processing system knows what role a word or concept is playing in a sentence. This approach is called **case grammar**. The ideas originally came from Fillmore (1968). The basic idea is that the semantic analysis of sentences involves figuring out what semantic role is being played by each word or concept in the sentence and computing sentence meaning based on those semantic roles. Two sample sentences illustrate this:

(12) The key will open the door.

(13) The janitor will open the door with the key.

Fillmore pointed out that syntactic aspects of sentences—which words serve as the subject, direct object, and so on—are irrelevant to sentence meaning. For example, in sentences 12 and 13 the word *key* plays different grammatical roles: subject of the sentence in 12 but object of the preposition in 13. For Fillmore, focusing on this difference misses a critical point for language. Regardless of its different grammatical roles, the key is doing exactly the same thing in both cases, playing the same semantic role of *instrument*. A purely syntactic analysis misses this, but a semantic analysis captures it perfectly.

9.5.1: Case Grammar

Fillmore's theory of case grammar proposes that sentence processing involves a semantic parsing that focuses on the semantic roles played by the content words in the sentences. These semantic roles are called *semantic cases*, or simply **case roles**. Thus, *door* is the recipient or patient of the action of *open* in sentences 12 and 13; *janitor* is the agent of *open*; *key* is the instrument; and so on. Stated simply, each content word plays a semantic role in the meaning of the sentence. That role is the word's semantic case.

9.5.2: Interaction of Syntax and Semantics

Note that semantic factors do not stand alone in language, just as syntactic factors are not independent of semantics. Syntax is more than just word and phrase order rules; it is a clue to understanding sentences. For example, O'Seaghdha's (1997) evidence shows separable effects of syntactic assignment and semantic integration of word meanings, with syntactic processes occurring before semantic integration. His results, based on RTs, are largely consistent with those in other studies (e.g., Peterson, Burgess, Dell, & Eberhard, 2001, on how we process

---

Analyzing Case Grammar (Figure 9.9)

The significant—indeed, critical—point about such a semantic parsing is that it relies on people's existing semantic and lexical knowledge, their knowledge of what kinds of things will open, who can perform the opening, and so on.

Interactive

Reconsider the *chase* sentence 11, "The policeman chased the burglar through the park," and three variations, thinking of the content words in terms of their semantic roles:

(14) The mouse chased the cat through the house.
(15) His insecurities chased him even in his sleep.
(16) *The book chased the flower.

Your lexical and semantic knowledge of *chase* is that some animate being does the chasing, the agent case. Some other thing is the recipient of the chasing, the patient, but that thing need not be animate, just capable of moving rapidly (e.g., you can chase a piece of paper being blown by the wind). On this analysis, it is clear that sentence 11 conforms to the normal situation stored in memory, so it is easy to comprehend. However, sentence 14 mismatches the typical state of affairs between mice and cats. Nonetheless, either of these creatures can serve as the required animate agent of the relation *chase*, so sentence 14 is sensible. Because of other semantic knowledge, you know that sentence 15 violates the literal meaning of *chase* but could still have a nonliteral, metaphorical meaning. But your semantic case analysis provides the reason sentence 16 is unacceptable. A book is inanimate, so it mismatches the required animate agent role for *chase*; book cannot play the role of agent for *chase*. Likewise, *flower* seems to violate the movable restriction on the patient case for *chase*.

idioms), including ERP studies of syntactic and semantic processing (Ainsworth-Darnell, Shulman, & Boland, 1998; Friederici, Hahne, & Mecklinger, 1996; Osterhout, Allen, McLaughlin, & Inoue, 2002). And, syntax in speech production is sensitive to a word's accessibility. Words that can be easily retrieved right now tend to appear earlier in a sentence.

SEMANTIC FOCUS Likewise, semantic factors refer to more than just word and phrase meanings because different syntactic devices can be clues to meaning. To anticipate just a bit, note how syntactic differences in the following sentences influence the semantic interpretation:

(17a) I'm going downtown with my sister at four o'clock.

(17b) It's at four o'clock that I'm going downtown with my sister.

(17c) It's my sister I'm going downtown with at four o'clock.

Sentences 17b and 17c differ subtly from 17a in the focus of the utterance. The focus of each sentence is different, so each means something slightly different. Imagine how inappropriate sentence 17c would be, for instance, as a response to the question "Did you say you're going downtown with your sister at three o'clock?" Our judgments about appropriateness make an important point: Our theories of language performance must be as sophisticated as our own knowledge of language. We are sensitive to the focus or highlighted aspects of sentences and subtleties of the ordering of clauses, so a theory of language must reflect this in a psychologically relevant way.

SEMANTICS CAN OVERPOWER SYNTAX Semantic features can do more than alter the syntax of sentences. Occasionally semantics can overpower syntax.

Conceptually Driven Processing of Language (Figure 9.10–Figure 9.11)

Let's focus on a classic study by Fillenbaum (1974).

> Fillenbaum presented several kinds of sentences and asked people to write paraphrases that preserved the original meaning. Ordinary *threat* sentences such as "Don't print that or I'll sue you" were then reordered into *perverse* threats, such as "Don't print that or I won't sue you." Regular *conjunctive* sentences such as "John got off the bus and went into the store" were then changed into *disordered* sentences, such as "John went into the store and got off the bus."

9.5.3: Evidence for the Semantic Grammar Approaches

A major prediction of *semantic grammar theory* can be stated in two parts. First, comprehenders begin to analyze a sentence immediately, as soon as the words are encountered. Second, this analysis assigns each word to a particular semantic case role, with each assignment contributing its part to overall sentence comprehension.

BILINGUALISM What happens if you know more than one language? How does cognition deal with that? Generally, when people are multilingual, cognition needs to keep the words, syntax, idioms, and the like for languages separate enough that they do not intrude on one another and intermix in the person's speech. At the same time, the words and phrases in one language need to map onto the same underlying ideas in the other languages. In general, human cognition is able to handle this task quite well. People who learn a second language early in life even show semantic priming effects across languages (Perea, Duñabeitia, & Carreiras, 2008), suggesting that the different words tap into the same underlying semantic knowledge base.

The need to keep two or more languages separate, but still be fluent in both, requires increased attention and cognitive control to mentally keep the languages from interfering with one another. This need for bilinguals to engage in such cognitive control spills over to other cognitive abilities that require cognitive control, such as showing larger negative priming effects, because bilinguals are more effective at suppressing salient but incorrect information (Treccani, Argyri, Sorace, & Della Salla, 2009).

Predictions of Semantic Grammar Theory (Figure 9.12)

Language is processed online, as it happens. This can sometimes lead people to initially make mistakes in parsing language, which cognition then needs to correct.

Interactive

As an example, read sentence 18:

(18) After the musician had bowed the piano was quickly taken off the stage.

Your analysis of this sentence proceeds easily and without disruption; it is a fairly straightforward sentence.

Thus, there seems to be an overall cognitive benefit to knowing more than one language. Bilinguals show superior performance on intelligence tests compared to otherwise similar monolinguals (Lambert, 1990), including doing better on tests of their primary language (Bialystok, 1988; van Hell & Dijkstra, 2002). Certain evidence exists that bilingualism can modestly protect people from cognitive declines associated with aging (Bialystok, Craik, & Luk, 2012). Yet, the evidence for a cognitive and intellectual advantage for bilingualism is mixed, with some studies finding no evidence for a difference (Ljungberg, Hansson, Andrés, Josefsson, & Nilsson, 2013; Paap & Greenberg, 2013; Ratiu & Azuma, 2015). Regardless, it is clear that bilingualism does not impair cognition and intelligence as was once thought.

Although bilingualism can have advantages for cognition, learning a second language as an adult is more difficult and requires more cognitive attention and effort. For example, some evidence suggests that when people are immersed in a second language they learned as an adult, their first language is actively suppressed (Linek, Kroll, & Sunderman, 2010), which is not observed with people who grew up speaking more than one language.

WRITING PROMPT

Meaning and Semantics

What kinds of meaning are captured by semantics? How do these differ from the meaning conveyed by the syntax of a language?

The response entered here will appear in the performance dashboard and can be viewed by your instructor.

Submit

9.6: Brain and Language

OBJECTIVE: Summarize the processes the brain uses to comprehend language

One of the most fruitful areas of research on the brain–cognition relation is work on language processing. So, in this section, we discuss aspects of language processing that have strong neural components. These include a consideration of people with intact brains, as well as the disruption of language processing in people who have suffered some sort of brain damage.

9.6.1: Language in the Intact Brain

With the advent of modern imaging methods, we have begun to learn an extraordinary amount about how the brain processes language from neurologically intact people. Consider a representative study looking at people's sensitivity to the syntactic structure of sentences. Osterhout and Holcomb (1992) presented sentences to people and recorded the changes in their brain wave patterns (ERPs) as they comprehended. In particular, they examined ERP patterns for sentences that violated syntactic or semantic expectations, comparing these with the patterns obtained with control sentences. When sentences ended in a semantically anomalous fashion ("The woman buttered her bread with socks"), a significant N400 ERP pattern was observed, much as reported in Kounios and Holcomb's (1992) study of semantic relatedness (see Figure 9.8). But when the sentence ended in a syntactically anomalous fashion ("The woman persuaded to catch up"), a strong **P600** pattern occurred (a positive electrical potential) 600 ms after the anomalous word *to* was seen (Figure 9.13).

Figure 9.13 ERP Patterns for Sentences Violating Syntactic or Semantic Expectations

Mean ERPs to syntactically acceptable sentences (solid curve) and syntactically anomalous sentences (dotted curve). The P600 component, illustrated as a downward dip in the dotted curve, shows the effect of detecting the syntactic anomaly. In this figure, positive changes go in the downward direction.

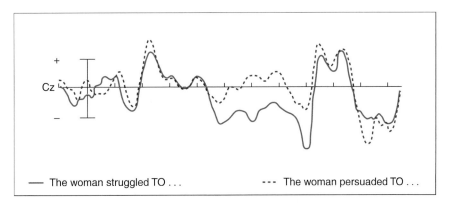

Research Techniques for Language Processing

A wealth of evidence illustrates the importance to cognitive science of such imaging and neuropsychological techniques and strongly suggests that the research on language processing will increasingly feature techniques such as imaging and ERP methods.

Interactive

Learning Language

As people acquire the meaning of words of a language, they are using both *experiential* (information derived from people's senses) and *distributional* (statistical occurrences of a word) knowledge to figure out word meanings and roles in the language (Andrews, Vigliocco, & Vinson, 2009). That is, people are using knowledge of what they actually experience during individual events, as well as how often and in which context various words and word combinations may occur. McCandliss, Posner, and Givon (1997) captured this basic idea as they taught people a new, miniature artificial language and recorded ERPs during learning. Early in training, words in the new language showed ERP patterns typical of nonsense material. People were focusing entirely on the experiential information of hearing the novel "words." However, after 5 weeks of training, the ERP patterns looked like those obtained with English words in which people had enough experience to develop a distributional understanding of the novel language to begin using that type of knowledge as well. Furthermore, left hemisphere frontal areas reacted to semantic aspects of the language (distributional), whereas posterior areas were sensitive to the visual characteristics of the words, the orthography (experiential). Thus, changes occur in the neural, and presumably cognitive, mechanisms involved as one learns a new language.

Syntactic Processing

Right Hemisphere Language

Individual Differences

This confirms the important and seemingly separate role of syntactic processing during language comprehension.

9.6.2: Aphasia

A large literature exists on brain-related disorders of language, based on people who through the misfortune of illness or brain injury have lost the ability to use language. Formal studies of such disorders date back to the mid-1800s, although records dating back to 3500 B.C. mention language loss caused by brain injury (see McCarthy & Warrington, 1990).

The disruption of language caused by a brain-related disorder is called **aphasia**. Aphasia is always the product of some physical injury to the brain sustained either in an accident or a blow to the head or in diseases and medical syndromes such as stroke. A major goal in neurology is to understand the aphasic syndromes more completely so that people who suffer from aphasia may be helped more effectively. From the standpoint of cognitive neuroscience, the language disruptions of aphasic patients can also help us understand language and its neurological basis.

Although there are many kinds of aphasias, with great variety in their effects and severity, three basic forms are the most common:

1. Broca's aphasia
2. Wernicke's aphasia
3. Conduction aphasia

BROCA'S APHASIA As described by Kertesz (1982), **Broca's aphasia** is characterized by severe difficulties in

Table 9.3 Brain-Related Disruptions of Language and Cognition

Table 9.3 provides a list and short explanation of these disruptions and some others you have already encountered.

Disorder	Disruption of
Language Related	
Broca's aphasia	Speech production, syntactic features
Wernicke's aphasia	Comprehension, semantic features
Conduction aphasia	Repetition of words and sentences
Anomia (anomic aphasia)	Word finding, either lexical or semantic
Pure word deafness	Perceptual or semantic processing of auditory word comprehension
Alexia	Reading, recognition of printed letters or words
Agraphia	Writing
Other Symbolic Related	
Acalculia	Mathematical abilities, retrieval or rule-based procedures
Perception, Movement Related	
Agnosia	Visual object recognition
Prosopagnosia	(Visual) face recognition
Apraxia	Voluntary action or skilled motor movement

producing speech. It is also called *expressive* or *production aphasia*. Patients with Broca's aphasia show speech that is hesitant, effortful, and phonemically distorted. Aside from stock phrases such as "I don't know," such patients generally respond to questions with one-word answers. If words are strung together, there are few if any grammatical markers, such as bound morphemes like *-ing*, *-ed*, and *-ly*. In less severe cases, the aphasia may be limited to more complex aspects of language production, such as the production of verb inflections (Faroqi-Shah & Thompson, 2007). Interestingly, such patients typically show less (or even no) impairment of comprehension for both spoken and written language.

French neurosurgeon Pierre Broca first described this syndrome in the 1860s and also identified the damaged area responsible for the disorder. The site of the brain damage, an area toward the rear of the left frontal lobe, is therefore called **Broca's area**. Broca's area lies adjacent to a major motor control center in the brain and is shown in Figure 9.14.

WERNICKE'S APHASIA Loosely speaking, the impairments in **Wernicke's aphasia** are the opposite of those in Broca's aphasia. In patients affected by Wernicke's aphasia, comprehension is impaired, as are repetition, naming, reading, and writing, but the syntactic aspects of speech are preserved. It is sometimes called *receptive* or *comprehension aphasia*. In this syndrome, there may be unrecognizable content words; recognizable but often inappropriate semantic substitutions; or *neologisms*, invented nonsense words. What is striking is that victims of this disorder are sometimes unaware of the aphasia.

German investigator Carl Wernicke identified this disorder, and the left-hemisphere region that is damaged, in 1874. This region is thus known as Wernicke's area. The area, toward the rear of the left temporal lobe, is adjacent to the auditory cortex, in the left temporal lobe, and is shown in Figure 9.15.

This is a very different area with very different abilities than Broca's area in the frontal lobe. Note also that this disorder demonstrates a *double dissociation*, a basic distinction at the level of brain organization between syntax and semantics (see Breedin & Saffran, 1999, for a case study showing loss of semantic knowledge but preserved syntactic performance).

CONDUCTION APHASIA Much less common than Broca's and Wernicke's aphasias, **conduction aphasia** is a narrower disruption of language ability. Both Broca's and Wernicke's areas seem to be intact in conduction aphasia, and people with conduction aphasia can understand and produce speech quite well. Their language impairment is an inability to repeat what they have just heard. In intuitive terms, the intact comprehension and production systems seem to have lost their normal connection or linkage. Indeed, the site of the brain lesion in conduction aphasia appears to be the primary pathway between Broca's and Wernicke's areas, called the *arcuate fasciculus* (Geschwind, 1970). Quite literally, the pathway between the comprehension and production areas is no longer able to conduct the linguistic message.

ANOMIA Another type of aphasia deserves brief mention here because it relates to the separation of the

Figure 9.14 Broca's Area

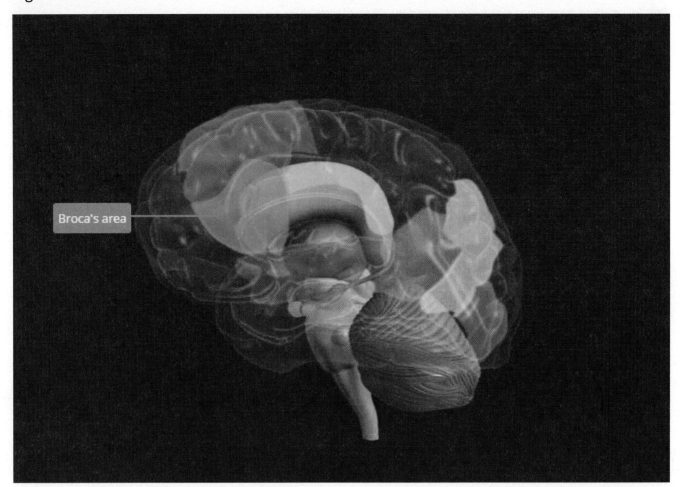

semantic and lexical systems discussed earlier. **Anomia**, or **anomic aphasia**, is a disruption of word finding, an impairment in the normal ability to retrieve a semantic concept and say its name. In anomia, some aspect of the normally automatic semantic or lexical components of retrieval has been damaged. Although moderate word-finding difficulty can result from damage almost anywhere in the left hemisphere, full-fledged anomia seems to involve damage especially in the left temporal lobe (Coughlan & Warrington, 1978; see McCarthy & Warrington, 1990, for details). There is a superficial similarity between anomia and the tip-of-the-tongue (TOT) effect. Several researchers (Geschwind, 1967; Goodglass, Kaplan, Weintraub, & Ackerman, 1976) have found no evidence among anomic patients of the partial knowledge that characterizes a TOT effect. Evidence also indicates that anomia can involve retrieval blockage only for the lexical component of retrieval, leaving semantic retrieval of the concept intact (Kay & Ellis, 1987). This finding, along with that of other cases (e.g., Ashcraft, 1993), suggests preserved semantic retrieval but a blockage in finding the lexical representation that corresponds to the already retrieved semantic concept.

OTHER APHASIAS As Table 9.3 shows, a variety of highly specific aphasias are also possible. Although most of these are quite rare, they nonetheless give evidence of the separability of several aspects of language performance. For instance, in **alexia** (or *dyslexia*), there is a disruption of reading without any necessary disruption of spoken language or aural comprehension. In **agraphia**, conversely, the patient is unable to write. Amazingly, a few reports describe patients with alexia but without agraphia—in other words, patients who can write but cannot read what they have just written (Benson & Geschwind, 1969). In **pure word deafness**, a patient cannot comprehend spoken language, although he or she is still able to read and produce written and spoken language.

There is documentation for even more specific forms of aphasia than these—for instance, difficulties in retrieval of verbs in written but not spoken language (Berndt & Haendiges, 2000) and difficulties in naming just visual stimuli, without either generalized visual *agnosia* or generalized anomia (Sitton, Mozer, & Farah, 2000). Thus, cognitive processing is quite complex, made of many different parts that need to work together seamlessly for language

Figure 9.15 Wernicke's Area

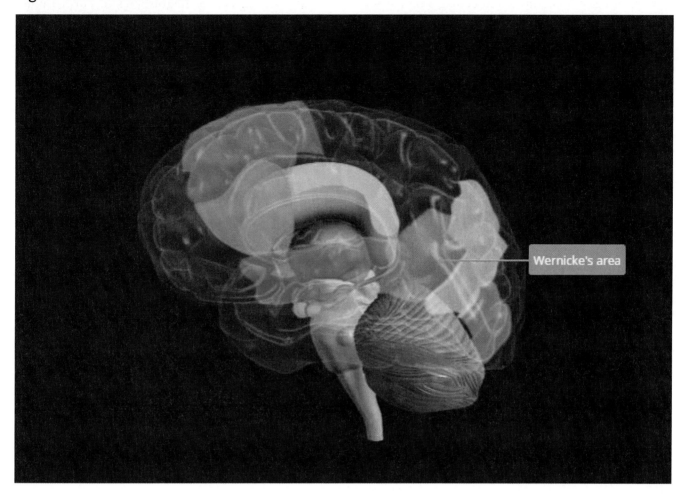

to work. When any one of those parts is faulty, unique language processing deficits can occur.

RIGHT HEMISPHERE DAMAGE Despite the fact that most of the aphasias discussed here involve processing in the left hemisphere of the cortex, there is also evidence of the right hemisphere's contribution to language comprehension and production. (See Beeman & Chiarello, 1998, for a useful overview of the complementary right- and left-hemisphere language processes.) Work by Beeman (1993, 1998) suggests that a problem of people with right

Table 9.4 Classic Impairments in Broca's and Wernicke's Aphasias

Review Table 9.4 for a listing of the typical impairments in both aphasias, including speech samples.

	Broca's Aphasia	Wernicke's Aphasia
Quality ofe Speech	Severely impaired; marked by extreme effort to generate speech, hesitant utterances, short (one-word) responses.	Little if any impairment; fluent speech productions, clear articulation, no hesitations.
Nature of Speech	Agrammatical; marked by loss of syntactic markers and inflections and use of simple noun and verb categories.	Neologistic; marked by invented words (neologisms) or semantically inappropriate substitutions; long strings of neologistic jargon.
Comprehension	Unimpaired compared with speech production. Word-finding difficulty caused by production difficulties.	Severely impaired; marked by lack of awareness that speech is incomprehensible; comprehension impaired also in nonverbal tasks (e.g., pointing).
Speech Samples	Experimenter asks the patient's address. "Oh dear. Um. Aah. O! O dear. Very-there-were-ave. avedeversher avenyer." (Correct address was Devonshire.)	Experimenter asks about the patient's work before hospitalization. "I wanna tell you this happened when happened when he rent. His-his kell come down here and is—he got ren something. It happened. In these ropliers were with him for hi-is friend—like was. And he roden all of these arranjen from the pedis on from iss pescid."

hemisphere damage is an inability to activate an appropriately diverse set of information from long-term memory from which inferences can be derived. In one study, after reading a text, people were given a lexical decision task. Some of the words in the task were related to inferences that needed to be drawn for comprehension. For example, read the following short text: "Then he went into the bathroom and discovered that he had left the bathtub water running. He had forgotten about it while watching the news. The mess took him a long time to mop up." After reading, you are then given a lexical decision probe word like *overflow*. Because the tub overflowing was not mentioned in the text, responding to *overflow* faster than baseline would be evidence of having inferred it based on reading the text itself. The results showed that normal controls responded 49 ms faster relative to neutral control words—they drew the appropriate inference. However, the patients with right hemisphere damage responded 148 ms more *slowly* to these words.

9.6.3: Generalizing from Cases of Brain Damage

Although it is a mistake to believe that our eventual understanding of language will be reducible to a catalog of biological and neurological processes (e.g., Mehler, Morton, & Jusczyk, 1984), knowledge of the neurological aspects of language is useful for something beyond the rehabilitation and treatment of aphasia. What do studies of such abnormal brain processes tell us about normal cerebral functioning and language?

Well, for one, the different patterns of behavioral impairments in Broca's and Wernicke's aphasias, stemming from different physical structures in the brain, imply that these two physical structures are responsible for different aspects of linguistic skill. Furthermore, these selective impairments reinforce the notion that syntax and semantics are two separable but interactive aspects of normal language (e.g., O'Seaghdha, 1997; Osterhout & Holcomb, 1992). The double dissociations show that different independent modules govern comprehension and speech production. Other dissociations indicate yet more independent modules of processing, such as separate ones corresponding to reading and writing.

An intriguing inference from such studies is that the specialized regions signal an innate, biological basis for language—that the human nervous system is specifically adapted to learn and use language, as opposed to simply being able to do so. Several theorists have gone so far as to discuss possible evolutionary mechanisms responsible for lateralization, hemispheric specialization, the dissociation of syntax and semantics revealed by Broca's and Wernicke's aphasias, and even cognition in general (Corballis, 1989; Geary, 1992; Lewontin, 1990). These are fascinating lines of reasoning on the nature of language and cognition as represented in the brain.

> **WRITING PROMPT**
>
> **Brain Components in Language Processing**
>
> What components of the brain are specialized for language use? What happens to language processing when different parts are damaged?
>
> The response entered here will appear in the performance dashboard and can be viewed by your instructor.
>
> Submit

Summary: Language

9.1 Linguistic Universals and Functions

- Language is our shared symbolic system of communication, unlike naturally occurring animal communication. True language involves a set of characteristics, linguistic universals, that emphasize the arbitrary connections between symbols and referents, the meaningfulness of the symbols, and our reliance on rules for generating and comprehending language.
- Three traditional levels of analysis—phonology, syntax, and semantics—are joined by two others in psycholinguistics, the levels of conceptual knowledge and beliefs. Linguists focus on an idealized language competence as they study language, but psycholinguists are also concerned with language performance. Therefore, the final two levels of analysis take on greater importance as we investigate language users and their behavior.
- To some degree, we can use people's linguistic intuitions, their linguistic competence, to discover what is known about language; language performance, on the other hand, is also affected by memory lapses and the like.
- The Sapir-Whorf linguistic relativity hypothesis claims that language controls or determines thought, making it impossible to think of an idea if there was no word for it in the language. The weak version of

this hypothesis is generally accepted now; language exerts an influence on thought by making it more difficult to think of an idea without having a word to name or express it.

9.2 Phonology

- Phonology is the study of the sounds of language. Spoken words consist of phonemes, the smallest units of sound that speakers of a language can distinguish. Surprisingly, a range of physically different sounds are classified as the same phoneme; we tolerate some variation in the sounds we categorize as "the same," called categorical perception.
- Categorical perception is particularly important in the study of speech recognition because the phonemes in a word exhibit coarticulation, overlapping effects among successive phonemes, such that an initial sound is influenced by the sounds that follow and the later sounds are influenced by what came before.
- Speech recognition relies heavily on conceptually driven processes. This includes our knowledge of the sentence and word context, our estimates of how we would produce the sounds ourselves, and our knowledge of what the words in our language are.
- As an illustration of embodied cognition, according to the motor theory of speech perception, part of the way that people go about understanding spoken language may be to try to mentally simulate those sounds as if they were actually being spoken.

9.3 Syntax

- Syntax involves the ordering of words and phrases in sentence structure and features such as active versus passive voice. Chomsky's theory of language is a heavily syntactic scheme with two sets of syntactic rules. Phrase structure rules are used to generate a deep structure representation of a sentence, and then transformational rules convert the deep structure into the surface structure, the string of words that makes up the sentence.
- There are a variety of syntactic clues to the meaning of a sentence, so an understanding of syntax is necessary to psycholinguists. On the other hand, psycholinguistics has developed its own theories of language, at least in part because of linguists' relative neglect of semantic and performance characteristics.
- Studies of how we plan and execute sentences reveal a highly interactive set of processes, rather than a strictly sequential sequence. We pause, delay, and rearrange sentences as a function of planning and memory-related factors like accessibility and working memory load.
- Meaning in language is conveyed not only by word order (syntax) but also by the changes in pitch (prosody) of the phonemes that are spoken. Prosody can convey information such as the nature of an utterance (e.g., the difference between a statement and a question) or which concepts the speaker wishes to emphasize.

9.4 Lexical Factors

- Semantic factors in language can sometimes override syntactic and phonological effects. The study of semantics breaks words down into morphemes, the smallest meaningful units in language; *cars* contains the free morpheme *car* and the bound morpheme *-s* signifying a "plural."
- Speech errors that people make can be used to help reveal the processes by which language is produced. These speech errors follow regularities that are likely to be produced by otherwise consistent and stable cognitive processes.

9.5 Semantics

- As the study of language comprehension has matured, the dominant approach to semantics claims that we perform a semantic parsing of sentences, assigning words to their appropriate semantic case roles as we hear or read.
- Garden path sentences, in which later phrases indicate an error in interpretation, have provided rich information about how syntax and semantics are processed online during comprehension and how we recover from comprehension errors.
- The cognitive processing consequences of bilingualism are not well understood. Whereas some evidence suggests that knowing more than one language can improve intelligence, other studies find no evidence of a benefit. Regardless, there appears to be no evidence that multilingualism compromises cognition and intelligence as was once thought.

9.6 Brain and Language

- Extensive evidence from studies with brain-damaged people and more modern work using imaging and ERP methods reveals several functional and anatomical dissociations in language ability.
- ERP methods show how language development unfolds over time, the time course of syntactic processing, the lateralized processing of language between the left and right hemispheres, and the strategies people use to comprehend language.

- The syntactic and articulatory aspects of language seem centered in Broca's area, in the left frontal lobe, whereas comprehension aspects are focused more on Wernicke's area, in the posterior left hemisphere junction of the temporal and parietal lobes.
- The study of these and other deficits, such as anomia and right hemispheric damage, converges with evidence from imaging and ERP studies to illustrate how various aspects of language performance act as separable, distinct components within the overall broad ability to produce and comprehend language.

SHARED WRITING

Language

The production and comprehension of language involves many levels of analysis, processing, and mental representation. Based on this, how is it that people sometimes misspeak, mistype, mishear, or misread things in everyday experience?

A minimum number of characters is required to post and earn points. After posting, your response can be viewed by your class and instructor, and you can participate in the class discussion.

Post   0 characters | 140 minimum

Chapter 10
Comprehension

Learning Objectives

10.1: Analyze the conceptual level of language analysis

10.2: Explain the measures that can be used to improve reading comprehension

10.3: Assess how current research integrates multiple factors in understanding reading comprehension

10.4: Evaluate the cognitive effects of conversational interactions

Comprehension—this is a concept that requires a bit of explanation. What does the word *comprehension* mean? Basically, the expanded meaning here includes both the fundamental language processes and the additional processes we use when comprehending realistic extended tracts of language, say a passage in a book or a connected, coherent conversation, or even a perceived event. How do we comprehend and understand? What do we *do* when we read, understand, and remember connected sentences? By taking a larger unit than isolated sentences of analysis, we confront a host of issues central to communication and to cognitive psychology. And by confronting Miller's (1973) highest two levels of analysis, conceptual knowledge and beliefs, we address the important issues Miller (1977) described as the "distant bridge that may someday need to be crossed." In short, it is time to cross the bridge.

10.1: Conceptual and Rule Knowledge

OBJECTIVE: Analyze the conceptual level of language analysis

Earlier in the course, we discussed the first three levels of language analysis: the phonological, syntactic, and lexical and semantic levels. Now we will dig into Miller's (1977) fourth and fifth levels: the conceptual and belief levels.

10.1.1: Comprehension Research

Much of the traditional evidence about comprehension relies on people's linguistic intuitions, their (leisurely) judgments about the acceptability of sentences, or simple measures of recall and accuracy. The Sachs (1967) study is a classic example of early comprehension research with a straightforward conclusion. Recall that as people were reading a passage, they were interrupted and tested on a target sentence, either 0, 80, or 160 syllables after the end of the target. Their recognition of the sentence was very accurate at the immediate interval. But beyond that, they were accurate only at rejecting the choice that changed the sentence meaning. So people could not accurately discriminate between a verbatim target sentence and a paraphrase: If the choice preserved the original meaning, then people mistakenly "recognized" it. Clearly, these results showed that memory for meaningful passages does not retain verbatim sentences for long but does retain meaning quite well.

10.1.2: Online Comprehension Tasks

As work on comprehension developed, researchers needed a task that measures comprehension as it happens, or an **online comprehension task**. Such tasks involve the same approach discussed throughout this course: Find a dynamic, time- or action-based task that yields measurements of the

Conceptual and Belief Levels of Language Analysis

The conceptual and belief levels aim more squarely at issues of comprehension and understanding.

Interactive

Here's the sentence Miller uses to illustrate these conceptual and belief levels:

(1) Mary and John saw the mountains while they were flying to California.

If this sentence were spoken aloud, your comprehension would begin with phonological processes, translating the stream of sounds into words. Your syntactic knowledge would parse the sentence into phrases and assist the semantic level of analysis as you determined the case role for each important word: *Mary* and *John* are the agents of *see*, the word *mountains* is assigned the *patient* or *recipient* role, *they* is the agent of *fly* in the second main clause, and so on.

---

underlying mental processes as they occur. Contrast performance in a variety of conditions, pitting factors against each other to see how they affect comprehension speed or difficulty. Then draw conclusions about the underlying mental processes, based on the performance measures.

WRITTEN LANGUAGE A common assessment of cognition during comprehension involves reading times. These can be gathered by using *eye movement data*, or having people control the presentation of text by pressing a button to advance to the next word, clause, or sentence. We can then analyze reading times for these individual components and draw inferences about online comprehension. In general, language that a person is prepared for is read faster, whereas those aspects that require a large involvement of mental resources result in longer reading times.

10.1.3: Metacomprehension

Reading a passage and having some understanding of what you have read does not mean you have actually learned something or will remember it later. Yet, we need to use our **metacomprehension** abilities (e.g., Dunlosky & Lipko, 2007) to monitor how well we understand and will remember information later.

10.1.4: Comprehension as Mental Structure Building

The mental representations that are created during comprehension are built up and updated in systematic ways. Specially, theoretical approaches, such as Gernsbacher's

Methods to Assess Online Comprehension
There are three other ways commonly used to assess online comprehension.

> **Interactive**
>
> **Sample Stimuli and Test Words**
>
> **Think-Aloud Verbal Protocol**
>
> Another way to assess online comprehension is the think-aloud verbal protocol method (e.g., Magliano, Trabasso, & Graesser, 1999). In this method, people are asked to verbalize their thoughts as they read a passage of text. The *verbal protocols* can then be analyzed later to assess the conscious thoughts people were having as they read, including the following:
>
> - How do they link up a current portion of text with events that occurred earlier?
> - Were they making predictions about what would happen next?
> - Did they notice an inconsistency in the text?
>
> The data generated from think-aloud verbal protocols can provide insight into the aspects of a text that might be fruitful candidates for further research. For example, this information can be used to focus investigation of which aspects of a text will yield interesting reading time data, or which kind of information to test using a probe task.
>
> **Neural Imaging**

(1990) **structure-building** framework, assume that language is essentially a set of instructions for how to build one's understanding of the world. This concerns how to start, how to update it, and how to remove irrelevant information.

A convenient way to organize thinking about comprehension is to use Gernsbacher's (1990) structure-building framework.

10.1.5: Levels of Comprehension
Comprehension is a complex process involving several different levels.

One of the ways of characterizing these different levels is the van Dijk and Kintsch's (1983) levels of representation theory. As a reminder, at one level is the **surface form**. This is our verbatim mental representation of the exact words and syntax used. At an intermediate level is the **propositional textbase**, which captures the basic idea units present in a text. Finally, there is the level of the **situation model** (Johnson-Laird, 1983; Van Dijk & Kintsch, 1983; Zwaan & Radvansky, 1998), which is a mental representation that serves as a simulation of a real or possible world as described by a text.

These three levels of comprehension involve different cognitive processes that operate at different time scales, with surface-level cognitive processing occurring more rapidly than situation-model–level processing (Huang & Gordon, 2011).

Measures of Metacomprehension

Metacomprehension is important because it can influence how much we may study information later, and to which information we devote our time. To this end, we need ways to measure or evaluate metacomprehension.

> **Interactive**
>
> A popular measure of metacomprehension is *judgments of learning (JOLs)* (Arbuckle & Cuddy, 1969). These are estimates people are asked to make of how well they feel they have learned some material they just read. Research on JOLs typically compares people's estimates of how well they have learned information with how they actually do. Unfortunately, in many cases, the relationship between JOLs and actual performance is quite low—in other words, people are typically not very good at estimating whether they have learned something or not. Consequently, when you plan your studying, say for an upcoming exam, you may not spend the time you need on certain material because you think you know it better than you actually do. Your test performance would improve if you could better monitor what you have and have not learned.
>
> 1 of 4   [Previous] [Next]

As you read, try to understand how comprehension may depend on these different levels. For example, research on order of mention in establishing discourse reference will depend on the surface form. Work on bridging inferences requires processing at the textbase level. Finally, work showing how people monitor various aspects of experience involves the situation model.

WRITING PROMPT

Cognitive Processes Resulting in Comprehension

How can the process of comprehension be broken down, both methodologically, in terms of how to test it, and theoretically, in terms of the levels of cognitive processes that result in comprehension?

 The response entered here will appear in the performance dashboard and can be viewed by your instructor.

[Submit]

10.2: Reading

OBJECTIVE: Explain the measures that can be used to improve reading comprehension

For years, the standard way to study reading was for people to read a passage of text and then take a memory test, such as a multiple-choice or recall test. Such tasks certainly

Gernsbacher's Structure-Building Framework (Figure 10.1)

The basic theme is that comprehension is a process of building mental structures. Laying a foundation, mapping information onto the structure, and shifting to new structures are the three principal components, whereas enhancement and suppression modify the relative salience of the structures created.

> **Laying a Foundation**
> As we read, we begin to build a mental structure that captures the meaning of a sentence. A foundation is initiated as the sentence begins and typically is built around the first mentioned character or idea. This is equivalent to saying that sentence 3 is about Dave and studying:
>
> (3) Dave was studying hard for his statistics midterm.

have face validity; they test memory for the text, because much of our reading is for learning and remembering what we read.

But this approach is lacking because it does not gather online measures of comprehension; it measures only what people remember *after* reading (not that this is not important, there's just more to what's going on). Earlier, you saw a graph of the hypothetical activation levels for concepts in a set of sentences. We would like to know directly how concepts vary in their activation levels across a passage because that tells us a great deal about online reading comprehension. A multiple-choice test is too blunt to give us such answers.

10.2.1: Gaze Duration

In reading research that assesses gaze duration, the equipment used is an **eye tracker**, a camera- and computer-based apparatus that records eye movements and the exact words that are fixated on during reading.

In this system of tracking, continuous recording of at least one eye, while tracking head position, enables the system to determine exactly what you are looking at on the computer screen. So, the machine records the duration of the eyes' gaze as they scan across lines of text (this system has other purposes in addition to reading, such as evaluating the usefulness of web pages). Typically, people

Table 10.1 Summary of Gernsbacher's Structure-Building Framework

Process /Control Mechanism	Explanation/ Function
Laying a foundation	
Mapping information	
Shifting	
Enhancement	
Suppression	

Increase the activation of coherent, related information.

Start Over

---

simply see a passage of text on the screen, and the eye tracker records the eye movements and durations as people read the words in the passage. In this task, the researcher knows which word is being processed on a moment-by-moment basis and how long the eyes dwell on each word. So, **gaze duration** is a prime measure of what is happening when people read (see Kambe, Duffy, Clifton, & Rayner, 2003, and Rayner, 1998, for thorough discussions of alternatives). Time-based eye movement data provide a window on the process of comprehension and reading.

MAKING EYE MOVEMENTS The eyes move in rapid sweeps—*saccades*—and then stop to focus on a word—*fixations*. Fixations in reading (English) last about 200 ms to 250 ms, and the average saccade size is from 7 to 9 letter spaces. However, as Figure 10.2 shows, there is considerable variability in these measures (Rayner, 1998).

IMMEDIACY AND EYE-MIND ASSUMPTIONS Two assumptions that have guided much of the work using eye movements are the immediacy and the eye-mind assumptions (Just & Carpenter, 1980, 1987, 1992). The **immediacy assumption** states that readers try to interpret each content word of a text as they encounter that word. In other words, we do not wait until we take in a group of words, say in a phrase, before we start to process them. Instead, we begin interpreting and comprehending as soon as we encounter a word. The **eye–mind assumption** is the idea that the pattern of eye movements directly reflects the complexity of the underlying cognitive processes.

Figure 10.2 Variability in Saccade and Fixation Measures

Frequency distributions for fixation durations (top) in ms, and saccade length (bottom) in number of character spaces.

SOURCE: From Rayner (1998).

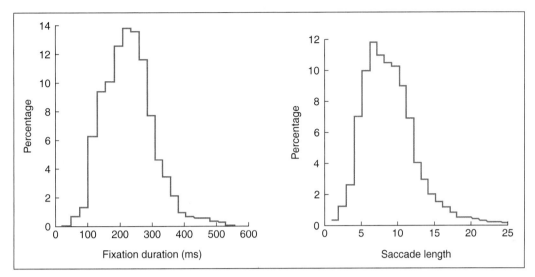

Although these assumptions are robust, they do have some limitations. For example, eye gazes often take in more than one word, depending on the length of the words, the size of the text fonts, and the span of the perceptual beam. So, there is not always a direct one-to-one relationship between an eye fixation and the words being processed. Moreover, eye gazes do not reflect only the processing of the current word; they can also reflect some spillover processing of previous words and some anticipatory processing of upcoming words (Kliegl, Nuthmann, & Engbert, 2006). Despite these limits, eye tracking is still a powerful tool for the reading researcher.

Also, outside of reading, you can use eye gaze to understand other aspects of comprehension, such as the understanding of spoken language. For example, if you listen to and follow a set of directions, say to pick up and move an object from one place to another, your eye movements track the spoken instructions very closely. As you hear, "Put the apple in the box," your eyes fixate immediately on those objects in the visual scene (Spivey, Tanenhaus, Eberhard, & Sedivy, 2002; see Crosby, Monin, & Richardson, 2008, for an application of eye tracking to social cognition).

10.2.2: Basic Online Reading Effects

An example of online reading research examined regressive eye movements back to a portion of text that had been read earlier. Just (1976) was interested in such eye movements when the referents in the sentence could not be immediately determined: If an initial assignment of a character to a case role was wrong, then what happened? Was there a regressive eye movement back to the correct referent? People read sentences such as 6 and 7, and eye movements were monitored:

(6) The tenant complained to his landlord about the leaky roof. The next day, he went to the attic to get his luggage.

(7) The tenant complained to his landlord about the leaky roof. The next day, he went to the attic to repair the damage.

In sentence 6, when *luggage* was encountered, eye movements bounced up immediately to the word *tenant*. In sentence 7, eye movements bounced up to *landlord*. These eye movements provided evidence of the underlying mental processes of finding antecedents and determining case roles.

Another study provides a demonstration of the detail afforded by eye trackers. Look at Figure 10.4, taken from Just and Carpenter (1987; see also Just & Carpenter, 1980). You see two sentences taken from a larger passage. Above the words are two numbers. The top number indicates the order in which people fixated on the elements in the sentence; 1 to 9 in the first sentence and 1 to 21 in the second. The number below is the gaze duration (in milliseconds). So, for example, the initial word in sentence 1, *Flywheels*, was fixated on for 1,566 ms. The next word, *are*, was fixated

Figure 10.3 The Pattern of Fixations

Eye tracking gives us gaze durations on a word-by-word basis.

SOURCE: Just, M. A., & Carpenter, P. A. (1987), *The psychology of reading and language comprehension*, Boston, MA: Allyn & Bacon. Figure 2.1 (p. 27). Credited source: Buswell, G. T. (1937). *How adults read*, Chicago, IL: Chicago University Press, Plates II and IV, pp. 6–7.

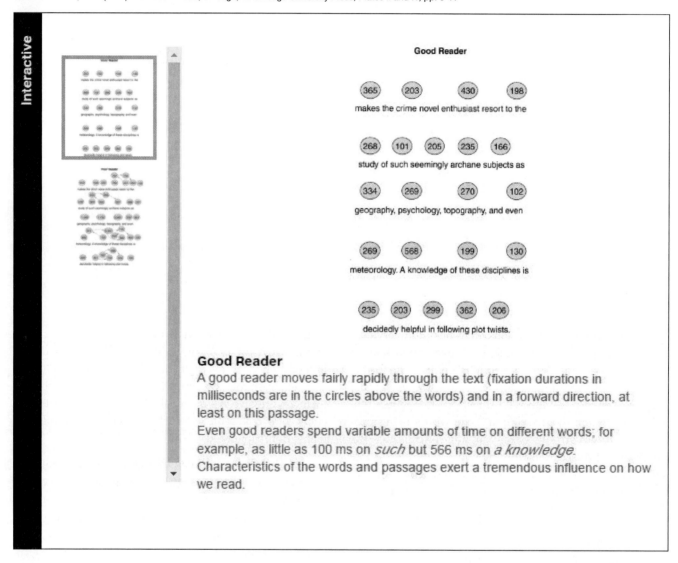

on only 267 ms. The fourth word, *of*, was not fixated on at all by this person, so neither a gaze number nor time is presented there. This is not unusual because many function words, such as *of* and *the*, are so automatically processed that they may not need to be fixated on (Angele & Rayner, 2013). In fact, you can raerragne the lteters in wodrs, usch as is odne in this snetence, and people have little trouble extracting the meaning. There is some disruption to reading, but not as much as if different letters were substituted for correct letters (Rayner, White, Johnson, & Liversedge, 2006). So, reading does not *require* a strict adherence to the printed form.

In the Just and Carpenter study, these passages were technical writing, in which a new concept, such as a flywheel, is introduced, defined, and explained. The average reading rate was about 225 words per minute, slower than for simpler material, such as newspaper stories or novels. At a general level, every content word was fixated. According to Just and Carpenter, this is the norm for all kinds of text. In fact, about 85% of the content words are fixated. Short function words, however, like *the* or *of* often tend not to be fixated; Rayner and Duffy (1988) estimate that function words are fixated only about 35% of the time. Readers also tend to skip some content words if the passage is very simple for them (say, a children's story given to an adult), if they are skimming or speed reading, or if a word is very predictable (Rayner & Well, 1996).

Figure 10.4 Eye Fixations of a College Student Reading a Scientific Passage

Gazes within each sentence are sequentially numbered above the fixated words with the durations (in milliseconds) indicated below the sequence numbers.

SOURCE: From Just & Carpenter (1980).

```
   1     2   3   4    5      6        7      8    9
 1566  267 400  83  267    617      767    450  450
 Flywheels are one of the oldest mechanical devices known to man.

   1       2        3         4     5      6        7
  400    616      517       684   250    317      617
 Every internal- combustion engine     contains a small

   8      9     10    11       12      13
 1116   367   467   483      450     383
 flywheel that converts the jerky motion of the pistons into the

  14    15   16   17      18    19   20   21
 284   383  317  283     533   50  366  566
 smooth flow of  energy that powers the drive shaft.
```

As noted already, gaze durations are quite variable, and the duration of a saccade is about 100 ms, followed by a fixation of 200 to 250 ms. These estimates come from situations in which the viewer is merely gazing out on a scene. In reading studies, however, people do not move their eyes as far, averaging 2 degrees of angle versus 5 degrees in scene perception. Hence, saccades during reading are shorter—they take about 30 ms. Although word fixations may be brief, readers often make repeated fixations on the *same* word. In some studies, successive fixation times are summed together. Alternatively, investigators report both the first-pass fixations and total fixation duration.

10.2.3: Benefits of Online Reading

A strength of online reading measures is that they provide evidence at two levels of comprehension. First, there are word-level processes operating at the surface form level. These are crucial to an understanding of reading. For instance, several studies attest to the early use of syntactic features of a sentence when we comprehend not just major syntactic characteristics such as phrase boundaries but even characteristics such as subject–verb agreement (Pearlmutter, Garnsey, & Bock, 1999) and pronoun gender (McDonald & MacWhinney, 1995). Reichle, Pollatsek, Fisher, and Rayner (1998) provide an account of such word-level processes with their E-Z Reader models of eye movement control in reading (for a review of several computational models of reading, see Norris, 2013).

Second, reading time measures can also be used to examine larger, macroscopic processes, such as at the textbase and situation model levels. We will hold off a discussion of situation model processing until the next section. At the textbase level, Table 10.2 presents Just and Carpenter's (1980) analysis of the "flywheel" passage. To the left of each line is a category label; each sector was categorized as to its role in the overall paragraph structure. To the right are two columns of numbers—observed gaze durations for a group of people and estimated durations—based on the "READER" model's predictions. For example, the 1,921 ms observed for sector 1 is the sum of the separate gaze durations for that sector (averaged across people). Different kinds of sectors take different amounts of time. For instance, definition sectors have more difficult words and are longer than other sector types, so they show longer gaze durations. Even a casual examination of the observed and predicted scores shows that the model does a good job of predicting reading times.

Generally, an analysis of reading times needs to account for several surface form and textbase factors that are tied to the text itself. Reading time is strongly influenced by word length, with words that are composed of more letters or syllables taking longer to read than shorter words. Also, word frequency plays a vital role, with infrequent words resulting in longer reading times as the reader needs to engage in extra mental effort to retrieve this lexical information from memory.

MODEL ARCHITECTURE AND PROCESSES So, just how does the mind go about reading? How can we address this issue in a systematic way? Figure 10.5 illustrates the

Table 10.2 Sector-by-Sector Analysis of "Flywheel" Passage

Category	Sector	Gaze Duration (ms) Observed	Gaze Duration (ms) Estimated
Topic	Flywheels are one of the oldest mechanical devices	1,921	1,999
Topic	known to man.	478	680
Expansion	Every internal-combustion engine contains a small flywheel	2,316	2,398
Expansion	that converts the jerky motion of the pistons into the smooth flow of energy	2,477	2,807
Expansion	that powers the drive shaft.	1,056	1,264
Cause	The greater the mass of a flywheel and the faster it spins,	2,143	2,304
Consequence	the more energy can be stored in it.	1,270	1,536
Subtopic	But its maximum spinning speed is limited by the strength of the material	2,400	2,553
Subtopic	it is made from.	615	780
Expansion	If it spins too fast for its mass,	1,414	1,502
Expansion	any flywheel will fly apart.	1,200	1,304
Definition	One type of flywheel consists of round sandwiches of fiberglass and rubber	2,746	3,064
Expansion	providing the maximum possible storage of energy	1,799	1,870
Expansion	when the wheel is confined in a small space	1,522	1,448
Detail	as in an automobile.	769	718
Definition	Another type, the "superflywheel," consists of a series of rimless spokes.	2,938	2,830
Expansion	This flywheel stores the maximum energy	1,416	1,596
Detail	when space is unlimited.	1,289	1,252

SOURCE: From Just & Carpenter (1980).

architecture and processes of the Just and Carpenter (1980, 1987, 1992) model.

Note that several elements are already familiar. For instance:

- Working memory is where different types of knowledge—visual, lexical, syntactic, semantic, and so on—are combined. Not surprisingly, working memory capacity is important in reading comprehension (e.g., Kaakinen, Hyona, & Keenan, 2003).
- Long-term memory contains a wide variety of knowledge used during reading. Each of these types of knowledge can match the current contents of working memory and update or alter those contents. In simple terms, what you know combines with what you have already read and understood. Together, these elements permit comprehension of what you are reading.
- Finally, wrap-up is an integrative process that occurs at the end of a sentence or clause. During wrap-up, readers tie up any loose ends. For instance, any remaining inconsistencies or uncertainties about reference are resolved here.

10.2.4: Factors That Affect Reading

An in-depth description of all the variables that influence reading is not possible here—there are simply too many. Here is just a brief list of some of the important factors:

- The effects of word frequency, syntactic structure, and context (Altmann, Garnham, & Dennis, 1992; Inhoff, 1984; Juhasz & Rayner, 2003; Schilling, Rayner, & Chumbley, 1998)
- The effects of sentence context on word identification (Paul, Kellas, Martin, & Clark, 1992; Schustack, Ehrlich, & Rayner, 1987; Simpson, Casteel, Peterson, & Burgess, 1989), including ERP work showing how rapidly we resolve anaphoric references (Van Berkum, Brown, & Hagoort, 1999)
- The effects of ambiguity (Frazier & Rayner, 1990; Rayner & Frazier, 1989) and figurative language (Frisson & Pickering, 1999)
- The effects of topic, plausibility, and thematic structure on reading (O'Brien & Myers, 1987; Pickering & Traxler, 1998; Rayner, Warren, Juhasz, & Liversedge, 2004; Speer & Clifton, 1998; Taraban & McClelland,

Variables That Affect Reading Times

Serial position is also an important factor. The further along a person is in a passage, the more of a foundation there is from which to build mental structures, thereby making comprehension easier and faster.

Variables That Increase Reading Times

- *Surface form effects:* sweep of the eyes to start a new line, sentence wrap-up, number of syllables, low frequency or new word, unusual spelling patterns
- *Textbase effects:* integration of information (after clauses, sentences, sectors, etc.), topic word, new argument, other error recovery, reference and inference processes, difficulty of passage/topic

Variables That Decrease Reading Times

1988), especially the relatedness of successive paragraphs and the presence of an informative introductory paragraph (Lorch, Lorch, & Matthews, 1985) or title (Wiley & Rayner, 2000)
- The effects of scripted knowledge on word recognition and comprehension (Sharkey & Mitchell, 1985)
- The effects of discourse structure on the understanding of reference (Malt, 1985; Murphy, 1985) and the resolution of ambiguity (Vu, Kellas, Metcalf, & Herman, 2000)

In addition, even phonology plays an important role in reading comprehension, such as research showing that phonological information is activated as rapidly as semantic knowledge in silent reading (Lee, Rayner, & Pollatsek, 1999; Rayner, Pollatsek, & Binder, 1998),

especially for readers of lower skill levels who rely more on print-to-sound-to-meaning processes than a direct print-to-meaning route (Jared, Levy, & Rayner, 1999).

WRITING PROMPT

Basic Mental Processes in Reading

What are the basic mental processes revealed by cognitive studies of reading? What are several different aspects of reading that can be pulled out of data as people read?

 The response entered here will appear in the performance dashboard and can be viewed by your instructor.

Submit

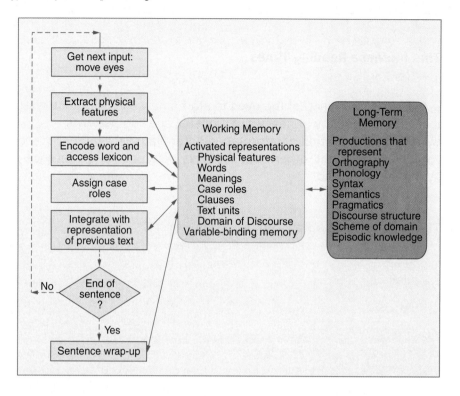

Figure 10.5 The Just and Carpenter (1980) Model

This model shows the major structures and processes that operate during reading. Solid lines represent the pathways of information flow; the dashed line shows the typical sequence of processing.

10.3: Reference, Situation Models, and Events

OBJECTIVE: Assess how current research integrates multiple factors in understanding reading comprehension

Although the cognitive mechanisms and processes involved in comprehension at the surface form and text-base levels are critically important, they are not the goal of comprehension. A person who has successfully comprehended something that has been read has derived not only an adequate representation of the text itself. This person also understands the circumstances being described—the reference of the text.

10.3.1: Reference

Reference involves finding the connections between elements in a passage of text, finding the words that refer to other concepts in the sentence. In sentence 3 from earlier, "Dave was studying hard for his statistics midterm," the word *his* refers to *Dave*. In this situation *Dave* is the **antecedent** of *his* because *Dave* comes before the pronoun. And the act of using a pronoun or possessive later is **anaphoric reference**. So, **reference** is the linguistic process of alluding to a concept by using another name. Commonly we use pronouns or synonyms to refer to the antecedent, although there are other types of reference. For example, using a person's name would be a form of *identity reference* in that it refers to a previous instance of using that person's name.

Reference is as common in language as any other feature we can identify. Part of reference is that it reduces redundancy and repetition. Contrast a normal passage such as 8a with 8b to see how boring and repetitive language would be without synonyms, pronouns, and so on.

(8a) Mike went to the pool to swim some laps. After his workout, he went to his psychology class. The professor asked him to summarize the chapter that he'd assigned the class to read.

(8b) Mike went to the pool to swim some laps. After Mike swam some laps, Mike went to Mike's psychology class. The professor of Mike's psychology class asked Mike to summarize the chapter that Mike's psychology professor had assigned Mike's psychology class to read.

This repetition of identity reference can be detrimental to comprehension. Research has shown a **repeated name penalty**, an increase in reading times when a direct reference is used again (e.g., the person's name) compared to when a pronoun is used (e.g., Almor, 1999; Gordon & Chan, 1995; Gordon & Scearce, 1995). That said, when we produce language, if there is more than one character being discussed or present in a situation, people are less likely to use indirect references, such as pronouns, and are more likely to use a direct reference, such as a person's name (Arnold & Griffin, 2007). This may be more acceptable under these circumstances because there may be some ambiguity as to who is being referenced.

SIMPLE REFERENCE Simple reference picks out an entity or entities in a more or less direct manner. That said, there are several ways of doing this. In natural discourse, different kinds of reference can occur.

Consider three simple forms of reference:

(9) I saw a convertible yesterday. The convertible was red.

(10) I saw a convertible yesterday. The car was red.

(11) I saw a convertible yesterday. It was red.

In sentence 9 the reference is so direct that it requires no inference on the part of the listener; this is *identity reference*, using the definite article *the* to refer back to a previously introduced concept, a *convertible*. *Synonym reference* requires that you consider whether the second word is an adequate synonym for the first, as in sentence 10; can *a convertible* also be referred to as *the car*? *Pronoun reference* requires similar reference and inference steps. In sentence 11, pronoun reference can refer only to the *convertible*, because it is the only concept in the earlier phrase that can be equated with *it*. That is, in English, the word *it* must refer to an ungendered concept, just as *he* must refer to a male, and so on. In some languages, the nouns have gender and pronouns must agree with the gender of the noun; translated literally from French we get, "Here is the Eiffel Tower. She is beautiful."

INFLUENCE OF ANTECEDENTS Another important idea is that there is some evidence that the order in which antecedents are encountered influences the likelihood that they will be linked to later reference. Two effects of this type are the **advantage of first mention** and the **advantage of clause recency**. In the advantage of first mention, characters and ideas that were mentioned first have a special significance. For example, in a study by Gernsbacher and Hargreaves (1988), after reading a sentence such as "Tina gathered the kindling as Lisa set up the tent," people responded faster to *Tina* than to *Lisa* (see Figure 10.6).

Table 10.3 Types of Reference and Implication
Clark's (1977) useful list is shown in Table 10.3.

Direct Reference
Identity. Michelle bought a computer. The computer was on sale.
Synonym. Michelle bought a computer. The machine was on sale.
Pronoun. Michelle bought a computer. It was on sale for 20% off.
Set membership. I talked to two people today. Michelle said she had just bought a computer.
Epithet. Michelle bought a computer. The stupid thing doesn't work.

Indirect Reference by Association
Necessary parts. Eric bought a used car. The tires were badly worn.
Probable parts. Eric bought a used car. The radio doesn't work.
Inducible parts. Eric bought a used car. The salesperson gave him a good price.

Indirect Reference by Characterization
Necessary roles. I taught my class yesterday. The time I started was 1:30.
Optional roles. I taught my class yesterday. The chalk tray was empty.

Other
Reasons. Rick asked a question in class. He hoped to impress the professor.
Causes. Rick answered a question in class. The professor had called on him.
Consequences. Rick asked a question in class. The professor was impressed.
Concurrences. Rick asked a question in class. Vicki tried to impress the professor too.

Figure 10.6 Advantage of First Mention
Mean response time to names that had appeared in the studied sentences when the name was the first- or second-mentioned participant and when the name played the agent or patient case role in the sentence.

SOURCE: Data from Gernsbacher & Hargreaves (1988).

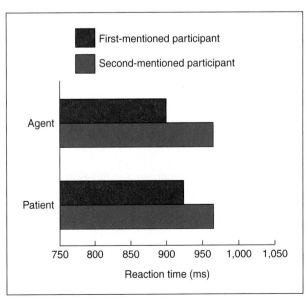

This effect involves remembering which entity was first and which was second. Patients with bilateral hippocampal damage do not show this effect (Kurczek, Brown-Schmidt, & Duff, 2013), suggesting we need some kind of memory for information order to perform normally.

ADVANTAGE OF CLAUSE RECENCY Conversely, there is also a time, at the end of the sentence, when the most recent character has an advantage—this is the *advantage of clause recency*. Again, for the sentence "Tina gathered the kindling as Lisa set up the tent," if you are probed immediately after it, Lisa has a slight advantage due to recency, but this advantage is shortlived, showing an advantage at about 50 ms to 60 ms, but disappearing by 150 ms (Gernsbacher, Hargreaves, & Beeman, 1989).

Other work has found that even the articles used can influence reference. Definite articles (such as *the*) convey given information and make sentences seem more coherent and sensible as compared to when indefinite articles (e.g., *a*, *an*, and *some*) are used (Robertson et al., 2000; see Table 10.4 for sample sentences), and sentences with definite articles are remembered better later (Haviland & Clark, 1974).

For Gernsbacher (1997), *the* is a cue for discourse coherence, enabling us to map information more efficiently and accurately. In one study (Robertson et al., 2000), people read sentences, followed by a recognition test (to make sure people tried to comprehend the sentences). Overall, sentences using *the* showed greater evidence of coherence than those with the indefinite *a*, *an*, and *some*. More important, this study tested people using fMRI and measured their levels of activity of different brain regions.

As Figure 10.7 shows, sentences that used the definite article showed greater activation than those with indefinite articles. Moreover, these activations were greater in the right hemisphere than the left, whereas more commonly the left hemisphere is implicated in language processing (e.g., Polk & Farah, 2002). Thus, the right hemisphere is particularly involved in establishing coherence in language comprehension.

Figure 10.7 Greater Activation in Sentences Using Definite Article

Activation levels for sentences presented with definite versus indefinite articles (levels in the figure are difference scores, showing how much greater the activations for definite than indefinite article sentences were), for seven left and right hemisphere locations in the brain.

SOURCE: From Robertson et al. (2000).

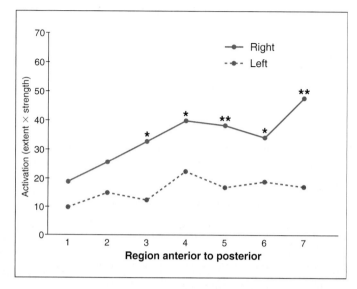

10.3.2: Situation Models

In constructing and using a situation model (Johnson-Laird, 1983; van Dijk & Kintsch, 1983; Zwaan & Radvansky, 1998), a person combines information that is available from the text, along with prior semantic and episodic knowledge, to create a mental simulation of the events being described. This is a *situation model*, a mental representation that serves as a simulation of a real or possible world as described by a text. The important idea is that comprehension is a search after meaning (Graesser, Singer, & Trabasso, 1994). Although comprehension does use some passive activation of semantic and episodic memories (e.g., McKoon & Ratcliff, 1992), we also actively build situation models that elaborate on the causal structure of the event a person is trying to understand. To improve our understanding of situation models, we examine three processes in the use of situation models in comprehension:

1. The use of inferences to elaborate on the information provided by a text
2. The influence of language structure, namely grammatical aspects, on situation model construction
3. The updating of the situation model as shifts in the described events are encountered

INFERENCE MAKING Instead of specifying everything for comprehension, we rely on people to know the

Table 10.4 Sample Sentences with Indefinite and Definite Articles

Indefinite	Definite
A grandmother sat at a table.	The grandmother sat at the table.
A child played in a backyard.	The child played in the backyard.
Some rain began to pour down.	The rain began to pour down.
An elderly woman led some others outside.	The elderly woman led the others outside.

meanings of words and about syntactic devices that structure our discourse, and to share our general conceptual knowledge of the world (e.g., to know that swimming laps can be a workout, or that professors assign chapters for their students to read; you remember a similar point from an earlier module, in discussing scripts). In fact, if you do specify everything exactly, you are breaking an important conversational rule and people will be unhappy with you.

> **Prove It**
>
> **Inferences and Language Comprehension**
> People spontaneously draw inferences as they comprehend language. These inferences are then incorporated into the situation models of what is being heard or read. Therefore, people frequently misremember information as having been heard or read, when in fact it was not. For this Prove It section, a list of sentences follows. Along with each sentence is an inference that people are likely to make (in parentheses). Read these sentences to a group of volunteers. Then, after you have read all the sentences, give your volunteers some distractor task, such as having them solve math problems for 3 to 5 minutes. When the distractor period is over, have your volunteers try to recall the sentences. You should find that people will likely report the inferences they made while they were comprehending. That is, they will "recall" more information than you actually read to them. These inferences—false memories, in a sense—are now part of their memory.
>
> 1. The housewife spoke to the manager about the increased meat prices. (complained)

Implication and Inference (Figure 10.8)

Let's turn to how comprehenders flesh out some missing information: implication and inference.

> **Interactive**
>
> In **implication**, there is an intended reference in a sentence or utterance, but it is not mentioned explicitly. The intention here is on the part of the speaker (or writer), who implies but does not state some conclusion or connection. In a sense, implication is in the mind of the speaker. If the listener (reader) draws the appropriate conclusion or connection during comprehension, then we say that the listener has drawn an inference. Thus, **inference** is the process by which the comprehender draws connections between concepts, determines the referents of words and ideas, and derives conclusions from a message. Implication is something that language producers do, and drawing inferences is something comprehenders do. If your professor says in class, "The next exam is on Wednesday, and it covers a lot of material," he or she is implying something about the difficulty of the exam, but is leaving it up to the students to draw that inference.

2. The paratrooper leaped out of the door. (jump out of a plane/helicopter)
3. The cute girl told her mother she wanted a drink. (asked)
4. The weatherman told the people of the approaching tornado. (warned)
5. The karate champion hit the cement block. (broke)
6. The absent-minded professor didn't have his car keys. (lost or forgot)
7. The safecracker put the match to the fuse. (lit)
8. The hungry python caught the mouse. (ate)
9. The man dropped the delicate glass pitcher. (broke)
10. The clumsy chemist had acid on his coat. (spilled)
11. The barnacle clung to the sides. (ship)
12. Dennis sat in Santa's chair and asked for an elephant. (lap)

GRAMMATICAL ASPECT When thinking about language comprehension in terms of the situation model, one can view language as being a set of instructions for how comprehenders should build a situation model in their minds (Zwaan, 1999). So, the things we say suggest to other people the elements of an event that should be part of their understanding. If you were to say to your friends, "I left my backpack at the library," this would tell them how to build their situation model and what to put in it. First, there is a spatial-temporal framework for that event. The place is the library, and the thing that was done (leaving the backpack there) was done in the past. Second, there is an entity involved, your backpack, and it is (or was) located within the library. Moreover, you may infer some unspoken information, such as that your backpack may be at a table on the second floor (where you usually study), and perhaps even create a mental image derived from this situation model of those circumstances.

Given that language is a set of instructions for understanding described circumstances, then saying different things, even when the differences seem subtle, can result in different understandings. A good illustration of this is *grammatical aspect*, which is how an event action or state is conveyed by a verb, as in the idea of a verb tense. For example, the sentence "Sam walked to the store" conveys the perfective aspect, and the sentence "Sam was walking to the store" contains an imperfective aspect. For the perfective aspect, the action *walk* is described as having already been completed. For the imperfective aspect, the action is described as ongoing. As such, you would create two different situation models for these two sentences. For the first, you might create a situation model in which Sam is in the store; for the second, you would create a situation model in which Sam is walking along a sidewalk on his way to the store (e.g., Anderson, Matlock, & Spivey, 2013; Magliano & Schleich, 2000).

A study by Madden and Zwaan (2003) offers an example of the influence of verb aspect on cognition. Researchers presented people with a series of pictures to view. Some of these pictures involved an ongoing action (e.g., a person driving), whereas other pictures involved a completed action (e.g., a driver having arrived home). After viewing each picture, people were given sentences to read and verify whether those sentences were satisfactory descriptions of the preceding picture. The sentences contained either a perfective verb ("The man made the fire") or an imperfective verb ("The man was making the fire"). People were asked to make these responses as quickly as possible. Madden and Zwaan found that for the completed event pictures, people responded faster to the perfective than to the imperfective sentences. Thus, the state of the described event as it is mentally represented is captured, to some degree, in the language used in a description.

SITUATION MODEL CREATION A simple type of inference making is what Clark (1977) termed a **bridging inference**, which is a process of constructing a connection between concepts. A bridging inference binds two units of language together. For example, determining that a reference like the epithet *the stupid thing* refers to the same entity as a *computer* is a bridging inference. It builds a connection between these two forms of reference, indicating that they refer to the same discourse entity. In bridging inference, the language producer uses reference to indicate the intended kinds of implications. For their part, comprehenders interpret the statement in the same fashion, computing the references and drawing the inferences needed. When the implication and inferences are intended, they are called **authorized inferences**. Alternatively, unintended implications and inferences are called **unauthorized inferences**, as when I say, "Your hair looks pretty today," and you respond, "So you think it was ugly yesterday?" (see also McKoon & Ratcliff, 1986).

The examples in Table 10.3 make it clear that the bridges we need to build for comprehension vary in their complexity, from simple and direct to difficult and remote.

Even on intuitive grounds, consider how the following sentences differ in the ease of comprehension:

(12) Marge went into her office. It was very dirty.

(13) Marge went into her office. The floor was very dirty.

(14) Marge went into her office. The African violet had bloomed.

Whereas sentence 12 is a simple case of pronoun reference, sentence 13 refers back to *office* with the word *floor*. Because an office necessarily has a floor, it is clear the implication in sentence 13 is that it was Marge's office floor that was dirty. One property you retrieve from semantic memory is that an office has a floor. Thus, if you comprehend that the office floor was dirty, you draw this inference. But it is an even longer chain of inference to draw the inference in sentence 14 that Marge has an African violet in her office; a floor is necessary, but an African violet is not. Overall, the integration of this semantic knowledge with the information in the text is part and parcel of creating a situation model.

It seems likely that the structure of concepts in semantic memory activation influences the ease with which information is inferred during situation model construction (e.g., Cook & Myers, 2004). So, more predictable pieces of information are processed faster (McKoon & Ratcliff, 1989; O'Brien, Plewes, & Albrecht, 1990). Marge's office necessarily has a floor as well as a desk, a chair, some shelves, and so on. It is conceivable that it has some plants, but that fact is optional enough that sentence 14 would take more time to comprehend.

INFLUENCES OF EMBODIED COGNITION Further evidence that people are drawing on their semantic knowledge, and that this knowledge has an embodied character, was found in a study by Zwaan, Stanfield, and Yaxley (2002). People read short descriptions of situations and then were shown pictures of objects. The task was to indicate whether the pictured object had been in the description they read. The critical manipulation was whether the picture either matched or mismatched the perceptual characteristics of the object in the description. For example, the critical sentence could be either "The ranger saw the eagle in the sky" or "The ranger saw the eagle in its nest" followed by a picture of either an eagle with its wings outstretched or perched (see Figure 10.9).

Zwaan et al. found that people responded faster when the picture matched the described state. Even though they saw an eagle in both pictures, the eagle with its wings outstretched "matched" the "eagle in the sky" description better, so people responded faster (see also Connell & Lynott,

Figure 10.9 Examples of Pictures of an Eagle in Flight or on a Perch

2009, for the activation of perceptual color information). Thus, people seemed to be activating perceptual qualities of objects during comprehension. Conversely, if people view a picture before reading, their response times are faster if the sentence matches the picture than if it doesn't (Wassenburg & Zwaan, 2010).

SPEECH ACT People are also aware of the intended consequence of someone saying something, called a **speech act** (Searle, 1969). For example, if you ask your roommate to turn down the stereo, the *speech* itself is the set of words you say, but the *speech act* is your intention, getting your roommate to let you study for an upcoming exam. Not only do people spontaneously derive the implied speech acts of what other people say, but they may misremember what was said in terms of the speech act itself. For instance, in a study by Holtgraves (2008b), people read a series of short vignettes, some of which conveyed speech acts. For example, suppose people read the following story: "Gloria showed up at the office wearing a new coat. When her coworker Heather saw it, she said to her, 'Gloria, I like your new coat.'" The last sentence conveys the speech act of complimenting Gloria. What Holtgraves found was that people were more likely to mistakenly remember that they had read "I'd like to compliment you on your new coat," an utterance that describes the speech act. However, a different group of people read a different version of the story, in which the last two sentences were: "When her coworker Heather saw it, she said *to* her friend Stacy: 'I like her new coat.'" In this condition, people were less likely to misremember having read, "I'd like to compliment her on her new coat." Because there was no actual complimenting speech act to Gloria, people did not store this information in memory and so did not make the memory error. Interestingly, the processing of speech act information appears to be dominated more by right hemisphere than left hemisphere (Holtgraves, 2012), consistent with the idea that understanding a speech act involves going beyond the actual words themselves.

Realize that we do not automatically and spontaneously draw all possible inferences while we read. Although some inferences are directly and typically drawn, such as simple and straightforward references, others are more complex and may not be drawn, and possibly should not be drawn. If you did draw such inferences, your cognitive resources would be quickly overwhelmed (Singer, Graesser, & Trabasso, 1994). For example, when you read a sentence like 12 or 13, you are likely not to draw an inference that Marge decided to clean her office, although if the next sentence in the story said that, you would certainly understand it. Most of the inferences you make are *backward* inferences. You are trying to understand what has already been described and how it all goes together. You make *forward* inferences—trying to predict what will happen next—under much rarer circumstances (Millis & Graesser, 1994).

Moreover, the creation of the situation model needs to account for constraints of embodiment. For example, in a study by de Vega, Robertson, Glenberg, Kaschak, and Rinck (2004; see also Radvansky, Zwaan, Federico, & Franklin, 1998), people were asked to read a series of passages. Embedded in those passages were critical sentences that described two actions that a character was doing, either at the same time or in sequence. If a person was described as doing two things that require the same parts of the body, such as "While chopping wood with his large axe, he painted the fence white," reading times were slower, as if readers were trying to figure out how this could be done. However, reading times were faster when either different parts of the body were being used, such as "While whistling a lively folk melody, he painted the fence white," or the actions were done in sequence, such as "After chopping wood with his large axe, he painted the fence white." Thus, we account for the limits of our human bodies, and the way actions happen in time, to help us comprehend what we are reading.

INDIVIDUAL DIFFERENCES Interestingly, reference and inference processes depend significantly on individual characteristics of the reader, particularly on the reader's skill. For instance, Long and De Ley (2000) found that less skilled readers resolve ambiguous pronouns just as well as more skilled readers, but they do so only when they are integrating meanings together. The more skilled readers resolve the pronouns earlier, probably when they first encounter a pronoun.

Several studies have also examined inferences as a function of working memory capacity (e.g., Fletcher & Bloom, 1988). One study by Singer, Andrusiak, Reisdorf, and Black (1992) explored individual differences in bridging as a function of working memory capacity and vocabulary knowledge. The gist of this work is that the greater your working memory capacity and vocabulary size, the greater the likelihood that information necessary for an inference will still be in working memory and can be used (see also Long, Oppy, & Seely, 1997; Miyake, Just, & Carpenter, 1994).

Evidence for individual differences in comprehension has also revealed itself in neurological measures. In a study by Virtue, van den Broek, and Linderholm (2006), people read sentences that had causal constraints that were either weak (e.g., As he arrived at the bus stop, he saw his bus was already five blocks away) or strong (As he arrived at the bus stop, he saw his bus was just pulling away). During reading, the researchers presented lexical decision probes that corresponded to likely inference that the readers might make (e.g., *run*, in this case). Of significance, this presentation was done to the left and right hemispheres by presenting the words on either the right or left half (respectively) of the computer screen. The data showed that the right hemisphere was more involved in generating remote associations (associated concepts that are not closely semantically related to the concepts in the sentences). Moreover, people with high working memory capacity activated fewer remote associations than low-span people. Essentially, people with a high working memory span were more focused in the amount of knowledge they activated during comprehension.

UPDATING Situations that we experience or read about often are in a state of flux. Things are always changing, and the events may differ from one moment to the next. Thus, the cognitive processes involved in comprehension must be able to shift the current understanding to adapt to these ongoing changes.

Situation Model Updating (Figure 10.10 and Figure 10.11)

Several **updating** processes alter a person's situation model in the face of information about how the situation has changed.

> **Interactive**
>
> To provide a framework for understanding how these changes can occur, we will use Zwaan's Event Indexing Model (Zwaan, Langston, & Graesser, 1995; Zwaan, Magliano, & Graesser, 1995; Zwaan & Radvansky, 1998). Per this theory, people actively monitor multiple event dimensions during reading to assess whether there has been a meaningful change along any of them.
>
> In the original version of the theory, five dimensions were proposed: space, time, entity, intentionality (goals), and causality. When there is a disruption along any one of these dimensions, people update their situation models, and this updating process takes time. For example, a break along the space and time dimensions could happen if a story protagonist were to move to a new location or there were a jump in time (e.g., a week later . . .). Similarly, if a new character were introduced into a story, the person would need to update the entity dimension, and so on. The mental processing involved in updating any of these dimensions appears to operate largely independently of the others (Curiel & Radvansky, 2014), suggesting that people are actively tracking multiple dimensions of experience when trying to understand stories. Further research has shown that people monitor more than just these five dimensions. For example, people also track emotional information (e.g., Komeda & Kusumi, 2006).

10.3.3: Events

Our discussion of comprehension up to now has largely focused on language comprehension of either written or spoken language, and this reflects the thrust of research in this area. However, this is not the only type of comprehension that people can engage in. We also comprehend events that we see or are involved in.

> **WRITING PROMPT**
>
> **Comprehension and the Real or Possible World**
>
> What is the relationship between comprehension and the real or possible world that is being comprehended? Which the cognitive representations and processes are critical for this?
>
> ▶ The response entered here will appear in the performance dashboard and can be viewed by your instructor.
>
> Submit

10.4: Conversation and Gesture

OBJECTIVE: Evaluate the cognitive effects of conversational interactions

We turn now to the comprehension of conversation and gesture. We focus on *conversation*—normal, everyday language interactions, such as an ordinary talk among friends. The issues we consider, however, apply to all kinds of linguistic interactions: how professors lecture and students comprehend, how people converse on the telephone, how an interviewer and a job applicant talk, how we reason and argue with one another (Rips, 1998), and so on. Furthermore, we look at how we expand on what we say by moving our hands about, making gestures, and by examining the cognitive role of these gestures.

Comprehending Events (Figure 10.12)

To better understand event comprehension, it helps to look at comics, video, and interactive experiences.

> **Interactive**
>
> **Comics**
>
> Narrative events often are thought of as being written or spoken descriptions, and certainly some narratives are. However, we encounter many different narratives in the world. A common type of narrative is comics. This includes comic strips, comic books, and graphic novels. What is interesting is that these narratives convey information visually as well as verbally in a printed format. How do they compare to written and spoken narratives in terms of the cognitive processes that are used to understand and remember them? The answer is: quite well.
>
> Like written narratives, comics have a structure to convey information, much like the syntactic structure of language. In fact, Cohn (2013b) has developed a theory of Visual Narrative Grammar he applies to comics and other visual narratives. For example, agents typically appear prior to patients in comics (Cohn & Paczynski, 2013). Thus, there is a mental syntax, or set of rules, that we follow to comprehend visual narratives of this type. That said, unique aspects of processing and comprehension occur when a person reads comics, particularly on a rather complicated page in a graphic novel (Cohn 2013a).
>
> **Video**
>
> **Interactive Experiences**

During a conversation, speakers develop a rhythm as each person takes successive turns speaking. Nonverbal interaction can occur during a turn, such as when a listener nods to indicate attention or agreement.

10.4.1: The Structure of Conversations

Conversations have a form and structure. They are not chaotic and random and we expect them to go certain ways. These expectations are part of the cognitive process of how conversations emerge. This section covers the structure of conversations and the implications for our understanding of cognition.

10.4.2: Cognitive Conversational Characteristics

Conversations are structured by cognitive factors. We focus on three: the conversational rules we follow, the issue of topic maintenance, and the online theories of conversational partners.

CONVERSATIONAL RULES Grice (1975; see also Norman & Rumelhart, 1975) suggested four **conversational rules** or maxims, rules that govern our conversational interactions with others, all derived from the **cooperative principle**. This is the idea that each participant in a conversation implicitly assumes all speakers are following the rules and that each contribution to the

Characteristics of Conversations

Let's examine two characteristics of conversations, the issues of turn taking and social roles, to get started and introduce some of the more cognitive effects of interest.

> **Taking Turns**
>
> Conversations are structured by a variety of cognitive and social variables and rules governing "the what and how" of our contributions. To begin, we take turns. Typically, there is little overlap between participants' utterances. Generally, two people speak simultaneously only at the change of turns, when one speaker is finishing and the other is beginning. In fact, interchanges in conversation often come in an *adjacency pair*, a pair of turns that sets the stage for another part of the conversation. For instance, if Ann wants to ask Betty a question, there can be an adjacency pair of utterances in which Ann sets the stage for the actual question:
>
> **Ann:** Oh, there's one thing I wanted to ask you.
>
> **Betty:** mhm
>
> **Ann:** In the village, they've got some of those . . . rings. . . . Would you like one? (From Svartvik & Quirk, 1980, cited in Clark, 1994.)
>
> The neutral "mhm" is both an indication of attention and a signal that Ann can go ahead and ask the question (Duncan, 1972).
>
> The rules we follow for turn taking are straightforward (Sacks, Schegloff, & Jefferson, 1974). First, the current speaker oversees selection of the next speaker. This is often done by directing a comment or question toward another participant ("What do you think about that, Fred?"). Second, if the first rule is not used, then anyone can become the current speaker. Third, if no one else takes the turn, the current speaker may continue but is not obliged to.

conversation is a sincere, appropriate contribution. In a sense, we enter a pact with our conversational partner, pledging to abide by certain rules and adopting certain conventions to make our conversations manageable and understandable (Brennan & Clark, 1996; Wilkes-Gibbs & Clark, 1992). This includes issues of syntax, where we choose syntactic structures that mention important discourse focus information early in our sentences (Ferreira & Dell, 2000) and use syntactic structures that are less ambiguous (Haywood, Pickering, & Branigan, 2005). We use intonation and prosody that help disambiguate an otherwise ambiguous syntactic form (Clifton, Carlson, & Frazier, 2006); we decide on word choice, as in situations when two conversational partners settle on a mutually acceptable term for referring to some object (Metzing & Brennan, 2003; Shintel & Keysar, 2007); and we often use gestures to amplify or disambiguate our speech (Goldin-Meadow, 1997; Kelly, Barr, Church, & Lynch, 1999; Özyürek, 2002).

A simple example or two should help you understand the point of these maxims. When a speaker violates or seems to violate a maxim, the listener assumes there is a reason and may not detect a violation (Engelhardt, Bailey, & Ferreira, 2006). That is, the listener assumes that the speaker is following the overarching cooperative principle and so must have intended the remark as something else, may be sarcasm or may be a nonliteral meaning (Kumon-Nakamura, Glucksberg, & Brown, 1995). As an

Table 10.5 Grice's (1975) Conversational Maxims, with Two Additional Rules

As Table 10.5 shows, the four maxims specify in detail how to follow the cooperative principle. (Two further rules have been added to the list for purposes that will become clear in a moment.)

The Cooperative Principle	
Relevance	Your utterances should be relevant to the discourse (e.g., stay on topic; don't make statements about things that others are not interested in).
Quantity	As needed, provide as much information as is necessary (e.g., don't give too much information; don't go beyond or give short shrift to what you know; don't give too much information).
Quality	Have what you say be truthful (e.g., don't give misleading information; don't lie; don't exaggerate).
Manner and Tone	Aim for clarity (e.g., avoid saying things that are unnecessarily ambiguous or obscure); keep it brief, but polite, and don't interrupt someone else.
Two Additional Rules	
Relations with Conversational Partner	Infer and respond to partner's knowledge and beliefs (e.g., tailor contributions to partner's level; correct misunderstandings).
Rule Violations	Signal or mark intentional violations of rules (e.g., use linguistic or pragmatic markers [stress, gestures]; use blatant violations; signal the reason for the violation). From Grice (1975); see also Norman and Rumelhart (1975).

example, imagine studying in the library when your friend asks:

(15) Can I borrow a pencil?

This is a straightforward speech act, a simple request you could respond to directly. But if you had just lent a pencil to your friend, and he said,

(16) Can I borrow a pencil with lead in it?

the question means something different. Assuming your friend was being cooperative, you now must figure out why he broke the quantity maxim about overspecifying: All pencils have lead in them, and mentioning the lead is a violation of a rule. You infer that it was a deliberate violation, where the friend's authorized implication can be expressed as "The pencil you lent me doesn't have any lead in it, so would you please lend me one I *can* use?" In general, people are adept at decoding speech acts and knowing what others are trying to achieve by what they say (Holtgraves, 2008a).

TOPIC MAINTENANCE We also follow the conversational rules in terms of **topic maintenance**, making our contributions relevant to the topic and sticking to it. Topic maintenance depends on two processes: comprehension of the speaker's remark and expansion, contributing something new to the topic.

Schank (1977; see also Litman & Allen, 1987) provides an analysis of topic maintenance and topic shift, including a consideration of what is and is not a permissible response, called simply a *move*, after one speaker's turn is over. The basic idea is that the listener comprehends the speaker's comment and stores it in memory. As in reading, the listener must infer what the speaker's main point was or what the discourse focus was. If the speaker, Ben, says,

(17) I bought a new car in Baltimore yesterday,

then Ed, his conversational partner, needs to infer Ben's main point and expand on that in his reply. Thus, sentence 18 is legal because it apparently responds to the speaker's authorized implication, whereas sentence 19* is probably not a legal move (denoted by the * sign):

(18) Ed: Really? I thought you said you couldn't afford a car.

(19*) Ed: I bought a new shirt yesterday.

Sentence 18 intersects with two main elements for sentence 17, "BUY" and "CAR," so it is probably an acceptable expansion. Sentence 19* intersects with "BUY," but the other common concept seems to be the time case role "YESTERDAY," an insufficient basis for most expansions. Thus, in general a participant's responsibility is to infer the speaker's focus and expand on it in an appropriate way. That is the relevance maxim: Sticking to the topic means you must infer it correctly. Ed seems to have failed to draw the correct inference.

On the other hand, may be Ed did comprehend Ben's statement correctly. If so, then he deliberately violated the relevance maxim in sentence 19*. But it is such a blatant violation that it suggests some other motive. Ed may be expressing disinterest in what Ben did or may be saying indirectly that he thinks Ben is bragging. And if Ed suspects Ben is telling a lie, then he makes his remark even more blatant, as in sentence 20:

(20) Yeah, and I had lunch with the Queen of England.

ONLINE THEORIES DURING CONVERSATION A final point involves the theories we develop of our conversational partners, something called **theory of mind**. The most obvious one is a **direct theory**. This is the mental model of what the conversational partner knows and is interested in, what the partner is like. We tailor our speech so we are not being too complex or too simplistic, and so we are not talking about something of no interest to the listener. Some clear examples of this involve adult–child speech, where a child's smaller vocabulary and knowledge prompt adults to modify and simplify their utterances in several ways (DePaulo & Bonvillian, 1978; Snow, 1972; Snow & Ferguson, 1977). But sensitivity to the partner's knowledge and interests is present to some degree in all conversations—although not perfectly, of course. We do not talk in our college classes the way we would to a group of second graders, nor do we launch into conversations with bank tellers about our research. Horton and Gerrig (2002) call this "audience design," which is awareness of the need to design our speech to the characteristics of our audience (e.g., Lockridge & Brennan, 2002). Alternatively, if we do not know much about another, we may assume that person knows what we know (Nickerson, 2001) and then revise our direct theory as we observe how well that person follows our remarks (Clark & Krych, 2004).

Audience design has implications beyond conversations. When we tell stories, we modify what we tell our listeners based on who they are and our social relationship to them. This retelling is not the same as recall. We modify the information we report to fit the social situation. In retelling stories, we often exaggerate some parts, minimize others, add information that was not there originally, and leave some bits out, all to suit our audience and the broader message we are trying to convey (Marsh, 2007).

There is another layer of theories during a conversation, an interpersonal level related to "face management," or public image (Holtgraves, 1994, 1998). Let's call this the **second-order theory**. This second-order theory is an evaluation of the other participant's direct theory: what you think the other participant believes about you.

Examples of Direct and Second-Order Theories

Let's develop examples of these two theories to illustrate their importance.

Imagine that you are registering for classes next semester and say to your friend Frank that you have decided to take Psychology of Personality. What would your reaction be if Frank responded to you with these statements?

(21) Why would you want to take that? It's just a bunch of experiments with rats, isn't it?

(22) Yeah, I'm taking Wilson's class next term too. John told me he's going to assign some books he thinks I'll really like.

(23) Maybe you shouldn't take Wilson's class next term. Don't you have to be pretty smart to do all that reading?

10.4.3: Empirical Effects in Conversation

Let's conclude with some evidence about the conversational effects we have been discussing. One of the most commonly investigated aspects of conversation involves **indirect requests**, such as when we ask someone to do something ("Close the window"; "Tell me what time it is") by an indirect and presumably politer statement ("It's drafty in here"; "Excuse me, but do you have the correct time?").

> **Prove It**
>
> ### Politeness Ethic in Conversational Requests
> One of the best student demonstration projects we have ever graded was a test of the politeness ethic in conversational requests. On five randomly selected days, the student sat next to a stranger on the bus, turned, and asked, "Excuse me, but do you have the correct time?" All five strangers answered her. On five other randomly selected days, she said to a stranger, "Tell me what time it is," not in an unpleasant tone, but merely in a direct fashion; none of the strangers answered. Devise other situations in which you violate the politeness ethic or other conversational rules and note people's reactions. If you do it properly, you will learn about the rules of conversation. But be careful that it doesn't turn into a demonstration project on aggression. Do the same thing again, but this time with a close friend or family. You will see how necessary some polite forms are with strangers and how inappropriate they are with people you know well.

INDIRECT REQUESTS Clark (1979) reported an impressive investigation of indirect requests. The study involved telephone calls to some 950 merchants in the San Francisco area in which the caller asked a question that the merchant normally would be expected to deal with on the phone (e.g., "What time do you close?" "Do you take credit cards?" "How much does something cost?"). The caller would write down a verbatim record of the call immediately after hanging up. A typical conversational interaction was as follows:

> (24) Merchant: "Hello, Scoma's Restaurant."
>
> Caller: "Hello. Do you accept any credit cards?"
>
> Merchant: "Yes we do; we even accept Carte Blanche."

Of course, the caller's question here was indirect: "Yes" isn't an acceptable answer to "Do you accept any credit cards?" because the authorized implication of the question was "What credit cards do you take?" Merchants almost always responded to the authorized implication rather than merely to the literal question. Furthermore, they tailored their answers to be informative while not saying more than necessary (obeying the second rule, on quantity), as in "We accept *only* Visa and MasterCard," or "We accept *all* major credit cards." Such responses are both informative and brief.

Such research has been extended to include not only indirect requests but also a variety of indirect statements and replies to questions. For instance, Holtgraves (1994) examined comprehension speed for indirect requests as a function of whether the speaker was of higher status than the listener (e.g., boss and employee) or whether they were of equal status (two employees). People read a short scenario (e.g., getting a conference room ready for a board of directors meeting), which concluded with one of two kinds of indirect statements. *Conventional statements* were normal indirect requests, such as "Could you go fill the water glasses?" *Negative state remarks* were more indirect, merely stating a negative situation and only indirectly implying that the listener should do something (e.g., "The water glasses seem to be empty."). People showed no effects of status when comprehending regular indirect requests; it did not matter whether it was a peer or the boss who said, "Could you go fill the water glasses?" But comprehension time increased significantly with negative state remarks made by peers. In other words, when the boss says, "The water glasses seem to be empty," we comprehend the conventional indirect request easily. But when a peer says it, we need additional time to comprehend.

INDIRECT REPLIES Holtgraves (1998) also focused on indirect replies, especially the idea of making a "face-saving" reply. His participants read a description of a situation, such as

> (25) Nick and Paul are taking the same history class. Students in this class must give a 20-minute presentation to the class on some topic.

They then read a sentence that gave positive (26) or negative (27) information about Nick's presentation or a sentence that was neutral (28):

> (26) Nick gave his presentation and it was excellent. He decides to ask Paul what he thought of it: "What did you think of my presentation?"
>
> (27) Nick gave his presentation and it was truly terrible. He decides to ask Paul what he thought of it: "What did you think of my presentation?"
>
> (28) Nick gave his presentation and then decided to ask Paul what he thought of it: "What did you think of my presentation?"

If you were Paul and faced the prospect of telling Nick that his presentation was awful, wouldn't you look for some face-saving response? This is exactly how people responded when they comprehended Paul's responses. In the excuse condition, Paul says,

(29) It's hard to give a good presentation,

in effect giving Nick a face-saving excuse for his poor performance. Another possible move is to change the topic, to avoid embarrassing Nick, as in

(30) I hope I win the lottery tonight.

Holtgraves (1998) collected several measures of comprehension, including overall comprehension time for the critical sentences 29 and 30. The comprehension times, shown in Figure 10.13 (from Experiment 2), were very clear. When people had heard positive information—the talk was excellent—it took them a long time to comprehend either the excuse (29) or topic change (30) response. But having heard negative information—the talk was terrible—was nearly the same as having heard nothing about the talk. People comprehended the excuse or topic change responses more rapidly, and there was no major difference between no information and negative information. People interpreted the violations of the relevance maxim as attempts to save face and avoid embarrassment.

Another role of indirect speech is to provide some element of *plausible deniability* (Lee & Pinker, 2010). Indirect requests and responses may carry implicit requests and replies that would be inappropriate otherwise. For example, suppose you are told by a not-so-savory relative of the defendant of a trial "I hear you're the foreman of the jury. It's an important responsibility. You have a wife and kids. We know you'll do the right thing." Thus, while the literal wording is acceptable, the implied meaning (a threat) is not. Thus, indirect speech can play many roles in human communication.

10.4.4: Metaphors and Idioms

There is a great deal of language that is not intended to be interpreted literally. Instead, the expectation is for people to read the intended meaning *into* what is being said. So, although language can be viewed as instructions for how to create one's situation models, these instructions are not always straightforward and direct.

Figure 10.13 Comprehension Times from Holtgraves's (1998) Study

Participants read settings in which negative information, positive information, or neutral information was offered about a character, followed by a conversational move in which the speaker made an excuse for the character or changed the topic. In both cases, it took longer to comprehend the remark when positive information about the character had just been encountered.

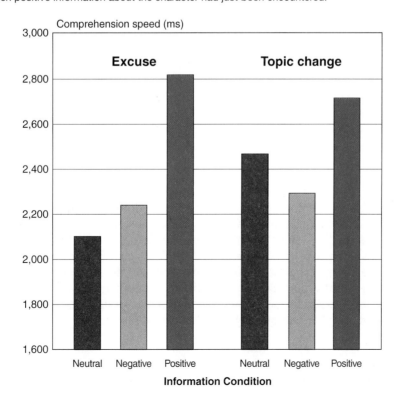

10.4.5: Gesture

When we speak, we not only move our lips, tongues, throats, and so on, we may also move our arms and hands. This movement, or **gesture**, is done to facilitate communication to listeners and excludes sign language and non-communicative mannerisms, such as touching one's hands to one's face (McNeill, 1992). That is, if they are not *beat gestures* that mark out important words or ideas, the movements we make with our hands communicate information to augment the words we are using. For example, when describing your route to school, you may make gestures to convey information about turns you made, obstructions you encountered, speeders you saw, and so on.

WRITING PROMPT

Cognitive Effects of Conversational Interactions

What information is actually present in a conversation that allows people to understand one another? What information is inferred from cognition during a conversation to make the interchange more efficient?

The response entered here will appear in the performance dashboard and can be viewed by your instructor.

Submit

Two Additional Forms of Language

Thus, we consider two additional forms of language in which what is said literally is not what is intended. These are metaphors and idioms.

Interactive

Metaphors

A *metaphor* is an expression using language to compare one thing with another, or to use language and knowledge of one domain to understand another domain. So, when you hear people say, "My job is my jail," they typically do no literally mean that they are physically imprisoned by their employment. Instead, they are comparing their job to the confining, restrictive, and inescapable experience of being in jail. This is using language to make an **analogy**, much as we have seen analogical thinking with other types of cognition. It is just much more explicit and out in the open here.

The standard way of thinking about how people cognitively process metaphors is that they first derive the literal meaning of a sentence. This process results in an interpretation that is literally false and may not even make sense. After drawing the literal interpretation, people compare it with what is going on in the current context in which the utterance was produced. At this point, it becomes clear that a literal interpretation does not work. So, people then try to derive a nonliteral meaning, using the metaphor as the basis of an analogy.

Psycholinguistically, although there may be cases in which people go through the steps outlined above, it appears that most of the time people are deriving the literal meaning of the metaphor while simultaneously, in parallel, obtaining the metaphoric meaning (Glucksberg, Gildea, & Bookin, 1982). This may involve an actual retrieval of a metaphor from memory if it has been encountered before or deriving a metaphorical interpretation by making a mental analogy using the literal circumstances of the language and comparing it with the set of circumstances confronting the person. This derivation of metaphoric analogy appears to be largely automatic and cannot be disrupted to lead people to interpret a metaphor only literally (Glucksberg, 2003).

Idioms

Importance of Gesture

Although gesturing may serve a more holistic function when we use it while speaking, when we gesture in the absence of spoken speech, it takes on more linguistic characteristics, much as sign languages do (Goldin-Meadow, 2006).

> **Interactive**
>
> **Involves Strong Social Component**
>
> Work by researchers such as Bavelas, Gerwing, Sutton, and Prevost (2008) shows that people even gesture when they are talking on the telephone and know the person on the other end cannot see them (although they gesture less than when they can see the other person). So, there must be something cognitively important about gesturing that facilitates language production in conversation. It should also be noted that people gesture less when they are speaking into a tape recorder (Bavelas et al., 2008) or when listeners do not appear to be attending to what they are saying (Jacobs & Garnham, 2007). So, this impact of gesture has a strong social component and is not purely psycholinguistic.
>
> **Helps Learning and Thinking**
>
> **Communicates Embodied Information**

Summary: Comprehension

10.1 Conceptual and Rule Knowledge

- A variety of online tasks have been devised to investigate comprehension, such as tasks involving reading times, the use of probes, think-aloud verbal protocols, and neuroimaging evidence.
- Successful comprehension is best achieved when people can self-monitor what they are and are not learning through judgments of learning. However, these judgments are often poor estimates of how much has actually been learned. Estimates can be improved by delaying these judgments, rereading, and providing summaries of the material.
- Comprehension involves processing at many levels, including the surface form, propositional textbase, and situation model levels. Evidence of processing at each of these levels can be derived across many different aspects of understanding.

10.2 Reading

- Tremendous progress has been made in understanding the mental processes of reading, largely by using the online measures of comprehension, such as reading times and gaze durations.

- Modern models of reading make predictions about comprehension based on a variety of factors; for instance, word frequency and recency in the passage influence surface form and textbase processing, respectively.
- Online measures of language comprehension provide a unique window into human cognition. Using these sorts of measures, we can gain moment-to-moment insights not only into the effectiveness of processing, but also into the very contents of people's minds.

10.3 Reference, Situation Models, and Events

- Reference in language involves the idea of bridging together different elements of a statement. The source of knowledge that permits speakers to include reference in their messages and listeners to infer the basis for those bridges is not just our knowledge of syntax and word meanings, but the entirety of semantic memory and much top-down processing.
- Situation models are created by combining information from the language itself as well as inferences people draw based on their prior semantic and episodic knowledge. This inference making is at the heart of comprehension and reflects information in the language, such as grammatical aspect.
- The capacity and operation of working memory are known to be important factors in understanding individual differences in reading comprehension.
- Situation model updating occurs when people detect a meaningful change in an event dimension. This updating process is cognitively effortful, resulting in increases in reading times and brain activity.
- Comprehension occurs not only for language that people hear or read, but also for other aspects of experience, including narrative comics, videos, and interactive experiences.

10.4 Conversation and Gesture

- Conversations follow a largely implicit set of conversational rules. Some of these involve turn taking and social status and conventions, but many more govern the nature or topic of participants' contributions. Topic shifts involve selecting some part of a person's utterance to form the basis for a new contribution but then adding some new information. Schank's work is a particularly important analysis of this process of topic shifting.
- Participants in a conversation develop theories of mind of the other speakers, such as direct theory, as well as theories of what the other speakers think of them, called second-order theories. When we converse, we tailor our contributions to these theories and follow a set of conversational rules, the unspoken contract between conversational partners. When a rule is violated intentionally, usually to make some other point (e.g., sarcasm), we mark our violation so that its apparent illegality as a conversational move is noticed and understood.
- Empirical work on conversational interaction often tests general notions about direct theories, the politeness rule, or indirect requests. Although we sometimes attempt to manipulate another person's direct theory of us, research also shows that the initially planned utterance usually is from a very egocentric perspective, whereas later adjustments may take the other person's perspective into account. The standard, everyday use of nonliteral language is also seen with the use of metaphors and idioms.
- Gestures made during conversation are a way that simulated spatial and action information can be communicated. Making gestures is part of the social act of conversation, although it may sometimes occur when our partner cannot see us, as when we are talking on the telephone. Gesture can serve as a working memory aid and help people solve problems.

SHARED WRITING

Comprehension in Written and Spoken Language

A lot of mental processes and representations are needed for successful comprehension. Given these, what everyday circumstances might make comprehension more difficult?

▶ A minimum number of characters is required to post and earn points. After posting, your response can be viewed by your class and instructor, and you can participate in the class discussion.

Post   0 characters | 140 minimum

Chapter 11
Reasoning and Decision Making

Learning Objectives

11.1: Differentiate between the forms of syllogistic and conditional reasoning

11.2: Assess the decision-making process

11.3: Explain the fundamental heuristics people use to make decisions

11.4: Assess how decisions are influenced by the context in which they are encountered

11.5: Explain the impact of adaptive thinking on heuristics

11.6: Evaluate the range of approaches to human decision making

11.7: Analyze the limitations in reasoning

A clear picture of cognitive psychology would be incomplete without a consideration of the slower, more deliberate kinds of thinking that we cover in this module. Specifically, how do we reason? How do we make decisions under conditions of uncertainty? These are the topics to which we now turn. A general thread running through reasoning and decision-making research is that we are often biased and overly influenced by our general world knowledge. These influences affect how we reason and make decisions. A second pervasive thread is our tendency to search for evidence that confirms our decisions, beliefs, and hypotheses far more than is logical. In general, we are much less skeptical than we ought to be.

We will begin by examining two classic forms of reasoning before moving to decision making about the likelihood of events for which relevant information in memory is generally lacking or insufficient. The strategies people use to make these judgments are interesting because they reveal a variety of rules of thumb or shortcuts on which people rely. These shortcuts work well sometimes, but can also lead us to distortions and biases. Overall, this research provides convincing examples of the uncertainty of human reasoning and the often-surprising inaccuracies in our knowledge.

11.1: Formal Logic and Reasoning

OBJECTIVE: Differentiate between the forms of syllogistic and conditional reasoning

At some point during their college careers—often in a course on logic—many students are exposed to the classic forms of reasoning. The two forms on which we will focus here are *syllogistic* and *conditional reasoning*. However, there are others, such as relational reasoning (e.g., Goodwin & Johnson-Laird, 2005). Overall, people are not particularly good at solving such problems when they are presented in an abstract form. They do better when the problems are presented in terms of concrete, real-world concepts, but still fall short. That said, in some situations, our world knowledge almost prevents us from seeing the "pure" (i.e., logical) answers (e.g., Markovits & Potvin, 2001), and errors are made. In short, people are not particularly logical thinkers.

When reading about these formal logic problems, do not let yourself be lulled into the feeling that these are just abstract, academic thought problems. The basic form of these problems has the potential to underlie much of our everyday thinking and decision making. For example,

275

Sudoku puzzles are built on logical, not mathematical, reasoning even though they involve numbers (Khemlani & Johnson-Laird, 2012). When a doctor is trying to determine a diagnosis, a detective is trying to solve a crime, or you are trying to figure out whether someone is drinking illegally, the principles of formal logic are at work.

11.1.1: Categorical Syllogisms

A **categorical syllogism** is a three-statement logical form, with the first two parts stating the premises taken to be true, and the third stating a conclusion based on those premises.

BIASES OF SYLLOGISTIC REASONING Let's look at some of the biases that have been observed when people try to reason syllogistically. First, there is the *figural effect*, which is a bias to arrive at conclusions in which the terms are in the same order as in the figure of the premises (Johnson-Laird, 1983). For example, if people learn that *"Some A are B"* and *"All B are C"* there is a bias to conclude that *"Some A are C."* Whereas if people learn that *"Some B are A"* and *"All C are B,"* there is a bias to conclude that *"Some C are A."* So, people are drawing conclusions not based on a rational assessment of the information, but simply based on the *order* in which the terms appear in the sentences.

Another difficulty or confusion people have is the **belief bias**, which is illustrated by the following example:

(1d) All poodles are animals.

All animals are wild.

Therefore, all poodles are wild.

The difficulty is that, although it seems to fly in the face of your world knowledge, the conclusion is logically true. If we assume that the premises are true, then the conclusion must be. However, the truth of the premises is

---

Syllogism Premises and Conclusions

The goal of syllogistic reasoning is to understand how the premises combine to yield logically true conclusions (if any).

Syllogisms may be presented in an abstract form using just letters, such as

(1a) All A are B.

All B are C.

Therefore, all A are C.

In this example, the two premises state a relation between the elements *A*, *B*, and *C*. "*All A are B*" states that the set *A* is a subset of the group *B*, that *A* is included in the set *B*. The third statement is the conclusion. By applying the rules of categorical reasoning, we can determine that the conclusion *"All A are C"* is true, so the conclusion follows logically from the premises.

Inserting words into the syllogism can help you understand this. For instance,

(1b) All poodles are dogs.

All dogs are animals.

Therefore, all poodles are animals.

another issue. Because the conclusion follows from the premises, the syllogism is valid. Of course, it is easy to think of counterexamples in the real world of poodles; hardly any poodles are wild, after all (Feldman, 1992). In this case, the premise "All animals are wild" violates your world knowledge. However, that is beside the point here. The point is to assess the validity of the conclusion given the premises.

The belief bias is the human tendency to ignore the logical form of an argument and focus instead on prior knowledge (Evans, Barston, & Pollard, 1983; see also Copeland, Gunawan, & Bies-Hernandez, 2011, for a similar source credibility effect). For the rules of syllogistic reasoning, the *truth* of the premises is *separate* from the *validity* of the logical argument. What matters is that the conclusion does or does not validly follow from the premises. So, applying syllogistic reasoning to real-world problems is at least a two-step process. First, determine whether the syllogism itself is valid; second, if the syllogism is valid, determine the truth of the premises.

Now consider another example:

(2a*) All A are B.

Some B are C.

Therefore, some A are C.

Note that in formal logic, *some* means "at least one and possibly all," although some people have trouble understanding this (Schmidt & Thompson, 2008). That said, not all the errors people make when solving syllogisms arise from relying on *Gricean maxims* of what is "likely meant" by the premises (Newstead, 1995). Try inserting words into this example to see whether the conclusion is correct.

(2b*) All polar bears are animals.

Some animals are white.

Therefore, some polar bears are white.

Euler Circle Illustrations for Three Categorical Syllogisms (Figure 11.1)

Euler circles can help you determine whether a syllogism is valid.

1. All A are B.
   All B are C.
   Therefore,
   all A are C.

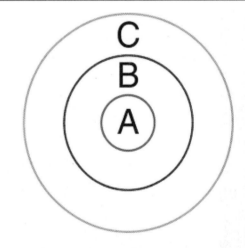

The "All–All" Form

In the above illustration, the "All–All" form shows that it is necessarily true that "All *A* are *C*." The circles, which represent the class of things *A*, *B*, and *C*, are nested such that *A* is a subset of *B* and *B* is a subset of *C*. There is no other way to represent the premises except by concentric circles (when *A* and *B* are identical, their boundaries overlap completely, and the diagram shows one circle labeled both *A* and *B*).

Despite the idea that word substitutions can lead to an empirically correct statement, this syllogism is invalid because the two premises do not invariably lead to a valid conclusion. The entire form of the syllogism is invalid (the reason for the asterisk). The incorrectness of the conclusion in (2a) stems from the qualifier *some*. Although the conclusion may be true in the world here, when you use another concrete example, you can see that this conclusion is not necessarily the case for *all* examples, as shown by the following:

(2c*) *All polar bears are animals.*

Some animals are brown.

Therefore, some polar bears are brown.

11.1.2: Theories of Syllogistic Reasoning

There are several theories of how people succeed in mentally solving syllogisms (or fail to do so). Following Khemlani and Johnson-Laird (2012), we discuss three basic types of theories: heuristics, rules, and mental models.

Heuristics are general rules of thumb that often provide the correct answer, but are not guaranteed to do so. The idea behind **heuristic theories** of reasoning is that people solve logical problems not by using well-reasoned steps, using instead shortcuts based on the nature of the problem itself. For example, according to the **atmosphere**

Mental Model Theories

Mental model theory (Johnson-Laird, 1983, 2013) has suggested that people create mental simulations of the circumstances derived from the premises. People then develop various conclusions from those mental models. Many problems are difficult because they require people to construct multiple mental models to derive the appropriate conclusion or to determine that no valid conclusion can be drawn because there are mutually incompatible circumstances that apply to those premises.

> To illustrate this process, assume that people are given the syllogism 2a. For the first premise, *All A are B*, people might create a mental model with tokens standing for A that are all associated with Bs, such as:
>
> A B
>
> A B
>
> Then, when people get the next premise, *Some B are C*, they might augment their mental model so that it has some Bs associated with C, and some not, such as:
>
> A B C
>
> A B
>   C

heuristic (Woodworth & Sells, 1935), people favor conclusions that match the premises. So if there is a *not* in the premises, then there is a bias to favor a conclusion that also has a *not* in it (a similar heuristic is the *matching heuristic*; Wetherick & Gilhooly, 1995). Another heuristic is **illicit conversion** (Chapman & Chapman, 1959) in which people inappropriately swap the terms of a statement, such as mentally converting "All *B* are *A*" to "All *A* are *B*." Chater and Oaksford (1999) have proposed a **probability heuristics theory** in which people make decisions using a collection of heuristics that use the information value of the premises. Regardless of the particular theory, all capture the finding that people sometimes do not put forth the kind of thinking needed to assess the validity of logical conclusions, but instead rely on superficial characteristics of the problem itself. This is the opposite of what most of us think of as rational thinking, but that is what many of us do.

Another group of theories are **mental rules theories** of human reasoning that assume that people are using rules, similar to the types of mental rules people use when processing language, to assess the validity of various logical conclusions (e.g., Rips, 1994). These mental rules are thought to capture an otherwise sound mental logic. As long as these rules are allowed to run and function without interruption, people can come to valid logical conclusions. Problems arise because either people are distracted and apply their mental resources elsewhere, or, more often in experiments on mental logic, problems stem from difficulties in converting the language of the problem into the appropriate mental logic form.

A final group of theories assumes that when people are given the premises of a syllogism, they create mental representations, called **mental models**, of the states of affairs they describe. An externally written analogue to this process would the Euler circles described earlier. However, Euler circles cannot capture what people are doing mentally because, often, this method must be explicitly taught. It is not something we do naturally or spontaneously.

11.1.3: Conditional Reasoning

CONDITIONAL REASONING: IF *P* THEN *Q* Conditional reasoning is a second kind of logical reasoning. Conditional reasoning problems contain two parts, a *conditional clause*, a statement that expresses a relationship (*if P then Q*), followed by some *evidence* pertaining to the conditional clause (*p*, for example). **Conditional reasoning** involves *a logical determination of whether the evidence supports, refutes, or is irrelevant to the stated if–then relationship*.

The *conditional* in these problems is the *if–then* statement. The *if* and *then* clauses are respectively known as the **antecedent** and the **consequent**. The *if* states a possible cause, and the *then* states an effect of that possible cause. So, *if P then Q* means "if *P* is true, then *Q* is true"; for example, if it rains (*P*) then the streets will be wet (*Q*). So far so good.

After the *if–then*, you get a second statement, some evidence about the truth or falsity of one part of the *if–then* relationship. The aim is to take this evidence and decide what follows logically from it. In other words, is the conditional *if–then* statement true or false given this evidence, or is the evidence irrelevant to it?

The evidence that follows an *if P, then Q* conditional results in one of the four possible outcomes. For "If it rains, the streets will be wet," the four possibilities are

P: That is, *P* is true, it's raining.

not P: That is, *P* is not true, it's not raining.

Q: That is, *Q* is true, the streets are wet.

not Q: That is, *Q* is not true, the streets are not wet.

Putting this together yields four possibilities:

Conditional	If *P*, then *Q*.	If *P*, then *Q*.	If *P*, then *Q*.	If *P*, then *Q*.
Evidence	*P*.	Not *P*.	*Q*.	Not *Q*.
Conclusion	Therefore, *Q*.	(No conclusion)	(No conclusion)	Not *P*.

For a conditional *if–then* statement, if the antecedent *P* is true, then its consequence (the *consequent*) *Q* must be true. As an expanded example, consider the following example, working out for all four possibilities in Table 11.1:

Table 11.1 Conditional Reasoning

Form	Name	Example
If *P*, then *Q*. Evidence: *P*. Therefore, *Q*.	Modus ponens: affirming the Antecedent (valid inference)	If I am a freshman, I must take Intro to Logic. Evidence: I am a freshman. Therefore, I must take Intro to Logic.
If *P*, then *Q*. Evidence: not *P*. *Therefore, not *Q*.	Denying the Antecedent (invalid inference)	If I am a freshman, I must take Intro to Logic. Evidence: I am not a freshman. *Therefore, I do not have to take Intro to Logic.
If *P*, then *Q*. Evidence: *Q*. *Therefore, *P*.	Affirming the Consequent (invalid inference)	If I am a freshman, I have to take Intro to Logic. Evidence: I have to take Intro to Logic. *Therefore, I am a freshman.
If *P*, then *Q*. Evidence: not *Q*. Therefore, not *P*.	Modus tollens: denying the consequent (valid inference)	If I am a freshman, I have to take Intro to Logic. Evidence: I do not have to take Intro to Logic. Therefore, I am not a freshman.

If I am a freshman, then I must take Intro to Logic.

VALID ARGUMENTS As the Table 11.1 shows, only two possibilities lead to a valid conclusion. In the first one, when given evidence that *P* is true, "I am a freshman," then the consequent *Q* must be true, "I do have to take Intro to Logic." This is *affirming the antecedent* (saying that the antecedent is true). The classic name for this is *modus ponens*. Likewise, if the evidence is that *Q* is not true, "I do not have to take Intro to Logic," it must therefore be that *P* is not true, "*I am not a freshman*." This is *denying the consequent* (saying that the consequent is not true) is called *modus tollens*.

In comparison, the other two arguments do not lead to a valid conclusion. By denying the antecedent you cannot conclude that the consequent is false; similarly, by affirming the consequent you cannot conclude that the antecedent is true. Continuing with the college registration example, "If I am a freshman, then I must take Intro to Logic." Denying the antecedent with "I am not a freshman" does not lead to the conclusion that "I do not have to take Intro to Logic." It could be that two groups must take Intro to Logic (e.g., all freshmen and all transfer students). So, not being a freshman doesn't necessarily mean you don't have to take Intro to Logic. Likewise, affirming the consequent with "I must take Intro to Logic" does not permit the conclusion that "I'm a freshman"; you might be a transfer student.

EVIDENCE ON CONDITIONAL REASONING Generally, people are good at inferring the validity of a

---

Sample Stories and Tests from Rader and Sloutsky (2002)

About 60% of people (incorrectly) "recognized" the affirm-the-consequent conclusion as having been in the story, not that different from the 61% who (correctly) recognized the *modus ponens* conclusion. So, both kinds of conclusions are routinely drawn as we read, even though one of them is incorrect (see Bonnefon & Hilton, 2004, for how the desirability of the consequent influences our predictions about the truth of the antecedent).

SOURCE: From Rader & Sloutsky (2002), Tables 2 and 4 (pp. 61, 65).

Interactive

Participants read four sentences. Sentence 1 was the same for both groups, but Sentence 2 differed between Version A and Version B. Sentence 3 was the same for both groups, but Sentence 4 differed for the two groups, containing either an inference or no inference. In other words, one quarter of the participants saw Version A and an Inference, one quarter saw Version A and No Inference, one quarter saw Version B and an Inference, and one quarter saw Version B and No Inference.

1. Frank woke up on his couch after taking a long nap and realized that he didn't know what time it was.

2. (Version A) He thought that if it was cold outside, then it was night.

   OR

   (Version B) He thought that if it was night, then it was cold outside.

3. Still feeling sleepy, Frank arose to open a window.

4. (Inference condition) He discovered that it was cold outside.

   OR

   (No Inference condition) He wondered whether it was cold outside.

consequent given evidence that the antecedent is true (affirming the antecedent, *modus ponens*). Rips and Marcus (1977) found that 100% of their sample drew this conclusion. The other valid inference, denying the consequent (*modus tollens*) is more difficult. Only 57% in Rips and Marcus's study drew this conclusion. Let's look at the mistakes people make by dividing conditional reasoning errors in into three broad categories, those involving the form of the problem, the search for evidence, and memory-related phenomena.

FORM ERRORS People sometimes draw invalid conclusions by using one of the two invalid forms, either denying the antecedent or affirming the consequent. Rader and Sloutsky (2002) found that we commonly draw such invalid conclusions when comprehending discourse. They gave people short scenarios that contained an *if–then* conditional, then tested recognition memory for either words or ideas in the stories (i.e., was this information in the story?).

Another form error is subtler. People tend to reverse the elements in the *if* and *then* and then go on to evaluate the evidence against the now-reversed conditional. This is called *illicit conversion*. So, given *If P, then Q* and evidence Q, people tend to switch the conditional to *If Q, then P*. They then decide that the evidence Q implies that P is true. This is wrong because the order of P and Q is meaningful. The *if* often specifies some cause, and the *then* specifies an effect. So, we cannot draw correct conclusions if we reverse the roles of the cause (P) for some effect (Q). For example, the statement "If the clock says 9:15, then you are late for work" cannot be validly converted to "If you are late for work, then the clock says 9:15."

SEARCH ERRORS A second kind of error involves a search for evidence. People often rely on a first impression or on the first example—the first mental model—that comes to mind (Evans, Handley, Harper, & Johnson-Laird, 1999). Unfortunately, this is often a *search for positive evidence*, also called the **confirmation bias**, where we often (inappropriately) seek *only* information that confirms a conclusion we have already drawn or a belief we already have. Knowing an outcome and the conditions that might lead to it causes people to overestimate the likelihood of that outcome. Because people find it easier to draw backward causal inferences, they mistakenly think that the prior events are more likely to lead to the actual outcome (Koriat, Fiedler, & Bjork, 2006), which is an analogue to the hindsight bias discussed later in this module.

As a demonstration, consider a classic study (reported in Wason & Johnson-Laird, 1972) shown at the top of Figure 11.2, the Wason card problem. Four cards are visible and each card has a letter on one side and a number on the other. The task is to pick the card or cards you would turn over to gather conclusive evidence on the following rule:

If a card has a vowel on one side, then it has an even number on the other side.

Think about this statement and decide how you would test the rule before you continue reading.

Of the people tested, 33% turned over only the *E* card, a correct choice conforming to *modus ponens* (affirming the antecedent). However, a thorough test of the rule's validity requires that another card be turned over (the rule might be rephrased "Only if a card has a vowel on one side will it have an even number on the other side"). Only 4% of the people turned over the correct combination to check on this, the *E* card (*modus ponens*) and the 7 card (*modus tollens*). Turning over the 7, which might yield negative evidence (*not Q*), was rarely considered. Instead, people preferred turning over the *E* and the 4 card: 46% of the people did this. Note two points. First, turning over the 4 is invalidly affirming the consequent (the rule doesn't say anything about the other side of a consonant; it could be

Figure 11.2 The Wason Card Problem

At the top of the illustration are the four cards in the Wason card problem. Which card or cards would you turn over to obtain conclusive evidence about the following rule: A card with a vowel on it will have an even number on the other side? At the bottom of the illustration are four envelopes. Which envelopes would you turn over to detect postal cheaters, under the rule that an unsealed envelope can be stamped with the less expensive stamp?

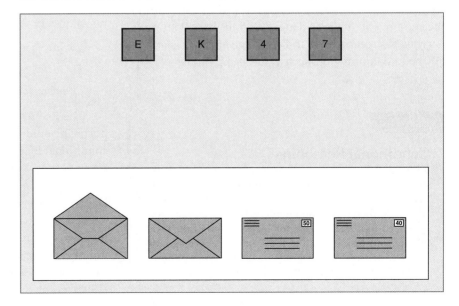

odd or an even). Second, turning over the *E* is a search for positive evidence, the (tentative) "yes" conclusion that *P* is true. Our bias is either to stop after turning over the *E* (positive evidence) or to continue searching for more positive evidence (turning over the 4). The poor performance on this task has spawned a great deal of work to understand the cognitive processes that give rise to it, including creating computational models (Klauer, Stahl, & Erdfelder, 2007) that suggest that people are not considering the cards one by one. Rather, they are looking at the configurations of the cards.

In a different situation, however, Johnson-Laird, Legrenzi, and Legrenzi (1972) found that 21 of 24 people made both of the correct choices when the situation was made more concrete. In that study, people were trying to find cheaters on the postal regulations, where unsealed envelopes could be mailed with a less expensive stamp than sealed envelopes.

Think about this. What *if–then* rule is being tested? Because either a sealed or an unsealed envelope could be mailed with a more expensive stamp, the rule must be

If the envelope is sealed, then it must carry the expensive stamp.

When asked to detect cheaters, people turned over not only the sealed envelope (*modus ponens*) but also the one with the cheaper stamp, that is, the *modus tollens* choice. Because the people were not postal workers, it is clear that the situation's concreteness helped them search for negative evidence; in the process, they showed valid conditional reasoning.

Although making the information more concrete can improve performance, other ways of making the information more "naturalistic" can impede reasoning. For example, people are more likely to make errors if more intense emotions are involved (Blanchette & Richards, 2004). If people are given the premise "If there is danger, then one feels nervous," they will be more likely to invalidly affirm the consequent "there is danger" if told that "Betty feels nervous" as compared to if they are given more emotionally neutral information, such as the stamp problem.

Prove It

Conditional Reasoning

Several things can come into play to affect how well people reason, such as whether the problem is naturalistic or abstract. Set up a version of the Wason card task, then devise some interesting variations. You will need four index cards for each version of the task. Set up your task so each of the four cards corresponds to each of the four response alternatives, namely *Modus Ponens* (affirming the antecedent), *Modus Tollens* (denying the consequent), affirming the consequent, and denying the antecedent. You might try an abstract version, such as the original card layout used by Wason or something similar (e.g., if the card is red on one side, it will be green on the other side), and then some more naturalistic version that you come up with (e.g., the "freshmen registering for class" example from earlier, or one testing the rule "If a person is drinking alcohol, then he or she is at least 21 years old").

After you have made your cards, lay them down in front of your volunteers, tell them the rule they are verifying, then have them tell you which card(s) they would choose to turn over.

You might also want to try some variations based on other things you have learned about memory and cognition. For example, what would happen if people were put in a dual-task, divided-attention situation? What would happen if you made the memory load a verbal load versus a visual/spatial load? Does emotional salience play a role? What if you prime different aspects of a person's semantic memory? The possibilities are endless.

MEMORY-RELATED ERRORS The third category of errors involves memory limitations. As noted earlier, Johnson-Laird has suggested that we reason by constructing *mental models*, mental representations of meanings of the terms in reasoning problems (Johnson-Laird & Byrne, 2002; Johnson-Laird, Byrne, & Schaeken, 1992). It is difficult to flesh out a set of meanings in abstract conditional reasoning problems of the "If *P* then *Q*" variety, but it is far easier in concrete, meaningful problems such as

If it was foggy, then the match was canceled.
It was foggy.
Therefore, the match was canceled.

Furthermore, if additional terms appear in the problem, additional mental models must be derived, two additional ones in the following case:

If it was foggy, then the match was canceled.
The match was not canceled.
Therefore, it was not foggy.

The use of mental models during reasoning is supported by evidence showing that the time to respond to a conditional logical problem is a function of the amount of effort needed to mentally construct and manipulate relevant mental models (Vergauwe, Gauffroy, Morsanyi, Dagry, & Barrouilet, 2013). When additional models are needed, the load on working memory mounts and can interfere with reasoning. The same is true when the phrasing of the problem places a greater load on comprehension (Thompson & Byrne, 2002). And finally, Evans et al. (1999) point out that if a conclusion matches the first mental model derived from the problem, it is particularly easy to accept the conclusion, leading to fallacies or errors in reasoning.

11.1.4: Hypothesis Testing

Part of the importance of conditional reasoning is its connection to scientific hypothesis testing. Consider a typical experimental hypothesis:

If theory A *is true, then data resembling* X *should be found in the experiment.*

If data resembling X are indeed found, there is a strong tendency to conclude that theory A must be true: It affirms the consequent. What's wrong with this? Well, affirming the consequent in an error in logical reasoning and could lead to mistakenly concluding that, based on this evidence, the antecedent is true. This error is seductive. Of course, it might be true that theory A is correct. But it is also possible that theory A is incorrect and that some other (correct) theory would also predict data X. People are strongly biased to draw conclusions based on what they know to be true about premises and not by what could be false about them (Espino, Santamaria, & Byrne, 2009).

Because of the invalidity of affirming the consequent, and because we want to test hypotheses, our experiments test a *different* hypothesis than "Theory A is correct." As you learned (or will learn) in statistics, we test the null hypothesis in hopes that the evidence will be inconsistent with the predicted null outcome. Note the form of such a test:

If the null hypothesis is true (if P*), then there will be no effect of the variable (then* Q*).*

If we obtain evidence that there *is* an effect of the variable, then we have evidence that the consequent is not true. We can then conclude, via denying the consequent (*modus tollens*), that the antecedent is not true. In other words, we reject the null hypothesis, deciding that it is false. Although people can make a variety of errors in such situations, especially when the *if–then* relationship is more complex (Cummins, Lubart, Alksnis, & Rist, 1991), the typical mistake is to search for positive, confirming evidence (Klayman & Ha, 1989). In similar vein, another strategy simply is to make a judgment as to the relevance or strength of the arguments and base a decision on that (e.g., Medin, Coley, Storms, & Hayes, 2003; Rips, 2001).

Overall, people have a bias to adopt what they view as the best explanation of a cause–effect relationship based on the plausibility of the various elements of the situation. When this requires that they place greater emphasis on confirmatory evidence they do so, but when it involves placing greater emphasis on disconfirming evidence, they also do that (Koslowski, Marasia, Vermeylen, & Hendrix, 2013). Moreover, the degree of emphasis people place on positive and negative evidence also reflects the source of that information, whether it comes randomly or from a helpful communicator (Voorspoels, Navarro, Perfors, Ransom, & Storms, 2015).

Thus, we have the capacity to use positive and negative evidence appropriately. However, the emphasis that we place on this information varies greatly with how plausible we think the evidence is and the circumstances surrounding how we learn about it. For example, suppose you are trying to learn what triggers rashes on some patients' skin. If you hear evidence about the weather, you may reject this as implausible, unless you later learn that a bacterium is responsive to exposure to sunlight. Similarly, you are less likely to place emphasis on weather information if you think you are getting random facts about the situation compared to if you are getting the information from a cooperative and knowledgeable source.

> **WRITING PROMPT**
>
> **Human Reasoning**
>
> How is human reasoning like formal proper reasoning, and how does it differ? What can people do to make their thinking more logical?
>
> ▶ The response entered here will appear in the performance dashboard and can be viewed by your instructor.
>
> Submit

11.2: Decisions

OBJECTIVE: Assess the decision-making process

How do we make decisions? How do we choose among several alternatives, say, on a multiple-choice test, or decide which of several apartments is best? What role does the information stored in memory play, and how certain are the decisions we make based on that information?

In a sense, we have been studying decision making all along in this course, although the decisions often were fairly simple—for example, deciding "yes" or "no" in semantic or lexical decision tasks. At its base, *decision making* can be viewed as a search for evidence where the ultimate decision depends on some criterion or rule for evaluating the evidence. A search may turn up either positive or negative evidence. How we make decisions as a function of such evidence and how we evaluate the evidence itself are at the heart of decision making and reasoning.

The bulk of the research on reasoning and decision making investigates processes that are slow and deliberate. In many ways, the research on such decisions is similar to the area of problem solving, in which there is a clear connotation of slow, deliberate processing. One aspect of this similarity relates to *familiarity*. The domains of reasoning and problem solving we investigate are not well known or understood by people, or they involve material that is not highly familiar. Another similarity involves the idea of *uncertainty*. There is often no certain answer to the problems or at least no good way of deciding whether a particular solution is the correct approach. Despite this, taking a principled and careful look at how people make decisions can provide enormous benefits and guide further research. For example, there are some findings that people who are convicted of crimes and sentenced to jail show principled differences in how they make decisions, and possibly that parts of their brains are working less effectively, thereby leading to these decision-making differences (Yechiam et al., 2008).

People make decisions and base their reasoning on a variety of strategies, some good and some not. This is also a characteristic of much problem solving. Because of these similarities, investigations of such strategies often are impossible to categorize clearly as reasoning on the one hand or problem solving on the other.

A rich source of evidence about human reasoning is gambling.

11.2.1: Algorithms and Heuristics

Reasoning and decision making can be done in different ways. There is no single approach that people take when trying to make sense of the world and what they should do.

Answer these questions:

1. If you toss a fair coin, what is the probability of getting heads?
2. If you toss a fair coin, what is the probability of getting heads two times in a row?

Most people know that the probability of tossing heads is .50—that's a simple 50/50 situation. But you may be hazy about the correct algorithm to apply to the second problem, two heads in a row. It is fairly simple and requires two basic probability statements.

First, the chance of getting heads once is 50/50, or stated as a probability, .50.

Second, the probability of any particular sequence of independent events (like coin tosses) is the basic probability of the event (.50) multiplied by itself once for each event in the sequence, that is $.50 \times .50$ for two heads in a row. More formally, the formula is $p(e)^2$, the basic probability of the event $p(e)$ raised to the nth power, where n is the number of events in the sequence. So the answer here is .25 for two heads, $.50^2$, or simply $.50 \times .50$.

People tend to be poor at probability questions. In a survey of an undergraduate class that used those questions, 89% got question 1 correct, but only 42% got the second question right; 37% said that the correct answer was .50, the same as in question 1 (Ashcraft, 1989). It is clear these people did not know the formula.

Likewise, people generally don't know the algorithm for answering this kind of question:

1. If each of 10 people at a business meeting shakes hands (once) with each other person, then how many handshakes are exchanged?

If you do not know how to compute this, then make an estimate—and then introspect for a moment on how you came up with that estimate. If you guessed, then what guided your guess? The algorithm for this problem is also fairly easy. If N is the number of people, then the number of handshakes is $N \times (N - 1)/2$; for 10 people, that's $10 \times 9/2 = 45$ handshakes.

Notice that the algorithm provides a systematic and orderly procedure that is guaranteed to yield the correct result (assuming you do the math correctly). Algorithmic methods, in all these settings, follow the *normative model*, the method provided by mathematics and probability. Heuristics, in contrast, seem very human. They are not necessarily systematic or orderly and rely heavily on educated guessing. This is referred to as the *descriptive model*, just a description of how the question was answered. A large part of the research on reasoning and decision making looks at how different the normative and descriptive models are, meaning how people diverge from the normative method. Under certain circumstances, or for particular kinds of questions, heuristics seem prone to distortions,

Two General Approaches in Decision Making

In many reasoning and problem-solving settings, two general approaches can be taken, an algorithmic approach and a heuristic approach.

Algorithm

An **algorithm** is a specific solution procedure, often detailed and complex, that is guaranteed to furnish the correct answer if it is followed correctly; for example, a formula. We are familiar with algorithms largely through our schoolwork in arithmetic and mathematics. We all learned an algorithmic approach to complex multiplication, a set of rules for applying operations in certain orders to arrive at the correct answer. If the rules are applied correctly, the algorithm provides the correct answer.

Heuristic

inaccuracies, and omissions—the descriptive model diverges from what's normative or "right."

One reason people use heuristics, rather than algorithms, to make decisions is because heuristics can be used more quickly and easily. This is an advantage when decisions need to be made rapidly. Generally, people do not exert the effort to do more deliberate reasoning when quick and dirty processes suffice, unless they encounter a difficulty in using their more intuitive judgments (Alter, Oppenheimer, Epley, & Eyre, 2007). A great deal of data show the sorts of heuristics people use and the mistakes they make when making decisions, and that people are consistent in their decision making in terms of the number of options they consider (Galotti, Wiener, & Tandler, 2014). That said, at this point we lack practical advice for how to train people to reason more effectively (Milkman, Chugh, & Bazerman, 2009).

FAST AND SLOW DECISIONS An idea that has appeared in recent years is that sometimes people are better at making decisions if they make them quickly, relying on their unconscious understanding, rather than deliberating and overthinking them (Gladwell, 2005). Although there is some evidence for this, especially if people are novices in a domain (Dijksterhuis, Bos, van der Leij, & van Baaren, 2009), a great deal of evidence contradicts this idea. For example, it is worse for people who are experts (Dijksterhuis et al., 2009), whom you would think would have a great deal of prior knowledge to draw on quickly. Under certain circumstances, or for particular kinds of questions, heuristics seem prone to distortions, inaccuracies, and omissions—the descriptive model diverges from what is normative or "right." There is also evidence that rapid decisions are no better than well-thought-out ones (Newell, Wong, Cheung, & Rakow, 2009).

Review: Algorithms and Heuristics

_____ Provides a systematic and orderly procedure that is guaranteed to yield the correct result

_____ Not necessarily systematic or orderly and relies heavily on educated guessing

_____ Follows the normative model, the method provided by mathematics and probability

_____ Follows the descriptive model, just a description of how the question was answered

_____ Can be used more quickly and easily

_____ Takes time and has to be followed in orderly fashion

Start Over   Check Answers

11.3: Classic Heuristics, Biases, and Fallacies

OBJECTIVE: Explain the fundamental heuristics people use to make decisions

By far the most influential work done on decision making and heuristics has been the classic work of Tversky and Kahneman (1973, 1974, 1980; Kahneman, Slovic, & Tversky, 1982; Kahneman & Tversky, 1972, 1973; Shafir & Tversky, 1992). Tversky and Kahneman's work on heuristics and fallacies has had an impact in such diverse areas as law, medicine, and business. Most prominently, it has affected the field of economics, and Daniel Kahneman received the Nobel Prize in Economics in 2002 (see Kahneman, 2003a, for a first-person account of that work, and 2003b for his personal history of the collaboration with the late Amos Tversky; see Kahneman & Tversky, 2000, for a compendium of chapters on this approach). In a way, it isn't surprising that this topic is of interest to many different fields—after all, think of how many situations and settings involve decision making. The Nobel Prize signifies more than just the relevance of the topic; it indicates noteworthy achievement in tackling and explaining a large and important set of ideas, how humans reason and make decisions.

Kahneman and Tversky's research focused extensively on a set of heuristics and biases that appears to characterize everyday decision making about uncertain events. In some of the situations they studied, these researchers found an algorithm can be applied to arrive at a correct answer. Many of these situations involve probabilistic reasoning. Knowledge of the algorithms doesn't necessarily mean that people understand them, can use them spontaneously, or can recognize when to apply them. Indeed, Kahneman and Tversky found that a sample of graduate students in psychology, all of whom had been exposed to statistical algorithms, did well on simple problems but still relied on a heuristic when given more complex situations. That said, some studies have shown good transfer and improved reasoning after relevant training (Agnoli, 1991; Agnoli & Krantz, 1989; Fong & Nisbett, 1991; Lehman, Lempert, & Nisbett, 1988). Other situations that have been studied involve estimates of likelihood when precise probabilities cannot be assigned or have not been supplied, although elements of statistical and probabilistic reasoning are still appropriate, and in situations contrasting verbal and numerical descriptions (for instance, "rain is likely" versus "there's a 70% chance of rain"; Windschitl & Weber, 1999). Finally, some settings involve very uncertain or even impossible situations; for example, asking people to predict the outcome of a hypothetical event, such as the outcome of World War II if Germany had developed the atomic bomb before the United States.

11.3.1: The Representativeness Heuristic

In this section, we begin by covering a series of classic heuristics that have been studied in research on decision making, including the representativeness, availability, simulation, and undoing heuristics. We also cover a newer approach that provides other ways of thinking about human decision making.

THE LAWS OF LARGE AND SMALL NUMBERS The representativeness heuristic embodies a bias of **insensitivity to sample size**. When people reason, they fail to account for the size of the sample or group on which the event is based. They seem to believe that both small and large samples should be equally similar to the population from which they were drawn. In other words, people believe in the law of small numbers. Now the *law of large numbers*—that a large sample is more representative of its

---

Examples of Classic Heuristics in Decision Making

As you read, try to develop your own examples of situations that are like the given ones. You will be surprised at how often we use heuristics in everyday judgments and decision making. Refer to Appendix at the end of the module for more details on these situations.

Interactive

If you toss a coin six times in a row, which of the following outcomes is more likely: HHHTTT or HHTHTT? Most of us would say—and quite rapidly at that—that the second alternative is more likely. But if you stop and think about it, you realize that each of these is *exactly* as likely as the other, because each is one of the possible ways six coin tosses can occur (the total number of outcomes is 2^6, i.e., 64 sequences of heads and tails). We tend to think of the alternating pattern HHTHTT as a representative of a whole class in which most outcomes have alternations between heads and tails. The thinking here, illogical but understandable, is that a random process ought to look random (e.g., Burns & Corpus, 2004), and if it doesn't, there might be some underlying systematic process (Hahn & Warren, 2009). The sequence of three heads then three tails looks nonrandom and so seems less likely. Likewise, because the likelihood for six tosses is three heads and three tails (in the long run), almost any sequence with three of each will appear more representative than sequences with more of one outcome than the other. (See Pollatsek, Konold, Well, & Lima, 1984, and Nickerson, 2002, for evidence on people's beliefs about random sampling processes, and the ability to produce and perceive randomness.) The mistake we make indicates how we reason in similar situations (the math for this problem is explained in the Appendix at the end of the module).

Note: The "^" symbol between numbers indicates superscript.

Biases in the Representativeness Heuristic

Let's take a look at several of the biases that stem from the representativeness heuristic.

> **Ignoring Base Rates (Ignoring Prior Odds) (Based on Johnson & Finke, 1985)**
>
> **Questions:**
>
> (a) Why are more fighter pilots first-born than second-born sons?
>
> (b) Why do more apartment building fires occur on the first three floors than the second three floors?
>
> (c) In football, do more interceptions occur on first down or second down?
>
> *The bias:*
>
> In all three questions, people tend to ignore base rates. To answer the questions correctly, we should consider: (a) How many first-born versus second-born sons are there? (b) How many apartment buildings even have more than three floors? (c) How many times do teams make first downs and compared to second downs?

Base Rates and Stereotypes

Gambler's Fallacy

---

population—is true. But people erroneously believe that there is also a **law of small numbers** (see also Bar-Hillel, 1980), but that is not the case.

STEREOTYPES Another bias resulting from the representativeness heuristic affects our reasoning about other people. Kahneman and Tversky (1973) reported evidence on estimations based on personality descriptions. They had people estimate the probability that a described person was a member of one or another profession. They found that estimations were influenced by the similarity of a description to a widely held **stereotype**. Consider first the situation: 100 people are in a room, 70 of them lawyers, 30 of them engineers. Given this situation, answer the following question:

1. A person named Bill was randomly selected from this roomful of 100 people. What is the likelihood that Bill is a lawyer?

Simple probability tells us that the chances of selecting a lawyer are .70. Consider two other situations. There are still the same 70 lawyers and 30 engineers. But now you are given a description of two randomly selected people and are asked "What is the likelihood that this person is an engineer?":

2. "Dick is a 30-year-old man. He is married with no children. A man of high ability and high motivation, he promises to be quite successful in his field. He is well liked by his colleagues."

3. "Jack is a 45-year-old man. He is married and has four children. He is generally conservative, careful, and ambitious. He shows no interest in political and social issues and spends most of his free time on his many hobbies, which include home carpentry, sailing, and mathematical puzzles."

Here, people did *not* judge the probabilities to be the same as the prior odds. Instead, they assumed that the personality descriptions contained relevant information and adjusted their estimates accordingly. Particularly, people responded that the probability was close to .50, that is, about a 50-50 chance that Dick and Jack were engineers. Description 3 resembles the stereotype for engineers. People de-emphasized the prior odds (.30) and based their judgments on the description that seemed representative of engineers, allowing an influence of the stereotype. Description 2 was written to be uninformative about Dick's profession. And yet, people still changed their estimates. People tend to view any evidence as a basis for changing and, they hope, improving their prediction (Fischhoff & Bar-Hillel, 1984; Griffin & Tversky, 1992).

According to the correct way to deal with these problems, the normative model is to assess the usefulness or relevance of the additional information to decide how much weight to give it; its usefulness or "diagnosticity." Then the estimate is adjusted by the appropriate weighting. This is *Bayes' theorem*, which states that estimates should be based on two kinds of information, the base rate of the event and the "likelihood ratio," which is an assessment of the usefulness of the new information.

11.3.2: The Availability Heuristic

What proportion of medical doctors are women? What proportion of U.S. households own a wireless router, a VR headset, or a tablet computer? How much safer are you in a commercial airliner than in a private car, or vice versa? Questions such as these ask you to estimate the frequency or probability of real-world events, even though you are unlikely to have more than a few shreds of relevant information stored in memory. Short of doing the fact-finding necessary to know the real answers, how do we make such

---

Aspects of the Availability Heuristic

In short, when people make estimates of likelihood, their estimates are influenced by the ease with which they can remember relevant examples.

> **Interactive**
>
> Because our judgments are based on what we can remember easily, any factor that leads to storing information in memory can influence our judgments. If reasonably accurate and undistorted information is in memory, then the availability heuristic does a good job. But if it contains information that is inaccurate, incomplete, or influenced by factors other than objective frequency, then it biases and distorts our reasoning. As a simple example, if your friend's Volvo needs repeated trips to the mechanic, you may view Volvos as unreliable. The availability heuristic has biased your judgment.

estimates? The simplest way of making estimates is to try to recall relevant information from memory. Event frequency is coded in memory (Brown & Siegler, 1992; Hasher & Zacks, 1984), perhaps automatically. If the retrieval of examples is easy, we infer that the event must be frequent or common. If the retrieval is difficult, then we estimate that it must not be frequent. Interestingly, frequency estimates affect your eventual judgments about the information: If it is repeated often enough, even false statements become "truer" (Brown & Nix, 1996). Frequency is related to the second heuristic that Tversky and Kahneman (1973) discussed, the **availability heuristic** in which people estimate the likelihood of events based on how easily examples come to mind. "Ease of retrieval" is what the term *availability* means here.[1]

11.3.3: The Simulation Heuristic

A variation on the availability heuristic is the **simulation heuristic**. Here people predict future events or are asked to imagine a different outcome of an event or action. The simulation involves a mental construction or imagining of outcomes, a forecasting of how some event will turn out or how it might have turned out under other circumstances. The ease with which these outcomes can be imagined is critical. If a sequence of events can easily be imagined, then the events are viewed as likely. Alternatively, if it is difficult to construct a plausible scenario, the hypothetical outcome is viewed as unlikely. An example of this was given earlier, when you were asked to imagine possible outcomes if Germany had developed the atomic bomb before the United States. Given the role of the atomic bomb in ending World War II, people would give that far more weight than if they were asked about the development of some other device, say a long-range bomber or submarine.

Here is another alternative by Kahneman and Tversky (1982, p. 203) in which imagining a different outcome may be difficult:

> Mr. Crane and Mr. Tees were scheduled to leave the airport on different flights, at the same time. They traveled from town in the same limousine, were caught in a traffic jam, and arrived at the airport 30 minutes after the scheduled departure time of their flights. Mr. Crane is told that his flight left on time. Mr. Tees is told that his flight was delayed and just left 5 minutes ago. Who is more upset, Mr. Crane or Mr. Tees?

As you would expect, almost everyone decides that Mr. Tees is more upset; in Kahneman and Tversky's study, 96% of the people made this judgment. However, from an objective standpoint, Mr. Crane and Mr. Tees are in identical positions: Both missed their planes, and because of the traffic jam, both expected to do so. The reason Mr. Tees is viewed as being more upset is that it was more "possible," in some sense, for him to have caught his flight. So it is easier to imagine an outcome in which the limousine arrives a few minutes earlier than it is to imagine one in which it arrives a half hour earlier. As such, we believe the traveler who "nearly caught his flight" will be more upset.

11.3.4: Elimination by Aspects

Another important heuristic that people use when coming to a decision is **elimination by aspects** (Tversky, 1972). Often the choices that you are presented with have multiple features or aspects. For example, when deciding which used car to buy, you may look at several features of several cars, such as price, gas mileage, sound system, and so on. Similarly, if you are trying to decide which apartment to rent, you may also consider different features of several alternatives, such as rent, location, size, pool, and so on.

A rational way to approach these sorts of situations would be to look at all the relevant features for all the options, tally up their various values, and then select the option that provides the best fit for the person across all available features. This approach is problematic because there often is a large number of options and features to consider. Imagine looking over and test-driving every single used car in town, or imagine visiting every single apartment that is available for rent in town prior to making a decision. What a lot of time and work!

To short-circuit this process, people can use the elimination by aspects heuristic. Basically, people go through the various features, one at a time, starting with those of most importance to them. Each option that does not meet some criterion for a given feature is then dropped from consideration. This process is repeated feature by feature until the number of options is drawn down to a manageable size or leads to a single option.

As an example of how this would work, if you were looking for a used car, you might first eliminate any cars that are too expensive or too cheap. Then, you might rule out any that are the wrong style (e.g., you don't want a convertible), followed by eliminating any with poor gas mileage, and so on. This would continue until you had the number of used cars down to a small number. Similarly, you might narrow down the set of apartments to look at by first eliminating those that have rents that are too high or too low. Then you would rule out those in parts of town where you do not want to live, those that are too high in a building, those that are part of complexes, and so on, until you

---

[1] In Module 8, *availability* meant whether some information was stored in memory, and *accessibility* referred to whether the information could be retrieved. Clearly, *availability* in the Kahneman and Tversky sense is referring to *accessibility* in the memory retrieval sense. Kahneman (2003a) has acknowledged that his original choice of terms was confusing in this sense and now refers to this as the *accessibility heuristic*. We continue to use the original term *availability*, however, to be consistent with the 50-year history of its usage in the decision-making literature.

get the number of apartments to consider down to a small number. Though this process will often get you to the ideal or close to ideal choice given what is available, it is not guaranteed to. It is possible that the best choice for you, given all the features to consider, might get eliminated early on.

11.3.5: The Undoing Heuristic

A more complete example of the simulation heuristic is illustrated in the simulation stories below, including the **undoing** of an outcome by changing what led up to it. This is called **counterfactual reasoning**, when a line of reasoning deliberately contradicts the facts in a "what if" kind of way (e.g., "What would have happened if Germany had developed the bomb first?"; see Mandel & Lehman, 1996; Roese, 1997, 1999; Spellman & Mandel, 1999). This is the process of judging that some event "nearly happened," "could have occurred," "might have happened if only," and so on.

TYPICAL BIAS Other factors have also been implicated in counterfactual reasoning. Byrne and McEleney (2000) suggested that people focus on actions, not failures to act, when they undo events. They had people read a story about Joe and Paul. Joe got an offer to trade his stock in Company B for stock in Company A, which he did, although he ultimately lost money on Company A. Paul got a comparable offer to trade his stock in Company A, but he decided not to take the offer. Staying with Company A, he ultimately lost the same amount of money as Joe did, even though they both started with the same amount. Despite the equal loss by both, 87.5% of the people claimed that Joe, the one who acted, would feel worse about his decision ("If only I hadn't traded my stock"). In a

---

Stories for the Simulation Heuristic (Table 11.2A–Table 11.2D)

Read the stories now and decide how you would complete the "if only" phrase before continuing.

SOURCE: From Kahneman & Tversky (1982).

1.	Mr. Jones was 47 years old, the father of three, and a successful banking executive. His wife had been ill at home for several months.
2a.	On the day of the accident, Mr. Jones left his office at the regular time. He sometimes left early to take care of home chores at his wife's request, but this was not necessary on that day. Mr. Jones did not drive home by his regular route. The day was exceptionally clear, and Mr. Jones told his friends at the office that he would drive along the shore to enjoy the view.
3.	The accident occurred at a major intersection. The light turned amber as Mr. Jones approached. Witnesses noted that he braked hard to stop at the crossing, although he could easily have gone through. His family recognized this as a common occurrence in Mr. Jones's driving. As he began to cross after the light changed, a light truck charged into the intersection at top speed and rammed Mr. Jones's car from the left. Mr. Jones was killed instantly.
4a.	It was later ascertained that the truck was driven by a teenage boy who was under the influence of drugs.
5.	As commonly happens in such situations, the Jones family and their friends often thought and often said, "If only . . ." during the days that followed the accident. How did they continue this thought? Please write one or more likely completions.

Route Version

companion article, McCloy and Byrne (2000) developed a scenario in which a character is late for an appointment, because of either controllable or uncontrollable factors—the character stopped to buy a hamburger or was delayed because a tree had fallen in the street. People undid the controllable factors far more frequently, although they distinguished among delaying factors in terms of interpersonal and social norms of how acceptable or polite the factor was; stopping to visit his parents on the way to the appointment was viewed less negatively than stopping for the hamburger.

A puzzle in the Mr. Jones story was that people seldom focused on the actual cause of the accident, the teenage boy. Kahneman and Tversky speculate that this was caused by a focus rule: We tend to maintain properties of the main object or focus of the story unless a different focus is provided. In support of this, when people read a version of the story that focused more on the boy, they were more likely to undo the boy's actions. However, undoing Mr. Jones's behavior—having him take his normal route home—essentially claims that Mr. Jones was responsible for the accident because it was his behavior that people altered. It is a case of "blaming the victim." This is especially common when there was an unusual event in the story that could have been altered via a downhill change. For instance, Goldinger, Kleider, Azuma, and Beike (2003) offer clear examples:

> Paul normally leaves work at 5:30 and drives directly home. One day, while following this routine, Paul is broadsided by a driver who violated a stop sign and receives serious injuries. (p. 81)

When people consider how much compensation Paul should receive for his injuries and how much punishment is appropriate for the other driver, they examine Paul's behavior closely. In this scenario, however, Paul tends not to be blamed for the accident. But if the scenario is changed to:

> Paul, feeling restless at work, leaves early to see a movie. . . . Paul is broadsided by a driver . . .

and so forth, then we tend to view him as less deserving of compensation. In addition, in an echo of the social norms result of McCloy and Byrne (2000), Goldinger et al. (2003) pointed out the following: *If* Paul receives an emergency call to return home and then is broadsided, "the accident now appears exceptionally tragic, and compensation awarded to him increases" (p. 81). This reversal, observed by Miller and McFarland (1986), depends on how free Paul was to choose what to do and how socially acceptable his choices were.

Related to this idea, people are more willing to say that an action was intentional when they are told it will be done in the future than if it has already happened in the past (Burns, Caruso, & Bartels, 2012). Essentially, people are perceived as having more control over future actions than past actions, even if the actions are identical. Also, regarding the future, people are more realistic when they are planning how they would do things differently in the future as compared to how they would change things in the past, even when what would need to be changed remains largely the same (Ferrante, Girotto, Stragà, & Walsh, 2013).

HINDSIGHT BIAS The simulation heuristic provides a nice explanation of **hindsight bias** (Fischhoff, 1975), the after-the-fact judgment that some event was very predictable, even though it wasn't. This is the "I knew it all along" effect. In thinking about the now-finished event, the scenario under which that event could have happened is easy to imagine—after all, it just happened (Sanna & Schwartz, 2006). The connection between the initial situation and the outcome is very available after the fact. This availability makes other possible connections seem less plausible than they otherwise would (e.g., Hell, Gigerenzer, Gauggel, Mall, & Muller, 1988; Hoch & Loewenstein, 1989). Hindsight bias can even distort our perceptual memories. Gray, Beilock, and Carr (2007) report that batters misremember how well they thought they would hit a baseball when they have been hitting well as compared to when they have struggled. That is, the current success with batting causes the player to experience hindsight bias by misremembering the batting as being better than it actually was.

Interestingly, hindsight bias even influences memory for events. People routinely "remember" their original position to be more consistent with their final decision than it really was (Erdfelder & Buchner, 1998; Holyoak & Simon, 1999) and even reconstruct story elements so that they are more consistent with the final outcome. As a demonstration of this, Carli (1999) had two groups of people read a story (about Pam and Peter in Experiment 1, and Barbara and Jack in Experiment 2) with no ending (control group), a happy ending (Jack proposes marriage to Barbara), or a tragic ending (Jack rapes Barbara). The stories had information that was consistent with both scenarios; for example, "Barbara met many men at parties" and "Pam wanted a family very much." After finishing the stories, people were questioned about the story events. The groups that heard either one of the stories with an ending agreed far more than the control group that they would have predicted that ending all along—but of course the endings were completely different. As a follow-up, in a later memory test, both groups mistakenly "remembered" information that had not been in the story but was consistent with the ending they had read. (See Harley, Carlsen, & Loftus, 2004, and Mather, Shafir, & Johnson, 2000, for an extension of hindsight bias to identifying pictures, important for eyewitness testimony situations, and remembering blind dates, respectively.)

In many ways, the simulation heuristic, which leads to hindsight bias, may come closer to what people mean by such terms as *everyday thinking* and *contemplating* than anything else we have covered so far. For example:

> If I stop for a cup of coffee, I might miss my bus.
>
> If I hadn't been so busy yesterday, I would have remembered to go to the ATM for money.
>
> In looking back, I guess I could have predicted that waiting until senior year to take statistics was a bad idea.

Such thinking often is the reason we decide to do one thing versus another; we think through the possible outcomes of different actions, then base our decision on the most favorable one. Thus, *mental simulation*, taking certain input conditions then forecasting possible outcomes, is an important way to understand cognitive processes related to planning. A general warning here is important, based on studies of how people generate possible outcomes of future events (Hoch, 1984, 1985). If you begin planning by thinking only of the desirable outcomes, you may blind yourself to possible undesirable outcomes. By starting with positives, you become more confident that your plan will have a good outcome. Overly optimistic predictions at the outset bias our ability to imagine negative outcomes and inflate our view of the likelihood of a positive outcome. ("Hey, what could go wrong if I wait until next week to start my term paper?"; see Petrusic & Baranski, 2003, on how confidence affects decision making.)

WRITING PROMPT

Algorithms Versus Heuristics

What is the fundamental difference between algorithms and heuristics? What are the advantages of each?

The response entered here will appear in the performance dashboard and can be viewed by your instructor.

Submit

11.4: Framing and Risky Decisions

OBJECTIVE: Assess how decisions are influenced by the context in which they are encountered

Not only are decisions made under conditions of uncertainty, they may also be made with an element of risk. The consequences of these decisions can come with costs and benefits. A *cost* is a negative outcome for a decision maker, whereas a *benefit* is a positive outcome. For example, imagine that people are asked to decide whether to take some cold medicine to relieve their symptoms. If they take it, the benefit is that they will alleviate many of their symptoms, but the cost is that they will feel sleepy. Alternatively, if they do not take the medicine, then they will have their cold symptoms in full force, but they will be more alert. So, this is a decision with costs and benefits. We make decisions by weighing the outcome of various choices and the degree to which we personally value different outcomes.

If people were rational decision makers, they would pick the option that leads to the best outcome for themselves, right? Sure. However, as you have already seen, people are not rational when making decisions that involve some element of risk. What influences the choice a person makes is a function of the *framing* of a situation, in terms of what information the person places an emphasis on.

Another way that people are influenced by assessing the amount of risk is by being attracted to 100% values. So if a treatment is described as being 100% effective on 70% of pathogens, it is preferred by people over one that is 70% effective on 100% of pathogens, even though the two have equivalent effectiveness (Li & Chapman, 2009). Thus, this illustrates again how people use heuristics to make decisions about probability and risk, even when they can calculate those values quickly and easily.

11.4.1: Risk Aversion and Seeking

The example of framing illustrates an important aspect of how people make decisions that involve uncertainty and risk, as well as the gains and losses that can follow from a decision. The choices that people make depending on the framing of a problem, and the emphasis on either whether something is gained (people will live) or something is lost (people will die), can be captured by a couple of principles. The first is that people tend to show *risk aversion for gains*. So when they are thinking about what they will gain, they have a bias to avoid making risky decisions. Thus, people prefer Program A for problem version 1 (Two Versions of a Problem Illustrating the Framing of a Choice) because the framing of the problem is for a gain (how many people will be saved). Version 1 presents the number of people saved as a sure thing.

The second principle is that people also tend to show risk seeking for losses, so when they are thinking about what they will lose, they have a bias to make risky decisions to avoid a loss. Thus, people prefer Program B for problem version 2 because the framing of the problem is for a loss (how many people will die). Version 2 presents the number of people who will die as a probability or a risk. Because risk aversion for gains and risk seeking for losses is a heuristic, it doesn't take much cognitive effort, and this basic pattern is observed even

Two Versions of a Problem Illustrating the Framing of a Choice

> **Version 1**
>
> Imagine that the country is preparing for the outbreak of an unusual disease expected to kill 6,000 people. Two alternative programs to combat the disease have been proposed, and only one of them can be implemented. Assume that the exact scientific estimate of the consequences of the programs are as follows:
>
> If Program A is adopted, 2,000 people will be saved.
>
> If Program B is adopted, there is 1/3 probability that 6,000 people will be saved, and 2/3 probability that no people will be saved.
>
> Which of the two programs would you favor?

when people are placed under a working memory load, such as having to do a secondary task in addition to the primary one of making a decision (Whitney, Rinehart, & Hinson, 2008).

11.4.2: Outcome Magnitude

If people were cold and rational, they would be able to see the true value of their various choices. But even the value that we place on something is colored by the context or framing in which the thing is embedded. Consider these two problems:

1. Imagine that you are about to purchase a jacket for $80. While you are looking at yourself in the mirror on the salesroom floor, another customer tells you that the same jacket is on sale for $60 at another store 20 minutes away. Would you make the trip to the other store?

2. Imagine that you are about to purchase a jacket for $500. While you are looking at yourself in the mirror on the salesroom floor, another customer tells you that the same jacket is on sale for $480 at another store 20 minutes away. Would you make the trip to the other store?

Overall, people are more likely to say that they would make the trip across town for the $80 jacket, but not for the $500 jacket, even though it is to save the exact same $20! Tversky and Kahneman (1981) also appealed to the idea of framing to explain this thinking. Essentially, the initial cost of the jacket sets up the framework to make a decision. It is an anchor from which people will adjust their assessments. People are more likely to make the trip across town for the $80 jacket because the $20 is a much larger proportion of the cost, compared to the $500 jacket. It is also noteworthy that people are more willing to spend extra effort on the jacket that is less valuable compared to the more valuable jacket.

Another example of framing and decision making has to do with whether options are added to or subtracted from something. For example, Biswas (2009) reported that people were more willing to pay for options on a car if the initial state of the car was with all the options present, and they had to select the ones that they did not want, compared to if none of the options were present and they needed to indicate which ones they wanted to add. Thus, the initial state of the car, in terms of the options, set the framework from which they based the other decisions.

Overall, though these framing effects are persistent across many people and situations, steps can be taken to lessen their impact. For example, if the impact of a decision is presented graphically, such as by presenting circles to represent people, and coloring the ones corresponding to people who will die, then framing effects can be attenuated or eliminated (Garcia-Retamero & Dhami, 2013). That said, it is important to keep in mind that images can be distorted to bias decisions as well.

WRITING PROMPT

Context in Decision Making

What does it mean that decision making is influenced by how questions are framed and the context in which the decision is placed?

The response entered here will appear in the performance dashboard and can be viewed by your instructor.

Submit

11.5: Adaptive Thinking and "Fast and Frugal" Heuristics

OBJECTIVE: Explain the impact of adaptive thinking on heuristics

Another way of thinking about heuristics during decision making is as *adaptive thinking*. The biggest proponent of this adaptive thinking view is Gigerenzer (e.g., 1996; but see Dougherty, Franco-Watkins, & Thomas, 2008, and Hilbig, 2010, for arguments against some fast and frugal heuristics). For Gigerenzer, it is a mistake to assume that the correct answer to any decision-making problem must be the normative answer supplied by classic probability theory. Instead, it is important to assess how well people's heuristics actually do in guiding behavior. People use heuristics not just because of memory limitations, incomplete algorithm knowledge, and so forth, but because they work; because they are adaptive in the sense of leading to successful behavior.[2] People use heuristics because they are tractable and robust (Gigerenzer, 2008). *Tractable* means that people can mentally track everything they need to use the heuristic, in contrast to algorithms that require people to track (often) much larger amounts of information. Calling heuristics *robust* means that they provide reasonable answers under a wide range of circumstances.

Gigerenzer (2008) provides some compelling illustrations of the value of heuristics. He notes that the 1990 Nobel Prize winner in economics, Harry Markowitz, derived an algorithm for maximizing the allocation of funds into various financial assets. However, even Markowitz did not use this principle for his own retirement savings, but instead used a simple $1/N$ heuristic (where N is the number of possible funds that the investments could go into). Based on this heuristic, your retirement funds are equally distributed across the number of available funds. Pretty simple, huh? This is the heuristic used by about half of the people in the real world. Although many financial wizards sneer at this strategy, when it was pitted against 12 different optimal return strategies, the $1/N$ heuristic beat them all. Bravo, heuristics!

Another example Gigerenzer (2008) gives deals with organ donation. He notes that the rate of organ donation is only 28% in the United States but is 99% in France. His explanation has to do with a go-with-the-default heuristic. In the United States, the default is that people do not donate their organs when they die. They need to try to be organ donors (so that other people may live), for example, by filling out a form or having the option checked on a driver's license. In the absence of this, the default is "not an organ donor." On the other hand, the default in France is to be an organ donor.

Part of the appeal of Gigerenzer's approach is that it is "positive"—it doesn't emphasize errors or deviations from some norm, but searches for the usefulness of heuristics. A second appealing feature is that the **"fast and frugal" heuristics** are simple. Third, the approach seems more open to input from general cognitive principles. For example, it acknowledges and considers memory limitations, incomplete knowledge, time limitations, and so on, as well as the general cognitive processes you have been studying. There is also the idea that human cognition is tuned to

---

[2]This is "adaptive" in an evolutionary sense, an explicit argument in Gigerenzer's approach. If a biological factor is adaptive, in evolutionary terms, it leads to greater success of the individual, hence greater spread of the feature into succeeding generations. In similar fashion, an adaptive heuristic will be successful, so it will be used more widely. Gigerenzer also advocates devoting more attention to the reasoning processes that are used, as opposed to the heavy focus on how answers deviate from the normative model.

Table 11.3 Fast and Frugal Heuristics

SOURCE: Adapted from Gigerenzer (2008).

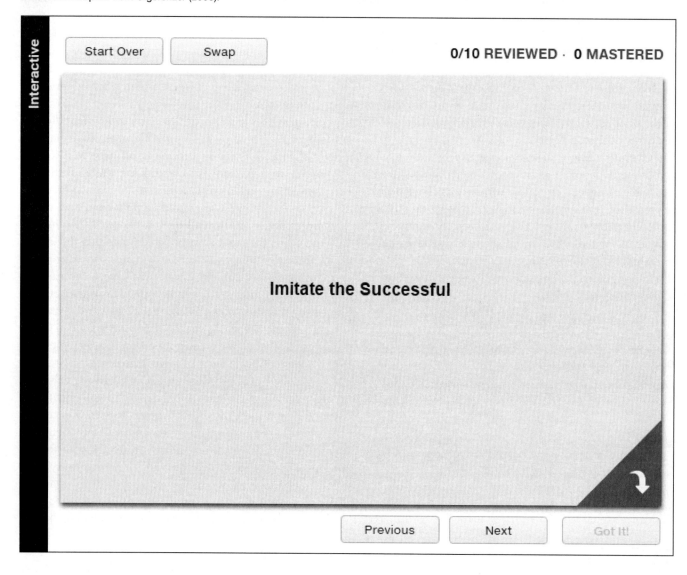

environmental statistics, and there is often a need to make assumptions based on small numbers of experiences (Hahn & Warren, 2009). So, if six coin flips in a row end in heads, although this is a completely acceptable outcome from a random process, it may also reflect some causal systematicity in the environment, such as a weighted coin.

This approach has generated several heuristics that differ in some ways from those outlined by Kahneman and Tversky and others.

11.5.1: Some Fast and Frugal Heuristics in Detail

Here we will look at some of these fast and frugal heuristics in more detail to give you a better understanding of how they would work. These are included in the following section—namely *satisficing*, *recognition*, and *take the best*—to better illustrate this approach. It should also be noted that although these heuristics can be useful, people sometimes become overly dependent on one, such as persisting with one when it has become clear that it is no longer optimal and that a switch in strategies would be better (Bröder & Schiffer, 2006).

The adaptive thinking approach does have its detractors. For instance, Newell and Shanks (2004) have questioned whether the simple recognition heuristic is as powerful as Goldstein and Gigerenzer (2002) claimed, showing how recognition is discounted when other cues of higher validity are presented. Similarly, Newell and Shanks (2003) showed that the "take the best" heuristic is used especially when the mental "cost" (e.g., in terms of time) of information is high. However, when the cost is low (for instance, when it takes very little time to retrieve additional information), people seem to rely on that heuristic less frequently.

Adaptive Thinking Heuristics (Figure 11.3)

Here are some of the adaptive fast and frugal heuristics that people may use in their decision making. Each is used in different circumstances to help people arrive at a satisfactory decision in a short period of time.

Satisficing Heuristic

An important adaptive thinking heuristic that has been on the scene for some time is Simon's (1979) **satisficing heuristic**. This principle is that we make a decision by taking the first solution that satisfies some criterion we may have—it is the "good enough" heuristic. For example, if you are looking for a place to eat dinner while traveling, rather than checking out every single eatery in town, you simply pick the first one that satisfies your criterion, such as "cheap fast-food place." People can use this heuristic to make reasonably optimal decisions.

The Recognition Heuristic

"Take the Best" Heuristic

11.5.2: The Ongoing Debate

Gigerenzer and Goldstein (1996; also Gigerenzer, 1996, and Goldstein & Gigerenzer, 2002) present considerable modeling and survey data to show how these heuristics do a good job of making decisions. Other supportive work has begun to appear as well (for instance, Burns, 2004, in a challenge to the well-known Gilovich, Vallone, & Tversky, 1985, paper on the "hot hand" in basketball).

Linda and the Conjunction Fallacy

Tversky and Kahneman (1983) had people read the Linda problem and then complete the rankings of the eight alternatives. They found that people endorse the compound alternative "bank teller and active in the feminist movement" as more likely than either "bank teller" or "active in the feminist movement." Such a judgment, from a purely probabilistic standpoint, is odd. It particularly illustrates what is known as the **conjunction fallacy**, the mistaken belief that a compound outcome of two characteristics is more likely than either one of the characteristics by itself. According to strict probability theory, this should be impossible—making up some numbers to illustrate, if the chances are .20 that Linda is a bank teller, and .30 that she is active in the feminist movement, then the conjunction of those two characteristics should never be larger than .20. In fact, in stripped-down form, the probability ought to be .06; that is,

In a room with 100 people, 20 are bank tellers, 80 are something else. Furthermore, 30 of the people are feminists, and 70 are not. What is the probability of randomly selecting someone who is both a bank teller and a feminist?

Table 11.4 Review List of Heuristics, Biases, and Fallacies

Heuristic	Biases and Fallacies
	Insensitivity to sample size Belief in the law of small numbers Stereotype bias, belief bias, confirmation bias
	Familiarity bias, salience or vividness bias
	Overly optimistic predictions inflate our confidence and prevent thinking of possible negative outcomes
	Bias to undo unusual event, bias to focus on action, bias to focus on controllable events, hindsight bias Blaming the victim
	Familiarity bias
	Reliance on ignorance or lack of knowledge

Undoing/counterfactual reasoning

Start Over

And yet people routinely say that the compound "bank teller and feminist" is more likely (has a higher rank) than the simpler probability that Linda is a bank teller. Considerable significance has attached to this fallacy in Kahneman and Tversky's work—it comes close to epitomizing the errors, the departures from the normative model, found in human reasoning.

WRITING PROMPT

Use of Fast and Frugal Heuristics

What are examples of everyday decisions in which people use fast and frugal heuristics? Why is the use of these advantageous compared to making more detailed lines of thought?

▶ The response entered here will appear in the performance dashboard and can be viewed by your instructor.

Submit

11.6: Other Explanations

OBJECTIVE: Evaluate the range of approaches to human decision making

Considering humans from a broader perspective than simply "lousy probabilists" reveals other reasonable interpretations of these rankings and judgments in decision making. Indeed, Moldoveanu and Langer (2002) supply a whole variety of explanations to justify this choice. For instance, many people treat the statement "Linda is a bank teller and is active in the feminist movement" not as a conjunction of two characteristics but as a *conditional probability*—in other words, as if it said, "*Given* that Linda is a bank teller, what is the likelihood that she is active in the feminist movement?" The probabilities for a conditional probability are very different than those for a conjunction.

The Linda Problem

Let's consider a final example to illustrate the use of heuristics to make decisions.

> Linda is 31 years old, single, outspoken, and very bright. She majored in philosophy. As a student, she was deeply concerned with issues of discrimination and social justice and also participated in antinuclear demonstrations.
>
> Now rank the following options in terms of the probability of their describing Linda. Give a ranking of 1 to the most likely option and a ranking of 8 to the least likely option.

A second reason for these interpretations involves the idea of how consistent personality tends to be—in other words, whatever job a person may end up in, there still should be some consistency in the person's personality, something like "once an activist, always an activist." Such a comment also exemplifies an idea on conversational rules and the cooperative principle. If you were supposed to judge the Linda problem solely on the probabilities of being a bank teller and being active in the feminist movement, then there is no communicative need to supply all of the background information on the character. In other words, "Why did you tell me about Linda's activism as a college student if I was supposed to ignore it?" The fact that personality information was supplied tells people that it is important and should be factored into the answer. So people rather naturally take that into account and come up with a plausible scenario—"OK, maybe she ended up as a bank teller, but she can still be a social activist by being active in the feminist movement." In short, they may be developing a personality-based model of Linda.

More recently, there have been some attempts to address how reasoning and decision making account for people's understanding of the causal structures of situations when making decisions (e.g., Garcia-Retamero, Wallin, & Dieckmann, 2007; Hagmayer & Sloman, 2009; Kim, Yopchick, & de Kwaadsteniet, 2008; Kynski & Tenebaum, 2007), as opposed to the more statistical approach of Kahneman and Tversky, and Gigerenzer's adaptive thinking approach. For example (see Kynski & Tenebaum, 2007), using standard base rate information, there is the finding that people who use sunscreen are more likely to develop skin cancer than those who do not, which runs counter to most people's intuitive judgments. This is because most people who don't use sunscreen are not out in the sun to begin with. It is important to note that using sunscreen does not cause skin cancer; rather, sunbathing does. When the problem is recast with this additional information, we see that among those people who sunbathe (a smaller group), those who use sunscreen are less likely to get skin cancer. Thus, when people are provided with the appropriate causal structure, their decisions processes can be very good.

This approach can also explain why people make some of the errors that they do. For example, Figure 11.4 illustrates the causal model that many people have of personality and career choice, and how it relates to the lawyer–engineer problem. When statistical principles are applied to this model, then the results are similar to the

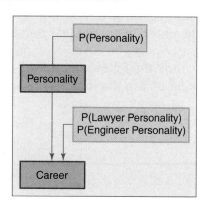

Figure 11.4 Causal Model Used by Kynski & Tenebaum (2007)

An illustration of the concept that people use ideas of causal information and relations to help them make decisions. In this case, people are assuming there is a causal relationship between a person's personality and his or her career choice.

Example Illustrating the Bayesian Theorem

OK, if that doesn't clear things up, let's go over a more concrete example.

Let's set A to be the probability of being audited by the IRS, and let's set B to be the probability of making more than $500,000 a year. What we want to know is what the probability is that a person will be audited if he or she makes over $500,000 a year, $P(A|B)$. For the sake of argument, let's say that the probability of being audited by the IRS, $P(A)$, is 5% or .05 (we are making up numbers here). Moreover, let's assume that the probability of making over $500,000 a year, $P(B)$, is 1% or .01. Finally, let's suppose that the probability of a person who makes more than $500,000 a year being audited by the IRS, $P(B|A)$, is 10% or .10. So, armed with this information, we can derive the following:

Probability of an IRS audit if you make over $500,000 = (.10*.05) / .01, which is .05 or 5%.

estimates provided by people. That is, people account for the causes that produce various outcomes, as well as their combined influence, then make their decisions.

11.6.1: Bayesian Theories

Another way of approaching human decision making is to take as a starting point that human decision making is Bayesian, that is, it follows Bayes' theorem. This probability theorem accounts for the various base rates of different circumstances that contribute to a given situation. Bayes' theorem can be stated formulaically as:

$P(A|B) = (P(B|A) * P(A)) / P(B)$

In English, we would read this as "The probability of *A* given *B* is equal to the probability of *B* given *A* times the probability of *A*, all divided by the probability of *B*." Is this clear?

11.6.2: Quantum Theory

More recently, cognitive psychologists have been using theoretical principles from quantum mechanics in the field of particle physics to address issues of human decision making. OK . . . wait . . . what? How does this even begin to make sense? It is true that the cognitive processes involved in human minds and brains are very different from quarks and such, and these cognitive scientists acknowledge the difference. Yet these researchers are attracted to and find useful the methods for analyzing probabilistic states and outcomes, which quantum mechanics and human decision making have in common. Thus, although the underlying processes may be different, the ways of thinking about and analyzing them may be analogous, and so warrant the use of an approach from a very different field. This application of quantum theory

Principles of Quantum Theory

Several principles of quantum theory apply to accounts of human decision making. These include the complementarity, uncertainty, and superimposition principles.

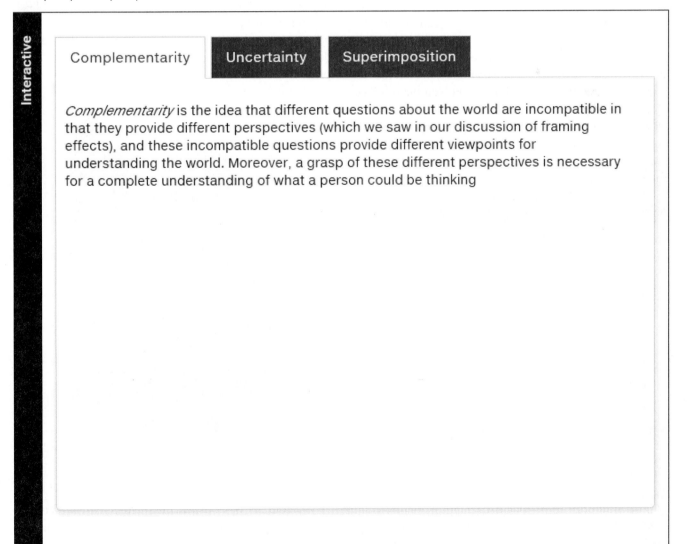

Complementarity is the idea that different questions about the world are incompatible in that they provide different perspectives (which we saw in our discussion of framing effects), and these incompatible questions provide different viewpoints for understanding the world. Moreover, a grasp of these different perspectives is necessary for a complete understanding of what a person could be thinking

has also been extended to studies of episodic memory (Brainerd, Wang, Reyna, & Nakamura, 2015) and semantic memory (Aerts, Sozzo, & Veloz, 2015).

If you remember from our discussion of sensory and perceptual processes, a heavy emphasis is placed on signal detection theory. Signal detection theory arose from work in the field of communications: signals traveling over wires or radio waves. Psychologists found a way to take that approach and reapply it to the study of cognition. A similar sort of thing is going on here.

Essentially, quantum theories of decision making take the probabilistic mathematical framework of quantum theory and apply it to cases of human decision making (Bruza, Wang, & Busemeyer, 2015; Busemeyer & Wang, 2015; Moreira, & Wichert, 2016).

Bruza et al. (2015) provide an account of the Linda problem from a quantum theory view. Specifically, a quantum cognition approach focuses on the complementarity of Linda's being a bank teller and a feminist. First, a person decides the probability that Linda is a bank teller (let's say the person decides that Linda *is* a teller). Then, from this point, a person decides whether Linda is a feminist. This is no longer an independent decision based on what was decided beforehand, even though traditional probability theory treats these decisions as independent. Because they are not independent in the mind of the decision maker, it is possible for the subjective probability to increase from that point given that Linda's characteristics match that of a feminist. The quantum theory approach can even account for phenomena such as preference reversals, which are a challenge for more traditional accounts of human decision making (Yukalov & Somette, 2015).

> **WRITING PROMPT**
>
> **Other Approaches to Decision Making**
>
> What are some of the advantages of the other approaches to understanding human decision making? What are some real-world settings in which knowing how people make decisions would be helpful?
>
> ▶ The response entered here will appear in the performance dashboard and can be viewed by your instructor.
>
> Submit

11.7: Limitations in Reasoning

OBJECTIVE: Analyze the limitations in reasoning

A central fact in studies of decision making and reasoning bears repeating. We use heuristics because of limitations in our cognition. We have **bounded rationality** and can only process so much information at a time. Moreover, there are limitations in our knowledge, both of relevant facts and of relevant algorithms. You had to estimate the number of handshakes in an earlier question because you didn't know the pertinent algorithm. In addition, there are limitations in the reasoner, sometimes as ordinary as unwillingness to make the effort needed but sometimes in working memory.

11.7.1: Limited Domain Knowledge

Everyday examples of how limited knowledge affects decision making and reasoning are abundant. Kempton (1986), for example, looked at reasoning based on analogies, particularly how we develop analogies based on known events and situations to reason about unknown or poorly understood domains. Focusing on a mechanical device, Kempton studied people's understanding of home heating, particularly their understanding of a furnace thermostat. The results indicated that some people's (incorrect) mental model is that a thermostat works like a water faucet: Turn it up higher to get a faster flow of heat. Likewise, many people's behavior suggests that they believe that the call button on an elevator works like a doorbell: If the elevator doesn't arrive reasonably soon, press the button again (see Shafir & Tversky, 1992).

We often are asked to make decisions, sometimes explicitly, about things that we may not have complete knowledge about.

It should be obvious that our mental models—our cognitive representations that we use during reasoning and decision making—can vary from true and complete knowledge (*expertise*, in other words) all the way down to no knowledge or information at all, *ignorance*. (People's awareness that they do not know something is quite interesting itself; see Gentner & Collins, 1981, and Glucksberg & McCloskey, 1981.) The most interesting situation to study is when knowledge is incomplete or inaccurate. Indeed, the fact that we are so concerned with how people estimate under uncertainty, and with their errors in reasoning, implies that complete and certain knowledge usually is not available to people.

NAIVE PHYSICS Some of the most intriguing research has been in **naive physics**, which is people's conceptions of the physical world, in particular, their understanding of the principles of motion.

A compelling aspect of the effects observed in naive physics studies is that the motion of physical objects is not a rarefied, unusual kind of knowledge. We have countless opportunities in our everyday experience to witness the behavior of objects in motion and to derive an understanding of the principles of motion. Anyone who has ever thrown a ball has had such opportunities. And yet the mental model we derive from that experience is flawed and is different from the perceptual–motor model that governs throwing a ball (Krist, Fieberg, & Wilkening, 1993; Schwartz & Black, 1999).

A second compelling aspect to the research concerns the nature of the mental model itself. People's erroneous understanding of bodies in motion is amazingly similar to the so-called *impetus theory*, which states that setting an object in motion puts some impetus or "movement force" into the object, with the impetus slowly dissipating across time (e.g., Catrambone, Jones, Jonides, & Seifert, 1995; see

---

Stimuli Used by McCloskey (Figure 11.5A–Figure 11.5C)

Take a look at the figures depicting several of the problems. The remaining section will be more meaningful to you if you spend a few moments working through the problems before reading further.

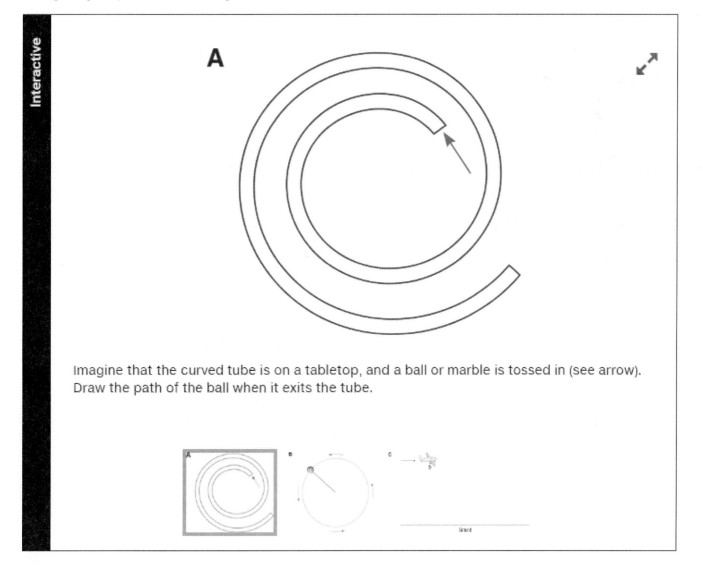

Imagine that the curved tube is on a tabletop, and a ball or marble is tossed in (see arrow). Draw the path of the ball when it exits the tube.

Cooke & Breedin, 1994, and Hubbard, 1996, on the notion of impetus). The punchline here is that the impetus theory was the accepted explanation of motion during the 13th and 14th centuries, a view abandoned by physics when Newton's laws of gravity and motion were advanced some 300 years ago. The correct mental model, basically, is that a body in motion continues in a straight line unless some other force, such as gravity, is applied. If some other force is applied, then that force combines with the continuing straight-line movement. Thus, when the ball leaves the tube, or when the string breaks, the ball moves in a straight line. No "curved force" continues to act on it because no such thing as "impetus" has been given to it. Likewise, the horizontal movement of the ball dropped from the airplane continues until the ball hits the ground. This movement is augmented by a downward movement caused by gravity; the ball accelerates vertically as it continues its previous horizontal motion. (If you demonstrated a naive belief in impetus, you might take some consolation in the fact that, across recorded history, people have believed in impetus theory longer than they have in Newton's laws.)

EXPERIENCE AND KNOWLEDGE Only a little research addresses the nature of our experience and the kind of information we derive from it. In some of the naive physics problems, especially the plane problem, an optical illusion seems partly responsible for the "straight down belief" (McCloskey, 1983; Rohrer, 2003). Beyond that, some inattentiveness on the reasoner's part, or perhaps difficulty in profiting from real-world feedback, may also account for part of the inaccuracy. To be sure, some of the difficulties we experience involve the difficulty of the problems themselves (Proffitt, Kaiser, & Whelan, 1990). For example, Medin and Edelson (1988) found that, depending on the structure and complexity of the problem, people may use base rate information appropriately, may use it inappropriately, or may ignore it entirely. We do know that instruction and training influence reasoning. Taking a physics class improves your knowledge of the rules of motion, but it doesn't completely eliminate the misbeliefs (Donley & Ashcraft, 1992; Kozhevnikov & Hegarty, 2001). Likewise, instruction in statistics, probability, and hypothesis testing improves your ability to reason accurately in those domains (Agnoli & Krantz,

---

Prove It

Naive Physics

Having people complete the diagrams on the naive physics beliefs is an almost fail-safe project. Interview your participants on their reasons for drawing the pathways they drew (be sure you understand the correct pathways before you try explaining the principles behind them). Also, come up with some new diagrams or problems to test other aspects of people's intuitive understanding of motion and gravity, for instance, the puzzling "water-level" problem (when a glass of fluid is tilted sideways, people—even those with presumed expertise—draw the line representing the water as perpendicular to the glass, rather than parallel to the horizon; for example, Hecht & Proffitt, 1995; Vasta & Liben, 1996).

It would be interesting to know how other groups of people respond. For example, in Donley and Ashcraft's (1992) article, the professors in the physics department performed essentially perfectly, whereas professors in other departments were no more accurate than students in the undergraduate physics sequence. It is not clear whether children would perform less accurately than adults; after all, adults perform badly. On the other hand, perhaps children will report more interesting reasons for their beliefs. Some adults complete the cliff-and-ball problem with a "straight out, then straight down" pathway, a pathway called the Road Runner effect, from the cartoon character. It would be interesting to know whether children (or adults) appeal to that in justifying their answers.

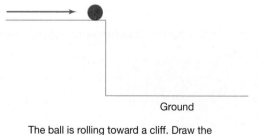

The ball is rolling toward a cliff. Draw the path of the ball as it goes over the edge.

If the two balls are dropped at the same time, do they hit the ground at the same time or at different times?

Figure 11.6 Response Choices for a Study by Lawson (2006)

In this study by Lawson, people needed to select the correct location of a bicycle's frame, pedals, and chain.

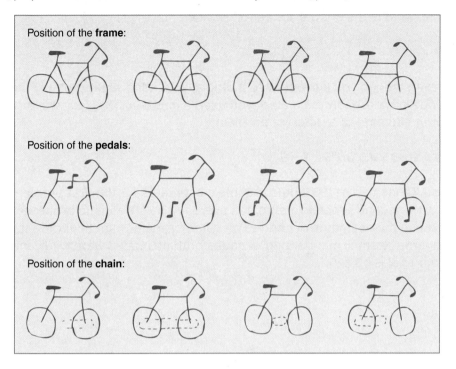

1989; Fong & Nisbett, 1991; Lehman et al., 1988). In short, acquiring a fuller knowledge of the domain is an important part of making more accurate decisions.

TYPES OF REASONING The types of reasoning errors we have been encountering here don't just apply to physics and statistics problems, but even extend to everyday objects. Lawson (2006) gave people partial drawings of bicycles, such as those shown in Figure 11.6, and asked people to select the drawings from a set of options that correctly depict the proper position of the bicycle frame, pedals, and chain. She found that, although bicycles are very familiar objects to most people, an average of 39% errors were made in their selections. Performance was better for expert cyclists, although even they made 15% errors. So, from this we can learn that even when people have a great deal of familiarity with a common object, they may still have trouble reasoning about the structure of that object and how it works.

11.7.2: Limitations in Processing Resources

Several studies attest to the role of processing resources in adequate decision making and problem solving.

Cherniak (1984; see also Tversky & Kahneman, 1983) has studied the *prototypicality heuristic*, a strategy in which we generate examples to reason out an answer rather than follow the correct, logical procedures of deductive reasoning. The heuristic is useful in the sense that it reduces people's errors when they are working under time constraints. However, it depends on a limitation, possibly of working memory, possibly of time in which to perform the task, or possibly in the willingness to do the slow, effortful work of following the algorithmic procedure.

WRITING PROMPT

Studying Errors in Decision Making

What is the value of studying the types of errors people make when making decisions? How can this knowledge be used to help people make better decisions for themselves?

 The response entered here will appear in the performance dashboard and can be viewed by your instructor.

Submit

Examples Illustrating Limitations in Processing Resources (Figure 11.7 and Figure 11.8)

> As an example, what is the answer to the following?
>
> $8 \times 7 \times 6 \times 5 \times 4 \times 3 \times 2 \times 1$
>
> Be honest—even though you know how to multiply, you didn't really multiply out all those values, did you? You probably estimated; you used a heuristic. One way we know this is by comparing your estimate to a different problem,
>
> $1 \times 2 \times 3 \times 4 \times 5 \times 6 \times 7 \times 8$
>
> In Tversky and Kahneman's (1983) data, people's estimates for the first problem averaged 2,250, but for the second problem, estimates averaged 512. Heuristic processing was clearly involved, because the estimates depended on whether the arithmetically identical sequences began with a large or small number and because both estimates were wildly inaccurate: The correct answer to 8! is 40,320.

Summary: Reasoning and Decision Making

11.1 Formal Logic and Reasoning

- Human reasoning is not especially logical, as shown in formal syllogistic and conditional reasoning problems. In syllogism tasks, conditional reasoning *if–then* problems, and hypothesis testing, people often fail to search for negative evidence. Instead, they frequently look for positive evidence, called confirmation bias, and are often influenced, both positively and negatively, by semantic knowledge. When a more skeptical attitude is adopted, and when the reasoning involves more concrete concepts, reasoning accuracy improves.
- There are a number of theories of how humans deal with logical reasoning. Heuristic theories assume that people largely use shortcuts based on the information in the premises. Probability heuristic theory assumes that people reason using subjective estimates of the likelihood of various premises. Mental rules theories assume that people use sets of implicit rules to reason, similar to the rules people may use when comprehending and producing language. Finally, mental model theory assumes that people create mental simulations of the circumstances described by the premises to reason.

11.2 Decisions

- People make decisions in many different ways. Sometimes they use algorithms, following a normative model, that are guaranteed to produce the correct answer. However, because they do not have enough time or the

- issues are not well defined, they may use heuristics or rules of thumb, following a descriptive model.
- Although there has been some suggestion in the popular press that people make better decisions when they make their selections quickly, research shows that slow, deliberative thinking produces better decisions far more often.

11.3 Classic Heuristics, Biases, and Fallacies

- Kahneman and Tversky investigated important heuristics used in circumstances when people reason about uncertain events. The representativeness heuristic guides people to judge outcomes as likely if they seem representative of the type of event being evaluated; for instance, a random-looking sequence of coin tosses is judged more representative of coin toss outcomes almost regardless of the true probabilities involved. Included among the reasoning effects predicted by this heuristic are various stereotyping results.
- In the availability heuristic, people judge the likelihood of events by how easily they can remember examples or instances. These judgments therefore can be biased by any other factor that affects memory, such as salience or vividness.
- In the simulation heuristic, people forecast or predict how some outcome could have been different. These forecasts are influenced by how easily the alternative outcomes can be imagined. Interestingly, when people complete "if only" statements, the changes they include tend to normalize the original situation by removing an unusual event and substituting a more common one. Such normalizations can be affected by the *focus* of the situation.
- With elimination by aspects, people can take a very large set of options and quickly reduce it down to a manageable size. People drop out options that do not meet various criteria that are set as important by a person.

11.4 Framing and Risky Decisions

- The choices that people make can be influenced by whether the choices are framed as either costs or benefits. In general, people are risk aversive for gains, meaning that they will go with the safer choice when they focus on what can be gained. However, they are risk seeking for losses in that they will be more likely to make a riskier choice if they focus on what will be lost.
- People are also somewhat blind at times to the absolute magnitude of a change, such as the cost of an item, or the amount saved. Instead, people focus more on the relative or proportional size of an outcome.

11.5 Adaptive Thinking and "Fast and Frugal" Heuristics

- Work on "fast and frugal" heuristics reveals that simple one-reason decision-making heuristics often do a very good job. These heuristics come from the adaptive thinking approach to decision making, the source of a current debate about the basis for reasoning under uncertainty.

11.6 Other Explanations

- Recent work on causal reasoning suggests that people use more normative reasoning than has otherwise been suspected if it is assumed that people have a reasonably correct causal model of the situation they are trying to make decisions about. Some of these newer approaches to human decision make use of ideas derived from the Bayes' theorem, whereas others use ideas developed from the field of quantum physics.

11.7 Limitations in Reasoning

- In everyday reasoning, we rely on mental models of the device or event to make our judgments. These mental models sometimes are quite inaccurate. In the best-known research, people's mental models of physical motion lead them to incorrect predictions (e.g., the trajectory of a ball dropped from an airplane).
- Ongoing research is focused on the kinds of limitations that lead to incorrect reasoning and decision making, including limited domain knowledge and limitations in working memory processes.

Appendix: Algorithms for Coin Tosses and Hospital Births

Coin Tosses

To begin with the obvious, the probability of a head on one coin toss is .50. Flipping a coin twice and keeping track of the possible sequences yields a .25 probability for each of the four possibilities HH, HT, TH, TT. In general, when the simple event has a probability of .50, the number of possibilities for a sequence of n events is 2 raised to the nth power. Thus, the number of distinct sequences for six coin tosses is 2^6, a total of 64 possibilities.

Two of the 64 possibilities are pure sequences, HHHHHH and TTTTTT. Two more are double sequences, HHHTTT and TTTHHH. All the remaining 60 possibilities involve either or both of the following characteristics: more of one outcome (e.g., heads) than the other and at least one alternation between the two outcomes at a position *other than* halfway through the sequence. Thus, the probability of a pure sequence is 2/64, as is the probability of a double sequence. Getting any one of the other 60 possibilities has a likelihood of 1/64. However, getting a "random-like" outcome—that is, any outcome other than straight or double—has a probability of 60/64.

Hospital Births

Many statistics texts contain tables of the binomial distribution, the best way to understand the hospital births example. Because most of these tables go up only to a sample size of 20, we will use a revised hospital example, comparing hospitals with three versus nine births per day (note that the 1:3 ratio is the same as the original example, 15:45). The probabilities for the original example are more extreme than these, but they will be in the same direction.

Just as with coin tosses, we are dealing with an event whose basic probability is .50, the likelihood that a newborn infant is male (ignoring the fact that male births are actually slightly more common than 50%). What is the probability that, in three births, all three will be boys? According to the binomial tables (Table 11.A), this probability is .1250. This is the probability that on any randomly selected day, the three-birth hospital will have all boys, $p = .1250$. Across the 365 days in a year, we expect an average of 45.625 such days (365 × .1250).

The temptation now is to consider the likelihood of exactly three boys in the nine-birth hospital. But this is not the relevant comparison. The relevant comparison to the all boys probability in the three-birth hospital would be all boys in the nine-birth hospital. This puts the comparison on the same footing as the original problem, 60% as the "extreme" cutoff.

The probability of exactly nine boys out of nine births is .0020, two chances in a thousand. For a whole year, we expect only 0.73 such days (365 × .0020). Now it should be clearer. The criterion of "extreme," all boys, is much more likely in the smaller sample than in the larger one, $p = .1250$ versus .0020. Multiplied out, the prediction is 45 days for the small hospital and .70 days for the large one.

By extension, and using the appropriate binomial values, the 15-birth hospital should have about 111 days per year with 60% or more boys, contrasted with 42 such days per year for the 45-birth hospital.

SHARED WRITING

Reasoning and Decision Making

Given the ways that people can and do make decisions, provide an example of an everyday choice, how people may typically reason to bring themselves to a poor decision, and what they can do to bring themselves to a better decision.

A minimum number of characters is required to post and earn points. After posting, your response can be viewed by your class and instructor, and you can participate in the class discussion.

Post   0 characters | 140 minimum

Table 11.A Binomial Probabilities for Exact Number of Relevant Outcomes, Where the Simple Probability of the Outcome Is .50

n = 3		n = 9		n = 15		n = 45	
			Exact Number of Relevant Outcomes:				
0	.1250	0	.0020	9	.1527	27	.0488
1	.3750	1	.0176	10	.0916	28	.0314
2	.3750	2	.0703	11	.0417	29	.0184
3	.1250	3	.1641	12	.0139	30	.0098
		4	.2461	13	.0032	31	.0047
		5	.2461	14	.0005	32	.0021
		6	.1641	15	.0000	33	.0008
		7	.0703	etc.		34	.0003
		8	.0176			35	.0001
		9	.0020			36	.0000

Chapter 12
Problem Solving

Learning Objectives

12.1: Explain the role of verbal protocol in studying problem solving

12.2: Analyze the approach adopted by cognitive psychologists to study the problem-solving process

12.3: Describe the work done by Gestalt psychologists to study problem solving

12.4: Compare insight and analogy as methods of problem solving

12.5: Outline the means–end analysis method of problem solving

12.6: Summarize a set of problem-solving techniques

A favorite example of "problem solving in action" is the following true story:

> When I (M. H. A.) was a graduate student, I attended a departmental colloquium at which a candidate for a faculty position was to present his research. As he started his talk, he realized that his first slide was projected too low on the screen. A flurry of activity around the projector ensued, one professor asking out loud, "Does anyone have a book or something?" Someone volunteered a book, the professor tried it, but it was too thick; the slide image was now too high. "No, this one's too big. Anyone got a thinner one?" he continued. After several more seconds of hurried searching for something thinner, another professor finally exclaimed, "Well, for Pete's sake, I don't believe this!" He marched over to the projector, grabbed the book, opened it halfway, and put it under the projector. He looked around the lecture hall and shook his head, saying, "I can't believe it. A roomful of PhDs, and no one knows how to open a book!"

This module examines the slow and deliberate cognitive processing called **problem solving**, which involves a goal-directed sequence of steps, the cognitive operations used to work through a problem, and which can be broken down into a collection of subgoals. As with decision making and reasoning, problem solving studies a person who is confronted with a difficult, time-consuming task: A problem is presented, the solution is not immediately obvious, and the person often is uncertain what to do next. We are interested in all aspects of the person's activities, from initial understanding of the problem to the steps that lead to a final solution, and, in some cases, how a person decides that a problem has been solved. Our interest needs no further justification than this: We confront countless problems in our daily lives, problems that are important for us to figure out and solve. We rely on our wits in these situations. We attempt to solve problems by mentally analyzing the situation, devising a plan of action, then carrying out that plan. Therefore, the mental processing involved in problem solving is, by definition, part of cognitive psychology.

Let's start with a simple "recreational" problem (Anderson, 1993). It will take you a minute or two at most to solve it, even if you lose patience with brain teasers very quickly. VanLehn's (1989) 9-year-old child seemed to understand it completely in about 20 seconds and solved it out loud in about 2 minutes:

> Three men want to cross a river. They find a boat, but it is a very small boat. It will only hold 200 pounds. The men are named Large, Medium, and Small. Large weighs 200 pounds, Medium weighs 120 pounds, and Small weighs 80 pounds. How can they all get across? They might have to make several trips in the boat.
>
> (VanLehn, 1989, p. 532)

Why should we be interested in such recreational problems? The answer is straightforward. As is typical of all scientific disciplines, cognitive science studies the simple before the complex, searching simpler settings to find basic principles that generalize to more complex settings. After all, not all everyday problems are tremendously complex. Figuring out how to prop up a slide projector is not of earthshaking significance (well, it probably was to the fellow interviewing for the job). In either case, the reasoning is that we often see large-scale issues and important processes more clearly when they are embedded in simple situations. Indeed, one aspect of problem solving you will read about, functional fixedness, accounts for why a roomful of PhDs didn't think about opening the book to make it thinner. Functional fixedness was discovered with a simple recreational problem.

12.1: Studying Problem Solving

OBJECTIVE: Explain the role of verbal protocol in studying problem solving

Unlike many areas of cognitive psychology, the study of significant problem solving requires us to examine a lengthy sample of behavior, often up to 20 or 30 minutes of activity. A major kind of data in problem solving is the *verbal protocol*, the transcription and analysis of people's verbalizations as they solve the problem.

Without a doubt, the use of verbal protocols influenced many opinions about problem solving. In fact, the status of verbal reports as data is still a topic of some debate (see Dunlosky & Hertzog, 2001; Ericsson & Simon, 1980, 1993; Fleck & Weisberg, 2004; and Russo, Johnson, & Stephens, 1989, for a range of views). That said, the ability to use other methods to investigate the cognitive processes involved in problem solving is always attractive. For example, Thomas (2013) used an eye tracker to find that people who made eye movements consistent with a problem solution were more likely to eventually solve it.

We will begin our coverage of problem solving by first defining the basic terms and ideas that pervade much of the research in this area. After that we will look at the classic problem-solving research of the Gestalt psychologists. The Gestalt movement coexisted with behaviorism early in the 20th century but never achieved the central status that behaviorism did. In retrospect, however, it was an important influence on cognitive psychology.

By the way, possibly more than in any material you have read so far, it is important in this module for you to spend some time working through the examples and problems. Hints usually accompany the problems, and the solutions are presented in the text. Many of the insights of the problem-solving literature pertain to conscious, strategic activities you will discover on your own as you work through the sample problems. Furthermore, simply by working the examples provided here, you may improve your own problem-solving skills.

12.2: Basics of Problem Solving

OBJECTIVE: Analyze the approach adopted by cognitive psychologists to study the problem-solving process

Cognitive psychology has adopted a *reductionistic approach* to the study of problem solving. That is, cognitive psychologists work to break the process of problem solving down into its various components and processes, and then understand how they work together. For instance, Newell and Simon's analysis of a cryptarithmetic problem (1972) is a microscopic analysis and interpretation of every statement made by one person as he solved a problem, all 2,186 words and 20 or so minutes of problem-solving activity. In Newell and Simon's (1972) description, "A person is confronted with a *problem* when he wants something and does not know immediately what series of actions he can perform to get it" (p. 72). The "something" can be renamed for more general use as a **goal**, the desired end point of the problem-solving activity.

12.2.1: Characteristics of Problem Solving

Problem solving consists of goal-directed activity, moving from some initial configuration or state through a series of intermediate steps until finally the overall goal has been reached: an adequate or correct solution. The difficulty is determining which intermediate states are on a correct pathway ("Will step A get me to step B or not?") and in devising operations or moves that achieve those intermediate states ("How do I get to step B from here?").

An intuitive illustration of such a nested solution structure is presented in Figure 12.1, a possible solution route to the locked-car problem. Note that during the solution, the first two plans led to barriers or blocks, thus requiring that another plan be devised. The problem solver finally decided on another plan, breaking a window to get into the locked car. This decision is followed by a sequence of related acts: the search for some heavy object that will break a window, the decision as to which

Four Aspects of Problem Solving

Let's start by listing several characteristics that define what is and is not a genuine instance of problem solving.

Interactive

Goal Directedness

The overall activity we are examining is directed toward achieving some goal or purpose. As such, we exclude daydreaming, for instance; it is mental, but it is not goal directed. Alternatively, if you have locked your keys in your car, both physical and mental activity are going on. The goal-directed nature of those activities, your repeated attempts to get into the locked car, makes this an instance of true problem solving.

Sequence of Operations

Cognitive Operations

Subgoal Decomposition

window to break, and so forth. Each of these decisions is a subgoal nested within the larger subgoal of breaking into the car, itself a subgoal in the overall solution structure.

12.2.2: A Vocabulary of Problem Solving

In this section, we present four characteristics that define what qualifies as problem solving from a cognitive science perspective. Many important ideas are embedded in these four points, however. Let's examine some of these points further, looking now toward an expanded vocabulary of problem solving, a set of terms we use to describe and understand how people solve problems.

THE PROBLEM SPACE The term **problem space** is critical. Anderson (1985) defines it as the various states or conditions that are possible. More concretely, the problem space includes the initial, intermediate, and goal states of the problem. It also contains the problem solver's knowledge at each of these steps, both knowledge that is currently being applied and knowledge that could be retrieved from memory and applied. Any available external devices, objects, or resources can also be included in the description of the problem space. Thus, a difficult arithmetic problem that must be completed mentally has a different problem space than the same problem as completed with pencil and paper.

To illustrate, VanLehn (1989) describes one man's error in the "three men and a rowboat" problem. The man focused only on the arithmetic of the problem and said essentially, "400 pounds of people, 200 pounds per trip, it'll take two trips of the boat." When he was reminded that the boat couldn't row itself back to the original side, he adopted a different problem space.

Figure 12.1 A Representation of Part of the Problem Space for Getting Into a Locked Car

Note the barriers encountered under plans A and B.

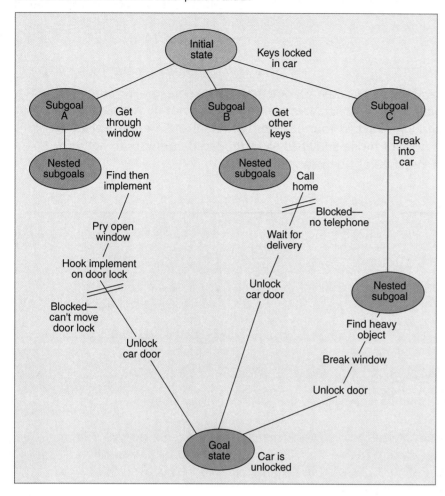

terminal node at the bottom. For "wide open" problems, each branch may need to be searched until a dead end is encountered. For other problems, information may be inferred that permits a restriction in the branches that are searched (the shaded area of the figure). Clearly, if the search space can be limited by pruning off the dead-end branches, then problem-solving efficiency increases.

THE LOOK AHEAD When problem solving, people are more likely to be successful if they have a plan about how they are going to solve the problem. In problem-solving research, this is called *lookahead* (VanLehn, 1989) and refers to how many steps in the future a problem solver is considering.

An example of failing to plan would be, when cooking, not warming up the oven while doing other preliminary steps, but turning on the oven at the point when something needs to be placed in it. Another example of lookahead failure would be cutting off a tree branch while sitting on it. Research has shown that people often do not adequately look ahead to the consequences of their actions when problem solving. Because they are more likely to be

In some problem contexts, we can speak of problem solving as a search of the problem space or, metaphorically, a search of the solution tree, in which each branch and twig represents a possible pathway from the initial state of the problem. For problems that are "wide open," that is, those with many possibilities that must be checked, there may be no alternative but to start searching the problem space, node by node, until some barrier or block is reached. As often as not, however, there is information in the problem that permits us to restrict the search space to a manageable size. Metaphorically, this information permits us to *prune* the search tree.

A general depiction of this is in Figure 12.2.

The initial state of the problem is the top node, and the goal state is some

Figure 12.2 A General Diagram of a Problem Space

This figure illustrates various branches of the space. Often a hint or an inference can prune the search tree, restricting the search to just one portion; this idea is represented by the shaded area of the figure. Note that, in most problems, the problem space tree is much larger, so the beneficial effect of pruning is far greater.

SOURCE: Adapted from Wickelgren (1974).

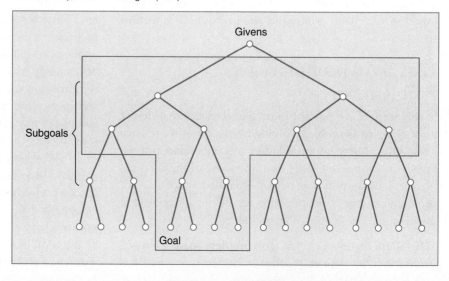

impulsive in making a move that initially seems to bring them closest to the solution, their efforts often do not result in success (Ormerod, MacGregor, Chronicle, Dewald, & Chu, 2013).

THE OPERATORS **Operators** are the set of legal moves or actions that can be done during problem solution. The term *legal* means permissible in the rules of the problem. For example, an illegal operator in the three men in a rowboat problem is having the men swim across the river or loading the boat with too heavy a load. An illegal operator in getting the keys out of your locked car is teleporting into the car. You just can't do that.

For transformation problems (Greeno, 1978), applying an operator transforms the problem into a new or revised state from which further work can be done. In general, a legal operator moves you from one node to the next along some connecting pathway in the search space. For instance, in solving algebraic equations, one transformation operator is "move the unknowns to the left." Thus, for the equation $2X + 7 = X + 10$, applying the operator would move the single X to the left of the equal sign by subtracting X from both sides of the equation.

Often, constraints within the problem prevent us from applying certain operators. The inability to operate many of the car's normal functions when on the outside is a constraint that makes it harder to solve the problem of getting your keys out of it when you have locked them inside. In algebra, by contrast, constraints are imposed by the rules of algebra; for example, you can't subtract X from one side of the equation without subtracting it from the other side, too.

THE GOAL The *goal* is the ultimate destination or solution to the problem. For recreational problems in particular, the goal is typically stated explicitly in the problem.

Sample Recreational Problems (Figure 12.3)

This section provides several recreational problems for you to work through to give you a better feel for and understanding of the problem-solving process. These include the Buddhist monk, drinking glasses, and six pennies problems.

Interactive

| Buddhist Monk | Drinking Glasses | Six Pennies |

The Buddhist monk problem in the interactive exercise below is distressingly vague in its specification of the goal (as are the problems "write a term paper that will earn you an **A**," "write a computer program that does *X* in as economical and elegant a fashion as possible," and so on). Here is the problem:

> One morning, exactly at sunrise, a Buddhist monk began to climb a tall mountain. The narrow path, no more than a foot or two wide, spiraled around the mountain to a glittering temple at the summit. The monk ascended the path at varying rates of speed, stopping many times along the way to rest and to eat the dried fruit he carried with him. He reached the temple shortly before sunset.
>
> After several days of fasting and meditation, he began his journey back along the same path, starting at sunrise and again walking at variable speeds with many pauses along the way. His average descending speed was, of course, greater than his average climbing speed.
>
> Show that there is a spot along the path that the monk will occupy on both trips at precisely the same time of day.

Hint: Although the problem seems to ask for a quantitative solution, think of a way of representing the problem using visual imagery.

The DONALD + GERALD Problem

This is a *cryptarithmetic* problem in which letters of the alphabet have been substituted for the digits in an addition problem.

> Your task is to reverse the substitutions—to figure out which digits go with which letters to yield a correct addition problem. The restriction is that the digits and letters must be in one-to-one correspondence (only one digit per letter and vice versa).
>
> Plan on spending 15 minutes or so, about the amount of time it takes people on their first attempt. Make notes on paper as you work so you can go back later to retrace and analyze your attempt to solve the problem. (Incidentally, this is the cryptarithmetic problem Newell and Simon's single person worked on.)
>
> ```
>   DONALD    (Hint: D = 5)
> + GERALD
> _____
>   ROBERT
> ```

Given that recreational problems usually present an explicit and complete specification of the initial and goal states, these are **well-defined problems**. In such problems, solutions involve progressing through the legal intermediate states, by means of known operators, until the goal is reached. In contrast, for **ill-defined problems**, the states, the operators, or both may be only vaguely specified.

DONALD + GERALD EXAMPLE OF RECREATIONAL PROBLEM Let's consider one recreational problem to pin down some of these terms and ideas.

Greeno (1978) calls this process a **constructive search**. Rather than blindly assigning digits and trying them out, people usually draw inferences from the other columns and use those to limit the possible values the letters can take. This approach is typical in arrangement problems, the third of Greeno's categories, in which some combination of the given components must be found that satisfies the constraints in the problem. In other kinds of arrangement problems, say anagrams, a constructive search heuristic would be to look for spelling patterns and form candidate words from those familiar units. The opposite approach, sometimes known as **generate and test**, merely uses some scheme to generate all possible arrangements, then tests those one by one to determine whether the problem solution has been found.

A related aspect of problem solving here (it can be postponed, but your solution will be more organized if it is done now) is quite general and almost constitutes good advice rather than an essential feature of performance. Some mechanism or system for keeping track of the information you know about the letters is needed, if only to prevent you from forgetting inferences you have already

Intermediate Steps for Tracking Known Values

The interactive below presents a compressed verbal protocol of the solution to the DONALD problem, which you might want to compare with your own solution pathway.

State	Known Values	Reasons and Statements From Protocol
1 5ONAL5 GERAL5 ROBERØ	Ø123456789 T D R is odd	Because D is 5, then T = Ø, and carry a 1 to the next column. So the first column is 5 + something = odd because L + L + 1 = R will make R odd.
	G is less than 5 R is odd and greater than 5	R must be bigger than 5 because less than 5 would yield a two-digit sum in the D + G column and there would be an extra column in the answer. G is less than 5.

drawn. Indeed, such an external memory aid can go a long way toward making your problem solving more efficient. In some instances, it may even help you generalize from one problem variant to another (as in the next example, the Tower of Hanoi problem).

WRITING PROMPT

Steps and Components of Problem Solving

Think of a problem you have encountered in your own life, and discuss how the steps and components of problem solving did or could go into solving that problem.

12.3: Gestalt Psychology and Problem Solving

OBJECTIVE: Describe the work done by Gestalt psychologists to study problem solving

Now that we have covered some of the basic terminology of research on problem solving, let's step back and look at some of the early research on the cognitive psychology of problem solving. As we noted previously, **Gestalt** is a German word that translates poorly into English; the one-word translations "whole," "shape," and "field" fail to capture what the term means. Roughly speaking, a Gestalt is a whole pattern, a form, or a configuration. It is a cohesive grouping, a perspective from which the entire field can be seen. A variety of translations have been used, but none ever caught on, which prompted Boring (1950) to remark that Gestalt psychology "suffered from its name." So, we use the German term *Gestalt*, rather than an inadequate translation. In perception, the Gestalt principles show that humans tend to perceive and deal with integrated, cohesive wholes.

12.3.1: Early Gestalt Research

The connection between Gestalt psychologists and problem solving is best explained by an anecdote (see Boring, 1950, pp. 595–597). In 1913, Wolfgang Köhler, a German psychologist, went to the Spanish island of Tenerife to study "the psychology of anthropoid apes" (p. 596). Trapped there by the outbreak of World War I, Köhler experimented with visual discrimination among several animal species. In the course of this research, he began to apply Gestalt principles to animal perception. His ultimate conclusion was that animals do not perceive individual elements in a stimulus, but that they perceive relations among stimuli. Furthermore, "Köhler also observed that the perception of relations is a mark of intelligence, and he called the sudden perception of useful or proper relations *insight*" (p. 596).

Still stranded on the island, Köhler continued to examine "insight learning." He presented problems to chimpanzees and searched for evidence of genuine problem solving in their behavior. By far, the most famous of his subjects was a chimpanzee named Sultan (Köhler, 1927). In a simple demonstration, Sultan was able to use a long pole to reach through the bars of his cage and get a bunch of bananas. Köhler made the situation more difficult by giving Sultan two shorter poles, neither of which was long enough to reach the bananas. After failing to get the bananas, and sulking in his cage for a while, Sultan (as the story goes) suddenly went over to the poles and put one inside the end of the other, thus creating one pole that was long enough to reach the bananas.

For Sultan, this is problem solving because the behavior is goal directed (to get the bananas), there was a sequence of operations (the pole parts needed to be put together before reaching for the bananas), and there were cognitive operations (Sultan needed to think about the situation). For Sultan, the problem space involved the confines of his cage, the distance of the bananas from him, and the pole segments. The operations were all of the things he could do inside his cage, and with the poles.

Köhler found this to be an apt demonstration of *insight*, a sudden solution to a problem by means of an insightful discovery. In another situation, Sultan discovered how to stand on a box to reach a banana that was otherwise too high to reach. In yet another, he discovered how to get a banana that was just out of reach through the cage bars: He walked *away* from the banana, out a distant door, and around the cage. All these problem solutions seemed to illustrate Sultan's perception of relations and the importance of insight in problem solving.

Grande builds a three-box structure to reach the bananas, while Sultan watches from the ground. *Insight*, sometimes referred to as an "Aha" experience, was the term Köhler used for the sudden perception of useful relations among objects during problem solving.

12.3.2: Difficulties in Problem Solving

Other Gestalt psychologists, most notably Duncker and Luchins, pursued research with humans. Two major contributions of this work are essentially the two sides of the problem-solving coin. One involved a set of negative effects related to rigidity or difficulty in problem solving; the other, insight and creativity during problem solving.

FUNCTIONAL FIXEDNESS Two articles on functional fixedness, one by Maier (1931) and one by Duncker (1945), identify and define this difficulty. **Functional fixedness** is a tendency to use objects and concepts in the problem environment in only their customary and usual ways.

It is probably not surprising that problem solvers experience functional fixedness. After all, we comprehend the problem situation by means of our world knowledge, along with whatever procedural knowledge we have that might

Problems Demonstrating Functional Fixedness (Figure 12.4)

Sometimes problems seem like they should be easy, but then they turn out to be difficult. This difficulty may stem from thinking about things in a typical or standard way. The following functional fixedness problems illustrate two ways in which research on problem solving has addressed this issue.

Interactive

| The Two-String Problem | **The Candle Problem** |

Maier (1931), for instance, had people work on the two-string problem. Two strings are suspended from the ceiling, and the goal is to tie them together. The problem is that the strings are too far apart for a person to hold one, reach the other, then tie them together. Also available are several other objects, including a chair, some paper, and a pair of pliers. Even standing on the chair does not get the person close enough to the two strings.

In Maier's results, only 39% of the people came up with the correct solution during a 10-minute period. The solution (if you haven't tried solving the problem, do so now) involves using an object in the room in a novel way. A correct solution is to tie the pliers to one string, swing it like a pendulum, then catch it while holding the other string. Thus, the functional fixedness in this situation was failing to think of the pliers in any but their customary function; people were fixed on the normal use for pliers and failed to appreciate how they could be used as a weight for a pendulum.

be relevant. When you find "PLIERS" in semantic memory, the most accessible properties involve the normal use for pliers. Far down on your list would be characteristics related to their weight or aspects of their shape that would enable you to tie a string to them. Likewise, "BOX" probably is stored in semantic memory in terms of "container" meanings—that a box can hold things, that you put things into a box—and not in terms of "platform or support" meanings (see Greenspan, 1986, for evidence on retrieval of central and peripheral properties). Simply from the standpoint of routine retrieval from memory, then, we can understand why people experience functional fixedness.

Problem solving in situations that elicit functional fixedness typically involves people failing to notice some minor or obscure feature, but one which is critical to solving the problem. This would be the weight of the pliers in the two-string problem and the rigidity of the box bottom (when inverted) in the candle problem. Problem solving (and creativity) can involve noticing and applying such obscure features. McCaffrey (2012) asked people to assess a series of objects to address the questions "Can this be decomposed further?" and "Does this description imply a use?" Afterward, people had to solve a series of problems that typically elicit functional fixedness. People who had just practiced assessing various parts of objects were more likely to notice an otherwise obscure feature of objects and solve the problems.

NEGATIVE SET A related difficulty in problem solving is **negative set** (or simply *set effects*). This is a bias or tendency to solve problems in a particular way, using a single specific approach, even when a different approach might be

The Water Jug Problem (Tables 12.1A–12.1B)

A classic demonstration of set effects comes from the water jug problem, studied by Luchins (1942). In this problem, you are given three jugs, each of a different capacity, and are to measure out a quantity of water using just the three jugs.

Problem	Capacity of Jug A	Capacity of Jug B	Capacity of Jug C	Desired Quantity
1	5 cups	40 cups	18 cups	28 cups
2	21 cups	127 cups	3 cups	100 cups

Measuring Quantities with Water Jugs

As a simple illustration, consider the first problem in above table. You need to measure out 28 cups of water and can use containers that hold 5, 40, and 18 cups (jugs A, B, and C). The solution is to fill A twice, then fill C once, each time pouring the contents into a destination jug. This approach is an addition solution because you add the quantities together.

For the second problem, a subtraction solution is appropriate: Fill B (127), subtract jug C from it twice (-3, -3), then subtract jug A (-21), yielding 100.

more productive. The term *set* is a rough translation of the original German term **Einstellung**, which means something like "approach" or "orientation."

As the slide projector problem in the introduction suggests, functional fixedness and negative set are common occurrences. The occurrence of mental set in problem solving can result from people focusing their attention on information that is consistent with an initial solution attempt for a problem, at the expense of other possible solutions. For example, eye tracking data reveal that people look more often at information consistent with their first solution attempt than at other information (Bilalić, McLeod, & Gobet, 2010). Possibly because we eventually find an adequate solution to our everyday problems despite the negative set or without overcoming our functional fixedness (e.g., eventually locating a thinner book), we are less aware of these difficulties in our problem-solving behavior. The classic demonstrations, however, illustrate dramatically how rigid such behavior can be and how barriers to successful problem solving can arise.

WRITING PROMPT

Gestalt Psychologist and Problem-Solving Research

What important contributions did Gestalt psychologists make to research in problem solving? How do these insights align with their other views on psychology in general?

 The response entered here will appear in the performance dashboard and can be viewed by your instructor.

12.4: Insight and Analogy

OBJECTIVE: Compare insight and analogy as methods of problem solving

On a more positive side of problem solving are the topics of insight and problem solving by analogy. These are ways that people arrive at a solution, either more consciously or more unconsciously, by using the knowledge that they have of the problem domain.

12.4.1: Insight

Insight is a deep, useful understanding of the nature of something, especially a difficult problem. We often include the idea that insight occurs suddenly—the "Aha!" reaction—possibly because a novel approach to the problem is taken or a novel interpretation is made (Sternberg, 1996), or even just because you have overcome an impasse (for research on the various sources of difficulty in insight problems, see Chronicle, MacGregor, & Ormerod, 2004, and Kershaw & Ohlsson, 2004).

Sometimes, the necessary insight for solving a problem comes from an analogy: An already-solved problem is similar to a current one, so the old solution can be adapted to the new situation. The historical example of this is the story of Archimedes, the Greek scientist who had to determine whether the king's crown was solid gold or whether some silver had been mixed with the gold. Archimedes knew the weights of both gold and silver per unit of volume but could not imagine how to measure the volume of the crown. As the anecdote goes, he stepped into his bath

---

Insight Problems (Figures 12.5 and 12.6)

Puzzle over the insight problems provided here for a moment to see whether you have a sudden "Aha!" experience when you realize how to solve the problems.

Chain Links

A woman has four pieces of chain. Each piece is made up of three links. She wants to join the pieces into a single closed ring of chain. To open a link costs 2 cents and to close a link costs 3 cents. She has only 15 cents. How does she do it?

Hint: You don't have to open a link on each piece of chain.

Four Trees

Prisoner's Escape

Bronze Coin

Nine Dots

Bowling Pins

one day and noticed how the water level rose as he sank into the water. He then realized the solution to his problem. The volume of the crown could be determined by immersing it in water and measuring how much water it displaced. Excited by his insight, he then jumped from the bath and ran naked through the streets, shouting, "Eureka! I have found it!"

Metcalfe and Wiebe (1987; also Metcalfe, 1986) studied how people solved such problems and compared that with how they solved algebra and other routine problems. They found two interesting results. First, people were rather accurate in predicting whether they would be successful in solving routine problems but not in predicting success with insight problems. Second, solutions to the insight problems seemed to come suddenly, almost without warning. This result is shown in Figure 12.7.

As they worked through the problems, people were interrupted and asked to indicate how "warm" they were, that is, how close they felt they were to finding the solution. For routine algebra problems, "warmth" ratings grew steadily as people worked through the problems, reflecting their feeling of getting closer and closer to the solution. But there was little or no such increase for the insight problems even 15 seconds before the solution was found.

LETTING GO AND IMPROVING INSIGHT Although these results support the idea that insight arrives suddenly, insight problems can be thought of in simpler terms, say overcoming functional fixedness or negative set (as in prisoner's escape and nine-dot), taking a different perspective (bronze coin), and the like (Smith, 1995).

A neuroimaging study by Kounios et al. (2006) provides some support for this idea. Using both electroencephalographic and functional magnetic resonance imaging (fMRI) recordings, they found increased cortical activity centered on the frontal lobes (particularly the anterior cingulate cortex, B.A.s 24, 32, and 33) when people produced insight solutions as compared to normal problem solving. Kounios et al.'s theory is that this part of the frontal lobe suppresses the irrelevant information (an attentional process) that tends to dominate a person's thinking up to that point. Suppression of these dominant thoughts allows more weakly activated ideas, such as those remote associations drawn by the right hemisphere, to come to the fore, possibly providing the solution to a problem. In other words, part of a person's thought processes is working on the problem along with the steps that are being worked on at the forefront of consciousness (which are going nowhere). When these dead-end thoughts are moved aside, alternative solutions can then present themselves.

This release from irrelevant modes of thinking, seen in the neuroimaging data, can be extended to a process called incubation. With **incubation**, when people have difficulty solving a problem, they may stop working on it for a while. Then at some point, the solution or key to a solution may present itself to them (Sio & Ormerod, 2009). Although this can work at times, it appears that incubation is most useful when people have originally been provided with misleading information, by either others or themselves, that steers them away from the correct solution. During incubation, the representations for these misleading ideas lose strength, so that later the more successful alternatives can then present themselves (Vul & Pashler, 2007). Interestingly, insight is more likely to occur when people are mind wandering while doing relatively undemanding tasks, as compared to doing a demanding task, simply resting, or not taking a break at all (Baird et al., 2012).

In some circumstances, insight may mean that we have drawn a critical inference that leads to a solution; for example, there is more than one way to divide a rope in half (Wickelgren, 1974). Weisberg (1995) reports that some people solve insight problems like those mentioned earlier without any of the sudden restructuring or understanding that supposedly accompanies insight.

Other evidence, however, suggests that verbalization can interfere with insight, can disrupt

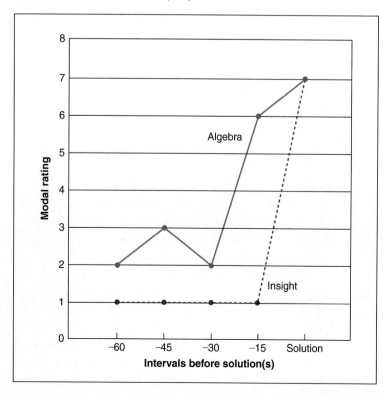

Figure 12.7 Modal (Most Frequent) Warmth Rating in the Four Time Periods Leading Up to a Problem Solution

SOURCE: Data from Metcalfe & Wiebe (1987).

"nonreportable processes that are critical to achieving insight solutions" (Schooler, Ohlsson, & Brooks, 1993, p. 166). Furthermore, being unable to report the restructuring that accompanies insight, or the actual insight itself, may be more common in insight situations than we realize. For instance, Siegler and Stern (1998; see also Siegler, 2000) conducted a study of second-graders solving arithmetic problems, then reporting verbally on their solutions. There was the regular computational, noninsightful way to solve the problems, which the second-graders followed, but also a shortcut that represented an insight (e.g., for a problem like 18 + 24 − 24, simply state 18). Almost 90% of the sample discovered the insight for solving such problems, as shown by the dramatic decrease in their solution times from around 12 seconds for the computational method to a mean of 2.7 seconds with the shortcut. However, the children were unaware of their discovery when questioned about how they had solved the problems. Within another five trials, however, 80% of the children's verbal reports indicated that they were aware of their discovery.

Sometimes we can take a long time to figure out a problem, but with effort, the passage of time, and insight, we can reach sometimes stunning and novel conclusions.

12.4.2: Analogy

In general, an *analogy* is a relationship between two similar situations, problems, or concepts. Understanding an analogy means putting the two situations into some kind of alignment so that the similarities and differences are made apparent (Gentner & Markman, 1997).

Take a simple example, the analogy "MERCHANT : SELL :: CUSTOMER : _____." Here, you must figure out the structure for the first pair of terms and then project or map that structure onto the second part of the analogy. Because "SELL" is the critical activity of "MERCHANT," the critical activity relationship is then mapped onto "CUSTOMER," and retrieval from memory yields "BUY."

Researchers argue that analogies provide excellent, widely applicable methods for solving problems. That is, if you are confronted with a difficult problem, a useful heuristic is to find a similar or related situation and build an analogy from it to the current problem. Such reasoning and problem solving may help us understand a variety of situations, such as how students should be taught in school, how people adopt professional role models, and how we empathize with others (Holyoak & Thagard, 1997; Kolodner, 1997). Furthermore, it has long been held that important scientific ideas, breakthroughs, and explanations often depend on finding analogies—for instance, that neurotransmitters fit into the receptor sites of a neuron much the way a key fits into a lock (see Gentner & Markman, 1997, for a description of reasoning by analogy in Kepler's discovery of the laws of planetary motion).

Curiously, analogical problem solving is better when people receive the information by hearing about it rather than reading it (Markman, Taylor, & Gentner, 2007), perhaps reflecting the more natural use of spoken over written language. Also, analogies are more effective if the analogy source is more abstract, rather than concrete (Day, Motz, & Goldstone, 2015), perhaps making it easier to apply to a new situation. Finally, people are more likely to use analogies to solve problems if they have had an opportunity to sleep (Monaghan et al., 2015). As with insight, perhaps this period of rest allows for remote associations to be made, leading people to use analogies.

ANALOGY PROBLEMS Although insight involves the seemingly sudden awareness of the solution to a problem, often as a result of unconscious processes, analogy involves the use of the solutions to prior problems that have a similar, underlying, abstract structure. This use of analogy can be conscious and explicit or, like insight, unconscious and implicit. Although people can certainly use analogies, especially when they are explicitly pointed out, they do not use them as often as would benefit them.

The radiation problem is interesting for a variety of reasons, including the fact that it is rather ill defined and thus comparable to many problems in the real world.

Different Analogy Problems

Gick and Holyoak (1980) had people read the parade problem, a somewhat different army fortress story, or no story at all. They then asked them to read and solve a second problem, the classic Duncker (1945) radiation problem.

> **The Parade Problem**
>
> A dictator ruled over a small country. He ruled from a strong fortress. The fortress was located in the center of the country, surrounded by numerous towns, villages, and farms. Like spokes on a wheel, many roads radiated out from the fortress. As part of a celebration of his glorious grab of power, this dictator demanded from one of his generals a large, over-the-top parade of his military might. The general's troops were assembled for the march in the morning, the day of the anniversary, at the end of one of these roads that led up to the dictator's fortress. At that time, the general was given a report by one of his captains that brought him up short. As commanded by the dictator, this parade needed to be far more spectacular and impressive than any other parade that had ever been seen in the land (or else). The dictator demanded that everyone in every region of the country see and hear his army at the same time. Given this demand, it seemed nearly impossible for the general to have the whole country see the parade as requested.

Duncker's participants produced two general approaches that led to dead ends: trying to avoid contact between the ray and nearby tissue, and trying to change the sensitivity of surrounding tissue to the effects of the ray. But the third approach, reducing the intensity of the rays, was more productive, especially if an analogy from some other, better understood situation was available.

GICK AND HOLYOAK'S (1980) RESULTS Gick and Holyoak (1980) used the radiation problem to study analogy. In fact, we have just simulated one of their experiments here by having you read the parade story first and then the radiation problem. In case you didn't notice, there are strong similarities between the problems, suggesting that the parade story can be used to develop an analogy for the radiation problem.

Gick and Holyoak found that 49% of people who first solved the parade problem realized it could be used as an analogy for the radiation problem. A different initial story, in which armies are attacking a fortress, provided a stronger hint about the radiation problem. Fully 76% of these participants used the attack analogy in solving the radiation problem. In contrast, only 8% of the control group, which merely attempted to solve the radiation problem, came up with the dispersion solution (i.e., multiple pathways).

When Gick and Holyoak provided a strong hint, telling people that the attack solution might be helpful as they worked on the radiation problem, 92% of them used the analogy, and most found it "very helpful." In contrast, only 20% of the people in the no-hint group produced the dispersion solution, even though they too had read the attack dispersion story. In short, only 20% spontaneously

Summary of Gick and Holyoak's (1980) Results
The simulation below summarizes Gick and Holyoak's results concerning the influence of analogical processing on problem solving.
SOURCE: From Kahneman & Tversky (1982)

Group	Order of Stories	Percentage of People Who Used the Analogy on the Radiation Problem
Group A	Parade, radiation	49%
Group B	Attack dispersion, radiation	76%
Group C	No story, radiation	8%

Study 1 (Experiment II originally; after Gick & Holyoak, Table 10)
People in groups A and B are given a general hint that their solution to one of the earlier stories may be useful in solving the radiation problem.

noticed and used the analogous relationship between the problems.

MULTICONSTRAINT THEORY Holyoak and Thagard (1997) proposed a theory of analogical reasoning and problem solving, based on such results. The theory, called the **multiconstraint theory**, predicts how people use analogies in problem solving and what factors govern the analogies people construct.

A final point is that most of the work on analogy, like many studies of problem solving, has focused on the conscious, explicit use of analogies. However, some evidence suggests that people may use analogies in a more unconscious, implicit manner as well. In a study by Day and Gentner (2007), people read given pairs of texts. When the events described by the second text were analogous to those described in the first (in terms of their relational structure), people read the second text faster. That is, people were able to use their unconscious knowledge of the event structure from the first text to help them understand the second text. When asked, people showed no awareness of this relationship between the two texts. So, in some sense, by having people read the first text, the relational structure of the event was primed, and this made the processing of the second text easier.

12.4.3: Neurocognition in Analogy and Insight

Some exciting work has been reported on the cognitive neuropsychology of analogical reasoning and insight. Wharton et al. (2000) identified brain regions associated

Factors Governing Analogy (Figure 12.8)

In particular, the multiconstraint theory states that people are constrained by three factors when they try to use or develop analogies.

Problem Similarity | Problem Structure | Purpose of the Analogy

The first factor is *problem similarity*. There must be some degree of similarity between the already-understood situation, the *source* domain, and the current problem, the *target* domain. In the parade story, for example, the fortress and troops are similar to the tumor and rays. Similarity between source and target has been shown to be important. Chen, Mo, and Honomichl (2004), for example, found that similarities from well-known folktales to new problems were especially important for finding problem solutions, even if participants did not report remembering the folktale. Alternatively, Novick (1988) found that novices focus especially on similarities, even when they are only superficial, which can interfere with performance.

with the mapping process in analogical reasoning. In their study, people saw a source picture of geometric shapes, followed by a target picture. They had to judge whether the target picture was an *analog pattern*—whether it had the same system of relations as the source picture. In the control condition, they judged whether the target was literally the same as the source. See Figure 12.9 for sample stimuli. In the top stimulus, the correct target preserves both the *spatial relations* in the source (a shape in all four quadrants) and the *object relations* (the patterned figures on the main diagonal are the same shape, and the shapes on the minor diagonal are different). Response times to analogy trials were in the 1,400 to 1,500 ms range and approximately 900 to 1,000 ms in the literal condition; accuracy was at or above 90% in both kinds of trials.

But the stunning result came from positron emission tomography (PET) scan images that were taken. Wharton et al. found significant activation in the medial frontal cortex, left prefrontal cortex, and left inferior parietal cortex.

In contrast, Bowden and Beeman (1998; Beeman & Bowden, 2000) found a significant role for *right* hemisphere processing in solving insight problems. Before reading further, try this demonstration:

> What one word can form a compound word or phrase with each of the following: Palm Shoe House?
>
> What one word can form a compound word or phrase with each of the following: Pie Luck Belly?

People were given such word triples—called "compound remote associates"—and had to think of a fourth word that combines with each of the three initial words to yield a familiar word pair. On many trials, people fail to find an immediate solution and end up spending considerable time working on the problem. They also report that when

Figure 12.9 A Depiction of Analogy Condition and Literal Condition Trials

The first column shows the source stimuli, the second shows the correct choice, and the third and fourth show incorrect choices for the stated reasons.

SOURCE: From Wharton et al. (2000), Figure 2, p. 179.

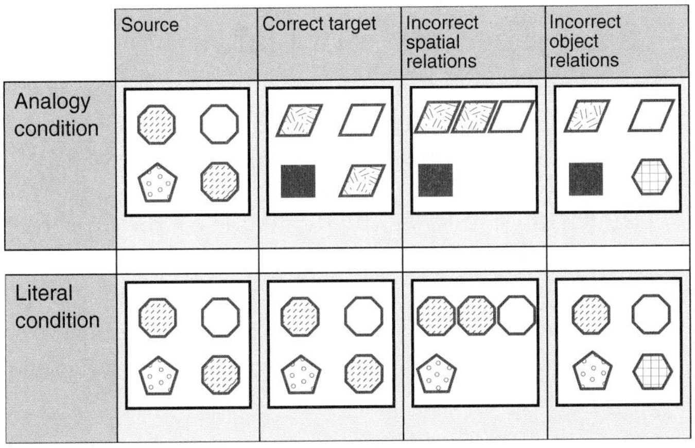

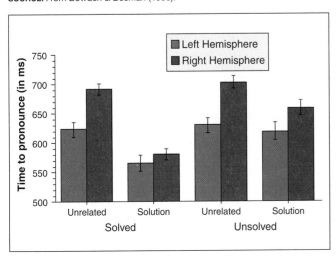

they finally solve the problem, the solution came to them as an insight—an "Aha!" type of solution.

In the Bowden and Beeman (1998) study, people saw the problems and then after 15 seconds were asked to name a new word that appeared on the screen (if they solved the problem before the 15-second period was up, they were given the word immediately). When the target word was unrelated to the three words seen before (e.g., *planet*), there was the typical effect—that targets presented to the right visual field, hence the left hemisphere of the brain, were named faster than those presented to the left visual field–right hemisphere.

Figure 12.10 Time to Pronounce the Target Word for Solved and Unsolved Trials

Bars labeled LH refer to target words presented to the right visual field, left hemisphere of the brain; RH means left visual field, right hemisphere. The figure shows priming effects for solution words, especially in the right hemisphere.

SOURCE: From Bowden & Beeman (1998).

But when the target was the word that solved the insight problem (*tree* in the first problem, *pot* in the second one), there was a significant priming effect. As shown in Figure 12.10, time to pronounce was shorter for the solution words than for the unrelated words. And the priming effect—the drop-off from "unrelated" to "solution"—was greater for targets presented to the right hemisphere than to the left (in other words, presented to the left visual field so going first to the right hemisphere).

Putting it differently, semantic priming in the right hemisphere was more prominent than in the left hemisphere for these problems: People were faster to name *pot* when it was presented to the right hemisphere, presumably because it had been primed by the initial three words. As the authors noted, these results fit nicely with other results concerning the role in language comprehension that the right hemisphere plays, especially the part having to do with drawing inferences (Bowden, Beeman, & Gernsbacher, 1995).

WRITING PROMPT

Improving Problem Solving

Given the problems people can sometimes have during problem solving, what steps can they take to improve based on what you have read in this section?

▶ The response entered here will appear in the performance dashboard and can be viewed by your instructor.

Submit

12.5: Means–End Analysis

OBJECTIVE: Outline the means–end analysis method of problem solving

Several problem-solving heuristics have been discovered and investigated. You have already read about analogy, and the final section of the module illustrates several others. But in terms of overall significance, no other heuristic comes close to means–end analysis. This formed the basis for Newell and Simon's groundbreaking work (1972), including their very first presentation of the information-processing framework in 1956 (on the "day cognitive psychology was born"). Because it shaped the entire area and the theories devised to account for problem solving, it deserves special attention.

12.5.1: The Basics of Means–End Analysis

With the **means–end analysis** approach, a problem is solved by repeatedly determining the difference between the current state and the goal or subgoal state, then finding and applying an operator that reduces this difference. Means–end analysis nearly always implies the use of subgoals because achieving the goal state usually involves the intermediate steps of achieving several subgoals along the way.

The basic notions of a means–end analysis can be summarized in a sequence of five steps:

1. Set up a goal or subgoal.
2. Look for a difference between the current state and the goal or subgoal state.
3. Look for an operator that will reduce or eliminate this difference. One such operator is the setting of a new subgoal.
4. Apply the operator.
5. Apply steps 2 through 4 repeatedly until all subgoals and the final goal are achieved.

At an intuitive level, means–end analysis and subgoals are familiar and represent "normal" problem solving. If you have to write a term paper for class, you break the overall goal down into a series of subgoals: Select a topic, find relevant material, read and understand the material, and so on. Each of these may contain its own sub-goals.

12.5.2: The Tower of Hanoi

One of the most thoroughly investigated problems is the Tower of Hanoi problem. This problem shows clearly the strengths and limitations of the means–end approach.

THE THREE-DISK VERSION Work on the Tower of Hanoi problem carefully, using the three-disk version in the simulation below. Try to keep track of your solution so you will understand how it demonstrates the usefulness of a means–end analysis. So that you will become familiar

---

The Seven-Step Solution for the Tower of Hanoi Problem (Figures 12.11A–12.11D)
Having done the three-disk version of the problem, consider your solution in terms of subgoals and means–end analysis.

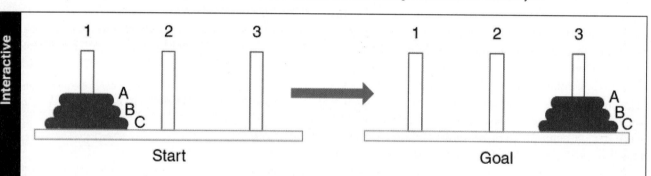

Goal of the Problem: The goal of the problem is to move all three disks from peg 1 to peg 3 so that C is on the bottom, B is in the middle, and A is on top. You may move only one disk at a time, and only to another peg; you may not place a larger disk on top of a smaller one.

Your goal, as stated in the Tower of Hanoi problem, is to move the ABC stack of disks from peg 1 to peg 3. Applying the means–end analysis, your first step sets up this goal.

with the problem and be able to reflect on your solution, do it several times again after you have solved it. See whether you can become skilled at solving the three-disk problem by remembering your solution and being able to generate it repeatedly. (By the way, an excellent heuristic for this problem is to solve it physically; draw the pegs on a piece of paper and move three coins of different sizes around to find the solution.)

THE FOUR-DISK VERSION After you have done the problem several times, solving it becomes easy. You come to see how each disk must move to get C on 3, then B, and finally A. Spend some time now on the same problem but use four disks instead of three. Don't work on this version blindly, however. Think of it as a variation on the three-disk problem, where parts of the new solution are "old." As a hint, try renaming the pegs as the source peg, the stack peg, and the destination peg. Furthermore, think of the seven moves not as seven discrete steps but as a single chunk, "moving a pyramid of three disks," which should help you see the relationships between the problems more clearly (Simon, 1975). According to Catrambone (1996), almost any label attached to a sequence of moves probably will help you remember the sequence better.

What did you discover as you solved the four-disk problem? Most people come to realize that the four-disk problem has two three-disk problems embedded in it, separated by the bridging move of D to 3. That is, to free D so it can move to peg 3 you must first move the top three disks out of the way, moving a "pyramid of three disks," getting D to peg 3 on the eighth move. Then the ABC pyramid has to move again to get them on top of D—another seven moves. Moving the disks entails the same order of moves as in the simpler problem, although the pegs take on different functions: For the four-disk problem, peg 2 serves as the destination for the first half of the solution, then as the source for the last half.

The Four-Disk Tower of Hanoi Problem, With Solution (Figures 12.12A–12.12C)

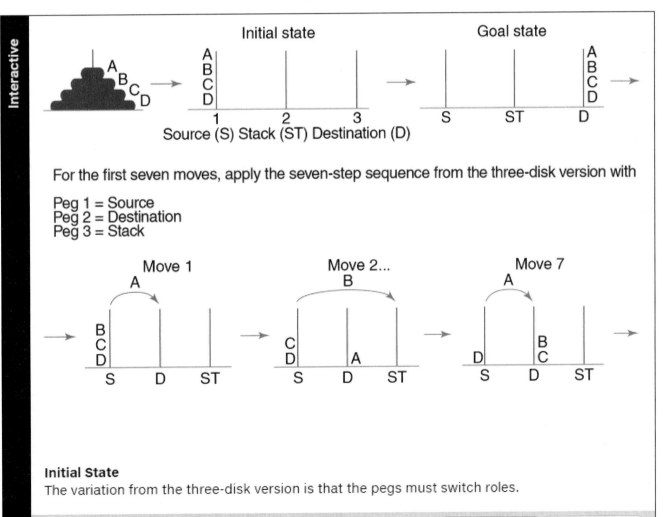

Initial State
The variation from the three-disk version is that the pegs must switch roles.

Because the three-disk solution is embedded in the four-disk problem—and, likewise, the four-disk solution is embedded in the five-disk problem—this is known as a recursive problem, where *recursive* simply means that simpler components are embedded in the more difficult versions.

12.5.3: General Problem Solver

Means–end analysis was an early focus of research on problem solving, largely because of work by Newell, Shaw, and Simon (1958; Ernst & Newell, 1969; Newell & Simon, 1972). Their computer simulation was called **general problem solver (GPS)**. This program was the first genuine computer simulation of problem-solving behavior. It was a general-purpose, problem-solving program, not limited to just one kind of problem but widely applicable to a large class of problems in which means–end analysis was appropriate.

Newell and Simon ran their simulation on various logical proofs, on the missionary–cannibal problem presented below, on the Tower of Hanoi problem, and on many other problems to demonstrate its generality. Notice the critical analogy here. Newell and Simon drew an analogy between the way computer programs solve problems and the way humans do: Human mental processes are of a symbolic nature, so the computer's manipulation of symbols is a fruitful analogy to those processes. This was a stunningly provocative and useful analogy for the science of cognition.

The Missionary–Cannibal Problem

Three missionaries and three cannibals are on one side of a river and need to cross to the other side. The only means of crossing is a boat, and the boat can hold only two people at a time. Devise a set of moves that will transport all six people across the river, bearing in mind the following constraint: The number of cannibals can never exceed the number of missionaries in any location, for the obvious reason. Remember that someone must row the boat back across each time.

Hint: At one point in your solution, you will have to send more people back to the original side than you just sent over to the destination.

PRODUCTION SYSTEMS An important characteristic of GPS was its formulation as a **production system** model, essentially the first such model proposed in psychology. A **production** is a pair of statements, called either a *condition–action* pair or an *if–then* pair. In such a scheme, if the production's conditions are satisfied, the action part of the pair takes place. In the GPS application to the Tower of Hanoi, three sample productions might be

1. If the destination peg is clear and the largest disk is free, then move the largest disk to the destination peg.
2. If the largest disk is not free, then set up a subgoal to free it.
3. If a subgoal to free the largest disk is set up and a smaller disk is on it, then move the smaller disk to the stack peg.

Such an analysis suggests a very "planful" solution by GPS: Setting up a goal and subgoals that achieve the goal sounds exactly like what we call planning. And indeed, such planning characterizes both people's and GPS's solutions to problems, not just the Tower of Hanoi but all kinds of transformation problems. GPS had what amounted to a planning mechanism, a mechanism that abstracted the essential features of situations and goals, then devised a plan that would produce a problem-solving sequence of moves. Provided with such a mechanism and the particular representational system necessary to encode the problem and the legal operators, GPS yielded an output that resembles the solution pathways taken by human problem solvers.

> **Prove It**
>
> The problems you have been solving throughout the module can be used without change to demonstrate the principles of problem solving. Here are some interesting contrasts and effects that you might want to test.
>
> Compare either the time or number of moves people make in learning and mastering the Tower of Hanoi problem when the pegs are labeled *1*, *2*, and *3* and when they are labeled *source*, *stack*, and *destination*. Try drawing the pegs in a triangular pattern rather than in a left-to-right display to see whether that makes the "stack" peg idea more salient. Compare how long it takes to master the problem when your participants learn to do it by moving three coins around on paper and when they keep track of their moves mentally.

LIMITATIONS OF GENERAL PROBLEM SOLVER Later investigators working with the general principles of GPS found some cases when the model did not do a good job of characterizing human problem solving. Consider the missionary–cannibal problem presented earlier; the solution pathway is presented in Figure 12.13.

The problem is difficult, most people find, at step 6, where the only legal move is to return one missionary and one cannibal back to the original side of the river. Having just brought two missionaries over, this return trip seems to be moving away from the overall goal. That is, returning one missionary and one cannibal seems to be incorrect because it appears to increase the distance to the goal: It is the only return trip that moves two characters back to the original side. Despite this being the only available move (other than returning the same two missionaries who just came over), people have difficulty in selecting this move (Thomas, 1974).

Figure 12.13 An Illustration of the Steps Needed to Solve the Missionary–Cannibal Problem

The left half of each box is the "start side" of the river, and the right half is the "destination side." The numbers and letters next to the arrows represent who is traveling on the boat.

SOURCE: Based on Glass & Holyoak (1986).

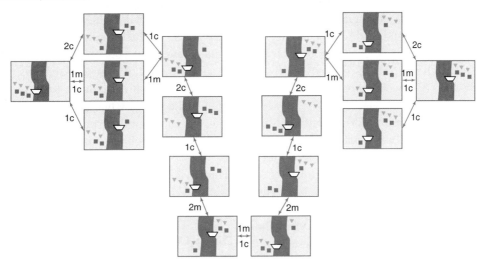

GPS did not have this difficulty because sending one missionary and one cannibal back was consistent with its immediate subgoal. On the other hand, at step 10, GPS is trying to fulfill its subgoal of getting the last cannibal to the destination side and seemingly can't let go of this subgoal. People, however, realize that this subgoal should be abandoned: Anyone can row back over to bring the last cannibal across and, in the process, finish the problem (Greeno, 1974). GPS was simply too rigid in its application of the means–end heuristic, however: It tried to bring the last cannibal across and then send the boat back again.

BEYOND GENERAL PROBLEM SOLVER Newell and Simon's GPS model, and models based on it, often provided a good description of human problem-solving performance (Atwood & Polson, 1976) and offered a set of predictions against which new experimental results could be compared (Greeno, 1974). Despite some limitations (Hayes & Simon, 1974), the model demonstrated the importance of means–end analysis for an understanding of human problem solving.

WRITING PROMPT

Applications of Means–End Analysis and Cognition

How do the issues involved with means–end analysis and its applications inform us about the step-by-step nature of cognition? How does this slower, more deliberative type of processing compare with other topics covered in this course?

 The response entered here will appear in the performance dashboard and can be viewed by your instructor.

Submit

12.6: Improving Your Problem Solving

OBJECTIVE: Summarize a set of problem-solving techniques

Sprinkled throughout the module have been hints and suggestions about how to improve your problem solving. Some of these are based on empirical research and some on intuitions that people have had about problem solving.

Let's close the module by pulling these hints and suggestions together and offering a few new ones. Here is a list of these suggestions for improving problem solving:

- Increase your domain knowledge.
- Automate some components of the problem-solving solution.
- Follow a systematic plan.
- Draw inferences.
- Develop subgoals.
- Work backward.
- Search for contradictions.
- Search for relations between problems.
- Find a different problem representation.
- Stay calm.
- If all else fails, try practice.

12.6.1: Increase Your Domain Knowledge

In thinking about what makes problems difficult, Simon suggests that the likeliest factor is **domain knowledge**, what one knows about the topic. Not surprisingly, a person who has only limited knowledge or familiarity with a topic is less able to solve problems efficiently in that domain (but see Wiley, 1998, on some disadvantages of too much domain knowledge). In contrast, extensive domain knowledge leads to expertise, a fascinating topic in its own right (see Ericsson & Charness, 1994, and Medin, Lynch, Coley, & Atran, 1997, for example).

Much of the research supporting this comes from Simon's work with chess (Chase & Simon, 1973; Gobet & Simon, 1996; see also Reeves & Weisberg, 1993). In several studies of chess masters, an important but not surprising result was obtained: Chess masters need only a glimpse of the arrangement of chess pieces to remember the arrangement, far beyond what novices or players of moderate skill can do. This advantage holds, however, only when the pieces are in legal locations (i.e., sensible within the context of a real game of chess). When the locations of the pieces are random, then there is no advantage for the skilled players. This advantage of expertise in remembering legal board positions is attributed to experts' more skilled perceptual encoding of the board, literally more efficient eye movements and fixations while looking at the board (Reingold, Charness, Pomplun, & Stampe, 2001).

12.6.2: Automate Some Components of the Problem-Solving Solution

A second connection also exists between the question "What makes problems difficult?" and the topics you have already studied. Kotovsky, Hayes, and Simon (1985) tested adults on various forms of the Tower of Hanoi problem and on problem *isomorphs*, problems with the same form but different details. Their results showed that a heavy working memory load was a serious impediment to successful problem solving: If a person had to hold three or four nested subgoals in working memory all at once, performance deteriorated.

Thus, a solution to this memory load problem was to *automate* the rules that govern moves, just as you were supposed to master and automate the seven-step sequence in the Tower of Hanoi. This frees working memory to be used for higher-level subgoals (Carlson, Khoo, Yaure, & Schneider, 1990). This is the same reasoning you encountered early in the course, where automatic processing uses few if any of the limited conscious resources of working memory.

12.6.3: Follow a Systematic Plan

Especially in long multistep problems, it is important to follow a *systematic plan* (Bransford & Stein, 1993; Polya, 1957). Although this seems straightforward, people do not always generate plans when solving problems, although doing so can dramatically improve performance (Delany, Ericsson, & Knowles, 2004). A plan helps you keep track of what you have done or tried and keeps you focused on the overall goal or subgoals you are working on. For example, on DONALD + GERALD, you need to devise a way to keep track of which digits you have used, which letters remain, and what you know about them. If nothing else, developing and following a plan helps you avoid redoing what you have already done. Keep in mind that people often make errors when planning how long a task will take, but can plan their time better if they break the task down into the problem subgoals, estimate the time needed for each of those, and then add those times together (Forsyth & Burt, 2008).

12.6.4: Draw Inferences and Develop Subgoals

Wickelgren's (1974) advice is to *draw inferences* from the givens, the terms, and the expressions in a problem before working on the problem itself.

It can also help you abandon a misleading representation of the problem and find one more suitable to solving the problem (Simon, 1995).

Beware of *unwarranted inferences*, the kinds of restrictions we place on ourselves that may lead to dead ends. For instance, for the nine-dot problem, an unwarranted inference is that you must stay within the boundaries of the nine dots.

Wickelgren also recommends a *subgoal heuristic* for problem solving, that is, breaking a large problem into separate subgoals. This is the heart of the means–end approach. There is a different slant to the subgoal approach, however, that bears mention. Sometimes in our real-world problem solving, there is only a vaguely specified goal and, as often as not, even more vaguely specified subgoals. How do you know when you have achieved a subgoal, say when the subgoal is "find enough articles on a particular topic to write a term paper that will earn an A"?

Simon's (1979) *satisficing* heuristic is important here; satisficing is a heuristic in which we find a solution to a goal or subgoal that is satisfactory although not necessarily the best possible one. For some problems, the term paper problem included, an initial satisfactory solution to subgoals may give you additional insight for further refinement of your solution. For instance, as you begin to write your rough draft, you realize there are gaps in your information. What seemed originally to be a satisfactory solution to the subgoal of finding references turns out to be insufficient, so you can recycle back to that subgoal to improve your solution. You might only discover this deficiency by going ahead and working on that next subgoal, the rough draft.

Two Trains and Fifteen Pennies Problems

If you do this appropriately, it can often save you from wasting time on blind alleys, as in the two trains and 15 pennies problems.

Interactive

Two Trains | **Fifteen Pennies**

Two train stations are 50 miles apart. At 2 P.M. one Saturday afternoon, the trains start toward each other, one from each station. Just as the trains pull out of the stations, a bird springs into the air in front of the first train and flies ahead to the front of the second train. When the bird reaches the second train it turns back and flies toward the first train. The bird continues to do this until the trains meet.
If both trains travel at the rate of 25 miles per hour and the bird flies at 100 miles per hour, how many miles will the bird fly before the trains meet?

Hint: Don't think about how far the bird is flying; think of how far the trains will travel and how long that will take.

12.6.5: Work Backward and Search for Contradictions

Another heuristic is *working backward*, in which a well-specified goal may permit a tracing of the solution pathway in reverse order, thus working back to the givens. The 15 pennies problem is an illustration, a problem that is best solved by working backward. Many math and algebra proofs can also be worked backward or in a combination of forward and backward methods.

In problems that ask "Is it possible to?" or "Is there a way that?" you should *search for contradictions* in the givens or goal state. Wickelgren uses the following illustration: Is there an integer x that satisfies the equation $x^2 + 1 = 0$? A simple algebraic operation, subtracting 1 from both sides, yields $x^2 = -1$, which contradicts the known property that any squared number is positive. This heuristic can also be helpful in multiple-choice exams. That is, maybe some of the alternatives contradict some idea or fact in the question or some fact you learned in the course. Either will enable you to rule out those choices immediately.

12.6.6: Search for Relations Among Problems

In *searching for relations* among problems, you actively consider how the current problem may resemble one you have already solved or know about. The four- and more disk Tower of Hanoi problems are examples of this, as are situations in which you search for an analogy (Bassok & Holyoak, 1989; Ross, 1987). Don't become impatient. Bowden (1985) discovered that people often found and used information from related problems, but only if sufficient time was allowed for them to do so.

> **Prove It**
>
> Take a paper and draw 16 dots as shown in the below figure.
>
> **Figure 12.14** The 16 Dots Problem
>
>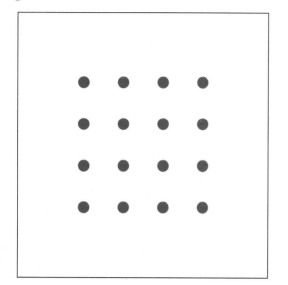
>
> Now, without lifting your pencil, join all 16 dots with six straight lines.

12.6.7: Find a Different Problem Representation

Another heuristic involves the more general issue of *problem representation*, or how you choose to represent and think about the problem you are working on. Often, when you get stuck on a problem, it is useful to go back to the beginning and *reformulate* or *reconceptualize* it. For instance, as you discovered in the Buddhist monk problem, a quantitative representation of the situation is unproductive. Return to the beginning and try to consider other ways to think about the situation, such as a visual imagery approach, especially a mental movie that includes action. In the Buddhist monk problem, superimposing two such mental movies permits you to see him walking up and down at the same time, thus yielding the solution. Likewise, animated diagrams, with arrows moving in toward a point of convergence, helped participants solve the radiation problem in Pedone, Hummel, and Holyoak's (2001) study, as compared to either static diagrams or a series of diagrams showing intermediate points in problem solution (see also Reed & Hoffman, 2004).

For other kinds of problems, try a numerical representation, including working the problem out with some examples, or a physical representation, using objects, scratch paper, and so forth. Simon (1995) makes a compelling point that one representation of a problem may highlight a particular feature while masking or obscuring a different, possibly important feature. According to Ahlum-Heath and DiVesta (1986), verbalizing your thinking also helps in the initial stages of problem solving.

Earlier, it was suggested that you can master the Tower of Hanoi problem more easily if you use three coins of different sizes. This is more than just good advice.

> Here's an example with patient H. M., who had profound anterograde amnesia. H. M. was unable to form new explicit long-term memories but performed normally when implicit learning was tested. The major result you read about earlier was the mirror tracing study: In a mirror tracing study, H. M. showed normal learning curves on this task, despite not remembering the task from day to day. Interestingly, H. M. was also tested on the Tower of Hanoi task, and he learned it as well as anyone (although he had no explicit memory of ever having done it before). The important ingredient here is the motor aspect of the tower problem: Learning a set of motor responses, even a complex sequence, relies on implicit memory. Thus, working the Tower of Hanoi manually by moving real disks or coins around should enable you to learn how to solve the problem from both an explicit and an implicit basis.

12.6.8: If All Else Fails, Try Practice

Finally, for problems we encounter in classroom settings, from algebra or physics problems up through such vague

Becoming an effective problem solver requires practice to strengthen certain knowledge, as these chess players exhibit.

problems as writing a term paper and studying effectively for an exam, a final heuristic should help. It is well known in psychology; even Ebbinghaus recommended it. If you want to be good at problem solving, *practice* problem solving. Practice within a particular knowledge domain strengthens that knowledge, pushes the problem-solving components closer to an automatic basis, and gives you a deeper understanding of the domain. Although it isn't flashy, practice is a major component of skilled problem solving and of gaining expertise in any area (Ericsson & Charness, 1994).

In Ericsson and Charness's (1994) review, people routinely believe that stunning talent and amazing accomplishments result from inherited, genetic, or "interior" explanations, when the explanation usually is dedicated, regular, long-term practice. This relationship between practice and performance level is seen in an analysis of practice and expertise data by Ericsson, Krampe, and Tesch-Römer (1993) shown in Figure 12.15.

As can be seen, the people who had higher levels of expertise also were the ones who engaged in more practice. So, practice is important to becoming an expert. However, it is unclear whether there is also some innate characteristic such as motivation, interest, or talent that could also be driving those people to practice more. Regardless, if you want to become highly skilled at something, your elementary school clarinet teacher was right—you really do need to practice.

Figure 12.15 Illustration of the Relationship Between Amount of Practice Over the Course of Years and the Level of Expertise

SOURCE: From Ericsson, Krampe, & Tesch-Römer (1993).

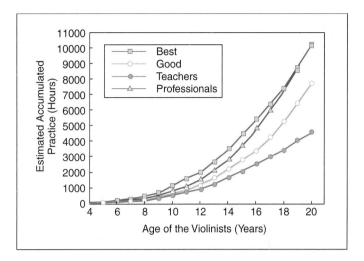

WRITING PROMPT

Real-World Applications of Problem-Solving Strategies

For each of the different steps that you can take to improve your problem solving, provide an example of a real-world, everyday problem to which it could be applied.

▶ The response entered here will appear in the performance dashboard and can be viewed by your instructor.

Submit

Summary: Problem Solving

12.1 Studying Problem Solving

- Unlike many topics of study in cognitive psychology, problem solving is slower and more deliberative. As such, verbal protocols of what people are thinking, which are normally actively avoided, can be used to gain insight into the steps that people take to arrive at a solution.
- Cognitive psychology often studies how people solve "recreational" problems, simple brain teasers, as a way of understanding problem solving. A common kind of data collected is the verbal protocol, a transcription of the person's verbalizations as the problem is being solved.

12.2 Basics of Problem Solving

- We are solving a problem when our behavior is goal directed and involves a sequence of cognitive steps or stages. The sequence involves separate cognitive operations, where each goal or subgoal can be decomposed into separate, smaller subgoals. The overall problem, including our knowledge, is called the problem space, within which we use lookahead, apply operators, draw inferences, and conduct a constructive search for moves that bring us closer to the goal.

12.3 Gestalt Psychology and Problem Solving

- The early Gestalt psychologists studied problem solving and discovered two major barriers to successful performance: functional fixedness and negative set. Köhler also studied chimpanzees and found evidence for insight during problem solving.

12.4 Insight and Analogy

- Insight is a deep understanding of a situation or problem, often thought to occur suddenly and without warning. Although there is some debate on the

- nature of insight, insights may be discovered and used unconsciously and only later be available to consciousness.
- Reasoning by analogy is a complex kind of problem solving in which relationships in one situation are mapped onto another. People are better at developing analogies if given a useful source problem and an explicit hint that the problem might be used in solving a target problem. Holyoak and Thagard's multiconstraint theory of analogical problem solving claims that we work under three constraints as we develop analogies: constraints related to the similarity of the source and target domains, the structure of the problems, and our purposes or goals in developing the analogies.
- Some evidence suggests a particularly important role for the left frontal and parietal lobes in solving problems by analogy and a right hemisphere role in insight problems involving semantic priming.

12.5 Means–End Analysis

- The best-known heuristic for problem solving is means–end analysis, in which the problem solver cycles between determining the difference between the current state and the goal or subgoal state and applying legal operators to reduce that difference. The importance of subgoals is revealed most clearly in problems such as the Tower of Hanoi.
- Newell and Simon's general problem solver (GPS) was the earliest cognitive theory of problem solving, implemented as a computer simulation. Studying GPS and comparing its performance with human problem solving shows the importance of means–end analysis.

12.6 Improving Your Problem Solving

- The set of recommendations for improving your problem solving includes increasing your knowledge of the domain, automaticity of components in problem solving, developing and following a plan, and not becoming anxious. Several special-purpose heuristics are also listed, including the mundane yet important advice about practice.

SHARED WRITING

Problem Solving

What is an everyday example of a problem that person needs to solve? In the language and theories of cognitive psychology, what goes into the way a typical person might solve that problem?

> A minimum number of characters is required to post and earn points. After posting, your response can be viewed by your class and instructor, and you can participate in the class discussion.

Post   0 characters | 140 minimum

Chapter 13
Emotion

Learning Objectives

13.1: Describe how emotions are processed in the brain

13.2: Contrast the effect of emotion on perception and attention

13.3: Explain how emotion influences memory

13.4: Summarize the impact of emotion on language processing

13.5: Analyze the influence of emotion on decision making

Although the computer metaphor has been a dominant guide for years in cognitive psychology, people are much more than computational machines. One of the big differences between us and our silicon-based creations is in the realm of emotions. We have them, and they don't. Although some might describe much of the research in cognitive psychology as "cold cognition" because it does not take people's emotions into account, there is more and more new research doing just that (see Mather & Sutherland, 2011, for a nice overview of major issues). What has been found is that emotion can influence cognition in a variety of complex but systematic ways.

This module presents an overview of various ways that emotion affects cognition. In a sense, this module serves to recapitulate many of the topics already covered in the course. We start with a consideration of how emotion can influence seemingly basic perceptual and attention processes. We then look at how emotion influences memory, making it better in some cases and worse in others. After this we consider roles of emotion in language, followed by some coverage of how emotion can influence our ability to make decisions and solve problems.

13.1: What Is Emotion?

OBJECTIVE: Describe how emotions are processed in the brain

Although we all have an intuitive sense about what emotions are, we need to go beyond that here and offer a formal definition to work with. Consider **emotion** to be both the state of mind a person is in at a particular moment, as well as the physiological response a person is experiencing at that time (in terms of heart rate, pupillary dilation, neurotransmitter release, and so on). There are other similar, related terms that mean somewhat different things, such as *affect, mood,* and *arousal,* but we leave these aside for now. Our purpose here is to look at some basic ideas about how emotion can influence and interact with other aspects of cognition, such as perception, attention, and memory.

There are numerous ways of dividing up and classifying different types of emotion. Also, it certainly is the case that we all have a lot of variety and subtlety in the different types of emotions we experience. However, for our purposes, we employ a simple approach of looking at two dimensions of emotion and use these to guide our coverage of how emotion influences cognition.

Many of the cognitive phenomena that we discuss can be understood in terms of where an emotion is along these two dimensions—that is, closer to the positive or negative end of the **valence** dimension, or closer to the low or high end of the **intensity** dimension.

13.1.1: Neurological Underpinnings

Emotion has both physical and mental components. It is certainly a visceral experience. We feel it in our

The Important Neurological Structures for Emotion Processing

The two structures that capture the back-and-forth influence of emotions on cognition and vice versa are the amygdala and the prefrontal cortex (Dolcos, Iordan, & Dolcos, 2011).

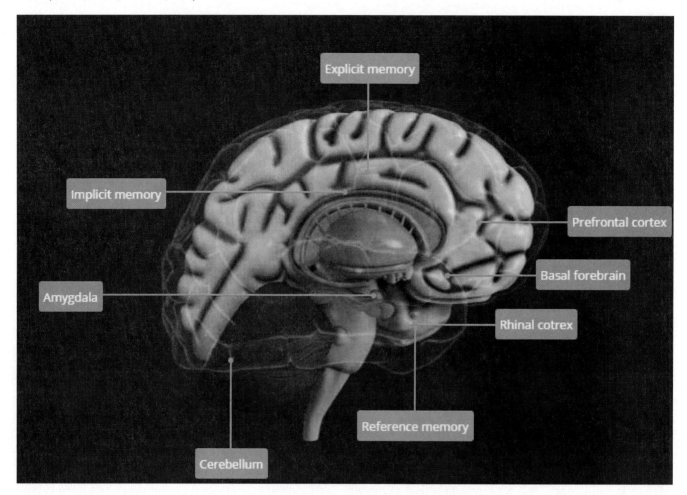

bodies—our hearts race, our breathing speeds up. It also triggers a different kind of mental experience than basic kinds of thought. As a mental experience, it has the power to influence brain and cognition in meaningful ways. In the brain, emotional experience is associated with several structures and areas, but we focus on two here that are most relevant for understanding how cognition and emotion interact.

AMYGDALA The *amygdala* is an almond-shaped structure located next to the hippocampus. The amygdala is critically involved in more instinctual emotions that are important for survival, such as the experience of fear (e.g., Davis, 1997; LeDoux, 2000) and is more active when a person is in an emotional state. The amygdala receives sensory information from various parts of the brain, allowing for a fast emotional response to environmental conditions. This is particularly true for biologically related emotions related to fear (e.g., seeing a snake), as compared with socially related emotions (e.g., seeing a happy family) (Sakaki, Niki, & Mather, 2012).

One interesting aspect of the amygdala is that there are very few neural synapses in the chain between the olfactory receptors of your sense of smell and the amygdala, which is why odors can be strongly associated with emotions (Herz & Engen, 1996).

The amygdala in turn sends its signals to the hypothalamus and brain stem, which help regulate the body's arousal state, as well as to areas tied to cognitive processes, such as the prefrontal cortex (attention) and the hippocampus (memory). Thus, emotional responses have the potential to directly influence the context and processing of thoughts in ways that differ from when we are not in an emotionally aroused state.

PREFRONTAL CORTEX In addition to the amygdala, another region of the brain that is important in emotional processing is the **ventromedial prefrontal cortex** (B.A. 10). This part of the brain is involved in the identification and interpretation of emotional stimuli and responses, and the integration of that emotional interpretation with the surrounding context (Roy, Shohamy, & Wager, 2012), as well as the regulation and control of those experiences.

The famous case of *Phineas Gage* illustrates the important role of the prefrontal cortex in emotion regulation. Specifically, Gage suffered a massive trauma that destroyed a large portion of his frontal lobe. This occurred in 1848 in Vermont when Gage was working on a blasting crew for a railroad. He was using a tamping rod to press some blasting power into a hole that had been drilled into a rock formation. The powder accidentally ignited, shooting the rod through Gage's skull, destroying part of his brain. Miraculously, Gage survived. After the accident, however, people claimed that Gage was no longer Gage. One of the biggest changes was that he was less able to control his emotions and would impulsively act and express himself in ways that were inconsistent with who he had been before the accident. This was because the part of his brain in the frontal lobe responsible for controlling and regulating emotional responses and behaviors was seriously damaged in the accident.

| WRITING PROMPT |

Emotion and Cognition

In what ways are emotions processed like other aspects of cognition, and in what ways are they processed differently? How often do you think emotion is involved with cognition, and why?

▶ The response entered here will appear in the performance dashboard and can be viewed by your instructor.

Submit

13.2: Emotion and Perception

OBJECTIVE: Contrast the effect of emotion on perception and attention

We turn now to the question of how emotion can influence cognition. We start with perception and attention, just as we started out this text. Then we move on to topics in memory, language, and decision making and problem solving.

Our first question seems very simple— Can emotion influence even basic perceptual processes?

Some evidence suggests that the answer is "yes," emotions *can* meaningfully influence perception, making some things easier to perceive than others.

Thus, the emotional content of an item can influence the accuracy with which it is processed. Looking more deeply, we can see that emotion appears to actually increase the amount of neural activity in perceptual brain areas, such as the *occipital* and *occipital-parietal cortex* areas (Taylor, Liberzon, & Koeppe, 2000).

However, it is not the case that there is a uniform boost in the ability to perceive emotional items. Instead, what appears to be happening is that there is an increase in the ability to process broad, global, or general characteristics of the threatening item, but a decline in the ability to perceive details (Bocanegra & Zeelenberg, 2011). That is, when emotions are aroused by a threatening stimulus, the perceptual processes in our cognitive system direct processing efforts to knowing generally what and where that emotional something is, so that we can be safe from it. In doing so, the cognition system is, in some sense, making sure that it does not waste its limited energies and resources on details that are probably trivial, thereby increasing the probability of survival.

In addition to influencing the ease with which a person recognizes different objects and entities in the world, emotion can alter the subjective perception of the world when the situation has characteristics that arouse emotions in the person.

13.2.1: Emotional Guidance of Attention

In addition to the effects of emotion on perceptual processes, there is some evidence that emotion can influence attention. A moment's reflection will reveal that this basic idea is not all that surprising. If something elicits our emotions, we are more likely to pay attention to it, whatever that thing is. What the work on emotion and attention has shown is that this influence can be more extensive and unconscious than a casual moment's reflection might reveal. Let's start by looking at a very basic form of attention, namely the *orienting reflex*.

Figure 13.1 Zeelenberg, Wagenmakers, and Rotteveel (2006) Study

For instance, people generally recognize things faster if they are emotionally meaningful (such as identifying briefly flashed words like *death* and *love*) than if they are not (Zeelenberg, Wagenmakers, & Rotteveel, 2006). This can be seen in Figure 13.1.

SOURCE: Data from Zeelenberg et al. (2006).

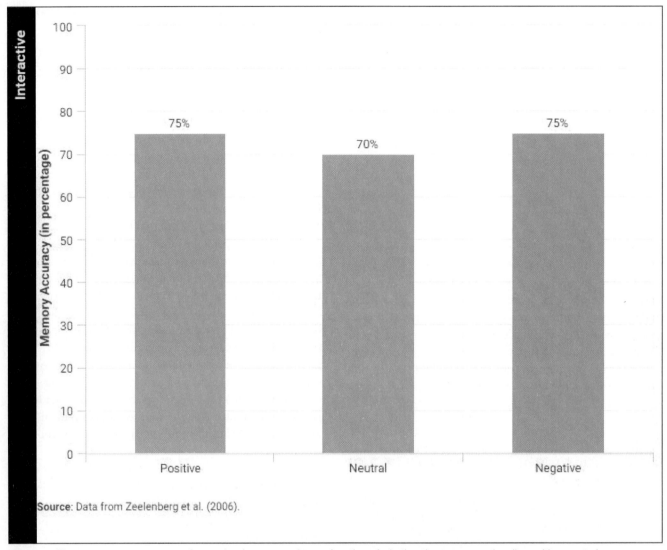

Source: Data from Zeelenberg et al. (2006).

This figure illustrates memory accuracy for previously seen words as a function of whether they were emotionally positive, neutral, or emotionally negative. Memory is better for emotionally charged items.

Cowan (1995) noted that the kinds of stimuli that can trigger the orienting reflex boil down to two basic categories:

1. Stimuli that are significant for the organism (the rock thrown toward your head)
2. Stimuli that are novel or unexpected

The important point here is that what is significant to you often has some relation to emotions, either positive or negative. It is vital to keep in mind that emotion and attention use some of the same neural components. These include structures such as the amygdala, portions of the frontal lobes, and the anterior cingulate cortex (B. A.s 24, 32, and 33).

Thus, emotion can affect the direction of attention (Vuilleumeir, 2005). For example, people are more likely to direct their attention to emotionally arousing stimuli, such as a seeing a snake in the grass. Attention can also influence how you feel about things. For example, people can develop a negative emotional response toward things that they try to ignore (Fenske & Raymond, 2006).

Embodied Perception (Figure 13.2)

Take the example of the height of a balcony on a building. When you are standing on the balcony looking down, how high off the ground do you perceive yourself to be?

Stefanucci and Storbeck (2009) Study: This graph shows the estimated height of a balcony as a function of whether a person was emotionally aroused (by viewing pictures) or not. Emotional arousal results in greater height estimations. How did people who were emotionally aroused estimate how high the balcony was?

A study by Stefanucci and Storbeck (2009) found that people who were more emotionally aroused (by viewing a series of emotionally arousing pictures), either positively or negatively, experienced the height of the balcony as greater (see the figure above).

1 of 4

13.2.2: Visual Search

One of the tasks in cognitive psychology that has been used extensively for the study of the operation of attention is *visual search*—trying to find an object in a display of irrelevant distractors. Research using this task has found that emotions, particularly negative emotions such as fear, can influence the visual search processes (e.g., Öhman, Flykt, & Esteves, 2001).

The inclusion of emotion-eliciting stimuli in a display, such as spiders or fearful faces, facilitates the direction of attention to such objects during visual search.

In some sense, this seems like a form of attention capture, with attention being preferentially moved to more fearful stimuli. This shifting of attention to emotional items seems to be directed more by the amygdala than by emotional control processes in the frontal lobe

Emotional stimuli have a way of capturing our attention.

During visual search, it is easier to direct our attention to emotionally charged items, such as a spider.

(Vuilleumeir & Huang, 2009). Also, it may have more to do with an increase in attentional resources because even the processing of nonemotional targets in a visual search task is facilitated when emotions are triggered (Becker, 2009).

Looking more deeply at emotion and the processing of visual stimuli, several investigators have tested the effects of emotional stimuli on the attentional blink phenomenon. As a reminder, if two stimuli are presented very rapidly in sequence, we sometimes miss the second one, as if our attentional mechanism had "blinked" for a brief moment while the second one was present. Researchers have now found that the length of this attentional blink is attenuated if the stimulus that occurs during that critical "moment" is emotionally loaded.

As an example, there is a reduced or absent attentional blink for emotion-eliciting words, such as *whore*, compared to emotionally neutral words, such as *veiled* (Anderson, 2005). Thus, emotional relevance and intensity can override other standard operating features of the cognitive system.

13.2.3: Emotional Stroop

Emotional processing can even influence attention and cognition in ways that might seem irrelevant at first. An example of this is the **emotional Stroop task** (see Williams, Matthews, & MacLeod, 1996, for a review).

Prove It! Stroop Task

Now, let's take a look at another list. Time yourself again as you say aloud the *color* of the word (not reading the word itself) as quickly as possible.

List 2
Blue
Purple
Red
Green
Purple
Green

In the emotional Stroop task, words are presented in different colors, and people are asked to name those colors. However, rather than having color words on the list, the critical comparison has to do with words that elicit an emotional response in a person, such as *spider*, as compared to more neutral words, such as *spade*.

What is typically found with this task is that people name the color of the word more slowly if the word is emotional for them. For example, a person who has a deathly fear of spiders would be slower to say "green" if *spider* were printed in green ink, whereas a person without this fear would not show this effect. The explanation is actually very straightforward.

Even though the person is supposed to focus on the color of ink, reading the word and accessing its meaning happens automatically. If the word is related to an emotional stressor for the person, it intrudes on the person's cognitive processing, and thus takes away resources from the other cognitive processes necessary for focusing on and naming the ink color. Color naming is slowed down for the emotional word condition, but not for the neutral word condition. This emotional Stroop task has been used to study numerous psychopathologies, including depression (e.g., Mitterschiffthaler et al., 2008), anxiety (e.g., Dresler, Mériau, Heekeren, & van der Meer, 2009), and post traumatic stress disorder (e.g., Cisler et al., 2011).

13.2.4: Emotion and Self-Control

An important function for attention is to guide thoughts and behavior. To some degree this is automatic, but to some degree it is under conscious control. What role does emotion play in this self-control? One would intuitively think that expressing emotions leads to less self-control; however, there is some evidence that *suppression* of your emotions can lead to attentional control problems.

An example of this is a study by Inzlicht and Gutsell (2007) in which people watched an emotional movie. Researchers asked them either to simply watch the movie (the control condition) or to suppress their emotions while viewing the film (the experimental suppression condition). After this, participants did a traditional color-word Stroop task. The results, as can be seen in Figure 13.3, were that people who had suppressed their emotions had a harder time doing the Stroop task.

The explanation was that suppressing one's emotions drained resources from the error monitoring aspect of attentional control, which is guided by the anterior cingulate cortex (B.A.s 24, 25, and 32). When these attentional resources were drained by the effort to control one's emotions over a long period of time, fewer resources were available for doing other tasks.

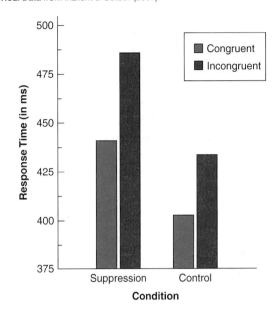

Figure 13.3 Response Times to Stroop Color Words With Suppressed and Non-Suppressed Emotions
SOURCE: Data from Inzlicht & Gutsell (2007).

So, overall, clear evidence indicates that emotional content and responses can influence even basic perceptual and attention processes. This suggests that emotion serves as a primary and fundamental force in guiding and influencing cognition.

WRITING PROMPT

Impact of Emotion on Perception and Attention

In what ways can emotionally relevant things in the environment take control of our perceptual and attention processes? Why should this be the case? What is an example of a personal experience in which something in the environment (such as a spider or snake) captured your attention?

▶ The response entered here will appear in the performance dashboard and can be viewed by your instructor.

Submit

13.3: Emotion and Memory

OBJECTIVE: Explain how emotion influences memory

One of the aspects of cognition where emotion has its largest effects is with memory. The influence of emotion on memory is somewhat complicated. In some ways and under certain circumstances, emotion can make memories

Influence of Emotion on Memory

Let's take a systematic look at these different influences and try to clarify when your emotions help or hinder what you remember from an event.

> **Interactive**
>
> The influence of emotion on memory is not a simple and direct process. In general, the amygdala is more involved in implicit aspects of memory, whereas the prefrontal cortex is more involved in explicit memory processes. This differential influence of these areas on memory can be seen in patients with brain damage. For example, Bechera et al. (1995) reported three case studies of patients: one with bilateral damage to the amygdala, one with bilateral damage to the hippocampus, and one with bilateral damage to both. These patients were exposed to a stimulus, either a tone or a visual image, that was paired with an aversive stimulus (i.e., a loud boat horn), a standard classical conditioning study. The results were fascinating. For the patients with damage to the amygdala, there was no evidence of classical conditioning when the tone or image was shown again; they had not learned the pairing with the loud boat horn. The results were very different when the damage was to the hippocampus, a structure that the prefrontal lobe relies on for conscious memories. In this case, the patients showed normal classical conditioning of the tone (image)-to-horn pairing—but they had no conscious memory for either the tone or the visual image. In other words, with damage to the amygdala, there was no implicit learning, and with damage to the hippocampus, there was no explicit learning. Thus, the different aspects of emotion processing in the brain directly influence different types of learning and memory.

Prove It

Emotional Memory

The influence of emotion on memory is a powerful one. This can easily be demonstrated by giving people lists of words and asking them to remember them later.

Here is a sample list of words: *hate, happy, joy, anger, disgust, fear, thrill, love, sad, pleased, date, caddy, boy, ranger, digest, near, drill, live, had, pleated.*

This list is 20 words long to make sure that it exceeds your participants' working memory capacity and that they are using long-term memory to retrieve the items.

Instruction:
Read the list of words, in a random order, to your participants. Read these words aloud at a rate of about one per second, using a metronome or a watch to help your pacing.

After you have read all of the words, give people a distractor task, such as solving three-digit math problems (e.g., 284 + 923 = ?) for 2 minutes.

This will displace any memory for the words in the list that still may be in working memory. After the 2 minutes are up, give your participants a sheet of paper and have them recall as many words from the original list of 20 as they can remember.

Observation:
What you should find is that people will remember more of the emotional words than the nonemotional words. If you want, try mixing things up by creating word lists of your own that compare emotionality to

> other things that can influence memory, such as word frequency, generation effects, or von Restorff effect.
>
> Emotional words: *hate, happy, joy, anger, disgust, fear, thrill, love, sad, pleased*
> Nonemotional words: *date, caddy, boy, ranger, digest, near, drill, live, had, pleated*

more durable, but in other ways or under different circumstances, emotion can have the opposite effect on memory.

13.3.1: Making Memory Better

If you think about your own life and the things you remember, one of the things you quickly notice is that your emotional experiences are often the most memorable. This can include negative memories, such as losing a close family member or friend, an emotional breakup, or being in a serious car accident, as well as positive memories, such as a new birth in the family, a marriage, or landing a greatly desired job. Thus, emotion can have clear benefits for memory. Emotional information is generally remembered better than more neutral information (see Kensinger, 2009, for a review), even when emotional words, such as *happy, disgust,* or *sad*, are compared to more neutral words, such as *hammer, digest,* or *fad*. Part of what influences memory is the quality of emotion—rather than the *valence* of the emotion (such as whether it is positive or negative)—and how intense the emotional experience is (Talarico, LaBar, & Rubin, 2004). In other words, what is important about the influence of emotion on memory is how strong the emotion is, not whether you feel good or bad about the event. That said, some memories may not be more intense but still may be remembered well if the information is viewed as being disgusting as compared to being fearful (Chapman, Johannes, Poppenk, Moscovitch, & Anderson, 2013).

The idea that people remember emotional information better than neutral information has also been demonstrated in laboratory work. For example, people remember emotionally arousing pictures better than neutral ones (Bradley, Greenwald, Petry, & Lang, 1992), particularly the details of negative images (Kensinger, Garoff-Eaton, & Schacter, 2006). People also remember emotional utterances better than neutral ones (Armony, Chochol, Fecteau, & Belin, 2007). Work using functional magnetic resonance imaging (fMRI) scanning, such as that by Dolcos, Labar, and Cabeza (2005) and Kensinger and Corkin (2004; see also Kensinger, 2007), has shown that the superior memory for emotional memories appears to reflect the involvement of the amygdala and medial temporal-lobe structures, such as the hippocampus, with the amygdala–hippocampus network being more important for emotional intensity, and a hippocampal–frontal lobe network being more important for emotional valence (whether the emotion is positive or negative, happy or sad).

EMOTION EFFECTS AFTER THE FACT It is important to note that the emotional benefit of memory does not need to be present at the time the event is originally experienced. In a fascinating study that exploits the phenomenon of reconsolidation, Finn and Roediger (2011) had people learn a set of English–Swahili vocabulary pairs (essentially the English translations of Swahili words). What was so interesting about this study is that after the initial learning phase was over, people were asked to recall the word pairs. As they recalled each pair, they were shown either a blank screen, or a neutral or an emotional picture. Then, after another period of time, they were asked to recall the word pairs again. On this final test, memory was better for words that had been followed by the emotional pictures, as can be seen in Figure 13.4.

It seemed that when the word pairs were retrieved the first time, they were in a labile and fluid state, not yet firmly in memory. If an emotional picture was shown with the pairs, the person's emotional response to the

Figure 13.4 Finn and Roediger (2011) Study Using English–Swahili Vocabulary Pairs

This figure illustrates the finding that recall memory for English–Swahili word pairs was better when people were looking at an emotional picture compared to looking at either a blank screen or a neutral picture.

SOURCE: Data from Finn & Roediger (2011).

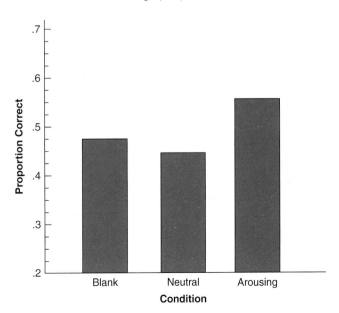

picture was incorporated into the memory trace and stored with it during the reconsolidation process. Thus, emotional responses can be incorporated into memory traces after the event has occurred if we experience an emotion when we are remembering the event. Note that Finn and Roediger did not find this benefit when people merely restudied the information, but only when they actively retrieved it, as would be expected in a reconsolidation process.

WHY DOES EMOTION HELP MEMORY? There are probably several reasons that, together, help make emotional memories easier to remember. First, emotional events are likely to be things that are important to us. Consequently, people are apt to devote more attention to processing that information relative to something that is more emotionally neutral. Part of this is driven by the recruitment of the amygdala, which is a critical brain structure for processing emotions. There is also some evidence that emotionally charged memories appear to benefit more from the process of memory consolidation offered by sleep compared to emotionally neutral memories (Hu, Stylos-Allan, & Walker, 2006).

Emotion may also help memory because emotional information is more distinctive. Much of what we encounter in our day-to-day lives does not elicit much in the way of a strong emotional reaction. Thus, truly emotional information is more likely to be distinctive, resulting in a kind of a **von Restorff effect**, which orients an increase in attention

---

Mood-Based Memories

There are a variety of ways in which emotion can influence memory. These depend on both the emotional content of the memories as well as your emotional state.

Mood-Congruent Memories

With **mood-congruent memories**, your emotional state at the time makes knowledge in long-term memory that is more consistent with that mood more available. So, when you are in a happy mood, you tend to think more about things that make you happy, such as rainbows, butterflies, and home, but when you are sad, you are more likely to think of things that make you sad, such as death, taxes, and arguments.

Most of what is going on with mood-congruent memories happens below consciousness, where you are not aware of it. It is a kind of priming. This can be seen in the fact that mood-congruent priming can be observed when people are doing a lexical decision task and are unaware that the influence of their emotional state on their performance is being tested (Olafson & Ferraro, 2001).

Mood-Dependent Memories

Studies of Flashbulb Memories

When surprising and important events happen, we form stronger detailed memories of when we learned about them. Sometimes these memories remain accurate, but sometimes they can have errors.

> Brown and Kulik (1977) examined the flashbulb memories of college students for the assassination of President Kennedy in 1963. People were asked to recall their own particular circumstances when they heard news of the event, not whether they remembered the event itself. The data showed an increase in the amount of recallable detailed information (see also Mahmood, Manier, & Hirst, 2004; and Winograd & Killinger, 1983; for flashbulb memories of emotional but not surprising events, such as good friends dying of AIDS, President Nixon's resignation, and the Apollo moon landings). Emotional experience is critical to the formation of flashbulb memories. Although no one has been able to measure brain activity when a person learns of news that would produce a false memory, fMRI scans of people remembering a flashbulb memory event, such as the terrorist attacks of September 11, 2001, show increased amygdala activity during retrieval (Sharot, Martorella, Delgado, & Phelps, 2007), highlighting the involvement of emotional experience in these memories. Note that whereas most flashbulb memories that have been studied are for negative events, there have been some studies of flashbulb memories for positive events, such as getting an invitation to join a fraternity or sorority (Kraha & Boals, 2014).

toward the emotion-eliciting stimulus (e.g., Talmi & Garry, 2012; Talmi, Schimmack, Paterson, & Moscovitch, 2007).

EMOTIONAL CONTEXT Emotions influence memory beyond just the fact than emotional memories are better. Another way that your emotions influence memory involves the kind of information that becomes activated or primed in long-term memory.

FLASHBULB MEMORIES We often seem to have—or believe we have—extremely accurate and very detailed memories of particular events, especially when the events were surprising and highly emotional. These are often called **flashbulb memories**.

So, the question is: Is memory good, even flashbulb-quality good, or is it widely subject to the sins of misattribution, suggestibility, bias, and the rest? The circumstances Conway et al. (1994) isolated as important for forming flashbulb memories—high level of importance, high affective response to the event—should also characterize memories of traumatic events, exactly those that are in dispute in cases of repressed and recovered memories.

13.3.2: Making Memory Worse

When you are in an excited emotional state, you certainly remember some things really, really well. So, it is clear there are ways in which emotions make memories better. However, when your emotions are running high, there also happen to be a lot of things that get missed, and so there are ways that emotions make memories less complete. In this section, we cover some ideas about how the accuracy of memory is tied to the intensity of your memories, and then discuss some ways in which memory is harmed by emotions.

INTENSITY AND MEMORY Emotions can be described in terms of how intense they are—that is, how much you are emotionally aroused. This level of arousal is systematically related to how much is remembered. In a general sense, memory follows what is known as the **Yerkes-Dodson law**, which is shown in Figure 13.5.

Figure 13.5 Illustration of the Yerkes-Dodson Law
Performance is poor at low levels of intensity, increases as intensity becomes greater, and then decreases for high levels of intensity.

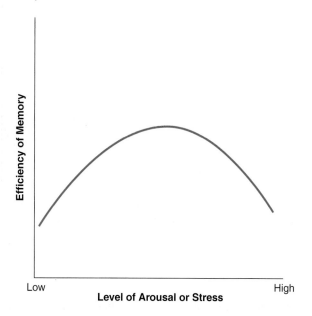

According to the Yerkes-Dodson law, when you are in a low arousal state, such as when you are tired or bored, your memory for information is not that good. However, as your arousal level goes up, so does your memory performance—but only up to a point. That point is an optimal level of arousal that allows the most learning to occur. Beyond that, things change. At even higher levels of arousal your memory starts to decline. You are too agitated and excited, and the amount of information you can adequately remember goes down. As you can see, at high levels of arousal, emotions can make memory worse.

Although the Yerkes-Dodson law accurately captures what is going on with emotional arousal in a general sense, the actual situation is a bit more complex. At higher levels of arousal, memory does not decline for everything—there are some things for which memory actually continues to improve. This is captured by the **Easterbrook hypothesis**, which states that at higher levels of emotional arousal there is a narrowing of attention onto whatever is eliciting the emotions in a person. *That* information can be thought of as being at the center of the event. The more irrelevant details of the event are less emotionally arousing, however, and can be thought of as the peripheral information. According to the Easterbrook hypothesis, when people are in more highly emotion-arousing situations, their attention narrows in on the central information and away from the peripheral information. Consequently, with high levels of emotional intensity, memory for the central details continues to get better, but memory for peripheral details declines. Because there are more peripheral than central details, overall memory is getting worse at high levels of emotional arousal, but things that are most important might be remembered really well.

So, as you can see, emotional information can both make memories better and make memories worse, depending on which aspect of an event is focused on. Memory for central information from an event is heightened by emotion, but memory for peripheral information is harmed by emotion. This differential influence of emotion on memory is exaggerated as time goes on during the consolidation process, particularly the consolidation that occurs during sleep (e.g., Payne & Kensinger, 2010), although this increase in emotional memory is more likely when people expect their memory to be tested (Cunningham, Chambers, & Payne, 2014). The neural processing that occurs during rapid eye movement (REM) sleep seems to be particularly important for the consolidation of emotional information in memory because of the increase in cortisol levels during this time, which reinstates the emotional experience.

WRITING PROMPT

The Role of Emotions on Memory

In what ways do emotions improve memories? Can you think of experiences from your own life that became more memorable because of the emotions you felt?

 The response entered here will appear in the performance dashboard and can be viewed by your instructor.

Submit

Studies Supporting the Easterbrook Hypothesis

The following memory studies support the Easterbrook hypothesis.

Interactive

| Tunnel Memories | **Weapon Focus Effect** |

In a study by Safer, Christiansen, Autry, and Österlund (1998), they showed people a series of pictures. Some of these depicted emotionally arousing scenes, such as a bloody body on the hood of a car, and others showed more emotionally neutral scenes, such as a picture of a woman in a park. On the later memory test, performance was better for central aspects of the scene in the emotionally arousing pictures compared to the more emotionally neutral pictures. Safer et al. called these **tunnel memories**—it was as if people were narrowing or tunneling their vision in on the emotional central details, causing them to be remembered better. There was no tunneling on the neutral scenes, however: Memory performance was more expansive and inclusive for the peripheral details in the emotionally neutral than the arousing scenes.

13.4: Emotion and Language

OBJECTIVE: Summarize the impact of emotion on language processing

Emotion can also influence language processes in cognition and is a quality that is communicated by language. In this section we cover the transmission and impact of emotional information. This includes emotional tone that is carried by the speech signal, as well as the tracking of emotional information in a situation model and how that influences the cognitive processing of other event dimensions.

13.4.1: Prosody

As noted in an earlier module, language is not delivered in a monotone, but involves changes in pitch. This change is called *prosody*. Prosody conveys meaning above and beyond the words used. The sentences "Those are my shoes." and "Those are my shoes?" consist of the same words but are spoken using different prosodies. One conveys a statement and the other a question, often with a rise in pitch at the end of the sentence. One other aspect of meaning that prosody can convey is the speaker's emotional tone. For example, imagine the sentence "Those are my shoes." spoken in happy, sad, and angry tones of voice, and notice the different prosodies used to convey these emotional meanings.

Recall from our earlier discussions of language that, in most people, most language-processing components are lateralized in the left hemisphere. For example, this is the location of Broca's and Wernicke's areas. However, not all language processing is strongly left lateralized, and the processing of emotional prosody is one example of this. Several studies, including research involving fMRI imaging, have found that the brain is right lateralized for the processing of emotional prosody (e.g., Buchanan et al., 2000). This is further supported by evidence from people with damage to the right hemisphere. Patients with right hemisphere damage often have difficulty deriving emotional information from linguistic prosody (e.g., Pell, 1999).

Note that although damage to the right hemisphere can impede the processing of linguistic prosody, damage to other brain structures involved in emotion processing, such as the amygdala, does not (Adolphs & Tranel, 1999). In other words, the processing of emotional prosody by the right hemisphere does not depend on emotional input from the amygdala; rather it involves a determination of emotional content from the pitch information in speech that is heard. Overall, although prosody is a deeply linguistic characteristic, this meaning is processed more by the right hemisphere than the left.

Hearing language that conveys characteristics of emotion does not mean that a listener will be convinced that the person speaking is experiencing that particular emotion. Sometimes, we know that people are only play-acting and that the emotion expressed is not one that the person is actually feeling. What people need to do is take the emotional information conveyed by the prosody, along with other characteristics of the language, to determine whether the emotion is genuinely being experienced or not. Evidence exists that people have some success at doing this and that different regions of the brain are involved in such detection (e.g., Drolet, Schubotz, & Fischer, 2012). For example, if vocal expressions of emotion are taken from the radio and played to people, the detection of genuine emotion is above chance, particularly for emotions such as anger and sadness. Moreover, as measured by fMRI recordings, there is greater activity in the medial prefrontal cortex and the temporal-parietal junction. The latter is often found to be important in perspective taking. Thus, people are able to use information in the language signal to determine not only the emotion that is being expressed by the prosody of the spoken words, but also the genuineness of that emotion.

13.4.2: Words and Situations

Let's shift our focus at this point away from *how* language is said to *what* is said, and how emotional processing influences language. First, and most obviously, some words are about emotion, such as *happy, sad,* and *angry*. We have specific words that aim to capture and communicate the experience of emotion, and the fact that such words are common in the language attests to the fact that such ideas are readily transmitted and received. We are quite facile at processing emotional information.

EMOTIONS AND EVENTS Thinking further about emotion and how it factors into our understanding of events, emotion can be experienced in two ways. One way is by the person who is feeling the emotion. This is the mental and physiological state of the person depending on the emotion being experienced. The second is by people other than the person experiencing the emotion. For example, some external person might see the emotion-experiencer smiling, scowling, blushing, and so on. These external manifestations of emotion are available to viewers. When comprehending language that involves emotion, how emotion is represented in the situation model may reflect, in part, whose perspective is emphasized by the language: the person experiencing or the one observing the emotion. These different ways of thinking about and processing emotional information may require different kinds of cognitive processes.

In a study by Oosterwijk et al. (2012), people read individual sentences that conveyed emotional information. A given sentence expressed emotion from either an internal perspective of the experiencer, such as "Hot embarrassment came over her," or the external perspective that some other person would see, "His nose wrinkled in disgust." People were simply asked to judge the sensibility of the sentences as fast and as accurately as possible (some non sensible sentences were included as well). The researchers found that people showed a processing cost when they had to switch from an internal emotional focus to an external one (or vice versa) from one sentence to the next, but not when the nature of the emotional focus stayed the same across the sentences. This shows that we mentally represent emotional information in our understanding of situations and events depending on the perspective we take, and the kind of experience that needs to be captured in the situation model.

Why would it be important to track emotional information in a story, other than the additional experience one might get from reading about a story character's emotional state? It is important to keep in mind that emotions do not exist in a vacuum, but are human reactions to experienced events. More specifically, emotions can be seen as being strongly tied to people's goals. For example, negative emotions come from having one's goals impeded or blocked, such as not being able to buy a bicycle or not having one's parents continue to be healthy. Likewise, positive emotions come from having one's goals achieved,

Processing Emotional Information

There is some evidence that people actual spend less time looking at emotional words compared to neutral words while reading, even when factors such as word frequency are taken into account (Scott, O'Donnell, & Sereno, 2012).

> **Interactive**
>
> There is some suggestion that many words in the language communicate some emotional character, even if they are abstract words. Words can vary along several dimensions, which can then influence cognition in meaningful ways. For example, words can be classified along the concrete–abstract dimension. *Concrete words*—words that refer to some physical entity, such as "ball"—are often remembered better than *abstract words*, such as "truth," because people can generate mental images of concrete words. For this reason, people can exploit dual encoding and form mental images of the concrete words, boosting their memory performance.
>
> The temptation at that point is to say that abstract words are more poorly remembered because they do not have any perceptual/embodied/grounded qualities. However, a closer look reveals that this is not the case. Abstract words *do* convey embodied information, but more in terms of emotional connotations than the more perceptual qualities of concrete words (e.g., Altarriba, Bauer, & Benvenuto, 1999). For example, words such as *courage* and *revenge* are abstract words that convey emotional qualities. If people are asked to process abstract words, those with higher emotional content are processed faster than neutral words (Kousta, Vigliocco, Vinson, Andrews, & Campo, 2010). Thus, emotional but abstract words can show an advantage in cognition. This benefit is due to the fact that people can take advantage of the embodied nature of emotion to facilitate their processing of these words.
>
> 1 of 2   Previous   Next

such as getting to go on a hot date or winning the lottery. Goals are important for language comprehension because they help provide an understanding of the motivations and causes of narrative events as the circumstances unfold.

WRITING PROMPT

The Impact of Emotions on Understanding Stories

How important is the experience of emotion, or knowledge of the experience of emotion, to how well you understand stories? What would stories be like without emotional information?

 The response entered here will appear in the performance dashboard and can be viewed by your instructor.

Submit

13.5: Emotion and Decision Making

OBJECTIVE: Analyze the influence of emotion on decision making

Decision making is another area of cognition that is affected by emotion. The effectiveness and quality of the decisions a person makes in a heightened emotional state differ from those made in a more relaxed and neutral state. Note that the terms *emotion*, *stress*, and *pressure* have different meanings, but for our current purposes, we gloss over some of these.

13.5.1: Stress Impairs Performance

It is clear to anyone who has ever been under emotional stress, such as being anxious, that decision-making and problem-solving performance can decline under these conditions. Essentially, when people experience anxiety, they tend to crowd their working memory with irrelevant thoughts about whatever it is they are anxious about. For example, people who are math anxious (i.e., they avoid doing math problems, taking math classes, exploring careers that use a lot of math) do worse on math problems because their working memory capacity is consumed by off-topic thoughts that stem from their math anxiety (Ashcraft & Krause, 2007). These thoughts detract from their limited capacity to devote to the problems, and their performance suffers. This anxiety may also disrupt the mental representations people hold in memory, making them less precise, thereby decreasing performance (Maloney, Ansari, & Fugelsang, 2011).

Sometimes we choke under pressure as when our performance deviates from what normally do, such as missing a shot at a critical time in a game.

People who are threatened also experience physiological changes due to the stress they are under (e.g., Kassam, Koslov, & Mendes, 2009). Specifically, there is a decrease in cardiovascular efficiency, reducing blood flow to the body and brain, causing the body to slow down. This makes thinking, including decision making, less effective. Part of the reason for this is that when a person is in a threatened state, it may be adaptive (under certain circumstances) to be more immobile and for the body to be prepared for some damage. Although there may be some survival advantages to this in the wild, in the circumstances of our everyday lives, this is a maladaptive response. It hinders our ability to make decisions that would be to our advantage.

CHOKING UNDER PRESSURE We have all been there. The pressure is on. We have done well in this situation before. It comes time to step up and do it again. And we choke (*rats!*). When people become anxious because of external pressures, their performance can decline.

As a process becomes increasingly practiced, it becomes more and more automatized, allowing people to do it better and more quickly. Thus, as people gain expertise, they become more fluid in the task as its components become more and more unconscious. For the most part, this is a good thing. However, it can also cause problems, especially when the pressure is on.

In a pressure situation, what skilled people should do is allow their unconscious cognitive processes to play themselves out. However, sometimes these people start to consciously think about what they are doing and how they are doing it. This is the classic case of when a person, such as a basketball player, experiences **choking under pressure**. What can happen under these circumstances is that the athlete's conscious thought processes about what he or she is doing begin to intrude on and compete with more unconscious and automatic cognitive processes.

In a classic demonstration of this, Beilock and Carr (2001) tested two groups of people in a golf-putting situation: expert golf players and golf novices. At first Beilock and Carr had people just putt to gain a sense of how well they did normally. Then they placed these people under pressure, asking them to consciously focus on their putting. What happened, not surprisingly, was that novices did better when they focused on accuracy. Surprisingly, however, golf experts did worse when they focused on their putting. This is because their conscious thoughts disrupted their normally automatic performance on this task. In fact, people who are experts actually did better if they focused on speed (Beilock, Bertenthal, McCoy, & Carr, 2004).

STEREOTYPE THREAT The influence of emotion and stress on cognition can come from several different sources. Some sources that people often do not think about are the social and ethnic groups we belong to and identify with, along with the cultural stereotypes of these groups that we carry around with us. When these stereotypes convey a negative view of our abilities in a typical domain, there can be an unconscious mental activation of this knowledge. This activation may happen even when doing something as simple as indicating gender or ethnic group when beginning to fill out a questionnaire. **Stereotype threat** occurs when this unconscious activation of a negative stereotype leads a person to perform worse on a task than he

Types of Pressure

Two different types of pressure can influence performance (DeCaro, Thomas, Albert, & Beilock, 2011).

> **Outcome-Based Pressure**
>
> **Monitoring Pressure**
>
> The other type of pressure situation is **monitoring pressure** in which people focus too much attention on the task and how they are doing it. For example, if a person is trying to make a free throw to win the game, he may start focusing on how he is standing, how he is holding the ball, how his arm is moving through the shot, and so on. This type of pressure is a result of conscious working memory processes disrupting more automatic processes, where any conflicts between the two disrupt performance. That is, more automatic processes that work well with largely unconscious control are disturbed when conscious thoughts in working memory jostle and conflict with those automatic processes.

or she would otherwise. This can come about in any way that orients a person to identify with the stereotyped group, such as indicating membership in that group on a form, or being placed in a situation in which the person feels like a minority (e.g., Murphy, Steele, & Gross, 2007). For example, asking people to indicate their gender before taking tests in science or engineering can lead women to do worse on such tests because of a (mistaken) cultural stereotype that women are not as good at these tasks (or lead Whites to do worse when exposed to the stereotype that Asians do better).

Stereotype threat can influence the types of choices and decisions a person makes (Carr & Steele, 2010). For example, people do worse at solving math problems resulting from stereotype threat if they belong to a group for which the stereotype claims that they should not be good at math (e.g., Beilock, Rydell, & McConnell, 2007).

As with the research on choking under pressure, stereotype threat lowers performance because a person's thoughts, and consequently working memory capacity, are consumed to some degree by counterproductive ideas related to the stereotype. This is the case even if that effort is oriented toward suppressing the stereotype-related thoughts (Schmader, 2010). As a result, less mental capacity is available for doing the task at hand, and performance suffers.

Note that priming can also lead to an increase in performance. For example, a study by Lang and Lang (2010) asked students to do a series of verbal analogy problems. Before actually doing the task, the researchers instructed one group of students to imagine a person who is successful at solving problems and write down several abilities such a person would have, the personality traits of such a person, and how such a person would feel just before starting to solve a problem. Compared to a control group that did not do this, people who spent time imagining what it was like to be such a successful person did better on the analogy test. Thus, it is possible to attenuate test anxiety as well as magnify it.

13.5.2: Stress Improves Performance

Emotion, stress, and pressure can decrease performance, but there are times when they can actually increase performance. There is bad stress and good stress. Performance can improve when the stress being experienced is viewed as challenging and exciting rather than threatening (e.g., Brooks, 2014; Kassam et al., 2009). Under such circumstances, there is an increase in cardiovascular efficiency, and more blood and oxygen are delivered to the body and brain, allowing the body to function at a heightened level. So, people perform better under these conditions.

As an example, consider a study by LePine, LePine, and Jackson (2004). The researchers assessed people for their stress level at school, along with other factors such as exhaustion and academic aptitude, and their performance as measured by their grade point average. Importantly, stress was identified as being either negative stress, where school work was viewed as a negative stressor and a threat, or as positive stress, where school work was viewed as a positive stressor and a challenge. Their results showed that when there was negative stress, performance was negatively affected—grades were lower. However, when school work was viewed as a challenge, performance was positively affected—students had higher grades.

In addition to physiological changes that can facilitate performance, there are also more cognitive changes where pressure can improve performance. For example, if people acquired a skill in a way that involved more implicit unconscious processes, such as learning to putt under dual-task conditions where attention is divided (Masters, 1992), then most of the cognitive processes involved in the task were unconscious. People trained under more explicit conscious conditions did worse under pressure. However, people who attended divided-attention training and then demonstrated more implicit learning, actually improved performance under pressure. To be sure, the implicit learning people had slower learning and had worse overall accuracy—but they experienced less disruption, compared to their baseline, when they were put under pressure. Thus, the overall effect of making a task more unconscious is to insulate it from the disruption of being in stressful situations. This is why it is valuable for some professions to constantly drill and practice their skills, in a variety of settings, to the point that they are automatic and unconscious. This way, when people need these skills for real, say in a genuine emergency, they can execute the skills without succumbing to the detrimental effects of pressure.

WRITING PROMPT

Impact of Emotions on Decision Making

You are not the calm, rational decision maker that you sometimes think that you are. What are some examples where your emotions have had a big influence on the decisions that you made? Was the outcome good or bad?

The response entered here will appear in the performance dashboard and can be viewed by your instructor.

Submit

Summary: Emotion

13.1 What Is Emotion?

- Emotion is a characteristic of human thought that varies in valence and intensity.
- Emotional experiences have meaningful influences on cognition.
- Two important neurological structures critical for emotion processing are the amygdala and the prefrontal cortex. The amygdala is important for the experience of emotion, and the prefrontal lobes are important for the control of emotional responses.

13.2 Emotion and Perception

- Emotion can influence perceptual processing by channeling cognitive processing resources toward those stimuli in the environment that are more emotional.

- The estimation of some perceptual qualities, such as size or height, can be influenced by whether an emotional state is aroused.
- Emotional responses can guide where attention is directed, even to the extent of sending attention to nominally irrelevant information in the environment, which can produce a processing cost, as in the emotional Stroop effect.
- The experience of intense emotions can consume cognitive resources, such as attention, to the point that that there is less available for other tasks and performance shows a deficit.

13.3 Emotion and Memory

- There is a general benefit for remembering emotional over neutral information. However, people who have suffered brain damage to areas critical for emotion processing, such as the amygdala, do not exhibit such a benefit.

- The memory benefit for emotional information comes from numerous sources, including its importance, its distinctiveness, the amount of attention and rehearsal it is given, and the superior benefit it seems to derive from neural consolidation processes.

- Emotion can serve as a form of contextual information that can facilitate some memory processes, as with mood-congruent and mood-dependent memories.

- The highly detailed memories we have for very emotional events, and for the contexts in which we learn about such events, are called flashbulb memories. Flashbulb memories, although not perfect and prone to some forgetting, appear to be more durable than normal everyday memories.

- Although emotion can improve some types of memories, it can also impose a cost on later memory. Specifically, at high levels of emotional intensity, attention is captured by central information at the cost of processing of peripheral information. Consequently, memory for more peripheral information is worse under conditions of higher emotional intensity. The weapon focus effect is an example of the kind of memory impairment that can occur with high levels of emotion.

13.4 Emotion and Language

- A great deal of emotional meaning is conveyed by the prosody of spoken utterances. Brain-damaged people, particularly those with damage to the right hemisphere, may be able to understand the words spoken to them but often lack the ability to process the prosody of what is being said, and so miss out on emotional cues in the speech stream.

- Emotion is infused in a lot of language processing, even beyond auditory cues such as prosody. Some words, even abstract words, carry emotional meaning in their lexical entries, thereby influencing how people process that information.

- People actively track and simulate emotional experiences in the situation models that are created during comprehension. There is evidence that people show processing costs of changes in emotional quality of a character, and in changes in the internal or external experience of an emotion.

13.5 Emotion and Decision Making

- When the source of stress is viewed as a threat, there can be declines in performance, causing people to choke under pressure.

- Choking under pressure can occur when the pressure is outcome based and the focus is on the outcome of the task, or monitoring based and the focus is on how the task is done. These different types of pressure arise under various kinds of situations.

- Some forms of pressure may be more subtle and unconscious, such as when a person experiences stereotype threat. This occurs when people are reminded of their membership in a group that cultural stereotypes suggest will not do well, and so their performance goes down.

- Stress can also improve performance when it involves activities that are highly over practiced, or when the stress is viewed as a challenge rather than a threat.

SHARED WRITING

Cognition and Emotion

Emotion pervades and influences cognition at all levels. Select a topic from anywhere in the text and describe how emotion might improve or impair that process.

A minimum number of characters is required to post and earn points. After posting, your response can be viewed by your class and instructor, and you can participate in the class discussion.

Post   0 characters | 140 minimum

Chapter 14
Cognitive Development in Infants and Children

Learning Objectives

14.1: Explain the lifespan perspective on human development

14.2: Outline the neurological changes that shape cognition from birth throughout childhood

14.3: Explain the changes in cognitive processes that occur during childhood

14.4: Describe developmental changes in memory

14.5: Summarize the process of language development

14.6: Outline the cognitive components of developing math skills

14.7: Compare the various approaches to the cognitive development of decision making

Not only do people differ from each other, they differ even within themselves—not just from moment to moment, but from year to year. Who you are today is different from who you were 10 years ago, and from who you will be 10 years from now. Part of this difference comes from the various experiences you have had and will have. Another part comes from your age. Systematic and regular changes that occur in memory and cognition can be identified as a function of a person's age.

In this module, we look at the influence of human development on cognition, concentrating on infancy and childhood.

14.1: A Lifespan Perspective

OBJECTIVE: Explain the lifespan perspective on human development

When we talk about development in psychology, the initial temptation is to view this only as what children do when they are growing up to become adults. Although this is certainly part of what development is about, it is not the perspective taken here. Instead, we advocate what is known as a **lifespan perspective** on human development. This view takes the position that there is no such thing as a nondeveloping person. We are all of us, in some way or another, going through a process of development as we age. Who we are at one age differs from who we were or will become at some other age. In short, the thinking of people of different ages, anywhere along the lifespan, is influenced by whatever their age may be. Ideally, for our purposes, studying this influence would involve looking at some aspect of cognition, such as semantic memory, and seeing how it changes across a person's lifetime.

Ideally, we would want to assess people across time to see how their lives change, an approach called a **longitudinal study**. Although this approach provides some of the more reliable data, it can be very time consuming and expensive, so only relatively few studies use it. Because longitudinal studies can take a long time, most developmental studies look at different people that are at different ages at the time of testing, and compare performance in the two groups, an approach known as a **cross-sectional study**. Data from these types of studies can be gathered quickly, but they do not show developmental processes as clearly because such studies involve looking at different people who have had different experiences.

However, we cheat a bit for the sake of clarity in getting the basic ideas across. What we do is divide people into three basic age categories:

1. A standard adult category. This can be considered to be anyone from age 18 to 60—and of course, our tests here are often based on college students.
2. The category of infants and children. This can be considered to be anyone from birth to age 17.
3. The category of older adults. These are people age 60 and older.

Of course, each of these categories spans a large range of ages, with lots of variations—think of how different a 2-year-old's language is from that of a 16-year-old! Not surprisingly, researchers in cognitive development typically take into account much smaller slices within these categories and make distinctions that we gloss over from time to time here. For example, someone studying cognitive development would make a distinction between someone who is 5 and someone who is 17.

In this module, we first look at cognition with regard to infants and children, using adults as the basis for comparison. We start with various neurological structures that are important for cognition, how well developed they are at birth, and how they change and mature as a person ages. After this we look at changes across childhood in perception and attention, memory, language acquisition, learning numbers and arithmetic, and decision making and problem solving.

14.2: Neurological Changes

OBJECTIVE: Outline the neurological changes that shape cognition from birth throughout childhood

In this section we look at various neurological structures that are important for cognition, how well developed they are at birth, and how they change and mature as a person ages.

Neurological Changes that Shape Cognition

> At birth, some of the neurological structures that are important for fairly primitive types of cognition are reasonably well developed, such as the thalamus and some medial temporal structures. The thalamus coordinates neural signals, and the medial temporal structures provide the basis for forming new memories. The basics of visual and auditory processing are also in place at birth. These include the eyes, the cochlea of the ear, as well as the neural structures needed for our senses of touch, taste, smell, and so on. That is, the fundamental neural hardware is in place. This then allows infants to begin learning about the environment and its structure. Note that this does not mean that the sensory systems are all working in an adult manner, but that the basics of the systems are established. These sensory systems need to develop, and this takes some time.

Because these various parts of the brain are important for more complex and conscious processes, such as conscious memory processing, it is to be expected that their later development places some limits on early cognition. These limits have implications across the range of cognitive processes. As you will see, there are developmental trends in the areas of perception and attention, memory, language acquisition, learning numbers and arithmetic, and decision making and problem solving.

> **WRITING PROMPT**
>
> **Neurobiological Changes that Impact Cognition**
>
> Given the changes that occur in infants and children as they mature, what specific changes in neurobiology produce changes in cognition? Have you noticed any of these changes in any infants or children you know?
>
> ▶ The response entered here will appear in the performance dashboard and can be viewed by your instructor.
>
> Submit

14.3: Perception and Attention

OBJECTIVE: Explain the changes in cognitive processes that occur during childhood

At birth, the infant has some basic perceptual abilities. For example, newborn infants prefer the sound of their mother's voice because it is familiar—they have been listening to it for quite some time before being born. Thus, audition is relatively well developed at birth, but then continues to develop and improve with experience. Other perceptual systems, not as developed in the newborn, also improve and mature during the first year of life. For example, the feature detectors in the visual cortex need to mature, and infants seem motivated to understand the visual world around them. This may be part of the reason why infants prefer to look at high-contrast images, such as patterns of black-and-white lines or shapes, rather than lower contrast images made of gray figures (e.g., Fantz, 1963). By 3 months or so, babies show visual preferences for a picture of a face over a non face, for their own mother over a stranger, and—most fascinating—for pictures of attractive faces rather than unattractive faces (Ramsey-Rennels & Langlois, 2007). In general, visual acuity increases as infants mature, reaching adult levels by about 1 year of age (Teller, Morse, Borton, & Regal, 1974).

14.3.1: Perceptual Memory

Given these characteristics, perhaps it is not surprising that even newborn infants can perceive and distinguish what might seem like complex perceptual qualities. For example, even 2-day-old infants can distinguish biological motion (point-light walkers) from other types of movement (Simion, Regolin, & Bulf, 2008). So, although perceptual systems do undergo some developmental changes, the basic elements of the systems appear to be present at birth. However, more complex perceptual processes, such as the abilities to use perceptual abstractions like geons to identify caricatures and abstract images of objects (such as cameras, chairs, or a slice of pizza) do not mature until a child is closer to 2 years of age (Smith, 2003).

Work with infants shows that visual sensory memory (i.e., iconic memory) is well developed by at least 6 months. Specifically, in a study by Blaser and Kaldy (2010), 6-month-old infants were shown a brief display with 2, 4, 6, 8, or 10 stars. Two of the stars then disappeared, and then in a third display reappeared—but in the third display, one of the stars changed color. The researchers monitored the infants' eye movements to see whether there was preferential looking at the one that had changed, thereby indicating some iconic memory. The results showed that the infants demonstrated preferential looking when there were 2, 4, or 6 items, but not with more. This is consistent with what has been reported in studies such as Sperling's (1960), that adults can extract a similar but limited number of items under whole report conditions. Thus, iconic memory appears to operate at an adultlike level even at a very young age.

BASIC NUMBER SENSE In terms of more complex cognition, one of the issues that emerges in cognitive development is that of basic number and quantitative principles, such as being able to perceptually identify and compare things based on magnitude or quantity. This is essentially a matter of knowing and dealing with how many things are present in a display or scene and reasoning with that information—part of what might be called **basic number sense**.

14.3.2: Attention Processes

Attention is another cognitive process that is present even in very young children. Take the example of visual search. As a reminder, in adults, visual search patterns can show evidence of either a *feature search*, in which there is a pop-out effect, with no increase in response time as display size increases, or a *conjunction search*, in which the target is identified by a unique combination of features, with an increase in response time as display size increases. A study by Gerhardstein and Rovee-Collier (2002) explored these patterns in young children. To make the task work with children, the researchers trained them to touch a computer screen whenever a certain cartoon character (e.g., a purple dinosaur) appeared. In the feature condition, none of the distractor items shared visual features with the character, whereas in the conjunction condition, they shared color or

Primary Systems for Handling Basic Number Sense

Some evidence suggests that infants, like many mammals, have two primary systems for handling basic number sense: a precise system for a very small number of things, and an approximate system for larger numbers of things (Feigenson, Dehaene, & Spelke, 2004).

Interactive

| Precise System | **Approximate Number System** |

The first system, the precise system, can be seen even in 10-month-old infants. In one demonstration, infants watched as one cracker was hidden inside a container, and then as one plus another cracker were hidden in a second container (they could not see into the containers, however). When they were then given a choice (the infants were allowed to crawl to the containers), the infants chose the container with 2 crackers, almost as if they had added 1 + 1 = 2. And they did the same when they saw 2 and 3 crackers hidden in containers—they crawled to the one with 3 crackers. If the number of crackers exceeded 3, however, the infants chose randomly, so it appears that this precise system is severely limited to very small numbers during infancy, and remains so until the child starts learning to count.

shape features. The results of this study, shown in Figure 14.1, found that children produced the same patterns of response times as are seen with adults. That is, children showed a pop-out effect in the feature search condition, and increasing response times with display size in the conjunction search condition. Thus, as early as 18 months old, some of the basic attentional processes seen in adults are also observed in young children.

Another task used to assess the attention performance of children is the Stroop color word task (in which people name the color that a word is printed in rather than the color word itself). What is found is that younger children generally show greater evidence of cognitive disruption on this task, and this level of disruption decreases as children grow older (Comalli, Wapner, & Werner, 1962). However, as shown in Figure 14.2, this seems to be tied to literacy (Schiller, 1966). Specifically, the Stroop color word task shows no evidence of disruption in first-graders. This is not particularly surprising, of course; first-graders are not very fluent readers, so they do not automatically read the color word itself. Thus, they can easily name the color of the ink that the word is printed in. However, by second grade, reading has become more automatic, and children start showing a disruption in performance. This disruption then grows somewhat smaller as children get older and they gain greater control over their attentional processes—but again, the size of the disruption effect is relatively small. Overall, this shows that even young children have the basic attentional processes shown by adults. Also, very interestingly, this indicates that tasks such as reading can become automatized very quickly with routine practice, such as the practice in reading that children receive in school.

Figure 14.1 Results of Gerhardstein and Rovee-Collier's (2002) Study

Both 24- and 36-month-old children show evidence of a pop-out effect for feature searches and an increase in response time for conjunction searches.

SOURCE: From Gerhardstein & Rovee-Collier (2002).

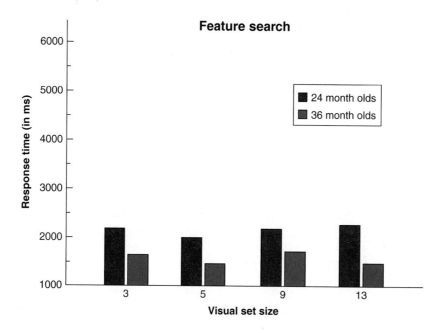

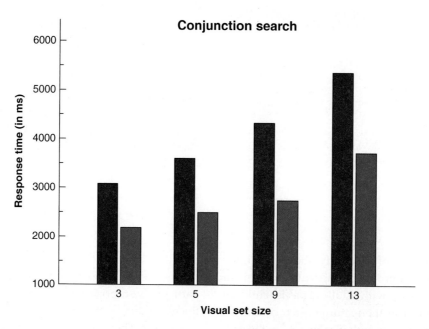

Attentional control shows shifts in processing throughout childhood, even for relatively older children. For example, take the case of task-unrelated thoughts that occur when a person is trying focus on a specific task. We can identify two different sources of distraction that can lead a person's thoughts off-task. The first are external sources of distraction, such as hearing noises in the environment. The second are internally generated, unrelated thoughts, as observed in cases of mind wandering. In a comparison of adolescents and young adults (college students), Stawarczyk, Majerus, Catale, and D'Argembeau (2014) found that although both groups were prone to both types of task-irrelevant thoughts, there were developmental differences. Specifically, adolescents were more prone to external distractions, whereas young adults were more prone to mind wandering. Thus, we can see that developmental differences exist in how people allocate their attention. At younger ages, some aspects of attention are more externally focused, whereas as we grow older there is a shift to more internal concerns.

Figure 14.2 Time-Ratio Score on Stroop Color Words

Notice that there is no color-word interference for the first-graders who have not yet learned to read. Also, there is more attentional control after literacy is achieved as children grow older.

SOURCE: From Schiller (1966).

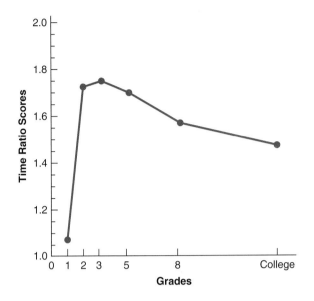

WRITING PROMPT

Developing Basic Number Sense

Given what is known about children, and their development of basic number sense, how does this number sense influence their ability to interact with the world? When would you expect them to have problems and when would you expect them to do well?

The response entered here will appear in the performance dashboard and can be viewed by your instructor.

Submit

14.4: Memory Development

OBJECTIVE: Describe developmental changes in memory

If you are like most people, your very first memory comes from when you were somewhere between 2 and 4 years old, although this can be influenced by gender and culture (Fitzgerald, 2010). This effect is called *infantile amnesia*. It certainly isn't the kind of "real" amnesia in which there is a catastrophic forgetting due to brain damage. That said, it does seem true that infantile amnesia means there are no memories from this early period of life. One conclusion you might be tempted to draw from this observation is that people do not form memories until later in life. However, this is not the case.

Even before the ability to effectively store and retrieve episodic memories is in place, the ability to form other memories, for instance, nondeclarative procedural memories such as learning how to write, is available to small children. In the following sections, we will address types of memories that are reasonably well developed at birth, followed by a discussion of memories that show clear developmental improvements.

Although infantile amnesia is generally confined to a period of life that ends somewhere around the ages of 2 to 4, it is not the case that memory is suddenly normal after that point. Memory continues to develop throughout childhood, and memory for events that we experience gets better and better over time. In addition to infantile amnesia, *childhood amnesia* comes into play during a period of life in which memory is still poorer than in adulthood. In this period of life, up to around age 7, there are fewer autobiographical memories than would be expected (Bauer & Larkina, 2014).

14.4.1: Memory Systems Present at Birth

When we are first born, there are some kinds of memory that are reasonably well developed. For example, nondeclarative memory is largely established. Think about all the things that a baby learns—and then remembers how to do—in its first year, such as how to roll over, sit up, crawl, sip from a cup, and even begin to make deliberate speech sounds, the rudimentary beginnings of language. There is a lot of memory processing going on then, although much of it is unconscious or implicit (and certainly nonverbal, to a very large degree).

There is even evidence of nondeclarative memory for the mother's voice prior to birth. In a study by Kisilevsky et al. (2003), recordings of the mother and another woman reading a poem were played to babies in the womb (about 38 weeks' gestation). There was also a control condition with no voices. The results showed that fetal heart rate increased to the sound of the mother's voice, but decreased to the sound of another woman's voice. The babies in the womb had memory for the sound of their own mother's voice. Of course, despite how well developed nondeclarative memory is at birth, a large number of nondeclarative skills are learned and mastered during later childhood.

14.4.2: Memory Systems that Improve With Age

In terms of declarative memories, short-term/working memory, and other kinds of more complex memories, infants are clearly less developed. Semantic memory develops sooner than episodic memory. Children will know what a ball is before they form memories of playing with a ball that will last long periods of time. Infants and children learn basic routines relatively early on and often cling to them, preferring them over changes. This is why children may prefer to watch the same television show or movie, or read the same book, over and over.

As semantic memory is developing, children are learning concepts and developing categories, scripts, and schemas. At 3 or 4 months old, infants are able to use basic categories, such as *dog* or *cat* (e.g., Quinn, Eimas, & Rosencrantz, 1993), and more fine-grained and elaborate category knowledge develops as they grow older. For example, a younger child might call all animals "doggie," a process known as **overextension**, and then later make a more adult distinction among dogs, cats, horses, and other animals.

EPISODIC MEMORY As noted, episodic memory, the other major type of declarative memory, seems to lag behind semantic memory in its development. This is no doubt why people cannot consciously remember events from early childhood—episodic memory was not fully functional that early. However, it is not the case that infants and young children have no episodic memory skills at all.

Again, although young infants do have a rudimentary form of episodic memory, it is just that—rudimentary. Episodic memory is underdeveloped at birth but shows rapid improvements over time. This can be seen in Figure 14.3,

---

Use of Cues in Episodic Memory

One of the important components of episodic memory is the ability to use context to help remember something. This is because context is serving as a retrieval cue, as in encoding specificity. Even young infants show that they can use these types of cues.

In a study by Butler and Rovee-Collier (1989), 3-month-old infants were placed in a crib with a ribbon attached to their ankle. This ribbon was also tied to a mobile hanging over the babies' head, so that the mobile would move when the babies kicked. Of course, when the babies figured this out, they kicked more to make the mobile move (finally, some control over something fun!). This method of assessing infant knowledge is called **conjugate reinforcement**. The important part of this study is that when the infants were first learning about the mobile and their kicking, they were in a crib lined with a padded bumper that had a certain pattern on the cloth. Five days later, the babies were retested, either with the same patterned bumper pad, or with a pad that had a different pattern.

Figure 14.3 Forgetting Curves for Children Ranging in Age from 2 to 18 Months Old

SOURCE: From Hartshorn et al. (1998).

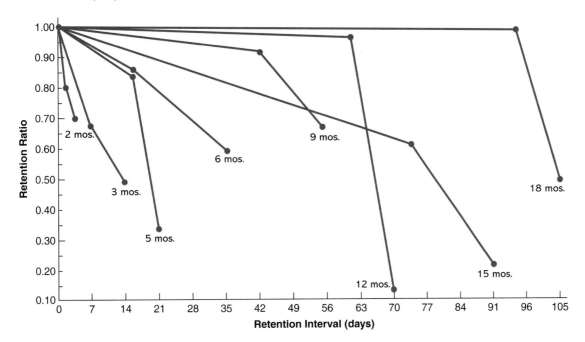

which shows how long children can remember information as they age.

As can be seen, there is an increase in the amount of time a memory can be retained as the child grows older. At some point episodic memory does improve, and the concept of the self develops enough that autobiographical memory emerges—people begin to more reliably recall events from their lives in a conscious fashion and start weaving their life story (Nelson & Fivush, 2004). In general, as children grow older, their ability to structure and organize new information improves, thereby increasing the probability that they will be able to form new, useful episodic memories.

SEMANTIC MEMORY As children grow older, their thinking is—well—childlike. Clearly, a big part of this is because their knowledge base is small, and their understanding of the world is just starting to develop, in terms of their categories, concepts, scripts, and schemas. However, part of it is also how much information they can handle at once, and their ability to process that information. Short-term/working memory is a component of memory that shows tremendous growth throughout childhood (e.g., Case, 1972).

Evidence for the development of short-term memory can be seen in the work by Renée Baillargeon (e.g., Baillargeon, 1986, 1987; Baillargeon, Spelke, & Wasserman, 1985). This research explored the idea of **object permanence** in infants, which is the ability to remember the presence of an object when it has been removed from view. You may remember this concept from studying about the Swiss psychologist Jean Piaget in course work on developmental psychology. Very early in life, infants are not at all surprised when an object disappears from their view, and they don't search for it. Then, around 6 months of age, they are surprised when it disappears and will search for it (e.g., when watching a toy train disappear into a tunnel, the baby looks at the other end of the tunnel). However, infants do not appear to retain strong memories for the features of an object, such as its shape, but only a memory that an object should be there (Kibbe & Leslie, 2011). This suggests that infants can track rudimentary information about objects in the world. Then, of course, as children grow older, these memory representations can become more detailed.

Throughout childhood, several factors can influence the development of semantic memory. Some of these are rather obvious, such as the exposure of a child to information. The more things children have been exposed to, the more opportunities they have to expand on their semantic knowledge base. Another factor that can impact semantic memory is the physical fitness of children. It has been shown that children who are physically fit, such as by engaging in more aerobic exercise, are better able to learn new information (Raine et al., 2013). So, gym and recess can be important to learning. Children's memories, and how permanent their learning becomes, also appear to be more dependent on sleep (Stickgold, 2013).

Information Handling in Children

Children simply cannot think about and handle as much information as adults. But why is this the case?

> **Interactive**
>
> Part of the reason younger children have lower memory-span scores is because they cannot think and speak as quickly as older children. Because they are processing information more slowly, they cannot get the information out fast enough, and they forget things, leading to lower memory-span scores (Hulme, Thompson, Muir, & Lawrence, 1984; Kail, 1991, 1997).
>
> 1 of 3 — Previous | Next

WORKING MEMORY When we measure the working memory capacity of children, we find that it is not as large as that of adults, although the older the child, the more adultlike it is.

WRITING PROMPT

Changes in Memory

A lot of things change in an infant and a child's ability to remember. What kinds of things would you expect a person to remember from these periods of life by the time adulthood is reached, and what would you expect to be forgotten?

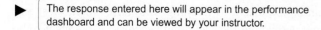
The response entered here will appear in the performance dashboard and can be viewed by your instructor.

Submit

14.5: Language Acquisition

OBJECTIVE: Summarize the process of language development

Although developmental change is clear and obvious when we look at the cognitive processes of perception, attention, and memory, it is absolutely stunning—even overwhelming—when we consider the development of language processing. Stop and think about this for a minute.

When people are first born, they know virtually nothing about the language they will grow up to speak (remember—we seem to be familiar with mom's speaking voice), although there might be some predisposition to learn a language. This may come from language-specific aspects of the human brain and/or from the propensity of the nervous system to extract and make use of regularities, called *invariances*, in the speech streams that an infant and child might hear (e.g., Gogate & Hollich, 2010).

Then, within a small number of years, children gain a strong mastery of that language, with apparent ease. If you have taken a foreign language course in high school or college, you know how difficult this can be when you are older. Thus, whatever is going on in the minds of children when they acquire their language (or in some cases, languages) must be an amazing thing—or at least it seems so from an adult's perspective. What we do in this section of the module is first outline the basic stages involved in language development. Then we highlight some findings that occur along the way that help illustrate this fantastic time of cognitive development.

Note that although much of our focus is on spoken language, there are similar patterns for children who acquire manually signed language, such as American Sign Language (ASL).

14.5.1: Stages of Language Acquisition

There is a set of stages that all children go through as they acquire language, no matter what culture they are raised in, and no matter what language they end up speaking—although of course there are individual differences in the rate at which different children proceed through the stages (Fernald & Marchman, 2006). These stages of language acquisition are defined by the *mean length of utterance*, or *MLU*, that typifies the average number of words in a child's speech.

Stages of Language Development
These stages are the babbling, single-word, two-word, and multiword stages.

Babbling Stage
At the beginning of language development is the **babbling stage**. During this time, an infant starts out by making vocalizations that are not crying, but that are the rudiments of what will be spoken language. This begins with cooing and similar sounds. By 6 months, the baby works up to repeating syllables, such as *ba-ba*, *da-da*, and *ma-ma*. Note that most babies will say *da-da* before they say *ma-ma*—not because they like Dad better, but because it is simply easier to articulate. Children will babble in the presence of other people, as well as by themselves in their cribs when no one is around, as if to practice saying things. They may also go through a period of babbling nonsense that has the prosody (up and down pitch) of language, but without any words. During this time, parents, who want to be able to communicate better with their children, are desperate for their language to emerge and are very excited by the first words (Dad thinks, "Oh, let her first word be *da-da*!").

Then there is the topic of categorical perception of phonemes. That is, people place boundaries in the speech signal to identify different phonemes, but where a person places a boundary varies as a function of what language they speak. The establishment of these boundaries occurs early on during language development. Essentially, at first, infants can recognize and identify all of the phonemes in all languages. However, with continued exposure to their own home language, by 8 months old infants are starting to hone in on the phonemes of their own language (and, note, losing sensitivity to phonemes that aren't in the home language). This process is nearly complete by about 12 months, it seems. It is at that point children start showing categorical perception patterns nearing those of the adults in their language community.

Table 14.1 Review: Stages of Language Acquisition

Stage	Description
Babbling	This is a prelinguistic stage from birth to 18 months. It involves crying, vocal play, and babbling.
Single-word	This stage is typically from 12 to 18 months of age (and beyond). It involves holophrasic speech in which a single word is used to convey a complete idea or sentence.
Two-word	This stage is typically at its peak around 18 months of age. It involves using basic syntactic forms, a focus on developing an understanding of word meaning, and an expansion of a child's vocabulary.
Multiword	This stage typically emerges between 18 and 24 months of age and grows more complex as a person grows older. It involves an increase in the complexity of syntactic understanding and usage, and longer and longer verbal expressions.

Check Your Understanding

TRAJECTORY OF LANGUAGE DEVELOPMENT Basically, there is a developmental trajectory from the simplest form of language to the most complex. Note that the ages given at which children reach these different stages are approximated averages, and there is always some overlap in the stages during periods of transition, rather than a clear boundary from one stage to the next. This can be seen in the plot shown in Figure 14.4, in which performance at different "stages" overlaps one another.

Also, as noted by looking at Figure 14.5, some children acquire language more quickly and some more slowly, depending on a number of factors.

Although this is not a deterministic outcome of future performance, *some* evidence suggests that this may be linked to other kinds of cognitive development (e.g., Marchman & Fernald, 2008).

14.5.2: Competence and Performance

We have been talking about the stages that children go through during language acquisition, in which their linguistic abilities are becoming more and more sophisticated. Although true, it would be inaccurate to think that what children are saying accurately reflects *all* they know about language. A general rule of thumb is that children almost always know more than their parents think they know (right?). What we observe when children are speaking is their **linguistic performance**—that is, the language that they demonstrate and can actually use. For instance, we speak of a child's active vocabulary, words a child will produce.

However, children often know more about language than this. Children have **linguistic competence**, an

Figure 14.4 Different Stages of Language Development

Illustrated here is the overlap that occurs in the different stages, which emphasizes there is not a discrete transition from one stage to the next as a child matures.

SOURCE: From Sachs (1983).

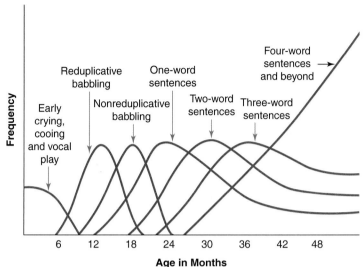

Figure 14.5 Mean Length of Utterance (in Morphemes)

Shown here is the mean length of utterance across chronological age of the three children tested by Brown (1973). Notice how the three children reach the various levels of linguistic complexity at very different ages, even though each child falls within the normal range.

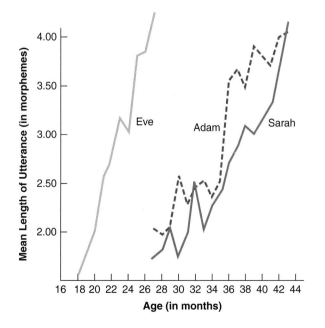

implicit understanding and knowledge of language that they cannot yet use, but that they can identify as being a proper part of the language. This implicit understanding is somewhat like when a child misspells a word and knows that it is misspelled, but doesn't know how to spell it correctly. The child has an implicit and linguistic competence understanding of the correct spelling, but lacks the explicit and linguistic performance ability to get it spelled right. (Adults have the same kind of linguistic competence; you can usually tell whether a sentence is ungrammatical, but that doesn't mean you can articulate the grammatical rule that is being violated. Your knowledge is implicit—the sentence just sounds wrong—but you can't say why explicitly.) Moreover, sometimes ideas that children wish to express, but that are beyond their linguistic performance level, can be captured by the gestures they are making while talking (Broaders & Goldin-Meadow, 2010). So, children may know more than they can say, having better linguistic competence and knowledge in memory than their linguistic performance reveals.

14.5.3: Learning New Words

One of the most obvious things that children do during language development is learn new words—the word explosion you just read about. Children hear new words and incorporate them into their vocabulary, and all is good. Sounds simple enough. But of course it isn't. How can children know what a new word possibly means? When a child hears the word *cup*, how does he or she know it refers to the object we call a cup, and not to any of the many other objects in the environment, the actions of any of the people in the room, someone's mental state, or something else?

LEARNING VERB TENSE Along with learning the names of objects, another important process in language development is learning the syntactic rules for the language.

Language processing is not just about individual words and sentences, of course. It is about communicating all kinds of ideas, simple and complex. What about language comprehension for stories, and processing at the situation model level? Even at this level there is developmental progress. For instance, a study by Bohn-Gettler, Rapp, van den Broek, Kendeou, and White (2011) compared 12-year-old children and adults in their online reading and processing of narratives stories, looking at the updating of situation models along various dimensions. Interestingly, the children and adults were similar in their updating of causal information, suggesting that for 12-year-olds, causality is an important, salient dimension. But only the adults showed evidence of online updating of spatial and temporal frameworks, new characters, and new character goals. Thus, there must be further development across adolescence in our abilities to

Understanding and Learning New Words

Children need some way to be able to pick out the referent of a word to know what object, action, or situation it refers to. Moreover, some words come in many forms, such as singular and plural nouns, or present-tense and past-tense verbs. How do children come to know the rules for how these different word forms work?

> **Interactive**
>
> One of the ways that adults help children learn their language is the specialized way that they talk to them. Adults often raise the pitch of their voice, slow their speech down, and exaggerate what they are saying in what was formerly called **motherese**, but is now called **child directed speech**, because people other than mothers use it when talking to infants and toddlers (some people also use this with foreigners and older adults, often insulting those people). Adults use child directed speech to get the attention of the child, and it seems to help the child orient to what the adults are saying. By one researcher's count, there are as many as 160 different ways in which adults can adjust their typical speech patterns as they talk to infants (e.g., Snow, 1972). Infants who listen to this speech get the idea that the language is meant for them—and that if they pay attention, they might learn something.
>
> 1 of 4

understand and track this kind of information as we read, comprehend, and process stories and text in situation model terms.

WRITING PROMPT

Language and Semantic Memory Development

How much of language development in children is due to language development per se, and how much do you think is due to semantic memory development? That is, how much of it could be due to small children not understanding enough about the world?

 The response entered here will appear in the performance dashboard and can be viewed by your instructor.

Submit

14.6: Learning Numbers and Arithmetic

OBJECTIVE: Outline the cognitive components of developing math skills

A major part of a child's early schooling involves numbers and arithmetic—one of the classic "three *R*s" of "reading and (w)riting and 'rithmetic," as the old expression goes. Children often start learning some simple addition in kindergarten, and addition is a major topic in most first-grade classrooms. But this builds on a substantial foundation of knowledge that the child has already acquired prior to formal education, particularly knowledge of magnitude and counting.

Syntactic Rules for the Language

To get a feel for this type of linguistic development, let's look at the acquisition of one rule, the past tense of verbs.

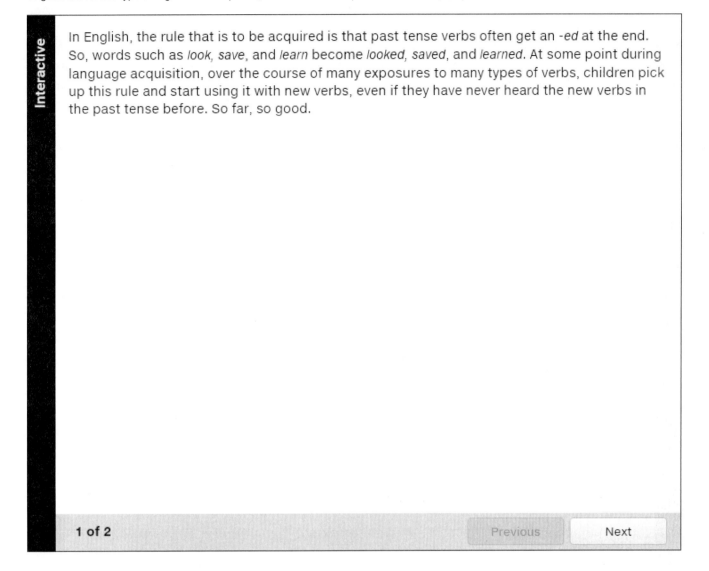

In English, the rule that is to be acquired is that past tense verbs often get an *-ed* at the end. So, words such as *look, save,* and *learn* become *looked, saved,* and *learned*. At some point during language acquisition, over the course of many exposures to many types of verbs, children pick up this rule and start using it with new verbs, even if they have never heard the new verbs in the past tense before. So far, so good.

14.6.1: Numerical Magnitude

As you read earlier, even infants show sensitivity to numerical *magnitude*, being able to discriminate two quantities of objects if the ratio between the quantities is large enough (say 1:2, for very young infants).

14.6.2: Counting

Another important line of research on very young children involves *counting*. We are not talking about learning the number names and reciting the count string, although doing so is important. Instead, we are talking about learning *how* to count objects; this is something that adults take for granted until they actually watch a young child do it.

Gelman and Gallistel (1978) claimed that these three principles explain counting:

- The *stable order principle*, using the count words in the same fixed order (one, two three, four, etc.)
- The *one-to-one principle*, that one count word gets attached to only one object being counted
- The *cardinality principle*, that the last count word used denotes the total number of things being counted

Children often violate one or more of these principles as they struggle to learn how to count. They may recount an already counted object, skip a number or an object in the sequence, or touch an object repeatedly, giving it extra count words in the process. As an example of the latter, one of the author's children, at age 3, counted a set of 3 glass turtles while touching them in sequence, saying "1, 2, 3, 4, 5," with the third turtle touched on counts 3, 4, and 5. Wagner and Walters (1982) noted that children

Young Infants' General Cognitive Sensitivity to Numerical Magnitude

However, the evidence shows that infants as young as 5 months also show rudimentary knowledge of addition- and subtraction-like relationships.

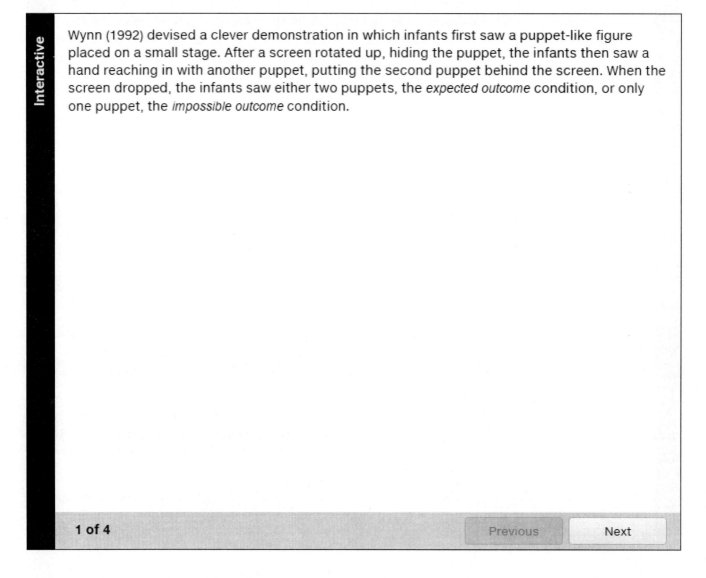

Wynn (1992) devised a clever demonstration in which infants first saw a puppet-like figure placed on a small stage. After a screen rotated up, hiding the puppet, the infants then saw a hand reaching in with another puppet, putting the second puppet behind the screen. When the screen dropped, the infants saw either two puppets, the *expected outcome* condition, or only one puppet, the *impossible outcome* condition.

often "recycled" some number sequences; in trying to finish counting a set of seven tokens, one said, "1, 2, 1, 2, 1, 2, 3" (p. 143).

14.6.3: Arithmetic

Counting is often not fully mastered until age 5 or 6, about when most children go to kindergarten. And it is in kindergarten that most children begin learning arithmetic, starting with addition. Given that counting has been their major numerical activity for several years, it is not surprising that their first attempts at addition involve counting.

Groen and Parkman (1972) conducted an early cognitive study on arithmetic. They tested both first-graders and college students on simple addition problems, simple one-digit plus one-digit problems known as the addition facts (e.g., 2 + 3, 4 + 5, 8 + 7). The results of first-graders' reaction times showed that the children were relying primarily on a counting procedure to solve the problems, a procedure called "counting on." For the problem 3 + 2, you take the larger number 3, then count up by ones twice, that is, by the smaller number 2. So first-graders were solving 3 + 2 = 5 by mentally counting "3 – 4, 5." The time it took them to increment up by ones was about 400 ms, which is 4/10ths of a second for each increment of one. The college students, in comparison, had incrementing times of only 20 ms, far too fast to represent true counting. Later work (e.g., Ashcraft & Battaglia, 1978; Campbell, 1987) suggested that adults were actually retrieving most of their addition (and multiplication) facts from memory, instead of relying on counting procedures.

But if first-graders rely heavily on counting when they add, and if adults are mostly retrieving answers from memory, then when do children switch from counting to retrieval? Considerable research has been devoted to questions like this, looking at both developmental and educational variables that make a difference in the cognitive processes used by children. It appears that children in the third grade are in the process of switching from counting solutions in addition to relying on memory retrieval (e.g., Ashcraft & Fierman, 1982), although some children lag behind in this switch. Interestingly, children who persist in counting when they add single-digit numbers are often later diagnosed as mathematically disabled and show other deficits as they struggle to learn more complex math concepts (e.g., Geary, Brown, & Samaranayake, 1991). And interestingly, some newer work has found that children's ability to estimate magnitudes, for example, to estimate where a number belongs on a 0–100 number line, is strongly related to their later math abilities and achievement (Moore & Ashcraft, 2015; Siegler & Booth, 2004).

WRITING PROMPT

Learning Math Skills as Children Versus Adults

Both children and adults are frequently presented with a need to do math problems in their daily lives. How do children and adults differ in their approaches to these problems? How would you modify a basic math problem you might encounter so that a child could better understand and solve it?

The response entered here will appear in the performance dashboard and can be viewed by your instructor.

Submit

14.7: Decision Making and Problem Solving

OBJECTIVE: Compare the various approaches to the cognitive development of decision making

For the topic of children's reasoning, no discussion would be complete without first covering the work of Jean Piaget, who is also credited with being one of the prime movers of the cognitive revolution itself. After Piaget, we discuss some more recent developments in theories of children's reasoning and decision making. Of course, outside of these larger theories there is evidence for how developments in other cognitive processes lead to changes in the ability to reason. For example, researchers have found that increases in working memory capacity seen as children grow older are associated with increased ability to do more complex forms of formal reasoning (Santamaría, Tse, Moreno-Rios, & García-Madruga, 2013). This is because more complex forms of reasoning require a person to coordinate larger and larger sets of information. As the ability to handle more information increases with better working memory capacity, performance on complex reasoning tasks also improves.

14.7.1: Piaget

Jean Piaget (1896–1980), the famous Swiss developmental psychologist, proposed that children go through a series of stages during cognitive development, and that these stages occur in a fixed order. At each stage, children acquire an ability to think about the world in a more complex and sophisticated way, but in a way that is characteristic of that particular stage. Ultimately, in the final stage of development, we achieve the ability to reason abstractly and to consider hypothetical situations, abilities Piaget considered to be at the highest level of thought. The stages he proposed were the sensorimotor stage, the preoperational stage, the concrete operations stage, and the formal operations stage (for Piaget, "operational" meant something very much like "logical"). We consider each of these in turn.

SENSORIMOTOR STAGE The first stage of thought is the **sensorimotor stage**, lasting from birth to around age 2. Here, children's thought and reasoning is governed by what they can see and do themselves.

PREOPERATIONAL STAGE The second of Piaget's stages is the **preoperational stage**, lasting from 2 to 7 years of age, in which children's thinking largely focuses on physical objects and how they are used.

CONCRETE OPERATIONS STAGE The third stage that children enter in Piaget's theory is the **concrete operations stage**, lasting from 7 to 11 years of age, in which children's thinking is becoming more reasoned and logical, and numbers are more easily managed, although this is largely applied to concrete objects. To the degree that children at this stage are able to reason abstractly, they are thinking about objects that are not present in the environment, or imagining different ways to use objects. It is at this stage that children can start using abstract thought forms such as *transitivity* (e.g., if $A < B$ and $B < C$, then $A < C$), but (in Piaget's view), only if the A, B, and C are tied to concrete things (for instance, $A < B$ expressed as "dogs are smaller than horses").

Another hallmark of this stage of reasoning is the ability to take perspectives other than one's own. It is at this time that children can reflect on another person's state of mind, how it might be different from their own,

Cognitive Development During the Sensorimotor Stage

When children are in the sensorimotor stage, their thinking is very embodied, relying on what they can sense and what their bodies can do. The following offers examples of such basic thinking on behavior.

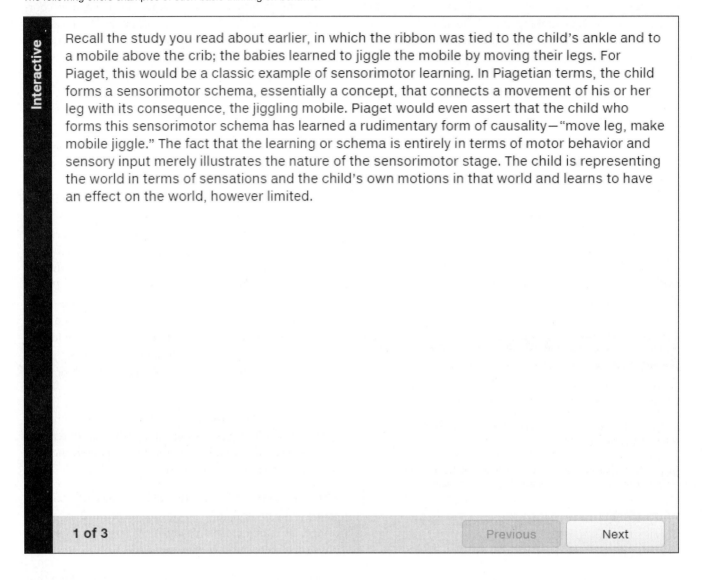

and how that person might think about and view the world.

FORMAL OPERATIONS STAGE The fourth stage in Piaget's theory is the **formal operations stage**, beginning somewhere around age 11, in which children are now able to think abstractly and hypothetically about the world. At this stage of cognitive development children acquire the ability to think about concepts that are not strongly grounded in real-world objects or interactions; in other words, A can be less than B without representing A and B as dogs and horses. Moreover, children at this stage are able imagine how things might be in the future, or how things might have been now if events had happened differently in the past. In this stage of cognitive development, scientific thinking becomes possible. This is how adults (are supposed to) think.

14.7.2: Vygotsky

Aside from Piaget, there have been other approaches to understanding cognitive development. One of the more salient of these was by Lev Vygotsky (1896–1934). Whereas Piaget emphasized the cross-cultural, innate processes that most children go through, Vygotsky stressed the influence of individual differences and the role of various experiences a person must have, such as those from one's culture. Also, whereas Piaget emphasized the learning children experience as they actively interact with the world, Vygotsky focused on the learning children experience as they interact socially with other people. Thus, for Vygotsky, the role of language and, in particular, the inner dialogue a person creates is critically important for development. Children learn through their social interactions, which involve talking with other people.

Cognitive Development During the Preoperational Stage

When children are in the preoperational stage, their thinking is very concrete, relying on the physical properties of objects in the world beyond themselves. The following offers examples of this more complex type of thinking on behavior.

> **Interactive**
>
> It is during this time that motor skills continue to develop, such as learning to throw a ball, ride a bicycle, and so on. There is a great deal of playing and pretending with objects at this stage, primarily with toys. Language is also exploding at this time, and children are acquiring knowledge of the names of the objects around them. During the later portions of this stage, children start asking a large number of "why" questions, as their reasoning develops, to try to understand the world around them and how it works.
>
> 1 of 3   Previous   Next

For Vygotsky, a critical idea was that children learn by doing things, and acquiring knowledge, in what he called their **zone of proximal development**. That is, Vygotsky claimed we learn—and grow cognitively—when we attempt tasks that challenge us just a bit, when we work just a little beyond our current levels of ability and knowledge. By reaching for these slightly more complex cognitive representations and processes, children eventually come to more and more closely approximate the mental functioning of adults.

14.7.3: Bruner

An alternative approach to cognition is the view advocated by Jerome Bruner (1915–2016), one of the pivotal figures as cognitive psychology emerged out of behaviorism. Bruner took an approach much closer to the information-processing orientation. In his view, children gain knowledge by building on the memories and processes that they have previously acquired, using earlier learned knowledge as scaffolding for learning new information.

Like Piaget, Bruner supported the idea that children go through stages of thinking and reasoning, although his stages had a different quality to them.

For Bruner, the stages of thinking and reasoning were as follows:

- **Enactive representation** (action based)
- **Iconic representation** (image based)
- **Symbolic representation** (language based)

So, children start out thinking about things in terms of how they interact with the world, in a sensorimotor

Review: Piaget's Cognitive Developmental Stages

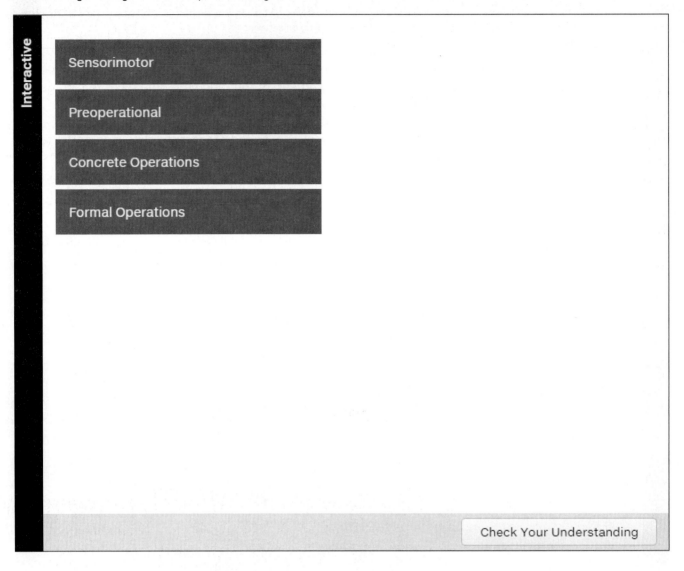

fashion. Then they move to a stage of representing information internally in terms of mental images of the various concepts they are dealing with. Note that Bruner still placed a strong emphasis on perceptual qualities at this stage.

Finally, children develop the ability to think more abstractly at the symbolic representation stage with the acquisition of language and its cognitive qualities, such as syntax. It is important to note that these stages of representation differ from those of Piaget in that children do not move from one stage to another. Instead, Bruner believed they maintain the ability to process at one stage, but acquire the ability to represent and process information at another stage and can also translate information from one representational stage to another.

This approach is in line with some aspects of embodied views of cognition in that more complex forms of thinking are based on more sensorimotor representations and processes.

14.7.4: Children as Rational Constructivists

More recent ideas about the cognitive development of decision making in children has focused on principles of reasoning and inference making. These are brought together as the **rational constructivist view** of cognitive development (Xu & Kushnir, 2013).

In terms of reasoning and rational thought, some of the basic principles that underlie adult's logical, rational thinking may also be behind children's development in terms of understanding the world and its structure. From this perspective, children are naive scientists, accounting for what they already know about the world and integrating that with what they experience.

In some sense, our minds keep track of the statistical and probabilistic aspects of our experiences (compare to our discussion of the creation of mental categories in semantic memory). Infants and children are sensitive to the probabilities of things happening in the world. That is, from their experiences, they know the likelihood that the birds that they see will be sparrows rather than penguins, which actions they take are more or less likely to get their parents' attention, or the likelihood that something they find is edible or not. This has been shown in laboratory studies (see Xu & Kushnir, 2013) in which children see different-colored Ping-Pong balls put into a box. In one case, 80% of the balls might be red, and 20% white. Infants were not surprised if an experimenter pulled out five balls from the box and four of them were red, but were surprised if the person pulled out five white balls and one red one. This is because the infants were sensitive to outcome likelihoods.

The other aspect of this theory is that children are constructing inferences about the world based on the knowledge that they gain from it. This is where the hypothesis testing part of their being like little scientists comes in. The rationale is that children have theories and hypotheses about how the world works, and they put these ideas to the test by interacting with the world. Children generally prefer things from which they can learn something new.

For example, young children are typically more likely to want to play with mini Slinkies rather than plain cubes because they can learn more from using the Slinkies. By interacting with more novel objects, children can gain further information about the world, and make better inferences and predictions about future situations. Even very young infants demonstrate a preference for novelty in learning. They are more likely to spend more time looking at something new as compared to something old. Infants also can learn more from something new.

WRITING PROMPT

Comparing Theories of Cognitive Development

Various theories of cognitive development have different strengths and weaknesses. Select a problem that children are presented with during their school-age years and how the different theories would account for how they would come to solve that problem.

> The response entered here will appear in the performance dashboard and can be viewed by your instructor.

Submit

Summary: Cognitive Development in Infants and Children

14.1 A Lifespan Perspective

- People are in a constant, continuous state of development. It would be ideal to use longitudinal data to assess this development. However, much of our coverage here involves cross-sectional studies that compare either infants and children with adults.

14.2 Neurological Changes

- Many neurological structures that shape cognition are well developed at birth, such as the thalamus, some parts of the cortex, and the sensory systems. During the course of development, the nervous system goes through a process of pruning unnecessary neural connections. Also, structures such as the frontal lobes and the hippocampus have a longer developmental progression to go through.

14.3 Perception and Attention

- Infants and children show evidence of many of the same perceptual and attention abilities as found in adults. Some of these are present with a fairly high degree of sophistication, such as identifying biological movement. They may not be as effective early on, but most show improvement across development.

- Infants can process small amounts of perceptual information, with the span of apprehension growing as a person continues to develop. This is also evident in the basic number senses available to people, even at a very young age.

- In terms of attention, children show patterns of attentional processes similar to those of adults, although there is a general increase in processing speed with age. This includes such basic attentional processes as visual search and Stroop interference, although the latter requires some degree of literacy to be observed.

14.4 Memory Development

- Although the declarative memory processes are not fully developed at birth, as evidenced by such phenomena as infantile and childhood amnesia, many nondeclarative memory processes are functioning well and can even be demonstrated in children before birth.

- As children grow older, their semantic memories become increasingly complex and detailed. These provide children with more and more sophisticated categories, schemas, and scripts.

- Episodic and autobiographical memories develop more slowly in children, although the rudiments are present at birth. For example, there appears to be some influence of context on memory as evidenced by work using conjugate reinforcement with ribbons tied to babies' ankles and to mobiles.

- Short-term memory starts out relatively impoverished, with little evidence of object permanence in young infants, but improves more and more with age, increasing in capacity and control.

14.5 Language Acquisition

- Children progress through a regular series of stages of language development, from simple babbling, through single-word, two-word, multiword, and more complex discourse in the period of a few short years.
- As children learn the rules of a language they may overregularize words (e.g., *goed* instead of *went*), even if they used the proper form earlier. This shows that children are acquiring the unconscious, implicit rules that guide language use and they are on their way to mastering the language.
- Even middle school children are progressing through a process of language development as they may continue to process language at the situation model level in a manner qualitatively different from that of adults.

14.6 Learning Numbers and Arithmetic

- Children goes through a developmental progression as they grasp the ideas of numbers and how they can be manipulated. This starts with a basic understanding of various numerical magnitudes.
- As children grow older, they develop the skills needed to count. This requires that they acquire the principles of stable order, one-to-one, and cardinality.

- As children enter the early school years, they gain the ability to do mathematical operations, such as addition and subtraction. This process moves from more basic counting methods to more sophisticated and faster direct retrieval processes. This is linked to the practice children have with mathematical concepts.

14.7 Decision Making and Problem Solving

- Early work on thinking and reasoning in children was oriented around Piaget's stages of development, namely the sensorimotor stage, the preoperational stage, the concrete operations stage, and the formal operations stage.
- Other cognitive developmental theories of thinking and reasoning include Vygotsky's culture- and language-based view, Bruner's information-processing view of more and more abstract stages of thinking, and more recent ideas that view children as rational constructivists.

SHARED WRITING

Cognitive Development

We all have gone through a process of cognitive development that has led to where we are now. Based on your own experiences, and what you have learned about cognitive development, what recommendations can you suggest that will help children be more successful in terms of what they learn, and how they think about the world?

▶ A minimum number of characters is required to post and earn points. After posting, your response can be viewed by your class and instructor, and you can participate in the class discussion.

Post   0 characters | 140 minimum

Chapter 15
Cognitive Aging

 Learning Objectives

15.1: Describe how aging affects the neurological processes associated with cognition

15.2: Identify the aspects of cognitive processes that change throughout adulthood

15.3: Summarize the changes associated with aging and memory

15.4: Outline the changes in language processing in older adults

15.5: Explain the age-related changes to complex thinking tasks

15.6: Contrast the effects of emotional information processing among older adults

We have looked at cognitive development in infants and children. Now we jump to the other end of the age spectrum to older adults. We are all growing older. Though we all have progressed through infancy and childhood, most of us have not yet reached old age.

> **What is in store for us when we cross into that part of life? How will our ability to think and remember be affected?**

This module addresses various changes that accompany cognitive aging. You will see there are several neurological and cognitive changes. Some of these have a negative effect on thinking and memory. Certain cognitive abilities peak early, say when people are in their 20s. One example of this would be some working memory span abilities, such as the number of digits that people can hold in mind at once. However, other cognitive abilities show either a neutral or positive effect. Some aspects of cognition continue to improve well into old age. For example, semantic and verbal memory capabilities, such as vocabulary knowledge, continue to grow and expand, peaking very late in life (Hartshorne & Germine, 2015).

Keep in mind that certain changes that occur are not a result of unavoidable shifts in our brains and thought, but in our attitudes. The better your attitude, the more successful you will be when you reach old age (e.g., Robertson, Savva, King-Kallimanis, & Kenny, 2015). So, try to keep a positive frame of mind about what you can do as you grow older. If you feel that your thinking and memory will get worse as you age, they will be worse. However, if you believe that you can be mentally sharp as you grow older, you will do better. Note that we are concerned here with changes in cognition due to the natural aging process—just by getting old. We are not so concerned about changes that occur because of age-related health conditions, such as Alzheimer's disease, although these are important topics.

In this module, we will cover issues of neurological change with aging, as well as the influence of the natural aging process on perception, attention, memory, language processing, reasoning, decision making, and problem solving. Finally, we will consider issues related to older adults' processing of emotional information and its influence on cognition.

15.1: Neurological and Cognitive Changes in Older Adults

OBJECTIVE: Describe how aging affects the neurological processes associated with cognition

People are always in a state of development. This continues throughout adulthood and into old age. Although we treat aging as if people are stable and consistent cognitively from 20 to 65, this simply is not the case. However, we focus on older adults here because the changes that do occur can become particularly pronounced during this time. First, we consider issues related to the challenges that occur when testing older adults and assessing the consequences of different ways of deriving these estimates of cognitive change with aging. Then we focus on neurological changes that accompany the natural aging process.

15.1.1: Cognitive Aging Studies

It is important to keep in mind how older and younger adults are compared to one another in cognitive studies. There is a vast array of questions that can be asked about aging and how that affects thinking. Each of these must take into consideration some of the challenges that cognitive aging research presents.

Although there is an advantage to longitudinal studies, they are more difficult to do, are more expensive, have a problem of people dropping out, and take much longer to produce meaningful results. Still, some important insight into cognitive aging can be gained. As one example, it is well known that episodic memory shows more dramatic declines with aging, the declines appear much earlier in cross-sectional studies (as early as in people in their 20s), whereas longitudinal studies show preserved and more consistent performance until quite old ages (e.g., Nyberg, Lövdén, Riklund, Lindenberger, & Bäckman, 2012). So, what initially looked like an age-related decline in cross-sectional studies might turn out to be an artifact of testing different groups of people. That said, it could be that part of this age-related perseveration is due to participants becoming familiar with the tests being given (Salthouse, 2014).

Studies for Assessing Cognitive Aging

Not all ways of assessing cognitive aging are the same. Each has its advantages and disadvantages. Which one a researcher chooses depends on the resources available.

Interactive

| Cross-Sectional Study | **Longitudinal Study** |

The simplest, most straightforward, and common way of scientifically studying cognitive aging is to use a *cross-sectional study* in which a group of younger adults is compared to a group of older adults. These studies are often done because they are easier to do, are less expensive, and can produce results fairly quickly. However, a major problem with them is that any observed age-related differences may be due to the natural aging process, or they may be because of other differences between the groups. For example, there may be cohort effects reflecting the fact that the older and younger adults grew up in quite different cultural and educational circumstances, which may influence the processes being measured.

15.1.2: Age-Related Neurological Changes

There are several age-related neurological changes that can affect cognition. Some are basic changes that occur in the nervous system, whereas others occur in more targeted areas.

Many cognitive changes come with aging, some for the worse and some for the better.

15.1.3: Neurological Preservation

In the face of all the age-related changes in the nervous system, there are some aspects that remain preserved. For example, although some older adults show substantial declines in neurological structures over time, not all do. In terms of neurological preservation, people with better neurological structures in aging often have a greater cortical thickness, a larger hippocampus, more preserved caudate nucleus, and more preserved tracts of white matter (myelinated neurons). Such preserved neurological structures help maintain cognitive functioning. Overall, what dictates whether we experience a greater or lesser decline is a function of a combination of genetic and experience factors—nature and nurture (Nyberg et al., 2012). By some estimates, about half of the variance associated with how much neural and cognitive decline or preservation a person experiences during aging can be attributed to genetic factors (McClearn et al., 1997).

Neurological Changes in Older Adults

Decline in the Speed of Neural Processing

One of the most basic changes that accompany the normal aging process is a slowdown in the speed with which neurons can fire. As a result, older adults can take longer to do some cognitive processes than younger adults, with more complex tasks showing a greater slowdown (Myerson, Ferraro, Hale, & Lima, 1992; Myerson, Hale, Wagstaff, Poon, & Smith, 1990). This is why older adults may have a greater following distance when they drive. They know, at least at an implicit, unconscious level, that it takes them longer to react if they suddenly need to brake or veer to the side.

The decline in the speed of neural processing has consequences for many mental processes. For example, when cognitive processes proceed more slowly, it is more likely that some tasks will not be able to be completed in time. Also, there will be greater forgetting with the more time that has passed, leading to poorer memory performance. Finally, complex trains of mental thought, in which people must keep track of intermediate steps in their thinking, will be derailed as they are forgotten because more time is needed to think through all of the steps (Salthouse, 1996).

Decline in the Frontal Lobes

Decline in the Dopamine System

Decline in the Operation of the Temporal Lobes

Decrease in Lateralization of Brain Hemispheres

Also keep in mind that the brain is not stable and set. Its plasticity enables it to change and adapt to new situations and demands, including those produced by aging (Park & McDonough, 2013). That is, stronger and more preserved aspects of a person's neurological and cognitive system can work to compensate for declines in other areas. Just because one thing shows some decline does not mean that something else cannot pick up the slack.

These different patterns of age-related neurological preservation highlight a fundamental aspect about the psychology of aging. Specifically, although there is some variability among people when they are children and young adults, it is much greater among older adults on just about any cognitive measure. Part of this has to do with the range of ages being studied. For example, with younger adults, the people characteristically studied are college students within a very narrow range of ages, typically between 18 and 22. In comparison, in studies of cognitive aging, the range of people in the older adult groups is from a much broader range of ages, such as from 65 to 80. Thus, some of the increased variability that is seen in aging studies is due to this. That said, even within narrow bands of aging there is still increased variability among older adults. That is, the degree of differences between older adults from each other increases as people progress through old age.

15.1.4: Successful Cognitive Aging

Fortunately, there are many ways to keep your cognition working optimally as you age. Perhaps the best advice is to be active (Hertzog, Kramer, Wilson, & Lindenberger, 2008). This starts with being *physically* active. Get out there and do things. Better physical health corresponds to better cognitive health.

Second, be *intellectually* active. Consistently pursue mental challenges, whether they are working puzzles, reading books, attending the theater, or doing scientific research. Keep your mind active.

Finally, you should be *socially* active. People who are more successful in their cognitive aging are socially active with other people. This does not mean that you need hundreds of friends that you see every month. It just means that you should get out there and interact with people. Keep in mind that as people age, their social circle shrinks from a large number of casual social relationships to focusing on a smaller number of more intimate, and rewarding, relationships of family and a smaller circle of friends.

Another aspect of how people live their lives that can lead to more successful cognitive aging is how large a person's "reserve" may be (Stern, 2009). For example, the greater your knowledge base entering old age, the better off your starting point. So, when cognitive aging starts to occur, even though there may be some declines, people starting with a larger reserve will be better able to cope with and manage the changes. So, try to expose yourselves to wide varieties of experiences. You will enjoy life more, and you will have more to fall back on as you grow older.

WRITING PROMPT

Steps for Aging Well

As you age, what changes do you expect to see in your own life, and what specific steps do you plan to take to make sure that your life develops in a way that keeps you more content?

▶ The response entered here will appear in the performance dashboard and can be viewed by your instructor.

Submit

15.2: Perception and Attention in Older Adults

OBJECTIVE: Identify the aspects of cognitive processes that change throughout adulthood

Now, having noted that there are neurological changes that accompany the natural aging process, what consequences do these changes have for different aspects of cognition?

VISION PROBLEMS In terms of vision, there are some noticeable changes in the ability to focus on objects in the world. In particular, the lens, the structure in the eye that helps focus the image on the retina, grows stiffer with age. As a result, it loses flexibility to the point that it can no longer change its shape to allow a person to focus on some objects, particularly close objects. As a result, usually at some time in their 40s, adults start finding themselves needing to move print farther and farther from themselves to read it. This is a condition called **presbyopia**, in which a person's lenses can no longer make the adjustments that they used to be able to do. Soon, the person will need to get either reading glasses or bifocals to make up for this declining ability. Note that this vision problem is unrelated to the state of a person's vision before this, in terms of whether corrective lenses were needed, because it is due to a different problem. Most corrective lenses people receive when they are younger are to deal with an improper eye shape (i.e., being nearsighted or farsighted) or problems with the cornea (having astigmatism). The age-related vision issues leading to reading glasses are due to the lens and are unavoidable as the lens hardens.

AUDITORY PROBLEMS In addition to vision problems, there are also problems with audition as people grow older, often eventually leading to a need to have hearing aids to compensate for changes in hearing abilities. This age-related change in hearing ability is called **presbycusis**. This is due to damage to or a loss in the hair cells in the cochlea of the inner ear—a loss that can be exacerbated by exposure to loud noises, such as from construction equipment, jet engines, or too loud music played over headphones when a person is younger (you've been warned). One of the consequences of this age-related hearing loss is a decline in the ability to hear high-pitched sounds, a decline that can begin in early adulthood. For example, there is a cell phone ringtone that is a high-pitched tone that typically only younger adults can hear (and is annoying for those who do hear it). As people grow older, it becomes more difficult to hear it, and to hear higher pitched voices, such as women's and children's voices. Also, it becomes harder to distinguish speech sounds that use higher pitches, such as the difference between "s" and "th" sounds.

15.2.1: Attention and Aging

In terms of attention, there are some clear age-related changes associated with age-related declines in the frontal lobes and other brain structures.

Consequences of Neurological Changes on Attention

Overall, there appears to be a decline in the ability of older adults to control attention as effectively as younger adults do.

Interactive

There is a decline in the size of the orienting reflex in older adults (e.g., Knight, 1984), so when something surprising happens in the environment, such as a sudden, unexpected loud noise, it will be less likely to grab their attention.

15.2.2: Attention Preservation and Improvement

Although there are some age-related declines in attention, the news is not all bad. For example, there are no meaningful differences between younger and older adults in terms of attention capture, such as by abrupt onsets of objects in a display (Kramer, Hahn, Irwin, & Theeuwes, 1999). Also, although older adults show greater slowdowns in visual conjunction searches, they are similar to younger adults in showing pop-out effects on visual feature searches (Plude & Doussard-Roosevelt, 1989; Zacks & Zacks, 1993).

One area of attention in which older adults outperform younger adults is seen with the phenomenon of *mind wandering*, in which people experience a lapse in attentional awareness away from the primary task toward more internally generated thoughts and ideas. Again, the classic example of this is getting to the bottom of the page and not remembering what you just read. If older adults have broad problems maintaining their attention, then one would expect them to mind wander more. However, this does not happen. Instead, they appear to mind wander *less* than younger adults (Carriere, Cheyne, Solman, & Smilek, 2010; Giambra, 1977, 1989; Krawietz, Tamplin, & Radvansky, 2012; Smallwood et al., 2004).

As an example of age-related improvement in mind wandering, in a study by Krawietz et al., younger and older adults were asked to read the first five chapters of Tolstoy's *War and Peace*, one sentence at a time, on a computer. At random intervals, people were stopped and asked whether they were mind wandering. The investigators found that the younger adults reported mind wandering 48% of the time, but the older adults reported mind wandering only 29% of the time. It is not just the case that the older adults were, for some reason, saying that they were mind wandering less often, perhaps to not look so bad. When the participants were tested for material that they had just read before being interrupted, the younger and older adults showed a similar pattern of accuracy, being more accurate when they said they were not mind wandering, and less accurate when they said they were. This age-related difference is also supported by more objective measures, such as recording of eye movements (Frank, Nara, Zavagnin, Touron, & Kane, 2015).

> **WRITING PROMPT**
>
> **Controlling Attention Through the Aging Process**
>
> The ability to perceive and control attention clearly changes in some ways as we age. How can you use what you have learned here to improve your interactions with older adults?

> The response entered here will appear in the performance dashboard and can be viewed by your instructor.
>
> Submit

15.3: Memory in Older Adults

OBJECTIVE: Summarize the changes associated with aging and memory

Perhaps the aspect of cognition that is most strongly associated with worry and concern among older adults is memory. According to the prevalent stereotype in our culture, older adults have poorer memories than do younger adults. What we see in this section is that although there is some truth to this, there are also aspects of human memory that are fairly well preserved as people age, and there are even characteristics that might improve. In general, the degree of success older adults experience with everyday memory tasks is related to the health of various underlying neurological components, such as the degree of cortical thickness (Bailey, Zacks, Hambrick, Zacks, Head, Kurby, & Sargent, 2013). Even in the face of global neurological changes, too often traditional views of memory and aging emphasize age-related declines and miss things that stay the same or actually improve (Hess, 2005).

With aging come changes in memory. Some aspects of memory become worse, whereas other aspects related to skill may stay preserved or even get better.

15.3.1: Short-Term Working Memory

For short-term/working memory, it is quite clear that older adults do worse on memory-span tasks compared to younger adults. That is, there is reduced capacity to actively hold on to information with old age (Craik & Byrd, 1982). Again, looking at cross-sectional data, short-term working memory span appears to peak around the age of

20 and decline through adulthood into old age (Brockmole & Logie, 2013). Older adults are not able to actively maintain as much information when they are thinking, which can then cause difficulties in more complex tasks that they may be asked to do, such as doing mental calculation or planning a new route somewhere.

Part of the reason for the reduction in effective working memory capacity may be due to declines in the ability to inhibit or suppress irrelevant information, information that remains in working memory and functionally reduces its size (Blair, Vadaga, Shuchat, & Li, 2011). Thus, continuing to remember too much information makes it more difficult for older adults to encode and use other information. Note that although these declines in working memory span are measurable and can have real-world consequences, there are not catastrophic reductions in working memory capacity. Older adults are still able to process and think deeply about many everyday things.

15.3.2: Episodic Long-Term Memory

When it comes to long-term memory, there are clear declines in episodic memory. Older adults often find it harder to recall and recognize newly learned episodic information, such as lists of words (e.g., Zelinski & Stewart, 1998). This is consistent with the idea that episodic memories are more fragile than other types of long-term memory, as is reflected by the fact that episodic memories are forgotten more quickly than are semantic and procedural memories. Note also that this memory system takes longer to develop in children. In other words, it is the last one to develop and the first one to be lost.

Part of the reason why older adults have trouble with episodic memory is that there is a decline in the ability to encode and use some kinds of contextual information. As you remember, context is a critical factor influencing the effectiveness of episodic memory. Because of this, older adults show smaller encoding specificity effects (Duchek, 1984; Luo & Craik, 2009) and smaller von Restorff effects (Bireta, Surprenant, & Neath, 2008). As a reminder, the *encoding specificity effect* is the finding that it is easier to remember something when we are in the same context during retrieval as we were in when we learned something, such as being in the same place. Older adults are measurably less likely to show this memory binding to a context.

Furthermore, the *von Restorff effect* is the finding that items that are unique in a context are remembered better, such as seeing the word *elephant* in a list of tools. Because older adults are not using context as effectively as younger adults, they are less influenced by the list context, and so do not show the memory benefit from having one of the items be different from the others. In other words, the different item does not show as great of an isolation effect for older adults as it does for younger adults.

RETRIEVAL INTERFERENCE Consistent with the fact that older adults may have some attentional selection problems, particularly when there is a need to suppress irrelevant information, they are also more susceptible to some forms of interference in long-term memory. *Retrieval interference* occurs when multiple memory traces compete during remembering, and a person has trouble selecting the appropriate one. The mental process of inhibition is helpful because related and irrelevant memory traces can have their activation reduced, thereby reducing the amount of experienced interference. If inhibition is less effective for older adults, then more interference will be seen in their attempt to remember.

An illustration of this increased interference with aging can be seen with the fan effect. Again, as a reminder, a *fan effect* is an increase in response time with an increase in the number of associations learned for a particular concept (Anderson, 1974). To better manage that retrieval interference, people need to suppress those memory traces that are related but irrelevant to one they are trying to retrieve. Because older adults are not as efficient at this, they experience more interference during memory retrieval and, therefore, show larger fan effects (Gerard, Zacks, Hasher, & Radvansky, 1991; Radvansky, Zacks, & Hasher, 1996; 2005). So, part of the problem with aging and memory is not that older adults are remembering too little; it is that they are remembering too much.

Although there are age-related changes in some aspects of episodic memory, especially in terms of *how much* is remembered, there do not appear to be major differences in the *quality* of what is remembered (Small, Dixon, Hultsch, & Herzog, 1999). For example, in terms of the organization and structure of information in episodic memory, there are no clear age differences. If anything, there is some evidence that the episodic memories that older adults have are *better* structured and organized than those of younger adults (Kahana & Wingfield, 2000). Moreover, and very importantly, if a person stays mentally active, then it may be possible for there to be little to no differences between older and younger adults' memories. A study by Shimamura, Berry, Mangels, Rusting, and Jurica (1995) showed no age-related memory impairments for things like proactive interference and prose memory for college professors.

15.3.3: Prospective Memory

Another area of memory in which declines in frontal lobe functioning and metamemory can cause difficulty is the domain of *prospective memory* (remembering to do something in the future) (e.g., Einstein & McDaniel, 1990; Smith & Bayen, 2004). That said, this depends in part on the kind of prospective memory that is being used. Specifically, age-related declines in prospective effectiveness are more pronounced for time-based prospective memory, in which a person must keep track of how much time has passed (Einstein, McDaniel, Richardson, Guynn, & Cunfer, 1995; Henry, MacLeod, Philips, & Crawford, 2004; Park, Hertzog, Kidder, Morrell, & Mayhorn, 1997). Thus, older adults might appear to need the aid of external cues or reminders to do better at remembering that it is time to take their medication again, to be at an appointment, or to take the pie out of the oven.

Interestingly, the picture reverses when testing is done in real-world settings where older adults actually tend to do better than younger adults compared to testing done in the laboratory (e.g., Henry et al., 2004; Kvavilashvili, Cockburn, & Kornbrot, 2013). Exactly why this is the case is not clear. It may be that older adults simply have better life skills in terms of managing their time and better strategies for reminding themselves of things than do younger adults. These life skills more than compensate for any cognitive declines that might be seen in more artificial laboratory tasks (e.g., Chasteen, Park, & Schwarz, 2001).

15.3.4: Semantic Long-Term Memory

Although there are age-related declines in episodic memory, this age difference is not observed with semantic memory (e.g., Rönnlund, Nyberg, Bäckman, & Nilsson, 2005; Spaniol, Madden, & Voss, 2006). For example, if younger and older adults are given a vocabulary test, a test of semantic memory, the older adults invariably do better on those tests. If you think about it, this preservation is a sensible outcome.

Preservation of Semantic Memories

The preservation of semantic memories in older adults takes many forms. In what follows, you can see examples of these different forms of preserved semantic memory.

Interactive

Older adults have been around for a lot longer than younger adults. They have had more experiences than younger adults, more exposure to the language, and so their semantic memories are going to be richer and more complex. Older adults do better on the vocabulary tests than younger adults (Ben-David, Erel, Goy, & Schneider, 2015) because they have had more opportunities to learn new words. This is not due to education getting worse in recent years, because, in many cases, the education of younger adults has been much better. It is simply due to the passage of time and being exposed to more of what life has to offer. There really is something to the idea of wisdom accumulating across age.

15.3.5: Metamemory

In terms of metamemory, because of declines in the effectiveness of their frontal lobes, older adults show some declines in more conscious processes, such as *source monitoring* (Hashtroudi, Johnson, & Chrosniak, 1990). Thus, they are more likely to make errors and misattribute information from one source to another, such as misremembering where they learned information, or whether they actually did something or just thought about doing it (such as taking medication). This has been confirmed by neuroimaging evidence. For example, event-related potentials (ERPs) recorded during source monitoring reveal that older adults do not exhibit the strong ERP waves consistent with memory discrimination shown in younger adults (Dywan, Segalowitz, & Webster, 1998). Because of this decline in source monitoring, it is also not surprising that older adults are more susceptible to *cryptomnesia*, or unconscious plagiarism (McCabe, Smith, & Parks, 2007), because they are not as effective at remembering whether an idea that they have is an old one or a new one.

Because older adults have relatively well-preserved semantic memories, but poorer metamemory abilities, they are more susceptible to the creation of some types of false memories, such as those seen with the DRM false memory effect (Norman & Schacter, 1997; Smith, Lozito, & Bayen, 2005). What happens is that when hearing a related list of words (e.g., *bed, pillow*), the *critical lure* word (*sleep*) becomes activated in semantic memory, just as with younger adults. But, because older adults do not monitor their memory processes as effectively, they are more likely to later misattribute these memories (of "sleep") as having been encountered before, when in fact they were not (Dehon & Brédart, 2004). Similarly, older adults are more likely to have false memories of events that are script-consistent because they have well-preserved scripts and schemas in semantic memory (Gallo & Roediger, 2003; LaVoie & Malmstrom, 1998).

So, what does this mean for more practical situations—for example, when older adults are eyewitnesses to an accident or a crime? Well, overall, older adults are susceptible to misleading post event information, but no more so than younger adults. If there is a problem, it is that older adults have a tendency to be more confident in the errors that they do make (Dodson & Krueger, 2006). So, they are just as likely to make mistakes as younger adults, but they are really sure about those mistakes.

15.3.6: Age-Related Stereotypes

An important point to keep in mind about age-related deficits in memory—and this is a point that is too often missed or overlooked—is that part of the problem can be attributed to how older people think about themselves and their own memory abilities. That is, there may be some unconscious handicapping that goes on when older adults adopt or accept common cultural stereotypes about the elderly, such as stereotypes that suggest that older adults are forgetful and have worse memories (Bouazzaoui et al., 2016). When these stereotypes are not emphasized, or the material is something that interests the older adults more, then age differences can be reduced or eliminated (McGillivray, Murayama, & Castel, 2015).

As one example of this influence of age-related stereotypes, a study by Levy (1996) administered several memory tests to older adults. However, before taking these memory tests, the older adults were subliminally exposed to some primes that reflected either positive stereotypes about aging, such as "wisdom," "sage," or "guidance," or negative stereotypes, such as "senile," "dementia," and "decrepit." Given what you know about stereotype threat, you probably won't be surprised by the results Levy found. The older adults did better when they had been unconsciously exposed to the positive aging words, but worse when they had been exposed to the negative aging words. A critical result in the study was that the positive and negative aging words had no effect on the younger adults who were tested—their memory task performance remained the same regardless of the prime words. Thus, the results showed that how people judge their own memory abilities (e.g., "I'm becoming forgetful," or "I've learned a lot across the years") can actually influence memory performance (see Thomas & Dubois, 2011, for a similar finding when age-related stereotype threats are reduced).

WRITING PROMPT

Aging and Different Kinds of Memory

Two stereotypes about older adults are that they are less able to remember things about the world and that they are wiser than younger adults. How might each of these play out in the workplace, and what can be done to maximize the contributions of older adults?

 The response entered here will appear in the performance dashboard and can be viewed by your instructor.

Submit

15.4: Language Processing Changes

OBJECTIVE: Outline the changes in language processing in older adults

Because of the sensory and perceptual challenges facing older adults, there can be age-related problems with language processing in terms of just being able to grasp and decode the words that they hear and see. This clearly is going to compromise older adults' ability to use language.

However, beyond that there are relatively few age-related changes in language processing per se. For example, as was mentioned in the section on semantic memory, older adults generally have larger vocabularies.

15.4.1: Anaphoric and Syntactic Complexity

The difficulties that older adults do experience in language processing are often a result of processing difficulties in other cognitive domains, such as not being able to process information fast enough, or not being able to maintain sufficient amounts of information in working memory. One linguistic consequence of reduced working memory capacity is that older adults may be less able to coordinate information sufficiently for efficient processing. For example, in a study by Light and Capps (1986), people were asked to listen to a series of short stories. The experimental texts in this study had final sentences with a pronoun that referred to a story character that had been mentioned earlier. Light and Capps varied the distance in the text between the pronoun and the person to whom it referred, by varying the number of intervening sentences. The results of this study are shown in Table 15.1.

As you can see in the table, as the distance between a pronoun and its referent increased, people's responses became less accurate. This makes sense, of course, in that the farther back in a story the referent was, the farther back in memory a person would have to search to find that referent. But this effect was much stronger for the older adults here—for them, having to search back farther in the story for the referent was even more difficult. The explanation for this was that older adults have a reduced working memory capacity and hence had a harder time maintaining the pronoun's referents in working memory. Consequently, searching back in time for the referents was even more difficult for them.

Similarly, because of their reduced working memory capacities, older adults are also more likely to have difficulty with syntactically complex or ambiguous sentences (Kemper, 1987; Kemtes & Kemper, 1997). The need to maintain more information in working memory, to understand these more complex language structures, leads to processing problems in older adults. Beyond the reduced working memory capacity, there is also evidence that some of their language processing abilities may be compromised. For example, older adults may spend less time on new concepts introduced in a text (Stine, Cheung, & Henderson, 1995) and seem less likely to allocate the extra processing effort that is needed at sentence boundaries (Stine & Wingfield, 1990; Stine et al., 1995) where linguistic wrap-up processes normally occur.

15.4.2: Discourse Processing

That said, there is some evidence of cognitive difficulties with more complex discourse. For example, older adults sometimes have a greater difficulty selecting the more important points in a text (Cohen, 1979; Dixon, Simon, Nowak, & Hultsch, 1982; Kintsch & van Dijk, 1978; Meyer & Rice, 1981; Stine & Wingfield, 1990; but see Adams, Smith, Nyquist, & Perlmutter, 1997; Spilich, 1983; Tun, 1989). This may be because older adults have a harder time organizing information from complex texts (Smith, Rebok, Smith, Hall, & Alvin, 1983), again because of working memory limitations.

In addition to processing the information in a discourse, successful comprehension requires that a person make inferences about things not explicitly mentioned. We need to fill in the gaps.

15.4.3: Situation Model Processing

An important part of language comprehension at the situation model level is knowing or figuring out what parts of described events are important, and what parts are less important. The preserved schemata and scripts of older adults can help with this. However, sometimes importance is defined not strictly by the knowledge in semantic memory, but by the structure and circumstances of the current event. That is, people need to understand the causal structure of those events. For example, if you read that a story character is outside while it's raining, finding out that the character is standing under a bridge involves a causal spatial relation because the bridge can keep that person dry. You would not draw the same causal connection if you learned that the character was standing under a lamppost, for example. Younger and older adults process such causal linguistic information similarly (Radvansky, Copeland, & Zwaan, 2003).

In general, many of the age-related difficulties in language processing that are found are at the level of individual words, or the *propositional textbase* (the ideas actually

Table 15.1 Light and Capps (1986) Study of Processing Difficulties in Older Adults

Aging and the ease of accessing (in proportion correct) a prior antecedent in a text as a function of how far back in the text it was (i.e., zero, one, or two sentences back).

	YOUNG	OLD
Number of intervening sentences		
Zero	65.1	64.1
One	61.8	58.3
Number of intervening sentences		
Zero	64.9	63.4
Two	61.3	54.7

Studies Demonstrating Cognitive Difficulties in Discourse Processing

Typically, the inferences that we make are appropriate to the discourse we are processing. However, at other times, we can make mistakes, and we need to go back and correct the inferences that we made. This correction of inappropriate inferences appears to be influenced by aging.

> **Interactive**
>
> In a study by Hamm and Hasher (1992; see also Zacks, Hasher, Doren, Hamm, & Attig, 1987), people were given passages to read in which the early part of a passage led readers to make one inference. An example of such a passage of text that first invites one inference and then forces the reader to adopt another is given here (from Hamm & Hasher, 1992, p. 58):
>
>> Carol was not feeling well and decided to find out what was wrong. She called her friend who was a nurse to ask her for some advice. The friend told Carol what to do. Carol went into town and apprehensively entered the large building hoping to find an answer. She walked through the doors and took an elevator to the third floor. She found a book that seemed relevant to her problem. Carol then went to the main desk and checked out the book for two weeks so that she could read it at home. When she left the building, she saw it had started snowing hard and she hailed a taxi to take her home.
>
> In this case, the early part of the passage leads most readers to infer that Carol is entering a hospital. However, as the passage continues it becomes clear that she is not going to a hospital, but a library. As such, people need to modify the inappropriate inference that was originally made (hospital) to a more appropriate inference (library).
>
> The study found that while both younger and older adults made both the inappropriate and appropriate inferences, the older adults had a tendency to hang on to the inappropriate inference in addition to the later, correct inference. This may happen because older adults have difficult suppressing newly irrelevant information.
>
> 1 of 2   Previous   Next

present in a text, and not inferred). For example, older adults take longer to read propositionally more complex sentences that have more idea units in them (Hartley, Stojack, Mushaney, Annon, & Lee, 1994; Stine & Hindman, 1994; Stine & Wingfield, 1990). This suggests that they have some trouble breaking complex sentences down into their component parts, again implicating working memory problems, as well as the general slowdown in processing that occurs with aging.

In contrast, language processing at the situation model level appears to be relatively well preserved regardless of increasing age (Radvansky & Dijkstra, 2007). Younger and older adults appear to use mental models quite similarly (Radvansky, Gerard, Zacks, & Hasher, 1990). In some cases, although older adults show difficulties at lower levels, such as remembering verbatim or propositional idea information, their ability to remember information at higher levels, such as the situation model level, is unaffected; in fact, it may even be better than in younger adults (Radvansky, Zwaan, Curiel, & Copeland, 2001). This preserved memory at higher levels of thought is seen in more everyday tasks, such as remembering news events (Frieske & Park, 1999). Compared to younger adults, older adults show better memory for the content of news stories and the sources of those stories.

AGING EFFECTS ON READING Reading time data have also provided information about aging effects. In the module on infancy and childhood, it was noted that 12-year-old children seemed to update their situation models much as young adults do, but only on the dimension of causality. On the other hand, comparisons of younger and older adults have essentially found no differences at the mental model level in terms of updating, on all relevant dimensions (e.g., causality, space, time). That is, when tested, older adults appear to be as sensitive to these dimensions as younger adults, although their reading times are somewhat slower overall (Radvansky et al., 2001; Stine-Morrow, Gagne, Morrow, & DeWall, 2004).

Thus, as they read and understand a text, older adults appear to be updating their situation models in a manner equivalent to younger adults. As a reminder, in a study by Morrow, Greenspan, and Bower (1987), people memorized a map of a building, along with the objects in each of the rooms of that building. Afterward, people read texts that described the movement of story protagonists through the building. The results showed a spatial gradient of availability, in which knowledge of objects in the building varied as a function of how far those objects were from the story protagonist's current location—the farther away objects were, the less available they were in memory. In studies by Morrow, Leirer, Altieri, and Fitzsimmons (1994) and by Morrow, Stine-Morrow, Leirer, Andrassy, and Kahn (1997), younger and older adults engaged in a similar task. In both cases, the spatial gradient was observed, although the older adults had a larger gradient than the younger adults, perhaps because of memory search difficulties.

Similarly, as described in the module on memory and forgetting, there is a finding that walking through doorways causes forgetting. That is, moving from one event (as defined by a given location) to another event causes people to forget information, such as which items they are carrying. This basic finding also appears to be unaffected by the natural aging process (Radvansky, Pettijohn, & Kim, 2015). Thus, older adults appear to be tracking developing and changing events in a manner similar to younger adults.

WRITING PROMPT

Information Processing in Older Adults

Despite the typical stereotype that older adults have poorer memory than younger adults, they do appear to do just as well as younger adults at processing information at the situation model level. Given this, what kinds of tasks do you think that older adults would have more problems with, and what sort of tasks would they do as well as, if not better than, younger adults?

The response entered here will appear in the performance dashboard and can be viewed by your instructor.

Submit

15.5: Reasoning, Decision Making, and Problem Solving

OBJECTIVE: Explain the age-related changes to complex thinking tasks

Up to this point we have seen that older adults have some age-related difficulties with perception and some aspects of attention, their working memory capacities are somewhat reduced, and they have trouble with episodic memory. On the other hand, they have quite good semantic memories, and reasonably good language processing abilities, especially at deeper levels of comprehension. So how do these circumstances affect their ability to reason, make decisions, and solve problems?

15.5.1: Reasoning in Older Adults

In terms of logical reasoning, older adults (Fisk & Sharp, 2002; Gilinsky & Judd, 1994) do not show differences compared to younger adults on categorical syllogisms (e.g., All A are B, Some B are C, therefore . . .) when there are only one or two mental models to manage, but do have difficulty with categorical syllogisms that require three mental models. That is, when there are a large number of possibilities that a person needs to consider to come to a valid conclusion, then older adults are more likely to have trouble. This age difference in reasoning is likely due to the same explanation you just read about—the difficulty older adults may have with managing more information in their limited working memories. Because they are less able to keep more information actively in mind, they are less able to exhaust many possibilities.

Another thing seen as an age-related problem with formal logical reasoning is a greater belief bias effect. Again, the *belief bias* is the tendency for people to ignore the logical form of a syllogism, and make decisions about the correctness of a conclusion based on their prior knowledge or beliefs about the world. That is, older adults are more likely to draw conclusions dependent on their personal understanding of the world, and not based on the information given in the logical premises. As such, they are less likely to come to a valid conclusion when the information involves something that they already have extensive knowledge about.

15.5.2: Decision Making in Older Adults

There do not appear to be major age differences in decision making or the use of heuristics, for example, in the use of the recognition heuristic in which people make choices based on what they better recognize (Pachur, Mata, & Schooler, 2007). Some evidence suggests that older adults spend less time gathering information before making a decision, and use more "fast and frugal" heuristics, but with little to no drop-off in the quality of their decisions (Mata & Nunes, 2010; Mata, Schooler, & Reiskamp, 2007). It may also be that their decision making is somewhat more intuitive than deliberate (Queen & Hess, 2010). So, it may be that older adults become more efficient at knowing what kind of information to target, and more facile at using their heuristics to make decisions.

One difference between the decision making of younger adults and older adults is related to the kind of decision process being tested, essentially whether the decision situation involves independent or dependent sequences. For example, the probability of getting heads on any particular coin flip is independent of any previous or future coin flip. They all stand apart. However, deciding which move to make next on a chessboard definitely depends on previous moves and clearly influences future moves—that is, it is a dependent choice process. A study by Worthy, Gorlick, Pacheco, Schnyer, and Maddox (2011) directly compared these two types of choice situations in younger adults and older adults. The pattern of what they found (although not actual data) is summarized in Figure 15.1.

Figure 15.1 Worthy, Gorlick, Pacheco, Schnyer, and Maddox (2011) Study

Patterns of performance (but not actual data—because they were difference scales) of younger adults and older adults on reasoning tasks that involve either independent outcome sequences (such as coin flipping) or dependent outcome sequences (such as chess moves). The younger adults outperformed the older adults when the choices involved independent outcomes, but the older adults outperformed the younger adults when the choices involved dependent outcomes.

SOURCE: From Worthy et al. (2011).

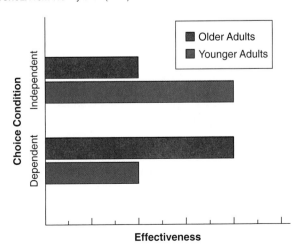

Basically, younger adults outperformed older adults on the independent decisions task, but older adults outperformed younger adults on the dependent decisions task. So, it appears that with aging there is a greater emphasis on the larger situation during decision making, and a greater emphasis on interrelations among decisions. Worthy et al. further suggest that this may be related to patterns of neurological activity that have been observed in other studies. Specifically, younger adults tend to show more activity in the ventral striatum (part of the basal ganglia, namely the nucleus accumbens and the olfactory tubercle), which is associated with the immediate rewards of a choice. In comparison, older adults tend to show more activity in the dorsolateral prefrontal cortex (B.A.s 9 and 46), which involves higher order behavioral regulation.

AGE-RELATED DIFFERENCES IN JUDGMENT AND DECISION MAKING In terms of decision framing and risky choices, there are effects of aging (Best & Charness, 2015). Specifically, older adults, compared to younger adults, are less likely to make risky choices when a choice is framed as a gain, such as being told how many people will live if a certain drug is taken. That is, the older adults are being more conservative in their choices. That noted, there also does not appear to be any age-related difference in the rate at which people tend to be risk avoidant when the choice is framed as a loss, such as being told how many people will die if they take the drug. So, there is some shift in the degree to which older adults are willing to take risks when they make these decisions, but only under certain circumstances, not universally.

Another area in which age-related differences in judgments are found involves the *hindsight bias*, in which people are more likely to misremember their past state of mind as being more consistent with their current state of mind. For example, after people learn something new about a topic, they then misremember themselves as having known more than they really knew at the earlier time; after your favorite player wins the game for your team, you forget your earlier misgivings and "remember" that you were certain your team was going to win. As it turns out, older adults show a greater hindsight bias effect than do younger adults (Bayen, Erdfelder, Bearden, & Lozito, 2006). This seems to occur because older adults have difficulty suppressing the more recent knowledge that they just learned, causing them to misattribute this knowledge to their earlier state of naiveté when, in fact, they did not have that information. This is evidence of how age-related declines in inhibitory processing can influence decision making.

15.5.3: Problem Solving in Older Adults

In terms of problem solving, there has been relatively little research done to date on standard cognitive psychology assessments of problem solving. Some of the work that has been done has not shown large age differences. For example, on the Tower of Hanoi problem using specific rules to move a stack of discs, there may be some planning differences for younger adults and older adults, with older adults not being able to plan out as far as younger adults because of their smaller working memory capacities. However, in terms of the number of moves and the success at solving the problem, there are apparently no age differences (Gilhooly, Philips, Wynn, Logie, & Della Salla, 1999).

There is some other evidence of age-related changes in problem solving. As a reminder, sometimes when people are solving a series of similar problems, they may fall into a certain way of solving the problem, a tendency called *negative set*. For example, with Luchins's (1942) water jug problem, people derived a complex way of measuring out a certain amount of liquid, and then were more likely to miss a simpler method when it was possible. Overall, when people have a certain way of approaching a problem, they may have trouble solving an easier problem because of their prior experience with more difficult solutions. This may slow down or impede the effort to find a path to a solution.

There is some evidence that older adults are more likely than younger adults to stick with prior problem solving strategies (Lemaire & Brun, 2014). That is, it may be that older adults have trouble suppressing a prior strategy of approaching a certain type of problem, and persist with less effective strategies. Thus, it may take them longer to adapt or find solutions to a problem that should be treated with a different approach from the one originally used. This may be why, in some cases, older adults may seem like they are resistant to change. It may not be from an unwillingness to consider other solutions, but from the simple fact of human nature that we fall back on our prior experiences of success and try to first solve problems, or deal with situations more generally, using those approaches.

15.6: Aging and Emotion

OBJECTIVE: Contrast the effects of emotional information processing among older adults

The experience of emotional events can lead to highly detailed memories for those events and the context in which that information was learned. These are called *flashbulb memories*, such as remembering where one was when the attacks of September 11, 2001, occurred. How is the creation of flashbulb memories affected by aging, especially given that older adults have trouble with episodic memories compared to younger adults? Although there has not been a great deal of study on this issue, it seems that older adults show flashbulb memory effects similar to those in younger adults (Berntsen & Rubin, 2006). Thus, when emotional information is encountered, the typical age-related memory deficits are not as strong as they would be otherwise.

That said, one last point to note about older adults is that they process emotional information differently than do younger adults. Specifically, older adults show an increase in the preference for cognitively processing positive emotional information. This is called the **positivity effect** and has often been presented in the context of *socioemotional selectivity theory* (e.g., Charles, Mather, & Carstensen, 2003). For this theory, age-related cognition is affected by changes in people's goals as a function of where they are in their lives. Thus for older adults there is a perspective shift because, in some sense, time is "running out." This altered perspective then leads older adults to focus more on emotional information, particularly emotionally positive information, and to avoid negative emotional information (e.g., Carstensen & Mikels, 2005; Hoyle & Sherrill, 2006). In a sense, older adults no longer want to be bothered by spending their time on the negative aspects of life, but want to move toward focusing on the positive aspects.

WRITING PROMPT

Problem Solving in Older Adults

How effective are older adults likely to be at solving problems that confront them in everyday life? What are examples of particular difficulties they may experience, and what are examples in which they would do as well as younger adults?

▶ The response entered here will appear in the performance dashboard and can be viewed by your instructor.

Submit

WRITING PROMPT

Emotional Processing in Older Adults

Given the changes in emotional processing as people grow older, how would you expect this to make your interactions with them different from those with younger adults? What kinds of everyday things do you think that they would find more or less interesting?

▶ The response entered here will appear in the performance dashboard and can be viewed by your instructor.

Submit

A Study on Positivity Effect (Figure 15.2)

This positivity effect is likely not strictly a function of chronological age. For example, a positivity effect is observed in younger adults who have some sort of serious illness (Fung, Lai, & Ng, 2001). Also, younger adults who are especially oriented toward their emotional states can also exhibit a positivity effect (Kennedy, Mather, & Carstensen, 2004; Löckenhoff & Carstensen, 2007; Reed & Carstensen, 2012).

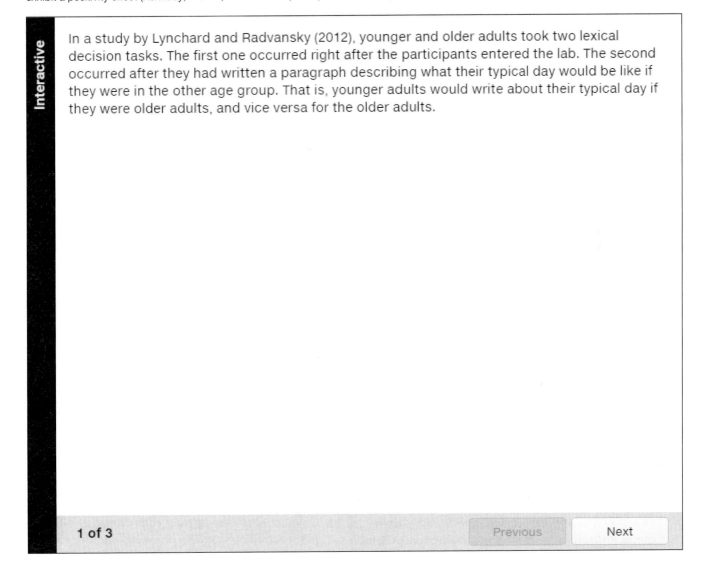

In a study by Lynchard and Radvansky (2012), younger and older adults took two lexical decision tasks. The first one occurred right after the participants entered the lab. The second occurred after they had written a paragraph describing what their typical day would be like if they were in the other age group. That is, younger adults would write about their typical day if they were older adults, and vice versa for the older adults.

Summary: Cognitive Aging

15.1 Neurological and Cognitive Changes in Older Adults

- Cognitive aging can be explored using either cross-sectional or longitudinal studies. Cross-sectional studies, which compare separate groups of younger and older adults, are easier to do, but are prone to cohort effects. Longitudinal studies, which compare the same people with themselves at different ages, are more robust assessments of cognitive aging, but are hard to do.

- Neurologically, older adults suffer from slower neural firings and declines in specific neural structures, such as the frontal lobes, temporal lobes, and the hippocampus. In addition, there appears to be a shift from more lateralized to less lateralized processing. Declines in the nervous system are not inevitable, but can be greatly attenuated in some cases. Being physically, intellectually, and socially active can provide great benefits in old age.

15.2 Perception and Attention in Older Adults

- Although there are some declines in the ability of the sensory organs, such as the eyes and the cochlea, to register and process information as efficiently as younger adults, the perceptual processes of older adults are very similar to those of younger adults.
- Similarly, although there are some issues with attentional control for older adults, particularly with simple perceptual tasks and tasks that involve the use of inhibitory processes, the basic patterns of data observed on attention tasks are similar in younger and older adults. There may also be a slowdown in the shift from controlled to automatic processes with practice. Mind wandering appears to improve with aging.

15.3 Memory in Older Adults

- The declines in short-term/working memory capacity that accompany aging are measurable, but not overly large.
- In terms of episodic memory, there are clear declines in the amount of newly learned information that can be recalled by older adults. However, whatever is recalled is well organized and structured. That said, this does not extend to an effective use of context in episodic memory as is seen with younger adults.
- Prospective memory appears to show clear declines, perhaps as a result of reduced frontal lobe activity. That said, these declines appear to be observed more with laboratory tasks than everyday tasks.
- An area of preservation with aging is semantic memory. Older adults have better developed semantic memories than younger adults, show normal priming effects, and can use their schemata and scripts to help compensate for declines in other areas.
- There are general declines in the metamemory abilities of older adults. This can be seen in age-related declines in source monitoring and some types of prospective memory. These declines can lead to an increase in some metamemory errors, such as increases in cryptomnesia, and the overconfident acceptance of false memories.
- Although some age-related changes in memory are unavoidable facts of the natural aging process, to some degree the memory experiences of older adults are due to the acceptance of unconscious stereotypes they carry around with them, and that are activated in their interactions with the world.

15.4 Language Processing Changes

- There are some problems for older adults in language processing due to sensory, perceptual, and working memory problems, but language processing per se seems to be relatively preserved.
- Older adults seem to perform in an equivalent manner to younger adults in terms of situation models; they update their models in terms of the same factors (e.g., time, space, causality), although they do tend to read more slowly. Difficulties older adults have at the level of situation models appear to be the result of limitations in working memory capacity, not true difficulties in understanding or comprehension at all.

15.5 Reasoning, Decision Making, and Problem Solving

- Aging affects the ability to engage in logical reasoning, per se, only for the most complex problems. However, older adults are more likely to be led astray by their prior beliefs.
- Older adults appear to use heuristics as well as younger adults and may actually use them more often, because older adults are less likely to gather background information before making a decision. Also, it may be the case that older adults use more relational information to make their decisions.
- The problem-solving process in older adults is qualitatively similar to that in younger adults in many ways, although older adults may be less able to plan out their future steps because of reduced working memory capacity. They may also be more likely to stick with a prior way of solving problems, even when a simpler solution path is available.

15.6 Aging and Emotion

- Older adults place a greater emphasis on emotionally positive information than do younger adults. This can even influence their unconscious cognitive processes. That said, this positivity bias can be an influence in anyone, depending on the perspective one takes. So, take a positive outlook on your life.

SHARED WRITING

Cognitive Aging

The population is aging. That is, a larger and larger proportion of the people are older adults. What recommendations can you suggest that will help people be more successful in their aging in terms of how they are able to think and remember?

▶ A minimum number of characters is required to post and earn points. After posting, your response can be viewed by your class and instructor, and you can participate in the class discussion.

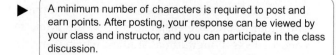

Glossary

Absent-mindedness Everyday memory failures in remembering information and intended activities, probably caused by sufficient attention or superficial, automatic processing during encoding (one of Schacter's seven sins of memory) (Ch. 8).

Accessibility The degree to which information can be retrieved from memory. A memory is said to be accessible if it is retrievable; memories that are not currently retrievable are said to have become inaccessible (contrast with *availability*) (Ch. 8).

Accuracy How correct a person is in the responses given; often quantified in terms of both number correct and number incorrect. Separately, or paired with response time, it can give an indication of cognitive functioning (Ch. 1).

Acetylcholine A neurotransmitter that may be involved in strengthening neural connections during long-term potentiation (Ch. 2).

Action potential The change in electrical charge that occurs when a neuron "fires." Neural firing follows the all-or-none principle, resulting in all action potentials being the same (Ch. 2).

Action slips Unintended, often automatic, actions that are inappropriate for the current situation (Ch. 4).

Activation How "active" the node is, how ready to fire. The mental activity of accessing and retrieving information from the network; if activation exceeds some threshold level, the concept "fires" and the activation spreads to related concepts (Ch. 7).

Actor See *agent*.

Ad hoc categories Categories that people can generate on the fly that have all of the qualities of more traditional categories, but that are based on situational circumstances (Ch. 7).

Advantage of clause recency The speedup of RT to information in the most recently processed clause (Ch. 10).

Advantage of first mention The speedup of RT to information mentioned first in the sentence (Ch. 10).

Affirm the antecedent (a valid inference): *Modus ponens*; evidence P for "if P then Q" allows you to conclude Q (Ch. 11).

Affirm the consequent (an invalid inference): Evidence Q for "if P then Q" does not allow you to conclude P (Ch. 11).

Agent (also actor) In the case grammar approach, the person who performs some action in a sentence is the agent, such as Bill in "Bill hit the ball with the bat" (see also *case grammar*) (Ch. 6).

Agnosia A disruption in the ability to recognize objects (Chs. 3, 9).

Agraphia A disruption in the ability to write, caused by a brain disorder or injury (Ch. 9).

Alertness and arousal The nervous system must be awake, responsive, and able to interact with the environment. Arousal is a necessary precondition for most cognitive processing. The system must be aroused to pay attention (Ch. 4).

Alexia A disruption in the ability to read or recognize printed letters or words, caused by a brain disorder or injury (Ch. 9).

Algorithm A specific rule or solution procedure that is certain to yield the correct answer if followed correctly (contrast with *heuristic*) (Ch. 11).

All-or-none principle The idea that either a neuron fires or it does not, with all action potentials being the same (Ch. 2).

Ambiguous Having more than one meaning, said both of words (e.g., *bank*) and sentences (e.g., "They are eating apples") (Ch. 9).

Amnesia Memory loss caused by brain damage or injury. Retrograde amnesia is loss of memory for information before the damage; anterograde amnesia is loss of memory for information after the damage (Ch. 8).

Amygdala An almond-shaped structure adjacent to one end of the hippocampus, often involved in emotion processing (Chs. 2, 13).

Analogy A relationship between two similar systems, problems, and so on; a heuristic in which a problem is solved by finding an analogy to a similar problem (Chs. 10, 12).

Anaphoric reference The act of using a pronoun or possessive (or synonym) to refer back to a previously mentioned concept (Ch. 10).

Anomia A disruption of word finding or retrieval, caused by a brain disorder or injury (Chs. 2, 7, 9).

Anomic aphasia See *anomia*.

Antecedent (in conditional reasoning) The *if* clause in standard conditional reasoning (*if–then*) tasks. In the statement "If it rains, then the picnic will be canceled," the antecedent is "If it rains" (Ch. 11).

Antecedent (in language) The concept to which a later word refers; for example, *he* refers to the antecedent *Bill* in "Bill said he was tired" (Ch. 10).

Anterograde amnesia Disruption in memory for events following the brain damage, usually a disruption in the storage of new information after brain damage (Ch. 8).

Aphasia A loss of some or all of previously intact language skills, caused by brain disorder or damage (Ch. 9).

Apperceptive agnosia A form of agnosia in which individual features cannot be integrated into a whole percept or pattern; a basic disruption in perceiving patterns (Ch. 3).

Arbitrariness One of Hockett's (1960a; 1960b) linguistic universals; the connections between linguistic units (sounds, words) and the concepts or meanings referred to by those units are entirely arbitrary; for example, it is arbitrary that we refer to a table by the linguistic unit *table* (Ch. 9).

Articulatory loop The part of the phonological loop involved in the active refreshing of information in the phonological store (Ch. 5).

Articulatory suppression effect The finding that people have poorer memory for a set of words if they are asked to say something while trying to remember the words (Ch. 5).

Association A general term referring to a connection or link between two elements. In classic behaviorism, an association was formed between a stimulus and a response (Chs. 2, 7, 8).

Associative agnosia A form of agnosia in which the individual can combine perceived features into a whole pattern but cannot associate the pattern with meaning, cannot link the perceived whole with stored knowledge about its identity (Ch. 3).

Associative interference Retrieval interference in memory that results in poorer performance (people are less accurate and/or slower) when information is associated with a target memory, but irrelevant. For example, it is harder to remember several facts about a person compared to just one (Ch. 8).

Atmosphere heuristic A logical reasoning heuristic in which people prefer to draw conclusions that use the same terms as the premises (e.g., if there is *some* in the premises, people prefer a conclusion that uses *some* (Ch. 11).

Attention The mental energy or resource necessary for completing mental processes, believed to be limited in quantity and under the control of some executive control mechanism (Ch. 4).

Attentional blink A brief slowdown in mental processing due to having processed another very recent event (Ch. 4).

Attention capture The spontaneous redirection of attention to stimuli in the world based on physical characteristics (Ch. 4).

Audition The sense of hearing (Ch. 3).

Auditory sensory memory (also *echoic memory*) The sensory memory system that encodes incoming auditory information and holds it briefly for further mental processing (Ch. 3).

Authorized inference Intended or correct. An implication of a speaker's statement is said to be authorized if the speaker intended the implication to be drawn; if the listener draws the intended inference, the inference is said to be authorized (contrast with *unauthorized inference*) (Ch. 10).

Autobiographical memory Memories of specific, personally experienced real-world information, such as of one's activities upon learning of the *Challenger* space shuttle disaster; the study of those memories (Ch. 6).

Automatic Occurring without conscious awareness or intention and consuming little, if any, of the available mental resources (Ch. 4).

Automaticity See *automatic*.

Availability (in memory research) Present in the memory system. Information is said to be available if it is currently stored in memory (contrast with *accessibility*) (Ch. 8).

Availability heuristic A decision-making heuristic in which we judge the frequency or probability of some event on the basis of how easily examples or instances can be recalled or remembered; thus the basis of this heuristic is ease of retrieval (Ch. 11).

Axon The long, extended portion of a neuron (Ch. 2).

Axon terminals The branchlike ending of the axon in the neuron, containing neurotransmitters (Ch. 2).

Babbling stage Stage of language acquisition, up to 18 months, in which there is a lot of vocal play, but no real words yet (Ch. 14).

Backward masking See *masking*.

Basic number sense Knowing and dealing with how many items are visible in a display (Ch. 14).

Behaviorism The movement or school of psychology in which the organism's observable behavior was the primary topic of interest; and the learning of new stimulus–response associations, whether by classical conditioning or by reinforcement principles, was deemed the most important kind of behavior to study (Ch. 1).

Belief bias The tendency to accept conclusions when reasoning is based on one's beliefs, rather than logical form (Chs. 11, 15).

Beliefs The fifth level of analysis of language, according to Miller, in which the listener's attitudes and beliefs about the speaker influence what is comprehended and remembered (Ch. 10).

Benefit See *facilitation*.

Beta movement Illusory movement that occurs when two or more pictures are viewed in rapid succession, as in a movie (Ch. 3).

Bias The tendency for knowledge, beliefs, and feelings to distort recollection of previous experiences and to affect current and future judgments of memory (one of Schacter's seven sins of memory) (Ch. 8).

Blocking The temporary retrieval failure or loss of access, such as the tip-of-the-tongue effect, in either episodic or semantic memory (one of Schacter's seven sins of memory) (Ch. 8).

BOLD signal Blood oxygenation level–dependent signal that informs the researcher about the relative level of activity in a region of the brain (Ch. 2).

Bottom-up processing See *data-driven processing*.

Boundary extension The finding that people tend to misremember more of a scene than was actually viewed, as if the boundaries of an image were extended farther out (Ch. 5).

Bounded rationality The idea that people can only think about so much information at one time when trying to reason and make decisions (Ch. 11).

Bridging inference Clark's (1977) term for the mental processes of reference, implication, and inference during language comprehension. Metaphorically, a bridge must be drawn from *he* back to *Gary* to comprehend the sentence "Gary pretended he wasn't interested" (Ch. 10).

Broca's aphasia A form of aphasia characterized by severe difficulties in producing spoken speech; that is, the speech is hesitant, effortful, and distorted phonemically (contrast with *Wernicke's aphasia*). The aphasia is caused by damage in Broca's area, a region of the cortex next to a major motor control center (Ch. 9).

Broca's area Site of brain damage associated with Broca's aphasia; located at the rear of the left frontal lobe, a region of the cortex next to a major motor control center (Ch. 9).

Brodmann's areas These are numbered areas of the cortex that were identified by an analysis of physical differences in different parts of the brain. These numbers are useful in locating general areas of the cortex (Ch. 2).

Brown–Peterson task A short-term memory task showing forgetting caused by proactive interference (Ch. 5).

Case grammar An approach in psycholinguistics in which the meaning of a sentence is determined by analyzing the semantic roles or cases played by different words, such as which word names the overall relationship and which names the agent or patient of the action. Other cases include time, location, and manner (Ch. 9).

Case role (also *semantic case*) One of the various semantic roles or functions of different words in a sentence (see also *case grammar*) (Ch. 9).

Categorical perception The perception of similar language sounds as being the same phoneme, despite the minor physical differences among them; for example, the classification of the initial sounds of *cool* and *keep* as both being the /k/ (hard *c*) phoneme, even though these initial sounds differ physically (Ch. 9).

Categorical syllogism A classic reasoning form composed of two premises and one conclusion in which the logical truth of the conclusion must be derived from the premises (Ch. 11).

Category A mental representation of a class of concepts (Ch. 7).

Category-specific deficit A disruption in which a person loses access to one semantic category of words or concepts while not losing others (Ch. 7).

Central executive In Baddeley's working memory model, the mechanism responsible for assessing the attentional needs of the different subsystems and furnishing attentional resources to those subsystems. Any executive or monitoring component of the memory system that is responsible for sequencing activities, keeping track of processes already completed, and diverting attention from one activity to another can be called an executive controller (Ch. 5).

Central tendency The idea that there is some mental core or center to the category where the best members will be found (Ch. 7).

Cerebral cortex See *neocortex*.

Cerebral hemispheres (*left* and *right*) The two major structures in the neocortex. In most people, the left cerebral hemisphere is especially responsible for language and other symbolic processing, and the right for nonverbal, perceptual processing (Ch. 2).

Cerebral lateralization The principle that different functions or actions within the brain tend to be localized in one or the other hemisphere. For instance, motor control of the left side of the body is lateralized in the right hemisphere of the brain (Ch. 2).

Change blindness The failure to notice changes in visual stimuli (e.g., photographs) when those changes occur during a saccade (Ch. 3).

Channel capacity An early analogy for the limited capacity of the human information-processing system (Ch. 1).

Characteristic feature In the Smith et al. (1974) model of semantic memory, characteristic features are the features or properties of a concept that are common but are not essential to the meaning of the concept; for example, "eats worms" may be characteristic of "BIRD," but the feature is not essential to the central meaning of the concept (contrast with *defining feature*) (Ch. 7).

Child directed speech Speech used when talking to children that is often higher in pitch, slower, and exaggerated (Ch. 14).

Choking under pressure Performance below typical performance levels as a result of emotional stress or pressure at the time of the action (Ch. 13).

Chunk A unit or grouping of information held in short-term memory (Ch. 5).

Circularity A major problem for the depth of processing concerns defining levels of rehearsal independently of retention scores (better memory was held as indicative of deeper processing; thus the depth of processing model could not fail to be consistent with the data) (Ch. 6).

Classic view of categorization The view that takes the position that people create and use categories based on a system of rules that define necessary and sufficient features (Ch. 7).

Closure A Gestalt grouping principle that is used by visual perception to close up gaps in a percept to help identify a whole object (Ch. 3).

Clustering The grouping together of related items during recall (e.g., recalling the words *apple, pear, banana, orange* together in a cluster, regardless of their order of presentation) (see also *organization*) (Ch. 6).

Coarticulation The simultaneous or overlapping articulation of two or more of the phonemes in a word (Ch. 9).

Cocktail party effect Selecting one message in a crowded, noisy environment (Ch. 4).

Cognition The collection of mental processes and activities used in perceiving, remembering, thinking, and understanding, and the act of using those processes (Ch. 1).

Cognitive neuroscience (a.k.a. *neurocognition/cognitive neuropsychology*) A hybrid term applied to the analysis of those handicaps in human cognitive functioning that result from brain injury and other neurophysiological effects on cognition (Ch. 2).

Cognitive psychology An objective, empirical discipline that favors an experimental approach. The major assumptions are (1) mental processes exist (vs. behaviorist perspective); (2) mental processes can be studied scientifically (vs. subjectivity of introspective method); (3) people are active information processors (vs. passive recipients of knowledge and experience) (Ch. 1).

Cognitive revolution The movement away from behaviorism to re-embrace the study of memory and mental activity (instinctive drift; language; perception; judgment; decision making) (Ch. 1).

Cognitive science A new term designating the study of cognition from the multiple standpoints of psychology, linguistics, computer science, and the neurosciences (Chs. 1, 2).

Common fate A Gestalt grouping principle that is used by visual perception to group together points that are moving together (Ch. 3).

Competence In linguistics, the internalized knowledge of language and its rules that fully fluent speakers of a language possess, uncontaminated by flaws in performance (contrast with *performance*) (Ch. 9).

Computer analogy Human information processing may be similar to the sequence of steps and operations in a computer program, similar to the flow of information from input to output when a computer processes information (Ch. 1).

Computerized axial tomography See *CT scan*.

Conceptual knowledge The fourth level of analysis of language in Miller's scheme, roughly equivalent to semantic memory (Ch. 10).

Conceptually driven processing (also *top-down processing*) Mental processing is said to be conceptually driven when it is guided and assisted by the knowledge already stored in memory (contrast with *data-driven processing*) (Chs. 1, 3).

Concrete operations stage The third Piagetian stage of development, from ages 7 to 11, in which children's thinking becomes more reasoned and logical, although there is still an emphasis on concrete objects (Ch. 14).

Conditional reasoning The form of reasoning in which the logical consequences of an *if–then* statement and some evidence are determined; for example, given "If it rains, then the picnic will be canceled," the phrase "It is raining" determines whether the picnic is canceled (Ch. 11).

Conduction aphasia A disruption of language in which the person is unable to repeat what has just been heard (Ch. 9).

Confirmation bias In reasoning, the tendency to search for evidence that confirms a conclusion (Ch. 11).

Conjugate reinforcement A methodology for testing memory in infants. The infant's leg is attached, via a ribbon, to a mobile, and when the infant kicks, the mobile moves. Various aspects of the environment can be manipulated to assess whether memory is retained for aspects of the testing environment by measuring the amount of kicking later (Ch. 14).

Conjunction fallacy The mistaken belief that a compound outcome of two characteristics is more likely than either one of the characteristics by itself (Ch. 11).

Connectionism See *connectionist*.

Connectionist (also *connectionism, neural net modeling, PDP modeling*) The terms refer to a recent development in cognitive theory, based on the notions that the several levels of knowledge necessary for performance can be represented as massive, interconnected networks; that performance consists of a high level of parallel processing among the several levels of knowledge; and that the basic building block of these interconnected networks is the simple connection between nodes stored in memory. For instance, perception of spoken speech involves several levels of knowledge, including knowledge of phonology, lexical information, syntax, and semantics. Processing at each level continually interacts with and influences processing at the other levels, in parallel. The connections in connectionist modeling are the network pathways both within and among the levels of knowledge (Chs. 2, 3, 7).

Connectionist models A computer-based technique for modeling complex systems. Knowledge is represented by the strength of the excitatory or inhibitory connections between massively interconnected nodes (Chs. 2, 7).

Conscious attention Awareness; a slower attentional mechanism especially influenced by top-down processing (Ch. 4).

Conscious processing Mental processing that is intentional, involves conscious awareness, and consumes mental resources (contrast with *automatic, automaticity*) (Ch. 4).

Consequent In conditional reasoning, the consequent is the *then* statement; in "If it rains, then the picnic will be canceled," the consequent is "then the picnic will be canceled" (Ch. 11).

Conservation A Piagetian milestone in cognitive development in which children learn that things can change their appearance but retain underlying qualities, such as quantity (Ch. 14).

Consolidation The more permanent establishment of memories in the neural architecture (Chs. 2, 6, 8).

Constructive search A problem-solving strategy of drawing problem-solving inferences based on what is known from other aspects of a problem (Ch.12).

Context The surrounding situation and its effect on cognition, including the concepts and ideas activated during comprehension (Chs. 3, 6, 9).

Contralaterality The principle that control of one side of the body is localized in the opposite-side cerebral hemisphere. The fact that the left hand, for instance, is largely under the control of the right cerebral hemisphere illustrates the principle of contralaterality (Ch. 2).

Controlled attention The deliberate, voluntary allocation of mental effort or concentration (Ch. 4).

Control processes The part of the standard or modal (Atkinson and Shiffrin, 1968, 1971) model of memory responsible for the active manipulation of information in short-term memory (Ch. 1).

Conversational rules The rules, largely tacit, that govern our participation in and contributions to conversations (Ch. 10).

Cooperative principle The most basic conversational postulate, stating that participants cooperate by sharing information in an honest, sincere, and appropriate fashion (Ch. 10).

Corpus callosum The fiber of neurons that connects the left and right cerebral hemispheres (Ch. 2).

Correlated attributes Features that tend to co-occur in various members of a category (Ch. 7).

Cost A response slower than baseline because of a misleading cue (Chs. 4, 7, 11).

Counterfactual reasoning (also *undoing*) A line of reasoning that deliberately contradicts the facts in a "what if" kind of way; in the simulation heuristic, the changing of details or events in a story to alter the (unfortunate or undesirable) outcome (Ch. 11).

Critical lure A word that is highly related to all the other words in a study list but that never was presented to the participant (but which the participant is likely to later report as part of the list) (Chs. 8, 15).

Cross-sectional study A study that compares different groups of people from different age groups. These studies are more prone to cohort effects due to different experiences of different age groups, rather than to developmental changes per se (Chs. 14, 15).

Cryptomnesia When a person unconsciously plagiarizes something heard or read, and because he or she has forgotten the source mistakenly thinks that it is an original idea (Chs. 6, 15).

CT scan Computed axial tomography; X-rays (of the head) that can be used to provide detailed 3-D information about the physical structure of the brain (Ch. 2).

Cued recall A form of recall in which the person is presented with part of the information as a cue to retrieve the rest of the information (Ch. 8).

Data-driven processing (also *bottom-up processing*) When mental processing of a stimulus is guided largely or exclusively by the features and elements in the pattern itself, this processing is described as being data-driven (contrast with *conceptually driven processing*) (Ch. 3).

Decay Simple loss of information across time, presumably caused by a fading process, especially in sensory memory; also, an older theory of forgetting from long-term memory (Chs. 3, 5, 8).

Declarative memory Long-term memory knowledge that can be retrieved and then reflected on consciously (see also *explicit memory*) (Ch. 6).

Deep structure In linguistics and psycholinguistics, the deep structure of a sentence is the meaning of the sentence; a deep structure is presumably the most basic and abstract level of representation of a sentence or idea (contrast with *surface structure*) (Ch. 9).

Default mode network (DMN) A collection of brain structures that tend to be more active when a person is at rest and not thinking about anything in particular. Hence, it is the part of the brain that is more active by default (Ch. 4).

Default value The common or ordinary value of some variable. In script theory, default value refers to an aspect of a story or scene that conforms to the typical or ordinary state of affairs; for instance, "MENU" is the default value that fills the slot in a script in which customers find out what can be ordered in a restaurant (Ch. 7).

Defining feature In Smith et al.'s (1974) theory of semantic memory, defining features are properties or features of a concept that are essential to the meaning of that concept; for instance, bearing live young is a defining feature of the concept "MAMMAL" (contrast with *characteristic feature*) (Ch. 7).

Dendrites The branching, input structures of the neuron (Ch. 2).

Deny the antecedent (an invalid inference) Evidence not *P* for "if *P* then *Q*" does not allow you to conclude not *Q* (Ch. 11).

Deny the consequent (a valid inference): *Modus tollens*; evidence not *Q* for "if *P* then *Q*" allows you to conclude not *P* (Ch. 11).

Depth of processing See *levels of processing*.

Direct stimulation The direct application of electrical current to the surface of the cortex. Used by Penfield and others to map the cortex, such as the sensory and motor homunculi (Ch. 2).

Direct theory In conversation, a direct theory is a person's appraisal of or informal theory about the other participant in the conversation, including information about that other person's knowledge, sophistication, and personal motives (contrast with *second-order theory*) (Ch. 10).

Displacement One of Hockett's (1960, 1960b) linguistic universals, referring to the fact that language permits us to talk about times other than the immediate present; language thus permits us to displace ourselves in time, by talking about the past, future, and so on (Ch. 9).

Dissociation Pattern of abilities and performance, especially among brain-damaged patients, revealing that one cognitive process can be disrupted while another remains intact. In a double dissociation, two patients show opposite patterns of disruption and preserved function, further evidence that the cognitive processes are functionally and anatomically separate (Chs. 2, 8).

Distributed practice This occurs when practice sessions are spaced out in time. This is a more effective way of encoding information (Ch. 6).

DMN See *default mode network*.

Domain knowledge A general term referring to one's knowledge of a specific domain or topic, especially in problem solving (Ch. 12).

Dorsal pathway The neural pathway across the top of the cortex, stemming from visual processing areas in the occipital lobe, primarily responsible for processing information about where things are in the world (Ch. 2).

Double dissociation See *doubly dissociated*.

Doubly dissociated Two mental processes are said to be doubly dissociated when a deficit in one of them, say due to brain damage, does not necessarily produce a deficit in the other process, and vice versa; for instance, a deficit in language comprehension due to brain damage (in Wernicke's area) does not necessarily produce a deficit in language production (in Broca's area), and vice versa (Chs. 2, 8, 9).

Downhill change In the simulation heuristic, an unusual or unexpected aspect of a story or situation that is changed to be more normal or customary. If a story character left work early and was involved in a car accident, a likely downhill change would be to normalize the unusual characteristic and substitute a more customary aspect, such as leaving work on time (Ch. 11).

Dual coding hypothesis According to Paivio (1971), concrete words can be encoded into memory twice, once as verbal symbols and once as image-based symbols, thus increasing the likelihood that they will be recalled or remembered (Chs. 6, 7).

Dual task method A method in which two tasks are performed simultaneously, such that the attentional and processing demands of one or both tasks can be assessed and varied. Dual task methodology is commonly used in studies of attention and attention-dependent mental processing (Chs. 4, 5).

Dysfluency Error, flaw, or irregularity in spoken speech (Ch. 9).

Early selection Selection or filtering based on early phases of perception (e.g., selection based on physical features of the message such as loudness or location; Broadbent's filter theory) (Ch. 4).

Easterbrook hypothesis Easterbrook's idea that at high levels of emotional arousal, there is a narrowing of attention that leads to better memory for details at the focus of an event, but poorer memory for details at the periphery of an event (Ch. 13).

Echoic memory See *auditory sensory memory*.

Ecological validity The hotly debated principle that research must resemble the situations and task demands that are characteristic of the real world rather than rely on artificial laboratory settings and tasks so that results will generalize to the real world; that is, will have ecological validity (Ch. 1).

EEG (electroencephalogram) Electrodes on a person's scalp pick up brain waves (Ch. 2).

Einstellung Another term for *negative set* (Ch. 12).

Elaborative rehearsal In the levels of processing framework, rehearsal that involves any rehearsal activity that processes a stimulus into the deeper, more meaningful levels of memory; any rehearsal that involves meaning, images, and other complex information from long-term memory (contrast with *maintenance rehearsal*) (Ch. 6).

Elimination by aspects A decision-making heuristic in which a person makes a choice by eliminating options that do not meet certain criteria (Ch 11).

Embodiment The way we think about and process information reflects the fact that we need to interact with the world using our bodies (Ch. 1).

Emergent properties The properties that emerge from collections of elements (e.g., neurons) working together to create a new process of property that the individual elements lack (e.g., reasoning) (Ch. 2).

Emotion Both the state of mind a person is in at a particular moment, as well as the physiological response a person is experiencing at that time (Ch. 13).

Emotional Stroop task A variant of the normal Stroop task in which potentially emotionally arousing words are used (in which people must name the colors they are printed in) to assess the degree to which processing is disrupted by this emotionally arousing content (Ch. 13).

Empiricism The philosophical position, originally from Aristotle, that advances observation and observation-derived data as the basis for all science (Ch. 1).

Enactive representation According to Bruner, an early, embodied, action-based stage of childhood thinking (Ch. 14).

Enactment effect The finding of improved memory for participant-performed tasks, relative to those that are not acted out (Ch. 6).

Encoding To input or take into memory, to convert to a usable mental form, to store into memory. We are said to encode auditory information into sensory memory; if that information is transferred to short-term memory, then it is said to have been encoded into STM (Ch. 1).

Encoding specificity Tulving's hypothesis that the specific nature of an item's encoding, including the entire context it was encoded in, determines how effectively the item can be retrieved (Chs. 6, 15).

Engle's controlled attention model A theory of working memory that assumes that working memory is essentially knowledge that is currently being thought about, along with the idea that working memory effectiveness is tied to the ability to control what information is and is not attended to by a person (Ch. 5).

Enhancement In Gernsbacher's theory, the boosting of concepts' levels of activation during comprehension (Ch. 10).

Episodic buffer The portion of working memory whereby information from different modalities and sources are bound together to form new episodic memories (Chs. 5, 6).

Episodic memory Tulving's term for the portion of long-term memory in which personally experienced information is stored (contrast with *semantic memory*) (Chs. 2, 6).

Erasure The masking or loss of information caused by subsequent presentation of another stimulus; usually in sensory memory (see also *masking*) (Ch. 3).

Event-related potentials (ERPs) Minute changes in electrical potentials in the brain, measured by EEG recording devices and related specifically to the presentation of a particular stimulus; the research technique used for determining neural correlates of cognitive activity (Chs. 2, 4, 7).

Executive control See *central executive*.

Exemplar theory A theoretical view of categorization that assumes that when people think about categories, they are mentally taking into account each experience, instance, or example of the various encounters that they have experienced with members of that category (Ch. 7).

Explanation-based theories Theoretical views of semantic categories that assume that people create mental categories as theories of the world to explain why things are the way they are (Ch. 7).

Explicit memory Long-term memory retrieval or performance that entails deliberate recollection or awareness (Ch. 6).

Explicit processing Involving conscious processing, conscious awareness that a task is being performed, and usually conscious awareness of the outcome of that performance (Ch. 4).

Eye-mind assumption The assumption that the eye normally remains fixated on a word as long as that word is being actively processed during reading (Ch. 10).

Eye tracker A device used to record eye movements and fixations (Ch. 10).

Facilitation (also *benefit*) Any positive or advantageous effect on processing, usually because of prior presentation of related information; in RT research, a speedup of RT due to related information (Chs. 4, 7, 11).

False alarm (also *false positive*) An error in a recognition task in which a response of "yes" is made to a new stimulus; any "yes" response in recognition when a "no" response is correct (Ch. 8).

False memory Memory of something that did not happen (Ch. 8).

False positive See *false alarm*.

Familiarity bias   In reasoning, the bias in the availability heuristic in which personal familiarity influences estimates of frequency, probability, and so on; judging events as more frequent or important just because they are more familiar in memory (Ch. 11).

Family resemblance   The idea that there is some set of features that is shared by many or most of a category's members, although all features may not be present in all members (Ch. 7).

Fan effect   An increase in response time for an increased number of associations with a concept on a study list (Chs. 8, 15).

"Fast and frugal" heuristics   A collection of heuristics in which people are using basic rules of thumb to quickly make decisions (Ch. 11).

Feature analysis   See *feature detection*.

Feature detection (also *feature analysis*)   A theoretical approach, most commonly in pattern recognition, in which stimuli (patterns) are identified by breaking them up into their constituent features (Ch. 3).

Feature list   See *semantic features*.

Feeling of knowing   An estimate of how likely it is that an item will be recognized on a later memory test (Ch. 6).

Figure-ground   A Gestalt grouping principle that is used by visual perception to segregate what part of the percept corresponds to an object, and what corresponds to the background behind it (Ch. 3).

Filtering (also *selecting*)   Especially in auditory perception, unwanted, unattended messages are filtered or screened out so that only the attended message is encoded into the central processing mechanism (e.g., Broadbent's filter theory) (Ch. 4).

Fixation   In visual perception, the pause during which the eye is almost stationary and is taking in visual information; also, the visual point on which the eyes focus during the fixation pause (see also *gaze duration*) (Chs. 3, 10).

Flashbulb memories   Memories of specific, emotionally salient events, reported subjectively to be as detailed and accurate as a photograph but now considered possibly to be more similar to normal, highly accurate memories (Chs. 13, 15).

Flexibility   The characteristic that enables the meaning of a language symbol to be changed and enables new symbols to be added to the language (Ch. 9).

Focal attention   Neisser's (1967) term for mental attention directed toward, for example, the contents of visual sensory memory and therefore responsible for transferring that information into short-term memory (Ch. 3).

Forgetting   Colloquially, losing information previously stored in memory. More technically, the term usually implies that the stored information is no longer in memory, that it is no longer available in the memory system (Chs. 6, 8).

Formal operations stage   The fourth Piagetian stage of development, beginning around age 11, in which children's thinking becomes able to handle and process abstract and hypothetical ideas (Ch. 14).

Fovea   The highly sensitive region of the retina responsible for precise, focused vision, composed largely of cones (Ch. 3).

Frame   In script theory, a slot or event in a stored script. In the restaurant script, for instance, there are frames for "How the customer gets the food" and "Who prepares the food" (Ch. 7).

Free recall   A recall task in which subjects may recall the list items in any order, regardless of their order of presentation (contrast with *serial recall*) (Chs. 5, 6).

Frontal lobe   Most forward part of cortex, and important for the control of thought and action (Ch. 2).

Functional fixedness   In problem solving, an inability to think of or consider any but the customary uses for objects and tools (Ch. 12).

Functionalism   The movement in psychology, closely associated with James, in which the functions of various mental and physical capacities were studied (contrast with *structuralism*) (Ch. 1).

Functional MRI (fMRI)   A use of MRI technology that provides online evidence about dynamic (functional) processes in the brain (Ch. 2).

GABA   An inhibitory neurotransmitter involved in weakening connections between neurons during learning (Ch. 2).

Garden path sentence   A sentence in which an early word or phrase tends to be misinterpreted and thus must be reinterpreted after the mistake is noticed; for example, "After the musician had played the piano was quickly taken off the stage" (Ch. 9).

Gaze duration   How long the eyes fixate on a specific word during reading, the principal measure of online comprehension during reading (Ch. 10).

General problem solver (GPS)   The first serious computer-based model of problem solving, by Newell, Shaw, and Simon (1958) (Ch. 12).

Generate and test   A problem-solving strategy of deriving possible solutions to a problem or subgoal and testing each one's effectiveness (Ch. 12).

Generation effect   The finding that information you generate or create yourself is better remembered compared to information you have only heard or read (Ch. 6).

Generativity   See *productivity*.

Geons   In Biederman's recognition by components (RBC) model, the basic primitives, the simple three-dimensional geometric forms in the human recognition system (Ch. 3).

Gestalt   A German term adopted into psychological terminology referring to an entire pattern, form, or configuration. The term always carries the connotation that decomposing a pattern into its components in some way loses the essential wholeness of the cohesive pattern (Chs. 3, 12).

Gesture   The movement of the hands and arms done to facilitate communication to listeners. This excludes sign language and noncommunicative mannerisms, such as touching one's hands to one's face (Ch. 10).

Given-new strategy   The idea that words and phrases that contain more accessible information (more active in memory), or given information, tend to occur earlier in sentences, whereas new information in a sentence tends to come later (Ch. 9).

Global–local distinction   When processing visual information, such as a scene or objects, one can focus attention on either the larger whole that the parts make up, or the smaller elements that make up the whole. In studies of global versus local processing, a larger letter (such as an *H*) may be composed of several smaller letters (such as *R*s) (Ch. 4).

Glutamate An excitatory neurotransmitter involved in strengthening connections between neurons during learning (Ch. 2).

Goal In problem solving, the end point or solution to the problem; the ending state toward which the problem-solving attempt is directed (Ch. 12).

Good continuation A Gestalt grouping principle that is used by visual perception to structure and organize together elements based on edges and trajectories so that a visually simple solution is obtained (Ch. 3).

Grammar In linguistics and psycholinguistics, a set of rules for forming the words or sentences in a language; optimally, the complete set of rules that characterizes a language, such that the rules generate only acceptable or legal sentences and do not generate any sentences that are unacceptable (Ch. 9).

Habituation A gradual reduction of the orienting response back to baseline (Ch. 4).

Hemineglect A disorder of attention in which half of the perceptual world, often the left, is neglected to some degree or cannot be attended to (Ch. 4).

Hemispheric specialization The principle that each cerebral hemisphere has specialized functions and abilities (Ch. 2).

HERA model Hemispheric Encoding/Retrieval Asymmetry model is an attempt by Tulving to explain patterns of neural activity data collected during episodic encoding and retrieval along with semantic retrieval. In general, the left hemisphere is more involved in episodic encoding and semantic retrieval, whereas the right hemisphere is more involved in episodic retrieval (Ch. 8).

Heuristic An informal "rule of thumb" method for solving problems, not necessarily guaranteed to solve the problem correctly but usually much faster or more tractable than the correct algorithm (Ch. 11).

Heuristic theory A theory of human mental reasoning in which people reach conclusions not based on logical steps, but based on convenient shortcuts and rules of thumb (Ch. 11).

Hidden units The level between the input and output units in a simple three-level connectionist model (Ch. 3).

Hindsight bias In reasoning, the bias or attitude that some completed event was very likely to have had just that outcome; the likelihood of misremembering a past state of mind as being more consistent with a current state of mind (Chs. 11, 15).

Hippocampus An internal brain structure, just internal to the temporal lobes, strongly implicated in the storing of new information into long-term memory (Chs. 2, 6).

Icon The contents of iconic (visual sensory) memory; the brief-duration visual image or record of a visual stimulus held in visual sensory memory (Ch. 3).

Iconic memory See *visual sensory memory*.

Iconic representation According to Bruner, a middle, perceptual, image-based stage of childhood thinking (Ch. 14).

Ill-defined problem A problem in which the initial, intermediate, or final goal state is poorly or vaguely defined or a problem in which the legal operators (moves) are not well specified (Ch. 12).

Illicit conversion A logical reasoning heuristic in which people inappropriately swap the terms in a premise, such as mentally converting "Some *B* are *A*" to "Some *A* are *C*" (Ch. 11).

Imagination inflation An increase in false memory for an event when the event has been imagined to have happened (Ch. 8).

Immediacy assumption The assumption that readers try to interpret each content word of a text as that word is encountered during reading (Ch. 10).

Implanted memories False memories that are placed in a person from some external source (Ch. 8).

Implication An unstated connection or conclusion that was nonetheless intended by a speaker (Ch. 10).

Implicit memory Long-term memory performance affected by prior experience with no necessary awareness of the influence (Chs. 6, 8).

Implicit processing Processing in which there is no necessary involvement of conscious awareness (Ch. 4).

Inattention blindness We sometimes fail to see an object we are looking at directly, even a highly visible one, because our attention is directed elsewhere (Ch. 3).

Incubation The time when a person stops actively and consciously thinking about a problem, allowing a solution to present itself (Ch. 12).

Independent and nonoverlapping stages The assumption in the strict information processing approach that the stages of processing are independent of one another in their functioning, and that they do not overlap in time. In other words, a stage begins its operations only when a previous stage has finished, and those operations are not changed by previous or subsequent stages (Ch. 1).

Indirect request A question or statement that is not intended to be taken literally but instead is a polite way of expressing the intended meaning; for example, "Do you have the time?" is an indirect way of asking "What time is it?" (Ch. 10).

Infantile amnesia The inability to remember early life events and very poor memory for your life at a very young age (Chs. 6, 14).

Inference Drawing a conclusion based on some statement, as in conversation or reading (Ch. 10).

Inhibition An active suppression of mental representations of salient but irrelevant information so that the activation level is reduced, perhaps below the resting baseline level (Chs. 4, 7).

Inhibition of return A process in which recently checked locations are mentally marked by attention as places that the search process would not return to (Ch. 4).

Input attention The basic processes of getting sensory information into the cognitive system (Ch. 4).

Input units The level analogous to the receptors in a simple three-level connectionist model (Ch. 3).

Insensitivity to sample size The particular bias in the representativeness heuristic. When people reason about events, they fail to take into account the size of the sample or group on which the event is based. They seem to believe that both small and large samples should be equally similar to the population from which they were drawn (the error is also referred to as the *law of small numbers*) (Ch. 11).

Insight Said to be an essential step in creativity and problem solving, though little, if any, research supports this notion empirically (Ch. 12).

Integration When memories from different experiences are combined they are integrated into a common memory trace. After this occurs, it is often difficult for the person to identify individual experiences (Ch. 8).

Intensity The relative strength of an emotional experience (Ch. 13).

Interference An explanation for "forgetting" of some target information in which related or recent information competes with or causes the loss of the target information (Chs. 3, 5, 8).

Introspection The largely abandoned method of investigation in which subjects look inward and describe their mental processes and thoughts; historically, the method of investigation promoted by Titchener (Ch. 1).

Involuntary memory Consists of autobiographical memories that come unbidden, often in response to some environmental cue, such as an odor (Ch. 6).

Isolation effect See *von Restorff effect*.

Judgments of learning (JOLs) A person makes a prediction, after studying some material, whether that information will be remembered on a later memory test (was it learned?) (Chs. 6, 10).

Just noticeable difference (JND) In psychophysics, the amount by which two stimuli must differ so that the difference can be perceived (Ch. 3).

Labor-in-vain effect An effect that occurs when people spend large amounts of time trying to learn information that is too far beyond their current level of knowledge, but end up with little to no new learning (Chs. 6, 10).

Lag In studies of mental processing, the number of intervening trials between a prime and a target (Ch. 7).

Language A shared symbolic system for communication (Ch. 9).

Late selection Selection or filtering based on the meaning and importance of information (e.g., selection based on semantic relevance) (Ch. 4).

Law of small numbers Mistaken belief that small samples will be representative of the population from which they are selected. (This is a mistake that people make. . . . it is not a real law) (Ch. 11).

Lesion Any damage to brain tissue, regardless of cause (e.g., from an accident, stroke, or surgery) (Ch. 2).

Levels of processing (also *depth of processing*) Craik and Lockhart's (1972) alternative to the standard three-component memory model. Information subjected only to maintenance rehearsal is not being processed more deeply into the meaning-based levels of the memory system and therefore tends not to be recalled or recognized as accurately as information subjected to elaborative rehearsal (Chs. 6, 8).

Lexical decision task A simple yes/no task in which subjects are timed as they decide whether the letter string being presented is a word; sometimes called simply the *word/nonword task* (Chs. 1, 7).

Lexical memory The mental lexicon or dictionary where our word knowledge (as distinct from conceptual knowledge) is stored (Ch. 7).

Lexicon See *mental lexicon*.

Lifespan perspective Taking the view that there is no such thing as a person that is not in some state of development. The entire lifespan is considered (Ch. 14).

Linguistic competence The level of language processing that a child can understand and identify (Ch. 14).

Linguistic performance The level of language processing that a child can demonstrate and use (Ch. 14).

Linguistic relativity hypothesis The hypothesis, credited to Whorf (1956), that one's language determines—or at least influences strongly—what one can think about (Ch. 9).

Linguistics The discipline that studies language as a formal system (Ch. 9).

Linguistic universals Features and characteristics that are universally true of all human languages (see also *displacement, productivity*) (Ch. 9).

Lobes (of the brain) Frontal, parietal, occipital, and temporal (Ch. 2).

Location The semantic case or argument in a proposition specifying the place or location of some event (Ch. 6).

Longitudinal study A study that looks at changes in cognition over time within a single group of people instead of comparing people from different age groups as in a cross-sectional study. This type of study provides a more sensitive and more accurate assessment of cognitive changes (Chs. 14, 15).

Long-term memory (LTM) The portion of the memory system responsible for holding information for more than a period of seconds or minutes; virtually permanent storage of information (Ch. 1).

Long-term potentiation (LTP) The temporary (days, weeks, or months) strengthening of connections between neurons as a temporary storage of memories prior to consolidation (Ch. 2).

LTM See *long-term memory*.

Magnetic resonance imaging (MRI) A medical scanning technology that reveals anatomical structure, especially of the brain (see also *functional MRI*) (Ch. 2).

Maintenance rehearsal In the levels of processing framework, rehearsal that merely repeats, recycles, or refreshes information at a particular level via repetition, without processing it to deeper, more meaningful levels of storage (contrast with *elaborative rehearsal*) (Ch. 6).

Mapping In Gernsbacher's theory, drawing the connections between words and their meanings to the overall meaning of the sentence; in general, the process of determining the connections between two sets of elements, including the relations in analogical problem solving (Ch. 10).

Masking An effect, often in perception experiments, in which a mask or pattern is presented very shortly after a stimulus and disrupts or even prevents the perception of the earlier stimulus (see also *erasure*) (Chs. 3, 7).

Massed practice   When information is studied as a part of one large session. This is a less effective means of memory encoding (Ch. 6).

Means–end analysis   A major heuristic in problem solving, assessing the distance between the current state and the goal or subgoal state, then applying some operator that reduces that distance (Ch. 12).

Memory   The mental processes of acquiring and retaining information for later retrieval; the mental storage system that enables these processes (Ch. 1).

Memory impairment   A specific interpretation of early eyewitness memory results in which a subsequent piece of information replaces a memory formed earlier, thus impairing memory of the original information (Ch. 8).

Mental lexicon   The mental dictionary of long-term memory; that is, the portion of long-term memory in which words and word meanings are stored (Chs. 7, 9).

Mental model   The mental representation of a situation or physical device; for example, a person's mental model of the physical motion of bodies or a person's mental model of a thermostat (Ch. 11).

Mental model theory   A theory of human mental reasoning in which people reach conclusions by creating mental simulations of the circumstances described in the premises (Ch. 11).

Mental rotation   Mental manipulation of a visual short-term memory code that reorients the imaged object in space (Ch. 5).

Mental rules theory   A theory of human mental reasoning in which people reach conclusions using rules or procedures about how to reach conclusions (Ch. 11).

Metacognition   Awareness and monitoring of one's own cognitive state or condition; knowledge about one's own cognitive processes and memory system (Chs. 1, 6).

Metacomprehension   The ability to monitor how well we are understanding and will remember information later (Ch. 10).

Metamemory   Knowledge about one's own memory system and its functioning (Ch. 6).

Method of loci   A classic mnemonic device in which the to-be-remembered items are mentally placed, one by one, into a set of prememorized locations, with retrieval consisting of a mental walk through the locations (Ch. 6).

Mind wandering   The situation in which a person's attention and thoughts wander from the current task to some other, inappropriate line of thought (Chs. 4, 15).

Mirror neurons   Neurons in the cortex specialized for planning and executing one's own movement, as well as simulating the movement of others that are being observed (Ch. 2).

Misattribution   Remembering a fact correctly from past experience but attributing it to an incorrect source or context (one of Schacter's seven sins of memory) (Ch 8).

Misinformation acceptance   The tendency to accept information presented after some critical event as being true of the original event itself; for example, accepting then reporting that a yield sign had appeared in an earlier description of a traffic accident (Ch. 8).

Misinformation effect   Incorrectly claiming to remember information that was not part of some original experience (Ch. 8).

Misleading information   Incorrect information about an event that is encountered after the event. This may then distort a person's memory for what actually happened in the original event (Ch. 8).

Mnemonic device   Any mental device or strategy that provides a useful rehearsal strategy for storing and remembering difficult material; see *method of loci*, for instance (Ch. 6).

Modality effect   In sensory memory research, the advantage in recall of the last few items in a list when those items have been presented orally rather than visually (Ch. 3).

Modal model of memory   The standard model of memory derived by Atkinson and Shiffrin (1968, 1971), which is made of three primary components: the sensory registers, the short-term store, and the long-term store (Ch. 1).

Monitoring pressure   Performance pressure occurring because of an emphasis on how a task is being done (Ch. 13).

Mood-congruent memories   The finding that it is easier to think of concepts, topics, ideas, or memories that are consistent with one's current mood (Ch. 13).

Mood-dependent memories   Similar to the idea of encoding specificity, this is the finding that it is easier to remember something when the mood at retrieval and encoding are the same or similar than if they are different (Ch. 13).

Morpheme   The smallest unit of meaning in language (Ch. 9).

Motherese   An older term for *child directed speech* (Ch. 14).

Motor cortex   The band of cortex at the back of the frontal lobe responsible for processing information about voluntary muscle movements throughout the body (Ch. 2).

Motor theory of speech perception   The idea that people perceive language, at least in part, by comparing the sounds that they are hearing with how they themselves would move their own vocal apparatus to make those sounds (Ch. 9).

MRI   See *magnetic resonance imaging*.

Multiconstraint theory   Holyoak and Thargard's (1997) multiconstraint theory of analogical problem solving claims that we work under three constraints as we develop analogies: constraints related to the similarity of the source and target domains (superficial similarities can be misleading), the structure of the problems (must be parallel mappings), and our purposes or goals in developing the analogies (Ch. 12).

Multiword stage   Stage of language development, from 18 to 24 months, in which multiple words are used but syntax is still somewhat simplified (Ch. 14).

Myelin sheath   The fatty coating on a neuron's axon that can facilitate neural communication (Ch. 2).

N400   ERP function associated with semantic anomaly (Ch. 9).

Naive physics   The study of people's misconceptions about the motion of physical objects, such as a ball rolling off a cliff (Ch. 11).

Naming The characteristic that human languages have names or labels for all the objects and concepts encountered by the speakers of the language (e.g., as opposed to most animal communication systems) (Ch. 9).

Negative priming Slower to respond to the target trials when they were preceded by irrelevant distractor primes compared to control trials where the ignored object on the prime trial was an unrelated item (Ch. 4).

Negative set In problem solving, a tendency to become accustomed to a single approach or way of thinking about a problem, making it difficult to recognize or generate alternative approaches (Chs. 12, 15).

Neocortex (also *cerebral cortex*) The top layer of the brain, newest (*neo-*) in terms of the evolution of the species, divided into left and right hemispheres; the locus of most higher-level mental processes (Ch. 2).

Network A structure for information stored in long-term semantic memory, assumed by several popular models of mental processing. In most network models, concepts are represented as nodes that are interconnected by means of links or pathways; activation is presumed to spread from concept to concept along these connecting pathways (Ch. 7).

Neural net modeling See *connectionist*.

Neurogenesis The creation of new neurons. This may be involved in memory formation even into adulthood (Ch. 2).

Neuron A specialized cell that conducts neural information through the nervous system; the basic building block of the nervous system (Ch. 2).

Neurotransmitter The chemical substance released into the synapse between two neurons, responsible for activating or inhibiting the next neuron in sequence (Ch. 2).

Node Especially in network models, a point or location in the long-term memory representation of knowledge; a concept or its representation in memory (Ch. 7).

Nodes of Ranvier Gaps along the myelin sheath that allow the action potential to jump from one point to another, thereby speeding neural communication (Ch. 2).

Nondeclarative memory See *implicit memory*.

Norepinephrine A neurotransmitter that is involved in the creation of new memories (Ch. 2).

Object See *patient*.

Object permanence The ability to remember the presence of an object when it is removed from view (Ch. 14).

Occipital lobe The lobe at the back of the brain that is most heavily involved in vision (Ch. 2).

Online comprehension task Task in which measurements of performance are obtained as comprehension takes place; *online* means happening and being measured right now (Ch. 10).

Onomatopoeia When a name is based on its referent sound (*buzz, hum, zoom*, etc.). Onomatopoeia is an exception to *arbitrariness* (Ch. 9).

Operator In problem solving, a legal move or operation that can occur during solution of a problem; the set of legal moves within some problem space (e.g., in algebra, one operator is "multiply both sides of the equation by the same number") (Ch. 12).

Organization Especially in studies of episodic long-term memory, the tendency to recall related words together, or the tendency to impose some form of grouping or clustering on information being stored in/retrieved from memory; related to chunking or grouping in short-term memory (Ch. 6).

Orienting reflex The reflexive redirection of attention that orients you toward the unexpected stimulus (Chs. 4, 13).

Outcome-based pressure Performance pressure occurring because of an emphasis on the final outcome of performance (Ch. 13).

Output units The level at which the input pattern has been categorized in a simple three-level connectionist model (Ch. 3).

Overextension Overextending a category to include more members than would be found in an adult version of the category, such as calling all men "daddy" (Ch. 14).

Overlearning The improved memory that results when a person continues to study material after it has already been memorized (Ch. 6).

Overregularizing The use of regular word forms, such as *goed*, for words that grammatically should have irregular forms, such as *went* (Ch. 14).

P600 ERP function associated with syntactic anomaly (Ch. 9).

Paired-associate learning A task in which pairs of items, respectively the stimulus and response terms, are to be learned, so that upon presentation of a stimulus, the response term can be recalled; a favorite learning task during the verbal learning period of human experimental psychology (Chs. 6, 8).

Pandemonium Selfridge's early model of letter identification (Ch. 3).

Parallel distributed processing (PDP) modeling See *connectionist*.

Parallel processing Any mental processing in which two or more processes or operations occur simultaneously (Ch. 1).

Parietal lobe Portion of the cortex on the top, behind the frontal lobe and in front of the occipital lobe. This part of the cortex is important for sensory processing, spatial processing, and working memory (Ch. 2).

Parse To divide or separate the words in a sentence into logical or meaningful groupings (Ch. 9).

Partial report condition An experimental condition in Sperling's (1960) research in which only a randomly selected portion of the entire stimulus display was to be reported (contrast with *whole report condition*) (Ch. 3).

Part-set cuing effect The finding that if you cue people with part of a list of words, they will have more difficulty recalling the rest of the set than if they had not been cued at all (Ch. 8).

Pathway In network representations in long-term memory, the connecting link between two concepts or nodes (Ch. 7).

Patient The object or recipient that receives the action in a sentence; one of the semantic cases in a case grammar approach (see also *case grammar*) (Ch. 6).

PDP modeling See *connectionist*.

Peg word mnemonic The mnemonic device in which a prememorized set of peg word connections is used to remember some new information; the peg words typically used are "One is a bun, Two is a shoe," and so on (Ch. 6).

Perception The process of interpreting and understanding sensory information; the act of sensing then interpreting that information (Ch. 3).

Perceptual symbols Symbolic representations used in memory and grounded in sensory and motor elements derived from experience (Ch. 7).

Performance Any observable behavior; in the context of linguistics, any behavior related to language (e.g., speech), influenced not only by linguistic factors but also by factors related to lapses in attention, memory, and so on (contrast with *competence*) (Ch. 9).

Persistence The tendency to remember facts or events, including traumatic memories, that one would rather forget; that is, failure to forget because of intrusive recollections and rumination (one of Schacter's seven sins of memory) (Ch. 8).

Phi phenomenon Illusory movement that occurs when two images are viewed in rapid succession in different points in space, as in a theater marquee or chasing Christmas lights (Ch. 3).

Phoneme A sound or set of sounds judged to be the same by speakers of a language (e.g., the initial sound in the words *cool* and *keep* for speakers of English). Note that because of categorical perception, we tend to judge some physically different sounds as the same and other different sounds as different, that is, belonging to a different phoneme category (Chs. 3, 9).

Phonemic competence One's basic knowledge of the phonology of the language (Ch. 9).

Phonological loop In Baddeley and Hitch's (1974) working memory model, the articulatory loop is the component responsible for recycling verbal material via rehearsal (Ch. 5).

Phonological similarity effect The finding that memory is poorer when people need to remember a set of words that are phonologically similar, compared to a set of words that are phonologically dissimilar (Ch. 5).

Phonological store The passive store component of the phonological loop (Ch. 5).

Phonology The study of the sounds of language, including how they are produced and how they are perceived (Ch. 9).

Phrase structure The underlying structure of a sentence in terms of the groupings of words into meaningful phrases, such as "[The young man] [ran quickly]" (Ch. 9).

PI See *proactive interference*.

Polysemy When a word in a language has multiple meanings (Ch. 9).

Pop-out effect In visual search, when a target item is highly discriminable from the distractor items, the attentional search mechanisms use a basic rapid attentional mechanism that operates in parallel across the visual field in a highly automatic fashion. The resulting visual search function is quite fast overall and shows only a minimal, if any, effect of set size (Ch. 4).

Positivity effect The finding that older adults place a greater emphasis on emotionally positive information (Ch. 15).

Positron emission tomography (PET) scan The technique that yields images of the functioning of the brain based on cerebral blood flow (Ch. 2).

Pragmatics The aspects of language that are "above and beyond" the words, so-called extralinguistic factors. For instance, part of our pragmatic knowledge of language rules includes the knowledge that the sentence, "Do you happen to know what time it is?" is actually an indirect request rather than a sentence to be taken literally (Ch. 10).

Preoperational stage The second Piagetian stage of development, from ages 2 to 7, in which children's thinking is largely focused on physical objects (Ch. 14).

Presbycusis The loss of hair cells in the cochlea that accompanies aging, and results in difficulty hearing high-pitched sounds. May be resolved with a hearing aid (Ch. 15).

Presbyopia The hardening of the lens of the eye that accompanies aging, and results in difficulty seeing close up. Often resolved with reading glasses (Ch. 15).

Primacy effect In recall performance, the elevated recall of the early positions of the list (contrast with *recency effect*) (Chs. 5, 6).

Prime The first stimulus in a prime–target pair, intended to exert some influence on the second stimulus (see also *priming*) (Ch. 7).

Priming Mental activation of a concept by some means, or the spread of that activation from one concept to another; also, the activation of some target information by action of a previously presented prime; sometimes loosely synonymous with the notion of accessing information in memory (Chs. 1, 4, 7).

Proactive interference (PI) Interference or difficulty, especially during recall, because of some previous activity, often the stimuli learned on some earlier list; any interference in which material presented at one time interferes with material presented later (Chs. 5, 8).

Probabilistic theories Theories that assume that categories in semantic memory are created by taking into account various probabilities and likelihoods across a person's experience. Prototype and exemplar theories are both probabilistic theories (Ch. 7).

Probability heuristics theory A theory of human mental reasoning in which people reach conclusions using subjective estimates of the probability of events (Ch. 11).

Problem of invariance In psycholinguistics, the problem that spoken sounds are not invariant, that they change depending on the preceding and following sounds in the word (Chs. 3, 9).

Problem solving Involves a goal-directed sequence of steps, the cognitive operations used to work through a problem, and can be broken down into a collection of subgoals (Ch. 12).

Problem space The initial, intermediate, and goal states of a problem, along with the problem solver's knowledge and any external resources that can be used to solve the problem (Ch. 12).

Processing fluency The ease with which something is processed or comes to mind (Ch. 8).

Process model A stage model designed to explain the several mental steps involved in performance of some task, usually implying that the stages occur sequentially and that they operate independently of one another (Chs. 1, 5).

Production A *production* is a simple *if–then* rule in models of memory processing, stating the conditions (*if*) necessary for some action (*then*) to be taken, whether that action is a physical response or a mental step or operation (Ch. 12).

Production system A *production system* is a large-scale model of some kind of performance or mental activity based on productions (Ch. 12).

Productivity (also *generativity*) One of Hockett's (1960a, 1960b) linguistic universals, referring to the rule-based nature of language, such that an infinite number of sentences can be generated or produced by applying the rules of the language (Ch. 9).

Propagation The movement of an action potential from the dendrites, through the soma, and down the axon (Ch. 2).

Property statements Simple statements in which the relationship being expressed is "X has the property or feature Y" (e.g., "A robin has wings") (Ch. 7).

Proposition A simple idea unit (Chs. 6, 7, 8).

Propositional textbase An intermediate level of representation that captures the basic idea units present in a text (Chs. 6, 10, 15).

Prosody The up and down pitch of an utterance that can convey emotional information (Chs. 9, 13).

Prosopagnosia Disruption in the ability to recognize faces (Chs. 3, 9).

Prospective memory Remembering to do something in the future; e.g., remembering to make a phone call tomorrow (Ch. 6).

Prototype The typical or average member of a category; the central or most representative member of a category. Note that a prototype may not exist for some categories, in which case the category's prototype would be some "average-like" combination of the various members (Ch. 7).

Prototype theory The probabilistic view that human categories are created by using a mental prototype (Ch. 7).

Proximity A Gestalt grouping principle that is used by visual perception to structure and organize together elements of a visual input that are nearer to one another in space (Ch. 3).

Psycholinguistics The study of language from the perspective of psychology; the study of language behavior and processes (Ch. 9).

Psychological essentialism The idea that people treat members of a category as if they have the same underlying, perhaps invisible, property or essence (Ch. 7).

Psychological refractory period A delay in a second decision or response cycle if it is required immediately after a preceding decision (Ch. 4).

Psychophysics The study of the relationship between physical stimuli and the perceived characteristics of those stimuli; the study of how perceptual experience differs from the physical stimulation that is being perceived (Ch. 3).

Pure word deafness Disruption of the perceptual or semantic processing in auditory word comprehension (Ch. 9).

Rational constructivist view A theory of cognitive development that assumes that children reason like little scientists trying to understand the world in which they find themselves (Ch. 14).

Recency effect In recall performance, the elevated recall of the last few items in a list, presumably because the items are stored in and retrieved from short-term memory (contrast with *primacy effect*) (Chs. 5, 6).

Recipient See *patient*.

Recoding Mentally transforming or translating a stimulus into another code or format; grouping items into larger units, as when recoding a written word into an acoustic–articulatory code (Ch. 5).

Recognition heuristic A heuristic in which you base a decision on whether you recognize the thing to be judged (Ch. 11).

Recognition task Any yes/no task in which subjects are asked to judge whether they have seen the stimulus before; more generally, any task asking for a simple yes/no (alternatively, true/false, same/different) response, often including a reaction time measurement (Chs. 5, 6, 7).

Reconsolidation When a memory is retrieved, this puts it in a plastic, malleable state where it can be changed before it is stored in memory again (Ch. 8).

Reconstructive memory The tendency in recall or recognition to include ideas or elements that were inferred or related to the original stimulus but were not part of the original stimulus itself (Ch. 7).

Reductionism The scientific approach in which a complex event or behavior is broken down into its constituents; the individual constituents are then studied individually (Ch. 1).

Reference In language, the allusion to or indirect mention of an element from elsewhere in the sentence or passage, as by using a pronoun or synonym (Ch. 10).

Region of proximal learning Information that is just beyond a person's current level of understanding (Chs. 6, 10).

Rehearsal The mental repetition or practicing of some to-be-learned material (Ch. 6).

Relation In case grammar, the central idea or relationship being asserted in a sentence or phrase. For instance, in "Bill hit the ball with the bat," the central relation is "HIT" (see also *case grammar*) (Ch. 6).

Relearning task An experimental task in which some material is learned, set aside for a period of time, then relearned to the same criterion in hopes that the relearning will take less time or effort to achieve the same level of accuracy; the task used by Ebbinghaus in his research on memory (Ch. 6).

Release from proactive interference The sudden reduction in proactive interference when the material to be learned is changed in some fashion, such as improved recall on a list of plant names after several trials involving animal names. The initial decline was caused by proactive interference, and the improvement on the last trial is caused by release from PI (Ch. 5).

Release from PI See *release from proactive interference*.

Reminiscence bump Superior memory than would otherwise be expected for life events around the age of 20, between the ages of 15 and 25 (Ch. 6).

Repeated name penalty   An increase in reading times when a direct reference is used again (e.g., the person's name) compared to when a pronoun is used (Ch. 10).

Repetition blindness   The tendency to not perceive a pattern, whether a word, a picture, or any other visual stimulus, when it is quickly repeated (Ch. 3).

Repetition priming   A priming effect caused by the exact repetition of a stimulus; often used in implicit memory tests (Chs. 6, 8).

Representation   A general term referring to the way information is stored in memory. The term always carries the connotation that we are interested in the format or organization of the information as it is stored (is the information stored in a semantic representation? a sound-based representation?) (Ch. 1).

Representational momentum   The phenomenon of misremembering the movement of an object farther along its path of travel than where it actually was when it was last seen (Ch. 5).

Representation of knowledge   See *representation*.

Representativeness heuristic   A reasoning heuristic in which we judge the likelihood of some event by deciding how representative that event seems to be of the larger group or population from which it was drawn (Ch. 11).

Repression   Intentional forgetting of painful or traumatic experiences, especially in Freudian theory (Ch. 8).

Response time (RT)   The elapsed time, usually measured in milliseconds, between some stimulus event and the subject's response to that event; a particularly common measure of performance in cognitive psychology (Ch. 1).

Retina   The layer of the eye covered with the rods and cones that initiate the process of visual sensation and perception (Ch. 3).

Retrieval   Accessing information stored in memory, whether or not that access involves conscious awareness (Chs. 1, 2).

Retrieval cue   Any cue, hint, or piece of information used to prompt retrieval of some target information (Ch. 6).

Retrieval-induced forgetting   Forgetting that occurs when one portion of a set of information is retrieved, but the remainder is not. The remaining part of the set is thought to be inhibited or suppressed, and so is less likely to be remembered (Ch. 8).

Retroactive interference (RI)   The interference from a recent event or experience that influences memory for an earlier event, such as trying to recall the items from list 1 but instead recalling the items from list 2 (Chs. 5, 8).

Retrograde amnesia   Loss of memories that preceded the brain damage (Ch. 8).

Rewrite rules   In a phrase structure grammar, the rules that specify the individual components of a phrase; for example, a noun phrase is rewritten as a determiner, an adjective, and a noun: $NP = D + N$ (Ch. 9).

RI   See *retroactive interference*.

Ribot's law   For retrograde amnesia, memory is more affected the closer it is to the time of the injury, with older memories being more likely to be preserved (Ch. 8).

RT   See *response time*.

Saccade   The voluntary sweeping of the eyes from one fixation point to another (Chs. 3, 10).

Salience (also *vividness*)   Source of bias in the availability heuristic in which a particularly notable or vivid memory influences judgments about the frequency or likelihood of such events (Ch. 11).

Sapir-Whorf linguistic relativity hypothesis   See *linguistic relativity hypothesis*.

Satisficing   Finding an acceptable or satisfactory solution to a problem, even though the solution may not be optimal (Chs. 11, 12).

Savings score   In a relearning task, the score showing how much was saved on second learning compared with original learning. For instance, if original learning took 10 trials and relearning took only 6 trials, then savings would be 40% (10 − 6)/10 (Ch. 6).

Schema   In Bartlett's (1932, p. 201) words, "an active organization of past reactions or past experiences"; a knowledge structure in memory (Ch. 7).

Schemata   Plural form of *schema*.

Script   Schank's (1977) term for a schema, a long-term memory representation of some complex event such as going to a restaurant (Ch. 7).

Second-order theory   In conversation, the informal theory we develop that expresses our knowledge of what the other participant knows about us, summarized by the phrase "what he/she thinks I know" (contrast with *direct theory*) (Ch. 10).

Selecting   See *filtering*.

Selective attention   The ability to attend to one source of information while ignoring or excluding other ongoing messages. (Ch. 4).

Self-reference effect   The finding that memory is generally better for information that is related to the self in some way (Ch. 6).

Semantic case (also *case role*)   In a case grammar approach, the particular case played by a word or concept is said to be that word's semantic case (see also *case grammar*) (Chs. 6, 9).

Semantic congruity effect   In the mental comparison task, reaction time is speeded or judgments are made easier when the basis for a judgment is congruent or similar to the stimuli being compared; for instance, a congruent condition would be "choose the smaller of second or minute," and an incongruent condition would be "choose the smaller of decade or century" (Ch. 3).

Semantic distance effect   See *semantic relatedness effect*.

Semantic features (also *feature list*)   Properties or characteristics stored in the mental representation of some concept, presumed by some theories to be accessed and evaluated in the process of making semantic judgments (Ch. 7).

Semanticity   One of Hockett's (1960a, 1960b) linguistic universals, expressing the fact that the elements of language convey meaning (Ch. 9).

Semantic memory   The long-term memory component in which general world knowledge is stored (contrast with *episodic memory*) (Chs. 2, 6, 7).

Semantic network   A structure for information stored in long-term semantic memory, assumed by several popular models of mental processing. In most network models, concepts are represented as nodes that are interconnected by means of links or pathways; activation is presumed to spread from concept to concept along these connecting pathways (Ch. 7).

Semantic priming When concepts are activated in memory, this activation spreads to semantic related concepts, making them easier to fully activate later if needed (Ch. 7).

Semantic relatedness effect In semantic memory tasks, reaction time is speeded up or judgments are made easier when the concepts are closer together in semantic distance, when they are more closely related. Note that the effect is reversed when the comparison is false; that is, RT is shorter for the comparison "A whale is a fish" than for "A whale is a bird" (Ch. 7).

Semantics The study of meaning (Ch. 9).

Sensation The reception of physical stimulation and encoding of it into the nervous system (Ch. 3).

Sensorimotor stage The first Piagetian stage of development, from birth to age 2, in which thinking is guided by what children can see and do themselves (Ch. 14).

Sensory cortex The band of cortex at the front of the parietal lobes responsible for processing sensory information from throughout the body (Ch. 2).

Sensory memory The initial mental storage system for sensory stimuli. There are presumably as many modalities of sensory memory as there are kinds of stimulation that we can sense (Ch. 1).

Sentence verification task A task in which subjects must respond true or false to simple sentences (Ch. 7).

Sequential stages of processing An assumption in most process models that the separate stages of processing occur in a fixed sequence, with no overlap of the stages (Ch. 1).

Serial exhaustive search A search process in which all possible elements are searched one by one before the decision is made, even if the target is found early in the search process (Ch. 5).

Serial position curve The display of accuracy in recall across the original positions in the to-be-learned list, often found to have a bowed shape, indicating lower recall in the middle of the list than in the initial or final position (Chs. 5, 6).

Serial processing Mental processing in which only one process or operation occurs at a time (Ch. 1).

Serial recall A recall task in which subjects must recall the list items in their original order of presentation (contrast with *free recall*) (Chs. 5, 6).

Shadowing task A task in which subjects hear a spoken message and must repeat the message out loud in a very short time; often used as one of the two tasks in a dual task method (Ch. 4).

Short-term memory (STM) The component of the human memory system that holds information for up to 20 s; the memory component where current and recently attended information is held; sometimes loosely equated with attention and consciousness (Chs. 1, 5).

Similarity A Gestalt grouping principle that is used by visual perception to structure and organize together elements of a visual input that have similar visual features, such as texture, color, brightness, and so forth (Ch. 3).

Simulation heuristic A reasoning heuristic in which we predict a future event or imagine a different outcome of a completed event; a forecasting of how some event will turn out or how it might have turned out under other circumstances (Ch. 11).

Single cell recording The measurement of firing rating from an individual neuron using an electrode implanted in vivo (Ch. 2).

Single-word stage Stage of language development, from 12 to 18 months, dominated by the use of single words, even to convey complex ideas (Ch. 14).

Situation model A memory representation of a real or possible-world situation; for example, of a situation described in a passage of text (Chs. 6, 8, 10).

SOA See *stimulus onset asynchrony*.

Soma The cell body of a neuron (Ch. 2).

Source memory Memory of the exact source of information (Ch. 8).

Source monitoring The ability to accurately remember the source of a memory, be it something you encountered in the world directly or learned indirectly from another source (Chs. 6, 15).

Span of apprehension (also *span of attention*) The number of simple elements (e.g., digits, letters) that can be heard and immediately reported in their correct order; a standard short-term memory task, common on standardized intelligence tests (Ch. 3).

Span of attention See *span of apprehension*.

Special populations Populations of people that have some consistent alteration of brain structure or activity, such as older adults or Korsakoff's patients (Ch. 2).

Speech act The intended consequence of an utterance. That is, what you are trying to accomplish when you say something (Ch. 10).

Split brain Refers to patients in whom the corpus callosum has been severed surgically and the resultant changes in their performance because of the surgery or, more generally, to research showing various specializations of the two cerebral hemispheres (Ch. 2).

Spotlight attention A rapid attentional mechanism operating in parallel and automatically across the visual field, especially for detecting simple visual features (Ch. 4).

Spreading activation The commonly assumed theoretical process by which long-term memory knowledge is accessed and retrieved. Some form of mental excitation or activation is believed to be passed or spread along the pathways that connect concepts in a memory network. When a concept has been activated, it has been retrieved or accessed within the memory representation. The process is loosely analogous to the spread of neural excitation in the brain (Ch. 7).

Stereotypes In reasoning, bias in judgments related to the typical characteristics of a profession, type of person, and so on (Ch. 11).

Stereotype threat Unconscious activation of negative stereotypes leads a person to perform worse on a task than he or she would otherwise (Ch. 13).

Sternberg task The short-term memory scanning task devised by Saul Sternberg (Ch. 5).

Stimulus onset asynchrony (SOA) In priming studies, the interval of time separating the prime and the target, usually a few hundred milliseconds (Ch. 7).

STM See *short-term memory*.

Story mnemonic A memory aid in which people construct a narrative story containing the material to help them remember it later (Ch. 6).

Stroop task Name the physical color of a color word (e.g., RED in the color blue → "blue") (Ch. 4).

Structuralism The approach, most closely identified with Titchener, in which the structure of the conscious mind—that is, the sensations, images, and feelings that are the elements of consciousness—was studied; the first major school of psychological thought, beginning in the late 1800s (contrast with *functionalism*) (Ch. 1).

Structure building The process of comprehension in Gernsbacher's theory of building a mental representation of the meaning of sentences (Ch. 10).

Subgoal In problem solving, an intermediate goal that must be achieved to reach a final goal (Ch. 12).

Subjective organization The grouping or organizing of items that are to be learned according to some scheme or basis devised by the subject (Ch. 6).

Suffix effect The inferior recall of the end of the list in the presence of an additional meaningful nonlist auditory stimulus (Ch. 3).

Suggestibility The tendency to incorporate information provided by others into our own recollection and memory representation (one of Schacter's seven sins of memory) (Ch. 8).

Suppression In Gernsbacher's theory, the active process of reducing the activation level of concepts no longer relevant to the meaning of a sentence (Ch. 10).

Surface form The level of representation in language comprehension that corresponds to a verbatim mental representation of the exact words and syntax used in a passage of text (Chs. 6, 10).

Surface structure In linguistics and psycholinguistics, the actual form of a sentence, whether written or spoken (contrast with *deep structure*); the literal string of words or sounds present in a sentence (Ch. 9).

Sustained attention See *vigilance*.

Symbolic distance effect The result, in symbolic comparison tasks, in which two relatively different stimuli (e.g., 1 and 8) are judged more rapidly than two relatively similar stimuli (e.g., 1 and 2) because of greater symbolic distance between 1 and 8 (Ch. 3).

Symbolic representation According to Bruner, a late, abstract, linguistic-based stage of childhood thinking (Ch. 14).

Synapse The junction of two neurons; the small gap between the terminal buttons of one neuron and the dendrites of another; as a verb, to form a junction with another neuron (Ch. 2).

Syntax The arrangement of words as elements in a sentence to show their relationship to one another; grammatical structure; the rules governing the order of words in a sentence (Ch. 9).

Tabula rasa Latin term meaning "blank slate." The term refers to a standard assumption of behaviorists that learning and experience write a record on the blank slate; in other words, the assumption that learning, as opposed to innate factors, is the most important factor in determining behavior (Ch. 1).

"Take the best" heuristic A heuristic in which you decide between alternatives based on the first useful information you retrieve about the alternatives (Ch. 11).

Target The second part of a prime–target stimulus (see *priming*); any concept or material that is designated as being of special interest (Ch. 7).

Task effects A second major difficulty for depth of processing (along with *circularity*). Different memory tasks revealed differences in memory performance that were incompatible with a unitary processing mechanism (Ch. 6).

Template A model or pattern. In theories of pattern recognition, a template is the pattern stored in memory against which incoming stimuli are compared to recognize the incoming patterns (Ch. 3).

Temporal lobe The lobe of the cortex on the sides, below the frontal and parietal lobes. This lobe is important for audition and memory (Ch. 2).

Thalamus ("inner room"; "inner chamber"; "gateway to the cortex") Major relay station from the sensory systems of the body to the cortex; almost all messages entering the cortex come through the thalamus (Ch. 2).

Theory of mind Theories we develop of our conversational partners (Ch. 10).

Three-eared man Procedure used to investigate echoic memory (Ch. 3).

Time In propositional or semantic case theories, the semantic case referring to when an event took place; for example, in the sentence "The car climbed the steep hill," Time = the past (Ch. 6).

Tip-of-the-tongue (TOT) effect Momentary retrieval failure, with the sense of being on the verge of retrieving the target concept (Chs. 6, 8).

Top-down processing See *conceptually driven processing*.

Topic maintenance Making conversational contributions relevant to the topic, sticking to the topic (Ch. 10).

Transcranial magnetic stimulation (TMS) The use of magnetic coils to affect the activity of targeted assemblies of neurons in the cortex to either increase their activation or give a person a temporary lesion (Ch. 2).

Transformational grammar Chomsky's theory of the structure of language, a combination of a phrase structure grammar and a set of transformational rules (Ch. 9).

Transformational rules In Chomsky's transformational grammar, the syntactic rules that transform an idea (a deep structure sentence) into its surface structure; for instance, rules that form a passive sentence or a negative sentence (Ch. 9).

Transience The tendency to lose access to information across time, whether through forgetting, interference, or retrieval failure (one of Schacter's seven sins of memory) (Ch. 8).

Trans-saccadic memory The memory system that is used across a series of eye movements to build up a more complete and stable understanding of the visual world (Ch. 3).

Tunnel memories At high levels of emotional arousal, better memories for the central details of an event, but poorer memories for peripheral details, as if a person is developing a kind of tunnel vision (Ch. 13).

Two-word stage Stage of language development, typically peaking around 18 months, dominated by two-word utterances (Ch. 14).

Typicality In semantic categories the degree to which items are viewed as typical, central members of a category; the central tendency of a category (Ch. 7).

Typicality effect In semantic memory research, the result that typical members of a category tend to be judged more rapidly than atypical members (Ch. 7).

Unauthorized inference Not intended, especially said of inferences drawn during a conversation (contrast with *authorized inference*) (Ch. 10).

Unconscious processing Mental processing outside of awareness (Ch. 1).

Undoing See *counterfactual reasoning*.

Unit See *chunk*.

Updating The process of altering a person's situation model in the face of information about how the situation has changed (Ch. 10).

Valence Whether an emotion is positive (e.g., happy) or negative (e.g., angry) (Ch. 13).

Ventral pathway The neural pathway along the bottom of the cortex, stemming from visual processing areas in the occipital lobe, primarily responsible for processing information about what things are in the world (Ch. 2).

Ventromedial prefontal cortex Part of the cortex involved in the identification and interpretation of emotional stimuli and responses (Ch. 13).

Verbal learning The branch of human experimental psychology, largely replaced by cognitive psychology in the late 1950s and early 1960s, investigating the learning and retention of "verbal," that is, language-based, stimuli; influenced directly by Ebbinghaus's methods and interests (Ch. 1).

Verbal protocol In studies of problem solving, a word-for-word transcription of what the subject said aloud during the problem-solving attempt (Chs. 1, 10, 12).

Vigilance (also *sustained attention*) The maintenance of attention for infrequent events over long periods of time (Ch. 4)

Visual attention Input attention (specifically as it relates to vision; typically associated with the spotlight metaphor) (Ch. 4).

Visual imagery The mental representation of visual information; the skill or ability to remember visual information (Ch. 6).

Visual persistence The perceptual phenomenon in which a visual stimulus still seems to be present even after its termination, usually a few hundred milliseconds to a few seconds (Ch. 3).

Visual search Search a spatial display of items (e.g., a set of 4, 8, or 16 letters) for the presence of a target (e.g., the letter K) (Chs. 4, 13).

Visual sensory memory (also *iconic memory*) The short-duration memory system specialized for holding visual information, lasting no more than about 250 ms to 500 ms (Ch. 3).

Visuo-spatial sketch pad The visual and perceptual component of Baddeley's working memory model (Ch. 5).

Vividness See *salience*.

Voicing Refers to whether the vocal cords begin to vibrate immediately with the obstruction of airflow (/b/ in *bat*) or whether the vibration is delayed until after the release of air (/p/ in *pat*) (Ch. 9).

von Restorff effect In a recall task, the elevated accuracy for an item that was noticeably different during list presentation; for instance, because it was written in a different color of ink (Chs. 6, 13, 15).

Weapon focus effect The finding that memory for peripheral details of an event are poorer when there is a salient, emotionally charged object present, such as a weapon, that is drawing attention and processing (Ch. 13).

Well-defined problem A problem in which the initial and final states and the legal operators are clearly specified (Ch. 12).

Wernicke's aphasia One of two common forms of aphasia in which the language disorder is characterized by a serious disruption of comprehension and the use of invented words as well as semantically inappropriate substitutions (contrast with *Broca's aphasia*). The aphasia is caused by damage in the region of the neocortex called *Wernicke's area* (Ch. 9).

Whole report condition Especially in Sperling's (1960) research, the condition in which the entire visual display was to be reported (contrast with *partial report condition*) (Ch. 3).

Word explosion The period of language development, from 18 months to age 5, in which there is an explosion in the number of words a child knows and can use (Ch. 14).

Word frequency effect Finding that frequent words in the language are processed more rapidly than infrequent words (Ch. 1).

Word stem completion task Complete word stems (e.g., D_ _ K) to form a complete word, generally the first word that comes to mind (e.g., DUCK) (Ch. 4).

Working memory The component, similar to short-term memory, in Baddeley and Hitch's (1974) theory in which verbal rehearsal and other conscious processing takes place; also, the component that contains the executive controller in charge of devoting conscious processing resources to the various other components in the memory system (Ch. 5).

Working memory span The amount of information a person can actively maintain in working memory at one time (Ch. 5).

Yerkes-Dodson law The inverted u-shaped function that shows that memory is best at moderate levels of emotional arousal, but poorer at low and high levels of arousal (Ch. 13).

Zone of proximal development The idea that children acquire information that is just beyond their current level of thinking (Ch. 14).

References

Abbot, V., Black, J. B., & Smith, E. E. (1985). The representation of scripts in memory. *Journal of Memory and Language, 24,* 179–199.

Abelson, R. P. (1981). Psychological status of the script concept. *American Psychologist, 36,* 715–729.

Abraham, W. C. (2006). Memory maintenance: The changing nature of neural mechanisms. *Current Directions in Psychological Science, 15,* 5–8.

Abrams, R. A., & Christ, S. E. (2003). Motion onset captures attention. *Psychological Science, 14,* 427–432.

Adams, C., Smith, M. C., Nyquist, L., & Perlmutter, M. (1997). Adult age-group differences in recall for the literal and interpretive meanings of narrative text. *The Journals of Gerontology, 52B,* P187–P195.

Addis, D. R., & Schacter, D. L. (2008). Constructive episodic simulation: Temporal distance and detail of past and future events modulate hippocampal engagement. *Hippocampus, 18*(2), 227–237.

Adolphs, R., & Tranel, D. (1999). Intact recognition of emotional prosody following amygdala damage. *Neuropsychologia, 37,* 1285–1292.

Aerts, D., Sozzo, S., & Veloz, T. (2015). Quantum structure of negation and conjunction in human thought. *Frontiers in Psychology, 6,* 1447.

Agnoli, F. (1991). Development of judgmental heuristics and logical reasoning: Training counteracts the representativeness heuristic. *Cognitive Development, 6,* 195–217.

Agnoli, F., & Krantz, D. H. (1989). Suppressing natural heuristics by formal instruction: The case of the conjunction fallacy. *Cognitive Psychology, 21,* 515–550.

Ahlum-Heath, M. E., & DiVesta, F. J. (1986). The effect of conscious controlled verbalization of a cognitive strategy on transfer in problem solving. *Memory & Cognition, 14,* 281–285.

Ainsworth-Darnell, K., Shulman, H. G., & Boland, J. E. (1998). Dissociating brain responses to syntactic and semantic anomalies: Evidence from event-related potentials. *Journal of Memory and Language, 38,* 112–130.

Albert, M. S., & Kaplan, E. (1980). Organic implications of neuropsychological deficits in the elderly. In L. W. Poon, J. L. Fozard, L. S. Cermak, D. Arenberg, & L. W. Thompson (Eds.), *New directions in memory and aging* (pp. 403–432). Hillsdale, NJ: Erlbaum.

Albrecht, J. E., & O'Brien, E. J. (1995). *Goal processing and the maintenance of global coherence.* Hillsdale, NJ: Erlbaum.

Allen, P. A., & Madden, D. J. (1990). Evidence for a parallel input serial analysis (PISA) model of word processing. *Journal of Experimental Psychology: Human Perception and Performance, 16,* 48–64.

Allen, R. J., Baddeley, A. D., & Hitch, G. J. (2006). Is the binding of visual features in working memory resource-demanding? *Journal of Experimental Psychology: General, 135,* 298–313.

Allport, A. (1989). Visual attention. In M. I. Posner (Ed.), *Foundations of cognitive science* (pp. 631–682). Cambridge, MA: Bradford.

Almor, A. (1999). Noun-phrase anaphora and focus: The informational load hypothesis. *Psychological Review, 106,* 748–765.

Altarriba, J., Bauer, L. M., & Benvenuto, C. (1999). Concreteness, context availability, and imageability ratings and word associations for abstract, concrete, and emotion words. *Behavior Research Methods, 31,* 578–602.

Alter, A. L., Oppenheimer, D. M., Epley, N., & Eyre, R. N. (2007). Overcoming intuition: Metacognitive difficulty activates analytic reasoning. *Journal of Experimental Psychology: General, 136,* 569–576.

Altmann, E. M., & Gray, W. D. (2002). Forgetting to remember: The functional relationship of decay and interference. *Psychological Science, 13,* 27–33.

Altmann, E. M., & Schunn, C. D. (2012). Decay versus interference: A new look at an old interaction. *Psychological Science, 23*(11), 1435–1437.

Altmann, G. T. M. (1998). Ambiguity in sentence processing. *Trends in Cognitive Sciences, 2,* 146–157.

Altmann, G. T. M., Garnham, A., & Dennis, Y. (1992). Avoiding the garden path: Eye movements in context. *Journal of Memory and Language, 31,* 685–712.

Altmann, G. T. M., & Steedman, M. (1988). Interaction with context during human sentence processing. *Cognition, 30,* 191–238.

Ambday, N., & Bharucha, J. (2009). Culture and the brain. *Current Directions in Psychological Science, 18,* 342–345.

Ambrosini, E., Sinigaglia, C., & Costantini, M. (2012). Tie my hands, tie my eyes. *Journal of Experimental Psychology: Human Perception and Performance, 38*(2), 263–266.

Anaki, D., & Henik, A. (2003). Is there a "strength effect" in automatic semantic priming? *Memory & Cognition, 31,* 262–272.

Anderson, A. K. (2005). Affective influences on the attentional dynamics supporting awareness. *Journal of Experimental Psychology: General, 134,* 258–281.

Anderson, B. (2011). There is no such thing as attention. *Frontiers in Psychology, 2*(246), 1–8.

Anderson, B. A., & Yantis, S. (2013). Persistence of value-driven attentional capture. *Journal of Experimental Psychology: Human Perception and Performance, 39*(1), 6–9.

Anderson, J. R. (1974). Retrieval of propositional information from long-term memory. *Cognitive Psychology, 6,* 451–474.

Anderson, J. R. (1980). *Cognitive psychology and its implications.* San Francisco, CA: Freeman.

Anderson, J. R. (1985). *Cognitive psychology and its implications* (2nd ed.). New York, NY: Freeman.

Anderson, J. R. (1993). Problem solving and learning. *American Psychologist, 48*, 35–44.

Anderson, J. R., & Bower, G. H. (1973). *Human associative memory*. Washington, DC: Winston & Sons.

Anderson, J. R., Qin, Y., Jung, K.-J., & Carter, C. S. (2007). Information-processing modules and their relative modality specificity. *Cognitive Psychology, 54*, 185–217.

Anderson, J. R., & Schooler, L. J. (1991). Reflections of the environment in memory. *Psychological Science, 2*, 396–408.

Anderson, M. C. (2003). Rethinking interference theory: Executive control and mechanisms of forgetting. *Journal of Memory and Language, 49*, 415–445.

Anderson, M. C., Bjork, E. L., & Bjork, R. A. (2000). Retrieval-induced forgetting: Evidence for a recall-specific mechanism. *Psychonomic Bulletin & Review, 7*, 522–530.

Anderson, R. J., Dewhurst, S. A., & Nash, R. A. (2012). Shared cognitive processes underlying past and future thinking: The impact of imagery and concurrent task demands on event specificity. *Journal of Experimental Psychology: Learning, Memory, and Cognition, 38*(2), 356–265.

Anderson, S. E., Matlock, T., & Spivey, M. (2013). Grammatical aspect and temporal distance in motion descriptions. *Frontiers in Psychology, 4*, 337.

Andrade, J. (1995). Learning during anaesthesia: A review. *British Journal of Psychology, 86*, 479–506.

Andres, M., Finocchiaro, C., Buiatti, M., & Piazza, M. (2015). Contribution of motor representations to action verb processing. *Cognition, 134*, 174–184.

Andrews, M., Vigliocco, G., & Vinson, D. (2009). Integrating experiential and distributional data to learn semantic representations. *Psychological Review, 116*(3), 463–498.

Andrews-Hanna, J. R. (2012). The brain's default network and its adaptive role in internal mentation. *The Neuroscientist, 18*(3), 251–270.

Angele, B., & Rayner, K. (2013). Processing the in the parafovea: Are articles skipped automatically? *Journal of Experimental Psychology: Learning, Memory, and Cognition, 39*(2), 649.

Arbuckle, T. Y., & Cuddy, L. L. (1969). Discrimination of item strength at time of presentation. *Journal of Experimental Psychology, 81*, 126–131.

Arbuckle, T. Y., Vanderleck, V. F., Harsany, M., & Lapidus, S. (1990). Adult age differences in memory in relation to availability and accessibility of knowledge-based schemas. *Journal of Experimental Psychology: Learning, Memory, and Cognition, 16*, 305–315.

Armony, J. L., Chochol, C., Fecteau, S., & Belin, P. (2007). Laugh (or cry) and you will be remembered. *Psychological Science, 18*, 1027–1029.

Armstrong, S. L., Gleitman, L. R., & Gleitman, H. (1983). What some concepts might not be. *Cognition, 13*, 263–308.

Arnold, J. E., & Griffin, Z. M. (2007). The effect of additional characters on choice of referring expression: Everyone counts. *Journal of Memory and Language, 56*, 521–536.

Arstila, V. (2012). Time slows down during accidents. *Frontiers in Psychology, 3*, 196.

Ashcraft, M. H. (1976). Priming and property dominance effects in semantic memory. *Memory & Cognition, 4*, 490–500.

Ashcraft, M. H. (1989). *Human memory and cognition*. Glenview, IL: Scott Foresman.

Ashcraft, M. H. (1993). A personal case history of transient anomia. *Brain and Language, 44*, 47–57.

Ashcraft, M. H. (1995). Cognitive psychology and simple arithmetic: A review and summary of new directions. *Mathematical Cognition, 1*, 3–34.

Ashcraft, M. H., & Battaglia, J. (1978). Cognitive arithmetic: Evidence for retrieval and decision processes in mental addition. *Journal of Experimental Psychology: Human Learning and Memory, 4*(5), 527–538.

Ashcraft, M. H., & Christy, K. S. (1995). The frequency of arithmetic facts in elementary texts: Addition and multiplication in grades 1–6. *Journal for Research in Mathematics Education, 26*, 396–421.

Ashcraft, M. H., & Fierman, B. A. (1982). Mental addition in third, fourth, and sixth graders. *Journal of Experimental Child Psychology, 33*(2), 216–234.

Ashcraft, M. H., Kellas, G., & Needham, S. (1975). Rehearsal and retrieval processes in free recall of categorized lists. *Memory & Cognition, 3*, 506–512.

Ashcraft, M. H., & Krause, J. A. (2007). Working memory, math performance, and math anxiety. *Psychonomic Bulletin & Review, 14*, 243–248.

Aslan, A., & Bäuml, K-H. (2011). Adaptive memory: Young children show enhanced retention of fitness-related information. *Cognition, 122*, 118–122.

Aslan, A., Bäuml, K-H., & Grundgeiger, T. (2007). The role of inhibitory processes in part-list cuing. *Journal of Experimental Psychology: Learning, Memory, and Cognition, 33*, 335–341.

Atance, C. M., & O'Neill, D. K. (2001). Episodic future thinking. *Trends in Cognitive Sciences, 5*(12), 533–539.

Atkins, S. A., & Reuter-Lorenz, P. A. (2008). False working memories? Semantic distortion in a mere 4 seconds. *Memory & Cognition, 36*, 74–81.

Atkinson, R. C., & Shiffrin, R. M. (1968). Human memory: A proposed system and its control processes. In W. K. Spence & J. T. Spence (Eds.), *The psychology of learning and motivation: Advances in research and theory* (Vol. 2, pp. 89–195). New York, NY: Academic Press.

Atkinson, R. C., & Shiffrin, R. M. (1971). The control of short-term memory. *Scientific American, 225*, 82–90.

Atwood, M. E., & Polson, P. (1976). A process model for water jug problems. *Cognitive Psychology, 8*, 191–216.

Au, J., Sheehan, E., Tsai, N., Duncan, G. J., Buschkuehl, M., & Jaeggi, S. M. (2015). Improving fluid intelligence with training on working memory: A meta-analysis. *Psychonomic Bulletin & Review, 22*(2), 366–377.

Averbach, E., & Coriell, A. S. (1961). Short-term memory in vision. *Bell System Technical Journal, 40*, 309–328. (Reprinted in *Readings in cognitive psychology*, by M. Coltheart, Ed., 1973, Toronto, Canada: Holt, Rinehart & Winston of Canada)

Averbach, E., & Sperling, G. (1961). Short term storage and information in vision. In C. Cherry (Ed.), *Information theory* (pp. 196–211). London, England: Butterworth.

Axmacher, N., Cohen, M. X., Fell, J., Haupt, S., Dümpelmann, M., Elger, C. E., et al. (2010). Intracranial EEG correlates of expectancy and memory formation in the human hippocampus and nucleus accumbens. *Neuron, 65,* 541–549.

Ayers, M. S., & Reder, L. M. (1998). A theoretical review of the misinformation effect: Predictions from an activation-based memory model. *Psychonomic Bulletin & Review, 5,* 1–21.

Ayers, T. J., Jonides, J., Reitman, J. S., Egan, J. C., & Howard, D. A. (1979). Differing suffix effects for the same physical suffix. *Journal of Experimental Psychology: Human Learning and Memory, 5,* 315–321.

Baars, B. J. (1986). *The cognitive revolution in psychology.* New York, NY: Guilford Press.

Bachoud-Levi, A. C., Dupoux, E., Cohen, L., & Mehler, J. (1998). Where is the length effect? A cross-linguistic study of speech production. *Journal of Memory and Language, 39,* 331–346.

Baddeley, A. D. (1966). Short-term memory for word sequences as a function of acoustic, semantic, and formal similarity. *The Quarterly Journal of Experimental Psychology, 18,* 302–309.

Baddeley, A. D. (1976). *The psychology of memory.* New York, NY: Basic Books.

Baddeley, A. D. (1978). The trouble with levels: A reexamination of Craik and Lockhart's framework for memory research. *Psychological Review, 85,* 139–152.

Baddeley, A. D. (1992a). Is working memory working? The Fifteenth Bartlett Lecture. *The Quarterly Journal of Experimental Psychology, 44A,* 1–31.

Baddeley, A. D. (1992b). Working memory. *Science, 255,* 556–559.

Baddeley, A. D. (2000a). The episodic buffer: A new component of working memory? *Trends in Cognitive Sciences, 4,* 417–423.

Baddeley, A. D. (2000b). The phonological loop and the irrelevant speech effect: Some comments on Neath (2000). *Psychonomic Bulletin & Review, 7,* 544–549.

Baddeley, A. D., & Hitch, G. (1974). Working memory. In G. H. Bower (Ed.), *The psychology of learning and motivation* (Vol. 8, pp. 47–89). New York, NY: Academic Press.

Baddeley, A. D., & Lieberman, K. (1980). Spatial working memory. In R. Nickerson (Ed.), *Attention and performance VIII.* Hillsdale, NJ: Erlbaum.

Baddeley, A. D., Thomson, N., & Buchanan, M. (1975). Word length and the structure of short-term memory. *Journal of Verbal Learning and Verbal Behavior, 14,* 575–589.

Baddeley, A. D., & Wilson, B. (1988). Comprehension and working memory: A single case neuropsychological study. *Journal of Memory and Language, 27,* 479–498.

Bahrick, H. P. (1983). The cognitive map of a city: 50 years of learning and memory. In G. H. Bower (Ed.), *The psychology of learning and motivation: Advances in research and theory* (Vol. 17, pp. 125–163). New York, NY: Academic Press.

Bahrick, H. P. (1984). Semantic memory content in permastore: Fifty years of memory for Spanish learned in school. *Journal of Experimental Psychology: General, 113,* 1–29.

Bahrick, H. P., Bahrick, L. E., Bahrick, A. S., & Bahrick, P. E. (1993). Maintenance of foreign language vocabulary and the spacing effect. *Psychological Science, 4,* 316–321.

Bahrick, H. P., Bahrick, P. C., & Wittlinger, R. P. (1975). Fifty years of memories for names and faces: A cross-sectional approach. *Journal of Experimental Psychology: General, 104,* 54–75.

Bahrick, H. P., & Hall, L. K. (1991). Lifetime maintenance of high school mathematics content. *Journal of Experimental Psychology: General, 120,* 20–33.

Bahrick, H. P., Hall, L. K., & Berger, S. A. (1996). Accuracy and distortion in memory for high school grades. *Psychological Science, 7,* 265–271.

Bailey, H. R., Zacks, J. M., Hambrick, D. Z., Zacks, R. T., Head, D., Kurby, C. A., & Sargent, J. Q. (2013). Medial temporal lobe volume predicts elders' everyday memory. *Psychological Science, 24*(7), 1113–1122.

Baillargeon, R. (1986). Representing the existence and the location of hidden objects: Object permanence in 6- and 8-month-old infants. *Cognition, 23,* 21–41.

Baillargeon, R. (1987). Object permanence in 3½- and 4½-month-old infants. *Developmental Psychology, 23,* 655–664.

Baillargeon, R., Spelke, E. S., & Wasserman, S. (1985). Object permanence in five-month-old infants. *Cognition, 20,* 191–208.

Baird, B., Smallwood, J., Mrazek, M. D., Kam, J. W., Franklin, M. S., & Schooler, J. W. (2012). Inspired by distraction mind wandering facilitates creative incubation. *Psychological Science, 23*(10), 1117–1122.

Balota, D. A, & Duchek, J. M. (1988). Age-related differences in lexical access, spreading activation, and simple pronunciation. *Psychology and Aging, 3,* 84–93.

Balota, D. A., & Paul, S. T. (1996). Summation of activation: Evidence from multiple primes that converge and diverge within semantic memory. *Journal of Experimental Psychology: Learning, Memory, and Cognition, 22,* 827–845.

Balota, D. A., & Yap, M. J. (2011). Moving beyond the mean in studies of mental chronometry: The power of response time distributional analyses. *Current Directions in Psychological Science, 20,* 160–166.

Banich, M. T. (1997). *Neuropsychology: The neural bases of mental function.* Boston, MA: Houghton Mifflin.

Banich, M. T. (2004). *Cognitive neuroscience and neuropsychology* (2nd ed.). Boston, MA: Houghton Mifflin.

Banich, M. T. (2009). Executive function: The search for an integrated account. *Current Directions in Psychological Science, 18,* 89–94.

Banks, W. P. (1977). Encoding and processing of symbolic information in comparative judgments. In G. H. Bower (Ed.), *The psychology of learning and motivation* (Vol. 11, pp. 101–159). New York, NY: Academic Press.

Banks, W. P., Clark, H. H., & Lucy, P. (1975). The locus of the semantic congruity effect in comparative judgments. *Journal of Experimental Psychology: Human Perception and Performance, 1,* 35–47.

Banks, W. P., Fujii, M., & Kayra-Stuart, F. (1976). Semantic congruity effects in comparative judgments of magnitude of digits. *Journal of Experimental Psychology: Human Perception and Performance, 2,* 435–447.

Bar-Hillel, M. (1980). What features make samples seem representative? *Journal of Experimental Psychology: Human Perception and Performance, 6,* 578–589.

Barnard, P. J., Scott, S., Taylor, J., May, J., & Knightley, W. (2004). Paying attention to meaning. *Psychological Science, 15,* 179–186.

Barner, D., Li, P., & Snedeker, J. (2010). Words as windows to thought: The case of object representation. *Current Directions in Psychological Science, 19*, 195–200.

Barnier, A. J. (2002). Posthypnotic amnesia for autobiographical episodes: A laboratory model for functional amnesia? *Psychological Science, 13*, 232–237.

Barron, E., Riby, L. M., Greer, J., & Smallwood, J. (2011). Absorbed in thought: The effect of mind wandering on the processing of relevant and irrelevant events. *Psychological Science, 22*, 596–601.

Barsalou, L. W. (1983). Ad hoc categories. *Memory & Cognition, 11*, 211–227.

Barsalou, L. W. (1999). Perceptual symbol systems. *Behavioral and Brain Sciences, 22*, 577–660.

Barshi, I., & Healy, A. F. (1993). Checklist procedures and the cost of automaticity. *Memory & Cognition, 21*, 496–505.

Bartlett, F. C. (1932). *Remembering: A study in experimental and social psychology.* London, England: Cambridge University Press.

Bassok, M., & Holyoak, K. H. (1989). Interdomain transfer between isomorphic topics in algebra and physics. *Journal of Experimental Psychology: Learning, Memory, and Cognition, 15*, 153–166.

Bassok, M., Pedigo, S. F., & Oskarsson, A. T. (2008). Priming addition facts with semantic relations. *Journal of Experimental Psychology: Learning, Memory, and Cognition, 34*, 343–352.

Bates, E., Masling, M., & Kintsch, W. (1978). Recognition memory for aspects of dialogue. *Journal of Experimental Psychology: Human Learning and Memory, 4*, 187–197.

Battig, W. F., & Montague, W. E. (1969). Category norms for verbal items in 56 categories: A replication and extension of the Connecticut category norms. *Journal of Experimental Psychology Monograph, 80*(3, Pt. 2), 1–46.

Bauer, P. J. (2007). Recall in infancy. *Current Directions in Psychological Science, 16*, 142–146.

Bauer, P. J., & Larkina, M. (2014). Childhood amnesia in the making: Different distributions of autobiographical memories in children and adults. *Journal of Experimental Psychology: General, 143*(2), 597–611.

Bäuml, K-H. T., & Samenieh, A. (2010). The two faces of memory retrieval. *Psychological Science, 21*, 793–795.

Bavelas, J., Gerwing, J., Sutton, C., & Prevost, D. (2008). Gesturing on the telephone: Independent effects of dialogue and visibility. *Journal of Memory and Language, 58*, 495–520.

Bayen, U. J., Erdfelder, E., Bearden, J. N., & Lozito, J. P. (2006). The interplay of memory and judgment processes in effects of aging on hindsight bias. *Journal of Experimental Psychology: Learning, Memory, and Cognition, 32*, 1003–1018.

Bechera, A., Tranel, D., Damasio, H., Adolphs, R., Rockland, C., & Damasio, A. R. (1995). Double dissociation of conditioning and declarative knowledge relative to the amygdala and hippocampus in humans. *Science, 269*, 1115–1118.

Becker, M. W. (2009). Panic search: Fear produces efficient visual search for nonthreatening objects. *Psychological Science, 20*, 435–437.

Beech, A., Powell, T., McWilliams, J., & Claridge, G. (1989). Evidence of reduced cognitive inhibition in schizophrenics. *British Journal of Psychology, 28*, 109–116.

Beeman, M. J. (1993). Semantic processing in the right hemisphere may contribute to drawing inferences from discourse. *Brain and Language, 44*, 80–120.

Beeman, M. J. (1998). Coarse semantic coding and discourse comprehension. In M. Beeman & C. Chiarello (Eds.), *Brain right hemisphere language comprehension: Perspectives from cognitive neuroscience* (pp. 255–284). Mahwah, NJ: Erlbaum.

Beeman, M. J., & Bowden, E. M. (2000). The right hemisphere maintains solution-related activation for yet-to-be-solved problems. *Memory & Cognition, 28*, 1231–1241.

Beeman, M. J., & Chiarello, C. (1998). Complementary right- and left-hemisphere language comprehension. *Current Directions in Psychological Science, 7*, 2–8.

Beilock, S. L. (2008). Math performance in stressful situations. *Current Directions in Psychological Science, 17*, 339–343.

Beilock, S. L., Bertenthal, B. I., McCoy, A. M., & Carr, T. H. (2004). Haste does not always make waste: Expertise, direction of attention, and speed versus accuracy in performing sensorimotor skills. *Psychonomic Bulletin & Review, 11*, 373–379.

Beilock, S. L., & Carr, T. H. (2001). On the fragility of skilled performance: What governs choking under pressure? *Journal of Experimental Psychology: Learning, Memory, and Cognition, 33*, 983–998.

Beilock, S. L., & DeCaro, M. S. (2007). From poor performance to success under stress: Working memory, strategy selection, and mathematical problem solving under pressure. *Journal of Experimental Psychology, 130*, 701–725.

Beilock, S. L., Kulp, C. A., Holt, L. E., & Carr, T. H. (2004). More on the fragility of performance: Choking under pressure in mathematical problem solving. *Journal of Experimental Psychology: General, 133*, 584–600.

Beilock, S. L., Rydell, R. J., & McConnell, A. R. (2007). Stereotype threat and working memory: Mechanisms, alleviation, and spillover. *Journal of Experimental Psychology: General, 136*, 256–276.

Bekoff, M., Allen, C., & Burghardt, G. (Eds.). (2002). *The cognitive animal: Empirical and theoretical perspectives on animal cognition.* Cambridge, MA: MIT Press.

Bellezza, F. S. (1992). Recall of congruent information in the self-reference task. *Bulletin of the Psychonomic Society, 30*, 275–278.

Belli, R. F. (1989). Influences of misleading postevent information: Misinformation interference and acceptance. *Journal of Experimental Psychology: General, 118*, 72–85.

Belli, R. F., Lindsay, D. S., Gales, M. S., & McCarthy, T. T. (1994). Memory impairment and source misattribution in postevent misinformation experiments with short retention intervals. *Memory & Cognition, 22*, 40–54.

Ben-David, B. M., Erel, H., Goy, H., & Schneider, B. A. (2015). "Older is always better": Age-related differences in vocabulary scores across 16 years. *Psychology and Aging, 30*(4), 856–862.

Benjamin, A. S., & Tullis, J. (2010). What makes distributed practice effective? *Cognitive Psychology, 61*(3), 228–247.

Benjamin, L. T., Jr., Durkin, M., Link, M., Vestal, M., & Acord, J. (1992). Wundt's American doctoral students. *American Psychologist, 47*, 123–131.

Benson, D. J., & Geschwind, N. (1969). The alexias. In P. Vincken & G. W. Bruyn (Eds.), *Handbook of clinical neurology* (Vol. 4, pp. 112–140). Amsterdam, The Netherlands: North-Holland.

Benz, S., Sellaro, R., Hommel, B., & Colzato, L. S. (2015). Music makes the world go round: The impact of musical training on non-musical cognitive functions—A review. *Frontiers in Psychology, 6,* 2023.

Berger, A., Henik, A., & Rafal, R. (2005). Competition between endogenous and exogenous orienting of visual attention. *Journal of Experimental Psychology: General, 134,* 207–221.

Berwick, R. C., Hauser, M., & Tattersall, I. (2013). Neanderthal language? Just-so stories take center stage. *Frontiers in Psychology, 4,* 671.

Berman, M. G., Jonides, J., & Kaplan, S. (2008). The cognitive benefits of interacting with nature. *Psychological Science, 19,* 1207–1212.

Berman, M. G., Jonides, J., & Lewis, R. L. (2009). In search of decay in verbal short-term memory. *Journal of Experimental Psychology: Learning, Memory, and Cognition, 35*(2), 317–333.

Berndt, R. S., & Haendiges, A. N. (2000). Grammatical class in word and sentence production: Evidence from an aphasic patient. *Journal of Memory and Language, 43,* 249–273.

Bernstein, D. M., & Loftus, E. F. (2009). How to tell if a particular memory is true or false. *Perspectives on Psychological Science, 4,* 370–374.

Berntsen, D. (2010). The unbidden past: Involuntary autobiographical memories as a basic mode of remembering. *Current Directions in Psychological Science, 19,* 138–142.

Berntsen, D., & Bohn, A. (2010). Remembering and forecasting: The relation between autobiographical memory and episodic future thinking. *Memory & Cognition, 38*(3), 265–278.

Berntsen, D., & Hall, N. M. (2004). The episodic nature of involuntary autobiographical memories. *Memory & Cognition, 32,* 789–803.

Berntsen, D., & Rubin, D. C. (2004). Cultural life scripts structure recall from autobiographical memory. *Memory & Cognition, 32,* 427–442.

Berntsen, D., & Rubin, D. C. (2006). Flashbulb memories and posttraumatic stress reactions across the life span: Age-related effects of the German occupation of Denmark during World War II. *Psychology and Aging, 21,* 127–139.

Bertsch, S., Pesta, B. J., Wiscott, R., & McDaniel, M. A. (2007). The generation effect: A meta-analytic review. *Memory & Cognition, 35,* 201–210.

Berwick, R. C., Hauser, M., & Tattersall, I. (2013). Neanderthal language? Just-so stories take center stage. *Frontiers in Psychology, 4,* 671.

Besner, D., & Stolz, J. A. (1999). What kind of attention modulates the Stroop effect? *Psychonomic Bulletin & Review, 6,* 99–104.

Best, R., & Charness, N. (2015). Age differences in the effect of framing on risky choice: A meta-analysis. *Psychology and Aging, 30*(3), 688–698.

Bialystok, E. (1988). Levels of bilingualism and levels of linguistic awareness. *Developmental Psychology, 24,* 560–567.

Bialystok, E., Craik, F. I., & Luk, G. (2012). Bilingualism: Consequences for mind and brain. *Trends in Cognitive Sciences, 16*(4), 240–250.

Biederman, I. (1987). Recognition by components: A theory of human image understanding. *Psychological Review, 94,* 115–147.

Biederman, I. (1990). Higher-level vision. In E. N. Osherson, S. M. Kosslyn, & J. M. Hollerbach (Eds.), *An invitation to cognitive science* (Vol. 2, pp. 41–72). Cambridge, MA: MIT Press.

Biederman, I., & Blickle, T. (1985). *The perception of objects with deleted contours*. Unpublished manuscript, State University of New York, Buffalo.

Biederman, I., Glass, A. L., & Stacy, E. W. (1973). Searching for objects in real world scenes. *Journal of Experimental Psychology, 97,* 22–27.

Biggs, A. T., Brockmole, J. R., & Witt, J. K. (2013). Armed and attentive: Holding a weapon can bias attentional priorities in scene viewing. *Attention, Perception, & Psychophysics, 75*(8), 1715–1724.

Bigler, E. D., Yeo, R. A., & Turkheimer, F. (1989). Introduction and overview. In E. D. Bigler, R. A. Yeo, & F. Turkheimer (Eds.), *Neuropsychological function and brain imaging* (p. 10). New York, NY: Plenum Press.

Bilalić, M., McLeod, P., & Gobet, F. (2008). Inflexibility of experts—Reality or myth? Quantifying the Einstellung effect in chess masters. *Cognitive Psychology, 56,* 73–102.

Bilalić, M., McLeod, P., & Gobet, F. (2010). The mechanism of the Einstellung (set) effect: A pervasive source of cognitive bias. *Psychological Science, 19,* 111–115.

Binder, K. S. (2003). Sentential and discourse topic effects on lexical ambiguity processing: An eye movement examination. *Memory & Cognition, 31,* 690–702.

Binder, K. S., & Rayner, K. (1998). Contextual strength does not modulate the subordinate bias effect: Evidence from eye fixations and self-paced reading. *Psychonomic Bulletin & Review, 5,* 271–276.

Bireta, T. J., Surprenant, A. M., & Neath, I. (2008). Age-related differences in the von Restorff isolation effect. *The Quarterly Journal of Experimental Psychology, 61,* 345–352.

Birmingham, E., Bischof, W. F., & Kingstone, A. (2008). Social attention and real world scenes: The roles of action, competition and social content. *The Quarterly Journal of Experimental Psychology, 61,* 986–998.

Bisiach, E., & Luzzatti, C. (1978). Unilateral neglect of representational space. *Cortex, 14,* 129–133.

Biswas, D. (2009). The effects of option framing on consumer choices: Making decisions in rational versus experiential processing modes. *Journal of Consumer Behaviour, 8*(5), 284–299.

Bjork, E. L., & Bjork, R. A. (2003). Intentional forgetting can increase, not decrease, residual influences of to-be-forgotten information. *Journal of Experimental Psychology: Learning, Memory, and Cognition, 29,* 524–531.

Bjorklund, D. F., Dukes, C., & Brown, R. D. (2009). The development of memory strategies. In M. L. Courage & N. Cowan (Eds.), *The development of memory in infancy and childhood* (pp. 145–175). New York, NY: Psychology Press.

Blair, M., Vadaga, K. K., Shuchat, J., & Li, K. Z. H. (2011). The role of age and inhibitory efficiency in working memory processing and storage components. *The Quarterly Journal of Experimental Psychology, 64,* 1157–1172.

Blakemore, C. (1977). *Mechanics of the mind*. Cambridge, England: Cambridge University Press.

Blanchette, I., & Richards, A. (2004). Reasoning about emotional and neutral materials. *Psychological Science, 15*, 745–752.

Blaser, E., & Kaldy, Z. (2010). Infants get five stars on iconic memory tests: A partial-report test of 6-month-old infant's iconic memory capacity. *Psychological Science, 21*, 1643–1645.

Bliss, T. V. P., & Collingridge, G. L. (1993). A synaptic model of memory: Long-term potentiation in the hippocampus. *Nature, 232*, 31–39.

Bliss, T. V. P., & Lomo, T. (1973). Long-lasting potentiation of synaptic transmission in the dentate area of the anaesthetized rabbit following stimulations of the preforant path. *The Journal of Physiology, 232*, 331–356.

Bloesch, E. K., Davoli, C. C., Roth, N., Brockmole, J. R., & Abrams, R. A. (2012). Watch this! Observed tool use affects perceived distance. *Psychonomic Bulletin & Review, 19*, 177–183.

Blom, J. P., & Gumperz, J. J. (1972). Social meaning in linguistic structure: Code-switching in Norway. In J. J. Gumperz & D. Hymes (Eds.), *Directions in sociolinguistics: The ethnography of communication* (pp. 407–434). New York, NY: Holt.

Bloom, F. E., & Lazerson, A. (1988). *Brain, mind, and behavior*. New York, NY: W. H. Freeman.

Bocanegra, B. R., & Zeelenberg, R. (2011). Emotion-induced trade-offs in spatiotemporal vision. *Journal of Experimental Psychology: General, 140*, 272–282.

Bock, J. K. (1982). Toward a cognitive psychology of syntax: Information processing contributions to sentence formulation. *Psychological Review, 89*, 1–47.

Bock, J. K. (1986). Meaning, sound, and syntax: Lexical priming in sentence production. *Journal of Experimental Psychology: Learning, Memory, and Cognition, 12*, 575–586.

Bock, K. (1995). Producing agreement. *Current Directions in Psychological Science, 4*, 56–61.

Bock, K. (1996). Language production: Methods and methodologies. *Psychonomic Bulletin & Review, 3*, 395–421.

Bock, K., & Griffin, Z. M. (2000). The persistence of structural priming: Transient activation or implicit learning? *Journal of Experimental Psychology: General, 129*, 177–192.

Bock, K., Irwin, D. E., Davidson, D. J., & Levelt, W. J. M. (2003). Minding the clock. *Journal of Memory and Language, 48*, 653–685.

Bock, K., & Miller, C. A. (1991). Broken agreement. *Cognitive Psychology, 23*, 45–93.

Bogg, T., & Lasecki, L. (2015). Reliable gains? Evidence for substantially underpowered designs in studies of working memory training transfer to fluid intelligence. *Frontiers in Psychology, 5*, 1589.

Bohn, M., Call, J., & Tomasello, M. (2015). Communication about absent entities in great apes and human infants. *Cognition, 145*, 63–72.

Bohn-Gettler, C. M., Rapp, D. N., van den Broek, P., Kendeou, P., & White, M. J. (2011). Adults' and children's monitoring of story events in the service of comprehension. *Memory & Cognition, 39*, 992–1011.

Bonebakker, A. E., Bonke, B., Klein, J., Wolters, G., Stijnen, T., Passchier, J., et al. (1996). Information processing during general anesthesia: Evidence for unconscious memory. *Memory & Cognition, 24*, 766–776.

Bonnefon, J. F., & Hilton, D. J. (2004). Consequential conditionals: Invited and suppressed inferences from valued outcomes. *Journal of Experimental Psychology: Learning, Memory, and Cognition, 30*, 28–37.

Boot, W. R., Blakely, D. P., & Simons, D. J. (2011). Do action video games improve perception and cognition? *Frontiers in Psychology, 2*, 226.

Borghi, A. M., Glenberg, A. M., & Kaschak, M. P. (2004). Putting words in perspective. *Memory & Cognition, 32*, 863–873.

Boring, E. G. (1950). *A history of experimental psychology* (2nd ed.). New York, NY: Appleton-Century-Crofts.

Boroditsky, L. (2001). Does language shape thought? Mandarin and English speakers' conceptions of time. *Cognitive Psychology, 43*, 1–22.

Boroditsky, L. (2011). How language shapes thought. *Scientific American, 304*, 63–65.

Botvinick, M. M., & Bylsma, L. M. (2006). Distraction and action slips in an everyday task: Evidence for a dynamic representation of task context. *Psychonomic Bulletin & Review, 12*, 1011–1017.

Bouazzaoui, B., Follenfant, A., Ric, F., Fay, S., Croizet, J. C., Atzeni, T., & Taconnat, L. (2016). Ageing-related stereotypes in memory: When the beliefs come true. *Memory, 24*(5), 659–668.

Bousfield, W. A. (1953). The occurrence of clustering in the recall of randomly arranged associates. *The Journal of General Psychology, 49*, 229–240.

Bousfield, W. A., & Sedgewick, C. H. W. (1944). An analysis of sequences of restricted associative responses. *The Journal of General Psychology, 30*, 149–165.

Bowden, E. M. (1985). Accessing relevant information during problem solving: Time constraints on search in the problem space. *Memory & Cognition, 13*, 280–286.

Bowden, E. M., & Beeman, M. J. (1998). Getting the right idea: Semantic activation in the right hemisphere may help solve insight problems. *Psychological Science, 9*, 435–440.

Bowden, E. M., Beeman, M., & Gernsbacher, M. A. (1995, March). *Two hemispheres are better than one: Drawing coherence inferences during story comprehension*. Paper presented at the annual meeting of the Cognitive Neuroscience Society, San Francisco, CA.

Bower, G. H. (1970). Analysis of a mnemonic device. *American Scientist, 58*, 496–510.

Bower, G. H. (1981). Mood and memory. *American Psychologist, 36*, 129–148.

Bower, G. H., Black, J. B., & Turner, T. J. (1979). Scripts in memory for text. *Cognitive Psychology, 11*, 177–220.

Bower, G. H., & Clark, M. C. (1969). Narrative stories as mediators for serial learning. *Psychonomic Science, 14*(4), 181–182.

Bower, G. H., Clark, M. C., Lesgold, A. M., & Winzenz, D. (1969). Hierarchical retrieval schemes in recall of categorical word lists. *Journal of Verbal Learning and Verbal Behavior, 8*, 323–343.

Bowers, J. S., & Davis, C. J. (2012). Bayesian just-so stories in psychology and neuroscience. *Psychological Bulletin, 138*(3), 389–414.

Bowers, J. S., & Jones, K. W. (2008). Detecting objects is easier than categorizing them. *The Quarterly Journal of Experimental Psychology, 61*, 552–557.

Bradley, M. M., Greenwald, M. K., Petry, M. C., & Lang, P. J. (1992). Remembering pictures: Pleasure and arousal in memory. *Journal of Experimental Psychology: Learning, Memory, and Cognition, 18*, 379–390.

Brady, T. F., Konkle, T., Alvarez, G. A., & Oliva, A. (2008). Visual long-term memory has a massive storage capacity for object details. *Proceedings of the National Academy of Sciences, 105*(38), 14325–14329.

Brainerd, C. J., Wang, Z., Reyna, V. F., & Nakamura, K. (2015). Episodic memory does not add up: Verbatim–gist superposition predicts violations of the additive law of probability. *Journal of Memory and Language, 84*, 224–245.

Branigan, H. P., Pickering, M. J., & Cleland, A. A. (1999). Syntactic priming in written production: Evidence for rapid decay. *Psychonomic Bulletin & Review, 6*, 635–640.

Branigan, H. P., Pickering, M. J., Stewart, A. J., & McLean, J. F. (2000). Syntactic priming in spoken production: Linguistic and temporal interference. *Memory & Cognition, 28*, 1297–1302.

Bransford, J. D. (1979). *Human cognition: Learning, understanding and remembering.* Belmont, CA: Wadsworth.

Bransford, J. D., & Franks, J. J. (1971). The abstraction of linguistic ideas. *Cognitive Psychology, 2*, 331–350.

Bransford, J. D., & Franks, J. J. (1972). The abstraction of linguistic ideas: A review. *Cognition: International Journal of Cognitive Psychology, 1*(2), 211–249.

Bransford, J. D., & Johnson, M. K. (1972). Contextual prerequisites for understanding: Some investigations of comprehension and recall. *Journal of Verbal Learning and Verbal Behavior, 11*, 717–726.

Bransford, J. D., & Stein, B. S. (1984). *The ideal problem solver.* New York, NY: Freeman.

Bransford, J. D., & Stein, B. S. (1993). *The ideal problem solver* (2nd ed.). New York, NY: Freeman.

Braver, T. S., Barch, D. M., Keys, B. A., Carter, C. S., Cohen, J. D., Kaye, J. A., et al. (2001). Context processing in older adults: Evidence for a theory relating cognitive control to neurobiology in healthy aging. *Journal of Experimental Psychology: General, 130*, 746–763.

Breedin, S. D., & Saffran, E. M. (1999). Sentence processing in the face of semantic loss: A case study. *Journal of Experimental Psychology: General, 128*, 547–562.

Bregman, A. S. (1990). *Auditory scene analysis: The perceptual organization of sound.* Cambridge, MA: MIT Press.

Bremner, J. D., Shobe, K. K., & Kihlstrom, J. F. (2000). False memories in women with self-reported childhood sexual abuse: An empirical study. *Psychological Science, 11*, 333–337.

Brendel, E., DeLucia, P. R., Hecht, H., Stacy, R. L., & Larsen, J. T. (2012). Threatening pictures induce shortened time-to-contact estimates. *Attention, Perception, & Psychophysics, 74*(5), 979–987.

Brennan, S. E., & Clark, H. H. (1996). Conceptual pacts and lexical choice in conversation. *Journal of Experimental Psychology: Learning, Memory, and Cognition, 22*, 1482–1493.

Breslin, C. W., & Safer, M. A. (2011). Effects of event valence on long-term memory for two baseball championship games. *Psychological Science, 22*, 1408–1412.

Bresnan, J. (1978). A realistic transformational grammar. In J. Bresnan, M. Halle, & G. Miller (Eds.), *Linguistic theory and psychological reality* (pp. 1–59). Cambridge, MA: MIT Press.

Bresnan, J., & Kaplan, R. M. (1982). Introduction: Grammars as mental representations of language. In J. Bresnan (Ed.), *The mental representation of grammatical relations* (pp. xvii–iii). Cambridge, MA: MIT Press.

Bridgeman, B. (1988). *The biology of behavior and mind.* New York, NY: Wiley.

Broadbent, D. E. (1952). Speaking and listening simultaneously. *Journal of Experimental Psychology, 43*, 267–273.

Broadbent, D. E. (1958). *Perception and communication.* London, England: Pergamon.

Broaders, S. C., Cook, S. W., Mitchell, Z., & Goldin-Meadow, S. (2007). Making children gesture brings out implicit knowledge and leads to learning. *Journal of Experimental Psychology: General, 136*, 539–550.

Broaders, S. C., & Goldin-Meadow, S. (2010). Truth is at hand: How gesture adds information during investigative interviews. *Psychological Science, 21*, 623–628.

Brockmole, J. R., & Logie, R. H. (2013). Age-related change in visual working memory: A study of 55,753 participants aged 8–75. *Frontiers in Psychology, 4*, 12.

Bröder, A., & Schiffer, S. (2006). Adaptive flexibility and maladaptive routines in selecting fast and frugal decision strategies. *Journal of Experimental Psychology: Learning, Memory, and Cognition, 32*, 904–918.

Broggin, E., Savazzi, S., & Marzi, C. A. (2012). Similar effects of visual perception and imagery on simple reaction time. *The Quarterly Journal of Experimental Psychology, 65*, 151–164.

Bronkhorst, A. W. (2015). The cocktail-party problem revisited: Early processing and selection of multi-talker speech. *Attention, Perception, & Psychophysics, 77*(5), 1465–1487.

Brooks, A. W. (2014). Get excited: Reappraising pre-performance anxiety as excitement. *Journal of Experimental Psychology: General, 143*(3), 1144–1158.

Brooks, J. O. III, & Watkins, M. J. (1990). Further evidence of the intricacy of memory span. *Journal of Experimental Psychology: Learning, Memory, and Cognition, 16*, 1134–1141.

Brooks, L. R. (1968). Spatial and verbal components of the act of recall. *Canadian Journal of Experimental Psychology, 22*, 349–368.

Brosch, T., Sander, D., Pourtois, G., & Scherer, K. R. (2008). Beyond fear: Rapid spatial orienting toward positive emotional stimuli. *Psychological Science, 19*, 362–370.

Brown, A. S. (1991). A review of the tip-of-the-tongue experience. *Psychological Bulletin, 109*, 204–223.

Brown, A. S. (1998). Transient global amnesia. *Psychonomic Bulletin & Review, 5*, 401–427.

Brown, A. S. (2002). Consolidation theory and retrograde amnesia in humans. *Psychonomic Bulletin & Review, 9*, 403–425.

Brown, A. S. (2004). The déjà vu illusion. *Current Directions in Psychological Science, 13*, 256–259.

Brown, A. S., & Halliday, H. E. (1991). Cryptomnesia and source memory difficulties. *American Journal of Psychology, 104*, 475–490.

Brown, A. S., & Murphy, D. R. (1989). Cryptomnesia: Delineating inadvertent plagiarism. *Journal of Experimental Psychology: Learning, Memory, and Cognition, 15*, 432–442.

Brown, A. S., Neblett, D. R., Jones, T. C., & Mitchell, D. B. (1991). Transfer of processing in repetition priming: Some inappropriate findings. *Journal of Experimental Psychology: Learning, Memory, and Cognition, 17*, 514–525.

Brown, A. S., & Nix, L. A. (1996). Turning lies into truths: Referential validation of falsehoods. *Journal of Experimental Psychology: Learning, Memory, and Cognition, 22*, 1088–1100.

Brown, G. D. A., Neath, I., & Chater, N. (2007). A temporal ratio model of memory. *Psychological Review, 114*, 539–576.

Brown, J. A. (1958). Some tests of the decay theory of immediate memory. *The Quarterly Journal of Experimental Psychology, 10*, 12–21.

Brown, N. R., & Siegler, R. S. (1992). The role of availability in the estimation of national populations. *Memory & Cognition, 20*, 406–412.

Brown, R. (1973). *A first language: The early stages.* Cambridge, MA: Harvard University Press.

Brown, R., & Ford, M. (1961). Address in American English. *Journal of Abnormal and Social Psychology, 62*, 375–385.

Brown, R., & Kulik, J. (1977). Flashbulb memories. *Cognition, 5*, 73–99.

Brown, R., & McNeill, D. (1966). The "tip-of-the-tongue" phenomenon. *Journal of Verbal Learning and Verbal Behavior, 5*, 325–337.

Bruner, J. S., Goodnow, J. J., & Austin, G. A. (1956). *A study of thinking.* New York, NY: Wiley.

Bruza, P. D., Wang, Z., & Busemeyer, J. R. (2015). Quantum cognition: A new theoretical approach to psychology. *Trends in Cognitive Sciences, 19*(7), 383–393.

Bryden, M. P. (1982). *Laterality: Functional asymmetry in the intact human brain.* New York, NY: Academic Press.

Buchanan, T. W., Lutz, K., Mirzazade, S., Specht, K., Shah, N. J., Zilles, K., et al. (2000). Recognition of emotional prosody and verbal components of spoken language: An fMRI study. *Cognitive Brain Research, 9*, 227–238.

Buchner, A., & Wippich, W. (2000). On the reliability of implicit and explicit memory measures. *Cognitive Psychology, 40*, 227–259.

Buckner, R. L. (1996). Beyond HERA: Contributions of specific prefrontal brain areas to long-term memory retrieval. *Psychonomic Bulletin & Review, 3*, 149–158.

Buckner, R. L., Andrews-Hanna, J. R., & Schacter, D. L. (2008). The brain's default network. *Annals of the New York Academy of Sciences, 1124*(1), 1–38.

Bundesen, C. (1990). A theory of visual attention. *Psychological Review, 97*, 523–547.

Bunting, M. F., Conway, A. R. A., & Heitz, R. P. (2004). Individual differences in the fan effect and working memory capacity. *Journal of Memory and Language, 51*, 604–622.

Burgess, G. C., Gray, J. R., Conway, A. R. A., & Braver, T. S. (2011). Neural mechanisms of interference control underlie the relationship between fluid intelligence and working memory span. *Journal of Experimental Psychology: General, 140*, 674–692.

Burgess, P. W., Scott, S. K., & Frith, C. D. (2003). The role of the rostral frontal cortex (area 10) in prospective memory: A lateral versus medial dissociation. *Neuropsychologia, 41*, 906–918.

Burgoon, E. M., Henderson, M. D., & Markman, A. B. (2013). There are many ways to see the forest for the trees: A tour guide for abstraction. *Perspectives on Psychological Science, 8*(5), 501–520.

Burke, D. M., MacKay, D. G., Worthley, J. S., & Wade, E. (1991). On the tip of the tongue: What causes word finding failures in young and older adults? *Journal of Memory and Language, 30*, 542–579.

Burns, B. D. (2004). Heuristics as beliefs and as behaviors: The adaptiveness of the "hot hand." *Cognitive Psychology, 48*, 295–331.

Burns, B. D., & Corpus, B. (2004). Randomness and inductions from streaks: "Gambler's fallacy" versus "hot hand." *Psychonomic Bulletin & Review, 11*, 179–184.

Burns, Z. C., Caruso, E. M., & Bartels, D. M. (2012). Predicting premeditation: Future behavior is seen as more intentional than past behavior. *Journal of Experimental Psychology: General, 141*(2), 227–232.

Burt, C. D. B. (1992). Reconstruction of the duration of autobiographical events. *Memory & Cognition, 20*, 124–132.

Burt, C. D. B., Kemp, S., & Conway, M. (2001). What happens if you retest autobiographical memory 10 years on? *Memory & Cognition, 29*, 127–136.

Busemeyer, J. R., & Wang, Z. (2015). What is quantum cognition, and how is it applied to psychology? *Current Directions in Psychological Science, 24*(3), 163–169.

Busey, T. A., & Parada, E. F. (2010). The nature of expertise in fingerprint examiners. *Psychonomic Bulletin & Review, 17*, 155–160.

Busey, T. A., Tunnicliff, J., Loftus, G. R., & Loftus, E. F. (2000). Accounts of the confidence–accuracy relation in recognition memory. *Psychonomic Bulletin & Review, 7*, 26–48.

Buswell, G. T. (1937). *How adults read.* Chicago, IL: Chicago University Press.

Butler, A. C. (2010). Repeated testing produces superior transfer of learning relative to repeated studying. *Journal of Experimental Psychology: Learning, Memory, and Cognition, 36*, 1118–1133.

Butler, C., Muhlert, N., & Zeman, A. (2010). Accelerated long-term forgetting. In S. Della Sala (Ed.), *Forgetting* (pp. 211–238). New York, NY: Psychology Press.

Butler, J., & Rovee-Collier, C. (1989). Contextual gating of memory retrieval. *Developmental Psychobiology, 22*, 533–552.

Byrne, R. M. J., & McEleney, A. (2000). Counterfactual thinking about actions and failures to act. *Journal of Experimental Psychology: Learning, Memory, and Cognition, 26*, 1318–1331.

Cabeza, R. (2002). Hemispheric asymmetry reduction in older adults: The HAROLD model. *Psychology and Aging, 17*, 85–100.

Campbell, J. I. (1987). Network interference and mental multiplication. *Journal of Experimental Psychology: Learning, Memory, and Cognition, 13*(1), 109–123.

Campbell, J. I. D., & Graham, D. J. (1985). Mental multiplication skill: Structure, process, and acquisition. *Canadian Journal of Psychology, 39*, 338–366.

Cantor, J., & Engle, R. W. (1993). Working-memory capacity as long-term memory activation: An individual differences approach. *Journal of Experimental Psychology: Learning, Memory, and Cognition, 19*, 1101–1114.

Carli, L. L. (1999). Cognitive reconstruction, hindsight, and reactions to victims and perpetrators. *Personality and Social Psychology Bulletin, 25*, 966–979.

Carlson, R. A., Khoo, B. H., Yaure, R. G., & Schneider, W. (1990). Acquisition of a problem-solving skill: Levels of organization and use of working memory. *Journal of Experimental Psychology: General, 119*, 193–214.

Carpenter, S. K., & Pashler, H. (2007). Testing beyond words: Using tests to enhance visuospatial map learning. *Psychonomic Bulletin & Review, 14*, 474–478.

Carr, P. B., & Steele, C. M. (2010). Stereotype threat affects financial decision making. *Psychological Science, 21*, 1411–1416.

Carr, T. H., McCauley, C., Sperber, R. D., & Parmalee, C. M. (1982). Words, pictures, and priming: On semantic activation, conscious identification, and the automaticity of information processing. *Journal of Experimental Psychology: Human Perception and Performance, 8*, 757–777.

Carriere, J. S. A., Cheyne, J. A., Solman, G. J. F., & Smilek, D. (2010). Age trends for failures of sustained attention. *Psychology and Aging, 25*, 569–574.

Carroll, D. W. (1986). *Psychology of language*. Pacific Grove, CA: Brooks/Cole.

Carstensen, L. L., & Mikels, J. A. (2005). At the intersection of emotion and cognition aging and the positivity effect. *Current Directions in Psychological Science, 14*, 117–121.

Casasanto, D. (2009). Embodiment of abstract concepts: Good and bad in right- and left-handers. *Journal of Experimental Psychology: General, 138*, 351–367.

Case, R. (1972). Validation of a neo-Piagetian mental capacity construct. *Journal of Experimental Child Psychology, 14*, 287–302.

Castel, A. D., Farb, N. A. S., & Craik, F. I. M. (2007). Memory for general and specific value information in younger and older adults: Measuring the limits of strategic control. *Memory & Cognition, 35*, 689–700.

Castel, A. D., McCabe, D. P., Roediger, H. L., & Heitman, J. L. (2007). The dark side of expertise: Domain-specific memory errors. *Psychological Science, 18*, 3–5.

Catrambone, R. (1996). Generalizing solution procedures learned from examples. *Journal of Experimental Psychology: Learning, Memory, and Cognition, 22*, 1020–1031.

Catrambone, R., Jones, C. M., Jonides, J., & Seifert, C. (1995). Reasoning about curvilinear motion: Using principles or analogy. *Memory & Cognition, 23*, 368–373.

Cavanagh, J. F., & Frank, M. J. (2014). Frontal theta as a mechanism for cognitive control. *Trends in Cognitive Sciences, 18*(8), 414–421.

Cave, K. R., & Bichot, N. P. (1999). Visuospatial attention: Beyond a spotlight model. *Psychonomic Bulletin & Review, 6*, 204–223.

Cave, K. R., & Kosslyn, S. M. (1989). Varieties of size-specific visual selection. *Journal of Experimental Psychology: General, 118*, 148–164.

Chaffin, R., & Imreh, G. (2002). Practicing perfection: Piano performance as expert memory. *Psychological Science, 13*, 342–349.

Chamberlain, R., & Wagemans, J. (2015). Visual arts training is linked to flexible attention to local and global levels of visual stimuli. *Acta Psychologica, 161*, 185–197.

Chambers, K. L., & Zaragoza, M. S. (2001). Intended and unintended effects of explicit warnings on eyewitness suggestibility: Evidence from source identification tests. *Memory & Cognition, 29*, 1120–1129.

Chan, J. C. K., Thomas, A. K., & Bulevich, J. B. (2009). Recalling a witnessed event increases eyewitness suggestibility. *Psychological Science, 20*, 66–73.

Chang, F., Dell, G. S., & Bock, J. K. (2006). Becoming syntactic. *Psychological Review, 113*, 234–272.

Chapman, H. A., Johannes, K., Poppenk, J. L., Moscovitch, M., & Anderson, A. K. (2013). Evidence for the differential salience of disgust and fear in episodic memory. *Journal of Experimental Psychology: General, 142*(4), 1100–1112.

Chapman, L. J., & Chapman, J. P. (1959). Atmosphere effect re-examined. *Journal of Experimental Psychology, 58*(3), 220–226.

Charles, S. T., Mather, M., & Carstensen, L. L. (2003). Aging and emotional memory: The forgettable nature of negative images for older adults. *Journal of Experimental Psychology: General, 132*, 310–324.

Chase, W. G., & Ericsson, K. A. (1982). Skill and working memory. In G. H. Bower (Ed.), *The psychology of learning and motivation* (Vol. 16, pp. 1–58). New York, NY: Academic Press.

Chase, W. G., & Simon, H. A. (1973). Perception in chess. *Cognitive Psychology, 4*, 55–81.

Chasteen, A. L., Park, D. C., & Schwarz, N. (2001). Implementation intentions and facilitation of prospective memory. *Psychological Science, 12*, 457–461.

Chater, N., & Oaksford, M. (1999). The probability heuristics model of syllogistic reasoning. *Cognitive Psychology, 38*, 191–258.

Chater, N., Tenenbaum, J. B., & Yuille, A. (2006). Probabilistic models of cognition: Conceptual foundations. *Trends in Cognitive Sciences, 10*(7), 287–291.

Chauvel, G., Maquestiaux, F., Hartley, A. A., Joubert, S., Didierjean, A., & Masters, R. S. W. (2012). Age effects shrink when motor learning is predominantly supported by nondeclarative, automatic memory processes: Evidence from golf putting. *The Quarterly Journal of Experimental Psychology, 65*, 25–38.

Chen, Y. (2007). Chinese and English speakers think about time differently? Failure of replicating Boroditsky (2001). *Cognition, 104*, 427–436.

Chen, Z. (2012). Object-based attention: A tutorial review. *Attention, Perception, & Psychophysics, 74*(5), 784–802.

Chen, Z., & Mo, L. (2004). Schema induction in problem solving: A multidimensional analysis. *Journal of Experimental Psychology: Learning, Memory, and Cognition, 30*, 583–600.

Chen, Z., Mo, L., & Honomichl, R. (2004). Having the memory of an elephant: Long-term retrieval and the use of analogues in problem solving. *Journal of Experimental Psychology: General, 133*, 415–433.

Cherniak, C. (1984). Prototypicality and deductive reasoning. *Journal of Verbal Learning and Verbal Behavior, 23*, 625–642.

Cherry, E. C. (1953). Some experiments on the recognition of speech, with one and with two ears. *Journal of the Acoustical Society of America, 25*, 975–979.

Cherry, E. C., & Taylor, W. K. (1954). Some further experiments on the recognition of speech with one and two ears. *Journal of the Acoustical Society of America, 26*, 554–559.

Chin-Parker, S., & Ross, B. H. (2004). Diagnosticity and prototypicality in category learning: A comparison of inference learning and classification learning. *Journal of Experimental Psychology: Learning, Memory, and Cognition, 30*, 216–226.

Chisholm, J. D., & Kingstone, A. (2015). Action video games and improved attentional control: Disentangling selection- and response-based processes. *Psychonomic Bulletin & Review, 22*(5), 1430–1436.

Chomsky, N. (1957). *Syntactic structures*. The Hague, The Netherlands: Mouton.

Chomsky, N. (1959). A review of Skinner's *Verbal Behavior. Language, 35*, 26–58.

Chomsky, N. (1965). *Aspects of a theory of syntax*. Cambridge, MA: Harvard University Press.

Chooi, W. T., & Thompson, L. A. (2012). Working memory training does not improve intelligence in healthy young adults. *Intelligence, 40*(6), 531–542.

Christianson, K., Hollingworth, A., Halliwell, J. F., & Ferreira, F. (2001). Thematic roles assigned along the garden path linger. *Cognitive Psychology, 42*, 368–407.

Christianson, S. (1989). Flashbulb memories: Special, but not so special. *Memory & Cognition, 17*, 435–443.

Chrobak, Q. M., & Zaragoza, M. S. (2008). Inventing stories: Forcing witnesses to fabricate entire fictitious events leads to freely reported false memories. *Psychonomic Bulletin & Review, 15*, 1190–1195.

Chronicle, E. P., MacGregor, J. N., & Ormerod, T. C. (2004). What makes an insight problem? The roles of heuristics, goal conception, and solution recoding in knowledge-lean problems. *Journal of Experimental Psychology: Learning, Memory, and Cognition, 30*, 14–27.

Chugani, H. T., Phelps, M. E., & Mazziotta, J. C. (1986). Positron emission tomography study of human brain functional development. *Annals of Neurology, 22*, 487–497.

Cisler, J. M., Wolitzky-Taylor, K. B., Adams, T. G., Babson, K. A., Badou, C. L., & Willems, J. L. (2011). The emotional Stroop task and posttraumatic stress disorder: A meta-analysis. *Clinical Psychology Review, 31*, 817–828.

Clancy, S. A., Schacter, D. L., McNally, R. J., & Pittman, R. K. (2000). False recognition in women reporting recovered memories of sexual abuse. *Psychological Science, 11*, 26–31.

Clapp, F. L. (1924). The number combinations: Their relative difficulty and frequency of their appearance in textbooks. *Research Bulletin No. 1*. Madison, WI: Bureau of Educational Research.

Clark, H. H. (1977). Bridging. In P. N. Johnson-Laird & P. C. Wason (Eds.), *Thinking: Readings in cognitive science* (pp. 411–420). Cambridge, England: Cambridge University Press.

Clark, H. H. (1979). Responding to indirect speech acts. *Cognitive Psychology, 11*, 430–477.

Clark, H. H. (1994). Discourse in production. In M. A. Gernsbacher (Ed.), *Handbook of psycholinguistics* (pp. 985–1021). San Diego, CA: Academic Press.

Clark, H. H., & Clark, E. V. (1977). *Psychology and language*. New York, NY: Harcourt Brace Jovanovich.

Clark, H. H., & Krych, M. A. (2004). Speaking while monitoring addressees for understanding. *Journal of Memory and Language, 50*, 62–81.

Clark, H. H., & Wasow, T. (1998). Repeating words in spontaneous speech. *Cognitive Psychology, 37*, 201–242.

Claus, B., & Kelter, S. (2006). Comprehending narratives containing flashbacks: Evidence for temporally organized representations. *Journal of Experimental Psychology: Learning, Memory, and Cognition, 32*, 1031–1044.

Cleary, A. M. (2008). Recognition memory, familiarity, and déjà vu experiences. *Current Directions in Psychological Science, 17*, 353–357.

Clifton, C., Jr., Carlson, K., & Frazier, L. (2006). Tracking the what and why of speakers' choices: Prosodic boundaries and the length of constituents. *Psychonomic Bulletin & Review, 13*, 854–861.

Clifton, C., & Frazier, L. (2004). Should given information come before new? Yes and no. *Memory & Cognition, 32*, 886–895.

Clifton, C., Jr., Traxler, M. J., Mohamed, M. T., Williams, R. S., Morris, R. K., & Rayner, K. (2003). The use of thematic role information in parsing: Syntactic processing autonomy revisited. *Journal of Memory and Language, 49*, 317–334.

Cohen, G. (1979). Language comprehension in old age. *Cognitive Psychology, 11*, 412–429.

Cohen, J. (1988). *Statistical power analysis for the behavioral sciences*. Hillsdale, NJ: Erlbaum.

Cohn, N. (2013a). Navigating comics: An empirical and theoretical approach to strategies of reading comic page layouts. *Frontiers in Psychology, 4*, 186.

Cohn, N. (2013b). Visual narrative structure. *Cognitive Science, 37*(3), 413–452.

Cohn, N., & Paczynski, M. (2013). Prediction, events, and the advantage of Agents: The processing of semantic roles in visual narrative. *Cognitive Psychology, 67*(3), 73–97.

Cokely, E. T., Kelley, C. M., & Gilchrist, A. L. (2006). Sources of individual differences in working memory capacity: Contributions of strategy to capacity. *Psychonomic Bulletin & Review, 13*, 991–997.

Colcombe, S. J., Kramer, A. F., Erickson, K. I., & Scalf, P. (2005). The implications of cortical recruitment and brain morphology for individual differences in inhibitory function in aging humans. *Psychology and Aging, 20*, 363–375.

Cole, G. G., & Kuhn, G. (2010). Attentional capture by object appearance and disappearance. *The Quarterly Journal of Experimental Psychology, 63*, 147–159.

Cole, S., Balcetis, E., & Dunning, D. (2013). Affective signals of threat increase perceived proximity. *Psychological Science, 24*(1), 34–40.

Colflesh, G. J. H., & Conway, A. R. A. (2007). Individual differences in working memory capacity and divided attention in dichotic listening. *Psychonomic Bulletin & Review, 14*, 699–703.

Colle, H. A., & Welsh, A. (1976). Acoustic masking in primary memory. *Journal of Verbal Learning and Verbal Behavior, 15*, 17–32.

Collins, A. M., & Loftus, E. F. (1975). A spreading-activation theory of semantic processing. *Psychological Review, 82*, 407–428.

Collins, A. M., & Quillian, M. R. (1969). Retrieval time from semantic memory. *Journal of Verbal Learning and Verbal Behavior, 8*, 240–247.

Collins, A. M., & Quillian, M. R. (1970). Does category size affect categorization time? *Journal of Verbal Learning and Verbal Behavior, 9*, 432–438.

Collins, A. M., & Quillian, M. R. (1972). How to make a language user. In E. Tulving & W. Donaldson (Eds.), *Organization of memory* (pp. 309–351). New York, NY: Academic Press.

Collins, K. A., Pillemer, D. B., Ivcevic, Z., & Gooze, R. A. (2007). Cultural scripts guide recall of intensely positive life events. *Memory & Cognition, 35*, 651–659.

Coltheart, M. (2013). How can functional neuroimaging inform cognitive theories? *Perspectives on Psychological Science, 8*(1), 98–103.

Comalli, P. E., Wapner, S., & Werner, H. (1962). Interference effects of Stroop color-word test in childhood, adulthood, and aging. *The Journal of Genetic Psychology: Research and Theory on Human Development, 100*, 47–53.

Coman, A., Manier, D., & Hirst, W. (2009). Forgetting the unforgettable through conversation: Socially shared retrieval-induced forgetting of September 11 memories. *Psychological Science, 20*, 627–633.

Connell, L., & Lynott, D. (2009). Is a bear white in the woods? Parallel representation of implied object color during language comprehension. *Psychonomic Bulletin & Review, 16*, 573–577.

Conrad, R., & Hull, A. (1964). Information, acoustic confusion, and memory span. *British Journal of Psychology, 55*, 75–84.

Conway, A. R. A., Cowan, N., & Bunting, M. F. (2001). The cocktail party phenomenon revisited: The importance of working memory capacity. *Psychonomic Bulletin & Review, 8*, 331–335.

Conway, M. A., Anderson, S. J., Larsen, S. F., Donnelly, C. M., McDaniel, M. A., McClelland, A. G. R., et al. (1994). The formation of flashbulb memories. *Memory & Cognition, 22*, 326–343.

Conway, M. A., Cohen, G., & Stanhope, N. (1991). On the very long-term retention of knowledge acquired through formal education: Twelve years of cognitive psychology. *Journal of Experimental Psychology: General, 120*, 395–409.

Conway, M. A., & Pleydell-Pearce, C. W. (2000). The construction of autobiographical memories in the self-memory system. *Psychological Review, 107*, 261–288.

Cook, A. E., & Myers, J. L. (2004). Processing discourse roles in scripted narratives: The influences of context and world knowledge. *Journal of Memory and Language, 50*, 268–288.

Cook, M. (1977). Gaze and mutual gaze in social encounters. *American Scientist, 65*, 328–333.

Cooke, N. J., & Breedin, S. D. (1994). Constructing naive theories of motion on the fly. *Memory & Cognition, 22*, 474–493.

Cooper, E. H., & Pantle, A. J. (1967). The total-time hypothesis in verbal learning. *Psychological Bulletin, 68*, 221–234.

Cooper, L. A., & Shepard, R. N. (1973). Chronometric studies of the rotation of mental images. In W. G. Chase (Ed.), *Visual information processing* (pp. 75–176). New York, NY: Academic Press.

Copeland, D. E., Gunawan, K., & Bies-Hernandez, N. J. (2011). Source credibility and syllogistic reasoning. *Memory & Cognition, 39*, 117–127.

Copeland, D. E., & Radvansky, G. A. (2001). Phonological similarity in working memory. *Memory & Cognition, 29*, 774–776.

Copeland, D. E., & Radvansky, G. A. (2004a). Working memory and syllogistic reasoning. *The Quarterly Journal of Experimental Psychology, 57A*, 1437–1457.

Copeland, D. E., & Radvansky, G. A. (2004b). Working memory span and situation model processing. *American Journal of Psychology, 117*, 191–213.

Copeland, D. E., Radvansky, G. A., & Goodwin, K. A. (2009). A novel study: Forgetting curves and the reminiscence bump. *Memory, 17*, 323–336.

Corballis, M. C. (1989). Laterality and human evolution. *Psychological Review, 96*, 492–505.

Corballis, M. C. (2004). The origins of modernity: Was autonomous speech a critical factor? *Psychological Review, 111*, 543–552.

Corder, M. (2004). Crippled but not crashed neural networks can help pilots land damaged planes. *Scientific American, 291*, 94–96.

Corley, M., Brocklehurst, P. H., & Moat, H. S. (2011). Error biases in inner and overt speech: Evidence from tongue twisters. *Journal of Experimental Psychology: Learning, Memory, and Cognition, 37*, 162–175.

Coughlan, A. K., & Warrington, E. K. (1978). Word comprehension and word retrieval in patients with localised cerebral lesions. *Brain, 101*, 163–185.

Coulson, S., Federmeier, K. D., Van Petten, C., & Kutas, M. (2005). Right hemisphere sensitivity to word- and sentence-level context: Evidence from event-related brain potentials. *Journal of Experimental Psychology: Learning, Memory, and Cognition, 31*, 127–147.

Courtney, S. M., Petit, L. Maisog, C. M., Ungerleider, L. G., & Haxby, J. V. (1998). An area specialized for spatial working memory in human frontal cortex. *Science, 279*, 1347–1351.

Cowan, N. (1995). *Attention and memory: An integrated framework*. New York, NY: Oxford University Press.

Cowan, N. (2010). The magical mystery four: How is working memory capacity limited, and why? *Current Directions in Psychological Science, 19*, 51–57.

Cowan, N. (2015). George Miller's magical number of immediate memory in retrospect: Observations on the faltering progression of science. *Psychological Review, 122*(3), 536–541.

Cowan, N., & Morey, C. C. (2007). How can dual-task working memory retention limits be investigated? *Psychological Science, 18,* 686–688.

Cowan, N., Wood, N. L., Wood, P. K., Keller, T. A., Nugent, L. D., & Keller, C. V. (1998). Two separate verbal processing rates contributing to short-term memory span. *Journal of Experimental Psychology: General, 127,* 141–160.

Craik, F. I. M., & Byrd, M. (1982). Aging and cognitive deficits: The role of attentional resources. In F. I. M. Craik & S. Trehub (Eds.), *Aging and cognitive processes* (pp. 191–211). New York, NY: Plenum Press.

Craik, F. I. M., Govoni, R., Naveh-Benjamin, M., & Anderson, N. D. (1996). The effects of divided attention on encoding and retrieval processes in human memory. *Journal of Experimental Psychology: General, 125,* 181–194.

Craik, F. I. M., & Lockhart, R. S. (1972). Levels of processing: A framework for memory research. *Journal of Verbal Learning and Verbal Behavior, 11,* 671–684.

Craik, F. I., & Tulving, E. (1975). Depth of processing and the retention of words in episodic memory. *Journal of Experimental Psychology: General, 104*(3), 268–294.

Craik, F. I., & Watkins, M. J. (1973). The role of rehearsal in short-term memory. *Journal of Verbal Learning and Verbal Behavior, 12*(6), 599–607.

Cree, G. S., & McRae, K. (2003). Analyzing the factors underlying the structure and computation of the meaning of *chipmunk, cherry, chisel, cheese,* and *cello* (and many other such concrete nouns). *Journal of Experimental Psychology: General, 132,* 163–201.

Crosby, J. R., Monin, B., & Richardson, D. (2008). Where do we look during potentially offensive behavior? *Psychological Science, 19,* 226–228.

Crovitz, H. F., & Shiffman, H. (1974). Frequency of episodic memories as a function of their age. *Bulletin of the Psychonomic Society, 4,* 517–518.

Crowder, R. G. (1970). The role of one's own voice in immediate memory. *Cognitive Psychology, 1,* 157–178.

Crowder, R. G. (1972). Visual and auditory memory. In J. F. Kavanaugh & I. G. Mattingly (Eds.), *Language by ear and by eye: The relationships between speech and reading* (pp. 251–276). Cambridge, MA: MIT Press.

Crowder, R. G., & Morton, J. (1969). Precategorical acoustic storage (PAS). *Perception & Psychophysics, 5,* 365–373.

Cummins, D. D., Lubart, T., Alksnis, O., & Rist, R. (1991). Conditional reasoning and causation. *Memory & Cognition, 19,* 274–282.

Cunningham, T. J., Chambers, A. M., & Payne, J. D. (2014). Prospection and emotional memory: How expectation affects emotional memory formation following sleep and wake. *Frontiers in Psychology, 5,* 862.

Curiel, J. M., & Radvansky, G. A. (2014). Spatial and character situation model updating. *Journal of Cognitive Psychology, 26*(2), 205–212.

Curran, T. (2000). Brain potentials of recollection and familiarity. *Memory & Cognition, 28,* 923–938.

Curtis, M. E., & Bharucha, J. J. (2009). Memory and musical expectation for tones in cultural context. *Music Perception, 26,* 365–375.

Cutica, I., & Bucciarelli, M. (2013). Cognitive change in learning from text: Gesturing enhances the construction of the text mental model. *Journal of Cognitive Psychology, 25*(2), 201–209.

Cutting, J. E. (2014). Event segmentation and seven types of narrative discontinuity in popular movies. *Acta Psychologica, 149,* 69–77.

Cutting, J. E., Brunick, K. L., & Candan, A. (2012). Perceiving event dynamics and parsing Hollywood films. *Journal of Experimental Psychology: Human Perception and Performance, 38*(6), 1476–1490.

Cutting, J., & Iricinschi, C. (2015). Re-presentations of space in Hollywood movies: An event-indexing analysis. *Cognitive Science, 39*(2), 434–456.

Dahan, D. (2010). The time course of interpretation in speech comprehension. *Current Directions in Psychological Science, 19,* 121–126.

Damasio, H., & Damasio, A. R. (1997). The lesion method in behavioral neurology and neuropsychology. In T. E. Feinberg & M. J. Farah (Eds.), *Behavioral neurology and neuropsychology* (pp. 69–82). New York, NY: McGraw-Hill.

D'Anastasio, R., Wroe, S., Tuniz, C., Mancini, L., Cesana, D. T., Dreossi, D., . . . Capasso, L. (2013). Micro-biomechanics of the Kebara 2 hyoid and its implications for speech in Neanderthals. *PloS one, 8*(12), e82261.

Daneman, M., & Carpenter, P. A. (1980). Individual differences in working memory and reading. *Journal of Verbal Learning and Verbal Behavior, 19,* 450–466.

Daneman, M., & Merikle, P. M. (1996). Working memory and language comprehension: A meta-analysis. *Psychonomic Bulletin & Review, 3,* 422–433.

D'Argembeau, A., Renaud, O., & Van der Linden, M. (2011). Frequency, characteristics and functions of future-oriented thoughts in daily life. *Applied Cognitive Psychology, 25*(1), 96–103.

Darwin, C. J., Turvey, M. T., & Crowder, R. G. (1972). An auditory analogue of the Sperling partial report procedure: Evidence for brief auditory storage. *Cognitive Psychology, 3,* 255–267.

Davachi, L., Mitchell, J. P., & Wagner, A. D. (2003). Multiple routes to memory: Distinct medial temporal lobe processes build item and source memories. *Proceedings of the National Academy of Sciences, 100,* 2157–2162.

Davelaar, E. J., Goshen-Gottstein, Y., Ashkenazi, A., Haarmann, H. J., & Usher, M. (2005). The demise of short-term memory revisited: Empirical and computational investigations of recency effects. *Psychological Review, 112,* 3–42.

Davis, M. (1997). Neurobiology of fear responses: The role of the amygdala. *Journal of Neuropsychiatry and Clinical Neurosciences, 9,* 382–402.

Davoli, C. C., Brockmole, J. R., & Goujon, A. (2012). A bias to detail: How hand position modulates visual learning and visual memory. *Memory & Cognition, 40,* 352–359.

Davoli, C. C., Brockmole, J. R., & Witt, J. K. (2012). Compressing perceived distance with remote tool-use: Real, imagined, and remembered. *Journal of Experimental Psychology: Learning, Memory, and Cognition, 38,* 80–89.

Davoli, C. C., Suszko, J. W., & Abrams, R. A. (2007). New objects can capture attention without a unique luminance transient. *Psychonomic Bulletin & Review, 14,* 338–343.

Day, S. B., & Gentner, D. (2007). Nonintentional analogical inference in text comprehension. *Memory & Cognition, 35,* 39–49.

Day, S. B., Motz, B. A., & Goldstone, R. L. (2015). The cognitive costs of context: The effects of concreteness and immersiveness in instructional examples. *Frontiers in Psychology, 6.*

DeCaro, M. S., Rotar, K. E., Kendra, M. S., & Beilock, S. L. (2010). Diagnosing and alleviating the impact of performance pressure on mathematical problem solving. *The Quarterly Journal of Experimental Psychology, 63,* 1619–1630.

DeCaro, M. S., Thomas, R. D., Albert, N. B., & Beilock, S. L. (2011). Choking under pressure: Multiple routes to skill failure. *Journal of Experimental Psychology: General, 140,* 390–406.

Dediu, D., & Levinson, S. C. (2013). On the antiquity of language: The reinterpretation of Neandertal linguistic capacities and its consequences. *Frontiers in Psychology, 4,* 397.

Deese, J. (1959). On the prediction of occurrence of particular verbal intrusions in immediate recall. *Journal of Experimental Psychology, 58,* 17–22.

de Hevia, M. D., & Spelke, E. S. (2010). Number-space mapping in human infants. *Psychological Science, 21,* 653–660.

Dehon, H., & Brédart, S. (2004). False memories: Young and older adults think of semantic associates at the same rate, but young adults are more successful at source monitoring. *Psychology and Aging, 19,* 191–197.

Delany, P. F., Ericsson, K. A., & Knowles, M. E. (2004). Immediate and sustained effects of planning in a problem-solving task. *Journal of Experimental Psychology: Learning, Memory, and Cognition, 30,* 1219–1234.

Dell, G. S. (1986). A spreading-activation theory of retrieval in sentence production. *Psychological Review, 93,* 283–321.

Dell, G. S., & Newman, J. E. (1980). Detecting phonemes in fluent speech. *Journal of Verbal Learning and Verbal Behavior, 20,* 611–629.

Dell'acqua, R., & Job, R. (1998). Is object recognition automatic? *Psychonomic Bulletin & Review, 5,* 496–503.

DePaulo, B. M., & Bonvillian, J. D. (1978). The effect on language development of the special characteristics of speech addressed to children. *Journal of Psycholinguistic Research, 7,* 189–211.

Descartes, R. (1972). *Treatise on man* (T. S. Hall, Trans.). Cambridge, MA: Harvard University Press. (Original work published 1637)

Desmarais, G., Dixon, M. J., & Roy, E. A. (2007). A role for action knowledge in visual object identification. *Memory & Cognition, 35,* 1712–1723.

Deutsch, G., Bourbon, T., Papanicolaou, A. C., & Eisenberg, H. M. (1988). Visuospatial tasks compared via activation of regional cerebral blood flow. *Neuropsychologia, 26,* 445–452.

Deutsch, J. A., & Deutsch, D. (1963). Attention: Some theoretical considerations. *Psychological Review, 70,* 80–90.

de Vega, M., León, I., & Díaz, J. M. (1996). The representation of changing emotions in reading comprehension. *Cognition and Emotion, 10,* 303–321.

de Vega, M., Robertson, D. A., Glenberg, A. M., Kaschak, M. P., & Rinck, M. (2004). On doing two things at once: Temporal constraints on action in language comprehension. *Memory & Cognition, 32,* 1033–1043

deWinstanley, P. A., & Bjork, E. L. (2004). Processing strategies and the generation effect: Implications for making a better reader. *Memory & Cognition, 32,* 945–955.

Dewitt, L. A., & Samuel, A. G. (1990). The role of knowledge-based expectations in music perception: Evidence from musical restoration. *Journal of Experimental Psychology: General, 119,* 123–144.

Dewsbury, D. A. (2000). Comparative cognition in the 1930s. *Psychonomic Bulletin & Review, 7,* 267–283.

Diamond, A., & Gilbert, J. (1989). Development as progressive inhibitory control of action: Retrieval of a contiguous object. *Cognitive Development, 4,* 223–249.

Dijksterhuis, A., Bos, M. W., van der Leij, A., & van Baaren, R. B. (2009). Predicting soccer matches after unconscious and conscious thought as a function of expertise. *Psychological Science, 20*(11), 1381–1387.

Di Pellegrino, G., Fadiga, L., Fogassi, L., Galese, V., & Rizzolatti, G. (1992). Understanding motor events: A neurophysiological study. *Experimental Brain Research, 91,* 176–180.

Ditman, T., & Kuperberg, G. R. (2005). A source-monitoring account of auditory verbal hallucinations in patients with schizophrenia. *Harvard Review of Psychiatry, 13,* 280–299.

Dixon, M. J., Smilek, D., & Merikle, P. M. (2004). Not all synesthetes are created equal: Projector versus associative synesthetes. *Cognitive, Affective, & Behavioral Neuroscience, 4,* 335–343.

Dixon, R. A., Simon, E. W., Nowak, C. A., & Hultsch, D. F. (1982). Text recall in adulthood as a function of level of information, input modality, and delay interval. *The Journals of Gerontology: Psychological Sciences, 37,* P358–P364.

Dodd, M. D., & MacLeod, C. M. (2004). False recognition without intentional learning. *Psychonomic Bulletin & Review, 11,* 137–142.

Dodd, M. D., Van der Stigchel, S., & Hollingworth, A. (2009). Inhibition of return and facilitation of return as a function of visual task. *Psychological Science, 20,* 333–339.

Dodson, C. S., & Krueger, L. E. (2006). I misremember it well: Why older adults are unreliable eyewitnesses. *Psychonomic Bulletin & Review, 13,* 770–775.

Dodson, C. S., & Schacter, D. L. (2002). When false recognition meets metacognition: The distinctiveness heuristic. *Journal of Memory and Language, 46,* 782–803.

Dolcos, F., Iordan, A. D., & Dolcos, S. (2011). Neural correlates of emotion–cognition interactions: A review of evidence from brain imaging investigations. *Journal of Cognitive Psychology, 23*(6), 669–694.

Dolcos, F., Labar, K. S., & Cabeza, R. (2005). Remembering one year later: Role of the amygdala and the medial temporal lobe memory system in retrieving emotional memories. *Proceedings of the National Academy of Sciences, 102,* 2626–2631.

Dolscheid, S., Shayan, S., Majid, A., & Casasanto, D. (2013). The thickness of musical pitch: Psychophysical evidence for linguistic relativity. *Psychological Science, 24*(5), 613–621.

Donchin, E. (1981). Surprise! . . . Surprise? *Psychophysiology, 18,* 493–513.

Donders, F. C. (1969). *Over de snelheid van psychische processen* [Speed of mental processes]. *Onderzoekingen gedann in het Psysiologish Laboratorium der Utrechtsche Hoogeschool* (W. G. Koster, Trans.). In W. G. Koster (Ed.), Attention and performance II. *Acta Psychologica, 30,* 412–431. (Original work published 1868)

Donley, R. D., & Ashcraft, M. H. (1992). The methodology of testing naive beliefs in the physics classroom. *Memory & Cognition, 20,* 381–391.

Dooling, D., & Lachman, R. (1971). Effects of comprehension on retention of prose. *Journal of Experimental Psychology, 88,* 216–222.

Dosher, B. A., & Ma, J-J. (1998). Output loss or rehearsal loop? Output-time versus pronunciation-time limits in immediate recall for forgetting-matched materials. *Journal of Experimental Psychology: Learning, Memory, and Cognition, 24,* 316–335.

Dougherty, M. R., Franco-Watkins, A. M., & Thomas, R. (2008). Psychological plausibility of the theory of probabilistic mental models and the fast and frugal heuristics. *Psychological Review, 115*(1), 199–213.

Drachman, D. A. (1978). Central cholinergic system and memory. In M. A. Lipton, A. D. Mascio, & K. F. Killam (Eds.), *Psychopharmacology: A generation of process.* New York, NY: Raven.

Dresler, T., Mériau, K., Heekeren, H. R., & van der Meer, E. (2009). Emotional Stroop task: Effect of word arousal and subject anxiety on emotional interference. *Psychological Research, 73,* 364–371.

Drolet, M., Schubotz, R. I., & Fischer, J. (2012). Authenticity affects the recognition of emotions in speech: Behavioral and fMRI evidence. *Cognitive, Affective, & Behavioral Neuroscience, 12,* 140–150.

Drosopoulos, S., Schulze, C., Fischer, S., & Born, J. (2007). Sleep's function in the spontaneous recovery and consolidation of memories. *Journal of Experimental Psychology: General, 136,* 169–183.

Du, F., Zhang, K., & Abrams, R. A. (2014). Hold the future, let the past go: Attention prefers the features of future targets. *Cognition, 131*(2), 205–215.

Duchek, J. M. (1984). Encoding and retrieval differences between young and old: The impact of attentional capacity usage. *Developmental Psychology, 20,* 1173–1180.

Dunbar, K., & MacLeod, C. M. (1984). A horse race of a different color: Stroop interference patterns with transformed words. *Journal of Experimental Psychology: Human Perception and Performance, 10,* 622–639.

Duncan, J., Bundesen, C., Olson, A., Humphreys, G., Chavda, S., & Shibuya, H. (1999). Systematic analysis of deficits in visual attention. *Journal of Experimental Psychology: General, 128,* 450–478.

Duncan, J., & Humphreys, G. (1989). Visual search and stimulus similarity. *Psychological Review, 96,* 433–458.

Duncan, S. (1972). Some signals and rules for taking speaking turns in conversations. *Journal of Personality and Social Psychology, 23,* 283–292.

Duncker, K. (1945). On problem solving. *Psychological Monographs, 58*(Whole no. 270).

Dunlosky, J., & Hertzog, C. (2001). Measuring strategy production during associative learning: The relative utility of concurrent versus retrospective reports. *Memory & Cognition, 29,* 247–253.

Dunlosky, J., & Lipko, C. (2007). Metacomprehension: A brief history and how to improve its accuracy. *Current Directions in Psychological Science, 16,* 228–232.

Dunlosky, J., & Nelson, T. O. (1994). Does the sensitivity of judgments of learning (JOLs) to the effects of various activities depend on when the JOLs occur? *Journal of Memory and Language, 33,* 545–565.

Dyer, F. C. (2002). When it pays to waggle. *Nature, 419,* 885–886.

Dywan, J., Segalowitz, S. J., & Webster, L. (1998). Source monitoring: ERP evidence for greater reactivity to nontarget information in older adults. *Brain and Cognition, 36,* 390–430.

Eakin, D. K., Schreiber, T. A., & Sergent-Marshall, S. (2003). Misinformation effects in eyewitness memory: The presence and absence of memory impairment as a function of warning and misinformation accessibility. *Journal of Experimental Psychology: Learning, Memory, and Cognition, 29,* 813–825.

Ebbinghaus, H. (1885/1913). *Memory: A contribution to experimental psychology* (H. A. Ruger & C. E. Bussenius, Trans.). New York, NY: Columbia University, Teacher's College. (Reprinted 1964, New York, NY: Dover).

Ebbinghaus, H. (1908). Abriss der Psychologie [Survey of psychology]. Leipzig: Veit & Comp. (Also cited as 1910.)

Edridge-Green, F. W. (1900). *Memory and its cultivation.* New York, NY: Appleton & Co.

Edwards, D., & Potter, J. (1993). Language and causation: A discursive action model of description and attribution. *Psychological Review, 100,* 23–41.

Egan, P., Carterette, E. C., & Thwing, E. J. (1954). Some factors affecting multichannel listening. *Journal of the Acoustic Society of America, 26,* 774–782.

Eichenbaum, H., & Fortin, N. (2003). Episodic memory and the hippocampus: It's about time. *Current Directions in Psychological Science, 12,* 53–57.

Eimas, P. D. (1975). Speech perception in early infancy. In L. B. Cohen & P. Salapatek (Eds.), *Infant perception: From sensation to cognition: Vol. II. Perception of space, speech, and sound* (pp. 193–231). New York, NY: Academic Press.

Einstein, G. O., & McDaniel, M. A. (1990). Normal aging and prospective memory. *Journal of Experimental Psychology: Learning, Memory, and Cognition, 16,* 717–726.

Einstein, G. O., & McDaniel, M. A. (2005). Prospective memory: Multiple retrieval processes. *Current Directions, 14,* 286–290.

Einstein, G. O., McDaniel, M. A., Richardson, S. L., Guynn, M. L., & Cunfer, A. R. (1995). Aging and prospective memory: Examining influences of self-initiated retrieval processes. *Journal of Experimental Psychology: Learning, Memory, and Cognition, 21,* 996–1007.

Ellis, H. D. (1983). The role of the right hemisphere in face perception. In A. W. Young (Ed.), *Functions of the right cerebral hemisphere* (pp. 33–64). New York, NY: Academic Press.

Emery, N. J. (2000). The eyes have it: The neuroethology, function and evolution of social gaze. *Neuroscience and Biobehavioral Review, 24*, 581–604.

Engelhardt, P. E., Bailey, K. G. D., & Ferreira, F. (2006). Do speakers and listeners observe the Gricean maxim of quantity? *Journal of Memory and Language, 54*, 554–573.

Engelkamp, J., & Dehn, D. M. (2000). Item and order information in subject-performed tasks and experimenter-performed tasks. *Journal of Experimental Psychology: Learning, Memory, and Cognition, 26*, 671–682.

Engle, R. W. (2001). What is working memory capacity? In H. L. Roediger, J. S. Nairne, I. Neath, & A. M. Suprenant (Eds.), *The nature of remembering: Essays in honor of Robert G. Crowder* (pp. 297–314). Washington, DC: American Psychological Association Press.

Engle, R. W. (2002). Working memory capacity as executive attention. *Current Directions in Psychological Science, 11*, 19–23.

Epstein, R., & Kanwisher, N. (1998). A cortical representation of the local visual area. *Nature, 392*, 598–601.

Erdelyi, M. H. (2010). The ups and downs of memory. *American Psychologist, 65*, 623–633.

Erdfelder, E., & Buchner, A. (1998). Decomposing the hindsight bias: A multinomial processing tree model for separating recollection and reconstruction in hindsight. *Journal of Experimental Psychology: Learning, Memory, and Cognition, 24*, 387–414.

Erickson, T. D., & Mattson, M. E. (1981). From words to meanings: A semantic illusion. *Journal of Verbal Learning and Verbal Behavior, 20*, 540–551.

Ericsson, K. A., & Charness, N. (1994). Expert performance: Its structure and acquisition. *American Psychologist, 49*, 725–747.

Ericsson, K. A., Delaney, P. F., Weaver, G., & Mahadevan, R. (2004). Uncovering the structure of a memorist's superior "basic" memory capacity. *Cognitive Psychology, 49*, 191–237.

Ericsson, K. A., Krampe, R. T., & Tesch-Römer, C. (1993). The role of deliberate practice in the acquisition of expert performance. *Psychological Review, 100*, 363–406.

Ericsson, K. A., & Simon, H. A. (1980). Verbal reports as data. *Psychological Review, 87*, 215–251.

Ericsson, K. A., & Simon, H. A. (1993). *Protocol analysis: Verbal reports as data* (Rev. ed.). Cambridge, MA: MIT Press.

Ernst, G. W., & Newell, A. (1969). *GPS: A case study in generality and problem solving*. New York, NY: Academic Press.

Enz, K. F., Pillemer, D. B., & Johnson, K. M. (2016). The relocation bump: Memories of middle adulthood are organized around residential moves. *Journal of Experimental Psychology: General, 145*, 935–940.

Espino, O., Santamaria, C., & Byrne, R. M. (2009). People think about what is true for conditionals, not what is false: Only true possibilities prime the comprehension of "if." *The Quarterly Journal of Experimental Psychology, 62*(6), 1072–1078.

Estes, Z., Verges, M., & Barsalou, L. W. (2008). Head up, foot down: Object words orient attention to the objects' typical location. *Psychological Science, 19*, 93–97.

Evans, J. St. B. T., Barston, J. L., & Pollard, P. (1983). On the conflict between logic and belief in syllogistic reasoning. *Memory & Cognition, 11*, 295–306.

Evans, J. St. B. T., Handley, S. J., Harper, C. N. J., & Johnson-Laird, P. H. (1999). Reasoning about necessity and possibility: A test of the mental model theory of deduction. *Journal of Experimental Psychology: Learning, Memory, and Cognition, 25*, 1495–1513.

Evans, J. St. B. T., Handley, S. J., Over, D. E., & Perham, N. (2002). Background beliefs in Bayesian inference. *Memory & Cognition, 30*, 179–190.

Fantz, R. L. (1963). Pattern vision in newborn infants. *Science, 140*, 296–297.

Farah, M. J., & McClelland, J. L. (1991). A computational model of semantic memory impairment: Modality specificity and emergent category specificity. *Journal of Experimental Psychology: General, 120*, 339–357.

Farias, A. R., Garrido, M. V., & Semin, G. R. (2016). Embodiment of abstract categories in space . . . grounding or mere compatibility effects? The case of politics. *Acta Psychologica, 166*, 49–53.

Faroqi-Shah, Y., & Thompson, C. K. (2007). Verb inflections in agrammatic aphasia: Encoding of tense features. *Journal of Memory and Language, 56*, 129–151.

Fausey, C. M., & Boroditsky, L. (2011). Who dunnit? Cross-linguistic differences in eye-witness memory. *Psychonomic Bulletin & Review, 18*, 150–157.

Fawcett, J. M. (2013). The production effect benefits performance in between-subject designs: A meta-analysis. *Acta Psychologica, 142*(1), 1–5.

Fazio, L. K., Barber, S. J., Rajaram, S., Ornstein, P. A., & Marsh, E. J. (2013). Creating illusions of knowledge: Learning errors that contradict prior knowledge. *Journal of Experimental Psychology: General, 142*(1), 1–5.

Fedorenko, E., Gibson, E., & Rohde, D. (2006). The nature of working memory capacity in sentence comprehension: Evidence against domain-specific working memory resources. *Journal of Memory and Language, 54*, 541–553.

Feigenson, L., Dehaene, S., & Spelke, E. (2004). Core systems of number. *Trends in Cognitive Sciences, 8*, 307–314.

Feldman, D. (1992). *When did wild poodles roam the earth? An Imponderables™ book*. New York, NY: HarperCollins.

Feldman, J. (2003). The simplicity principle in human concept learning. *Current Directions in Psychological Science, 12*, 227–232.

Feng, J., Spence, I., & Pratt, J. (2007). Playing an action video game reduces gender differences in spatial cognition. *Psychological Science, 18*, 850–855.

Fenn, K. M., & Hambrick, D. Z. (2013). What drives sleep-dependent memory consolidation: Greater gain or less loss? *Psychonomic Bulletin & Review, 20*(3), 501–506.

Fenske, M. J., & Raymond, J. E. (2006). Affective influences of selective attention. *Current Directions in Psychological Science, 15*, 312–316.

Ferguson, B., Graf, E., & Waxman, S. R. (2014). Infants use known verbs to learn novel nouns: Evidence from 15- and 19-month-olds. *Cognition, 131*(1), 139–146.

Fernald, A., & Marchman, V. A. (2006). Language learning in infancy. In M. J. Traxler & M. A. Gernsbacher (Eds.), *Handbook of psycholinguistics* (2nd ed., pp. 1027–1071). New York, NY: Elsevier.

Ferrante, D., Girotto, V., Stragà, M., & Walsh, C. (2013). Improving the past and the future: A temporal asymmetry in hypothetical thinking. *Journal of Experimental Psychology: General, 142*(1), 23–27.

Ferreira, F., Henderson, J. M., Anes, M. D., Weeks, P. A., Jr., & McFarlane, D. K. (1996). Effects of lexical frequency and syntactic complexity in spoken-language comprehension: Evidence from the auditory moving-window technique. *Journal of Experimental Psychology: Learning, Memory, and Cognition, 22,* 324–335.

Ferreira, F., & Swets, B. (2002). How incremental is language production? Evidence from the production of utterances requiring the computation of arithmetic sums. *Journal of Memory and Language, 46,* 57–84.

Ferreira, V. S. (1996). Is it better to give than to donate? Syntactic flexibility in language production. *Journal of Memory and Language, 35,* 724–755.

Ferreira, V. S., Bock, K., Wilson, M. P., & Cohen, N. J. (2008). Memory for syntax despite amnesia. *Psychological Science, 19*(9), 940–946.

Ferreira, V. S., & Dell, G. S. (2000). Effect of ambiguity and lexical availability on syntactic and lexical production. *Cognitive Psychology, 40,* 296–340.

Ferreira, V. S., & Firato, C. E. (2002). Proactive interference effects on sentence production. *Psychonomic Bulletin & Review, 9,* 795–800.

Ferreira, V. S., & Humphreys, K. R. (2001). Syntactic influences on lexical and morphological processing in language production. *Journal of Memory and Language, 44,* 52–80.

Fillenbaum, S. (1974). Pragmatic normalization: Further results for some conjunctive and disjunctive sentences. *Journal of Experimental Psychology, 102,* 574–578.

Fillmore, C. J. (1968). Toward a modern theory of case. In D. A. Reibel & S. A. Schane (Eds.), *Modern studies in English* (pp. 361–375). Englewood Cliffs, NJ: Prentice Hall.

Finke, R. A., & Freyd, J. J. (1985). Transformations of visual memory induced by implied motions of pattern elements. *Journal of Experimental Psychology: Learning, Memory, and Cognition, 11,* 780–794.

Finkenhauer, C., Luminet, O., Gisle, L., El-ahmadi, A., Van der Linden, M., & Philipott, P. (1998). Flashbulb memories and the underlying mechanisms of their formation: Toward an emotional-integrative model. *Memory & Cognition, 26,* 516–531.

Finlayson, N. J., & Grove, P. M. (2015). Visual search is influenced by 3D spatial layout. *Attention, Perception, & Psychophysics, 77*(7), 2322–2330.

Finn, B., & Roediger, H. L. (2011). Enhancing retention through reconsolidation: Negative emotional arousal following retrieval enhances later recall. *Psychological Science, 22,* 781–786.

Fischhoff, B. (1975). Hindsight does not equal foresight: The effect of outcome knowledge on judgment under uncertainty. *Journal of Experimental Psychology: Human Perception and Performance, 1,* 288–299.

Fischhoff, B., & Bar-Hillel, M. (1984). Diagnosticity and the base-rate effect. *Memory & Cognition, 12,* 402–410.

Fisk, J. E., & Sharp, C. (2002). Syllogistic reasoning and cognitive ageing. *The Quarterly Journal of Experimental Psychology, 55A,* 1273–1293.

Fitzgerald, J. M. (2010). Culture, gender, and the first memories of black and white American students. *Memory & Cognition, 38,* 785–796.

Fleck, J. I., & Weisberg, R. W. (2004). The use of verbal protocols as data: An analysis of insight in the candle problem. *Memory & Cognition, 32,* 990–1006.

Fletcher, C. R., & Bloom, C. P. (1988). Causal reasoning in the comprehension of simple narrative texts. *Journal of Memory and Language, 27,* 235–244.

Flusberg, S. J., & Boroditsky, L. (2011). Are things that are hard to physically move also hard to imagine moving? *Psychonomic Bulletin & Review, 18,* 158–164.

Fodor, J. A., & Garrett, M. (1966). Some reflections on competence and performance. In J. Lyons & R. J. Wales (Eds.), *Psycholinguistic papers* (pp. 135–154). Edinburgh, Scotland: Edinburgh University Press.

Fong, G. T., & Nisbett, R. E. (1991). Immediate and delayed transfer of training effects in statistical reasoning. *Journal of Experimental Psychology: General, 120,* 34–45.

Forgus, R. H., & Melamed, L. E. (1976). *Perception: A cognitive-stage approach.* New York, NY: McGraw-Hill.

Foroughi, C. K., Werner, N. E., Barragán, D., & Boehm-Davis, D. A. (2015). Interruptions disrupt reading comprehension. *Journal of Experimental Psychology: General, 144*(3), 704–709.

Forrin, N. D., MacLeod, C. M., & Ozubko, J. D. (2012). Widening the boundaries of the production effect. *Memory & Cognition, 40*(7), 1046–1055.

Förster, J. (2012). GLOMOsys: The how and why of global and local processing. *Current Directions in Psychological Science, 21*(1), 15–19.

Forsyth, D. K., & Burt, C. D. B. (2008). Allocating time to future tasks: The effect of task segmentation on planning fallacy bias. *Memory & Cognition, 36,* 791–798.

Foss, D. J., & Hakes, D. T. (1978). *Psycholinguistics: An introduction to the psychology of language.* Englewood Cliffs, NJ: Prentice Hall.

Franconeri, S. L., & Simons, D. J. (2003). Moving and looming stimuli capture attention. *Perception & Psychophysics, 65,* 999–1010.

Frank, D. J., Nara, B., Zavagnin, M., Touron, D. R., & Kane, M. J. (2015). Validating older adults' reports of less mind-wandering: An examination of eye movements and dispositional influences. *Psychology and Aging, 30*(2), 266–278.

Frank, M. J., & Badre, D. (2015). How cognitive theory guides neuroscience. *Cognition, 135,* 14–20.

Franklin, M. S., Broadway, J. M., Mrazek, M. D., Smallwood, J., & Schooler, J. W. (2013). Window to the wandering mind: Pupillometry of spontaneous thought while reading. *The Quarterly Journal of Experimental Psychology, 66*(12), 2289–2294.

Frazier, L., & Rayner, K. (1982). Making and correcting errors during sentence comprehension: Eye movements in the analysis of structurally ambiguous sentences. *Cognitive Psychology, 14*, 178–210.

Frazier, L., & Rayner, K. (1990). Taking on semantic commitments: Processing multiple meanings vs. multiple senses. *Journal of Memory and Language, 29*, 181–200.

Freedman, J. L., & Loftus, E. F. (1971). Retrieval of words from long-term memory. *Journal of Verbal Learning and Verbal Behavior, 10*, 107–115.

Freeman, J. E., & Ellis, J. A. (2003). The representation of delayed intentions: A prospective subject-performed task? *Journal of Experimental Psychology: Learning, Memory, and Cognition, 29*, 976–992.

Freud, S. (1899/1938). Childhood and concealing memories. In A. A. Brill (Ed.), *The basic writings of Sigmund Freud.* New York, NY: Modern Library.

Freud, S. (1953/1905). Three essays on the theory of sexuality. In J. Strachey (Ed.), *The standard edition of the complete psychological works of Sigmund Freud* (Vol. 7, pp. 135–423). London, England: Hogarth. (Original work published 1905)

Freyd, J. J., & Finke, R. A. (1984). Representational momentum. *Bulletin of the Psychonomic Society, 23*, 443–446.

Friederici, A. D., Hahne, A., & Mecklinger, A. (1996). Temporal structure of syntactic parsing: Early and late event-related brain potential effects. *Journal of Experimental Psychology: Learning, Memory, and Cognition, 22*, 1219–1248.

Friedman, A., & Polson, M. C. (1981). Hemispheres as independent resource systems: Limited-capacity processing and cerebral specialization. *Journal of Experimental Psychology: Human Perception and Performance, 7*, 1031–1058.

Friedman, N. P., Miyake, A., Young, S. E., DeFries, J. C., Corley, R. P., & Hewitt, J. K. (2008). Individual differences in executive functions are almost entirely genetic in origin. *Journal of Experimental Psychology: General, 137*, 201–225.

Friedrich, F. J., Henik, A., & Tzelgov, J. (1991). Automatic processes in lexical access and spreading activation. *Journal of Experimental Psychology: Human Perception and Performance, 17*, 792–806.

Frieske, D. A., & Park, D. C. (1999). Memory for news in young and old adults. *Psychology and Aging, 14*, 90–98.

Frings, C., Schneider, K. K., & Fox, E. (2015). The negative priming paradigm: An update and implications for selective attention. *Psychonomic Bulletin & Review, 22*(6), 1577–1597.

Frisson, S., & Pickering, M. J. (1999). The processing of metonymy: Evidence from eye movements. *Journal of Experimental Psychology: Learning, Memory, and Cognition, 25*, 1366–1383.

Frisson, S., & Wakefield, M. (2012). Psychological essentialist reasoning and perspective taking during reading: A donkey is not a zebra, but a plate can be a clock. *Memory & Cognition, 40*, 297–310.

Fromkin, V. A. (1971). The non-anomalous nature of anomalous utterances. *Language, 47*, 27–52.

Fromkin, V. A., & Rodman, R. (1974). *An introduction to language.* New York, NY: Holt, Rinehart & Winston.

Frost, P. (2000). The quality of false memory over time: Is memory for misinformation "remembered" or "known"? *Psychonomic Bulletin & Review, 7*, 531–536.

Fukuda, K., Vogel, E., Mayr, U., & Awh, E. (2010). Quality, not quantity: The relationship between fluid intelligence and working memory capacity. *Psychonomic Bulletin & Review, 17*, 673–679.

Fung, H. H., Lai, P., & Ng, R. (2001). Age differences in social preferences among Taiwanese and Mainland Chinese: The role of perceived time. *Psychology and Aging, 16*, 351–356.

Galanter, E. (1962). Contemporary psychophysics. In R. Brown, E. Galanter, E. H. Hess, & G. Mandler (Eds.), *New directions in psychology* (Vol. 1, pp. 87–156). New York, NY: Holt, Rinehart & Winston.

Galantucci, B., Fowler, C. A., & Turvey, M. T. (2006). The motor theory of speech perception reviewed. *Psychonomic Bulletin & Review, 13*, 361–377.

Gallo, D. A. (2010). False memories and fantastic beliefs: 15 years of the DRM illusion. *Memory & Cognition, 38*, 833–848.

Gallo, D. A., & Roediger, H. L. (2003). The effects of associations and aging on illusory recollection. *Memory & Cognition, 31*, 1036–1044.

Galotti, K. M., Wiener, H. J., & Tandler, J. M. (2014). Real-life decision making in college students. I: Consistency across specific decisions. *The American Journal of Psychology, 127*(1), 19–31.

Galton, F. (1879). Psychometric experiments. *Brain: A Journal of Neurology, II*, 149–162.

Galván, V. V., Vessal, R. S., & Golley, M. T. (2013). The effects of cell phone conversations on the attention and memory of bystanders. *PloS one, 8*(3), e58579.

Garcia-Retamero, R., & Dhami, M. K. (2013). On avoiding framing effects in experienced decision makers. *The Quarterly Journal of Experimental Psychology, 66*(4), 829–842.

Garcia-Retamero, R., Wallin, A., & Dieckmann, A. (2007). Does causal knowledge help us be faster and more frugal in our decisions? *Memory & Cognition, 35*, 1399–1409.

Gardner, H. (1985). *The mind's new science: A history of the cognitive revolution.* New York, NY: Basic Books.

Garrett, M. F. (1975). The analysis of sentence production. In G. H. Bower (Ed.), *The psychology of learning and memory* (Vol. 9, pp. 133–177). New York, NY: Academic Press.

Garry, M., Manning, C. G., Loftus, E. F., & Sherman, S. J. (1996). Imagination inflation: Imagining a childhood event inflates confidence that it occurred. *Psychonomic Bulletin & Review, 3*, 208–214.

Garry, M., & Polaschek, D. L. L. (2000). Imagination and memory. *Current Directions in Psychological Science, 9*, 6–10.

Gartus, A., Klemer, N., & Leder, H. (2015). The effects of visual context and individual differences on perception and evaluation of modern art and graffiti art. *Acta Psychologica, 156*, 64–76.

Gates, A. I. (1917). Recitation as a factor in memorizing. *Archives of Psychology, 40*, 104.

Gazzaniga, M. S. (1995). Principles of human brain organization derived from split-brain studies. *Neuron, 14*, 217–228.

Gazzaniga, M. S., Ivry, R. B., & Mangun, G. R. (1998). *Cognitive neuroscience: The biology of the mind*. New York, NY: Norton.

Gazzaniga, M. S., & Sperry, R. W. (1967). Language after section of the cerebral commissures. *Brain, 90*, 131–148.

Geary, D. C. (1992). Evolution of human cognition: Potential relationship to the ontogenetic development of behavior and cognition. *Evolution and Cognition, 1*, 93–100.

Geary, D. C. (1994). *Children's mathematical development: Research and practical applications*. Washington, DC: American Psychological Association.

Geary, D. C., Brown, S. C., & Samaranayake, V. A. (1991). Cognitive addition: A short longitudinal study of strategy choice and speed-of-processing differences in normal and mathematically disabled children. *Developmental Psychology, 27*(5), 787–797.

Gelman, R., & Gallistel, C. (1978). *Young children's understanding of numbers*. Cambridge, MA: Harvard University Press.

Gennari, S. P. (2004). Temporal references and temporal relations in sentence comprehension. *Journal of Experimental Psychology: Learning, Memory, and Cognition, 30*, 877–890.

Gentner, D., & Collins, A. (1981). Studies of inference from lack of knowledge. *Memory & Cognition, 9*, 434–443.

Gentner, D., & Markman, A. B. (1997). Structure mapping in analogy and similarity. *American Psychologist, 52*, 45–56.

Gerard, L. D., Zacks, R. T., Hasher, L., & Radvansky, G. A. (1991). Age deficits in retrieval: The fan effect. *The Journals of Gerontology, 46*, 131–136.

Gerbier, E., & Toppino, T. C. (2015). The effect of distributed practice: Neuroscience, cognition, and education. *Trends in Neuroscience and Education, 4*(3), 49–59.

Gerhardstein, P., & Rovee-Collier, C. (2002). The development of visual search in infants and very young children. *Journal of Experimental Child Psychology, 81*, 194–215.

Gernsbacher, M. A. (1990). *Language comprehension as structure building*. Hillsdale, NJ: Erlbaum.

Gernsbacher, M. A. (1997). Two decades of structure building. *Discourse Processes, 23*, 265–304.

Gernsbacher, M. A., & Faust, M. E. (1991). The mechanism of suppression: A component of general comprehension skill. *Journal of Experimental Psychology: Learning, Memory, & Cognition, 17*, 245–262.

Gernsbacher, M. A., Hallada, B. M., & Robertson, R. R. W. (1998). How automatically do readers infer fictional characters' emotional states? *Scientific Studies of Reading, 2*, 271–300.

Gernsbacher, M. A., & Hargreaves, D. (1988). Accessing sentence participants: The advantage of first mention. *Journal of Memory and Language, 27*, 699–717.

Gernsbacher, M. A., Hargreaves, D., & Beeman, M. (1989). Building and accessing clausal representations: The advantage of first mention versus the advantage of clause recency. *Journal of Memory and Language, 28*, 735–755.

Gernsbacher, M. A., Keysar, B., Robertson, R. R. W., & Werner, N. K. (2001). The role of suppression and enhancement in understanding metaphors. *Journal of Memory and Language, 45*, 433–450.

Gernsbacher, M. A., & Robertson, R. R. W. (1992). Knowledge activation vs. mapping when representing fictional characters' emotional states. *Language and Cognitive Processes, 7*, 337–353.

Gershkoff-Stowe, L., & Goldin-Medow, S. (2002). Is there a natural order for expressing semantic relations? *Cognitive Psychology, 45*, 375–412.

Geschwind, N. (1967). The varieties of naming errors. *Cortex, 3*, 97–112.

Geschwind, N. (1970). The organisation of language and the brain. *Science, 170*, 940–944.

Gevins, A., Smith, M. E., McEvoy, L., & Yu, D. (1997). High resolution EEG mapping of cortical activation related to working memory: Effects of task difficulty, type of processing, and practice. *Cerebral Cortex, 7*, 374–385.

Giambra, L. M. (1977). Adult male daydreaming across the life span: A replication, further analyses, and tentative norms based upon retrospective reports. *International Journal of Aging and Human Development, 8*, 197–228.

Giambra, L. M. (1989). Task-unrelated-thought frequency as a function of age: A laboratory study. *Psychology and Aging, 4*, 136–143.

Gibbs, R. W., & Nayak, N. P. (1989). Psycholinguistic studies on the syntactic behavior of idioms. *Cognitive Psychology, 21*(1), 100–138.

Gibbs, R. W., Nayak, N. P., & Cutting, C. (1989). How to kick the bucket and not decompose: Analyzability and idiom processing. *Journal of Memory and Language, 28*(5), 576–593.

Gibson, B. S., & Sztybel, P. (2014). The spatial semantics of symbolic attention control. *Current Directions in Psychological Science, 23*(4), 271–276.

Gick, M. L., & Holyoak, K. J. (1980). Analogical problem solving. *Cognitive Psychology, 12*, 306–355.

Gigerenzer, G. (1996). On narrow norms and vague heuristics: A reply to Kahneman and Tversky. *Psychological Review, 103*, 592–596.

Gigerenzer, G. (2004). Dread risk, September 11, and fatal traffic accidents. *Psychological Science, 15*, 286–287.

Gigerenzer, G. (2008). Why heuristics work. *Perspectives in Psychological Science, 3*, 20–29.

Gigerenzer, G., & Goldstein, D. G. (1996). Reasoning the fast and frugal way: Models of bounded rationality. *Psychological Review, 104*, 650–669.

Gilhooly, K. J., Philips, L. H., Wynn, V., Logie, R. H., & Della Sala, S. (1999). Planning processes and age in the five-disc tower of London task. *Thinking and Reasoning, 5*, 339–361.

Gilinsky, A. S., & Judd, B. B. (1994). Working memory and bias in reasoning across the life span. *Psychology and Aging, 9*, 356–371.

Gillihan, S. J., & Farah, M. J. (2005). Is the self special? A critical review of evidence from experimental psychology and cognitive neuroscience. *Psychological Bulletin, 131*, 76–97.

Gilovich, T., Vallone, R., & Tversky, A. (1985). The hot hand in basketball: On the misperception of random sequences. *Cognitive Psychology, 17*, 295–314.

Gladwell, M. (2005). *Blink: The power of thinking without thinking*. New York, NY: Little, Brown.

Glanzer, M. (1972). Storage mechanisms in recall. In G. H. Bower & J. T. Spence (Eds.), *The psychology of learning and motivation* (Vol. 5, pp. 129–193). New York, NY: Academic Press.

Glanzer, M., & Cunitz, A. R. (1966). Two storage mechanisms in free recall. *Journal of Verbal Learning and Verbal Behavior, 5,* 351–360.

Glass, A. L., & Holyoak, K. J. (1986). *Cognition* (2nd ed.). New York, NY: Random House.

Glaze, J. A. (1928). The association value of nonsense syllables. *The Journal of Genetic Psychology: Research and Theory on Human Development, 35,* 255–269.

Gleason, J. B. (1985). *Studying language development. The development of language.* Boston, MA: Allyn & Bacon/Longman.

Glenberg, A. M. (1976). Monotonic and nonmonotonic lag effects in paired-associate and recognition memory paradigms. *Journal of Verbal Learning and Verbal Behavior, 15*(1), 1–16.

Glenberg, A. M. (1979). Component-levels theory of the effects of spacing of repetitions on recall and recognition. *Memory & Cognition, 7*(2), 95–112.

Glenberg, A. (2011). Introduction to the mirror neuron forum. *Perspectives on Psychological Science, 6,* 363–368.

Glenberg, A., & Adams, F. (1978). Type I rehearsal and recognition. *Journal of Verbal Learning and Verbal Behavior, 17,* 455–464.

Glenberg, A. M., & Lehmann, T. S. (1980). Spacing repetitions over 1 week. *Memory & Cognition, 8,* 528–538.

Glenberg, A., Smith, S. M., & Green, C. (1977). Type I rehearsal: Maintenance and more. *Journal of Verbal Learning and Verbal Behavior, 11,* 403–416.

Glück, J., & Bluck, S. (2007). Looking back across the life span: A life story account of the reminiscence bump. *Memory & Cognition, 35,* 1928–1939.

Glucksberg, S. (2003). The psycholinguistics of metaphor. *Trends in Cognitive Sciences, 7*(2), 92–96.

Glucksberg, S., & Danks, J. H. (1975). *Experimental psycholinguistics: An introduction.* Hillsdale, NJ: Erlbaum.

Glucksberg, S., Gildea, P., & Bookin, H. B. (1982). On understanding nonliteral speech: Can people ignore metaphors? *Journal of Verbal Learning and Verbal Behavior, 21*(1), 85–98.

Glucksberg, S., & Keysar, B. (1990). Understanding metaphorical comparisons: Beyond similarity. *Psychological Review, 97,* 3–18.

Glucksberg, S., & McCloskey, M. (1981). Decisions about ignorance: Knowing that you don't know. *Journal of Experimental Psychology: Human Learning and Memory, 7,* 311–325.

Gobet, F., Johnston, S. J., Ferrufino, G., Johnston, M., Jones, M. B., Molyneux, . . . Weeden, L. (2015). "No level up!" No effects of video game specialization and expertise on cognitive performance. *Frontiers in Psychology, 5,* 1337.

Gobet, F., & Simon, H. A. (1996). Recall of random and distorted chess positions: Implications for the theory of expertise. *Memory & Cognition, 24,* 493–503.

Godden, D. B., & Baddeley, A. D. (1975). Context-dependent memory in two natural environments: On land and underwater. *British Journal of Psychology, 66,* 325–331.

Gogate, L. J., & Hollich, G. (2010). Invariance detection within an interactive system: A perceptual gateway to language development. *Psychological Review, 117,* 496–516.

Gold, P. E., Cahill, L., & Wenk, G. L. (2002). Ginkgo biloba: A cognitive enhancer? *Psychological Science in the Public Interest, 3,* 2–11.

Gold, P. E., Cahill, L., & Wenk, G. L. (2003). The lowdown on ginkgo biloba. *Scientific American, 288,* 86–92.

Goldin-Meadow, S. (1997). When gestures and words speak differently. *Psychological Science, 6,* 138–143.

Goldin-Meadow, S. (2006). Talking and thinking with our hands. *Current Directions in Psychological Science, 15,* 34–39.

Goldin-Meadow, S., & Beilock, S. L. (2010). Action's influence on thought: The case of gesture. *Perspectives on Psychological Science, 5,* 664–674.

Goldinger, S. D., Kleider, H. M., Azuma, T., & Beike, D. R. (2003). "Blaming the victim" under memory load. *Psychological Science, 14,* 81–85.

Goldstein, D. G., & Gigerenzer, G. (2002). Models of ecological rationality: The recognition heuristic. *Psychological Review, 109,* 75–90.

Goodglass, H., Kaplan, E., Weintraub, S., & Ackerman, N. (1976). The "tip-of-the-tongue" phenomenon in aphasia. *Cortex, 12,* 145–153.

Goodwin, D. W., Powell, B., Bremeer, D., Hoine, H., & Stern, J. (1969). Alcohol and recall: State-dependent effects in man. *Science, 163,* 2358–2360.

Goodwin, G. P., & Johnson-Laird, P. N. (2005). Reasoning about relations. *Psychological Review, 112,* 468–493.

Gordon, B., Hart, J., Jr., Boatman, D., & Lesser, R. P. (1997). Cortical stimulation (interference) during behavior. In T. E. Feinberg & M. J. Farah (Eds.), *Behavioral neurology and neuropsychology* (pp. 667–672). New York, NY: McGraw-Hill.

Gordon, P. C., & Chan, D. (1995). Pronouns, passives, and discourse coherence. *Journal of Memory and Language, 34,* 216–231.

Gordon, P. C., & Scearce, K. A. (1995). Pronominalization and discourse coherence, discourse structure and pronoun interpretation. *Memory & Cognition, 23,* 313–323.

Gordon, R. L., Fehd, H. M., & McCandliss, B. D. (2015). Does music training enhance literacy skills? A meta-analysis. *Frontiers in Psychology, 6.*

Goshen-Gottstein, Y., & Kempinsky, H. (2001). Probing memory with conceptual cues at multiple retention intervals: A comparison of forgetting rates on implicit and explicit tests. *Psychonomic Bulletin & Review, 8,* 139–146.

Graesser, A. C. (1981). *Prose comprehension beyond the word.* New York, NY: Springer.

Graesser, A. C., & Nakamura, G. V. (1982). The impact of a schema on comprehension and memory. In G. H. Bower (Ed.), *The psychology of learning and motivation* (pp. 59–109). New York, NY: Academic Press.

Graesser, A. C., Singer, M., & Trabasso, T. (1994). Constructing inferences during narrative text comprehension. *Psychological Review, 101,* 371–395.

Graf, P., & Schacter, D. L. (1987). Selective effects of interference on implicit and explicit memory for new associations.

Journal of Experimental Psychology: Learning, Memory, and Cognition, 13, 45–53.

Grafman, J., Pascual-Leone, A., Always, D., Nichelli, E., Gomez-Tortosa, E., & Hallett, M. (1994). Induction of a recall deficit by rapid-rate transcranial magnetic stimulation. Neuroreport, 5, 1157–1160.

Graham, E. R., & Burke, D. M. (2011). Aging increases inattention blindness to the gorilla in our midst. Psychology and Aging, 26, 162–166.

Gray, R., Beilock, S. L., & Carr, T. H. (2007). "As soon as the bat met the ball, I knew it was gone": Outcome prediction, hindsight bias, and the representation and control of action in expert and novice baseball players. Psychonomic Bulletin & Review, 14, 669–675.

Greenberg, J. H. (1978). Generalizations about numeral systems. In J. H. Greenberg (Ed.), Universals of human language: Vol. 3. Word structure (pp. 249–295). Stanford, CA: Stanford University Press.

Greenberg, S. N., Healy, A. F., Koriat, A., & Kreiner, H. (2004). The GO model: A reconsideration of the role of structural units in guiding and organizing text on line. Psychonomic Bulletin & Review, 11, 428–433.

Greene, R. L. (1986). Effects of intentionality and strategy on memory for frequency. Journal of Experimental Psychology: Learning, Memory, and Cognition, 12, 489–495.

Greeno, J. G. (1974). Hobbits and orcs: Acquisition of a sequential concept. Cognitive Psychology, 6, 270–292.

Greeno, J. G. (1978). Natures of problem-solving abilities. In W. K. Estes (Ed.), Handbook of learning and cognitive processes: Vol. 5. Human information processing (pp. 239–270). Hillsdale, NJ: Erlbaum.

Greenspan, S. L. (1986). Semantic flexibility and referential specificity of concrete nouns. Journal of Memory and Language, 25, 539–557.

Greenwald, A. G., Spangenberg, E. R., Pratkanis, A. R., & Eskenazi, J. (1991). Double-blind tests of subliminal self-help audiotapes. Psychological Science, 2, 119–122.

Grice, H. P. (1975). Logic and conversation. In P. Cole & J. L. Morgan (Eds.), Syntax and semantics: Vol. 3. Speech acts (pp. 41–58). New York, NY: Seminar Press.

Griffin, D., & Tversky, A. (1992). The weighing of evidence and the determinants of confidence. Cognitive Psychology, 24, 411–435.

Griffin, T. D., Wiley, J., & Thiede, K. W. (2008). Individual differences, rereading, and self-explanation: Concurrent processing and cue validity as constraints on metacomprehension accuracy. Memory & Cognition, 36, 93–103.

Griffin, Z. M. (2003). A reversed word length effect in coordinating the preparation and articulation of words in speaking. Psychonomic Bulletin & Review, 10, 603–609.

Griffin, Z. M., & Bock, K. (2000). What the eyes say about speaking. Psychological Science, 11, 274–279.

Griffiths, T. L., Steyvers, M., & Tenebaum, J. B. (2007). Topics in semantic representation. Psychological Review, 114, 211–244.

Grill-Spector, K., & Kanwisher, N. (2005). Visual recognition: As soon as you know it is there, you know what it is. Psychological Science, 16, 152–160.

Groen, G. J., & Parkman, J. M. (1972). A chronometric analysis of simple addition. Psychological Review, 79(4), 329–343.

Grossenbacher, P. G., & Lovelace, C. T. (2001). Mechanisms of synesthesia: Cognitive and physiological constraints. Trends in Cognitive Sciences, 5, 36–41.

Grossman, M., Smith, E. E., Koenig, P., Glosser, G., DeVita, C., Moore, P., & McMillan, C. (2002). The neural basis for categorization in semantic memory. Neuroimage, 17, 1549–1561.

Gruppuso, V., Lindsay, D. S., & Masson, M. E. J. (2007). I'd know that face anywhere! Psychonomic Bulletin & Review, 14, 1085–1089.

Grysman, A., Prabhakar, J., Anglin, S. M., & Hudson, J. A. (2015). Self-enhancement and the life script in future thinking across the lifespan. Memory, 23(5), 774–785.

Gupta, P., & Cohen, N. J. (2002). Theoretical and computational analysis of skill learning, repetition priming, and procedural memory. Psychological Review, 109, 401–448.

Gustafsson, B., & Wigstrom, H. (1988). Physiological mechanisms underlying long-term potentiation. Trends in Neuroscience, 11, 156–163.

Gutchess, A. H., Welsh, R. C., Boduroglu, A., & Park, D. C. (2006). Cultural differences in neural function associated with object processing. Cognitive, Affective, & Behavioral Neuroscience, 6, 102–109.

Gygax, P., Garnham, A., & Oakhill, J. (2004). Inferring characters' emotional states: Can readers infer specific emotions? Language and Cognitive Processes, 19, 613–639.

Haber, R. N., & Hershenson, M. (1973). The psychology of visual perception. New York, NY: Holt, Rinehart & Winston.

Habib, R., Nyberg, L., & Tulving, E. (2003). Hemispheric asymmetries of memory: The HERA model revisited. Trends in Cognitive Sciences, 7, 241–245.

Hagmayer, Y., & Sloman, S. A. (2009). Decision makers conceive of their choices as interventions. Journal of Experimental Psychology: General, 138, 22–38.

Hahn, U., & Warren, P. A. (2009). Perceptions of randomness: Why three heads are better than four. Psychological Review, 116, 454–461.

Haines, R. F. (1991). A breakdown in simultaneous information processing. In G. Obrecht & L. W. Stark (Eds.), Presbyopia research (pp. 171–175). New York, NY: Plenum Press.

Hall, J. F. (1971). Verbal learning and retention. Philadelphia, PA: Lippincott.

Hall, M. L., Ferreira, V. S., & Mayberry, R. I. (2015). Syntactic priming in American Sign Language. PLoS one, 10(3), e0119611.

Hambrick, D. Z. (2003). Why are some people more knowledgeable than others? A longitudinal study of knowledge acquisition. Memory & Cognition, 31, 902–917.

Hambrick, D. Z., & Engle, R. W. (2002). Effects on domain knowledge, working memory capacity, and age on cognitive performance: An investigation of the knowledge-is-power hypothesis. Cognitive Psychology, 44, 339–387.

Hambrick, D. Z., & Meinz, E. J. (2011). Limits on the predictive power of domain-specific experience and knowledge in skilled performance. Current Directions in Psychological Science, 20, 275–279.

Hamm, V. P., & Hasher, L. (1992). Age and the availability of inferences. *Psychology and Aging, 7*(1), 56–64.

Hanley, J. R., & Chapman, E. (2008). Partial knowledge in a tip-of-the-tongue state about two- and three-word proper names. *Psychonomic Bulletin & Review, 15,* 156–160.

Hannigan, S. L., & Reinitz, M. T. (2001). A demonstration and comparison of two types of inference-based memory errors. *Journal of Experimental Psychology: Learning, Memory, and Cognition, 27,* 931–940.

Hannigan, S. L., & Reinitz, M. T. (2003). Migration of objects and inferences across episodes. *Memory & Cognition, 31,* 434–444.

Hanson, C., & Hirst, W. (1988). Frequency encoding of token and type information. *Journal of Experimental Psychology: Learning, Memory, and Cognition, 14,* 289–297.

Harley, E. M., Carlsen, K. A., & Loftus, G. R. (2004). The "saw-it-all-along" effect: Demonstrations of visual hindsight bias. *Journal of Experimental Psychology: Learning, Memory, and Cognition, 30,* 960–968.

Harris, R. J., & Monaco, G. E. (1978). Psychology of pragmatic implication: Information processing between the lines. *Journal of Experimental Psychology, 107,* 1–22.

Harrison, T. L., Shipstead, Z., Hicks, K. L., Hambrick, D. Z., Redick, T. S., & Engle, R. W. (2013). Working memory training may increase working memory capacity but not fluid intelligence. *Psychological Science, 24*(12), 2409–2419.

Hart, J. T. (1965). Memory and the feeling-of-knowing experience. *Journal of Educational Psychology, 56,* 208–216.

Hartley, J. T., Stojack, C. C., Mushaney, T. J., Annon, T. A. K., & Lee, D. W. (1994). Reading speed and prose memory in older and younger adults. *Psychology and Aging, 9,* 216–223.

Hartshorn, K., Rovee-Collier, C., Gerhardstein, P., Bhatt, R. S., Wondoloski, T. L., Klein, P., et al. (1998). The ontogeny of long-term memory over the first year-and-a-half of life. *Developmental Psychobiology, 32,* 69–89.

Hartshorne, J. K., & Germine, L. T. (2015). When does cognitive functioning peak? The asynchronous rise and fall of different cognitive abilities across the life span. *Psychological Science, 26*(4), 433–443.

Hartsuiker, R. J., Anton-Mendez, I., & van Zee, M. (2001). Object attraction in subject-verb agreement construction. *Journal of Memory and Language, 45,* 546–572.

Hartsuiker, R. J., & Kolk, H. H. J. (2001). Error monitoring in speech production: A computational test of the perceptual loop theory. *Cognitive Psychology, 42,* 113–157.

Harvey, A. J., Kneller, W., & Campbell, A. C. (2013). The effects of alcohol intoxication on attention and memory for visual scenes. *Memory, 21*(8), 969–980.

Hasher, L., & Griffin, M. (1978). Reconstructive and reproductive processes in memory. *Journal of Experimental Psychology: Human Learning and Memory, 4,* 318–330.

Hasher, L., & Zacks, R. T. (1984). Automatic processing of fundamental information: The case of frequency of occurrence. *American Psychologist, 39,* 1372–1388.

Hasher, L., & Zacks, R. T. (1988). Working memory, comprehension, and aging: A review and a new view. In G. H. Bower (Ed.), *The psychology of learning and motivation* (Vol. 22, pp. 193–225). New York, NY: Academic Press.

Hashtroudi, S., Johnson, M. K., & Chrosniak, L. D. (1990). Aging and qualitative characteristics of memories for perceived and imagined complex events. *Psychology and Aging, 5,* 119–126.

Hasson, U., Furman, O., Clark, D., Dudai, Y., & Davachi, L. (2008). Enhanced intersubject correlations during movie viewing correlate with successful episodic encoding. *Neuron, 57*(3), 452–462.

Haviland, S. E., & Clark, H. H. (1974). What's new? Acquiring new information as a process in comprehension. *Journal of Verbal Learning and Verbal Behavior, 13,* 512–521.

Hayes, J. R., & Simon, H. A. (1974). Understanding written problem instructions. In L. W. Gregg (Ed.), *Knowledge and cognition* (pp. 167–200). Hillsdale, NJ: Erlbaum.

Haywood, S. L., Pickering, M. J., & Branigan, H. P. (2005). Do speakers avoid ambiguities during dialogue? *Psychological Science, 16,* 362–366.

Heaps, C. M., & Nash, M. (2001). Comparing recollective experience in true and false autobiographical memories. *Journal of Experimental Psychology: Learning, Memory, and Cognition, 27,* 920–930.

Hecht, H., & Proffitt, D. R. (1995). The price of expertise: Effects of experience on the water-level task. *Psychological Science, 6,* 90–95.

Heil, M., Rolke, B., & Pecchinenda, A. (2004). Automatic semantic activation is no myth: Semantic context effects on the N400 in the letter-search task in the absence of response time effects. *Psychological Science, 15,* 852–857.

Heimler, B., Pavani, F., Donk, M., & van Zoest, W. (2014). Stimulus-and goal-driven control of eye movements: Action videogame players are faster but not better. *Attention, Perception, & Psychophysics, 76*(8), 2398–2412.

Hell, W., Gigerenzer, G., Gauggel, S., Mall, M., & Muller, M. (1988). Hindsight bias: An interaction of automatic and motivational factors? *Memory & Cognition, 16,* 533–538.

Hellyer, S. (1962). Frequency of stimulus presentation and short-term decrement in recall. *Journal of Experimental Psychology, 64,* 650.

Helsabeck, F., Jr. (1975). Syllogistic reasoning: Generation of counterexamples. *Journal of Educational Psychology, 67,* 102–108.

Helton, W. S., & Russell, P. N. (2015). Rest is best: The role of rest and task interruptions on vigilance. *Cognition, 134,* 165–173.

Henderson, J. M., & Anes, M. D. (1994). Roles of object-file review and type priming in visual identification within and across eye fixations. *Journal of Experimental Psychology: Human Perception and Performance, 20,* 826–839.

Henderson, J. M., & Hollingworth, A. (2003). Global transsaccadic change in blindness during scene perception. *Psychological Science, 14,* 493–497.

Henkel, L. A. (2004). Erroneous memories arising from repeated attempts to remember. *Journal of Memory and Language, 50,* 26–46.

Henkel, L. A. (2014). Point-and-shoot memories the influence of taking photos on memory for a museum tour. *Psychological Science, 25*(2), 396–402.

Henry, J. D., MacLeod, M. S., Philips, L. H., & Crawford, J. R. (2004). A meta-analytic review of prospective memory and aging. *Psychology and Aging, 19,* 27–39.

Hertwig, R., Barron, G., Weber, E. U., & Erev, I. (2004). Decision from experience and the effect of rare events in risky choice. *Psychological Science, 15,* 534–539.

Hertzog, C., Kramer, A. F., Wilson, R. S., & Lindenberger, U. (2008). Enrichment effects on adult cognitive development: Can the functional capacity of older adults be preserved and enhanced? *Psychological Science in the Public Interest, 9*(1), 1–65.

Herz, R. S., & Engen, T. (1996). Odor memory: Review and analysis. *Psychonomic Bulletin & Review, 3,* 300–313.

Herz, R. S., & Schooler, J. W. (2002). A naturalistic study of autobiographical memories evoked by olfactory and visual cues: Testing the Proustian hypothesis. *American Journal of Psychology, 115,* 21–32.

Hess, T. M. (2005). Memory and aging in context. *Psychological Bulletin, 131,* 383–406.

Hicks, J. L., & Marsh, R. L. (2000). Toward specifying the attentional demands of recognition memory. *Journal of Experimental Psychology: Learning, Memory, and Cognition, 26,* 1483–1498.

Higgins, E. & Rayner, K. (2015). Transsaccadic processing: Stability, integration, and the potential role of remapping. *Attention, Perception, & Psychophysics, 77*(1), 3–27.

Higham, P. A., & Garrard, C. (2005). Not all errors are created equal: Metacognition and changing answers on multiple-choice tests. *Canadian Journal of Experimental Psychology, 59,* 28–34.

Hilbig, B. E. (2010). Reconsidering "evidence" for fast-and-frugal heuristics. *Psychonomic Bulletin & Review, 17*(6), 923–930.

Hildreth, E. C., & Ullman, S. (1989). The computational study of vision. In M. I. Posner (Ed.), *Foundations of cognitive science* (pp. 581–630). Cambridge, MA: Bradford.

Hinson, J. M., Jameson, T. L., & Whitney, P. (2003). Impulsive decision making and working memory. *Journal of Experimental Psychology: Learning, Memory, and Cognition, 29,* 298–306.

Hintzman, D. L. (1978). *The psychology of learning and memory.* San Francisco, CA: Freeman.

Hirshman, E., & Durante, R. (1992). Prime identification and semantic priming. *Journal of Experimental Psychology: Learning, Memory, and Cognition, 18,* 255–265.

Hirshman, E., Whelley, M. M., & Palij, M. (1989). An investigation of paradoxical memory effects. *Journal of Memory and Language, 28,* 594–609.

Hirst, W., & Kalmar, D. (1987). Characterizing attentional resources. *Journal of Experimental Psychology: General, 116,* 68–81.

Hoch, S. J. (1984). Availability and interference in predictive judgment. *Journal of Experimental Psychology: Learning, Memory, and Cognition, 10,* 649–662.

Hoch, S. J. (1985). Counterfactual reasoning and accuracy in predicting personal events. *Journal of Experimental Psychology: Learning, Memory, and Cognition, 11,* 719–731.

Hoch, S. J., & Loewenstein, G. F. (1989). Outcome feedback: Hindsight and information. *Journal of Experimental Psychology: Learning, Memory, and Cognition, 15,* 605–619.

Hochel, M., & Milán, E. G. (2008). Synaesthesia: The existing state of affairs. *Cognitive Neuropsychology, 25,* 93–117.

Hockett, C. F. (1960a). Logical considerations in the study of animal communication. In W. E. Lanyon & W. N. Tavolga (Eds.), *Animal sounds and communication* (pp. 392–430). Washington, DC: American Institute of Biological Sciences.

Hockett, C. F. (1960b). The origin of speech. *Scientific American, 203,* 89–96.

Hockett, C. F. (1966). The problem of universals in language. In J. H. Greenberg (Ed.), *Universals of language* (2nd ed., pp. 1–29). Cambridge, MA: MIT Press.

Hoffrage, U., Krauss, S., Martignon, L., & Gigerenzer, G. (2015). Natural frequencies improve Bayesian reasoning in simple and complex inference tasks. *Frontiers in Psychology, 6,* 1473.

Holcomb, P. J. (1993). Semantic priming and stimulus degradation: Implications for the role of the N400 in language processing. *Psychophysiology, 30,* 47–61.

Holcomb, P. J., Kounios, J., Anderson, J. E., & West, W. C. (1999). Dual-coding, context availability, and concreteness effects in sentence comprehension: An electrophysiological investigation. *Journal of Experimental Psychology: Learning, Memory, and Cognition, 25,* 721–742.

Holtgraves, T. (1994). Communication in context: Effects of speaker status on the comprehension of indirect requests. *Journal of Experimental Psychology: Learning, Memory, and Cognition, 20,* 1205–1218.

Holtgraves, T. (1998). Interpreting indirect replies. *Cognitive Psychology, 37,* 1–27.

Holtgraves, T. (2008a). Automatic intention recognition in conversation processing. *Journal of Memory and Language, 58,* 627–645.

Holtgraves, T. (2008b). Conversation, speech acts, and memory. *Memory & Cognition, 36,* 361–374.

Holtgraves, T. (2012). The role of the right hemisphere in speech act comprehension. *Brain and Language, 121*(1), 58–64.

Holyoak, K. J., & Simon, D. (1999). Bidirectional reasoning in decision making by constraint satisfaction. *Journal of Experimental Psychology: General, 128,* 3–31.

Holyoak, K. J., & Thagard, P. (1997). The analogical mind. *American Psychologist, 52,* 35–44.

Hopkins, W. D., Russell, J. L., & Cantalupo, C. (2007). Neuroanatomical correlates of handedness for tool use in chimpanzees (Pan troglodytes): Implication for theories on the evolution of language. *Psychological Science, 18,* 971–977.

Horner, A. J., Bisby, J. A., Wang, A., Bogus, K., & Burgess, N. (2016). The role of spatial boundaries in shaping long-term event representations. *Cognition, 154,* 151–164.

Horton, W. S., & Gerrig, R. J. (2002). Speakers' experiences and audience design: Knowing when and knowing how to adjust utterances to addressees. *Journal of Memory and Language, 47,* 589–606.

Hostetter, A. B., & Alibali, M. (2008). Visible embodiment: Gestures as simulated action. *Psychonomic Bulletin & Review, 15,* 495–514.

Hothersall, D. (1984). *History of psychology.* New York, NY: Random House.

Hothersall, D. (1985). *Psychology.* Columbus, OH: Merrill.

Howe, M. L., & Courage, M. L. (1993). On resolving the enigma of infantile amnesia. *Psychological Bulletin, 113*, 305–326.

Hoyle, R. H., & Sherrill, M. R. (2006). Future orientation in the self-system: Possible selves, self-regulation, and behavior. *Journal of Personality, 74*(6), 1673–1696.

Hu, P., Stylos-Allan, M., & Walker, M. P. (2006). Sleep facilitates consolidation of emotional declarative memory. *Psychological Science, 17*, 891–898.

Hu, Y., Ericsson, K. A., Yang, D., & Lu, C. (2009). Superior self-paced memorization of digits in spite of a normal digit span: The structure of a memorist's skill. *Journal of Experimental Psychology: Learning, Memory, and Cognition, 35*(6), 1426–1442.

Huang, T-R., & Grossberg, S. (2010). Cortical dynamics of contextually cued attentive visual learning and search: Spatial and object evidence accumulation. *Psychological Review, 117*, 1080–1112.

Huang, Y. T., & Gordon, P. C. (2011). Distinguishing the time course of lexical and discourse processes through context, coreference, and quantified expressions. *Journal of Experimental Psychology: Learning, Memory, and Cognition, 37*, 966–978.

Hubbard, E. M., & Ramachandran, V. S. (2005). Neurocognitive mechanisms of synesthesia. *Neuron, 48*, 509–520.

Hubbard, T. L. (1990). Cognitive representation of linear motion: Possible direction and gravity effects in judged displacement. *Memory & Cognition, 18*, 299–309.

Hubbard, T. L. (1995). Environmental invariants in the representation of motion: Implied dynamics and representational momentum, gravity, friction, and centripetal force. *Psychonomic Bulletin & Review, 2*, 322–338.

Hubbard, T. L. (1996). Representational momentum, centripetal force, and curvilinear impetus. *Journal of Experimental Psychology: Learning, Memory, and Cognition, 22*, 1049–1060.

Hubbard, T. L. (2005). Representational momentum and related displacements in spatial memory: A review of the findings. *Psychonomic Bulletin & Review, 12*, 822–851.

Hubbard, T. L., Hutchison, J. L., & Courtney, J. R. (2010). Boundary extension: Findings and theories. *The Quarterly Journal of Experimental Psychology, 63*, 1467–1494.

Hubel, D. H., & Wiesel, T. N. (1962). Receptive fields, binocular interaction, and functional architecture in the cat's visual cortex. *Journal of Physiology, 160*, 106–154.

Huettel, S. A., Song, A. W., & McCarthy, G. (2004). *Functional magnetic resonance imaging*. Sunderland, MA: Sinauer Associates.

Huff, M., Meitz, T. G., & Papenmeier, F. (2014). Changes in situation models modulate processes of event perception in audiovisual narratives. *Journal of Experimental Psychology: Learning, Memory, and Cognition, 40*(5), 1377–1388.

Hulme, C., Thomson, N., Muir, C., & Lawrence, A. (1984). Speech rate and the development of short-term memory span. *Journal of Experimental Child Psychology, 38*, 241–253.

Hunt, E., & Agnoli, F. (1991). The Whorfian hypothesis: A cognitive psychology perspective. *Psychological Review, 98*, 377–389.

Hunt, R. R., & Lamb, C. A. (2001). What causes the isolation effect? *Journal of Experimental Psychology: Learning, Memory, and Cognition, 27*, 1359–1366.

Hussain, Z., Sekuler, A. B., & Bennett, P. J. (2011). Superior identification of familiar visual patterns a year after learning. *Psychological Science, 22*, 724–730.

Hyman, I. E., Jr., Husband, T. H., & Billings, F. J. (1995). False memories of childhood experiences. *Applied Cognitive Psychology, 9*, 181–197.

Hyman, I. E., Jr., & Pentland, J. (1996). The role of mental imagery in the creation of false childhood memories. *Journal of Memory and Language, 35*, 101–117.

Inhoff, A. W. (1984). Two stages of word processing during eye fixations in the reading of prose. *Journal of Verbal Learning and Verbal Behavior, 23*, 612–624.

Intraub, H., & Richardson, M. (1989). Wide-angle memories of close-up scenes. *Journal of Experimental Psychology: Learning, Memory, and Cognition, 15*, 179–187.

Intriligator, J., & Cavanagh, P. (2001). The spatial resolution of visual attention. *Cognitive Psychology, 43*, 171–216.

Inzlicht, M., & Gutsell, J. N. (2007). Running on empty: Neural signals for self-control failure. *Psychological Science, 18*, 933–937.

Irwin, D. E. (1991). Information integration across saccadic eye movements. *Cognitive Psychology, 23*, 420–456.

Irwin, D. E. (1992). Memory for position and identity across eye movements. *Journal of Experimental Psychology: Learning, Memory, and Cognition, 18*, 307–317.

Irwin, D. E. (1996). Integrating information across saccadic eye movements. *Current Directions in Psychological Science, 5*, 94–100.

Irwin, D. E. (2014). Short-term memory across eye blinks. *Memory, 22*(8), 898–906.

Irwin, D. E., & Brockmole, J. R. (2004). Suppressing where but not what: The effect of saccades on dorsal- and ventral-stream visual processing. *Psychological Science, 15*, 467–473.

Irwin, D. E., Yantis, S., & Jonides, J. (1983). Evidence against visual integration across saccadic eye movements. *Perception & Psychophysics, 34*, 49–57.

Jackendoff, R. S. (1992). *Languages of the mind: Essays on mental representation*. Cambridge, MA: MIT Press.

Jacobs, N., & Garnham, A. (2007). The role of conversational hand gestures in a narrative task. *Journal of Memory and Language, 56*(2), 291–303.

Jacoby, L. L. (1991). A process dissociation framework: Separating automatic from intentional uses of memory. *Journal of Memory and Language, 30*, 513–541.

Jacoby, L. L., & Dallas, M. (1981). On the relationship between autobiographical memory and perceptual learning. *Journal of Experimental Psychology: General, 110*, 306–340.

Jacoby, L. L., Toth, J. P., & Yonelinas, A. P. (1993). Separating conscious and unconscious influences of memory: Measuring recollection. *Journal of Experimental Psychology: General, 122*, 139–154.

Jacoby, L. L., Woloshyn, V., & Kelley, C. (1989). Becoming famous without being recognized: Unconscious influences of memory produced by dividing attention. *Journal of Experimental Psychology: General, 118*, 115–125.

James, W. (1890). *The principles of psychology*. New York, NY: Dover.

January, D., & Kako, E. (2007). Re-evaluating evidence for linguistic relativity: Reply to Boroditsky (2001). *Cognition, 104,* 417–426.

Jared, D., Levy, B. A., & Rayner, K. (1999). The role of phonology in the activation of word meanings during readings: Evidence from proofreading and eye movements. *Journal of Experimental Psychology: General, 128,* 219–264.

Jefferies, E., Lambdon Ralph, M. A., & Baddeley, A. D. (2004). Automatic and controlled processing in sentence recall: The role of long-term and working memory. *Journal of Memory and Language, 51,* 623–643.

Jenkins, J. G., & Dallenbach, K. M. (1924). Obliviscence during sleep and waking. *American Journal of Psychology, 35,* 605–612.

Jessberger, S., & Gage, F. H. (2008). Stem-cell-associated structural and functional plasticity in the aging hippocampus. *Psychology and Aging, 23,* 684–691.

Jesse, A., & Massaro, D. W. (2010). Seeing a singer helps comprehension of the song's lyrics. *Psychonomic Bulletin & Review, 17,* 323–328.

Johansson, G. (1973). Visual perception of biological motion and a model for its analysis. *Perception & Psychophysics, 14*(2), 201–211.

Johnson, J. T., & Finke, R. A. (1985). The base-rate fallacy in the context of sequential categories. *Memory & Cognition, 13,* 63–73.

Johnson, M. K., Hashtroudi, S., & Lindsay, S. (1993). Source monitoring. *Psychological Bulletin, 114,* 3–28.

Johnson, M. R., Higgins, J. A., Norman, K. A., Sederberg, P. B., Smith, T. A., & Johnson, M. K. (2013). Foraging for thought: An inhibition-of-return-like effect resulting from directing attention within working memory. *Psychological Science, 24*(7), 1104–1112.

Johnson, N. F. (1970). The role of chunking and organization in the process of recall. In G. H. Bower (Ed.), *The psychology of learning and motivation* (Vol. 4, pp. 172–247). New York, NY: Academic Press.

Johnson, S. K., & Anderson, M. C. (2004). The role of inhibitory control in forgetting semantic knowledge. *Psychological Science, 15,* 448–453.

Johnson-Laird, P. N. (1983). *Mental models: Towards a cognitive science of language, inference and consciousness.* Cambridge, MA: Harvard University Press.

Johnson-Laird, P. N. (2013). Mental models and cognitive change. *Journal of Cognitive Psychology, 25*(2), 131–138.

Johnson-Laird, P. N., & Byrne, R. M. J. (2002). Conditionals: A theory of meaning, pragmatics, and inference. *Psychological Review, 109,* 646–678.

Johnson-Laird, P. N., Byrne, R. M. J., & Schaeken, W. (1992). Propositional reasoning by model. *Psychological Review, 99,* 418–439.

Johnson-Laird, P. N., Legrenzi, P., & Legrenzi, M. S. (1972). Reasoning and a sense of reality. *British Journal of Psychology, 63,* 395–400.

Johnston, J. C., McCann, R. S., & Remington, R. W. (1995). Chronometric evidence for two types of attention. *Psychological Science, 6,* 365–369.

Johnston, W. A., & Heinz, S. P. (1978). Flexibility and capacity demands of attention. *Journal of Experimental Psychology: General, 107,* 420–435.

Jones, D. M., Macken, W. J., & Nicholls, A. P. (2004). The phonological store of working memory: Is it phonological and is it a store? *Journal of Experimental Psychology: Learning, Memory, and Cognition, 30,* 656–674.

Jones, G. V. (1989). Back to Woodworth: Role of interlopers in the tip-of-the-tongue phenomenon. *Memory & Cognition, 17,* 69–76.

Jones, G. V., & Martin, M. (2006). Primacy of memory linkage in choice among valued objects. *Memory & Cognition, 34,* 1587–1597.

Jonides, J., & Jones, C. M. (1992). Direct coding for frequency of occurrence. *Journal of Experimental Psychology: Learning, Memory, and Cognition, 18,* 368–378.

Jonides, J., Lacey, S. C., & Nee, D. E. (2005). Processes of working memory in mind and brain. *Current Directions in Psychological Science, 14,* 2–5.

Jonides, J., Smith, E. E., Koeppe, R. A., Awh, E., Minoshima, S., & Mintun, M. A. (1993). Spatial working-memory in humans as revealed by PET. *Nature, 363,* 623–625.

Joye, Y., Pals, R., Steg, L., & Evans, B. L. (2013). New methods for assessing the fascinating nature of nature experiences. *PloS one, 8*(7), e65332.

Juhasz, B. J., & Rayner, K. (2003). Investigating the effects of a set of intercorrelated variables on eye fixation durations in reading. *Journal of Experimental Psychology: Learning, Memory, and Cognition, 29,* 1312–1318.

Just, M. A. (1976, May). *Research strategies in prose comprehension.* Paper presented at the meetings of the Midwestern Psychological Association, Chicago, IL.

Just, M. A., & Carpenter, P. A. (1980). A theory of reading: From eye fixations to comprehension. *Psychological Review, 87,* 329–354.

Just, M. A., & Carpenter, P. A. (1987). *The psychology of reading and language comprehension.* Boston, MA: Allyn & Bacon.

Just, M. A., & Carpenter, P. A. (1992). A capacity theory of comprehension. *Psychological Review, 99,* 122–149.

Just, M. A., Carpenter, P. A., Maguire, M., Diwadkar, V., & McMains, S. (2001). Mental rotation of objects retrieved from memory: A functional MRI study of spatial processing. *Journal of Experimental Psychology: General, 130,* 493–504.

Kaakinen, J. K., Hyona, J., & Keenan, J. M. (2003). How prior knowledge, WMC, and relevance of information affect eye fixations in expository text. *Journal of Experimental Psychology: Learning, Memory, and Cognition, 29,* 447–457.

Kahana, M. J., & Wingfield, A. (2000). A functional relation between learning and organization in free recall. *Psychonomic Bulletin & Review, 7,* 516–521.

Kahneman, D. (1968). Method, findings and theory in studies of visual masking. *Psychological Bulletin, 70,* 404–426.

Kahneman, D. (1973). *Attention and effort.* Englewood Cliffs, NJ: Prentice Hall.

Kahneman, D. (2003a). A perspective on judgment and choice: Mapping bounded rationality. *American Psychologist, 58,* 697–720.

Kahneman, D. (2003b). Experiences of collaborative research. *American Psychologist, 58,* 723–730.

Kahneman, D., Slovic, P., & Tversky, A. (Eds.). (1982). *Judgment under uncertainty: Heuristics and biases.* Cambridge, England: Cambridge University Press.

Kahneman, D., Triesman, A., & Gibbs, B. J. (1992). The reviewing of object files: Object-specific integration of information. *Cognitive Psychology, 24,* 175–219.

Kahneman, D., & Tversky, A. (1972). Subjective probability: A judgment of representativeness. *Cognitive Psychology, 3,* 430–454.

Kahneman, D., & Tversky, A. (1973). On the psychology of prediction. *Psychological Review, 80,* 237–251.

Kahneman, D., & Tversky, A. (1982). The simulation heuristic. In D. Kahneman, P. Slovic, & A. Tversky (Eds.), *Judgment under uncertainty: Heuristics and biases* (pp. 201–208). Cambridge, England: Cambridge University Press.

Kahneman, D., & Tversky, A. (Eds.). (2000). *Choice, values, and frames.* New York, NY: Cambridge University Press.

Kail, R. (1991). Processing time declines exponentially during childhood and adolescence. *Developmental Psychology, 27,* 259–266.

Kail, R. (1997). Processing time, imagery, and spatial memory. *Journal of Experimental Child Psychology, 64,* 67–78.

Kalénine, S., & Bonthoux, F. (2008). Object manipulability affects children's and adults' conceptual processing. *Psychonomic Bulletin & Review, 15,* 667–672.

Kambe, G., Duffy, S. A., Clifton, C., Jr., & Rayner, K. (2003). An eye-movement-contingent probe paradigm. *Psychonomic Bulletin & Review, 10,* 661–666.

Kamienkowski, J. E., Navajas, J., & Sigman, M. (2012). Eye movements blink the attentional blink. *Journal of Experimental Psychology: Human Perception and Performance, 38*(3), 555–560.

Kane, M. J., & Engle, R. W. (2000). Working memory capacity, proactive interference, and divided attention: Limits on long-term memory retrieval. *Journal of Experimental Psychology: Learning, Memory, and Cognition, 26,* 336–358.

Kane, M. J., & Engle, R. W. (2002). The role of prefrontal cortex in working-memory capacity, executive attention, and general fluid intelligence: An individual-differences perspective. *Psychonomic Bulletin & Review, 9,* 637–671.

Kane, M. J., & Engle, R. W. (2003). Working-memory capacity and the control of attention: The contributions of goal neglect, response competition and task set to Stroop interference. *Journal of Experimental Psychology: General, 132,* 47–70.

Kane, M. J., Hambrick, D. Z., Tuholski, S. W., Wilhelm, O., Payne, T. W., & Engle, R. W. (2004). The generality of working memory capacity: A latent-variable approach to verbal and visuospatial memory span and reasoning. *Journal of Experimental Psychology: General, 133,* 189–217.

Kane, M. J., & McVay, J. C. (2012). What mind wandering reveals about executive-control abilities and failures. *Current Directions in Psychological Science, 21*(5), 348–354.

Kanwisher, N. (1987). Repetition blindness: Type recognition without token individuation. *Cognition, 27,* 117–143.

Kanwisher, N. (1991). Repetition blindness and illusory conjunctions: Errors in binding visual types with visual tokens. *Journal of Experimental Psychology: Human Perception and Performance, 17,* 404–421.

Kanwisher, N., & Driver, J. (1992). Objects, attributes, and visual attention: Which, what, and where. *Psychological Science, 1,* 26–31.

Karpicke, J. D. (2012). Retrieval-based learning active retrieval promotes meaningful learning. *Current Directions in Psychological Science, 21*(3), 157–163.

Karpicke, J. D., & Roediger, H. L. (2007). Repeated during retrieval is the key to long-term retention. *Journal of Memory and Language, 57,* 151–162.

Kaschak, M. P., & Glenberg, A. M. (2000). Constructing meaning: The role of affordances and grammatical constructions in sentence comprehension. *Journal of Memory and Language, 43,* 508–529.

Kaschak, M. P., Kutta, T. J., & Schatschneider, C. (2011). Long-term cumulative structural priming persists for (at least) one week. *Memory & Cognition, 39*(3), 381–388.

Kassam, K. S., Gilbert, D. T., Swencionis, J. K., & Wilson, T. D. (2009). Misconceptions of memory: The Scooter Libby effect. *Psychological Science, 20,* 551–552.

Kassam, K. S., Koslov, K., & Mendes, W. B. (2009). Decisions under distress: Stress profiles influence anchoring and adjustment. *Psychological Science, 20,* 1394–1399.

Kay, J., & Ellis, A. (1987). A cognitive neuropsychological case study of anomia: Implications for psychological models of word retrieval. *Brain, 110,* 613–629.

Kelley, C. M., & Jacoby, L. L. (1996). Memory attributions: Remembering, knowing, and feeling of knowing. In L. M. Reder (Ed.), *Implicit memory and metacognition* (pp. 287–308). Hillsdale, NJ: Erlbaum.

Kelley, C. M., & Lindsay, D. S. (1993). Remembering mistaken for knowing: Ease of retrieval as a basis for confidence in answers to general knowledge questions. *Journal of Memory and Language, 32,* 1–24.

Kelley, M. R., & Nairne, J. S. (2001). von Restorff revisited: Isolation, generation, and memory of order. *Journal of Experimental Psychology: Learning, Memory, and Cognition, 27,* 54–66.

Kelly, S. D., Barr, D. J., Church, R. B., & Lynch, K. (1999). Offering a hand to pragmatic understanding: The role of speech and gesture in comprehension and memory. *Journal of Memory and Language, 40,* 577–592.

Kelly, S. D., Özyürek, A., & Maris, E. (2010). Two sides of the same coin: Speech and gesture mutually interact to enhance comprehension. *Psychological Science, 21,* 260–267.

Kemmerer, D., Castillo, J. G., Talavage, T., Patterson, S., & Wiley, C. (2008). Neuroanatomical distribution of five semantic components of verbs: Evidence from fMRI. *Brain and Language, 107,* 16–43.

Kempen, G., & Hoehkamp, E. (1987). An incremental procedural grammar for sentence formulation. *Cognitive Science, 11,* 201–258.

Kemper, S. (1987). Life-span changes in syntactic complexity. *The Journals of Gerontology: Psychological Sciences, 42,* P323–P328.

Kemper, S., & Thissen, D. (1981). Memory for the dimensions of requests. *Journal of Verbal Learning and Verbal Behavior, 20,* 552–563.

Kempton, W. (1986). Two theories of home heat control. *Cognitive Science, 10,* 75–90.

Kemtes, K. A., & Kemper, S. (1997). Younger and older adults' on-line processing of syntactically ambiguous sentences. *Psychology and Aging, 12,* 362–371.

Kennedy, Q., Mather, M., & Carstensen, L. L. (2004). The role of motivation in the age-related positivity effect in autobiographical memory. *Psychological Science, 15,* 208–214.

Kensinger, E. A. (2007). Negative emotion enhances memory accuracy: Behavioral and neuroimaging evidence. *Current Directions in Psychological Science, 16,* 213–218.

Kensinger, E. A. (2009). *Emotional memory across the adult lifespan.* New York, NY: Psychology Press.

Kensinger, E. A., & Corkin, S. (2004). Two routes to emotional memory: Distinct neural processes for valence and arousal. *Proceedings of the National Academy of Sciences, 101,* 3310–3315.

Kensinger, E. A., Garoff-Eaton, R. J., & Schacter, D. L. (2006). Memory for specific visual details can be enhanced by negative arousing content. *Journal of Memory and Language, 54,* 99–112.

Kensinger, E. A., & Schacter, D. L. (2006). When the Red Socks shocked the Yankees: Comparing negative and positive memories. *Psychonomic Bulletin & Review, 13,* 757–763.

Keppel, G., & Underwood, B. J. (1962). Proactive inhibition in short-term retention of single items. *Journal of Verbal Learning and Verbal Behavior, 1,* 153–161.

Kershaw, T. C., & Ohlsson, S. (2004). Multiple causes of difficulty in insight: The case of the nine-dot problem. *Journal of Experimental Psychology: Learning, Memory, and Cognition, 30,* 3–13.

Kertesz, A. (1982). Two case studies: Broca's and Wernicke's aphasia. In M. A. Arbib, D. Caplan, & J. C. Marshall (Eds.), *Neural models of language processes* (pp. 25–44). New York, NY: Academic Press.

Keysar, B., Shen, Y., Glucksberg, S., & Horton, W. S. (2000). Conventional language: How metaphorical is it? *Journal of Memory and Language, 43,* 576–593.

Khemlani, S., & Johnson-Laird, P. N. (2012). Theories of the syllogism: A meta-analysis. *Psychological Bulletin, 138*(3), 427–457.

Kibbe, M. M., & Leslie, A. M. (2011). What do infants remember when they forget? Location and identity in 6-month-olds' memory for objects. *Psychological Science, 22,* 1500–1505.

Kim, N. S., & Ahn, W. (2002). Clinical psychologists' theory-based representations of mental disorders predict their diagnostic reasoning and memory. *Journal of Experimental Psychology: General, 131,* 451–476.

Kim, N. S., Yopchick, J. E., & de Kwaadsteniet, L. (2008). Causal diversity effects in information seeking. *Psychonomic Bulletin & Review, 15,* 81–88.

Kinder, A., & Shanks, D. R. (2003). Neuropsychological dissociations between priming and recognition: A single-system connectionist account. *Psychological Review, 110,* 728–744.

Kingstone, A., Smilek, D., Ristic, J., Friesen, C. K., & Eastwood, J. D. (2003). Attention, researchers! It is time to take a look at the real world. *Current Directions in Psychological Science, 12,* 176–180.

Kintsch, W. (1974). *The representation of meaning in memory.* Hillsdale, NJ: Erlbaum.

Kintsch, W. (2000). Metaphor comprehension: A computational theory. *Psychonomic Bulletin & Review, 7,* 257–266.

Kintsch, W., & Bates, E. (1977). Recognition memory for statements from a classroom lecture. *Journal of Experimental Psychology: Human Learning and Memory, 3,* 150–159.

Kintsch, W., & van Dijk, T. A. (1978). Toward a model of text comprehension and production. *Psychological Review, 85,* 363–394.

Kintsch, W., Welsch, D., Schmalhofer, F., & Zimny, S. (1987). Sentence memory: A theoretical analysis. *Journal of Memory and Language, 29,* 133–159.

Kishiyama, M. M., Yonelinas, A. P., & Lazzara, M. M. (2004). The von Restorff effect in amnesia: The contribution of the hippocampal system to novelty-related memory enhancements. *Journal of Cognitive Neuroscience, 16,* 15–23.

Kisilevsky, B. S., Hains, S. M. J., Lee, K., Xie, X., Huang, H., Ye, H. H., et al. (2003). Effects of experience on fetal voice recognition. *Psychological Science, 14,* 220–224.

Kissler, J., Herbert, C., Peyk, P., & Junghofer, M. (2007). Buzzwords: Early cortical response to emotional words during reading. *Psychological Science, 18,* 475–480.

Klatzky, R. L. (1980). *Human memory: Structures and processes* (2nd ed.). San Francisco, CA: Freeman.

Klatzky, R. L., Pellegrino, J. W., McCloskey, B. P., & Doherty, S. (1989). Can you squeeze a tomato? The role of motor representations in semantic sensibility judgments. *Journal of Memory and Language, 28,* 56–77.

Klauer, K. C., Stahl, C., & Erdfelder, E. (2007). The abstract selection task: New data and an almost comprehensive model. *Journal of Experimental Psychology: Learning, Memory, and Cognition, 33,* 680–703.

Klayman, J., & Ha, Y.-W. (1989). Hypothesis testing in rule discovery: Strategy, structure, and content. *Journal of Experimental Psychology: Learning, Memory, and Cognition, 15,* 596–604.

Klein, D. V., & Murphy, G. L. (2002). Paper has been my ruin: Conceptual relations of polysemous senses. *Journal of Memory and Language, 47,* 548–570.

Klein, R. (2000). Inhibition of return. *Trends in Cognitive Sciences, 4,* 138–147.

Klein, S. B. (2012). A role for self-referential processing in tasks requiring participants to imagine survival on the savannah. *Journal of Experimental Psychology: Learning, Memory, and Cognition, 38*(5), 1234–1242.

Klein, S. B., Robertson, T. E., & Delton, A. W. (2010). Facing the future: Memory as an evolved system for planning future acts. *Memory & Cognition, 38,* 13–22.

Kliegl, R., Nuthmann, A., & Engbert, R. (2006). Tracking the mind during reading: The influence of past, present, and future words on fixation durations. *Journal of Experimental Psychology: General, 135,* 12–35.

Knight, R. T. (1984). Decreased response to novel stimuli after prefrontal lesions in man. *Electroencephalography and Clinical Neurophysiology: Evoked Potentials Section, 59,* 9–20.

Koechlin, E., Dehaene, S., & Mehler, J. (1998). Numerical transformations in five-month-old human infants. *Mathematical Cognition, 3,* 89–104.

Koehler, J. J., & Macchi, L. (2004). Thinking about low-probability events: An exemplar-cuing theory. *Psychological Science, 15,* 540–546.

Köhler, W. (1927). *The mentality of apes.* New York, NY: Harcourt, Brace.

Koivisto, M., & Revonsuo, A. (2007). How meaning shapes seeing. *Psychological Science, 18,* 845–849.

Kolers, P. A., & Roediger, H. L. III (1984). Procedures of mind. *Journal of Verbal Learning and Verbal Behavior, 23,* 425–449.

Kolodner, J. L. (1997). Educational implications of analogy: A view from case-based reasoning. *American Psychologist, 52,* 57–66.

Komeda, H., & Kusumi, T. (2006). The effect of a protagonist's emotional shift on situation model construction. *Memory & Cognition, 34,* 1548–1556.

Koriat, A. (1993). How do we know that we know? The accessibility model of the feeling of knowing. *Psychological Review, 100,* 609–639.

Koriat, A. (1995). Dissociating knowing and the feeling of knowing: Further evidence for the accessibility model. *Journal of Experimental Psychology: General, 124,* 311–333.

Koriat, A., Fiedler, K., & Bjork, R. A. (2006). Inflation of conditional predictions. *Journal of Experimental Psychology: General, 135,* 429–447.

Koriat, A., Levy-Sadot, R., Edry, E., & de Marcus, S. (2003). What do we know about what we cannot remember? Accessing the semantic attributes of words that cannot be recalled. *Journal of Experimental Psychology: Learning, Memory, and Cognition, 29,* 1095–1105.

Koriat, A., & Pearlman-Avnion, S. (2003). Memory organization of action events and its relationship to memory performance. *Journal of Experimental Psychology: General, 132,* 435–454.

Koriat, A., Sheffer, L., & Ma'ayan, H. (2002). Comparing objective and subjective learning curves: Judgments of learning exhibit increased underconfidence with practice. *Journal of Experimental Psychology: General, 131,* 147–162.

Koslowski, B., Marasia, J., Vermeylen, F., & Hendrix, V. (2013). A disconfirming strategy is not necessarily better than a confirming strategy. *American Journal of Psychology, 126*(3), 335–354.

Kosslyn, S. M., Alpert, N. M., Thompson, W. L., Maljkovic, V., Weise, S. B., Chabris, C. F., et al. (1993). Visual mental imagery activates topographically organized visual cortex: PET investigations. *Journal of Cognitive Neuroscience, 5,* 263–287.

Kotovsky, K., Hayes, J. R., & Simon, H. A. (1985). Why are some problems hard? Evidence from Tower of Hanoi. *Cognitive Psychology, 17,* 248–294.

Kounios, J. (1996). On the continuity of thought and the representation of knowledge: Electrophysiological and behavioral time-course measures reveal levels of structure in semantic memory. *Psychonomic Bulletin & Review, 3,* 265–286.

Kounios, J., Frymiare, J. L., Bowden, E. M., Fleck, J. I., Subramaniam, K., Parrish, T. B., et al. (2006). The prepared mind: Neural activity prior to problem presentation predicts subsequent solution by sudden insight. *Psychological Science, 17,* 882–890.

Kounios, J., & Holcomb, P. J. (1992). Structure and process in semantic memory: Evidence from event-related brain potentials and reaction times. *Journal of Experimental Psychology: General, 121,* 459–479.

Kounios, J., & Holcomb, P. J. (1994). Concreteness effects in semantic processing: ERP evidence supporting dual-coding theory. *Journal of Experimental Psychology: Learning, Memory, and Cognition, 20,* 804–823.

Kousta, S-T., Vigliocco, G., Vinson, D. P., Andrews, M., & Campo, E. D. (2010). The representation of abstract words: Why emotion matters. *Journal of Experimental Psychology: General, 140,* 14–34.

Kozhevnikov, M., & Hegarty, M. (2001). Impetus beliefs as default heuristics: Dissociation between explicit and implicit knowledge about motion. *Psychonomic Bulletin & Review, 8,* 439–453.

Kozhevnikov, M., Louchakova, O., Josipovic, Z., & Motes, M. A. (2009). The enhancement of visuospatial processing efficiency through Buddhist Deity Meditation. *Psychological Science, 20,* 645–653.

Kozlov, M. D., Hughes, R. W., & Jones, D. M. (2012). Gummed-up memory: Chewing gum impairs short-term recall. *The Quarterly Journal of Experimental Psychology, 65*(3), 501–513.

Kraha, A., & Boals, A. (2014). Why so negative? Positive flashbulb memories for a personal event. *Memory, 22*(4), 442–449.

Kraljik, T., Samuel, A. G., & Brennan, S. E. (2008). First impressions and last resorts: How listeners adjust to speaker variability. *Psychological Science, 19,* 332–338.

Kramer, A. F., Hahn, S., Irwin, D. E., & Theeuwes, J. (1999). Attention capture and aging: Implications for visual search performance and oculomotor control. *Psychology and Aging, 14,* 135–154.

Kramer, T. H., Buckout, R., & Eugenio, P. (1990). Weapon focus, arousal, and eyewitness memory. *Law and Human Behavior, 14,* 167–184.

Krawietz, S. A., Tamplin, A. K., & Radvansky, G. A. (2012). Aging and mind wandering during text comprehension. *Psychology and Aging, 27,* 951–958.

Krist, H., Fieberg, E. L., & Wilkening, F. (1993). Intuitive physics in action and judgment: The development of knowledge about projectile motion. *Journal of Experimental Psychology: Learning, Memory, and Cognition, 19,* 952–966.

Kubilius, J., Wagemans, J., & Op de Beeck, H. P. (2011). Emergence of perceptual gestalts in the human visual cortex: The case of the configural superiority effect. *Psychological Science, 22,* 1296–1303.

Kucera, H., & Francis, W. N. (1967). *Computational analysis of present day American English.* Providence, RI: Brown University Press.

Kumon-Nakamura, S., Glucksberg, S., & Brown, M. (1995). How about another piece of pie? The allusional pretense theory of discourse irony. *Journal of Experimental Psychology: General, 124,* 3–21.

Kunar, M A., Carter, R., Cohen, M., & Horowitz, T. S. (2008). Telephone conversation impairs sustained visual attention via a central bottleneck. *Psychonomic Bulletin & Review, 15*, 1135–1140.

Kurczek, J., Brown-Schmidt, S., & Duff, M. (2013). Hippocampal contributions to language: Evidence of referential processing deficits in amnesia. *Journal of Experimental Psychology: General, 142*(4), 1346–1354.

Kurtz, K. J., & Loewenstein, J. (2007). Converging on a new role for analogy in problem solving and retrieval: When two problems are better than one. *Memory & Cognition, 35*, 334–341.

Kurumada, C., Brown, M., Bibyk, S., Pontillo, D. F., & Tanenhaus, M. K. (2014). Is it or isn't it: Listeners make rapid use of prosody to infer speaker meanings. *Cognition, 133*(2), 335–342.

Kutas, M., & Hillyard, S. A. (1980). Reading senseless sentences: Brain potentials reflect semantic incongruity. *Science, 207*, 203–205.

Kvavilashvili, L., Cockburn, J., & Kornbrot, D. E. (2013). Prospective memory and ageing paradox with event-based tasks: A study of young, young-old, and old-old participants. *The Quarterly Journal of Experimental Psychology, 66*(5), 864–875.

Kynski, T. R., & Tenebaum, J. B. (2007). The role of causality in judgment under uncertainty. *Journal of Experimental Psychology: General, 136*, 430–450.

Lachman, R., Lachman, J. L., & Butterfield, E. C. (1979). *Cognitive psychology and information processing: An introduction.* Hillsdale, NJ: Erlbaum.

Lachter, J., Forster, K. I., & Ruthruff, E. (2004). Forty-five years after Broadbent (1958): Still no identification without attention. *Psychological Review, 111*, 880–913.

Lakens, D., Schneider, I. K., Jostmann, N. B., & Schubert, T. W. (2011). Telling things apart: The distance between response keys influences categorization times. *Psychological Science, 22*, 887–890.

Lambert, W. E. (1990). Issues in foreign language and second language education. *Proceedings of the Research Symposium on Limited English Proficient Students' Issues* (1st, Washington, DC, September 10–12).

Lane, S. M., & Zaragoza, M. S. (2007). A little elaboration goes a long way: The role of generation in eyewitness suggestibility. *Memory & Cognition, 35*, 1255–1266.

Lang, J. W. B., & Lang, J. (2010). Priming competence diminishes the link between cognitive test anxiety and test performance: Implications for the interpretation of test scores. *Psychological Science, 21*, 811–819.

Langner, R., & Eickhoff, S. B. (2013). Sustaining attention to simple tasks: A meta-analytic review of the neural mechanisms of vigilant attention. *Psychological Bulletin, 139*(4), 870–900.

Larsen, A., & Bundesen, C. (2009). Common mechanisms in apparent motion perception and visual pattern matching. *Scandinavian Journal of Psychology, 50*, 526–534.

Lashley, K. D. (1950). In search of the engram. *Symposia for the Society for Experimental Biology, 4*, 454–482.

Laszlo, S., & Federmeier, K. D. (2009). A beautiful day in the neighborhood: An event-related potential study of lexical relationships and prediction in context. *Journal of Memory and Language, 61*, 326–338.

Latham, A. J., Patston, L. L., & Tippett, L. J. (2013). The virtual brain: 30 years of video-game play and cognitive abilities. *Frontiers in Psychology, 4*, 629.

Lavaer, G. D, & Burke, D. M. (1993). Why do semantic priming effects increase in old age? A meta-analysis. *Psychology and Aging, 8*, 34–43.

Lavie, N. (2010). Attention, distraction, and cognitive control under load. *Current Directions in Psychological Science, 19*, 143–148.

LaVoie, D. J., & Malmstrom, T. (1998). False recognition effects in younger and older adults' memory for text passages. *The Journals of Gerontology: Psychological Sciences, 53B*, P255–P262.

Lawson, R. (2006). The science of cycology: Failures to understand how everyday objects work. *Memory & Cognition, 34*, 1667–1675.

Leahey, T. H. (1992). The mythical revolutions of American psychology. *American Psychologist, 47*, 308–318.

Leahey, T. H. (2000). *A history of psychology: Main currents in psychological thought* (3rd ed.). Englewood Cliffs, NJ: Prentice Hall.

Leahey, T. H. (2003). Herbert A. Smith Nobel Prize in Economic Sciences, 1978. *American Psychologist, 58*, 753–755.

LeDoux, J. E. (2000). Emotion circuits in the brain. *Annual Review of Neuroscience, 23*, 155–184.

Lee, C-L., & Federmeier, K. D. (2009). Wave-ering: An ERP study of syntactic and semantic context effects on ambiguity resolution for noun/verb homographs. *Journal of Memory and Language, 61*, 538–555.

Lee, H-W., Rayner, K., & Pollatsek, A. (1999). The time course of phonological, semantic, and orthographic coding in reading: Evidence from the fast-priming technique. *Psychonomic Bulletin & Review, 6*, 624–634.

Lee, J. J., & Pinker, S. (2010). Rationales for indirect speech: The theory of the strategic speaker. *Psychological Review, 117*, 785–807.

LeFevre, J., Bisanz, J., Daley, K. E., Buffone, L., Greenham, S. L., & Sadesky, G. S. (1996). Multiple routes to solution of single-digit multiplication problems. *Journal of Experimental Psychology: General, 125*, 284–306.

Lehman, D. R., Lempert, R. O., & Nisbett, R. E. (1988). The effects of graduate training on reasoning. *American Psychologist, 43*, 431–442.

Lemaire, P., & Brun, F. (2014). Effects of strategy sequences and response–stimulus intervals on children's strategy selection and strategy execution: A study in computational estimation. *Psychological Research, 78*(4), 506–519.

Leonard, C. J., Balestreri, A., & Luck, S. J. (2015). Interactions between space-based and feature-based attention. *Journal of Experimental Psychology: Human Perception and Performance, 41*(1), 11–16.

Leonesio, R. J., & Nelson, T. O. (1990). Do different metamemory judgments tap the same underlying aspects of memory? *Journal of Experimental Psychology: Learning, Memory, and Cognition, 16*, 464–470.

LePine, J. A., LePine, M. A., & Jackson, C. L. (2004). Challenge and hindrance stress: Relationships with exhaustion, motivation to learn, and learning performance. *Journal of Applied Psychology, 89*, 883–891.

Lerner, Y., Honey, C. J., Silbert, L. J., & Hasson, U. (2011). Topographic mapping of a hierarchy of temporal receptive windows using a narrated story. *The Journal of Neuroscience, 31*(8), 2906–2915.

Levy, B. (1996). Improving memory in old age through implicit self-stereotyping. *Journal of Personality and Social Psychology, 71*, 1092–1107.

Levy, D. A., Stark, C. E. L., & Squire, L. R. (2004). Intact conceptual priming in the absence of declarative memory. *Psychological Science, 15*, 680–686.

Levy, J., Pashler, H., & Boer, E. (2006). Central interference in driving: Is there any stopping the psychological refractory period? *Psychological Science, 17*, 228–235.

Lewandowsky, S., Stritzka, W. G. K., Oberauer, K., & Morales, M. (2005). Memory for fact, fiction, and misinformation. *Psychological Science, 16*, 190–195.

Lewis, J. L. (1970). Semantic processing of unattended messages using dichotic listening. *Journal of Experimental Psychology, 85*, 225–228.

Lewontin, R. C. (1990). The evolution of cognition. In D. N. Osherson & E. E. Smith (Eds.), *Thinking: An invitation to cognitive science* (Vol. 3, pp. 229–246). Cambridge, MA: MIT Press.

Li, M., & Chapman, G. B. (2009). "100% of anything looks good": The appeal of one hundred percent. *Psychonomic Bulletin & Review, 16*(1), 156–162.

Li, X., Schweickert, R., & Gandour, J. (2000). The phonological similarity effect in immediate recall: Positions of shared phonemes. *Memory & Cognition, 28*, 1116–1125.

Libby, L. K. (2003). Imagery perspective and source monitoring in imagination inflation. *Memory & Cognition, 31*, 1072–1081.

Liberman, A. M. (1957). Some results of research on speech perception. *Journal of the Acoustical Society of America, 29*, 117–123.

Liberman, A. M., Cooper, F. S., Shankweiler, D. P., & Studdert-Kennedy, M. (1967). Perception of the speech code. *Psychological Review, 74*, 431–461.

Liberman, A. M., Harris, K. S., Hoffman, H. S., & Griffith, B. C. (1957). The discrimination of speech sounds within and across phoneme boundaries. *Journal of Experimental Psychology, 54*, 358–368.

Liberman, A. M., & Mattingly, I. G. (1985). The motor theory of speech perception revised. *Cognition, 21*, 1–36.

Libkuman, T. M., Nichols-Whitehead, P., Griffith, J., & Thomas, R. (1999). Source of arousal and memory for detail. *Memory & Cognition, 27*, 166–190.

Light, L. L., & Anderson, P. A. (1983). Memory for scripts in young and older adults. *Memory & Cognition, 11*(5), 435–444.

Light, L. L., & Capps, J. L. (1986). Comprehension of pronouns in young and older adults. *Developmental Psychology, 22*, 580–585.

Lindberg, M. A. (1980). Is knowledge base development a necessary and sufficient condition for memory development? *Journal of Experimental Child Psychology, 30*, 401–410.

Linder, I., Echterhoff, G., Davidson, P. S. R., & Brand, M. (2010). Observation inflation: Your actions become mine. *Psychological Science, 21*, 1291–1299.

Lindsay, D. S., Allen, B. P., Chan, J. C. K., & Dahl, L. C. (2004). Eyewitness suggestibility and source similarity: Intrusions of details from one event into memory reports of another event. *Journal of Memory and Language, 50*, 96–111.

Lindsay, D. S., Hagen, L., Read, J. D., Wade, K. A., & Garry, M. (2004). True photographs and false memories. *Psychological Science, 15*, 149–154.

Lindsay, P. H., & Norman, D. A. (1977). *Human information processing: An introduction to psychology*. New York, NY: Academic Press.

Lindsley, J. R. (1975). Producing simple utterances: How far ahead do we plan? *Cognitive Psychology, 7*, 1–19.

Linek, J. A., Kroll, J. F., & Sunderman, G. (2010). Losing access to the native language while immersed in a second language: Evidence for the role of inhibition in second-language learning. *Psychological Science, 20*, 1507–1515.

Linton, M. (1975). Memory for real-world events. In D. A. Norman & D. E. Rumelhart (Eds.), *Explorations in cognition* (pp. 376–404). San Francisco, CA: Freeman.

Linton, M. (1978). Real world memory after six years: An in vivo study of very long term memory. In M. M. Gruneberg, P. E. Morris, & R. N. Sykes (Eds.), *Practical aspects of memory* (pp. 69–76). Orlando, FL: Academic Press.

Liszkowski, U., Schäfer, M., Carpenter, M., & Tomasello, M. (2009). Prelinguistic infants, but not chimpanzees, communicate about absent entities. *Psychological Science, 20*, 654–660.

Litman, D. J., & Allen, J. F. (1987). A plan recognition model for subdialogues in conversation. *Cognitive Science, 11*, 163–200.

Livesey, E. J., Harris, I. M., & Harris, J. A. (2009). Attentional changes during implicit learning: Signal validity protects target stimulus from the attentional blink. *Journal of Experimental Psychology: Learning, Memory, and Cognition, 35*, 408–422.

Ljungberg, J. K., Hansson, P., Andrés, P., Josefsson, M., & Nilsson, L. G. (2013). A longitudinal study of memory advantages in bilinguals. *PLoS one, 8*(9), e73029.

Lochner, M. J., & Trick, L. M. (2014). Multiple-object tracking while driving: The multiple-vehicle tracking task. *Attention, Perception, & Psychophysics, 76*(8), 2326–2345.

Löckenhoff, C. E., & Carstensen, L. L. (2007). Aging, emotion, and health-related decision strategies: Motivational manipulations can reduce age differences. *Psychology and Aging, 22*, 134–146.

Lockridge, C. B., & Brennan, S. E. (2002). Addressees' needs influence speakers' early syntactic choices. *Psychonomic Bulletin & Review, 9*, 550–557.

Loftus, E. F. (1971). Memory for intentions: The effect of presence of a cue and interpolated activity. *Psychonomic Science, 23*, 315–316.

Loftus, E. F. (1991). Made in memory: Distortions in recollection after misleading information. In G. H. Bower (Ed.), *The psychology of learning and motivation* (Vol. 27, pp. 187–215). New York, NY: Academic Press.

Loftus, E. F. (1993). The reality of repressed memories. *American Psychologist, 48*, 518–537.

Loftus, E. F. (2003). Make-believe memories. *American Psychologist, 58*, 867–873.

Loftus, E. F. (2004). Memories of things unseen. *Current Directions in Psychological Science, 13*, 145–147.

Loftus, E. F., & Coan, D. (1994). The construction of childhood memories. In D. Peters (Ed.), *The child witness in context: Cognitive, social, and legal perspectives*. New York, NY: Kluwer.

Loftus, E. F., Donders, K., Hoffman, H. G., & Schooler, J. W. (1989). Creating new memories that are quickly accessed and confidently held. *Memory & Cognition, 17*, 607–616.

Loftus, E. F., & Hoffman, H. G. (1989). Misinformation and memory: The creation of new memories. *Journal of Experimental Psychology: General, 118*, 100–104.

Loftus, E. F., & Ketcham, K. (1991). *Witness for the defense*. New York, NY: St. Martin's Press.

Loftus, E. F., & Palmer, J. C. (1974). Reconstruction of automobile destruction: An example of the interaction between language and memory. *Journal of Verbal Learning and Verbal Behavior, 13*, 585–589.

Loftus, G. R., & Hanna, A. M. (1989). The phenomenology of spatial integration: Data and models. *Cognitive Psychology, 21*, 363–397.

Loftus, G. R., & Irwin, D. E. (1998). On the relations among different measures of visible and informational persistence. *Cognitive Psychology, 35*, 135–199.

Loftus, G. R., & Loftus, E. F. (1974). The influence of one memory retrieval on a subsequent memory retrieval. *Memory & Cognition, 2*, 467–471.

Logan, G. D. (1988). Toward an instance theory of automatization. *Psychological Review, 95*, 492–527.

Logan, G. D. (1990). Repetition priming and automaticity: Common underlying mechanisms? *Cognitive Psychology, 22*, 1–35.

Logan, G. D. (2003). Executive control of thought and action: In search of the wild homunculus. *Current Directions in Psychological Science, 12*, 45–48.

Logan, G. D., & Crump, M. J. C. (2009). The left hand doesn't know what the right hand is doing: The disruptive effects of attention to the hands in skilled typewriting. *Psychological Science, 20*, 1296–1300.

Logan, G. D., & Etherton, J. L. (1994). What is learned during automatization? The role of attention in constructing an instance. *Journal of Experimental Psychology: Learning, Memory, and Cognition, 20*, 1022–1050.

Logan, G. D., & Klapp, S. T. (1991). Automatizing alphabet arithmetic: I. Is extended practice necessary to produce automaticity? *Journal of Experimental Psychology: Learning, Memory, and Cognition, 17*, 179–195.

Logie, R. H., Zucco, G., & Baddeley, A. D. (1990). Interference with visual short-term memory. *Acta Psychologica, 75*, 55–74.

Long, D. L., & De Ley, L. (2000). Implicit causality and discourse focus: The interaction of text and reader characteristics in pronoun resolution. *Journal of Memory and Language, 42*, 545–570.

Long, D. L., Golding, J., Graesser, A. C., & Clark, L. F. (1990). Inference generation during story comprehension: A comparison of goals, events, and states. In A. C. Graesser & G. H. Bower (Eds.), *The psychology of learning and motivation* (Vol. 25). New York, NY: Academic Press.

Long, D. L., Oppy, B. J., & Seely, M. R. (1997). Individual differences in readers' sentence- and text-level representations. *Journal of Memory and Language, 36*, 129–145.

Lorch, R. F., Jr., Lorch, E. P., & Matthews, P. D. (1985). On-line processing of the topic structure of a text. *Journal of Memory and Language, 24*, 350–362.

Lourenco, S. F., & Longo, M. R. (2010). General magnitude representation in human infants. *Psychological Science, 21*, 873–881.

Lozano, S. C., & Tversky, B. (2006). Communicative gestures facilitate problem solving for both communicators and recipients. *Journal of Memory and Language, 55*, 47–63.

Lu, S., & Zhou, K. (2005). Stimulus-driven attentional capture by equiluminent color change. *Psychonomic Bulletin & Review, 12*, 567–572.

Luchins, A. S. (1942). Mechanization in problem solving. *Psychological Monographs, 54*(Whole no. 248).

Luo, L., & Craik, F. I. M. (2009). Age differences in recollection: Specificity effects at retrieval. *Journal of Memory and Language, 60*, 421–436.

Luria, A. (1968). *The mind of a mnemonist*. New York, NY: Basic Books.

Lustig, C., & Hasher, L. (2001). Implicit memory is vulnerable to proactive interference. *Psychological Science, 12*, 408–412.

Lutz, M. E., & Radvansky, G. A. (1997). The fate of completed goal information. *Journal of Memory and Language, 36*, 293–310.

Lynchard, N. A., & Radvansky, G. A. (2012). Age-related perspectives and emotion processing. *Psychology and Aging, 27*, 934–939.

Maass, A., & Köhnken, G. (1989). Simulating the "weapon effect." *Law and Human Behavior, 13*, 397–408.

Mack, A. (2003). Inattentional blindness: Looking without seeing. *Current Directions in Psychological Science, 12*, 180–184.

Mack, A., & Rock, I. (1998). *Inattentional blindness*. Cambridge, MA: MIT Press.

Mackworth, N. H. (1948). The breakdown of vigilance during prolonged visual search. *The Quarterly Journal of Experimental Psychology, 1*, 6–21.

MacLean, K. A., Ferrer, E., Aichele, S. R., Bridwell, D. A., Zanesco, A. P., Jacobs, T. L., et al. (2010). Intensive meditation training improves perceptual discrimination and sustained attention. *Psychological Science, 21*, 829–839.

MacLeod, C. M. (1988). Forgotten but not gone: Savings for pictures and words in long-term memory. *Journal of Experimental Psychology: Learning, Memory, and Cognition, 14*, 195–212.

MacLeod, C. M. (1991). Half a century of research on the Stroop effect: An integrative review. *Psychological Bulletin, 109*, 163–203.

MacLeod, C. M. (1992). The Stroop task: The "gold standard" of attentional measures. *Journal of Experimental Psychology: General, 121*, 12–14.

MacLeod, C. M., Gopie, N., Hourihan, K. L., Neary, K. R., & Ozubko, J. D. (2010). The production effect: Delineation of a phenomenon. *Journal of Experimental Psychology. Learning, Memory, and Cognition, 36*, 671–685.

MacLeod, M. D., & Macrae, C. N. (2001). Gone but not forgotten: The transient nature of retrieval-induced forgetting. *Psychological Science, 12*, 148–152.

MacLeod, M. D., & Saunders, J. (2008). Retrieval inhibition and memory distortion. *Current Directions in Psychological Science, 17*, 26–30.

MacPherson, S. E., Phillips, L. H., & Sala, S. D. (2002). Age, executive functioning, and social decision making: A dorsolateral prefrontal theory of cognitive aging. *Psychology and Aging, 17*, 598–609.

MacQueen, G. M., Tipper, S. P., Young, L. T., Joffe, R. T., & Levitt, A. J. (2000). Impaired distractor inhibition on a selective attention task in unmedicated, depressed subjects. *Psychological Medicine, 30*, 557–564.

Madden, C. J., & Zwaan, R. A. (2003). How does verb aspect constrain event representations? *Memory & Cognition, 31*(5), 663–672.

Madden, D. J., Turkington, J. M., Provenzale, J. M., Denny, L. L., Langley, L. K., Hawk, T. C., et al. (2002). Aging and attentional guidance during visual search: Functional neuroanatomy by positron emission tomography. *Psychology and Aging, 17*, 24–43.

Madigan, S., & O'Hara, R. (1992). Short-term memory at the turn of the century: Mary Whiton Calkins's memory research. *American Psychologist, 47*, 170–174.

Magliano, J. P., Miller, J., & Zwaan, R. A. (2001). Indexing space and time in film understanding. *Applied Cognitive Psychology, 15*, 533–545.

Magliano, J. P., Radvansky, G. A., Forsythe, J. C., & Copeland, D. E. (2014). Event segmentation during first-person continuous events. *Journal of Cognitive Psychology, 26*(6), 649–661.

Magliano, J. P., & Schleich, M. C. (2000). Verb aspect and situation models. *Discourse Processes, 29*(2), 83–112.

Magliano, J. P., Taylor, H. A., & Kim, H. J. (2005). When goals collide: Monitoring the goals of multiple characters. *Memory & Cognition, 33*, 1357–1367.

Magliano, J. P., Trabasso, T., & Graesser, A. C. (1999). Strategic processing during comprehension. *Journal of Educational Psychology, 91*, 615–629.

Maglio, S. J., & Trope, Y. (2012). Disembodiment: Abstract construal attenuates the influence of contextual bodily state in judgment. *Journal of Experimental Psychology: General, 141*(2), 211–216.

Magno, E., & Allan, K. (2007). Self-reference during explicit memory retrieval: An event-related potential analysis. *Psychological Science, 18*, 672–677.

Mahmood, D., Manier, D., & Hirst, W. (2004). Memory for how one learned of multiple deaths from AIDS: Repeated exposure and distinctiveness. *Memory & Cognition, 32*, 125–134.

Maier, N. R. F. (1931). Reasoning in humans: II. The solution of a problem and its appearance in consciousness. *Journal of Comparative Psychology, 12*, 181–194.

Maki, R. H. (1989). Recognition of added and deleted details in scripts. *Memory & Cognition, 17*, 274–282.

Maloney, E. A., Ansari, D., & Fugelsang, J. A. (2011). The effect of mathematics anxiety on the processing of numerical magnitude. *The Quarterly Journal of Experimental Psychology, 64*, 10–16.

Malt, B. C. (1985). The role of discourse structure in understanding anaphora. *Journal of Memory and Language, 24*, 271–289.

Malt, B. C., Sloman, S. A., & Gennari, S. P. (2003). Universality and language specificity in object naming. *Journal of Memory and Language, 49*, 20–42.

Mandel, D. R., & Lehman, D. R. (1996). Counterfactual thinking and ascriptions of cause and preventability. *Journal of Personality and Social Psychology, 71*, 450–463.

Mandler, G. (1967). Organization and memory. In K. W. Spence & J. T. Spence (Eds.), *The psychology of learning and motivation* (Vol. 1, pp. 327–372). New York, NY: Academic Press.

Mandler, G. (1972). Organization and recognition. In E. Tulving & W. Donaldson (Eds.), *Organization of memory* (pp. 139–166). New York, NY: Academic Press.

Mandler, G. (2007). *A history of modern experimental psychology: From James and Wundt to cognitive science*. Cambridge, MA: MIT Press.

Mandler, J. M., & Mandler, G. (1969). The diaspora of experimental psychology: The Gestaltists and others. In D. Fleming & B. Bailyn (Eds.), *The intellectual migration: Europe and America, 1930–1960* (pp. 371–419). Cambridge, MA: Harvard University Press.

Manwell, L. A., Roberts, M. A., & Besner, D. (2004). Single letter coloring and spatial cuing eliminates a semantic contribution to the Stroop effect. *Psychonomic Bulletin & Review, 11*, 458–462.

Marcel, A. J. (1980). Conscious and preconscious recognition of polysemous words: Locating the selective effects of prior verbal context. In R. S. Nickerson (Ed.), *Attention and performance VIII* (pp. 435–457). Hillsdale, NJ: Erlbaum.

Marcel, A. J. (1983). Conscious and unconscious perception: Experiments on visual masking and word recognition. *Cognitive Psychology, 15*, 197–237.

Marchman, V. A., & Fernald, A. (2008). Speed of word recognition and vocabulary knowledge in infancy predict cognitive and language outcomes in later childhood. *Developmental Science, 11*, 1–14.

Marcus, G. F. (1996). Why do children say "breaked"? *Current Directions in Psychological Science, 5*, 81–85.

Marian, V., & Neisser, U. (2000). Language-dependent recall of autobiographical memories. *Journal of Experimental Psychology: General, 129*, 361–368.

Markman, A. B., Taylor, E., & Gentner, D. (2007). Auditory presentation leads to better analogical retrieval than written presentation. *Psychonomic Bulletin & Review, 14*, 1101–1106.

Markovits, H., & Doyon, C. (2004). Information processing and reasoning with premises that are empirically false: Interference, working memory, and processing speed. *Memory & Cognition, 32,* 592–601.

Markovits, H., & Potvin, F. (2001). Suppression of valid inferences and knowledge structures: The curious effect of producing alternative antecedents on reasoning with causal conditionals. *Memory & Cognition, 29,* 736–744.

Marler, P. (1967). Animal communication signals. *Science, 35,* 63–78.

Marsh, E. J. (2007). Retelling is not the same as recalling. *Current Directions in Psychological Science, 16,* 16–20.

Marsh, E. J., Meade, M. L., & Roediger, H. L. III (2003). Learning facts from fiction. *Journal of Memory and Language, 49,* 519–536.

Marsh, E. J., Roediger, H. L., Bjork, R. A., & Bjork, E. L. (2007). The memorial consequences of multiple-choice testing. *Psychonomic Bulletin & Review, 14,* 194–199.

Marsh, R. L., Cook, G. I., Meeks, J. T., Clark-Foos, A., & Hicks, J. L. (2007). Memory for intention-related material presented in a to-be-ignored channel. *Memory & Cognition, 35,* 1197–1204.

Marsh, R. L., Hicks, J. L., & Cook, G. I. (2005). On the relationship between effort toward an ongoing task and cue detection in event-based prospective memory. *Journal of Experimental Psychology: Learning, Memory, and Cognition, 31,* 68–75.

Marshall, P. J. (2009). Relating psychology and neuroscience. *Perspectives on Psychological Science, 4,* 113–125.

Marslen-Wilson, W. D., & Welsh, A. (1978). Processing interactions and lexical access during word recognition in continuous speech. *Cognitive Psychology, 30,* 509–517.

Martin, R. C. (2000). Contribution from the neuropsychology of language and memory to the development of cognitive theory. *Journal of Memory and Language, 43,* 149–156.

Martindale, C. (1991). *Cognitive psychology: A neural-network approach.* Pacific Grove, CA: Brooks/Cole.

Mason, R. A., Just, M. A., Keller, T. A., & Carpenter, P. A. (2003). Ambiguity in the brain: What brain imaging reveals about the processing of syntactically ambiguous sentences. *Journal of Experimental Psychology: Learning, Memory, and Cognition, 29,* 1319–1338.

Masson, M. E. J. (1984). Memory for the surface structure of sentences: Remembering with and without awareness. *Journal of Verbal Learning and Verbal Behavior, 23,* 579–592.

Masson, M. E. J. (1995). A distributed memory model of semantic priming. *Journal of Experimental Psychology: Learning, Memory, and Cognition, 21,* 3–23.

Masters, R. S. W. (1992). Knowledge, knerves and know-how: The role of explicit versus implicit knowledge in the breakdown of a complex motor skill under pressure. *British Journal of Psychology, 83,* 343–358.

Mata, R., & Nunes, L. (2010). When less is enough: Cognitive aging, information search, and decision quality in consumer choice. *Psychology and Aging, 25,* 289–298.

Mata, R., Schooler, L. J., & Reiskamp, J. (2007). The aging decision maker: Cognitive aging and the adaptive selection of decision strategies. *Psychology and Aging, 22,* 796–810.

Mather, M., Cacioppo, J. T., & Kanwisher, N. (2013). How fMRI can inform cognitive theories. *Perspectives on Psychological Science, 8*(1), 108–113.

Mather, M., Shafir, E., & Johnson, M. K. (2000). Misremembrance of options past: Source monitoring and choice. *Psychological Science, 11,* 132–138.

Mather, M., & Sutherland, M. R. (2011). Arousal-based competition in perception and memory. *Perspectives on Psychological Science, 6,* 114–133.

Mathis, K. M. (2002). Semantic interference from objects both in and out of scene context. *Journal of Experimental Psychology: Learning, Memory, and Cognition, 28,* 171–182.

Matlin, M. (1983). *Cognition.* New York, NY: Holt.

Maurer, D. (1997). Neonatal synaesthesia: Implications for the processing of speech and faces. In S. Baron-Cohen & J. E. Harrison (Eds.), *Synaesthesia: Classic and contemporary readings* (pp. 224–242). Malden, MA: Blackwell.

Mayr, S., & Buchner, A. (2006). Evidence for episodic retrieval of inadequate prime responses in auditory negative priming. *Journal of Experimental Psychology: Human Perception and Performance, 32,* 932–943.

Mazzoni, G., & Cornoldi, C. (1993). Strategies in study time allocation: Why is study time sometimes not effective? *Journal of Experimental Psychology: General, 122,* 47–60.

Mazzoni, G., & Memon, A. (2003). Imagination can create false autobiographical memories. *Psychological Science, 14,* 186–188.

McCabe, D. P. (2010). The influence of complex working memory span task administration methods on prediction of higher level cognition and metacognitive control of response times. *Memory & Cognition, 38,* 868–882.

McCabe, D. P., Smith, A. D., & Parks, C. M. (2007). The role of working memory capacity in reducing memory errors. *Memory & Cognition, 35,* 231–241.

McCabe, J. (2011). Metacognitive awareness of learning strategies in undergraduates. *Memory & Cognition, 39,* 462–476.

McCaffrey, T. (2012). Innovation relies on the obscure: A key to overcoming the classic problem of function fixedness. *Psychological Science, 23*(3), 215–218.

McCandliss, B. D., Posner, M. I., & Givon, T. (1997). Brain plasticity in learning visual words. *Cognitive Psychology, 33,* 88–110.

McCarley, J. S., Kramer, A. F., Wickens, C. D., Vidoni, E. D., & Boot, W. R. (2004). Visual skills in airport security screening. *Psychological Science, 15,* 302–306.

McCarthy, R. A., & Warrington, E. K. (1984). A two route model of speech production: Evidence from aphasia. *Brain, 107,* 463–485.

McCarthy, R. A., & Warrington, E. K. (1990). *Cognitive neuropsychology: A clinical introduction.* San Diego, CA: Academic Press.

McClearn, G. E., Johansson, B., Berg, S., Pedersen, N. L., Ahern, F., Petrill, S. A., & Plomin, R. (1997). Substantial genetic influence on cognitive abilities in twins 80 or more years old. *Science, 276*(5318), 1560–1563.

McClelland, J. L. (1979). On the time relations of mental processes: An examination of systems of processes in cascade. *Psychological Review, 86,* 287–330.

McClelland, J. L., & Elman, J. L. (1986). The TRACE model of speech perception. *Cognitive Psychology, 18*, 1–86.

McClelland, J. L., & Rumelhart, D. E. (1981). An interactive activation model of context effects in letter perception: Part 1. An account of basic findings. *Psychological Review, 88*, 375–407.

McClelland, J. L., Rumelhart, D. E., & Hinton, G. E. (1986). The appeal of parallel distributed processing. In D. E. Rumelhart, J. L. McClelland, & PDP Research Group (Eds.), *Parallel distributed processing* (Vol. 1, pp. 3–44). Cambridge, MA: MIT Press.

McCloskey, M. (1983). Naive theories of motion. In D. Gentner & A. L. Stevens (Eds.), *Mental models* (pp. 299–324). Hillsdale, NJ: Erlbaum.

McCloskey, M. (1992). Cognitive mechanisms in numerical processing: Evidence from acquired dyscalculia. *Cognition, 44*, 107–157.

McCloskey, M., Wible, C. G., & Cohen, N. J. (1988). Is there a special flashbulb-memory mechanism? *Journal of Experimental Psychology: General, 117*, 171–181.

McCloy, R., & Byrne, R. M. (2000). Counterfactual thinking about controllable events. *Memory & Cognition, 28*, 1071–1078.

McCrae, C. S., & Abrams, R. A. (2001). Age-related differences in object- and location-based inhibition of return of attention. *Psychology and Aging, 16*, 437–449.

McDaniel, M. A., Maier, S. F., & Einstein, G. O. (2002). "Brain-specific" nutrients: A memory cure? *Psychological Science in the Public Interest, 31*, 12–38.

McDaniel, M. A., Roediger, H. L., & McDermott, K. B. (2007). Generalized test-enhanced learning from the laboratory to the classroom. *Psychonomic Bulletin & Review, 14*, 200–206.

McDonald, J. L., & MacWhinney, B. (1995). The time course of anaphor resolution: Effects of implicit verb causality and gender. *Journal of Memory and Language, 34*, 543–566.

McGeoch, J. A. (1932). Forgetting and the law of disuse. *Psychological Review, 39*, 352–370.

McGillivray, S., Murayama, K., & Castel, A. D. (2015). Thirst for knowledge: The effects of curiosity and interest on memory in younger and older adults. *Psychology and Aging, 30*(4), 835–841.

McKoon, G., & Macfarland, T. (2002). Event templates in the lexical representations of verbs. *Cognitive Psychology, 45*, 1–44.

McKoon, G., & Ratcliff, R. (1986). Inferences about predictable events. *Journal of Experimental Psychology: Learning, Memory, and Cognition, 12*, 82–91.

McKoon, G., & Ratcliff, R. (1989). Inferences about contextually defined categories. *Journal of Experimental Psychology: Learning, Memory, and Cognition, 15*, 1134–1146.

McKoon, G., & Ratcliff, R. (1992). Inference during reading. *Psychological Review, 99*, 440–466.

McKoon, G., & Ratcliff, R. (2007). Interactions of meaning and syntax: Implications for models of sentence comprehension. *Journal of Memory and Language, 56*, 270–290.

McNamara, D. S., & McDaniel, M. A. (2004). Suppressing irrelevant information: Knowledge activation or inhibition? *Journal of Experimental Psychology: Learning, Memory, and Cognition, 30*, 465–482.

McNamara, T. P. (1992). Priming and constraints it places on theories of memory and retrieval. *Psychological Review, 99*, 650–662.

McNeill, D. (1992). *Hand and mind: What gestures reveal about thought*. Chicago, IL: University of Chicago Press.

McRae, K., & Boisvert, S. (1998). Automatic semantic similarity priming. *Journal of Experimental Psychology: Learning, Memory, and Cognition, 24*, 558–572.

McRae, K., de Sa, V. R., & Seidenberg, M. S. (1997). On the nature and scope of featural representations of word meaning. *Journal of Experimental Psychology: General, 126*, 99–130.

Meade, M. L., & Roediger, H. L. III. (2002). Explorations in the social contagion of memory. *Memory & Cognition, 30*, 995–1009.

Meade, M. L., Watson, J. M., Balota, D. A., & Roediger, H. L. (2007). The roles of spreading activation and retrieval mode in producing false recognition in the DRM paradigm. *Journal of Memory and Language, 56*, 305–320.

Medin, D. L. (1989). Concepts and conceptual structure. *American Psychologist, 44*, 1469–1481.

Medin, D. L., & Atran, S. (2004). The native mind: Biological categorization and reasoning in development and across cultures. *Psychological Review, 111*, 960–983.

Medin, D. L., Coley, J. D., Storms, G., & Hayes, B. K. (2003). A relevance theory of induction. *Psychonomic Bulletin & Review, 10*, 517–532.

Medin, D. L., & Edelson, S. M. (1988). Problem structure and the use of base-rate information from experience. *Journal of Experimental Psychology: General, 117*, 68–85.

Medin, D. L., Goldstone, R. L., & Gentner, D. (1993). Respects for similarity. *Psychological Review, 100*, 254–278.

Medin, D. L., Lynch, E. B., Coley, J. D., & Atran, S. (1997). Categorization and reasoning among tree experts: Do all roads lead to Rome? *Cognitive Psychology, 32*, 49–96.

Mehler, J., Morton, J., & Jusczyk, P. W. (1984). On reducing language to biology. *Cognitive Neuropsychology, 1*, 83–116.

Melton, A. W. (1963). Implications of short-term memory for a general theory of memory. *Journal of Verbal Learning and Verbal Behavior, 2*, 1–21.

Metcalfe, J. (1986). Feeling of knowing in memory and problem solving. *Journal of Experimental Psychology: Learning, Memory, and Cognition, 12*, 288–294.

Metcalfe, J. (2002). Is study time allocated selectively to a region of proximal learning? *Journal of Experimental Psychology: General, 131*, 349–363.

Metcalfe, J., & Finn, B. (2008). Evidence that judgments of learning are causally related to study choice. *Psychonomic Bulletin & Review, 15*, 174–179.

Metcalfe, J., & Kornell, N. (2003). The dynamics of learning and allocation of study time to a region of proximal learning. *Journal of Experimental Psychology: General, 132*, 530–542.

Metcalfe, J., & Wiebe, D. (1987). Intuition in insight and noninsight problem solving. *Memory & Cognition, 15*, 238–246.

Metcalfe, J., & Xu, J. (2016). People mind wander more during massed than spaced inductive learning. *Journal of Experimental Psychology: Learning, Memory, and Cognition, 42*, 978–984.

Metzing, C., & Brennan, S. E. (2003). When conceptual pacts are broken: Partner-specific effects on the comprehension of referring expressions. *Journal of Memory and Language, 49*, 201–213.

Meyer, A. S., & Bock, K. (1992). The tip-of-the-tongue phenomenon: Blocking or partial activation? *Memory & Cognition, 20*, 715–726.

Meyer, B. J. F., & Rice, G. E. (1981). Information recalled from prose by young, middle, and old adult readers. *Experimental Aging Research, 7*, 253–268.

Meyer, D. E., & Schvaneveldt, R. W. (1971). Facilitation in recognizing pairs of words: Evidence of a dependence between retrieval operations. *Journal of Experimental Psychology, 90*, 227–234.

Meyer, D. E., Schvaneveldt, R. W., & Ruddy, M. G. (1975). Loci of contextual effects on visual word-recognition. In P. M. A. Rabbitt & S. Dornic (Eds.), *Attention and performance* (Vol. 5, pp. 98–118). London, England: Academic Press.

Milham, M. P., Erickson, K. I., Banich, M. T., Kramer, A. F., Webb, A., Wszalek, T., et al. (2002). Attentional control in the aging brain: Insights from an fMRI study of the Stroop task. *Brain and Cognition, 49*, 277–296.

Milkman, K. L., Chugh, D., & Bazerman, M. H. (2009). How can decision making be improved? *Perspectives on Psychological Science, 4*(4), 379–383.

Miller, D. T., & McFarland, C. (1986). Counterfactual thinking and victim compensation: A test of norm theory. *Personality and Social Psychology Bulletin, 12*, 513–519.

Miller, G. A. (1956). The magical number seven, plus or minus two: Some limits on our capacity for processing information. *Psychological Review, 63*, 81–97.

Miller, G. A. (1973). Psychology and communication. In G. A. Miller (Ed.), *Communication, language, and meaning: Psychological perspectives* (pp. 3–12). New York, NY: Basic Books.

Miller, G. A. (1977). Practical and lexical knowledge. In P. N. Johnson-Laird & P. C. Wason (Eds.), *Thinking: Readings in cognitive science* (pp. 400–410). Cambridge, England: Cambridge University Press.

Miller, G. A., Galanter, E., & Pribram, K. H. (1960). *Plans and the structure of behavior*. New York, NY: Holt.

Miller, G. A., & Isard, S. (1963). Some perceptual consequences of linguistic rules. *Journal of Verbal Learning and Verbal Behavior, 2*, 217–228.

Millis, K. K., & Graesser, A. C. (1994). The time-course of constructing knowledge-based inferences for scientific texts. *Journal of Memory and Language, 33*, 583–599.

Millis, K. K., & Just, M. A. (1994). The influence of connectives on sentence comprehension. *Journal of Memory and Language, 33*, 128–147.

Mills, C. B., Innis, J., Westendorf, T., Owsianiecki, L., & McDonald, A. (2006). Effect of a synesthete's photisms on name recall. *Cortex, 42*, 155–163.

Milner, B., Corkin, S., & Teuber, H. L. (1968). Further analysis of the hippocampal amnesic syndrome: 14–year follow up study of H. M. *Neuropsychologia, 6*, 215–234.

Minsky, M. L. (1986). *The society of mind*. New York, NY: Simon & Schuster.

Mishkin, M., & Appenzeller, T. (1987). The anatomy of memory. *Scientific American, 256*, 80–89.

Mishkin, M., Ungerleider, L., & Macko, K. A. (1983). Object vision and spatial vision: Two cortical pathways. *Trends in Neurosciences, 6*, 414–417.

Mitchell, D. B. (2006). Nonconscious priming after 17 years. *Psychological Science, 17*, 925–929.

Mitchell, D. C., & Holmes, V. M. (1985). The role of specific information about the verb in parsing sentences with local structural ambiguity. *Journal of Memory and Language, 24*, 542–559.

Mitchell, K. J., & Zaragoza, M. S. (1996). Repeated exposure to suggestion and false memory: The role of contextual variability. *Journal of Memory and Language, 35*, 246–260.

Mitchell, K. J., & Zaragoza, M. S. (2001). Contextual overlap and eyewitness suggestibility. *Memory & Cognition, 29*, 616–626.

Mitterschiffthaler, M. T., Williams, S. C. R., Walsh, N. D., Cleare, A. J., Donaldson, C., Scott, J., et al. (2008). Neural basis of the emotional Stroop interference effect in major depression. *Psychological Medicine, 38*, 24–256.

Miyake, A., Friedman, N. P., Emerson, M. J., Witzki, A. H., Howerter, A., & Wager, T. D. (2000). The unity and diversity of executive functions and their contributions to complex "frontal lobe" tasks: A latent variable analysis. *Cognitive Psychology, 41*, 49–100.

Miyake, A., Just, M. A., & Carpenter, P. A. (1994). Working memory constraints on the resolution of lexical ambiguity: Maintaining multiple interpretations in neutral contexts. *Journal of Memory and Language, 33*, 175–202.

Miyake, A., & Shah, P. (Eds.). (1999). *Models of working memory: Mechanisms of active maintenance and executive control*. New York, NY: Cambridge University Press.

Moldoveanu, M., & Langer, E. (2002). False memories of the future: A critique of the applications of probabilistic reasoning to the study of cognitive processes. *Psychological Review, 109*, 358–375.

Monaghan, P., & Pollmann, S. (2003). Division of labor between the hemispheres for complex but not simple tasks: An implemented connectionist model. *Journal of Experimental Psychology: General, 132*, 379–399.

Monaghan, P., & Shillcock, R. (2004). Hemispheric asymmetries in cognitive modeling: Connectionist modeling of unilateral visual neglect. *Psychological Review, 111*, 283–308.

Monaghan, P., Sio, U. N., Lau, S. W., Woo, H. K., Linkenauger, S. A., & Ormerod, T. C. (2015). Sleep promotes analogical transfer in problem solving. *Cognition, 143*, 25–30.

Moore, A. B., Clark, B. A., & Kane, M. J. (2008). Who shalt not kill? Individual differences in working memory capacity, executive control, and moral judgment. *Psychological Science, 19*, 549–557.

Moore, A. M., & Ashcraft, M. H. (2015). Children's mathematical performance: Five cognitive tasks across five grades. *Journal of Experimental Child Psychology, 135*, 1–24.

Moray, N. (1959). Attention in dichotic listening: Affective cues and the influence of instructions. *The Quarterly Journal of Experimental Psychology, 11*, 56–60.

Moray, N., Bates, A., & Barnett, T. (1965). Experiments on the four-eared man. *The Journal of the Acoustical Society of America, 38*, 196–201.

Moreira, C., & Wichert, A. (2016). Quantum-like Bayesian networks for modeling decision making. *Frontiers in Psychology, 7*, 11.

Moreno, S., Bialystok, E., Barac, R., Schellenberg, E. G., Cepeda, N. J., & Chau, T. (2011). Short-term music training enhances verbal intelligence and executive function. *Psychological Science, 22,* 1425–1433.

Morey, C. C., & Cowan, N. (2004). When visual and verbal memories compete: Evidence of cross-domain limits in working memory. *Psychonomic Bulletin & Review, 11,* 296–301.

Morris, A. L., & Harris, C. L. (2002). Sentence context, word recognition, and repetition blindness. *Journal of Experimental Psychology: Learning, Memory, and Cognition, 28,* 962–982.

Morrow, D. G., Greenspan, S. L., & Bower, G. H. (1987). Accessibility and situation models in narrative comprehension. *Journal of Memory and Language, 26,* 165–187.

Morrow, D. G., Leirer, V. O., Altieri, P. A., & Fitzsimmons, C. (1994). Age differences in creating spatial models from narratives. *Language & Cognitive Processes, 9,* 203–220.

Morrow, D. G., Stine-Morrow, E. A. L., Leirer, V. O., Andrassy, J. M., & Kahn, J. (1997). The role of reader age and focus of attention in creating situation models from narratives. *The Journals of Gerontology, 52B,* P73–P80.

Morton, J. (1979). Facilitation in word recognition: Experiments causing change in the logogen models. In P. A. Kolers, M. E. Wrolstad, & H. Bouma (Eds.), *Processing of visible language* (Vol. 1, pp. 259–268). New York, NY: Plenum.

Moscovitch, M. (1979). Information processing and the cerebral hemispheres. In M. S. Gazzaniga (Ed.), *Handbook of behavioral neurobiology: Vol. 2. Neuropsychology* (pp. 379–446). New York, NY: Plenum.

Moshinsky, A., & Bar-Hillel, M. (2002). Where did 1850 happen first—in America or in Europe? A cognitive account for a historical bias. *Psychological Science, 13,* 20–26.

Most, S. B., Scholl, B. J., Clifford, E. R., & Simons, D. J. (2005). What you see is what you set: Sustained inattentional blindness and the capture of awareness. *Psychological Review, 112,* 217–242.

Moyer, R. S., & Bayer, R. H. (1976). Mental comparison and the symbolic distance effect. *Cognitive Psychology, 8,* 228–246.

Mueller, S. T., Seymour, T. L., Kieras, D. E., & Meyer, D. E. (2003). Theoretical implications of articulatory duration, phonological similarity, and phonological complexity in verbal working memory. *Journal of Experimental Psychology: Learning, Memory, and Cognition, 29,* 1353–1380.

Murdock, B. B., Jr. (1962). The serial position effect of free recall. *Journal of Experimental Psychology, 64,* 482–488.

Murphy, G. L. (1985). Processes of understanding anaphora. *Journal of Memory and Language, 24,* 290–303.

Murphy, G. L., Hampton, J. A., & Milovanovic, G. S. (2012). Semantic memory redux: An experimental test of hierarchical category representation. *Journal of Memory and Language, 67*(4), 521–539.

Murphy, G. L., & Medin, D. L. (1985). The role of theories in conceptual coherence. *Psychological Review, 92,* 289–316.

Murphy, M. C., Steele, C. M., & Gross, J. J. (2007). Signaling threat: How situational cues affect women in math, science, and engineering settings. *Psychological Science, 18,* 879–885.

Murray, D. J. (1967). The role of speech responses in short-term memory. *Canadian Journal of Psychology, 21,* 263–276.

Myers, E. B., Blumstein, S. E., Walsh, E., & Eliassen, J. (2009). Inferior frontal regions underlie the perception of phonetic category invariance. *Psychological Science, 20,* 895–903.

Myerson, J., Ferraro, F. R., Hale, S., & Lima, S. D. (1992). General slowing in semantic priming and word recognition. *Psychology and Aging, 7,* 257–270.

Myerson, J., Hale, S., Wagstaff, D., Poon, L. W., & Smith, G. A. (1990). The information loss model: A mathematical theory of age-related cognitive slowing. *Psychological Review, 97,* 475–487.

Nadel, L., & Zola-Morgan, S. (1984). Infantile amnesia: A neurobiological perspective. In M. Moscovitch (Ed.), *Infant memory*. New York, NY: Plenum Press.

Nairne, J. S., & Pandeirada, J. N. (2008). Adaptive memory remembering with a stone-age brain. *Current Directions in Psychological Science, 17*(4), 239–243.

Nairne, J. S., Pandeirada, N. S., & Thompson, S. R. (2008). Adaptive memory: The comparative value of survival processing. *Psychological Science, 19,* 176–180.

Nairne, J. S., Thompson, S. R., & Pandeirada, N. S. (2007). Adaptive memory: Survival processing enhances retention. *Journal of Experimental Psychology: Learning, Memory, and Cognition, 33,* 263–273.

Navon, D. (1977). Forest before trees: The precedence of global features in visual perception. *Cognitive Psychology, 9*(3), 353–383.

Navon, D. (1984). Resources: A theoretical soup stone? *Psychological Review, 91,* 216–234.

Neath, I., Surprenant, A. M., & Crowder, R. G. (1993). The context-dependent stimulus suffix effect. *Journal of Experimental Psychology: Learning, Memory & Cognition, 19,* 698–703.

Nee, D. E., & Jonides, J. (2008). Dissociable interference control processes in perception and memory. *Psychological Science, 19,* 490–500.

Neely, J. H. (1977). Semantic priming and retrieval from lexical memory: Roles of inhibitionless spreading activation and limited-capacity attention. *Journal of Experimental Psychology: General, 106,* 226–254.

Neill, W. T. (1977). Inhibition and facilitation processes in selective attention. *Journal of Experimental Psychology: Human Perception & Performance, 3,* 444–450.

Neill, W. T., Valdes, L. A., & Terry, K. M. (1995). Selective attention and the inhibitory control of cognition. In F. N. Dempster & C. J. Brainerd (Eds.), *New perspectives on interference and inhibition in cognition* (pp. 207–261). New York, NY: Academic Press.

Neisser, U. (1964). Visual search. *Scientific American, 210,* 94–102.

Neisser, U. (1967). *Cognitive psychology*. New York, NY: Appleton-Century-Crofts.

Neisser, U. (1976). *Cognition and reality*. San Francisco, CA: Freeman.

Neisser, U. (1981). John Dean's memory: A case study. *Cognition, 9,* 1–22.

Neisser, U. (1982). *Memory observed: Remembering in natural contexts*. San Francisco, CA: Freeman.

Neisser, U., Novick, R., & Lazar, R. (1963). Searching for ten targets simultaneously. *Perceptual and Motor Skills, 17,* 955–961.

Nelson, K. (1993). The psychological and social origins of autobiographical memory. *Psychological Science, 4*, 7–14.

Nelson, K., & Fivush, R. (2004). The emergence of autobiographical memory: A social cultural developmental theory. *Psychological Review, 111*, 486–511.

Nelson, T. O. (1978). Savings and forgetting from long-term memory. *Journal of Verbal Learning and Verbal Behavior, 10*, 568–576.

Nelson, T. O. (1985). Ebbinghaus's contribution to the measurement of retention: Savings during relearning. *Journal of Experimental Psychology: Learning, Memory, and Cognition, 11*, 472–479.

Nelson, T. O. (1988). Predictive accuracy of the feeling of knowing across different criterion tasks and across different subject populations and individuals. In M. Gruneberg, P. Morris, & R. Sykes (Eds.), *Practical aspects of memory: Current research and issues* (Vol. 1, pp. 190–196). New York, NY: Wiley.

Nelson, T. O. (1993). Judgments of learning and the allocation of study time. *Journal of Experimental Psychology: General, 122*, 269–273.

Nelson, T. O., & Leonesio, R. J. (1988). Allocation of self-paced study time and the "labor-in vain effect." *Journal of Experimental Psychology: Learning, Memory, and Cognition, 14*, 676–686.

Nelson, T. O., McSpadden, M., Fromme, K., & Marlatt, G. A. (1986). Effects of alcohol intoxication on metamemory and on retrieval from long-term memory. *Journal of Experimental Psychology: General, 115*, 247–254.

Neuschatz, J. S., Benoit, G. E., & Payne, D. G. (2003). Effective warnings in the Deese–Roediger–McDermott false-memory paradigm: The role of identifiability. *Journal of Experimental Psychology: Learning, Memory, and Cognition, 29*, 35–41.

Newell, A., Shaw, J. C., & Simon, H. A. (1958). Elements of a theory of human problem solving. *Psychological Review, 65*, 151–166.

Newell, A., & Simon, H. A. (1972). *Human problem solving*. Englewood Cliffs, NJ: Prentice Hall.

Newell, B. R., & Shanks, D. R. (2003). Take the best or look at the rest? Factors influencing "one-reason" decision making. *Journal of Experimental Psychology: Learning, Memory, and Cognition, 29*, 53–65.

Newell, B. R., & Shanks, D. R. (2004). On the role of recognition in decision making. *Journal of Experimental Psychology: Learning, Memory, and Cognition, 30*, 923–935.

Newell, B. R., Wong, K. Y., Cheung, J. C., & Rakow, T. (2009). Think, blink or sleep on it? The impact of modes of thought on complex decision making. *The Quarterly Journal of Experimental Psychology, 62*(4), 707–732.

Newstead, S. E. (1995). Gricean implicatures and syllogistic reasoning. *Journal of Memory and Language, 34*(5), 644–664.

Nickerson, R. S. (2001). The projective way of knowing: A useful heuristic that sometimes misleads. *Current Directions in Psychological Science, 10*, 168–172.

Nickerson, R. S. (2002). The production and perception of randomness. *Psychological Review, 109*, 330–357.

Nickerson, R. S., & Adams, M. J. (1979). Long-term memory for a common object. *Cognitive Psychology, 11*, 287–307.

Nisbett, R. E., & Masuda, T. (2003). Culture and point of view. *Proceedings of the National Academy of Sciences, 100*, 11163–11170.

Nisbett, R. E., Peng, K., Choi, I., & Norenzayan, A. (2001). Culture and systems of thought: Holistic versus analytic cognition. *Psychological Review, 108*, 291–310.

Noh, S. R., & Stine-Morrow, E. A. L. (2009). Age differences in tracking characters during narrative comprehension. *Memory & Cognition, 37*, 769–778.

Noice, H., & Noice, T. (1999). Long-term retention of theatrical roles. *Memory, 7*, 357–382.

Noice, H., & Noice, T. (2001). Learning dialogue with and without movement. *Memory & Cognition, 29*, 820–827.

Norman, D. A. (1981). Categorization of action slips. *Psychological Review, 88*, 1–15.

Norman, D. A. (1986). Reflections on cognition and parallel distributed processing. In J. L. McClelland, D. E. Rumelhart, & PDP Research Group (Eds.), *Parallel distributed processing: Explorations in the microstructure of cognition: Vol. 2. Psychological and biological models* (pp. 531–546). Cambridge, MA: MIT Press.

Norman, D. A., & Rumelhart, D. E. (Eds.). (1975). *Explorations in cognition*. San Francisco, CA: Freeman.

Norman, K. A., & Schacter, D. L. (1997). False recognition in younger and older adults: Exploring the characteristics of illusory memories. *Memory & Cognition, 25*, 838–848

Norris, D. (2013). Models of visual word recognition. *Trends in Cognitive Sciences, 17*(10), 517–524.

Norris, D., McQueen, J. M., Cutler, A., & Butterfield, S. (1997). The possible-word constraint in the segmentation of continuous speech. *Cognitive Psychology, 34*, 191–243.

Nosofsky, R. M. (1986). Attention, similarity, and the identification-categorization relationship. *Journal of Experimental Psychology: General, 115*, 39–57.

Novick, L. R. (1988). Analogical transfer, problem similarity, and expertise. *Journal of Experimental Psychology: Learning, Memory, and Cognition, 14*, 510–520.

Nyberg, L., Cabeza, R., & Tulving, E. (1996). PET studies of encoding and retrieval: The HERA model. *Psychonomic Bulletin & Review, 3*, 135–148.

Nyberg, L., Lövdén, M., Riklund, K., Lindenberger, U., & Bäckman, L. (2012). Memory aging and brain maintenance. *Trends in Cognitive Sciences, 16*(5), 292–305.

Nyberg, L., McIntosh, A. R., & Tulving, E. (1998). Functional brain imaging of episodic and semantic memory with positron emission tomography. *Journal of Molecular Medicine, 76*, 48–53.

Oberfeld, D., Hecht, H., & Gamer, M. (2010). Surface lightness influences perceived room height. *The Quarterly Journal of Experimental Psychology, 63*, 1999–2011.

O'Brien, E. J., Albrecht, J. E., Hakala, C. M., & Rizzella, M. L. (1995). Activation and suppression of antecedents during reinstatement. *Journal of Experimental Psychology: Learning, Memory, and Cognition, 21*, 626–634.

O'Brien, E. J., & Myers, J. L. (1987). The role of causal connections in the retrieval of text. *Memory & Cognition, 15*, 419–427.

O'Brien, E. J., Plewes, P. S., & Albrecht, J. E. (1990). Antecedent retrieval processes. *Journal of Experimental Psychology: Learning, Memory, and Cognition, 16*, 241–249.

Ofshe, R. J. (1992). Inadvertent hypnosis during interrogation: False confession due to dissociative state, misidentified multiple personality, and the satanic cult hypothesis. *International Journal of Clinical and Experimental Hypnosis, 40*, 125–156.

Öhman, A., Flykt, A., & Esteves, F. (2001). Emotion drives attention: Detecting the snake in the grass. *Journal of Experimental Psychology: General, 130*, 466–478.

Ojemann, G. A. (1982). Models of the brain organization for higher integrative functions derived with electrical stimulation techniques. *Human Neurobiology, 1*, 243–250.

Ojemann, G. A., & Creutzfeldt, O. D. (1987). Language in humans and animals: Contribution of brain stimulation and recording. In *Handbook of physiology: The nervous system* (Vol. 5). Bethesda, MD: American Physiological Society.

Olafson, K. M., & Ferraro, F. R. (2001). Effects of emotional state on lexical decision performance. *Brain and Cognition, 45*, 15–20.

Oosterwijk, S., Winkielman, P., Pecher, D., Zeelenberg, R., Rottveel, M., & Fischer, A. H. (2012). Mental states inside out: Switching costs for emotional and nonemotional sentences that differ in internal and external focus. *Memory & Cognition, 40*, 93–100.

Open Science Collaboration (2015). Estimating the reproducibility of psychological science. *Science, 349*(6251). doi: 10.1126/science.aac4716

Oppenheim, G. M., & Dell, G. S. (2010). Motor movement matters: The flexible abstractness of inner speech. *Memory & Cognition, 38*, 1147–1160.

Ormerod, T. C., MacGregor, J. N., Chronicle, E. P., Dewald, A. D., & Chu, Y. (2013). Act first, think later: The presence and absence of inferential planning in problem solving. *Memory & Cognition, 41*(7), 1096–1108.

O'Seaghdha, P. G. (1997). Conjoint and dissociable effects of syntactic and semantic context. *Journal of Experimental Psychology: Learning, Memory, and Cognition, 23*, 807–828.

Osterhout, L., Allen, M. D., McLaughlin, J., & Inoue, K. (2002). Brain potentials elicited by prose-embedded linguistic anomalies. *Memory & Cognition, 30*, 1304–1312.

Osterhout, L., & Holcomb, P. J. (1992). Event-related brain potentials elicited by syntactic anomaly. *Journal of Memory and Language, 31*, 785–806.

Özyürek, S. (2002). Do speakers design their cospeech gestures for their addressees? The effects of addressee location on representational gestures. *Journal of Memory and Language, 46*, 688–704.

Paap, K. R., & Greenberg, Z. I. (2013). There is no coherent evidence for a bilingual advantage in executive processing. *Cognitive Psychology, 66*(2), 232–258.

Pachur, T., Mata, R., & Schooler, L. J. (2007). Cognitive aging and the adaptive use of recognition in decision making. *Psychology and Aging, 24*, 901–915.

Paivio, A. (1971). *Imagery and verbal processes*. New York, NY: Holt.

Paivio, A. (1990). *Mental representations: A dual coding approach*. New York, NY: Oxford University Press.

Palermo, D. S. (1978). *Psychology of language*. Glenview, IL: Scott, Foresman.

Palmer, S. E. (1975). The effects of contextual scenes on the identification of objects. *Memory & Cognition, 3*, 519–526.

Park, D. C., Hertzog, C., Kidder, D. P., Morrell, R. W., & Mayhorn, C. B. (1997). Effect of age on event-based and time-based prospective memory. *Psychology and Aging, 12*, 314–327.

Park, D. C., & Huang, C-M. (2010). Culture wires the brain: A cognitive neuroscience perspective. *Perspectives on Psychological Science, 5*, 391–400.

Park, D. C., & McDonough, I. M. (2013). The dynamic aging mind revelations from functional neuroimaging research. *Perspectives on Psychological Science, 8*(1), 62–67.

Pashler, H. (1994). Dual-task interference in simple tasks: Data and theory. *Psychological Bulletin, 116*, 220–244.

Pashler, H., & Johnson, J. C. (1998). Attentional limitations in dual-task performance. In H. Pashler (Ed.), *Attention* (pp. 155–189). Hove, England: Psychology Press.

Pashler, H., Rohrer, D., Cepeda, N. J., & Carpenter, S. K. (2007). Enhancing learning and retarding forgetting: Choices and consequences. *Psychonomic Bulletin & Review, 14*, 187–193.

Pastötter, B., & Bäuml, K. H. T. (2014). Retrieval practice enhances new learning: The forward effect of testing. *Frontiers in Psychology, 5*, 3389.

Paul, S. T., Kellas, G., Martin, M., & Clark, M. B. (1992). Influence of contextual features on the activation of ambiguous word meanings. *Journal of Experimental Psychology: Learning, Memory, and Cognition, 18*, 703–717.

Payne, D. G., Toglia, M. P., & Anastasi, J. S. (1994). Recognition performance level and the magnitude of the misinformation effect in eyewitness memory. *Psychonomic Bulletin & Review, 1*, 376–382.

Payne, J. D., & Kensinger, E. A. (2010). Sleep's role in the consolidation of emotional episodic memories. *Psychological Science, 19*, 290–295.

Pearlmutter, N. J., Garnsey, S. M., & Bock, K. (1999). Agreement processes in sentence comprehension. *Journal of Memory and Language, 41*, 427–456.

Pecher, D., Zeelenberg, R., & Barsalou, L. W. (2003). Verifying different modality properties for concepts produces switching costs. *Psychological Science, 14*, 119–124.

Pedone, R., Hummel, J. E., & Holyoak, K. J. (2001). The use of diagrams in analogical problem solving. *Memory & Cognition, 29*, 214–221.

Pell, M. D. (1999). Fundamental frequency encoding of linguistic and emotional prosody by right hemisphere–damaged speakers. *Brain and Language, 69*, 161–192.

Penfield, W., & Jasper, H. H. (1954). *Epilepsy and the functional anatomy of the human brain*. Boston, MA: Little, Brown.

Penfield, W., & Milner, B. (1958). Memory deficit produced by bilateral lesions in the hippocampal zone. *Archives of Neurology and Psychiatry, 79*, 475–497.

Perea, M., Duñabeitia, J. A., & Carreiras, M. (2008). Masked associative/semantic priming effects across languages with highly proficient bilinguals. *Journal of Memory and Language, 58,* 916–930.

Peretz, I., Radeau, M., & Arguin, M. (2004). Two-way interactions between music and language: Evidence from priming recognition of tune and lyrics in familiar songs. *Memory & Cognition, 32,* 142–152.

Perfect, T. J., Andrade, J., & Eagan, I. (2011). Eye closure reduces the cross-modal memory impairment caused by auditory distraction. *Journal of Experimental Psychology: Learning, Memory, and Cognition, 37,* 1008–1013.

Perrachione, T. K., Fedorenko, E. G., Vinke, L., Gibson, E., & Dilley, L. C. (2013). Evidence for shared cognitive processing of pitch in music and language. *PloS one, 8*(8), e73372.

Peterson, L. R., & Peterson, M. J. (1959). Short-term retention of individual items. *Journal of Experimental Psychology, 58,* 193–198.

Peterson, R. R., Burgess, C., Dell, G. S., & Eberhard, K. M. (2001). Dissociation between syntactic and semantic processing during idiom comprehension. *Journal of Experimental Psychology: Learning, Memory, and Cognition, 27*(5), 1223.

Peterson, W. W., Birdsall, T. G., & Fox, W. C. (1954). The theory of signal detectability. *Institute of Radio Engineers Transactions, PGIT-4,* 171–212.

Petrusic, W. M., & Baranski, J. V. (2003). Judging confidence influences decision processing in comparative judgments. *Psychonomic Bulletin & Review, 10,* 177–183.

Pettijohn, K. A., & Radvansky, G. A. (2016a). Walking through doorways causes forgetting: Environmental effects. *Journal of Cognitive Psychology, 28*(3), 329–340.

Pettijohn, K. A., & Radvansky, G. A. (2016b). Walking through doorways causes forgetting: Event structure or updating disruption? *The Quarterly Journal of Experimental Psychology, 69*(11), 2119–2129.

Pettijohn, K. A., Thompson, A. N., Tamplin, A. K., Krawietz, S. A., & Radvansky, G. A. (2016). Event boundaries and memory improvement. *Cognition, 148,* 136–144.

Pexman, P. M., Holyk, G. G., & Monfils, M. (2003). Number-of-features effects and semantic processing. *Memory & Cognition, 31,* 842–855.

Pickering, M. J., & Traxler, M. J. (1998). Plausibility and recovery from garden paths: An eye-tracking study. *Journal of Experimental Psychology: Learning, Memory, and Cognition, 24,* 940–961.

Piercey, C. D., & Joordens, S. (2000). Turning an advantage into a disadvantage: Ambiguity effects in lexical decision versus reading tasks. *Memory & Cognition, 28,* 657–666.

Pinker, S. (1994). *The language instinct: How the mind creates language.* New York, NY: Morrow.

Pitt, M. A., & Samuel, A. G. (1995). Lexical and sublexical feedback in auditory word recognition. *Cognitive Psychology, 29,* 149–188.

Pittenger, C., & Kandel, E. R. (2003). In search of general mechanisms for long-lasting plasticity: Aplysia and the hippocampus. *Philosophical Transactions of the Royal Society of London, 358,* 757–763.

Plude, D. J., & Doussard-Roosevelt, J. A. (1989). Aging, selective attention, and feature integration. *Psychology and Aging, 4,* 98–105.

Poldrack, R. A., & Wagner, A. D. (2004). What can neuroimaging tell us about the mind? *Current Directions in Psychological Science, 13,* 177–181.

Poljac, E., de-Wit, L., & Wagemans, J. (2012). Perceptual wholes can reduce the conscious accessibility of their parts. *Cognition, 123*(2), 308–312.

Polk, T. A., & Farah, M. J. (2002). Functional MRI evidence for an abstract, not perceptual, word-form area. *Journal of Experimental Psychology: General, 131,* 65–72.

Pollack, I., & Pickett, J. M. (1964). Intelligibility of excerpts from fluent speech: Auditory vs. structural context. *Journal of Verbal Learning and Verbal Behavior, 3,* 79–84.

Pollatsek, A., Konold, C. E., Well, A. D., & Lima, S. D. (1984). Beliefs underlying random sampling. *Memory & Cognition, 12,* 395–401.

Polya, G. (1957). *How to solve it.* Garden City, NY: Doubleday/Anchor.

Posner, M. I., & Cohen, Y. (1984). Components of visual orienting. In H. Bouma & D. G. Bouwhuis (Eds.), *Attention and performance X* (pp. 531–556). Hillsdale, NJ: Erlbaum.

Posner, M. I., Kiesner, J., Thomas-Thrapp, L., McCandliss, B., Carr, T. H., & Rothbart, M. K. (1992, November). *Brain changes in the acquisition of literacy.* Paper presented at the meetings of the Psychonomic Society, St. Louis, MO.

Posner, M. I., Nissen, M. J., & Ogden, W. C. (1978). Attended and unattended processing modes: The role of set for spatial location. In H. L. Pick & I. J. Saltzman (Eds.), *Modes of perceiving and processing information* (pp. 137–157). Hillsdale, NJ: Erlbaum.

Posner, M. I., & Snyder, C. R. R. (1975). Facilitation and inhibition in the processing of signals. In P. M. A. Rabbitt & S. Dornic (Eds.), *Attention and performance V* (pp. 669–682). New York, NY: Academic Press.

Posner, M. I., Snyder, C. R. R., & Davidson, B. J. (1980). Attention and the detection of signals. *Journal of Experimental Psychology: General, 109,* 160–174.

Postman, L., & Underwood, B. J. (1973). Critical issues in interference theory. *Memory & Cognition, 1,* 19–40.

Pratt, J., Radulescu, P. V., Guo, R. M., & Abrams, R. A. (2010). It's alive! Animate motion captures visual attention. *Psychological Science, 21,* 1724–1730.

Price, R. H. (1987). *Principles of psychology.* Glenview, IL: Scott Foresman.

Prince, S. E., Tsukiura, T., & Cabeza, R. (2007). Distinguishing the neural correlates of episodic encoding and semantic retrieval. *Psychological Science, 18,* 144–151.

Pritchard, R. M. (1961). Stabilized images on the retina. *Scientific American, 204,* 72–78.

Proffitt, D. R. (2006). Embodied perception and the economy of action. *Perspectives on Psychological Science, 1,* 110–122.

Proffitt, D. R., Kaiser, M. K., & Whelan, S. M. (1990). Understanding wheel dynamics. *Cognitive Psychology, 22,* 342–373.

Protopapas, A., Archonti, A., & Skaloumbakas, C. (2007). Reading ability is negatively related to Stroop interference. *Cognitive Psychology, 54,* 251–282.

Provins, K. A. (1997). Handedness and speech: A critical reappraisal of the role of genetic and environmental factors in the cerebral lateralization of function. *Psychological Review, 104*, 554–571.

Queen, T. L., & Hess, T. M. (2010). Age differences in the effects of conscious and unconscious thought in decision making. *Psychology and Aging, 25*, 251–261.

Quillian, M. R. (1968). Semantic memory. In M. Minsky (Ed.), *Semantic information processing* (pp. 216–270). Cambridge, MA: MIT Press.

Quillian, M. R. (1969). The teachable language comprehender: A simulation program and theory of language. *Communications of the ACM, 12*, 459–476.

Quinn, P. C., Eimas, P. D., & Rosenkrantz, S. L. (1993). Evidence for representations of perceptually similar natural categories by 3-month-old and 4-month-old infants. *Perception, 22*, 463–475.

Rader, A. W., & Sloutsky, V. M. (2002). Processing of logically valid and logically invalid conditional inferences in discourse comprehension. *Journal of Experimental Psychology: Learning, Memory, and Cognition, 28*, 59–68.

Radvansky, G. A., & Copeland, D. E. (2006a). Memory retrieval and interference: Working memory issues. *Journal of Memory and Language, 55*, 33–46.

Radvansky, G. A., & Copeland, D. E. (2006b). Walking through doorways causes forgetting. *Memory & Cognition, 34*, 1150–1156.

Radvansky, G. A., Copeland, D. E., & Zwaan, R. A. (2003). Aging and functional spatial relations in comprehension and memory. *Psychology and Aging, 18*, 161–165.

Radvansky, G. A., & Dijkstra, K. (2007). Aging and situation model processing. *Psychonomic Bulletin & Review, 14*, 1027–1042.

Radvansky, G. A., Gerard, L. D., Zacks, R. T., & Hasher, L. (1990). Younger and older adults' use of mental models as representations for text materials. *Psychology and Aging, 5*, 209–214.

Radvansky, G. A., Gibson, B. S., & McNerney, M. W. (2011). Synesthesia and memory: Color congruency, von Restorff and false memory effects. *Journal of Experimental Psychology: Learning, Memory, and Cognition, 37*, 219–229.

Radvansky, G. A., Krawietz, S. A., & Tamplin, A. K. (2011). Walking through doorways causes forgetting: Further explorations. *The Quarterly Journal of Experimental Psychology, 64*, 1632–1645.

Radvansky, G. A., Pettijohn, K. A., & Kim, J. (2015). Walking through doorways causes forgetting: Younger and older adults. *Psychology and Aging, 30*(2), 259–265.

Radvansky, G. A., Spieler, D. H., & Zacks, R. T. (1993). Mental model organization. *Journal of Experimental Psychology: Learning, Memory, and Cognition, 19*, 95–114.

Radvansky, G. A., Tamplin, A. K., & Krawietz, S. A. (2010). Walking through doorways causes forgetting: Environmental integration. *Psychonomic Bulletin & Review, 17*, 900–904.

Radvansky, G. A., & Zacks, R. T. (1991). Mental models and the fan effect. *Journal of Experimental Psychology: Learning, Memory, and Cognition, 17*, 940–953.

Radvansky, G. A., Zacks, R. T., & Hasher, L. (1996). Fact retrieval in younger and older adults: The role of mental models. *Psychology and Aging, 11*(2), 258–271.

Radvansky, G. A., Zwaan, R. A., Curiel, J. M., & Copeland, D. E. (2001). Situation models and aging. *Psychology and Aging, 16*, 145–160.

Radvansky, G. A., Zwaan, R. A., Federico, T., & Franklin, N. (1998). Retrieval from temporally organized situation models. *Journal of Experimental Psychology: Learning, Memory, and Cognition, 24*, 1224–1237.

Rafal, R. D. (1997). Hemispatial neglect: Cognitive neuropsychological aspects. In T. E. Feinberg & M. J. Farah (Eds.), *Behavioral neurology and neuropsychology* (pp. 319–336). New York, NY: McGraw-Hill.

Raine, L. B., Lee, H. K., Saliba, B. J., Chaddock-Heyman, L., Hillman, C. H., & Kramer, A. F. (2013). The influence of childhood aerobic fitness on learning and memory. *PloS one, 8*(9), e72666.

Ramsey-Rennels, J. L., & Langlois, J. H. (2007). How infants perceive and process faces. In A. Slater & M. Lewis (Eds.), *Introduction to infant development* (pp. 191–232). Oxford, England: Oxford University Press.

Randall, J. G., Oswald, F. L., & Beier, M. E. (2014). Mind-wandering, cognition, and performance: A theory-driven meta-analysis of attention regulation. *Psychological Bulletin, 140*(6), 1411–1431.

Ranganath, C., & Pallar, K. A. (1999). Frontal brain activity during episodic and semantic retrieval: Insights from event-related potentials. *Journal of Cognitive Neuroscience, 11*, 598–609.

Rapp, B., & Goldrick, M. (2000). Discreteness and interactivity in spoken word production. *Psychological Review, 107*, 460–499.

Rasch, B., & Born, J. (2008). Reactivation and consolidation of memory during sleep. *Current Directions in Psychological Science, 17*, 188–192.

Rasmussen, A. S., & Berntsen, D. (2013). The reality of the past versus the ideality of the future: Emotional valence and functional differences between past and future mental time travel. *Memory & Cognition, 41*(2), 187–200.

Ratcliff, G., & Newcombe, F. (1982). Object recognition: Some deductions from the clinical evidence. In A. W. Ellis (Ed.), *Normality and pathology in cognitive functions* (p. 162). London, England: Academic Press.

Ratcliff, R., & McKoon, G. (1988). A retrieval theory of priming in memory. *Psychological Review, 95*, 385–408.

Ratiu, I., & Azuma, T. (2015). Working memory capacity: Is there a bilingual advantage? *Journal of Cognitive Psychology, 27*(1), 1–11.

Raven, J. C. (1941). Standardization of progressive matrices, 1938. *British Journal of Medical Psychology, 19*, 137–150.

Rawson, K. A., & Touron, D. R. (2015). Preservation of memory-based automaticity in reading for older adults. *Psychology and Aging, 30*(4), 809–823.

Rayner, K. (1998). Eye movements in reading and information processing: 20 years of research. *Psychological Bulletin, 124*, 372–422.

Rayner, K., Carlson, M., & Frazier, L. (1983). The interaction of syntax and semantics during sentence processing: Eye movements in the analysis of semantically biased sentences. *Journal of Verbal Learning and Verbal Behavior, 22*, 358–374.

Rayner, K., & Clifton, C., Jr. (2002). Language comprehension. In D. L. Medin (Ed.), *Steven's handbook of experimental psychology* (Vol. X, pp. 261–316). New York, NY: Wiley.

Rayner, K., & Duffy, S. A. (1988). On-line comprehension processes and eye movements in reading. In M. Daneman, G. E. MacKinnon, & T. G. Waller (Eds.), *Reading research: Advances in theory and practice* (pp. 13–66). New York, NY: Academic Press.

Rayner, K., & Frazier, L. (1989). Selection mechanisms in reading lexically ambiguous words. *Journal of Experimental Psychology: Learning, Memory, and Cognition, 15,* 779–790.

Rayner, K., Inhoff, A. W., Morrison, P. E., Slowiaczek, M. L., & Bertera, J. H. (1981). Masking of foveal and parafoveal vision during eye fixations in reading. *Journal of Experimental Psychology: Human Perception and Performance, 7,* 167–179.

Rayner, K., Pollatsek, A., & Binder, K. S. (1998). Phonological codes and eye movements in reading. *Journal of Experimental Psychology: Learning, Memory, and Cognition, 24,* 476–497.

Rayner, K., Warren, T., Juhasz, B. J., & Liversedge, S. P. (2004). The effect of plausibility on eye movements in reading. *Journal of Experimental Psychology: Learning, Memory, and Cognition, 30,* 1290–1301.

Rayner, K., & Well, A. D. (1996). Effects of contextual constraint on eye movements in reading: A further examination. *Psychonomic Bulletin & Review, 3,* 504–509.

Rayner, K., White, S. J., Johnson, R. L., & Liversedge, S. P. (2006). Raeding wrods with jubmled letters: There is a cost. *Psychological Science, 17,* 192–193.

Reason, J. (1990). *Human error.* New York, NY: Cambridge University Press.

Reder, L. M., & Kusbit, G. W. (1991). Locus of the Moses illusion: Imperfect encoding, retrieval, or match? *Journal of Memory and Language, 30,* 385–406.

Reed, A. E., & Carstensen, L. L. (2012). The theory behind the age-related positivity effect. *Frontiers in Psychology, 3,* 339.

Reed, S. K. (1992). *Cognition: Theory and applications* (3rd ed.). Pacific Grove, CA: Brooks/Cole.

Reed, S. K., & Hoffman, B. (2004). Use of temporal and spatial information in estimating event completion time. *Memory & Cognition, 32,* 271–282.

Reeves, L. M., & Weisberg, R. W. (1993). Abstract versus concrete information as the basis for transfer in problem solving: Comment on Fong and Nisbett (1991). *Journal of Experimental Psychology: General, 122,* 125–128.

Regev, M., Honey, C. J., Simony, E., & Hasson, U. (2013). Selective and invariant neural responses to spoken and written narratives. *The Journal of Neuroscience, 33*(40), 15978–15988.

Regier, T., Kay, P., & Khetarpal, N. (2009). Color naming and the shape of color space. *Language, 85*(4), 884–892.

Rehder, B. (2003). A causal-model theory of conceptual representation and categorization. *Journal of Experimental Psychology: Learning, Memory, and Cognition, 29,* 1141–1159.

Rehder, B., & Burnett, R. C. (2005). Feature inference and the causal structure of categories. *Cognitive Psychology, 50,* 264–314.

Reichle, E. D., Carpenter, P. A., & Just, M. A. (2000). The neural bases of strategy and skill in sentence–picture verification. *Cognitive Psychology, 40,* 261–295.

Reichle, E. D., Pollatsek, A., Fisher, D. L., & Rayner, K. (1998). Eye movements during mindless reading. *Psychological Science, 21,* 1300–1310.

Reichle, E. D., Reineberg, A. E., & Schooler, J. W. (2010). Toward a model of eye movement control in reading. *Psychological Review, 105,* 125–157.

Reingold, E. M., Charness, N., Pomplun, M., & Stampe, D. M. (2001). Visual span in expert chess players: Evidence from eye movements. *Psychological Science, 12,* 48–55.

Reiser, B. J., Black, J. B., & Abelson, R. P. (1985). Knowledge structures in the organization and retrieval of autobiographical memories. *Cognitive Psychology, 17,* 89–137.

Reuter-Lorenz, P. A., & Cappell, K. A. (2008). Neurocognitive aging and the compensation hypothesis. *Current Directions in Psychological Science, 17,* 177–182.

Reyna, V. F. (2004). How people make decisions that involve risk. *Current Directions in Psychological Science, 13,* 60–67.

Ribot, T. (1882). *Diseases of memory: An essay in the positive psychology.* New York, NY: Appleton.

Rich, A. N., & Mattingly, J. B. (2002). Anomalous perception in synaesthesia: A cognitive neuroscience perspective. *Nature Reviews: Neuroscience, 3,* 43–51.

Riley, K. P. (1989). Psychological interventions in Alzheimer's disease. In G. C. Gilmore, P. J. Whitehouse, & M. R. Wykle (Eds.), *Memory, aging and dementia.* New York, NY: Springer.

Rips, L. J. (1975). Inductive judgments about natural categories. *Journal of Verbal Learning and Verbal Behavior, 14,* 665–681.

Rips, L. J. (1989). Similarity, typicality, and categorization. In S. Vosniadou & A. Ortony (Eds.), *Similarity and analogical reasoning* (pp. 21–59). Cambridge, England: Cambridge University Press.

Rips, L. J. (1994). *The psychology of proof: Deductive reasoning in human thinking.* Cambridge, MA: MIT Press.

Rips, L. J. (1998). Reasoning and conversation. *Psychological Review, 105,* 411–441.

Rips, L. J. (2001). Two kinds of reasoning. *Psychological Science, 12,* 129–134.

Rips, L. J., & Collins, A. (1993). Categories and resemblance. *Journal of Experimental Psychology: General, 122,* 468–486.

Rips, L. J., & Marcus, S. L. (1977). Supposition and the analysis of conditional sentences. In M. A. Just & P. A. Carpenter (Eds.), *Cognitive processes in comprehension* (pp. 185–220). Hillsdale, NJ: Erlbaum.

Rips, L. J., Shoben, E. J., & Smith, E. E. (1973). Semantic distance and the verification of semantic relations. *Journal of Verbal Learning and Verbal Behavior, 12,* 1–20.

Roberson, D., Davies, I., & Davidoff, J. (2000). Color categories are not universal: Replications and new evidence from a Stone-Age culture. *Journal of Experimental Psychology: General, 129,* 369–398.

Robertson, D. A., Gernsbacher, M. A., Guidotti, S. J., Robertson, R. R. W., Irwin, W., Mock, B. J., et al. (2000). Functional neuroanatomy of the cognitive process of mapping during discourse comprehension. *Psychological Science, 11,* 255–260.

Robertson, D. A., Savva, G. M., King-Kallimanis, B. L., & Kenny, R. A. (2015). Negative perceptions of aging and decline in walking speed: A self-fulfilling prophecy. *PloS one, 10*(4), e0123260.

Robins, R. W., Gosling, S. D., & Craik, K. H. (1999). An empirical analysis of trends in psychology. *American Psychologist, 54,* 117–128.

Roediger, H. L. III. (1996). Memory illusions. *Journal of Memory and Language, 35,* 76–100.

Roediger, H. L. III, & Crowder, R. G. (1976). A serial position effect in recall of United States presidents. *Bulletin of the Psychonomic Society, 8*(4), 275–278.

Roediger, H. L. III, Jacoby, D., & McDermott, K. B. (1996). Misinformation effects in recall: Creating false memories through repeated retrieval. *Journal of Memory and Language, 35,* 300–318.

Roediger, H. L., & Karpicke, J. D. (2006). The power of resting memory: Basic research and implications for educational practice. *Perspectives on Psychological Science, 1,* 181–210.

Roediger, H. L. III, Marsh, E. J., & Lee, S. C. (2002). Kinds of memory. In D. Medin (Ed.), *Stevens' handbook of experimental psychology,* (3rd ed., Vol. 2, pp. 1–42). New York, NY: Wiley.

Roediger, H. L. III, & McDermott, K. B. (1995). Creating false memories: Remembering words not presented in lists. *Journal of Experimental Psychology: Learning, Memory, and Cognition, 21,* 803–814.

Roediger, H. L. III, & McDermott, K. B. (2000). Tricks of memory. *Current Directions in Psychological Science, 9,* 123–127.

Roediger, H. L. III, Meade, M. L., & Bergman, E. T. (2001). Social contagion of memory. *Psychonomic Bulletin & Review, 8,* 365–371.

Roediger, H. L. III, Stadler, M. L., Weldon, M. S., & Riegler, G. L. (1992). Direct comparison of two implicit memory tests: Word fragment and word stem completion. *Journal of Experimental Psychology: Learning, Memory, and Cognition, 18,* 1251–1269.

Roese, N. J. (1997). Counterfactual thinking. *Psychological Bulletin, 121,* 133–148.

Roese, N. J. (1999). Counterfactual thinking and decision making. *Psychonomic Bulletin & Review, 6,* 570–578.

Rogers, T. B., Kuiper, N. A., & Kirker, W. S. (1977). Self-reference and the encoding of personal information. *Journal of Personality and Social Psychology, 35,* 677–688.

Rogers, T. T., Lambon Ralph, M. A., Gerrard, P., Bozeat, S., McClelland, J. L., Hodges, J. R., et al. (2004). Structure and deterioration of semantic memory: A neuropsychological and computational investigation. *Psychological Review, 111,* 205–235.

Rohrer, D. (2003). The natural appearance of unnatural incline speed. *Memory & Cognition, 31,* 816–826.

Rönnlund, M., Nyberg, L., Bäckman, L., & Nilsson, L-G. (2005). Stability, growth, and decline in adult life span development of declarative memory: Cross-sectional and longitudinal data from a population-based study. *Psychology and Aging, 20,* 3–18.

Rosch, E. H. (1978). Principles of categorization. In E. H. Rosch & B. B. Lloyd (Eds.), *Cognition and categorization* (pp. 27–48). Hillsdale, NJ: Erlbaum.

Rosch, E. H., & Mervis, C. B. (1975). Family resemblances: Studies in the internal structure of categories. *Cognitive Psychology, 7,* 573–605.

Rosch-Heider, E. (1972). Universals in color naming and memory. *Journal of Experimental Psychology, 93,* 10–21.

Rosen, V. M., & Engle, R. W. (1997). The role of working memory capacity in retrieval. *Journal of Experimental Psychology: General, 126,* 211–227.

Rosler, F., Pechmann, T., Streb, J., Roder, B., & Hennighausen, E. (1998). Parsing of sentences in a language with varying word order: Word-by-word variations of processing demands are revealed by event-related brain potentials. *Journal of Memory and Language, 38,* 150–176.

Ross, B. H. (1987). This is like that: The use of earlier problems and the separation of similarity effects. *Journal of Experimental Psychology: Learning, Memory, and Cognition, 13,* 629–640.

Ross, B. H., & Murphy, G. L. (1999). Food for thought: Cross-classification and category organization in a complex real-world domain. *Cognitive Psychology, 38,* 495–553.

Ross, M., & Wang, Q. (2010). Why we remember and what we remember: Culture and autobiographical memory. *Perspectives on Psychological Science, 5,* 401–409.

Roy, M., Shohamy, D., & Wager, T. D. (2012). Ventromedial prefrontal-subcortical systems and the generation of affective meaning. *Trends in Cognitive Sciences, 16,* 147–156.

Rubin, D. C. (2007). A basic-systems model of episodic memory. *Perspectives on Psychological Science, 1,* 277–311.

Rubin, D. C., & Berntsen, D. (2003). Life scripts help to maintain autobiographical memories of highly positive, but not highly negative, events. *Memory & Cognition, 31,* 1–14.

Rubin, D. C., Rahhal, T. A., & Poon, L. W. (1998). Things learned in early adulthood are remembered best. *Memory & Cognition, 26,* 3–19.

Rubin, D. C., Schrauf, R. W., & Greenberg, D. L. (2003). Belief and recollection of autobiographical memories. *Memory & Cognition, 31,* 887–901.

Ruff, C. C., Kristkjánsson, Á., & Driver, J. (2007). Readout from iconic memory and selective spatial attention involve similar neural processes. *Psychological Science, 18,* 901–909.

Rugg, M. D., & Coles, M. G. H. (Eds.). (1995). *Electrophysiology of mind: Event-related brain potential and cognition.* New York, NY: Oxford University Press.

Rumelhart, D. E. (1989). The architecture of mind: A connectionist approach. In M. I. Posner (Ed.), *Foundations of cognitive science* (pp. 133–159). Cambridge, MA: MIT Press.

Rumelhart, D. E., Lindsay, P. H., & Norman, D. A. (1972). A process model for long-term memory. In E. Tulving & W. Donaldson (Eds.), *Organization of memory* (pp. 197–246). New York, NY: Academic Press.

Rumelhart, D. E., & McClelland, J. L. (1986). *Parallel distributed processing: Explorations in the microstructure of cognition: Vol. 1. Foundations.* Cambridge, MA: Bradford.

Rundus, D. (1971). Analysis of rehearsal processes in free recall. *Journal of Experimental Psychology, 89,* 63–77.

Rundus, D., & Atkinson, R. C. (1970). Rehearsal processes in free recall: A procedure for direct observation. *Journal of Verbal Learning and Verbal Behavior, 9,* 99–105.

Russo, J. E., Johnson, E. J., & Stephens, D. L. (1989). The validity of verbal protocols. *Memory & Cognition, 17,* 759–769.

Rypma, B., Prabhakaran, V., Desmond, J. E., & Gabriei, D. E. (2001). Age differences in prefrontal cortical activity in working memory. *Psychology and Aging, 16,* 371–384.

Sachs, J. S. (1967). Recognition memory for syntactic and semantic aspects of connected discourse. *Perception & Psychophysics, 2,* 437–442.

Sachs, J. (1983). Talking about the there and then: The emergence of displaced reference in parent-child discourse. In K. E. Nelson (Ed.), *Children's language* (Vol. 3, pp. 1–28). Hillsdale, NJ: Erlbaum.

Sacks, H., Schegloff, E. A., & Jefferson, G. (1974). A simplest systematics for the organization of turntaking for conversation. *Language, 50,* 696–735.

Sacks, O. (1970). *The man who mistook his wife for a hat.* New York, NY: Harper & Row.

Safer, M. A., Christiansen, S-A., Autry, M. W., & Österlund, K. (1998). Tunnel memory for traumatic events. *Applied Cognitive Psychology, 12,* 99–117.

Sakaki, M., Niki, K., & Mather, M. (2012). Beyond arousal and valence: The importance of the biological versus social relevance of emotional stimuli. *Cognitive Affective & Behavioral Neuroscience, 12,* 115–139.

Salame, P., & Baddeley, A. D. (1982). Disruption of short-term memory by unattended speech: Implications for the structure of working memory. *Journal of Verbal Learning and Verbal Behavior, 21,* 150–164.

Salthouse, T. A. (1984). Effects of age and skill in typing. *Journal of Experimental Psychology: General, 113,* 345–371.

Salthouse, T. A. (1996). The processing-speed theory of adult age differences in cognition. *Psychological Review, 103*(3), 403–427.

Salthouse, T. A. (2014). Why are there different age relations in cross-sectional and longitudinal comparisons of cognitive functioning? *Current Directions in Psychological Science, 23*(4), 252–256.

Salthouse, T. A., & Pink, J. E. (2008). Why is working memory related to fluid intelligence? *Psychonomic Bulletin & Review, 15,* 364–371.

Saltz, E., & Donnenwerth-Nolan, S. (1981). Does motoric imagery facilitate memory for sentences? A selective interference test. *Journal of Verbal Learning and Verbal Behavior, 20,* 322–332.

Samuel, A. G. (2001). Knowing a word affects the fundamental perception of the sounds within it. *Psychological Science, 12,* 348–351.

Samuel, D. (1999). *Memory: How we use it, lose it and can improve it.* New York, NY: New York University Press.

Sanbonmatsu, D. M., Strayer, D. L., Biondi, F., Behrends, A. A., & Moore, S. M. (2016). Cell-phone use diminishes self-awareness of impaired driving. *Psychonomic Bulletin & Review, 23*(2), 617–623.

Sanchez, C. A. (2012). Enhancing visuospatial performance through video game training to increase learning in visuospatial science domains. *Psychonomic Bulletin & Review, 19,* 58–65.

Sanchez, C. A., & Wiley, J. (2006). An examination of the seductive details effect in terms of working memory capacity. *Memory & Cognition, 34,* 344–355.

Sanna, L. J., & Schwartz, N. (2006). Human judgment: The case of the hindsight bias and its debiasing. *Current Directions in Psychological Science, 15,* 172–176.

Sanocki, T., Islam, M., Doyon, J. K., & Lee, C. (2015). Rapid scene perception with tragic consequences: Observers miss perceiving vulnerable road users, especially in crowded traffic scenes. *Attention, Perception, & Psychophysics, 77*(4), 1252–1262.

Santamaría, C., Tse, P. P., Moreno-Ríos, S., & García-Madruga, J. A. (2013). Deductive reasoning and metalogical knowledge in preadolescence: A mental model appraisal. *Journal of Cognitive Psychology, 25*(2), 192–200.

Sara, S. J. (2000). Retrieval and reconsolidation: Toward a neurobiology of remembering. *Learning and Memory, 7,* 73–84.

Sargent, J. Q., Zacks, J. M., Hambrick, D. Z., Zacks, R. T., Kurby, C. A., Bailey, H. R., . . . & Beck, T. M. (2013). Event segmentation ability uniquely predicts event memory. *Cognition, 129*(2), 241–255.

Sarter, M., Berntson, G. G., & Cacioppo, J. T. (1996). Brain imaging and cognitive neuroscience: Toward strong inference in attributing function to structure. *American Psychologist, 51,* 13–21.

Sattler, J. M. (1982). *Assessment of children's intellectual and special abilities* (2nd ed.). Boston, MA: Allyn & Bacon.

Sayette, M. A., Reichle, E. D., & Schooler, J. W. (2009). Lost in the sauce: The effects of alcohol on mind wandering. *Psychological Science, 20,* 747–752.

Sayette, M. A., Schooler, J. W., & Reichle, E. D. (2010). Out for a smoke: The impact of cigarette craving on zoning out during reading. *Psychological Science, 21,* 26–30.

Schab, F. R. (1990). Odors and the remembrance of things past. *Journal of Experimental Psychology: Learning, Memory, and Cognition, 16,* 648–655.

Schacter, D. L. (1987). Implicit memory: History and current status. *Journal of Experimental Psychology: Learning, Memory, and Cognition, 13,* 501–518.

Schacter, D. L. (1989). Memory. In M. I. Posner (Ed.), *Foundations of cognitive science* (pp. 683–725). Cambridge, MA: MIT Press.

Schacter, D. L. (1996). *Searching for memory.* New York, NY: Basic Books.

Schacter, D. L. (1999). The seven sins of memory: Insights from psychology and cognitive neuroscience. *American Psychologist, 54,* 182–203.

Schacter, D. L., & Addis, D. R. (2007). The cognitive neuroscience of constructive memory: Remembering the past and imagining the future. *Philosophical Transactions of the Royal Society of London B: Biological Sciences, 362*(1481), 773–786.

Schacter, D. L., & Badgaiyan, R. D. (2001). Neuroimaging of priming: New perspectives on implicit and explicit memory. *Current Directions in Psychological Science, 10,* 1–4.

Schank, R. C. (1977). Rules and topics in conversation. *Cognitive Science, 1,* 421–441.

Schank, R. C., & Abelson, R. P. (1977). *Scripts, plans, goals and understanding.* Hillsdale, NJ: Erlbaum.

Schiller, P. H. (1966). Developmental study of color-word interference. *Journal of Experimental Psychology, 72,* 105–108.

Schilling, H. E. H., Rayner, K., & Chumbley, J. I. (1998). Comparing naming, lexical decision, and eye fixation times: Word frequency effects and individual differences. *Memory & Cognition, 26,* 1270–1281.

Schmader, T. (2010). Stereotype threat deconstructed. *Current Directions in Psychological Science, 19,* 14–18.

Schmeichel, B. J. (2007). Attention control, memory updating, and emotion regularity temporarily reduce the capacity for executive control. *Journal of Experimental Psychology: General, 136,* 241–255.

Schmidt, J. R., & Thompson, V. A. (2008). "At least one" problem with "some" formal reasoning paradigms. *Memory & Cognition, 36,* 217–229.

Schmidt, S. R. (1985). Encoding and retrieval processes in the memory for conceptually distinctive events. *Journal of Experimental Psychology: Learning, Memory, and Cognition, 11,* 565–578.

Schmidt, S. R. (1994). Effects of humor on sentence memory. *Journal of Experimental Psychology: Learning, Memory, and Cognition, 20,* 953–967.

Schmidt, S. R. (2004). Autobiographical memories for the September 11th attacks: Reconstructive errors and emotional impairment of memory. *Memory & Cognition, 32,* 443–454.

Schmolck, H., Buffalo, E. A., & Squire, L. R. (2000). Memory distortions develop over time: Recollections of the O. J. Simpson trial verdict after 15 and 32 months. *Psychological Science, 11,* 39–45.

Schneider, V. I., Healy, A. F., & Bourne, L. E., Jr. (2002). What is learned under difficult conditions is hard to forget: Contextual interference effects in foreign vocabulary acquisition, retention, and transfer. *Journal of Memory and Language, 46,* 419–440.

Schneider, W., & Shiffrin. R. M. (1977). Controlled and automatic human information processing: I. Detection, search, and attention. *Psychological Review, 84,* 1–66.

Schnorr, J. A., & Atkinson, R. C. (1969). Repetition versus imagery instructions in the short- and long-term retention of paired associates. *Psychonomic Science, 15,* 183–184.

Schooler, J. W., Ohlsson, S., & Brooks, K. (1993). Thoughts beyond words: When language overshadows insight. *Journal of Experimental Psychology: General, 122,* 166–183.

Schrauf, R. W., & Rubin, D. C. (1998). Bilingual autobiographical memory in older adult immigrants: A test of cognitive explanations of the reminiscence bump and the linguistic encoding of memories. *Journal of Memory and Language, 39,* 437–457.

Schrauf, R. W., & Rubin, D. C. (2000). Internal languages of retrieval: The bilingual encoding of memories for the personal past. *Memory & Cognition, 28,* 616–623.

Schreiber, T. A., & Sergent, S. D. (1998). The role of commitment in producing misinformation effects in eyewitness memory. *Psychonomic Bulletin & Review, 5,* 443–448.

Schustack, M. W., Ehrlich, S. F., & Rayner, K. (1987). Local and global sources of contextual facilitation in reading. *Journal of Memory and Language, 26,* 322–340.

Schutzwohl, A. (1998). Surprise and schema strength. *Journal of Experimental Psychology: Learning, Memory, and Cognition, 24,* 1182–1199.

Schwartz, D. L., & Black, T. (1999). Inferences through imagined actions: Knowing by simulated doing. *Journal of Experimental Psychology: Learning, Memory, and Cognition, 25,* 116–136.

Schwarz, K. A., & Pfister, R. (2016). Scientific psychology in the 18th century: A historical rediscovery. *Perspectives on Psychological Science, 11*(3), 399–407.

Scott, G. G., O'Donnell, P. J., & Sereno, S. C. (2012). Emotion words affect eye fixations during reading. *Journal of Experimental Psychology: Learning, Memory, and Cognition, 38*(3), 783–792.

Scullin, M. K., & McDaniel, M. A. (2010). Remembering to execute a goal: Sleep on it! *Psychological Science, 21,* 1028–1035.

Seamon, J. G., Philbin, M. M., & Harrison, L. G. (2006). Do you remember proposing marriage to the Pepsi machine? False recollections from a campus walk. *Psychonomic Bulletin & Review, 13,* 752–756.

Searle, J. R. (1969). *Speech acts.* Cambridge, England: Cambridge University Press.

Sederberg, P. B., Schulze-Bonhage, A., Madsen, J. R., Bromfield, E. B., Litt, B., Brandt, A., et al. (2007). Gamma oscillations distinguish true from false memories. *Psychological Science, 18,* 927–932.

See, J. E., Howe, S. R., Warm, J. S., & Dember, W. N. (1995). Meta-analysis of the sensitivity decrement in vigilance. *Psychological Bulletin, 2,* 230–249.

Segel, E., & Boroditsky, L. (2011). Grammar in art. *Frontiers in Psychology, 1,* 244.

Sehulster, J. R. (1989). Content and temporal structure of autobiographical knowledge: Remembering twenty-five seasons at the Metropolitan Opera. *Memory & Cognition, 17,* 590–606.

Seidenberg, M. S. (1993). Connectionist models and cognitive theory. *Psychological Science, 4,* 228–235.

Seifert, C. M., Robertson, S. P., & Black, J. B. (1985). Types of inferences generated during reading. *Journal of Memory and Language, 24,* 405–422.

Selfridge, O. G. (1959). Pandemonium: A paradigm for learning. In D. V. Blake & A. M. Uttley (Eds.), *Proceedings of the symposium on the mechanisation of thought processes.* London, England: H. M. Stationery Office.

Senkfor, A. J., & Van Petten, C. (1998). Who said what? An event-related potential investigation of source and item memory. *Journal of Experimental Psychology: Learning, Memory, and Cognition, 24,* 1005–1025.

Serences, J. T., Shomstein, S., Leber, A. B., Golay, X., Egeth, H. E., & Yantis, S. (2005). Coordination of voluntary and stimulus-driven attentional control in human cortex. *Psychological Science, 16,* 114–122.

Sereno, S. C., Brewer, C. C., & O'Donnell, P. J. (2003). Context effects in word recognition: Evidence for early interactive processing. *Psychological Science, 14,* 328–333.

Shafir, E., & Tversky, A. (1992). Thinking through uncertainty: Nonconsequential reasoning and choice. *Cognitive Psychology, 24,* 449–474.

Shallice, T., Fletcher, P., & Dolan, R. (1998). The functional imaging of recall. In M. A. Conway, S. E. Gathercole, & C. Cornoldi (Eds.), *Theories of memory* (Vol. II, pp. 247–258). Hove, England: Psychology Press.

Shallice, T., & Warrington, E. K. (1970). Independent functioning of the verbal memory stores: A neuropsychological study. *The Quarterly Journal of Experimental Psychology, 22,* 261–273.

Shand, M. A. (1982). Sign-based short-term coding of American Sign Language signs and printed English words by congenitally deaf signers. *Cognitive Psychology, 14,* 1–12.

Sharkey, N. E., & Mitchell, D. C. (1985). Word recognition in a functional context: The use of scripts in reading. *Journal of Memory and Language, 24,* 253–270.

Sharot, T., Martorella, E. A., Delgado, M. R., & Phelps, E. A. (2007). How personal experience modulates the neural circuitry of memories of September 11. *Proceedings of the National Academy of Science, 104,* 389–394.

Shaw, J. S. III. (1996). Increases in eyewitness confidence resulting from postevent questioning. *Journal of Experimental Psychology: Applied, 2,* 126–146.

Shaw, J. S. III, Bjork, R. A., & Handal, A. (1995). Retrieval-induced forgetting in an eyewitness–memory paradigm. *Psychonomic Bulletin & Review, 2,* 249–253.

Shaywitz, S. E., Mody, M., & Shaywitz, B. A. (2006). Neural mechanisms in dyslexia. *Current Directions in Psychological Science, 15,* 278–281.

Shelton, A. L., & McNamara, T. P. (2001). Visual memories from nonvisual experiences. *Psychological Science, 12,* 343–347.

Shelton, J. T., Elliott, E. M., Matthews R. A., Hill, B. D., & Gouvier, W. D. (2010). The relationships of working memory, secondary memory, and general fluid intelligence: Working memory is special. *Journal of Experimental Psychology: Learning, Memory, and Cognition, 36,* 813–820.

Shepard, R. N., & Metzler, J. (1971). Mental rotation of three-dimensional objects. *Science, 171,* 701–703.

Sherman, G., & Visscher, P. K. (2002). Honeybee colonies achieve fitness through dancing. *Nature, 419,* 920–922.

Shiffrin, R. M., & Schneider, W. (1977). Controlled and automatic human information processing: II. Perceptual learning, automatic attending, and a general theory. *Psychological Review, 84,* 127–190.

Shimamura, A. P., Berry, J. M., Mangels, J. A., Rusting, C. L., & Jurica, P. J. (1995). Memory and cognitive abilities in university professors: Evidence for successful aging. *Psychological Science, 6*(5), 271–277.

Shintel, H., & Keysar, B. (2007). You said it before and you'll say it again: Expectations and consistency in communication. *Journal of Experimental Psychology: Learning, Memory, and Cognition, 33,* 357–369.

Shipstead, Z., Harrison, T. L., & Engle, R. W. (2015). Working memory capacity and the scope and control of attention. *Attention, Perception, & Psychophysics, 77*(6), 1863–1880.

Shors, T. J. (2014). The adult brain makes new neurons, and effortful learning keeps them alive. *Current Directions in Psychological Science, 23*(5), 311–318.

Siegler, R. S. (2000). Unconscious insights. *Current Directions in Psychological Science, 9,* 79–83.

Siegler, R. S., & Booth, J. L. (2004). Development of numerical estimation in young children. *Child Development, 75*(2), 428–444.

Siegler, R. S., & Stern, E. (1998). A microgenetic analysis of conscious and unconscious strategy discoveries. *Journal of Experimental Psychology: General, 127,* 377–397.

Simion, F., Regolin, L., & Bulf, H. (2008). A predisposition for biological motion in the newborn baby. *Proceedings of the National Academy of Sciences, 105,* 809–813.

Simon, H. A. (1975). The functional equivalence of problem solving skills. *Cognitive Psychology, 7,* 268–288.

Simon, H. A. (1979). *Models of thought.* New Haven, CT: Yale University Press.

Simon, H. A. (May, 1995). *Thinking in words, pictures, equations, numbers: How do we do it and what does it matter?* Invited address presented at the meeting of the Midwestern Psychological Association, Chicago, IL.

Simons, D. J., & Ambinder, M. S. (2005). Change blindness: Theory and consequences. *Current Directions in Psychological Science, 14,* 44–48.

Simpson, G. B. (1981). Meaning dominance and semantic context in the processing of lexical ambiguity. *Journal of Verbal Learning and Verbal Behavior, 20,* 120–136.

Simpson, G. B. (1984). Lexical ambiguity and its role in models of word recognition. *Psychological Bulletin, 96,* 316–340.

Simpson, G. B., Casteel, M. A., Peterson, R. R., & Burgess, C. (1989). Lexical and sentence context effects in word recognition. *Journal of Experimental Psychology: Learning, Memory, and Cognition, 15,* 88–97.

Singer, M. (1990). *Psychology of language: An introduction to sentence and discourse processes.* Hillsdale, NJ: Erlbaum.

Singer, M., Andrusiak, P., Reisdorf, P., & Black, N. L. (1992). Individual differences in bridging inference processes. *Memory & Cognition, 20,* 539–548.

Singer, M., Graesser, A. C., & Trabasso, T. (1994). Minimal or global inference during reading. *Journal of Memory and Language, 33,* 421–441.

Sio, U. N., & Ormerod, T. C. (2009). Does incubation enhance problem solving? A meta-analytic review. *Psychological Bulletin, 135*(1), 94–120.

Sitaran, N., Weingartner, H., Caine, E. D., & Gillin, J. C. (1978). Choline: Selective enhancement of serial learning and encoding of low imagery words in man. *Life Sciences, 22,* 1555–1560.

Sitton, M., Mozer, M. C., & Farah, M. J. (2000). Superadditive effects of multiple lesions in a connectionist architecture: Implications for the neuropsychology of optic aphasia. *Psychological Review, 107,* 709–734.

Skinner, B. F. (1957). *Verbal behavior.* New York, NY: Appleton-Century-Crofts.

Slamecka, N. J. (1968). An examination of trace storage in free recall. *Journal of Experimental Psychology, 4,* 504–513.

Slamecka, N. J. (1985). Ebbinghaus: Some associations. *Journal of Experimental Psychology: Learning, Memory, and Cognition, 11,* 414–435.

Slamecka, N. J., & Graf, P. (1978). The generation effect: Delineation of a phenomenon. *Journal of Experimental Psychology: Human Learning and Memory, 4,* 592–604.

Sloman, S. A. (1998). Categorical inference is not a tree: The myth of inheritance hierarchies. *Cognitive Psychology, 35,* 1–33.

Small, B. J., Dixon, R. A., Hultsch, D. F., & Hertzog, C. (1999). Longitudinal changes in quantitative and qualitative indicators of word and story recall in young-old and old-old adults. *The Journals of Gerontology: Psychological Sciences, 54B*, P107–P115.

Smallwood, J., Davies, J. B., Heim, D., Finnigan, F., Sudberry, M. V., O'Connor, R. C., et al. (2004). Subjective experience and the attentional lapse. Task engagement and disengagement during sustained attention. *Consciousness and Cognition, 4*, 657–690.

Smallwood, J., Fishman, D. J., & Schooler, J. W. (2007). Counting the cost of an absent mind: Mind wandering as an unrecognized influence on educational performance. *Psychonomic Bulletin & Review, 14*, 230–236.

Smallwood, J., McSpadden, M., & Schooler, J. W. (2007). The lights are on but no one's home: Meta-awareness and the decoupling of attention when the mind wanders. *Psychonomic Bulletin & Review, 14*, 527–533.

Smallwood, J., McSpadden, M., & Schooler, J. W. (2008). When attention matters: The curious incident of the wandering mind. *Memory & Cognition, 36*(6), 1144–1150.

Smallwood, J., & Schooler, J. W. (2006). The restless mind. *Psychological Bulletin, 132*, 946–958.

Smilek, D., Carriere, J. S. A., & Cheyne, J. A. (2010). Out of mind, out of sight: Eye blinking as indicator and embodiment of mind wandering. *Psychological Science, 21*, 786–789.

Smilek, D., Dixon, M. J., Cudahy, C., & Merikle, P. M. (2002). Synesthetic color experiences influence memory. *Psychological Science, 13*, 548–552.

Smith, D. A., & Graesser, A. C. (1981). Memory for actions in scripted activities as a function of typicality, retention interval, and retrieval task. *Memory & Cognition, 9*, 550–559.

Smith, E. E. (2000). Neural bases of human working memory. *Current Directions in Psychological Science, 9*, 45–49.

Smith, E. E., & Jonides, J. (1999). Storage and executive processes in the frontal lobes. *Science, 283*, 1657–1661.

Smith, E. E., Rips, L. J., & Shoben, E. J. (1974). Semantic memory and psychological semantics. In G. H. Bower (Ed.), *The psychology of learning and motivation* (Vol. 8, pp. 1–45). New York, NY: Academic Press.

Smith, J. D., Redford, J. S., Washburn, D. A., & Taglialatela, L. A. (2005). Specific-token effects in screening tasks: Possible implications for aviation security. *Journal of Experimental Psychology: Learning, Memory, and Cognition, 31*, 1171–1185.

Smith, L. (2003). Learning to recognize objects. *Psychological Science, 14*, 244–250.

Smith, M. C., Bentin, S., & Spalek, T. M. (2001). Attention of semantic activation during visual word recognition. *Journal of Experimental Psychology: Learning, Memory, and Cognition, 27*, 1289–1298.

Smith, R. E. (2003). The cost of remembering to remember in event-based prospective memory: Investigating the capacity demands of delayed intention performance. *Journal of Experimental Psychology: Learning, Memory, and Cognition, 29*, 347–361.

Smith, R. E., & Bayen, U. J. (2004). A multinomial model of event-based prospective memory. *Journal of Experimental Psychology: Learning, Memory, and Cognition, 30*, 756–777.

Smith, R. E., Lozito, J. P., & Bayen, U. J. (2005). Adult age differences in distinctive processing: The modality effect on false recall. *Psychology and Aging, 20*, 486–492.

Smith, S. M. (1995). Getting into and out of mental ruts: A theory of fixation, incubation, and insight. In R. J. Sternberg & J. E. Davidson (Eds.), *The nature of insight* (pp. 229–251). Cambridge, MA: MIT Press.

Smith, S. W., Rebok, G. W., Smith, W. R., Hall, S. E., & Alvin, M. (1983). Adult age differences in the use of story structure in delayed free recall. *Experimental Aging Research, 9*(3), 191–195.

Snow, C. (1972). Mother's speech to children learning language. *Child Development, 43*, 549–565.

Snow, C., & Ferguson, C. (Eds.). (1977). *Talking to children: Language input and acquisition.* Cambridge, England: Cambridge University Press.

Son, L. K. (2004). Spacing one's study: Evidence for a metacognitive control strategy. *Journal of Experimental Psychology: Learning, Memory, and Cognition, 30*, 601–604.

Son, L. K., & Metcalfe, J. (2000). Metacognitive and control strategies in study-time allocation. *Journal of Experimental Psychology: Learning, Memory, and Cognition, 26*, 204–221.

Song, H., & Schwarz, N. (2009). If it's difficult to pronounce, it must be risky: Fluency, familiarity, and risk perception. *Psychological Science, 20*, 135–138.

Spaniol, J., Madden, D. J., & Voss, A. (2006). A diffusion model analysis of adult age differences in episodic and semantic long-term memory retrieval. *Journal of Experimental Psychology: Learning, Memory, and Cognition, 32*, 101–117.

Speer, N. K., Zacks, J. M., & Reynolds, J. R. (2007). Human brain activity time-locked to narrative event boundaries. *Psychological Science, 18*, 449–455.

Speer, S. R., & Clifton, C., Jr. (1998). Plausibility and argument structure in sentence comprehension. *Memory & Cognition, 26*, 965–978.

Spelke, E., Hirst, W., & Neisser, U. (1976). Skills of divided attention. *Cognition, 4*, 215–230.

Spellman, B. A., & Busey, T. A. (2010). Emerging trends in psychology and law: An editorial overview. *Psychonomic Bulletin & Review, 17*, 141–142.

Spellman, B. A., & Holyoak, K. J. (1996). Pragmatics in analogical mapping. *Cognitive Psychology, 31*, 307–346.

Spellman, B. A., & Mandel, D. R. (1999). When possibility informs reality: Counterfactual thinking as a cue to causality. *Current Directions in Psychological Science, 8*, 120–123.

Spence, C., & Read, L. (2003). Speech shadowing while driving: On the difficulty of splitting attention between eye and ear. *Psychological Science, 14*, 251–256.

Sperling, G. (1960). The information available in brief visual presentations. *Psychological Monographs, 74*(Whole no. 48).

Sperling, G. (1963). A model for visual memory tasks. *Human Factors, 5*, 9–31.

Sperry, R. W. (1964). The great cerebral commissure. *Scientific American, 210*, 42–52.

Spieth, W., Curtis, J. F., & Webster, J. C. (1954). Responding to one of two simultaneous messages. *Journal of the Acoustical Society of America, 26*, 391–396.

Spilich, G. J. (1983). Life-span components of text processing: Structural and procedural differences. *Journal of Verbal Learning and Verbal Behavior, 22,* 231–244.

Spivey, M. J., Tanenhaus, M. K., Eberhard, K. M., & Sedivy, J. C. (2002). Eye movements and spoken language comprehension: Effects of visual context on syntactic ambiguity resolution. *Cognitive Psychology, 45,* 447–481.

Spreng, R. N., & Levine, B. (2006). The temporal distribution of past and future autobiographical events across the lifespan. *Memory & Cognition, 34*(8), 1644–1651.

Squire, L. R. (1986). Mechanisms of memory. *Science, 232*(4578), 1612–1619.

Squire, L. R. (1987). *Memory and brain.* New York, NY: Oxford University Press.

Squire, L. R. (1993). The organization of declarative and nondeclarative memory. In T. Ono, L. R. Squire, M. E. Raichle, D. I. Perrett, & M. Fukuda (Eds.), *Brain mechanisms of perception and memory: From neuron to behavior* (pp. 219–227). New York, NY: Oxford University Press.

Stadler, M. A., Roediger, H. L. III., & McDermott, K. B. (1999). Norms for word lists that create false memories. *Memory & Cognition, 27,* 494–500.

Stallings, L. M., MacDonald, M. C., & O'Seaghdha, P. G. (1998). Phrasal ordering constraints in sentence production: Phrase length and verb disposition in heavy-NP shift. *Journal of Memory and Language, 39,* 392–417.

Standing, L. (1973). Learning 10000 pictures. *The Quarterly Journal of Experimental Psychology, 25*(2), 207–222.

Stark, L., & Perfect, T. J. (2008). The effects of repeated idea elaboration on unconscious plagiarism. *Memory & Cognition, 36,* 65–73.

Stawarczyk, D., Majerus, S., Catale, C., & D'Argembeau, A. (2014). Relationships between mind-wandering and attentional control abilities in young adults and adolescents. *Acta Psychologica, 148,* 25–36.

Stazyk, E. H., Ashcraft, M. H., & Hamann, M. S. (1982). A network approach to simple multiplication. *Journal of Experimental Psychology: Learning, Memory, and Cognition, 17,* 355–376.

Stebbins, G. T., Carrillo, M. C., Dorfman, J., Dirksen, C., Desmond, J. E., Turner, D. A., et al. (2002). Aging effects on memory encoding in the frontal lobes. *Psychology and Aging, 17,* 44–55.

Steblay, N. M. (1992). A meta-analytic review of the weapon focus effect. *Law and Human Behavior, 16,* 413–424.

Stefanucci, J. K., & Storbeck, J. (2009). Don't look down: Emotional arousal elevates height perception. *Journal of Experimental Psychology: General, 131,* 131–145.

Stern, Y. (2009). Cognitive reserve. *Neuropsychologia, 47*(10), 2015–2028.

Sternberg, R. J. (1996). *Cognitive psychology.* Fort Worth, TX: Harcourt Brace.

Sternberg, S. (1966). High-speed scanning in human memory. *Science, 153,* 652–654.

Sternberg, S. (1969). The discovery of processing stages: Extensions of Donders's method. In W. G. Koster (Ed.), Attention and performance II. *Acta Psychologica, 30,* 276–315.

Sternberg, S. (1975). Memory scanning: New findings and current controversies. *The Quarterly Journal of Experimental Psychology, 27,* 1–32.

Stickgold, R. (2013). Parsing the role of sleep in memory processing. *Current Opinion in Neurobiology, 23*(5), 847–853.

Stickgold, R., & Walker, M. P. (2005). Memory consolidation and reconsolidation: What is the role of sleep? *Trends in Neurosciences, 28*(8), 408–415.

Stine, E. A. L., Cheung, H., & Henderson, D. (1995). Adult age differences in the on-line processing of new concepts in discourse. *Aging & Cognition, 2,* 1–18.

Stine, E. A. L., & Hindman, J. (1994). Age differences in reading time allocation for propositionally dense sentences. *Aging & Cognition, 1,* 2–16.

Stine, E. A. L., & Wingfield, A. (1990). How much do working memory deficits contribute to age differences in discourse memory? *European Journal of Cognitive Psychology, 2,* 289–304.

Stine-Morrow, E. A. L., Gagne, D. D., Morrow, D. G., & DeWall, B. H. (2004). Age differences in rereading. *Memory & Cognition, 32,* 696–710.

Storm, B. C. (2011). The benefit of forgetting in thinking and remembering. *Current Directions in Psychological Science, 20,* 291–295.

Stothart, C., Mitchum, A., & Yehnert, C. (2015). The attentional cost of receiving a cell phone notification. *Journal of Experimental Psychology: Human Perception and Performance, 41*(4), 893–897.

Strayer, D. L., & Drews, F. A. (2007). Cell-phone-induced driver distraction. *Current Directions in Psychological Science, 16,* 128–131.

Strayer, D. L., & Johnston, W. A. (2001). Driven to distraction: Dual-task studies of stimulated driving and conversing on a cellular phone. *Psychological Science, 12,* 462–466.

Stroop, J. R. (1935). Studies of interference in serial verbal reactions. *Journal of Experimental Psychology, 18,* 643–662.

Struiksma, M. E., Noordzij, M. L., & Postma, A. (2011). Embodied representation of the body contains veridical spatial information. *The Quarterly Journal of Experimental Psychology, 64,* 1124–1137.

Suh, S. Y., & Trabasso, T. (1993). Inferences during reading: Converging evidence from discourse analysis, talk-aloud protocols, and recognition priming. *Journal of Memory and Language, 32,* 279–300.

Sulin, R. A., & Dooling, D. J. (1974). Intrusion of a thematic idea in retention of prose. *Journal of Experimental Psychology, 103,* 255–262.

Sullivan, S. J., Mikels, J. A., & Carstensen, L. L. (2010). You never lose the ages you've been: Affective perspective taking in older adults. *Psychology and Aging, 25,* 229–234.

Sun, X., Punjabi, P. V., Greenberg, L. T., & Seamon, J. G. (2009). Does feigning amnesia impair subsequent recall? *Memory & Cognition, 37,* 81–89.

Svartvik, J., & Quirk, R. (Eds.). (1980). *A corpus of English conversation.* Lund, Sweden: CWK Gleerup.

Symons, C. S., & Johnson, B. T. (1997). The self-reference effect in memory: A meta-analysis. *Psychological Bulletin, 121,* 371–394.

Szpunar, K. K. (2010). Episodic future thought: An emerging concept. *Perspectives on Psychological Science, 5*(2), 142–162.

Szpunar, K. K., Moulton, S. T., & Schacter, D. L. (2013). Mind wandering and education: From the classroom to online learning. *Frontiers in Psychology, 4*, 495.

Szpunar, K. K., & Radvansky, G. A. (2016). Cognitive approaches to the study of episodic future thinking. *The Quarterly Journal of Experimental Psychology, 69*(2), 209–216.

Szpunar, K. K., & Schacter, D. L. (2013). Get real: Effects of repeated simulation and emotion on the perceived plausibility of future experiences. *Journal of Experimental Psychology: General, 142*(2), 323–327.

Tabor, W., & Hutchins, S. (2004). Evidence for self-organized sentence processing: Digging-in effects. *Journal of Experimental Psychology: Learning, Memory, and Cognition, 30*, 431–450.

Takahashi, M., Shimizu, H., Saito, S., & Tomoyori, H. (2006). One percent ability and ninety-nine percent perspiration: A study of a Japanese memorist. *Journal of Experimental Psychology: Learning, Memory, and Cognition, 32*, 1195–1200.

Talarico, J. M., LaBar, K. S., & Rubin, D. C. (2004). Emotional intensity predicts autobiographical memory experience. *Memory & Cognition, 32*, 1118–1132.

Talarico, J. M., & Rubin, D. C. (2003). Confidence, not consistency, characterizes flashbulb memories. *Psychological Science, 14*, 455–461.

Talmi, D., & Garry, L. M. (2012). Accounting for immediate emotional memory enhancement. *Journal of Memory and Language, 66*, 93–108.

Talmi, J. M., Schimmack, U., Paterson, T., & Moscovitch, M. (2007). The role of attention and relatedness in emotionally enhanced memory. *Emotion, 7*, 89–102.

Tamplin, A. K., Krawietz, S. A., Radvansky, G. A., & Copeland, D. E. (2013). Event memory and moving in a well-known environment. *Memory & Cognition, 41*(8), 1109–1121.

Tanaka, J. T., & Curran, T. (2001). A neural basis for expert object recognition. *Psychological Science, 12*, 43–47.

Tanenhaus, M. K., & Trueswell, J. C. (1995). Sentence comprehension. In J. Miller & P. Eiman (Eds.), *Handbook of perception and cognition: Speech, language, and communication* (2nd ed., Vol. 11, pp. 217–262). San Diego, CA: Academic Press.

Tang, Y., Zhang, W., Chen, K., Feng, S., Ji, Y., Shen, J., et al. (2006). Arithmetic processing in the brain shaped by cultures. *Proceedings of the National Academy of Sciences, 103*, 10775–10780.

Tanner, W. P., Jr., & Swets, J. A. (1954). A decision-making theory of visual detection. *Psychological Review, 61*, 401–409.

Taraban, R., & McClelland, J. L. (1988). Constituent attachment and thematic role assignment in sentence processing: Influences of content-based expectations. *Journal of Memory and Language, 27*, 597–632.

Taylor, S. F., Liberzon, I., & Koeppe, R. A. (2000). The effect of graded aversive stimuli on limbic and visual activation. *Neuropsychologia, 38*, 1415–1425.

Teller, D. Y., Morse, R., Borton, R., & Regal, D. (1974). Visual acuity for vertical and diagonal gratings in human infants. *Vision Research, 14*, 1433–1439.

Thapar, A., & Greene, R. L. (1994). Effects of level of processing on implicit and explicit tasks. *Journal of Experimental Psychology: Learning, Memory, and Cognition, 20*, 671–679.

Theeuwes, J., Kramer, A. F., Hahn, S., & Irwin, D. E. (1998). Our eyes do not always go where we want them to go: Capture of the eyes by new objects. *Psychological Science, 9*, 379–385.

Thiede, K. W. (1999). The importance of monitoring and self-regulation during multitrial learning. *Psychonomic Bulletin & Review, 6*, 662–667.

Thomas, A. K, Bulevich, J. B., & Loftus, E. F. (2003). Exploring the role of repetition and sensory elaboration in the imagination inflation effect. *Memory & Cognition, 31*, 630–640.

Thomas, A. K., & Dubois, S. J. (2011). Reducing the burden of stereotype threat eliminates age differences in memory distortion. *Psychological Science, 22*, 1515–1517.

Thomas, J. C., Jr. (1974). An analysis of behavior in the hobbits–orcs problem. *Cognitive Psychology, 6*, 257–269.

Thomas, L. E. (2013). Spatial working memory is necessary for actions to guide thought. *Journal of Experimental Psychology: Learning, Memory, and Cognition, 39*(6), 1974–1981.

Thomas, M. H., & Wang, A. Y. (1996). Learning by the keyword mnemonic: Looking for long-term benefits. *Journal of Experimental Psychology: Applied, 2*, 330–342.

Thompson, R. F. (1986). The neurobiology of learning and memory. *Science, 233*, 941–947.

Thompson, V. A., & Byrne, R. M. J. (2002). Reasoning counterfactually: Making inferences about things that didn't happen. *Journal of Experimental Psychology: Learning, Memory, and Cognition, 28*, 1154–1170.

Thomson, D. M., & Tulving, E. (1970). Associative encoding and retrieval: Weak and strong cues. *Journal of Experimental Psychology, 86*, 255–262.

Thorndike, E. L. (1914). *The psychology of learning*. New York, NY: Teachers College.

Tillman, B., & Dowling, W. J. (2007). Memory decreases for prose but not for poetry. *Memory & Cognition, 35*, 628–639.

Tipper, S. P. (1985). The negative priming effect: Inhibitory priming with to be ignored objects. *The Quarterly Journal of Experimental Psychology, 37A*, 571–590.

Tipper, S. P. (2010). From observation to action simulation: The role of attention, eye-gaze, emotion, and body state. *The Quarterly Journal of Experimental Psychology, 63*, 2081–2105.

Tolman, E. C. (1948). Cognitive maps in rats and men. *Psychological Review, 55*, 189–208.

Tourangeau, R., & Rips, L. (1991). Interpreting and evaluating metaphors. *Journal of Memory and Language, 30*, 452–472.

Townsend, J. T., & Wenger, M. J. (2004). The serial–parallel dilemma: A case study in a linkage of theory and method. *Psychonomic Bulletin & Review, 11*, 391–418.

Trabasso, T., & Bartolone, J. (2003). Story understanding and counterfactual reasoning. *Journal of Experimental Psychology: Learning, Memory, and Cognition, 29*, 904–923.

Trawley, S. L., Law, A. S., & Logie, R. H. (2011). Event-based prospective remembering in a virtual world. *The Quarterly Journal of Experimental Psychology, 64*, 2181–2193.

Treccani, B., Argyri, E., Sorace, A., & Della Salla, S. (2009). Spatial negative priming in bilingualism. *Psychonomic Bulletin & Review, 16*, 320–327.

Treisman, A. M. (1960). Contextual cues in selective listening. *The Quarterly Journal of Experimental Psychology, 12*, 242–248.

Treisman, A. M. (1964). Monitoring and storage of irrelevant messages in selective attention. *Journal of Verbal Learning and Verbal Behavior, 3*, 449–459.

Treisman, A. M. (1965). The effects of redundancy and familiarity on translating and repeating back a foreign and a native language. *British Journal of Psychology, 56*, 369–379.

Treisman, A. (1982). Perceptual grouping and attention in visual search for features and for objects. *Journal of Experimental Psychology: Human Perception and Performance, 8*, 194–214.

Treisman, A. (1988). Features and objects: The Fourteenth Bartlett Memorial Lecture. *The Quarterly Journal of Experimental Psychology, 40A*, 201–237.

Treisman, A. (1991). Search, similarity, and integration of features between and within dimensions. *Journal of Experimental Psychology: Human Perception and Performance, 17*, 652–676.

Treisman, A., & Gelade, G. (1980). A feature integration theory of attention. *Cognitive Psychology, 12*, 97–136.

Treisman, A. M., Russell, R., & Green, J. (1975). Brief visual storage of shape and movement. In P. M. A. Rabbitt & S. Dornic (Eds.), *Attention and performance* (Vol. 5, pp. 699–721). New York, NY: Academic Press.

Trigg, G. L., & Lerner, R. J. (Eds.). (1981). *Encyclopedia of physics*. Reading, MA: Addison-Wesley.

Tucker, M., & Ellis, R. (1998). On the relations between seen objects and components of potential actions. *Journal of Experimental Psychology: Human Perception and Performance, 24*, 830–846.

Tuholski, S. W., Engle, R. W., & Baylis, G. C. (2001). Individual differences in working memory capacity and enumeration. *Memory & Cognition, 29*, 484–492.

Tulving, E. (1962). Subjective organization in free recall of "unrelated" words. *Psychological Review, 69*, 344–354.

Tulving, E. (1972). Episodic and semantic memory. In E. Tulving & W. Donaldson (Eds.), *Organization of memory* (pp. 381–403). New York, NY: Academic Press.

Tulving, E. (1983). *Elements of episodic memory*. Oxford, England: Clarendon.

Tulving, E. (1989). Remembering and knowing the past. *American Scientist, 77*, 361–367.

Tulving, E. (1993). What is episodic memory? *Current Directions in Psychological Science, 2*, 67–70.

Tulving, E., & Pearlstone, Z. (1966). Availability versus accessibility of information in memory for words. *Journal of Verbal Learning and Verbal Behavior, 5*, 381–391.

Tulving, E., & Thompson, D. M. (1973). Encoding specificity and retrieval processes in episodic memory. *Psychological Review, 80*, 352–373.

Tun, P. A. (1989). Age differences in processing expository and narrative text. The *Journals of Gerontology: Psychological Sciences, 44*, P9–P15.

Turner, M. L., & Engle, R. W. (1989). Is working memory capacity task dependent? *Journal of Memory and Language, 28*, 127–154.

Tversky, A. (1972). Elimination by aspects: A theory of choice. *Psychological Review, 79*(4), 281–299.

Tversky, A., & Kahneman, D. (1973). Availability: A heuristic for judging frequency and probability. *Cognitive Psychology, 5*, 207–232.

Tversky, A., & Kahneman, D. (1974). Judgment under uncertainty: Heuristics and biases. *Science, 185*, 1124–1131.

Tversky, A., & Kahneman, D. (1980). Causal schemas in judgments under uncertainty. In M. Fishbein (Ed.), *Progress in social psychology* (Vol. 1, pp. 49–72). Hillsdale, NJ: Erlbaum.

Tversky, A., & Kahneman, D. (1981). The framing of decisions and the psychology of choice. *Science, 211*, 453-458.

Tversky, A., & Kahneman, D. (1983). Extensional versus intuitive reasoning: The conjunction fallacy in probability judgment. *Psychological Review, 90*, 293–315.

Tyler, L. K., Voice, J. K., & Moss, H. E. (2000). The interaction of meaning and sound in spoken word recognition. *Psychonomic Bulletin & Review, 7*, 320–326.

Underwood, B. J. (1957). Interference and forgetting. *Psychological Review, 64*, 49–60.

Underwood, B. J., Keppel, G., & Schulz, R. W. (1962). Studies of distributed practice: XXII. Some conditions which enhance retention. *Journal of Experimental Psychology, 64*, 112–129.

Underwood, B. J., & Schulz, R. W. (1960). *Meaningfulness and verbal learning*. Philadelphia, PA: Lippincott.

Ungerleider, L. G., & Haxby, J. V. (1994). "What" versus "where" in the human brain. *Current Opinion in Neurobiology, 4*, 157–165.

Unsworth, N. (2007). Individual differences in working memory capacity and episodic retrieval: The dynamics of delayed and continuous distractor free recall. *Journal of Experimental Psychology: Learning, Memory, and Cognition, 33*, 1020–1034.

Unsworth, N., & Engle, R. W. (2007). The nature of individual differences in working memory capacity: Active maintenance in primary memory and controlled search from secondary memory. *Psychological Review, 114*, 104–132.

Unsworth, N., Heitz, R. P., & Parks, N. A. (2008). The importance of temporal distinctiveness for forgetting over the short-term. *Psychological Science, 19*, 1078–1081.

Unsworth, N., & Spillers, G. J, (2010). Working memory capacity: Attention control, secondary memory, or both? A direct test of the dual component model. *Journal of Memory and Language, 62*, 392–406.

Unsworth, N., Spillers, G. J., & Brewer, G. A. (2012). Dynamics of context-dependent recall: An examination of internal and external context change. *Journal of Memory and Language, 66*, 1–16.

Vachon, F., Hughes, R. W., & Jones, D. M. (2012). Broken expectations: Violation of expectancies, not novelty, captures auditory attention. *Journal of Experimental Psychology: Learning, Memory, and Cognition, 38*, 164–177.

Vallar, G., & Baddeley, A. D. (1984). Fractionation of working memory: Neuropsychological evidence for a phonological short-term store. *Journal of Verbal Learning and Verbal Behavior, 23*, 151–161.

Vallée-Tourangeau, F., Euden, G., & Hearn, V. (2011). Einstellung defused: Interactivity and mental set. *The Quarterly Journal of Expermental Psychology, 64*, 1889–1895.

Van Berkum, J. J. A. (2008). Understanding sentences in context: What brain waves can tell us. *Current Directions in Psychological Science, 17*, 376–380.

Van Berkum, J. J. A., Brown, C. M., & Hagoort, P. (1999). Early referential context effects in sentence processing: Evidence from event-related brain potentials. *Journal of Memory and Language, 41*, 147–182.

Vandierendonck, A. (2016). A working memory system with distributed executive control. *Perspectives on Psychological Science, 11*(1), 74–100.

van Dijk, T. A., & Kintsch, W. (1983). *Strategies in discourse comprehension*. New York, NY: Academic Press.

van Elk, M., & Blanke, O. (2011). The relation between body semantics and spatial body representations. *Acta Psychologica, 138*, 347–358.

van Hell, J. G., & Dijkstra, T. (2002). Foreign language knowledge can influence native language performance in exclusively native contexts. *Psychonomic Bulletin & Review, 9*, 780–789.

VanLehn, K. (1989). Problem solving and cognitive skill acquisition. In M. I. Posner (Ed.), *Foundations of cognitive science* (pp. 527–579). Cambridge, MA: MIT Press.

Van Overschelde, J. P., Rawson, K. A., & Dunlosky, J. (2004). Category norms: An updated and expanded version of the Battig and Montague (1969) norms. *Journal of Memory and Language, 50*, 289–335.

Vasta, R., & Liben, L. S. (1996). The water-level task: An intriguing puzzle. *Current Directions in Psychological Science, 5*, 171–177.

Vecera, S. P., Behrmann, M., & McGoldrick, J. (2000). Selective attention to the parts of an object. *Psychonomic Bulletin & Review, 7*, 301–308.

Verfaille, K., & Y'dewalle, G. (1991). Representational momentum and event course anticipation in the perception of implied motions. *Journal of Experimental Psychology: Learning, Memory, and Cognition, 17*, 302–313.

Vergauwe, E., Barrouillet, P., & Camos, V. (2010). Do mental processes share a domain-general resource? *Psychological Science, 21*, 384–390.

Vergauwe, E., Gauffroy, C., Morsanyi, K., Dagry, I., & Barrouillet, P. (2013). Chronometric evidence for the dual-process mental model theory of conditional. *Journal of Cognitive Psychology, 25*(2), 174–182.

Verhaeghen, P. (2011). Aging and executive control: Reports of a demise greatly exaggerated. *Psychological Science, 20*, 174–180.

Verhaeghen, P., Cerella, J., & Basak, C. (2004). A working memory workout: How to expand the focus of serial attention from one to four items in 10 hours or less. *Journal of Experimental Psychology: Learning, Memory, and Cognition, 30*, 1322–1337.

Verhaeghen P., & Meersman, L. D. (1998). Aging and the Stroop effect: A meta-analysis. *Psychology and Aging, 13*, 120–126.

Virtue, S., van den Broek, P., & Linderholm, T. (2006). Hemispheric processing of inferences: The effects of textual constraint and working memory capacity. *Memory & Cognition, 34*, 1341–1354.

Vivas, A. B., Humphreys, G. W., & Fuentes, L. J. (2006). Abnormal inhibition of return: A review and new data on patients with parietal lobe damage. *Cognitive Neuropsychology, 23*, 1049–1064.

von Frisch, K. (1967). *The dance language and orientation of honeybees*. Cambridge, MA: Harvard University Press.

von Restorff, H. (1933). Über die Wirkung von Bereichsbildungen im Spurenfeld [On the effect of sphere formations in the trace field]. *Psychologische Forschung, 18*, 299–342.

Voorspoels, W., Navarro, D. J., Perfors, A., Ransom, K., & Storms, G. (2015). How do people learn from negative evidence? Non-monotonic generalizations and sampling assumptions in inductive reasoning. *Cognitive Psychology, 81*, 1–25.

Vredeveldt, A., Hitch, G. J., & Baddeley, A. D. (2011). Eyeclosure helps memory by reducing cognitive load and enhancing visualization. *Memory & Cognition, 39*, 1253–1263.

Vu, H., Kellas, G., Metcalf, K., & Herman, R. (2000). The influence of global discourse on lexical ambiguity resolution. *Memory & Cognition, 28*, 236–252.

Vuilleumeir, P. (2005). How brains beware: Neural mechanisms of emotional attention. *Trends in Cognitive Sciences, 9*, 585–594.

Vuilleumeir, P., & Huang, Y-M. (2009). Emotional attention: Uncovering the mechanisms of affective biases in perception. *Current Directions in Psychological Science, 18*, 148–152.

Vul, E., & Pashler, H. (2007). Incubation benefits only after people have been misdirected. *Memory & Cognition, 35*, 701–710.

Wade, K. A., Garry, M., Read, J. D., & Lindsay, S. (2002). A picture is worth a thousand lies: Using false photographs to create false childhood memories. *Psychonomic Bulletin & Review, 9*, 597–603.

Wagar, B. M., & Thagard, P. (2004). Spiking Phineas Gage: A neurocomputational theory of cognitive-affective integration in decision making. *Psychological Review, 111*, 67–79.

Wagemans, J., Elder, J. H., Kubovy, M., Palmer, S. E., Peterson, M. A., Singh, M., & von der Heydt, R. (2012). A century of Gestalt psychology in visual perception: I. Perceptual grouping and figure–ground organization. *Psychological Bulletin, 138*(6), 1172–1217.

Wagemans, J., Feldman, J., Gepshtein, S., Kimchi, R., Pomerantz, J. R., van der Helm, P. A., & van Leeuwen, C. (2012). A century of Gestalt psychology in visual perception: II. Conceptual and theoretical foundations. *Psychological Bulletin, 138*(6), 1218–1252.

Wagenaar, W. A. (1986). My memory: A study of autobiographical memory over six years. *Cognitive Psychology, 18*, 225–252.

Wagner, S. H., & Walters, J. (1982). A longitudinal analysis of early number concepts: From numbers to number. In G. Forman (Ed.), *Action and thought: From sensorimotor schemes to symbolic operations* (pp. 137–161). New York, NY: Academic Press.

Waltz, J. A., Knowlton, B. J., Holyoak, K. J., Boone, K. B., Mishkin, F. S., Santos, M. M., et al. (1999). A system for relational reasoning in human prefrontal cortex. *Psychological Science, 10*, 119–125.

Waltz, J. A., Lau, A., Grewal, S. K., & Holyoak, K. J. (2000). The role of working memory in analogical mapping. *Memory & Cognition, 28*, 1205–1212.

Ward, J. (2008). *The frog who croaked blue*. New York, NY: Routledge.

Waris, O., Soveri, A., & Line, M. (2015). Transfer after working memory updating. *PloS one, 10*(9), e0138734.

Warm, J. S. (1984). An introduction to vigilance. In J. S. Warm (Ed.), *Sustained attention in human performance* (pp. 1–14). Chichester, England: Wiley.

Warm, J. S., & Jerison, H. J. (1984). The psychophysics of vigilance. In J. S. Warm (Ed.), *Sustained attention in human performance* (pp. 15–60). Chichester, England: Wiley.

Warren, R. M. (1970). Perceptual restoration of missing speech sounds. *Science, 167*, 392–393.

Warren, R. M., & Warren, R. P. (1970). Auditory illusions and confusions. *Scientific American, 223*, 30–36.

Warrington, E. K., & James, M. (1988). Visual apperceptive agnosia: A clinico-anatomical study of three cases. *Cortex, 24*, 13–32.

Warrington, E. K., & McCarthy, R. (1983). Category specific access dysphasia. *Brain, 106*, 859–878.

Warrington, E. K., & Shallice, T. (1969). The selective impairment of auditory verbal short-term memory. *Brain, 92*, 885–896.

Warrington, E. K., & Shallice, T. (1984). Category specific semantic impairments. *Brain, 107*, 829–854.

Warrington, E. K., & Weiskrantz, L. (1970). The amnesic syndrome: Consolidation or retrieval? *Nature, 228*, 628–630.

Wason, P. C., & Johnson-Laird, P. N. (1972). *Psychology of reasoning: Structure and content*. Cambridge, MA: Harvard University Press.

Wassenburg, S. I., & Zwaan, R. A. (2010). Readers routinely represent implied object rotation: The role of visual experience. *The Quarterly Journal of Experimental Psychology, 63*, 1665–1670.

Watson, J. B. (1913). Psychology as the behaviorist sees it. *Psychological Review, 20*, 158–177.

Watson, J. M., McDermott, K. B., & Balota, D. A. (2004). Attempting to avoid false memories in the Deese–Roediger–McDermott paradigm: Assessing the combined influence of practice and warnings in young and old adults. *Memory & Cognition, 32*, 135–141.

Watson, R. I. (1968). *The great psychologists from Aristotle to Freud*. New York, NY: Lippincott.

Waugh, N. C., & Norman, D. A. (1965). Primary memory. *Psychological Review, 72*, 89–104.

Weaver, C. A. III. (1993). Do you need a "flash" to form a flashbulb memory? *Journal of Experimental Psychology: General, 122*, 39–46.

Weinstein, Y., Bugg, J. M., & Roediger, H. L. (2008). Can the survival recall advantage be explained by basic memory processes? *Memory & Cognition, 36*, 913–919.

Weisberg, R. (1995). Prolegomena to theories of insight in problem solving: A taxonomy of problems. In R. J. Sternberg & J. E. Davidson (Eds.), *The nature of insight* (pp. 157–196). Cambridge, MA: MIT Press.

Wells, G. L., Olson, E. A., & Charman, S. D. (2002). The confidence of eyewitnesses in their identifications from lineups. *Current Directions in Psychological Science, 11*, 151–154.

Wenger, M. J., & Payne, D. G. (1995). On the acquisition of mnemonic skill: Application of skilled memory theory. *Journal of Experimental Psychology: Applied, 1*, 194–215.

Werner, H. (1935). Studies on contour. *American Journal of Psychology, 47*, 40–64.

Wertheimer, M. (1912). *Experimentelle Studien über das Sehen von Bewegung* [Experimental studies on the seeing of motion]. *Zeitschrift für Psychologie und Physiologie der Sinnesorgane, 61*, 161–265.

West, R. F., & Stanovich, K. E. (1986). Robust effects of syntactic structure on visual word processing. *Memory & Cognition, 14*, 104–112.

Westmacott, R., & Moscovitch, M. (2003). The contribution of autobiographical significance to semantic memory. *Memory & Cognition, 31*, 761–774.

Wetherick, N. E., & Gilhooly, K. J. (1995). "Atmosphere," matching, and logic in syllogistic reasoning. *Current Psychology, 14*(3), 169–178.

Whaley, C. P. (1978). Word–nonword classification time. *Journal of Verbal Learning and Verbal Behavior, 17*, 143–154.

Wharton, C. M., Grafman, J., Flitman, S. S., Hansen, E. K., Brauner, J., Marks, A., et al. (2000). Toward neuroanatomical models of analogy: A positron emission tomography study of analogical mapping. *Cognitive Psychology, 40*, 173–197.

Whitney, P. (1998). *The psychology of language*. Boston, MA: Houghton Mifflin.

Whitney, P., Rinehart, C. A., & Hinson, J. M. (2008). Framing effects under cognitive load: The role of working memory in risky decisions. *Psychonomic Bulletin & Review, 15*(6), 1179–1184.

Whorf, B. L. (1956). Science and linguistics. In J. B. Carroll (Ed.), *Language, thought, and reality: Selected writings of Benjamin Lee Whorf* (pp. 207–219). Cambridge, MA: MIT Press.

Wickelgren, W. A. (1974). *How to solve problems*. San Francisco, CA: Freeman.

Wickens, D. D. (1972). Characteristics of word encoding. In A. W. Melton & E. Martin (Eds.), *Coding processes in human memory* (pp. 191–215). New York, NY: Winston.

Wickens, D. D., Born, D. G., & Allen, C. K. (1963). Proactive inhibition and item similarity in short-term memory. *Journal of Verbal Learning and Verbal Behavior, 2*, 440–445.

Wiener, E. L. (1984). Vigilance and inspection. In J. S. Warm (Ed.), *Sustained attention in human performance* (pp. 207–246). Chichester, England: Wiley.

Wiggs, C. L., Weisberg, J., & Martin, A. (1999). Neural correlates of semantic and episodic memory retrieval. *Neuropsychologia, 37*, 103–118.

Wiley, J. (1998). Expertise as mental set: The effects of domain knowledge in creative problem solving. *Memory & Cognition, 26*, 716–730.

Wiley, J., & Rayner, K. (2000). Effects of titles on the processing of text and lexically ambiguous words: Evidence from eye movements. *Memory & Cognition, 28,* 1011–1021.

Wilkes-Gibbs, D., & Clark, H. H. (1992). Coordinating beliefs in conversation. *Journal of Memory and Language, 31,* 183–194.

Willander, J., & Larsson, M. (2006). Smell your way back to childhood. *Psychonomic Bulletin & Review, 13,* 240–244.

Willander, J., & Larsson, M. (2007). Olfaction and emotion: The case of autobiographical memory. *Memory & Cognition, 35,* 1659–1663.

Willems, R. M., Hagoort, P., & Casasanto, D. (2010). Body-specific representations of action verbs: Neural evidence from right- and left-handers. *Psychological Science, 21,* 67–74.

Willems, R. M. Labruna, L., D'Espisito, M., Ivry, R., & Casasanto, D. (2011). A functional role for the motor system in language understanding: Evidence from theta-burst transcranial magnetic stimulation. *Psychological Science, 22,* 849–854.

Williams, J. M. G., Mathews, A., & MacLeod, C. (1996). The emotional Stroop task and psychopathology. *Psychological Bulletin, 120,* 3–24.

Williams, R. W., & Herrup, K. (1988). The control of neuron number. *Annual Review of Neuroscience, 11,* 423–453.

Williams, S. E., & Horst, J. S. (2014). Goodnight book: Sleep consolidation improves word learning via storybooks. *Frontiers in Psychology, 5,* 184.

Williamson, V. J., Baddeley, A. D., & Hitch, G. J. (2010). Musicians' and nonmusicians' shortterm memory for verbal and musical sequences: Comparing phonological similarity and pitch proximity. *Memory & Cognition, 38,* 163–175.

Wilson, M., & Emmorey, K. (2006). Comparing sign language and speech reveals a universal limit on short-term memory capacity. *Psychological Science, 17,* 682–683.

Wilson, M., & Fox, G. (2007). Working memory for language is not special: Evidence for an articulatory loop for novel stimuli. *Psychonomic Bulletin & Review, 14,* 470–473.

Wilson, M., & Knoblich, G. (2005). The case for motor involvement in perceiving conspecifics. *Psychological Bulletin, 131,* 460–473.

Wilson, M., & Wilson, T. P. (2005). An oscillator model of the timing of turn-taking. *Psychonomic Bulletin & Review, 12,* 957–968.

Windschitl, P. D., & Weber, E. U. (1999). The interpretation of "likely" depends on the context, but "70%" is 70%—right? The influence of associative processes on perceived certainty. *Journal of Experimental Psychology: Learning, Memory, and Cognition, 25,* 1514–1533.

Winer, G. A., Cottrell, J. E., Gregg, V., Fournier, J. S., & Bica, L. A. (2002). Fundamentally misunderstanding visual perception: Adults' belief in visual emissions. *American Psychologist, 57,* 417–424.

Wingfield, A., Goodglass, H., & Lindfield, K. C. (1997). Separating speed from automaticity in a patient with focal brain atrophy. *Psychological Science, 8,* 247–249.

Winograd, E., & Killinger, W. A., Jr. (1983). Relating age at encoding in early childhood to adult recall: Development of flashbulb memories. *Journal of Experimental Psychology: General, 112,* 413–422.

Witt, J. K. (2011). Action's effect on perception. *Current Directions in Psychological Science, 20,* 201–206.

Witt, J. K., Proffitt, D. R., & Epstein, W. (2005). Tool use affects perceived distance, but only when you intend to use it. *Journal of Experimental Psychology: Human Perception and Performance, 31,* 880–888.

Wixted, J. T. (2004). On common ground: Jost's (1897) law of forgetting and Ribot's (1881) law of retrograde amnesia. *Psychological Review, 111,* 864–879.

Wixted, J. T. (2005). A theory about why we forget what we once knew. *Current Directions in Psychological Science, 14,* 6–9.

Wixted, J. T., & Ebbesen, E. B. (1991). On the form of forgetting. *Psychological Science, 2,* 409–415.

Wood, N. L., & Cowan, N. (1995a). The cocktail party phenomenon revisited: Attention and memory in the classic selective listening procedure of Cherry (1953). *Journal of Experimental Psychology: General, 124,* 243–262.

Wood, N., & Cowan, N. (1995b). The cocktail party phenomenon revisited: How frequent are attention shifts to one's name in an irrelevant auditory channel? *Journal of Experimental Psychology: Learning, Memory, and Cognition, 21,* 255–260.

Woodworth, R. S., & Sells, S. B. (1935). An atmosphere effect in formal syllogistic reasoning. *Journal of Experimental Psychology, 18*(4), 451–460.

Worthy, D. A., Gorlick, M. A., Pacheco, J. L., Schnyer, D. M., & Maddox, W. T. (2011). With age comes wisdom: Decision making in younger and older adults. *Psychological Science, 22,* 1375–1380.

Wraga, M., Swaby, M., & Flynn, C. M. (2008). Passive tactile feedback facilitates mental rotation of handheld objects. *Memory & Cognition, 36,* 271–281.

Wright, A. A., & Roediger, H. L. III. (2003). Interference processes in monkey auditory list memory. *Psychonomic Bulletin & Review, 10,* 696–702.

Wright, O., Davies, I. R., & Franklin, A. (2015). Whorfian effects on colour memory are not reliable. *The Quarterly Journal of Experimental Psychology, 68*(4), 745–758.

Wurm, L. H. (2007). Danger and usefulness: An alternative framework for understanding rapid evaluation effects in perception. *Psychonomic Bulletin & Review, 14,* 1218–1225.

Wurm, L. H., & Seaman, S. R. (2008). Semantic effects in naming and perceptual identification but not in delayed naming: Implications for models and tasks. *Journal of Experimental Psychology: Learning, Memory, and Cognition, 34,* 381–398.

Wynn, K. (1992). Addition and subtraction by human infants. *Nature, 358*(6389), 749–750.

Xu, F., & Kushnir, T. (2013). Infants are rational constructivist learners. *Current Directions in Psychological Science, 22*(1), 28–32.

Yamauchi, T., & Markman, A. B. (2000). Inference using categories. *Journal of Experimental Psychology: Learning, Memory, and Cognition, 26,* 776–795.

Yang, S-J., & Beilock, S. L. (2011). Seeing and doing: Ability to act moderates orientation effects in object perception. *The Quarterly Journal of Experimental Psychology, 64,* 639–648.

Yantis, S. (2008). The neural basis of selective attention. *Current Directions in Psychological Science, 17,* 86–90.

Yantis, S., & Jonides, J. (1984). Abrupt visual onsets and selective attention: Evidence from visual search. *Journal of Experimental Psychology: Human Perception and Performance, 10*, 601–621.

Yaro, C., & Ward, J. (2007). Searching for Shereshevskii: What is superior about the memory of synaesthetes? *The Quarterly Journal of Experimental Psychology, 60*, 681–695.

Yates, M., Locker, L., Jr., & Simpson, G. B. (2003). Semantic and phonological influences on the processing of words and pseudohomophones. *Memory & Cognition, 31*, 856–866.

Yechiam, E., Kanz, J. E., Bechara, A., Stout, J. C., Busemeyer, J. R., Altmaier, E. M., et al. (2008). Neurocognitive deficits related to poor decision making in people behind bars. *Psychonomic Bulletin & Review, 15*, 44–51.

Yonelinas, A. P. (2002). The nature of recollection and familiarity: A review of 30 years of research. *Journal of Memory and Language, 46*, 441–517.

Yu, C., & Smith, L. B. (2011). What you learn is what you see: Using eye movements to study infant cross-situational word learning. *Developmental Science, 14*(2), 165–180.

Yuille, J. C., & Paivio, A. (1967). Latency of imaginal and verbal mediators as a function of stimulus and response concreteness-imagery. *Journal of Experimental Psychology, 75*, 540–544.

Yukalov, V. I., & Sornette, D. (2015). Preference reversal in quantum decision theory. *Frontiers in Psychology, 6*, 1538.

Zacks, J. L., & Zacks, R. T. (1993). Visual search times assessed without reaction times: A new method and an application to aging. *Journal of Experimental Psychology: Human Perception and Performance, 19*, 798–813.

Zacks, J. M., Braver, T. S., Sheridan, M. A., Donaldson, D. I., Snyder, A. Z., Ollinger, J. M., et al. (2001). Human brain activity time-locked to perceptual event boundaries. *Nature Neuroscience, 4*, 651–655.

Zacks, J. M., Speer, N. K., Swallow, K. M., Braver, T. S., & Reynolds, J. R. (2007). Event perception: A mind/brain perspective. *Psychological Bulletin, 133*, 273–293.

Zacks, R. T., Hasher, L., Doren, B., Hamm, V., & Attig, M. S. (1987). Encoding and memory of explicit and implicit information. *The Journals of Gerontology, 42*(4), 418-422.

Zaragoza, M. S., & Lane, S. M. (1994). Source misattributions and the suggestibility of eyewitness memory. *Journal of Experimental Psychology: Learning, Memory, and Cognition, 20*, 934–945.

Zaragoza, M. S., McCloskey, M., & Jamis, M. (1987). Misleading postevent information and recall of the original event: Further evidence against the memory impairment hypothesis. *Journal of Experimental Psychology: Learning, Memory, and Cognitio*n, *13*, 36–44.

Zaragoza, M. S., Payment, K. E., Ackil, J. K., Drivdahl, S. B., & Beck, M. (2001). Interviewing witnesses: Forced confabulation and confirmatory feedback increase false memories. *Psychological Science, 12*, 473–477.

Zaromb, F. M., Karpicke, J. D., & Roediger, H. L. (2010). Comprehension as a basis for metacognitive judgments: Effects of effort after meaning on recall and metacognition. *Journal of Experimental Psychology: Learning, Memory, and Cognition, 36*, 552–557.

Zbrodoff, N. J., & Logan, G. D. (1986). On the autonomy of mental processes: A case study of arithmetic. *Journal of Experimental Psychology: General, 115*, 118–130.

Zechmeister, E. B., & Shaughnessy, J. J. (1980). When you think that you know and when you think that you know but you don't. *Bulletin of the Psychonomic Society, 15*, 41–44.

Zeelenberg, R., Wagenmakers, E., & Rotteveel, M. (2006). The impact of emotion on perception: Bias or enhanced processing? *Psychological Science, 17*, 287–291.

Zelinski, E. M., & Stewart, S. T. (1998). Individual differences in 16-year memory changes. *Psychology and Aging, 13*, 622–630.

Zhang, W., & Luck, S. J. (2009). Sudden death and gradual decay in visual working memory. *Psychological Science, 20*, 423–428.

Zola-Morgan, S., Squire, L., & Amalral, D. G. (1986). Human amnesia and the medial temporal region: Enduring memory impairment following a bilateral lesion limited to field CA1 of the hippocampus. *Journal of Neuroscience, 6*, 2950–2967.

Zwaan, R. A. (1996). Processing narrative time shifts. *Journal of Experimental Psychology: Learning, Memory, and Cognition, 22*, 1196–1207.

Zwaan, R. A. (1999). Situation models: The mental leap into imagined worlds. *Current Directions in Psychological Science, 8*(1), 15–18.

Zwaan, R. A., Langston, M. C., & Graesser, A. C. (1995). The construction of situation models in narrative comprehension: An event-indexing model. *Psychological Science, 6*, 292–297.

Zwaan, R. A., Magliano, J. P., & Graesser, A. C. (1995). Dimensions of situation model construction in narrative comprehension. *Journal of Experimental Psychology: Learning, Memory, and Cognition, 21*, 386–397.

Zwaan, R. A., & Radvansky, G. A. (1998). Situation models in language comprehension and memory. *Psychological Bulletin, 123*, 162–185.

Zwaan, R. A., Stanfield, R. A., & Yaxley, R. H. (2002). Language comprehenders mentally represent the shapes of objects. *Psychological Science, 13*, 168–171.

Credits

Cover

P. Aeyaey/Fotolia

Chapter 1

P. 004 Neisser, U. (1967). Cognitive psychology. New York, NY: Appleton-Century-Crofts.

P. 008–009 Lachman, R., Lachman, J. L., & Butterfield, E. C. (1979). Cognitive psychology and information processing: An introduction. Hillsdale, NJ: Erlbaum.

P. 010 Chomsky, N. (1959). A review of Skinner's Verbal Behavior. Language, 35, 26–58.

P. 011 Gardner, H. (1985). The mind's new science: A history of the cognitive revolution. New York, NY: Basic Books.

P. 011 Norman, D. A. (1986). Reflections on cognition and parallel distributed processing. In J. L. McClelland, D. E. Rumelhart, & PDP Research Group (Eds.), Parallel distributed processing: Explorations in the microstructure of cognition: Vol. 2. Psychological and biological models (pp. 531–546). Cambridge, MA: MIT Press.

P. 013 Based on Campbell, J. I. D., & Graham, D. J. (1985). Mental multiplication skill: Structure, process, and acquisition. Canadian Journal of Psychology, 39, 338–366.

P. 014 Based on Glanzer, M., & Cunitz, A. R. (1966). Two storage mechanisms in free recall. Journal of Verbal Learning and Verbal Behavior, 5, 351–360.

P. 016 Based on Atkinson, R. C., & Shiffrin, R. M. (1971). The control of short-term memory. Scientific American, 225, 82–90. and Atkinson, R. C., & Shiffrin, R. M. (1968). Human memory: A proposed system and its control processes. In W. K. Spence & J. T. Spence (Eds.), The psychology of learning and motivation: Advances in research and theory (Vol. 2, pp. 89–195). New York: Academic Press.

P. 017 Based on Sternberg, S. (1969). The discovery of processing stages: Extensions of Donder's method. In W. G. Koster (Ed.), Attention and performance II. Acta Psychologica, 30, 276–315.

P. 020 Reed, S. K. (1992). Cognition: Theory and applications (3rd ed.). Pacific Grove, CA: Brooks/Cole.

P. 006 Bettmann/Getty Images

P. 007 Bettmann/Getty Images

P. 007 Antman Archives/The Image Works

P. 009 Burben/Shutterstock

P. 010 Bernal Revert/Alamy Stock Photo

Chapter 2

P. 039 Banich, M. T. (2004). Cognitive neuroscience and neuropsychology (2nd ed.). Boston, MA: Houghton Mifflin.

P. 026 Pearson Education

P. 027 Pearson Education

P. 027 Pearson Education

P. 029 Pearson Education

P. 031 Pearson Education

P. 032 Pearson Education

P. 039 Osterhout, L., & Holcomb, P. J. (1992). Event-related brain potentials elicited by syntactic anomaly. Journal of Memory and Language, 31, 785–806.

P. 042 McClelland, J. L., & Rumelhart, D. E. (1981). An interactive activation model of context effects in letter perception: Part 1. An account of basic findings. Psychological Review, 88, 375–407.

P. 040 Davachi, L., Mitchell, J. P., & Wagner, A. D. (2003). Multiple routes to memory: Distinct medial temporal lobe processes build item and source memories. Proceedings of the National Academy of Sciences, 100, 2157–2162.

P. 038 IxMaster/Shutterstock

P. 038 Patrick J. Lynch/Science Source

P. 032 Layland Masuda/Shutterstock

P. 039 Deco/Alamy Stock Photo

P. 036 Pearson Education

P. 025 Pearson Education

P. 035 Pearson Education

Chapter 3

P. 066 Osherson, Daniel N., Stephen M. Kosslyn, and John Hollerbach, eds., An Invitation to Cognitive Science, Volume 2: Visual Cognition and Action, figure 3.15: "Geons and the objects they may", © 1990 Massachusetts Institute of Technology, by permission of The MIT Press.

P. 045 E. C. Hildreth & S. Ullman, The Computational Study of Vision, 1988, Massachusetts Institute of Technology.

P. 046 Based on Galanter

P. 048 Banks, W. P. (1977). Encoding and processing of symbolic information in comparative judgments. In G. H. Bower (Ed.), The psychology of learning and motivation (Vol. 11, pp. 101–159). New York, NY: Academic Press.

P. 050 Based on Hothersall, D. (1984). History of psychology. New York: Random House.

P. 055 Data from Sperling, G. (1960). The information available in brief visual presentations. Psychological Monographs, 74 (Whole No. 498).

P. 058 Based on PSYCHOLOGY 2nd edition by Saundra K. Ciccarelli and J. Noland White.

P. 062 Neisser, U. (1964). Visual search. Scientific American, 210, 94–102.

P. 062 Morris, A. L., & Harris, C. L. (2002). Sentence context, word recognition, and repetition blindness. Journal of Experimental Psychology: Learning, Memory, and Cognition, 28, 962–982.

P. 065 Based on Selfridge, O. G. (1959). Pandemonium: A paradigm for learning. In The mechanisation of thought processes, National Physical Laboratory, London: H. M. Stationery Office.

P. 067 Biederman, I. (1987). Recognition by components: A theory of human image understanding. Psychological Review, 94, 115–147.

P. 065 Based on Rumelhart, D. E., & McClelland, J. L. (1986). Parallel distributed processing: Explorations in the microstructure of cognition: Vol. 1. Foundations. Cambridge, MA: Bradford.

P. 068 Biederman, I. (1987). Recognition by components: A theory of human image understanding. Psychological Review, 94, 115–147.

P. 069 Based on Tucker, M. & Ellis, R. (1998). On the relations of seen objects and components of potential actions. Journal of Experimental Psychology: Human Perception and Performance, 24, 830–846.

P. 070 Sacks, O. (1970). The man who mistook his wife for a hat. New York, NY: Harper & Row.

P. 070 Banich, M. T. (2004). Cognitive neuroscience and neuropsychology (2nd ed.). Boston, MA: Houghton Mifflin.

P. 072–073 Neisser, U. (1967). Cognitive psychology. New York, NY: Appleton-Century-Crofts.

P. 074 Based on Darwin, C. J., Turvey, M. T., & Crowder, R. G. (1972). An auditory analogue of the Sperling partial report procedure: Evidence for brief auditory storage. Cognitive Psychology, 3, 255–267.

P. 076 Based on Warren, R. M., & Warren, R. P. (1970). Auditory illusions and confusions. Scientific American, 223, 30–36.

P. 060 Adapted from Selfridge, O. G. (1959). Pandemonium: A paradigm for learning. In The mechanisation of thought processes, National Physical Laboratory, London: H. M. Stationery Office.

P. 072 Price, R.H., Principles of Psychology, 1st Ed., (c) 1987. Reprinted and Electronically reproduced by permission of Pearson Education, Inc., Upper Saddle River, New Jersey.

P. 053 Sputnik/Alamy Stock Photo

P. 058 Dorling Kindersley Limited

P. 054 Fesus Robert/Shutterstock

P. 061 Pearson Education

P. 047 Banks, W. P., Clark, H. H., & Lucy, P. (1975). The locus of the semantic congruity effect in comparative judgments. Journal of Experimental Psychology: Human Perception and Performance, 1, 35–47.

P. 050 Pearson Education

P. 050 Pearson Education

P. 073 Pearson Education

Chapter 4

P. 102 Barshi, I., & Healy, A. F. (1993). Checklist procedures and the cost of automaticity. Memory & Cognition, 21, 496–505.

P. 085 Posner, M. I., Snyder, C. R. R., & Davidson, B. J. (1980). Attention and the detection of signals. Journal of Experimental Psychology: General, 109, 160–174.

P. 087 Based on A Feature-Integration Theory of Attention by Anne M. Treisman and Arry Gelade, Cognitive Psychology 12, 97–136.

P. 091 Based on F. E. Bloom & A. Lazerson. Brain, Mind, and Behavior, 2nd ed.

P. 095 Based on Broadbent, D. E. (1958). Perception and communication. London: Pergamon

P. 096 Based on Lindsay, P. H., & Norman, D. A. (1977). Human information processing: An introduction to psychology. New York: Academic Press.

P. 101 Barshi, I., & Healy, A. F. (1993). Checklist procedures and the cost of automaticity. Memory & Cognition, 21, 496–505.

P. 090 Solso, R. L., Cognitive Psychology, 5th Ed., © 1998. Reprinted and Electronically reproduced by permission of Pearson Education, Inc., Upper Saddle River, New Jersey.

P. 079 Chris Madden/Alamy Stock Photo

P. 092 Ale Ventura/PhotoAlto/Alamy Stock Photo

P. 092 Asiseeit/Vetta/Getty Images

P. 100 Lennox McLendon/AP Images

P. 101 Ondrejschaumann/Fotolia

P. 103 Ariwasabi/Shutterstock

P. 088–089 Banich, M. T. (1997). Neuropsychology: The neural bases of mental function. Boston, MA: Houghton Mifflin.

Chapter 5

P. 115 Baddeley, A. D., & Hitch, G. (1974). Working memory. In G. H. Bower (Ed.), The psychology of learning and motivation (Vol. 8, pp. 47–89). New York, NY: Academic Press.

P. 108 Peterson, L. R., & Peterson, M. J. (1959). Short-term retention of individual items. Journal of Experimental Psychology, 58, 193–198.

P. 113 Based on Sternberg, S. (1969). The discovery of processing stages: Extensions of Donder's method. In W. G. Koster (Ed.), Attention and performance II. Acta Psychologica, 30, 276–315.

P. 116 Based on Baddeley, A. D. (2000a). The episodic buffer: A new component of working memory? Trends in Cognitive Sciences, 4, 417–423.

P. 120 Mental Rotation of Three-Dimensional Objects by Roger N. Shepard and Jacqueline Metzler. Copyright © 1971 by Roger N. Shepard and Jacqueline Metzler. Reprinted by permission of American Association for the Advancement of Science.

P. 126 Based on Daneman, M., & Carpenter, P. A. (1980). Individual differences in working memory and reading. Journal of Verbal Learning and Verbal Behavior, 19, 450–466.

P. 128 Rosen, V. M., & Engle, R. W. (1997). The role of working memory capacity in retrieval. Journal of Experimental Psychology: General, 126, 211–227.

P. 107 Thomas Del Brase/Photographer's Choice/Getty Images

P. 123 Larry Lilac/Alamy Stock Photo

P. 111 Glanzer, M., & Cunitz, A. R. (1966). Two storage mechanisms in free recall. Journal of Verbal Learning and Verbal Behavior, 5, 351–360.

P. 124 Based on Working Memory, by A. D. Baddeley and G. Hitch, in Gordon H. Bower (Ed.) The Psychology of Learning and Motivation, vol. 8.

P. 125 Based on Logie, R. H., Zucco, G., & Baddeley, A. D. (1990). Interference with visual short-term memory. Acta Psychologica, 75, 55–74.

Chapter 6

P. 132 Based on Squire and Zola-Morgan (1991). Source: Squire, L. R., & Zola-Morgan, S. (1991). The medial temporal lobe memory system. Science, 253, 1380–1386.

P. 138 Based on The von Restorff Effect in Amnesia: The Contribution of the Hippocampal System to Novelty-Related Memory Enhancements.

P. 145 Bower, G. H., Clark, M. C., Lesgold, A. M., & Winzenz, D. (1969). Hierarchical retrieval schemes in recall of categorical word lists. Journal of Verbal Learning and Verbal Behavior, 8, 323–343.

P. 145 Based on Bower, G. H., Clark, M. C., Lesgold, A. M., & Winzenz, D. (1969). Hierarchical retrieval schemes in recall of categorical word lists. Journal of Verbal Learning and Verbal Behavior, 8, 323–343.

P. 149 Based on A propositional theory of recognition memory, John R. Anderson and Gordon H. Bower (1972).

P. 152 Based on Kintsch, W., Welsch, D., Schmalhofer, F., & Zimny, S. (1990). Sentence memory: A theoretical analysis. Journal of Memory and language, 29(2), 133–159

P. 139 Rundus, D. (1971). Analysis of rehearsal processes in free recall. Journal of Experimental Psychology, 89, 63–77.

P. 136 Data from Ebbinghaus, H. (1885/1913). Memory: A contribution to experimental psychology (H. A. Ruger & C. E. Bussenius, Trans.). New York: Columbia University, Teacher's College. (Reprinted 1964, New York: Dover).

P. 147 Based on Godden, D. B., & Baddeley, A. D. (1975). Context-dependent memory in two natural environments: On land and underwater. British Journal of Psychology, 66, 325–331.

Chapter 7

P. 173 Meyer, D. E., & Schvaneveldt, R. W. (1971). Facilitation in recognizing pairs of words: Evidence of a dependence between retrieval operations. Journal of Experimental Psychology, 90, 227–234.

P. 173 Neely, J. H. (1977). Semantic priming and retrieval from lexical memory: Roles of inhibitionless spreading activation and limited-capacity attention. Journal of Experimental Psychology: General, 106, 226–254.

P. 180 Bransford, J. D. (1979). Human cognition: Learning, understanding and remembering. Belmont, CA: Wadsworth.

P. 175 Bartlett's (1932) "The War of the Ghosts", Author: Frederic Charles Bartlett, in source: Bartlett, F. C. (1932). Remembering: A study in experimental and social psychology. Reprinted with the permission of Cambridge University Press.

P. 176 Bartlett's (1932) "The War of the Ghosts", Author: Frederic Charles Bartlett, in source: Bartlett, F. C. (1932). Remembering: A study in experimental and social psychology. Reprinted with the permission of Cambridge University Press.

P. 178 Sulin, R. A., & Dooling, D. J. (1974). Intrusion of a thematic idea in retention of prose. Journal of Experimental Psychology, 103, 255–262.

P. 179 Hasher, L., & Griffin, M. (1978). Reconstructive and reproductive processes in memory. Journal of Experimental Psychology: Human Learning and Memory, 4, 318–330.

P. 179 Schank, R. C., & Abelson, R. P. (1977). Scripts, plans, goals and understanding. Hillsdale, NJ: Erlbaum.

P. 179 Abelson, R. P. (1981). Psychological status of the script concept. American Psychologist, 36, 715–729.

P. 181 Smith, D. A., & Graesser, A. C. (1981). Memory for actions in scripted activities as a function of typicality, retention interval, and retrieval task. Memory & Cognition, 9, 550–559.

P. 180 Based on Schank, R. C., & Abelson, R. P. (1977). Scripts, plans, goals and understanding. Hillsdale, NJ: Erlbaum.

P. 160 Bahrick, H. P., Bahrick, P. C., & Wittlinger, R. P. (1975). Fifty years of memories for names and faces: A cross-sectional approach. Journal of Experimental Psychology: General, 104, 54–75.

P. 170 Farah, M. J., & McClelland, J. L. (1991). A computational model of semantic memory impairment: Modality specificity and emergent category specificity. Journal of Experimental Psychology: General, 120, 339–357.

P. 162 Based on Smith, E. E., Shoben, E. J., & Rips, L. J. (1974). Structure and process in semantic memory: A featural model for semantic decisions. Psychological Review, 81, 214–241

P. 169 Based on Martindale, C. (1991). Cognitive psychology: A neural-network approach.

P. 174 Based on Neely, J. H. (1977). Semantic priming and retrieval from lexical memory: Roles of inhibitionless spreading activation and limited-capacity attention. Journal of Experimental Psychology: General, 106, 226–254.

P. 166 Kounios, J., & Holcomb, P. J. (1992). Structure and process in semantic memory: Evidence from event-related brain potentials and reaction times. Journal of Experimental Psychology: General, 121, 459–479.

P. 165 Kounios, J., & Holcomb, P. J. (1992). Structure and process in semantic memory: Evidence from event-related brain potentials and reaction times. Journal of Experimental Psychology: General, 121, 459–479.

P. 167 Kounios, J., & Holcomb, P. J. (1994). Concreteness effects in semantic processing: ERP evidence supporting dual-coding theory. Journal of Experimental Psychology: Learning, Memory, and Cognition, 20, 804–823.

Chapter 8

P. 188 Schacter, D. L. (1999). The seven sins of memory: Insights from psychology and cognitive neuroscience. American Psychologist, 54, 182–203.

P. 194 Bransford, J. D., & Stein, B. S. (1984). The ideal problem solver. New York, NY: Freeman.

P. 196 Roediger, H. L. III, & McDermott, K. B. (1995). Creating false memories: Remembering words not presented in lists. Journal of Experimental Psychology: Learning, Memory, and Cognition, 21, 803–814.

P. 198 Bransford, J. D., & Franks, J. J. (1971). The abstraction of linguistic ideas. Cognitive Psychology, 2, 331–350.

P. 203 Nickerson, R. S., & Adams, M. J. (1979). Long-term memory for a common object. Cognitive Psychology, 11, 287–307.

P. 205 Conway, M. A., Cohen, G., & Stanhope, N. (1991). On the very long-term retention of knowledge acquired through formal education: Twelve years of cognitive psychology. Journal of Experimental Psychology: General, 120, 395–409.

P. 200 Based on Reconstruction of automobile destruction: an example of the interaction between language and memory from Journal Of Verbal Learning And Verbal Behavior.

P. 192 Radvansky, G. A., & Zacks, R. T. (1991). Mental models and the fan effect. Journal of Experimental Psychology: Learning, Memory, and Cognition, 17, 940–953.

P. 209 Blakemore, C. (1977). Mechanics of the mind. Cambridge, England: Cambridge University Press.

P. 188 Based on Jenkins, J. G., & Dallenbach, K. M. (1924). Obliviscence during sleep and waking. The American Journal of Psychology, 35(4), 605–612.

P. 207 Dr. John Mazziotta et al./Neurology/Science Source

Chapter 9

P. 221 Liberman, A. M., Harris, K. S., Hoffman, H. S., & Griffith, B. C. (1957). The discrimination of speech sounds within and across phoneme boundaries. Journal of Experimental Psychology, 54, 358–368.

P. 243 Kertesz, A. (1982). Two case studies: Broca's and Wernicke's aphasia. In M. A. Arbib, D. Caplan, & J. C. Marshall (Eds.), Neural models of language processes (pp. 25–44). New York, NY: Academic Press.

P. 234 Simpson, G. B. (1981). Meaning dominance and semantic context in the processing of lexical ambiguity. Journal of Verbal Learning and Verbal Behavior, 20, 120–136.

P. 234 Holcomb, P. J., Kounios, J., Anderson, J. E., & West, W. C. (1999). Dual-coding, context availability, and concreteness effects in sentence comprehension: An electrophysiological investigation. Journal of Experimental Psychology: Learning, Memory, and Cognition, 25, 721–742.

P. 240 Coulson, S., Federmeier, K. D., Van Petten, C., & Kutas, M. (2005). Right hemisphere sensitivity to word- and sentencelevel context: Evidence from event-related brain potentials. Journal of Experimental Psychology: Learning, Memory, and Cognition, 31, 127–147.

P. 244 Beeman, M. J. (1993). Semantic processing in the right hemisphere may contribute to drawing inferences from discourse. Brain and Language, 44, 80–120.

P. 230 Fromkin, V. A. (1971). The non-anomalous nature of anomalous utterances. Language, 47, 27–52.

P. 214 Based on "Logical Considerations in the study of Animal Communication" by C. F. Hockett in Animal sounds and communication (pp. 392–430) ed. by W. E. Lanyon and W. N. Tavolga.

P. 219–220 Based on Glucksberg, S., & Danks, J. H. (1975). Experimental psycholinguistics: An introduction. Hillsdale, NJ: Erlbaum.

P. 214 Hockett, C. F. (1960a). Logical considerations in the study of animal communication. In W. E. Lanyon & W. N. Tavolga (Eds.), Animal sounds and communication (pp. 392–430). Washington, DC: American Institute of Biological Sciences.

P. 216–217 Glass, A. L., & Holyoak, K. J. (1986). Cognition (2nd ed.). New York, NY: Random House.

P. 218 Greenberg, J. H. (1978). Generalizations about numeral systems. In J. H. Greenberg (Ed.), Universals of human language: Vol. 3. Word structure (pp. 249–295). Stanford, CA: Stanford University Press.

P. 223–224 Miller, G. A., & Isard, S. (1963). Some perceptual consequences of linguistic rules. Journal of Verbal Learning and Verbal Behavior, 2, 217–228.

P. 224 Warren, R. M. (1970). Perceptual restoration of missing speech sounds. Science, 167, 392–393.

P. 229 Chomsky, N. (1957). Syntactic structures. The Hague, The Netherlands: Mouton.

P. 235 Holcomb, P. J., Kounios, J., Anderson, J. E., & West, W. C. (1999). Dual-coding, context availability, and concreteness effects in sentence comprehension: An electrophysiological investigation. Journal of Experimental Psychology: Learning, Memory, and Cognition, 25, 721–742.

P. 237 Fillenbaum, S. (1974). Pragmatic normalization: Further results for some conjunctive and disjunctive sentences. Journal of Experimental Psychology, 102, 574–578.

P. 239 Osterhout, L., & Holcomb, P. J. (1992). Event-related brain potentials elicited by syntactic anomaly. Journal of Memory and Language, 31, 785–806.

P. 220 Based on Fromkin, V. A., & Rodman, R. (1974). An introduction to language. New York: Holt, Rinehart & Winston.

P. 236 Based on Whitney, P. (1998). The psychology of language. Boston: Houghton Mifflin.

P. 238 Rayner, K., Carlson, M., & Frazier, L. (1983). The interaction of syntax and semantics during sentence processing: Eye movements in the analysis of semantically biased sentences. Journal of Verbal Learning and Verbal Behavior, 22, 358–374.

P. 225 Foss & Hakes, Psycholinguistics: An Introduction to the Psychology of Language., 1st Ed., (c) 1978. Reprinted and Electronically reproduced by permission of Pearson Education, Inc., Upper Saddle River, New Jersey.

P. 215 Elena Efimova/Shutterstock

P. 237 Pearson Education

P. 242 Pearson Education

P. 243 Pearson Education

P. 224 Based on Miller, G. A., & Isard, S. (1963). Some perceptual consequences of linguistic rules. Journal of Verbal Learning and Verbal Behavior, 2, 217–228

P. 227 Bernal Revert/Alamy Stock Photo

Chapter 10

P. 247 Miller, G. A. (1977). Practical and lexical knowledge. In P. N. Johnson-Laird & P. C. Wason (Eds.), Thinking: Readings in cognitive science (pp. 400–410). Cambridge, England: Cambridge University Press.

P. 259 Gernsbacher, M. A. (1990). Language comprehension as structure building. Hillsdale, NJ: Erlbaum.

P. 256 Just, M. A., & Carpenter, P. A. (1980). A theory of reading: From eye fixations to comprehension. Psychological Review, 87, 329–354.

P. 261 Based on Foroughi, C. K., Werner, N. E., Barragan, D., & Boehm-Davis, D. A. (2015). Interruptions disrupt reading comprehension. Journal of Experimental Psychology: General, 144, 704–709.

P. 252 Based on Language Comprehension as Structure Building by Morton Ann Gernsbacher, 1990 Psychology Press.

P. 259 Baased on Ashcraft's Fundamentals of Cognition, 1998, Addison Wesley. Adapted from a table in Clark, H. H. (1977). Bridging. In P. N. Johnson-Laird & P. C. Wason (Eds.), Thinking: Readings in Cognitive Science (pp. 141–420). Cambridge: Cambridge University Press.

P. 260 Based on Functional Neuroanatomy Of The Cognitive Process Of Mapping During Discourse Comprehension.

P. 261 Harris, R. J., & Monaco, G. E. (1978). Psychology of pragmatic implication: Information processing between the lines. Journal of Experimental Psychology, 107, 1–22.

P. 263–264 Thomas Holtgraves, Conversations Memory: Intentions, Politeness, and The Social Context In Language and Memory: Aspects of Knowledge Representation, Hanna Pishwa, 2006 Walter de Gruyter.

P. 264 de Vega, M., Robertson, D. A., Glenberg, A. M., Kaschak, M. P., & Rinck, M. (2004). On doing two things at once: Temporal constraints on action in language comprehension. Memory & Cognition, 32, 1033–1043.

P. 268 Based on Grice, H. P. (1975). Logic and conversation. In P. Cole & J. L. Morgan (Eds.), Syntax and se-mantics: Vol. 3. Speech acts (pp. 41–58). New York: Seminar Press.

P. 270 Holtgraves, T. (1998). Interpreting indirect replies. Cognitive Psychology, 37, 1–27.

P. 271 Lee, J. J., & Pinker, S. (2010). Rationales for indirect speech: The theory of the strategic speaker. Psychological Review, 117, 785–807.

P. 267 Svartvik, J., & Quirk, R. (Eds.). (1980). A corpus of English conversation. Lund, Sweden: CWK Gleerup.

P. 253 Rayner, K. (1998). Eye movements in reading and information processing: 20 years of research. Psychological Bulletin, 124, 372–422.

P. 254 Based on Just, M. A., & Carpenter, P. A. (1987), The psychology of reading and language comprehension. Boston: Allyn & Bacon. Buswell, G. T. (1937). How adults read. Chicago: Chicago University Press, Plates II and IV, pp. 6–7.

P. 255 Just, M. A., & Carpenter, P. A. (1980). A theory of reading: From eye fixations to comprehension. Psychological Review, 87, 329–354.

P. 258 Just, M. A., & Carpenter, P. A. (1980). A theory of reading: From eye fixations to comprehension. Psychological Review, 87, 329–354.

P. 259 Based on Language Comprehension as Structure Building by Morton Ann Gernsbacher, 1990 Psychology Press.

P. 265 Rinck, M., Hähnel, A., Bower, G. H., & Glowalla, U., The metrics of spatial situation models, In Journal of Experimental Psychology: Learning, Memory, and Cognition, 23(3), 622–637.

P. 265 Data from On doing two things at once: Temporal constraints on actions in language comprehension by Manuel De Vega, David A. Robertson, Arthur M. Glenberg, Michael P. Kaschak and Mike Rinck in Memory & Cognition 32(7), 1033–1043, 2004 Springer.

P. 266 Data from Aging and Mind Wandering During Text Comprehension by Gabriel A. Radvansky, Sabine A. Krawietz, and Andrea K. Tamplin in Psychology and Aging, Vol. 27, No. 4, 951–958.

P. 271 Based on Holtgraves, T. (1998). Interpreting indirect replies. Cognitive Psychology, 37, 1–27.

P. 263 Wollertz/Fotolia

P. 263 Searagen/Fotolia

Chapter 11

P. 279 Based on Cognitive Psychology: Memory, Language, and Thought, 2/e, by Darlene V. Howard.

P. 280 Rader, A. W., & Sloutsky, V. M. (2002). Processing of logically valid and logically invalid conditional inferences in discourse comprehension. Journal of Experimental Psychology: Learning, Memory, and Cognition, 28, 59–68.

P. 281 Wason, P. C., & Johnson-Laird, P. N. (1972). Psychology of reasoning: Structure and content. Cambridge, MA: Harvard University Press.

P. 284 Ashcraft, M. H. (1989). Human memory and cognition. Glenview, IL: Scott Foresman.

P. 287 Kahneman, D., & Tversky, A. (1972). Subjective probability: A judgment of representativeness. Cognitive Psychology, 3, 430–454.

P. 288 Based on Johnson, J. T., & Finke, R. A. (1985). The base-rate fallacy in the context of sequential categories. Memory & Cognition, 13, 63–73.

P. 288 Kahneman, D., & Tversky, A. (1973). On the psychology of prediction. Psychological Review, 80, 237–251.

P. 290 Kahneman, D., & Tversky, A. (1982). The simulation heuristic. In D. Kahneman, P. Slovic, & A. Tversky (Eds.), Judgment under uncertainty: Heuristics and biases (pp. 201–208). Cambridge, England: Cambridge University Press.

P. 291 "Stories for the Simulation Heuristic", Author: Daniel Kahneman, Author: Amos Tversky, in source: Kahneman, D., & Tversky, A. (1982). The simulation heuristic. In D. Kahneman, P. Slovic, & A. Tversky (Eds.), Judgment under uncertainty: Heuristics and biases (pp. 201–208). Reprinted with the permission of Cambridge University Press.

P. 292 Goldinger, S. D., Kleider, H. M., Azuma, T., & Beike, D. R. (2003). "Blaming the victim" under memory load. Psychological Science, 14, 81–85.

P. 294 Tversky, A., & Kahneman, D. (1981). The framing of decisions and the psychology of choice. Science, 211, 453–458.

P. 296 Gigerenzer, G. (2008). Why heuristics work. Perspectives in Psychological Science, 3, 20–29.

P. 299 Tversky, A., & Kahneman, D. (1983). Extensional versus intuitive reasoning: The conjunction fallacy in probability judgment. Psychological Review, 90, 293–315.

P. 306 Waltz, J. A., Knowlton, B. J., Holyoak, K. J., Boone, K. B., Mishkin, F. S., Santos, M. M., et al. (1999). A system for relational reasoning in human prefrontal cortex. Psychological Science, 10, 119–125.

P. 281 Based on Wason, P. C., & Johnson-Laird, P. N. (1972). Psychology of reasoning: Structure and content. Cambridge, MA: Harvard University Press.

P. 297 Based on Goldstein, D. G., & Gigerenzer, G. (2002). Models of ecological rationality: The recognition heuristic. Psychological Review, 109, 75–90.

P. 300 Kynski, T. R., & Tenebaum, J. B. (2007). The role of causality in judgment under uncertainty. Journal of Experimental Psychology: General, 136, 430–450.

P. 303 M. McCloskey, Naive Theories of Motion, 1982 National Institute of Education.

P. 305 The science of cycology: Failures to understand how everyday objects work by Rebecca Lawson. Copyright © 1669 by Rebecca Lawson. Reprinted by permission of Springer.

P. 306 Waltz, J. A., Knowlton, B. J., Holyoak, K. J., Boone, K. B., Mishkin, F. S., Santos, M. M., et al. (1999). A system for relational reasoning in human prefrontal cortex. Psychological Science, 10, 119–125.

P. 306 Waltz, J. A., Knowlton, B. J., Holyoak, K. J., Boone, K. B., Mishkin, F. S., Santos, M. M., et al. (1999). A system for relational reasoning in human prefrontal cortex. Psychological Science, 10, 119–125.

P. 302 EdBockStock/Shutterstock

P. 284 Siart/Shutterstock

Chapter 12

P. 332 AVAVA/Shutterstock

P. 321 Huntstock, Inc/Alamy Stock Photo

P. 316 Intelligenzprüfungen an Menschenaffen. By Wolfgang Kohler. Tafel IV. © 1921 Verlag Von Julius Springer in Berlin.

P. 329 Glass, A. L., & Holyoak, K. J. (1986). Cognition (2nd ed.). New York, NY: Random House.

P. 331 Based on Posner 1973, Cognition: An introduction.

P. 331 M. I. Posner, Cognition: An introduction, 1973 Scott Foresman.

P. 333 Ericsson, K. A., Krampe, R. T., & Tesch-RÖmer, C. (1993). The role of deliberate practice in the acquisition of expert performance. Psychological Review, 100, 363–406.

P. 309 VanLehn, K. (1989). Problem solving and cognitive skill acquisition. In M. I. Posner (Ed.), Foundations of cognitive science (pp. 527–579). Cambridge, MA: MIT Press.

P. 310 Newell, A., & Simon, H. A. (1972). Human problem solving. Englewood Cliffs, NJ: Prentice Hall.

P. 316 Boring, E. G. (1950). A history of experimental psychology (2nd ed.). New York, NY: Appleton-Century-Crofts.

P. 319 Metcalfe, J. (1986). Feeling of knowing in memory and problem solving. Journal of Experimental Psychology: Learning, Memory, and Cognition, 12, 288–294.

P. 319 Metcalfe, J. (1986). Feeling of knowing in memory and problem solving. Journal of Experimental Psychology: Learning, Memory, and Cognition, 12, 288–294.

P. 319 Metcalfe, J., & Wiebe, D. (1987). Intuition in insight and noninsight problem solving. Memory & Cognition, 15, 238–246.

P. 321 Schooler, J. W., Ohlsson, S., & Brooks, K. (1993). Thoughts beyond words: When language overshadows insight. Journal of Experimental Psychology: General, 122, 166–183.

P. 319 Based on Metcalfe, J. (1986). Feeling of knowing in memory and problem solving. Journal of Experimental Psychology: Learning, Memory, and Cognition, 12, 288–294 and Metcalfe, J., & Wiebe, D. (1987). Intuition in insight and noninsight problem solving. Memory & Cognition, 15, 238–246.

P. 320 Based on Metcalfe, J., & Wiebe, D. (1987). Intuition in insight and noninsight problem solving. Memory & Cognition, 15, 238–246.

P. 322 Based on Gick, M. L., & Holyoak, K. J. (1980). Analogical problem solving. Cognitive Psychology, 12, 306–355.

P. 323 Data from Gick, M. L., & Holyoak, K. J. (1980). Analogical problem solving. Cognitive Psychology, 12, 306–355.

P. 325 Based on Wharton, C. M., Grafman, J., Flitman, S. S., Hansen, E. K., Brauner, J., Marks, A., et al. (2000). Toward neuroanatomical models of analogy: A positron emission tomography study of analogical mapping. Cognitive Psychology, 40, 173–197.

P. 328 Based on The Missionary–Cannibal Problem.

P. 326 Based on The Tower of Hanoi problem.

P. 312 How to solve problems by W.A. Wickelgren. Copyright © 1995 by W.A. Wickelgren. Reprinted by permission of Dover Publications.

P. 322 Gick, M. L., & Holyoak, K. J. (1980). Analogical problem solving. Cognitive Psychology, 12, 306–355.

P. 317 Pearson Education

Chapter 13

P. 340 Asti Suwandi/Shutterstock

P. 339 Denkou Images/Alamy Stock Photo

P. 350 ZUMA Press, Inc./Alamy Stock Photo

P. 338 Data from The impact of emotion on perception: Bias or enhanced processing? in Psychological Science by Zeelenberg, Wagenmakers, & Rotteveel (2006).

P. 339 Data from Don't look down: Emotional arousal elevates height perception in Journal of Experimental Psychology: by Stefanucci & Storbeck (2009).

P. 341 Data from Running on empty: Neural signals for self-control failure in Psychological Science, by Inzlicht & Gutsell (2007).

P. 343 Data from Enhancing retention through reconsolidation: Negative emotional arousal following retrieval enhances later recall in Psychological Science, by Finn & Roediger (2011).

P. 346 Based on Yerkes-Dodson law.

P. 345 Masatomo Kuriya/Corbis Premium Historical/Getty Images

P. 340 Pearson Education

P. 336 Pearson Education

Chapter 14

P. 363 Iofoto/Shutterstock

P. 358 Data from The development of visual search in infants and very young children in Journal of Experimental Child Psychology by P. Gerhadstein; C. Rovee-Collier.

P. 359 Schiller, P. H. (1966). Developmental study of color-word interference. Journal of Experimental Psychology, 72, 105–108.

P. 361 Hartshorn, K., Rovee-Collier, C., Gerhardstein, P., Bhatt, R. S., Wondoloski, T. L., Klein, P., et al. (1998). The ontogeny of long-term memory over the first year-and-a-half of life. Developmental Psychobiology, 32, 69–89.

P. 365 Data from Talking about the there and then: The emergence of displaced reference in parent-child discourse in Children's language by Sachs, Psychology Press.

P. 365 Brown, R. (1973). A first language: The early stages. Cambridge, MA: Harvard University Press.

Chapter 15

P. 384 Light, L. L., & Capps, J. L. (1986). Comprehension of pronouns in young and older adults. Developmental Psychology, 22, 580–585.

P. 385 Hamm, V. P., & Hasher, L. (1992). Age and the availability of inferences. Psychology and Aging, 7(1), 56–64.

P. 387 Data from With Age Comes Wisdom: Decision-Making in Younger and Older Adults by Darrell A. Worthy, Marissa A. Gorlick, Jennifer L. Pacheco, David M. Schnyer, and W. Todd Maddox, Sage Publications.

P. 389 Data from Age-related perspectives and emotion processing. Psychology and Aging, by Lynchard, N. A., & Radvansky, G. A.

P. 377 Monkey Business Images/Shutterstock

P. 380 Diego Cervo/Shutterstock

Name Index

A
Abraham, W. C., 29, 30
Abrams, R. A., 68, 84
Ackerman, N., 242
Acord, J., 6
Adams, C., 284
Adams, F., 141
Adams, M. J., 203
Adams, T. G., 341
Addis, D. R., 156
Adolphs, R., 348
Aerts, D., 302
Agnoli, F., 287
Ahern, F., 377
Ahlum-Heath, M. E., 332
Aichele, S. R., 82
Ainsworth-Darnell, K., 237
Albert, N. B., 351
Albrecht, J. E., 263
Alksnis, O., 283
Allan, K., 153
Allen, C. K., 109, 217
Allen, J. F., 268
Allen, M. D., 237
Allen, P. A., 18
Allen, R. J., 121
Almor, A., 259
Alpert, N. M., 119
Alter, A. L., 285
Altieri, P. A., 386
Altmaier, E. M., 284
Altmann, E. M., 108
Altmann, G. T. M., 256
Alvarez, G. A., 144
Alvin, M., 384
Always, D., 40
Amalral, D. G., 208
Ambday, N., 35
Ambrosini, E., 68
Anaki, D., 172
Anderson, A. K., 340, 343
Anderson, B. A., 78, 84
Anderson, J. E., 235
Anderson, J. R., 18, 149, 190, 204, 309, 311, 381
Anderson, M. C., 170, 192
Anderson, N. D., 142
Anderson, R. J., 156
Anderson, S. E., 262
Anderson, S. J., 345
Andrade, J., 82
Andrassy, J. M., 386
Andrés, P., 239
Andrews-Hanna, J. R., 80
Andrusiak, P., 264
Anes, M. D., 57
Angele, B., 254
Anglin, S. M., 156
Annon, T. A. K., 385
Ansari, D., 350
Appenzeller, T., 28
Arguin, M., 171
Argyri, E., 238

Armony, J. L., 343
Arnold, J. E., 259
Ashcraft, M. H., 14, 170, 172, 242, 284, 304, 350, 368, 369
Ashkenazi, A., 139
Aslan, A., 194
Atance, C. M., 156
Atkins, S. A., 196
Atkinson, R. C., 16, 138, 139, 146
Atran, S., 159, 330
Atwood, M. E., 329
Atzeni, T., 383
Au, J., 127
Austin, G. A., 9, 181
Averbach, E., 54, 56
Awh, E., 116, 129
Ayers, M. S., 200
Ayers, T. J., 74
Azuma, T., 239, 292

B
Baars, B. J., 8
Babson, K. A., 341
Bäckman, L., 376, 377, 382
Baddeley, A. D., 105, 114, 115, 116, 117, 119, 121, 122, 125, 141, 144, 146, 147
Badgaiyan, R. D., 175
Badou, C. L., 341
Badre, D., 37
Bahrick, H. P., 153
Bailey, H. R., 380
Bailey, K. G. D., 267
Baillargeon, R., 361
Baird, B., 320
Balestreri, A., 80
Balota, D. A., 14, 196, 234
Banich, M. T., 29, 34, 39, 88, 89, 90, 117, 208
Banks, W. P., 47
Bar-Hillel, M., 288, 289
Barac, R., 126
Baranski, J. V., 293
Barber, S. J., 200
Barnard, P. J., 97
Barnett, T., 73
Barnier, A. J., 208
Barr, D. J., 267
Barrouillet, P., 117, 282
Barsalou, L. W., 84, 184
Barshi, I., 101
Barston, J. L., 277
Bartels, D. M., 292
Bartlett, F. C., 175, 176
Basak, C., 126
Bassok, M., 331
Bates, A., 73
Bates, E., 150
Battaglia, J., 368
Bauer, P. J., 359
Bäuml, K-H. T., 140, 192, 194
Bayen, U. J., 382, 383, 387

Bayer, R. H., 47
Baylis, G. C., 129
Bazerman, M. H., 285
Bearden, J. N., 387
Bechara, A., 284
Beck, T. M., 380
Becker, M. W., 340
Beeman, M. J., 243, 260, 324, 325
Behrends, A. A., 93
Behrmann, M., 98
Beier, M. E., 103
Beike, D. R., 292
Beilock, S. L., 68, 129, 292, 350, 351, 365
Bekoff, M., 217
Belin, P., 343
Bellezza, F. S., 143
Belli, R. F., 199
Benjamin, A. S., 140
Benjamin, L. T., 6
Bennett, P. J., 64
Benson, D. J., 242
Bentin, S., 173
Benz, S., 127
Berg, S., 377
Berger, A., 89
Berger, S. A., 153
Berman, M. G., 97, 108
Berndt, R. S., 242
Bernstein, D. M., 202
Berntsen, D., 154, 156, 388
Berntson, G. G., 25
Berry, J. M., 381
Bertenthal, B. I., 350
Bertera, J. H., 55
Bertsch, S., 143
Berwick, R. C., 217
Besner, D., 98
Best, R., 387
Bharucha, J., 35
Bhatt, R. S., 361
Bialystok, E., 126, 239
Bica, L. A., 51
Bichot, N. P., 86
Biederman, I., 66, 67, 71
Bies-Hernandez, N. J., 277
Biggs, A. T., 88
Bigler, E. D., 39
Bilaliæ, M., 318
Binder, K. S., 234, 257
Biondi, F., 93
Birdsall, T. G., 48
Bireta, T. J., 381
Birmingham, E., 84
Bisanz, J., 14
Bisby, J. A., 151
Bischof, W. F., 84
Bisiach, E., 91
Biswas, D., 295
Bjork, E. L., 140, 143, 192
Bjork, R. A., 140, 192, 281
Black, J. B., 180, 303

Black, N. L., 264
Black, T., 303
Blair, M., 381
Blakely, D. P., 88
Blakemore, C., 209
Blanchette, I., 282
Blanke, O., 168
Blaser, E., 356
Blickle, T., 67
Bliss, T. V. P., 29
Bloesch, E. K., 68
Bloom, C. P., 264
Bloom, F. E., 264
Blumstein, S. E., 223
Bocanegra, B. R., 337
Bock, J. K., 230, 231
Bock, K., 231, 255
Boduroglu, A., 62
Boer, E., 124
Bogg, T., 127
Bogus, K., 151
Bohn, A., 156
Bohn, M., 217
Bohn-Gettler, C. M., 365
Boisvert, S., 174
Boland, J. E., 237
Bonebakker, A. E., 82
Bonke, B., 82
Bonnefon, J. F., 280
Bonvillian, J. D., 269
Boot, W. R., 66, 88
Booth, J. L., 369
Boring, E. G., 316
Born, D. G., 109
Born, J., 29, 188
Boroditsky, L., 119
Borton, R., 356
Bos, M. W., 285
Bouazzaoui, B., 383
Bousfield, W. A., 9, 144
Bowden, E. M., 320, 324, 325, 331
Bower, G. H., 133, 135, 146, 149, 180, 386
Bowers, J. S., 67
Bradley, M. M., 343
Brady, T. F., 144
Brainerd, C. J., 302
Brandt, A., 203
Branigan, H. P., 231, 267
Bransford, J. D., 180, 192, 194, 196, 197, 198, 199, 330
Brauner, J., 323, 324, 325
Braver, T. S., 129
Brédart, S., 383
Breedin, S. D., 241, 304
Bremeer, D., 147
Brennan, S. E., 224, 267, 269
Brewer, C. C., 234
Brewer, G. A., 146
Bridgeman, B., 83
Bridwell, D. A., 82
Broadbent, D. E., 9, 94
Broaders, S. C., 365
Brocklehurst, P. H., 225
Brockmole, J. R., 68, 88, 381
Bröder, A., 296
Broggin, E., 119
Bromfield, E. B., 203
Brooks, A. W., 352
Brooks, J. O., 107

Brooks, K., 321
Brooks, L. R., 119
Brosch, T., 83
Brown, A. S., 147, 206, 210, 290
Brown, C. M., 256
Brown, G. D. A., 139
Brown, J. A., 105, 107
Brown, M., 267
Brown, N. R., 290
Brown, R., 365
Brown, S. C., 369
Brown-Schmidt, S., 260
Brun, F., 388
Bruner, J. S., 9, 181
Bruza, P. D., 302
Bryden, M. P., 34
Buchanan, M., 105
Buchanan, T. W., 348
Buchner, A., 96, 133, 292
Buckner, R. L., 80, 207
Buffone, L., 14
Bugg, J. M., 145
Bulf, H., 356
Bundesen, C., 60
Bunting, M. F., 127, 128, 190
Burgess, C., 236, 256
Burgess, G. C., 129
Burgess, N., 151
Burghardt, G., 217
Burgoon, E. M., 181
Burns, B. D., 297
Burns, Z. C., 292
Burt, C. D. B., 330
Buschkuehl, M., 127
Busemeyer, J. R., 284, 302
Busey, T. A., 20
Buswell, G. T., 254
Butler, A. C., 140
Butterfield, E. C., 8, 9, 11
Butterfield, S., 226
Byrd, M., 380
Byrne, R. M. J., 282, 283, 291, 292

C

Cabeza, R., 159, 207, 343
Cacioppo, J. T., 25, 37
Cahill, L., 134
Caine, E. D., 28
Call, J., 217
Camos, V., 117
Campbell, A. C., 88
Campbell, J. I. D., 13, 14, 17, 368
Cantalupo, C., 217
Cantor, J., 128
Capasso, L., 217
Cappell, K. A., 196
Capps, J. L., 384
Carli, L. L., 292
Carlsen, K. A., 267, 292
Carlson, R. A., 390
Carpenter, M., 217
Carpenter, P. A., 119, 125, 126, 252, 253, 254, 255, 256, 258, 264
Carpenter, S. K., 140
Carr, P. B., 351
Carr, T. H., 292, 350
Carreiras, M., 238
Carriere, J. S. A., 380

Carroll, D. W., 232
Carstensen, L. L., 388, 389
Carter, C. S., 18
Carter, R., 92
Carterette, E. C., 93
Caruso, E. M., 292
Case, R., 361
Casteel, M. A., 256
Castel, A. D., 383
Castillo, J. G., 37
Catale, C., 358
Catrambone, R., 303, 327
Cavanagh, J. F., 93
Cavanagh, P., 90
Cave, K. R., 67, 86
Cepeda, N. J., 126
Cerella, J., 126
Cesana, D. T., 217
Chabris, C. F., 119
Chaddock-Heyman, L., 361
Chaffin, R., 134
Chamberlain, R., 88
Chambers, A. M., 346
Chan, D., 259
Chapman, G. B., 293
Chapman, H. A., 343
Chapman, J. P., 279
Chapman, L. J., 279
Charles, S. T., 388
Charman, S. D., 201
Charness, N., 330, 333, 387
Chase, W. G., 107, 126, 330
Chasteen, A. L., 382
Chater, N., 139, 279
Chau, T., 126
Chen, K., 35
Cherniak, C., 305
Cherry, E. C., 93
Cheung, H., 384
Cheung, J. C., 285
Cheyne, J. A., 380
Chiarello, C., 243
Chin-Parker, S., 183
Chisholm, J. D., 88
Chochol, C., 343
Choi, I., 159
Chomsky, N., 10, 215, 218
Chooi, W. T., 127
Christ, S. E., 84
Christy, K. S., 14
Chronicle, E. P., 313, 319
Chrosniak, L. D., 383
Chu, Y., 313
Chugh, D., 285
Chumbley, J. I., 256
Church, R. B., 267
Cisler, J. M., 341
Clapp, F. L., 14
Claridge, G., 96
Clark, B. A., 128
Clark, D., 80
Clark, E. V., 231
Clark, H. H., 47, 231, 259, 260, 262, 267, 269, 270
Clark, L. F., 180
Clark, M. B., 256
Clark, M. C., 133, 135
Cleare, A. J., 341
Cleland, A. A., 231

Clifton, C., 231, 252, 256, 267
Cohen, G., 204, 205, 384
Cohen, J., 15
Cohen, M., 92
Cohen, N. J., 208, 231
Cohen, Y., 87
Cokely, E. T., 129
Cole, G. G., 84
Coles, M. G. H., 39
Coley, J. D., 283, 330
Colle, H. A., 117
Collingridge, G. L., 29
Collins, A. M., 163, 165, 303
Coltheart, M., 37
Colzato, L. S., 127
Comalli, P. E., 357
Connell, L., 263
Conrad, R., 117
Conway, A. R. A., 127, 128, 129, 190
Conway, M. A., 152, 204, 205, 345
Cook, A. E., 263
Cooke, N. J., 304
Cooper, F. S., 221, 223, 225
Cooper, L. A., 119
Copeland, D. E., 122, 126, 128, 190, 277, 384, 385
Corballis, M. C., 33, 213, 244
Corder, M., 66
Coriell, A. S., 56
Corkin, S., 208, 343
Corley, M., 225
Corley, R. P., 116
Costantini, M., 68
Cottrell, J. E., 51
Coughlan, A. K., 242
Courage, M. L., 153
Courtney, J. R., 120
Courtney, S. M., 116
Cowan, N., 84, 93, 105, 106, 125, 127, 338
Craik, F. I. M., 140, 142, 239, 380, 381
Craik, K. H., 2
Crawford, J. R., 382
Croizet, J. C., 383
Crosby, J. R., 253
Crovitz, H. F., 154
Crowder, R. G., 73, 74, 75, 139
Cudahy, C., 54
Cummins, D. D., 283
Cunfer, A. R., 382
Cunitz, A. R., 14, 110
Cunningham, T. J., 346
Curiel, J. M., 385
Curran, T., 67, 142
Curtis, J. F., 93
Cutler, A., 226

D

D'Anastasio, R., 217
D'Argembeau, A., 156, 358
Dagry, I., 282
Dahan, D., 224
Daley, K. E., 14
Dallenbach, K. M., 188
Daneman, M., 125, 126
Danks, J. H., 220
Darwin, C. J., 73, 74
Davachi, L., 80
Davelaar, E. J., 139
Davidson, B. J., 85
Davies, J. B., 380

Davis, M., 336
Davoli, C. C., 68, 84, 88
Day, S. B., 321, 323
de Kwaadsteniet, L., 299
De Ley, L., 264
de Marcus, S., 144
de Sa, V. R., 169
de Vega, M., 264
de-Wit, L., 57
DeCaro, M. S., 129, 351
Dediu, D., 217
Deese, J., 195
DeFries, J. C., 116
Dehaene, S., 357
Dehn, D. M., 144
Dehon, H., 383
Delany, P. F., 330
Dell, G. S., 224, 225, 236, 267
Dell'acqua, R., 67
Della Sala, S., 238, 388
Delton, A. W., 137, 156
Dember, W. N., 82
Dennis, Y., 256
DePaulo, B. M., 269
Descartes, R., 5
DeVita, C., 184
Dewald, A. D., 313
DeWall, B. H., 386
Dewhurst, S. A., 156
deWinstanley, P. A., 143
Dewitt, L. A., 224
Dewsbury, D. A., 8
Dhami, M. K., 295
Di Pelligrino, G., 36
Diamond, A., 96
Dieckmann, A., 299
Dijksterhuis, A., 285
Dijkstra, K., 385
Dijkstra, T., 239
Dilley, L. C., 231
DiVesta, F. J., 332
Diwadkar, V., 119
Dixon, M. J., 53, 54
Dixon, R. A., 381, 384
Dodd, M. D., 87
Dodson, C. S., 383
Dolan, R., 207
Dolcos, F., 336, 343
Dolcos, S., 336
Donaldson, C., 341
Donchin, E., 39
Donders, F. C., 14
Donders, K., 199
Donk, M., 88
Donley, R. D., 304
Donnelly, C. M., 345
Donnenwerth-Nolan, S., 144
Dosher, B. A., 105
Dougherty, M. R., 295
Doussard-Roosevelt, J. A., 380
Dowling, W. J., 150
Doyon, C., 128
Doyon, J. K., 84
Drachman, D. A., 28
Dreossi, D., 217
Dresler, T., 341
Drews, F. A., 92
Drolet, M., 348
Drosopoulos, S., 188
Dubois, S. J., 383

Duchek, J. M., 381
Dudai, Y., 80
Duff, M., 260
Duffy, S. A., 252, 254
Duñabeitia, J. A., 238
Duncan, G. J., 127
Duncker, K., 316, 322
Dunlosky, J., 248, 310
Durante, R., 174
Durkin, M., 6
Dyer, F. C., 216
Dywan, J., 383

E

Eastwood, J. D., 84
Ebbinghaus, H., 7, 14, 135
Eberhard, K. M., 236, 253
Edelson, S. M., 304
Edry, E., 144
Egan, J. C., 74
Egan, P., 93
Egeth, H. E., 84
Ehrlich, S. F., 256
Eichenbaum, H., 208
Eickhoff, S. B., 82
Eimas, P. D., 72, 360
Einstein, G. O., 134, 155, 382
Elder, J. H., 57
Eliassen, J., 223
Elliott, E. M., 129
Ellis, A., 170, 242
Ellis, J. A., 144
Ellis, R., 69
Elman, J. L., 224
Emerson, M. J., 116
Emery, N. J., 84
Emmorey, K., 106
Engbert, R., 253
Engelhardt, P. E., 267
Engelkamp, J., 144
Engen, T., 30, 336
Engle, R. W., 116, 121, 122, 125, 126, 127, 128, 129
Epley, N., 285
Epstein, R., 32
Epstein, W., 68
Erdfelder, E., 282, 292, 387
Erickson, T. D., 177
Ericsson, K. A., 30, 107, 126, 134, 310, 330, 333
Ernst, G. W., 328
Eskenazi, J., 134
Espino, O., 283
Estes, Z., 84
Esteves, F., 83, 339
Etherton, J. L., 97
Evans, B. L., 97
Evans, J. St. B. T., 277, 281, 282
Eyre, R. N., 285

F

Fadiga, L., 36
Fantz, R. L., 356
Farah, M. J., 143, 242, 260
Faroqi-Shah, Y., 241
Fausey, C. M., 119
Faust, M. E., 123, 234
Fawcett, J. M., 143
Fay, S., 383
Fazio, L. K., 200

Fecteau, S., 343
Federico, T., 264
Federmeier, K. D., 234
Fedorenko, E. G., 231
Fehd, H. M., 127
Feigenson, L., 357
Feldman, D., 277
Feldman, J., 57, 183
Feng, J., 88, 126
Feng, S., 35
Fenn, K. M., 136
Fenske, M. J., 338
Ferguson, C., 269
Fernald, A., 363, 364
Ferrante, D., 292
Ferreira, F., 267
Ferreira, V. S., 231, 232, 267
Ferrer, E., 82
Ferrufino, G., 88
Fieberg, E. L., 303
Fiedler, K., 281
Fierman, B. A., 369
Fillenbaum, S., 237
Fillmore, C. J., 235
Finke, R. A., 55, 120
Finlayson, N. J., 87
Finn, B., 343
Finnigan, F., 380
Fischer, A. H., 348
Fischer, J., 348
Fischer, S., 188
Fischhoff, B., 289, 292
Fisher, D. L., 255
Fisk, J. E., 386
Fitzgerald, J. M., 359
Fitzsimmons, C., 386
Fivush, R., 361
Fleck, J. I., 310, 320
Fletcher, C. R., 264
Fletcher, P., 207
Flitman, S. S., 323, 324, 325
Flusberg, S. J., 119
Flykt, A., 83, 339
Flynn, C. M., 119
Fodor, J. A., 229
Fogassi, L., 36
Follenfant, A., 383
Fong, G. T., 287, 305
Forgus, R. H., 71
Forrin, N. D., 143
Förster, J., 80
Forsyth, D. K., 330
Fortin, N., 208
Foss, D. J., 225
Fournier, J. S., 51
Fowler, C. A., 225
Fox, E., 96
Fox, G., 106
Fox, W. C., 48
Francis, W. N., 18
Franco-Watkins, A. M., 295
Franconeri, S. L., 84
Frank, D. J., 380
Frank, M. J., 37, 93
Franklin, M. S., 320
Franklin, N., 264
Franks, J. J., 196, 197, 198, 199
Frazier, L., 231, 256, 267
Freedman, J. L., 172
Freeman, J. E., 144

Freud, S., 153
Freyd, J. J., 55, 120
Friederici, A. D., 237
Friedman, A., 34
Friedman, N. P., 116
Friedrich, F. J., 174
Friesen, C. K., 84
Frieske, D. A., 385
Frings, C., 96
Frisson, S., 256
Fromkin, V. A., 220, 232
Fromme, K., 30
Frost, P., 203
Frymiare, J. L., 320
Fuentes, L. J., 87
Fugelsang, J. A., 350
Fujii, M., 47
Fukuda, K., 129
Fung, H. H., 289
Furman, O., 80

G

Gagne, D. D., 386
Galanter, E., 45
Galantucci, B., 225
Galese, V., 36
Gallistel, C., 367
Gallo, D. A., 195, 383
Galotti, K. M., 285
Galton, F., 154
Galván, V. V., 93
Gamer, M., 51
Gandour, J., 118
García-Madruga, J. A., 369
Garcia-Retamero, R., 295, 299
Gardner, H., 11
Garnham, A., 256
Garnsey, S. M., 255
Garoff-Eaton, R. J., 343
Garrard, C., 137
Garrett, M. F., 229, 232
Garry, L. M., 345
Gartus, A., 62
Gates, A. I., 140
Gauffroy, C., 282
Gauggel, S., 292
Gazzaniga, M. S., 34
Geary, D. C., 14, 244, 369
Gelade, G., 86
Gelman, R., 367
Gentner, D., 303, 321, 323
Gepshtein, S., 57
Gerard, L. D., 381, 385
Gerbier, E., 140
Gerhardstein, P., 356, 358, 361
Germine, L. T., 375
Gernsbacher, M. A., 123, 234, 249, 259, 260, 325
Gerrig, R. J., 269
Geschwind, N., 241, 242
Gevins, A., 117
Giambra, L. M., 380
Gibbs, B. J., 57
Gibson, B. S., 54
Gibson, E., 231
Gick, M. L., 322, 323
Gigerenzer, G., 292, 295, 296, 297
Gilbert, D. T., 352
Gilbert, J., 96
Gilchrist, A. L., 129

Gilhooly, K. J., 279, 388
Gilinsky, A. S., 386
Gillihan, S. J., 143
Gillin, J. C., 28
Gilovich, T., 297
Girotto, V., 292
Gladwell, M., 285
Glanzer, M., 14, 110
Glass, A. L., 67, 215, 216, 329
Glaze, J. A., 9
Glenberg, A. M., 36, 140, 141, 142, 215, 264
Glosser, G., 184
Glucksberg, S., 220, 267, 303
Gobet, F., 88, 318, 330
Godden, D. B., 146, 147
Gogate, L. J., 363
Golay, X., 84
Gold, P. E., 134
Goldin-Meadow, S., 273, 365
Golding, J., 180
Goldinger, S. D., 292
Goldrick, M., 224
Goldstein, D. G., 296, 297
Goldstone, R. L., 321
Golley, M. T., 93
Gomez-Tortosa, E., 40
Goodglass, H., 242
Goodnow, J. J., 9, 181
Goodwin, D. W., 147
Goodwin, G. P., 275
Gopie, N., 143
Gordon, P. C., 249, 259
Gordon, R. L., 127
Gorlick, M. A., 387
Goshen-Gottstein, Y., 139
Gosling, S. D., 2
Goujon, A., 88
Gouvier, W. D., 129
Govoni, R., 142
Graesser, A. C., 150, 180, 249, 250, 260, 264
Graf, P., 143, 209
Grafman, J., 40, 323, 324, 325
Graham, D. J., 13, 14, 17
Gray, J. R., 129
Gray, R., 292
Gray, W. D., 108
Green, C., 141, 142
Green, J., 55
Greenberg, S. N., 20
Greenberg, Z. I., 239
Greenham, S. L., 14
Greeno, J. G., 313, 314, 329
Greenspan, S. L., 317, 386
Greenwald, A. G., 134
Greenwald, M. K., 343
Gregg, V., 51
Grice, H. P., 266, 268
Griffin, D., 289
Griffin, M., 178
Griffin, Z. M., 231, 259
Griffith, B. C., 221, 223, 225
Grill-Spector, K., 67
Groen, G. J., 368
Gross, J. J., 351
Grossberg, S., 88
Grossenbacher, P. G., 53, 54
Grossman, M., 184
Grove, P. M., 87
Grundgeiger, T., 194
Grysman, A., 156

Guidotti, S. J., 260
Gunawan, K., 277
Guo, R. M., 84
Gupta, P., 208
Gustafsson, B., 29
Gutchess, A. H., 62
Gutsell, J. N., 341
Guynn, M. L., 382

H

Ha, Y.-W., 283
Haarmann, H. J., 139
Haber, R. N., 50
Habib, R., 207
Haendiges, A. N., 242
Hagmayer, Y., 299
Hagoort, P., 256
Hahn, S., 380
Hahn, U., 296
Hahne, A., 237
Haines, R. F., 52
Hains, S. M. J., 359
Hakes, D. T., 225
Hall, J. F., 7
Hall, L. K., 153
Hall, M. L., 231
Hall, N. M., 154
Hall, S. E., 384
Hallett, M., 40
Halliday, H. E., 147
Hamann, M. S., 14
Hambrick, D. Z., 127, 128, 129, 136, 167, 380
Hampton, J. A., 165
Handley, S. J., 281, 282
Hanna, A. M., 55
Hansen, E. K., 323, 324, 325
Hansson, P., 239
Hargreaves, D., 259, 260
Harley, E. M., 292
Harper, C. N. J., 281, 282
Harris, I. M., 97
Harris, J. A., 97
Harris, K. S., 221, 223, 225
Harris, R. J., 262
Harrison, T. L., 122, 127
Hartley, J. T., 385
Hartshorn, K., 361
Hartshorne, J. K., 375
Hartsuiker, R. J., 232
Harvey, A. J., 88
Hasher, L., 96, 178, 290, 381, 385
Hashtroudi, S., 383
Hasson, U., 80
Hauser, M., 217
Haviland, S. E., 260
Haxby, J. V., 36, 116
Hayes, B. K., 283
Hayes, J. R., 329, 330
Haywood, S. L., 267
Head, D., 380
Healy, A. F., 20, 101
Heaps, C. M., 203
Hecht, H., 51, 304
Heekeren, H. R., 341
Hegarty, M., 304
Heil, M., 175
Heim, D., 380
Heimler, B., 88
Heinz, S. P., 93
Heitz, R. P., 108, 128, 190

Hell, W., 292
Hellyer, S., 138, 139
Helton, W. S., 82
Henderson, D., 384
Henderson, J. M., 57
Henderson, M. D., 181
Hendrix, V., 283
Henik, A., 89, 172, 174
Henkel, L. A., 140
Henry, J. D., 382
Herman, R., 257
Herrup, K., 26
Hershenson, M., 50
Hertzog, C., 310, 378, 381, 382
Herz, R. S., 30, 154, 336
Hess, T. M., 380, 387
Hewitt, J. K., 116
Hicks, J. L., 142
Hicks, K. L., 127
Higgins, E., 56, 57
Higgins, J. A., 123
Higham, P. A., 137
Hilbig, B. E., 295
Hildreth, E. C., 45
Hill, B. D., 129
Hillman, C. H., 361
Hillyard, S. A., 39
Hilton, D. J., 280
Hindman, J., 385
Hinson, J. M., 294
Hinton, G. E., 169
Hintzman, D. L., 159
Hirshman, E., 174
Hirst, W., 101
Hitch, G. J., 115, 116, 117, 119, 121
Hoch, S. J., 292, 293
Hochel, M., 53
Hockett, C. F., 213
Hoffman, B., 332
Hoffman, H. G., 199, 200
Hoffman, H. S., 221, 223, 225
Hoine, H., 147
Holcomb, P. J., 39, 163, 165, 166, 235, 239, 244
Hollich, G., 363
Hollingworth, A., 87
Holmes, V. M., 257
Holtgraves, T., 263, 264, 268, 269, 270, 271
Holyk, G. G., 166
Holyoak, K. H., 331
Holyoak, K. J., 215, 216, 292, 321, 322, 323, 329, 332
Hommel, B., 127
Honey, C. J., 80
Hopkins, W. D., 217
Horner, A. J., 151
Horowitz, T. S., 92
Horton, W. S., 269
Hothersall, D., 5, 50
Hourihan, K. L., 143
Howard, D. A., 74
Howe, M. L., 153
Howe, S. R., 82
Howerter, A., 116
Hoyle, R. H., 388
Hu, P., 29, 384
Hu, Y., 134
Huang, C-M., 29
Huang, H., 359
Huang, T-R., 88
Huang, Y-M., 340

Huang, Y. T., 249
Hubbard, E. M., 53
Hubbard, T. L., 120, 304
Hudson, J. A., 156
Hughes, R. W., 84, 117
Hull, A., 117
Hultsch, D. F., 381, 384
Hummel, J. E., 332
Humphreys, G. W., 87
Humphreys, K. R., 232
Hussain, Z., 64
Hutchison, J. L., 120
Hyona, J., 256

I

Imreh, G., 134
Inhoff, A. W., 55, 256
Innis, J., 54
Inoue, K., 237
Intraub, H., 120
Intriligator, J., 90
Inzlicht, M., 341
Iordan, A. D., 336
Irwin, D. E., 55, 57, 380
Irwin, W., 260
Isard, S., 223, 224
Islam, M., 84

J

Jackson, C. L., 352
Jacobs, T. L., 82
Jacoby, L. L., 133
Jaeggi, S. M., 127
James, W., 8
Jamis, M., 199
Jared, D., 257
Jefferies, E., 122
Jenkins, J. G., 188
Jerison, H. J., 82
Jesse, A., 225
Ji, Y., 35
Job, R., 67
Joffe, R. T., 97
Johannes, K., 343
Johansson, B., 377
Johansson, G., 58
Johnson, B. T., 143
Johnson, E. J., 310
Johnson, J. C., 97
Johnson, M. K., 123, 292, 383
Johnson, M. R., 123
Johnson, R. L., 254
Johnson, S. K., 170
Johnson-Laird, P. N., 150, 249, 260, 275, 276, 278, 281, 282
Johnston, J. C., 88
Johnston, M., 88
Johnston, S. J., 88
Johnston, W. A., 92, 93
Jones, C. M., 303
Jones, D. M., 84, 117
Jones, G. V., 154
Jones, K. W., 67
Jones, M. B., 88
Jones, T. C., 210
Jonides, J., 57, 74, 84, 97, 108, 116, 119, 303
Joordens, S., 234
Josefsson, M., 239
Josipovic, Z., 126

Joye, Y., 97
Judd, B. B., 386
Juhasz, B. J., 256
Jung, K.-J., 18
Jurica, P. J., 381
Jusczyk, P. W., 244
Just, M. A., 119, 252, 253, 254, 255, 256, 258, 264

K

Kaakinen, J. K., 256
Kahana, M. J., 144, 381
Kahn, J., 386
Kahneman, D., 57, 97, 286, 288, 290, 294, 297
Kaiser, M. K., 304
Kaldy, Z., 356
Kalmar, D., 101
Kam, J. W., 320
Kambe, G., 252
Kamienkowski, J. E., 97
Kandel, E. R., 29
Kane, M. J., 116, 121, 122, 126, 127, 128, 129, 380
Kanwisher, N., 32, 37, 67
Kanz, J. E., 284
Kaplan, E., 242
Kaplan, S., 97
Karpicke, J. D., 139, 140
Kaschak, M. P., 215, 231, 264
Kassam, K. S., 350, 352
Kay, J., 170, 242
Kayra-Stuart, F., 47
Keenan, J. M., 256
Kellas, G., 256, 257
Keller, C. V., 105
Keller, T. A., 105
Kelley, C. M., 129, 201
Kelly, S. D., 267
Kemmerer, D., 37
Kemper, S., 384
Kempton, W., 302
Kemtes, K. A., 384
Kendeou, P., 365
Kennedy, Q., 389
Kenny, R. A., 375
Kensinger, E. A., 343, 346
Keppel, G., 108
Kershaw, T. C., 319
Kertesz, A., 240
Keysar, B., 267
Khemlani, S., 276, 278
Khoo, B. H., 390
Kibbe, M. M., 361
Kidder, D. P., 382
Kieras, D. E., 117
Kim, J., 386
Kim, N. S., 299
Kimchi, R., 57
King-Kallimanis, B. L., 375
Kingstone, A., 84, 88
Kintsch, W., 149, 150, 151, 152, 249, 260, 384
Kirker, W. S., 143
Kisilevsky, B. S., 359
Klatzky, R. L., 190
Klauer, K. C., 282
Klayman, J., 283
Kleider, H. M., 292
Klein, D. V., 234
Klein, J., 82
Klein, P., 361

Klein, R., 87
Klein, S. B., 137, 145, 156
Klemer, N., 62
Kliegl, R., 253
Kneller, W., 88
Knightley, W., 97
Knoblich, G., 68
Knowles, M. E., 330
Koenig, P., 184
Koeppe, R. A., 116, 337
Köhler, W., 316
Kolers, P. A., 209
Kolk, H. H. J., 232
Kolodner, J. L., 321
Konkle, T., 144
Koriat, A., 20, 137, 144, 281
Koslov, K., 350
Koslowski, B., 283
Kosslyn, S. M., 67, 119
Kotovsky, K., 330
Kounios, J., 163, 165, 166, 235, 320
Kozhevnikov, M., 126, 304
Kozlov, M. D., 117
Kraljik, T., 224
Kramer, A. F., 66, 361, 378, 380
Krampe, R. T., 333
Krantz, D. H., 287
Krause, J. A., 350
Krawietz, S. A., 151, 192, 380
Kreiner, H., 20
Krist, H., 303
Kroll, J. F., 239
Krueger, L. E., 383
Krych, M. A., 269
Kubilius, J., 57
Kubovy, M., 57
Kucera, H., 18
Kuhn, G., 84
Kuiper, N. A., 143
Kumon-Nakamura, S., 267
Kunar, M. A., 92
Kurby, C. A., 380
Kurczek, J., 260
Kurtz, K. J., 292
Kusbit, G. W., 191
Kushnir, T., 372, 373
Kutas, M., 39
Kutta, T. J., 231
Kynski, T. R., 299, 300

L

LaBar, K. S., 154, 343
Lacey, S. C., 116
Lachman, J. L., 8, 9, 11
Lachman, R., 8, 9, 11
Lai, P., 289
Lambdon Ralph, M. A., 122
Lambert, W. E., 239
Lang, J. W. B., 351
Lang, P. J., 343
Langer, E., 298
Langlois, J. H., 256
Langner, R., 82
Larkina, M., 359
Larsen, A., 60
Larsen, S. F., 345
Larsson, M., 154
Lasecki, L., 127
Lashley, K. D., 207
Laszlo, S., 234

Latham, A. J., 88
Lau, S. W., 321
LaVoie, D. J., 383
Law, A. S., 156
Lawson, R., 305
Lazerson, A., 264
Leahey, T. H., 6, 8, 11
Leber, A. B., 84
Leder, H., 62
LeDoux, J. E., 336
Lee, C-L., 84, 234
Lee, D. W., 385
Lee, H-W., 257
Lee, H. K., 361
Lee, J. J., 271
Lee, K., 359
Lee, S. C., 208
LeFevre, J., 14
Legrenzi, M. S., 282
Legrenzi, P., 282
Lehman, D. R., 287, 291, 305
Lehmann, T. S., 140
Leirer, V. O., 386
Lemaire, P., 388
Lempert, R. O., 287, 305
Leonard, C. J., 80
Leonesio, R. J., 137
LePine, J. A., 352
LePine, M. A., 352
Lerner, R. J., 54
Lerner, Y., 80
Leslie, A. M., 361
Levine, B., 156
Levinson, S. C., 217
Levitt, A. J., 97
Levy, B. A., 257, 383
Levy, D. A., 175
Levy, J., 124
Levy-Sadot, R., 144
Lewandowsky, S., 199
Lewis, R. L., 108
Lewontin, R. C., 244
Li, K. Z. H., 381
Li, M., 293
Li, X., 118
Liben, L. S., 304
Liberman, A. M., 221, 223, 225
Liberzon, I., 337
Lieberman, K., 115
Light, L. L., 384
Lindenberger, U., 376, 377, 378
Linderholm, T., 264
Lindfield, K. C., 242
Lindsay, D. S., 201
Line, M., 127
Linek, J. A., 239
Link, M., 6
Linkenauger, S. A., 321
Linton, M., 152
Lipko, C., 248
Liszkowski, U., 217
Litman, D. J., 268
Litt, B., 203
Liversedge, S. P., 254, 256
Livesey, E. J., 97
Ljungberg, J. K., 239
Lochner, M. J., 92
Löckenhoff, C. E., 389
Locker, L., 166
Lockhart, R. S., 140

Lockridge, C. B., 269
Loewenstein, G. F., 292
Loewenstein, J., 292
Loftus, E. F., 155, 160, 165, 172, 199, 200, 202
Loftus, G. R., 55, 172, 292
Logan, G. D., 18, 97, 101, 116, 210
Logie, R. H., 125, 156, 381, 388
Lomo, T., 29
Long, D. L., 180, 264
Lorch, E. P., 257
Lorch, R. F., 257
Louchakova, O., 126
Lövdén, M., 376, 377
Lovelace, C. T., 53, 54
Lozito, J. P., 383, 387
Lu, C., 134
Lu, S., 84
Lubart, T., 283
Luchins, A. S., 318, 388
Luck, S. J., 80
Lucy, P., 47
Luk, G., 239
Luo, L., 381
Luria, A., 54
Lutz, K., 348
Luzzatti, C., 91
Lynch, E. B., 330
Lynch, K., 267
Lynchard, N. A., 389
Lynott, D., 263

M

Ma, J-J., 105
Ma'ayan, H., 137
Macfarland, T., 233
MacGregor, J. N., 313, 319
Mack, A., 52
Macken, W. J., 117
Macko, K. A., 36
Mackworth, N. H., 82
MacLean, K. A., 82
MacLeod, C. M., 11, 143, 232
MacLeod, M. D., 192
MacLeod, M. S., 382
MacQueen, G. M., 97
Macrae, C. N., 192
MacWhinney, B., 255
Madden, C. J., 262
Madden, D. J., 18, 382
Maddox, W. T., 387
Madigan, S., 105
Madsen, J. R., 203
Magliano, J. P., 150, 249, 250, 262
Maglio, S. J., 68
Magno, E., 153
Maguire, M., 119
Maier, N. R. F., 316
Maier, S. F., 134
Maisog, C. M., 116
Majerus, S., 358
Maki, R. H., 180
Maljkovic, V., 119
Mall, M., 292
Malmstrom, T., 383
Maloney, E. A., 350
Malt, B. C., 257
Mancini, L., 217
Mandel, D. R., 291
Mandler, G., 5, 8
Mandler, J. M., 8

Mangels, J. A., 381
Manwell, L. A., 98
Marasia, J., 283
Marcel, A. J., 174
Marchman, V. A., 363, 364
Marcus, S. L., 281
Marian, V., 193
Markman, A. B., 181, 183, 321
Markovits, H., 128, 275
Marks, A., 323, 324, 325
Marlatt, G. A., 30
Marler, P., 216
Marsh, E. J., 140, 200, 208, 269
Marsh, R. L., 142
Marshall, P. J., 25, 37
Marslen-Wilson, W. D., 224
Martin, A., 207
Martin, M., 154, 256
Martin, R. C., 25
Martindale, C., 168, 169
Marzi, C. A., 119
Masling, M., 150
Massaro, D. W., 225
Masson, M. E. J., 82, 150, 169, 210
Masters, R. S. W., 352
Masuda, T., 62
Mata, R., 387
Mather, M., 37, 292, 335, 336, 388, 389
Mathis, K. M., 171
Matlock, T., 262
Matthews R. A., 129
Matthews, P. D., 257
Mattingly, I. G., 225
Mattingly, J. B., 53
Mattson, M. E., 177
Maurer, D., 54
May, J., 97
Mayberry, R. I., 231
Mayhorn, C. B., 382
Mayr, S., 96
Mayr, U., 129
McCabe, D. P., 129, 383
McCabe, J., 137
McCaffrey, T., 317
McCandliss, B. D., 127
McCann, R. S., 88
McCarley, J. S., 66
McCarthy, R. A., 115, 169, 242
McClearn, G. E., 377
McClelland, A. G. R., 345
McClelland, J. L., 42, 62, 114, 169, 224, 256
McCloskey, M., 25, 199, 303, 304
McCloy, R., 292
McConnell, A. R., 351
McCoy, A. M., 350
McDaniel, M. A., 134, 140, 143, 155, 156, 345, 382
McDermott, K. B., 140, 195, 196, 203
McDonald, A., 54
McDonald, J. L., 255
McDonough, I. M., 378
McEleney, A., 291
McEvoy, L., 117
McFarland, C., 292
McGeoch, J. A., 188
McGillivray, S., 383
McGoldrick, J., 98
McIntosh, A. R., 207
McKoon, G., 171, 233, 260, 262, 263
McLaughlin, J., 237

McLean, J. F., 231
McLeod, P., 318
McMains, S., 119
McMillan, C., 184
McNamara, T. P., 144, 171
McNeill, D., 292
McNerney, M. W., 54
McQueen, J. M., 226
McRae, K., 169, 174
McSpadden, M., 30
McWilliams, J., 96
Meade, M. L., 196
Mecklinger, A., 237
Medin, D. L., 159, 283, 304, 330
Mehler, J., 244
Meinz, E. J., 128
Melamed, L. E., 71
Melton, A. W., 134
Mendes, W. B., 350
Mériau, K., 341
Merikle, P. M., 53, 54, 126
Metcalf, K., 257
Metcalfe, J., 140, 319, 320
Metzing, C., 267
Metzler, J., 119
Meyer, B. J. F., 384
Meyer, D. E., 17, 117, 173
Mikels, J. A., 388
Milán, E. G., 53
Milkman, K. L., 285
Miller, D. T., 292
Miller, G. A., 105, 217, 223, 224, 247
Millis, K. K., 264
Mills, C. B., 54
Milner, B., 208
Milovanovic, G. S., 165
Minoshima, S., 116
Minsky, M. L., 37
Mintun, M. A., 116
Mirzazade, S., 348
Mishkin, M., 28, 36
Mitchell, D. B., 172, 210
Mitchell, D. C., 257
Mitchell, K. J., 202
Mitterschiffthaler, M. T., 341
Miyake, A., 116, 126, 264
Moat, H. S., 225
Mock, B. J., 260
Mody, M., 41
Moldoveanu, M., 298
Monaco, G. E., 262
Monaghan, P., 43, 321
Monfils, M., 166
Monin, B., 253
Moore, A. B., 128
Moore, A. M., 369
Moore, P., 184
Moore, S. M., 93
Morales, M., 199
Moray, N., 73
Moreira, C., 302
Moreno, S., 126
Moreno-Ríos, S., 369
Morey, C. C., 125
Morrell, R. W., 382
Morrison, P. E., 55
Morrow, D. G., 386
Morsanyi, K., 282
Morse, R., 356
Morton, J., 75, 210, 244

Moscovitch, M., 167, 343, 345
Moss, H. E., 225
Motes, M. A., 126
Motz, B. A., 321
Moyer, R. S., 47
Mozer, M. C., 242
Mrazek, M. D., 320
Mueller, S. T., 117
Muller, M., 292
Murayama, K., 383
Murdock, B. B., 110
Murphy, G. L., 165, 183, 234, 257
Murphy, M. C., 351
Murray, D. J., 117
Mushaney, T. J., 385
Myers, E. B., 223
Myers, J. L., 256, 263

N

Nadel, L., 153
Nairne, J. S., 145
Nakamura, G. V., 180
Nakamura, K., 302
Nara, B., 380
Nash, M., 203
Nash, R. A., 156
Navajas, J., 97
Navarro, D. J., 283
Naveh-Benjamin, M., 142
Navon, D., 80, 97
Neary, K. R., 143
Neath, I., 74, 139, 381
Neblett, D. R., 210
Nee, D. E., 116
Neely, J. H., 173, 174
Neill, W. T., 96
Neisser, U., 15, 54, 55, 56, 58, 72, 101, 193
Nelson, K., 153, 361
Nelson, T. O., 30, 137, 153
Newell, A., 325, 328
Newell, B. R., 285, 296
Newman, J. E., 224
Newstead, S. E., 277
Ng, R., 289
Nichelli, E., 40
Nicholls, A. P., 117
Nickerson, R. S., 203, 269
Niki, K., 336
Nilsson, L-G., 239, 382
Nisbett, R. E., 62, 159, 287, 305
Nissen, M. J., 85
Nix, L. A., 290
Noice, H., 144, 193
Noice, T., 144, 193
Noordzij, M. L., 168
Norenzayan, A., 159
Norman, D. A., 11, 107, 108, 138, 266, 268
Norman, K. A., 123, 383
Norris, D., 226, 255
Nowak, C. A., 384
Nugent, L. D., 105
Nunes, L., 387
Nuthmann, A., 253
Nyberg, L., 207, 376, 377, 382
Nyquist, L., 284

O

O'Brien, E. J., 256, 263
O'Connor, R. C., 380
O'Donnell, P. J., 234, 349
O'Hara, R., 105
O'Neill, D. K., 156
O'Seaghdha, P. G., 236, 244
Oaksford, M., 279
Oberauer, K., 199
Oberfeld, D., 51
Ogden, W. C., 85
Ohlsson, S., 319, 321
Öhman, A., 83, 339
Oliva, A., 144
Olson, E. A., 201
Oosterwijk, S., 348
Op de Beeck, H. P., 57
Oppenheim, G. M., 225
Oppenheimer, D. M., 285
Oppy, B. J., 264
Ormerod, T. C., 313, 319, 320, 321
Ornstein, P. A., 200
Osterhout, L., 39, 237, 239, 244
Oswald, F. L., 103
Owsianiecki, L., 54
Ozubko, J. D., 143
Özyürek, S., 267

P

Paap, K. R., 239
Pacheco, J. L., 387
Pachur, T., 387
Paivio, A., 146, 166
Palermo, D. S., 219, 229
Pallar, K. A., 207
Palmer, J. C., 160
Palmer, S. E., 57, 67
Pals, R., 97
Pandeirada, J. N., 145
Pandeirada, N. S., 145
Parada, E. F., 20
Park, D. C., 29, 62, 378, 382, 385
Parkman, J. M., 368
Parks, C. M., 383
Parks, N. A., 108
Parrish, T. B., 320
Pascual-Leone, A., 40
Pashler, H., 97, 124, 140, 320
Passchier, J., 82
Pastötter, B., 140
Paterson, T., 345
Patston, L. L., 88
Patterson, S., 37
Paul, S. T., 234, 256
Pavani, F., 88
Payne, D. G., 134
Payne, J. D., 346
Payne, T. W., 129
Pearlman-Avnion, S., 144
Pearlmutter, N. J., 255
Pecchinenda, A., 175
Pecher, D., 348
Pedersen, N. L., 377
Pedone, R., 332
Pell, M. D., 348
Penfield, W., 208
Peng, K., 159
Perea, M., 238
Peretz, I., 171
Perfors, A., 283
Perlmutter, M., 284
Perrachione, T. K., 231
Pesta, B. J., 143
Peterson, L. R., 105, 107, 108

Peterson, M. A., 57
Peterson, M. J., 105, 107, 108
Peterson, R. R., 236, 256
Peterson, W. W., 48
Petit, L., 116
Petrill, S. A., 377
Petrusic, W. M., 293
Petry, M. C., 343
Pettijohn, K. A., 151, 192, 386
Pexman, P. M., 166
Pfister, R., 5
Philips, L. H., 382, 388
Pickering, M. J., 231, 256, 267
Pickett, J. M., 223
Piercey, C. D., 234
Pink, J. E., 129
Pinker, S., 214, 271
Pitt, M. A., 224
Pittenger, C., 29
Plewes, P. S., 263
Pleydell-Pearce, C. W., 152
Plomin, R., 377
Plude, D. J., 380
Poldrack, R. A., 25
Poljac, E., 57
Polk, T. A., 260
Pollack, I., 223
Pollard, P., 277
Pollatsek, A., 255, 257
Pollmann, S., 43
Polson, M. C., 34
Polson, P., 329
Polya, G., 330
Pomer-antz, J. R., 57
Pomplun, M., 330
Poon, L. W., 153
Poppenk, J. L., 343
Posner, M. I., 85, 87, 97
Postma, A., 168
Postman, L., 190
Potvin, F., 275
Pourtois, G., 83
Powell, B., 147
Powell, T., 96
Prabhakar, J., 156
Pratkanis, A. R., 134
Pratt, J., 84, 88, 126
Price, R. H., 72
Prince, S. E., 159
Proffitt, D. R., 68, 304

Q

Qin, Y., 18
Queen, T. L., 387
Quillian, M. R., 161, 163
Quinn, P. C., 360

R

Radeau, M., 171
Rader, A. W., 280, 281
Radulescu, P. V., 84
Radvansky, G. A., 54, 122, 126, 128, 151, 156, 190, 191, 192, 264, 380, 381, 384, 385, 386, 389
Rafal, R. D., 89, 90
Rahhal, T. A., 153
Raine, L. B., 361
Rajaram, S., 200
Rakow, T., 285
Ramachandran, V. S., 53

Ramsey-Rennels, J. L., 256
Randall, J. G., 103
Ranganath, C., 207
Ransom, K., 283
Rapp, B., 224
Rapp, D. N., 365
Rasch, B., 29
Rasmussen, A. S., 156
Ratcliff, R., 171, 260, 262, 263
Ratiu, I., 239
Raymond, J. E., 338
Rayner, K., 55, 56, 57, 234, 252, 253, 254, 255, 256, 257
Read, L., 92
Reason, J., 102
Rebok, G. W., 384
Reder, L. M., 191, 200
Redford, J. S., 66
Redick, T. S., 127
Reed, A. E., 389
Reed, S. K., 20, 332
Reeves, L. M., 330
Regal, D., 356
Regev, M., 80
Regolin, L., 356
Rehder, B., 183
Reichle, E. D., 255
Reingold, E. M., 330
Reisdorf, P., 264
Reiskamp, J., 387
Reitman, J. S., 74
Remington, R. W., 88
Renaud, O., 156
Reuter-Lorenz, P. A., 196
Reyna, V. F., 302
Ric, F., 383
Rice, G. E., 384
Rich, A. N., 53
Richards, A., 282
Richardson, D., 253
Richardson, M., 120
Richardson, S. L., 382
Riklund, K., 376, 377
Riley, K. P., 29
Rinck, M., 264
Rinehart, C. A., 294
Rips, L. J., 163, 165, 265, 281, 283
Rist, R., 283
Ristic, J., 84
Rizzolatti, G., 36
Roberts, M. A., 98
Robertson, D. A., 260, 264, 375
Robertson, R. R. W., 260
Robertson, S. P., 303
Robertson, T. E., 137, 156
Robins, R. W., 2
Rock, I., 52
Rodman, R., 220
Roediger, H. L., 108, 139, 140, 145, 195, 196, 200, 203, 208, 209, 343, 383
Roese, N. J., 291
Rogers, T. B., 143
Rohrer, D., 304
Rolke, B., 175
Rönnlund, M., 382
Rosen, V. M., 125, 127, 128
Rosenkrantz, S. L., 360
Ross, B. H., 183, 331
Roth, N., 68
Rotteveel, M., 338

Rottveel, M., 348
Rovee-Collier, C., 356, 358, 361
Roy, M., 337
Rubin, D. C., 133, 153, 154, 193, 343, 388
Ruddy, M. G., 17, 173
Rugg, M. D., 39
Rumelhart, D. E., 42, 62, 169, 266, 268
Rundus, D., 139
Russell, J. L., 217
Russell, P. N., 82
Russell, R., 55
Russo, J. E., 310
Rusting, C. L., 381
Rydell, R. J., 351

S

Sachs, J. S., 149, 150, 247
Sacks, O., 70
Sadesky, G. S., 14
Saffran, E. M., 241
Sakaki, M., 336
Salame, P., 116
Saliba, B. J., 361
Salthouse, T. A., 129, 376
Saltz, E., 144
Samaranayake, V. A., 369
Samenieh, A., 192
Samuel, A. G., 224
Samuel, D., 28
Sanbonmatsu, D. M., 93
Sanchez, C. A., 126, 127
Sander, D., 83
Sanna, L. J., 292
Sanocki, T., 84
Santamaría, C., 283, 369
Sargent, J. Q., 380
Sarter, M., 25
Sattler, J. M., 106
Savazzi, S., 119
Savva, G. M., 375
Scearce, K. A., 259
Schab, F. R., 147
Schacter, D. L., 80, 82, 156, 175, 187, 208, 209, 343, 383
Schaeken, W., 282
Schäfer, M., 217
Schank, R. C., 268
Schatschneider, C., 231
Schellenberg, E. G., 126
Scherer, K. R., 83
Schiffer, S., 296
Schiller, P. H., 357, 359
Schilling, H. E. H., 256
Schimmack, U., 345
Schleich, M. C., 262
Schmader, T., 351
Schmalhofer, F., 151, 152
Schmeichel, B. J., 117
Schmidt, J. R., 277
Schmidt, S. R., 150
Schneider, K. K., 96
Schneider, W., 97, 390
Schnorr, J. A., 146
Schnyer, D. M., 387
Schooler, J. W., 154, 199, 320, 321
Schooler, L. J., 204, 387
Schrauf, R. W., 193
Schubotz, R. I., 348
Schulz, R. W., 190
Schulze, C., 188

Schulze-Bonhage, A., 203
Schunn, C. D., 108
Schustack, M. W., 256
Schvaneveldt, R. W., 17, 173
Schwartz, D. L., 303
Schwartz, N., 292
Schwarz, K. A., 5
Schwarz, N., 382
Schweickert, R., 118
Scott, G. G., 349
Scott, J., 341
Scott, S., 97
Scullin, M. K., 156
Seaman, S. R., 145
Searle, J. R., 263
Sederberg, P. B., 123, 203
Sedgewick, C. H. W., 9, 144
Sedivy, J. C., 253
See, J. E., 82
Seely, M. R., 264
Segalowitz, S. J., 383
Sehulster, J. R., 139, 152
Seidenberg, M. S., 169
Seifert, C. M., 303
Sekuler, A. B., 64
Selfridge, O. G., 59
Sellaro, R., 127
Sells, S. B., 279
Serences, J. T., 84
Sereno, S. C., 234, 349
Seymour, T. L., 117
Shafir, E., 286, 292
Shah, N. J., 348
Shah, P., 126
Shallice, T., 115, 169, 207
Shand, M. A., 119
Shanks, D. R., 296
Shankweiler, D. P., 221, 223, 225
Sharkey, N. E., 257
Sharp, C., 386
Shaw, J. C., 328
Shaywitz, B. A., 41
Shaywitz, S. E., 41
Sheehan, E., 127
Sheffer, L., 137
Shelton, A. L., 144
Shelton, J. T., 129
Shen, J., 35
Shepard, R. N., 119
Sherman, G., 216
Sherrill, M. R., 388
Shiffman, H., 154
Shiffrin, R. M., 16, 97, 138
Shimamura, A. P., 381
Shintel, H., 267
Shipstead, Z., 122, 127
Shoben, E. J., 163, 165
Shohamy, D., 337
Shomstein, S., 84
Shors, T. J., 30
Shuchat, J., 381
Shulman, H. G., 237
Siegler, R. S., 290, 321, 369
Sigman, M., 97
Silbert, L. J., 80
Simion, F., 356
Simon, D., 292
Simon, E. W., 384
Simon, H. A., 30, 310, 325, 327, 328, 329, 330, 330, 332

Simons, D. J., 84, 88
Simony, E., 80
Simpson, G. B., 18, 166, 234, 256
Singer, M., 260, 264
Singh, M., 57
Sinigaglia, C., 68
Sio, U. N., 320, 321
Sitaran, N., 28
Sitton, M., 242
Skinner, B. F., 10
Slamecka, N. J., 143, 193
Sloman, S. A., 299
Sloutsky, V. M., 280, 281
Slovic, P., 286
Slowiaczek, M. L., 55
Small, B. J., 381
Smallwood, J., 320, 380
Smilek, D., 53, 54, 84, 380
Smith, A. D., 383
Smith, E. E., 116, 119, 163, 165, 184
Smith, J. D., 66
Smith, L., 356
Smith, M. C., 173, 284
Smith, M. E., 117
Smith, R. E., 356, 382, 383
Smith, S. M., 141, 142, 320
Smith, S. W., 384
Smith, T. A., 123
Smith, W. R., 384
Snow, C., 269
Snyder, C. R. R., 85, 97
Solman, G. J. F., 380
Son, L. K., 137
Sorace, A., 238
Sornette, D., 302
Soveri, A., 127
Sozzo, S., 302
Spalek, T. M., 173
Spangenberg, E. R., 134
Spaniol, J., 382
Specht, K., 348
Speer, S. R., 256
Spelke, E. S., 101, 357, 361
Spellman, B. A., 20, 291
Spence, C., 92
Spence, I., 88, 126
Sperling, G., 54, 356
Sperry, R. W., 34
Spieler, D. H., 191
Spieth, W., 93
Spilich, G. J., 384
Spillers, G. J., 126, 146
Spivey, M. J., 253, 262
Spreng, R. N., 156
Squire, L. R., 28, 132, 175, 207, 208
Stacy, E. W., 67
Stadler, M. A., 196
Stahl, C., 282
Stampe, D. M., 330
Standing, L., 144
Stanfield, R. A., 262
Stanhope, N., 204, 205
Stanovich, K. E., 231
Stark, C. E. L., 175
Stawarczyk, D., 358
Stazyk, E. H., 14
Steele, C. M., 351
Steg, L., 97
Stein, B. S., 192, 194, 330
Stephens, D. L., 310

Stern, E., 321
Stern, J., 147
Stern, Y., 378
Sternberg, R. J., 319
Sternberg, S., 17, 111, 113
Stewart, A. J., 231
Stewart, S. T., 381
Stickgold, R., 136, 361
Stijnen, T., 82
Stine, E. A. L., 384, 385
Stine-Morrow, E. A. L., 386
Stojack, C. C., 385
Stolz, J. A., 98
Storm, B. C., 192
Storms, G., 283
Stout, J. C., 284
Stragà, M., 292
Strayer, D. L., 92, 93
Stritzka, W. G. K., 199
Stroop, J. R., 10, 11
Struiksma, M. E., 168
Studdert-Kennedy, M., 221, 223, 225
Stylos-Allan, M., 29, 384
Subramaniam, K., 320
Sudberry, M. V., 380
Sunderman, G., 239
Surprenant, A. M., 74, 381
Suszko, J. W., 84
Sutherland, M. R., 335
Swaby, M., 119
Swencionis, J. K., 352
Swets, J. A., 48
Symons, C. S., 143
Szpunar, K. K., 156

T

Taconnat, L., 383
Taglialatela, L. A., 66
Talarico, J. M., 154, 343
Talavage, T., 37
Talmi, D., 345
Talmi, J. M., 345
Tamplin, A. K., 151, 192, 380
Tanaka, J. T., 67
Tandler, J. M., 285
Tanenhaus, M. K., 253
Tang, Y., 35
Tanner, W. P., 48
Taraban, R., 256
Tattersall, I., 217
Taylor, E., 321
Taylor, J., 97
Taylor, S. F., 337
Taylor, W. K., 93
Teller, D. Y., 356
Tenebaum, J. B., 299, 300
Terry, K. M., 96
Tesch-Römer, C., 333
Teuber, H. L., 208
Thagard, P., 321, 323
Theeuwes, J., 380
Thomas, A. K., 383
Thomas, J. C., 328
Thomas, L. E., 310
Thomas, R. D., 295, 351
Thompson, A. N., 151
Thompson, C. K., 241
Thompson, D. M., 146
Thompson, L. A., 127
Thompson, R. F., 30

Thompson, S. R., 145
Thompson, V. A., 277, 282
Thompson, W. L., 119
Thomson, D. M., 192
Thomson, N., 105
Thorndike, E. L., 188
Thwing, E. J., 93
Tillman, B., 150
Tipper, S. P., 68, 95, 96, 97
Tippett, L. J., 88
Tolman, E. C., 48
Tomasello, M., 217
Toppino, T. C., 140
Toth, J. P., 133
Touron, D. R., 380
Townsend, J. T., 18
Trabasso, T., 260, 264
Tranel, D., 348
Trawley, S. L., 156
Traxler, M. J., 256
Treccani, B., 238
Treisman, A., 86
Treisman, A. M., 55, 94, 224
Trick, L. M., 92
Treisman, A., 57
Trigg, G. L., 54
Trope, Y., 68
Tsai, N., 127
Tse, P. P., 369
Tsukiura, T., 159
Tucker, M., 69
Tuholski, S. W., 129
Tullis, J., 140
Tulving, E., 24, 133, 140, 144, 146, 192, 206, 207
Tun, P. A., 384
Tuniz, C., 217
Turkheimer, F., 39
Turner, M. L., 125
Turner, T. J., 180
Turvey, M. T., 73, 74, 225
Tversky, A., 286, 288, 289, 290, 294, 297
Tyler, L. K., 225
Tzelgov, J., 174

U

Ullman, S., 45
Underwood, B. J., 108, 190
Ungerleider, L. G., 36, 116
Unsworth, N., 108, 126, 128, 146
Usher, M., 139

V

Vachon, F., 84
Vadaga, K. K., 381
Valdes, L. A., 96
Vallar, G., 115
Vallone, R., 297
van Baaren, R. B., 285
Van Berkum, J. J. A., 256
van den Broek, P., 264, 365
van der Helm, P. A., 57
van der Leij, A., 285
Van der Linden, M., 156
van der Meer, E., 341
Van der Stigchel, S., 87
van Dijk, T. A., 151, 249, 260, 384
van Elk, M., 168
van Hell, J. G., 239

van Leeuwen, C., 57
van Zoest, W., 88
Vandierendonck, A., 117
VanLehn, K., 309, 311, 312
Vasta, R., 304
Vecera, S. P., 98
Veloz, T., 302
Verfaille, K., 120
Vergauwe, E., 117, 282
Verges, M., 84
Verhaeghen, P., 126
Vermeylen, F., 283
Vessal, R. S., 93
Vestal, M., 6
Vidoni, E. D., 66
Vinke, L., 231
Virtue, S., 264
Visscher, P. K., 216
Vivas, A. B., 87
Vogel, E., 129
Voice, J. K., 225
von der Heydt, R., 57
von Frisch, K., 216
Voorspoels, W., 283
Voss, A., 382
Vredeveldt, A., 117
Vu, H., 257
Vuilleumeir, P., 338, 340
Vul, E., 320

W

Wagemans, J., 57, 88
Wagenaar, W. A., 152
Wagenmakers, E., 338
Wager, T. D., 116, 337
Wagner, A. D., 25
Wagner, S. H., 367
Walker, M. P., 29, 136, 384
Wallin, A., 299
Walsh, C., 292
Walsh, E., 223
Walsh, N. D., 341
Walters, J., 367
Wang, A., 151
Wang, Z., 302
Wapner, S., 357
Ward, J., 54
Waris, O., 127
Warm, J. S., 82
Warren, P. A., 296
Warren, R. M., 99, 224
Warren, T., 256
Warrington, E. K., 115, 169, 242
Washburn, D. A., 66
Wason, P. C., 281
Wassenburg, S. I., 263
Wasserman, S., 361
Watkins, M. J., 107, 140
Watson, J. B., 8, 10
Watson, J. M., 196
Waugh, N. C., 107, 108, 138
Weber, E. U., 287

Webster, J. C., 93
Webster, L., 383
Weeden, L., 88
Weingartner, H., 28
Weinstein, Y., 145
Weintraub, S., 242
Weisberg, J., 207
Weisberg, R. W., 310, 320, 330
Weise, S. B., 119
Weiskrantz, L., 115
Well, A. D., 254
Wells, G. L., 201
Welsch, D., 151, 152
Welsh, A., 117, 224
Welsh, R. C., 62
Wenger, M. J., 18, 134
Wenk, G. L., 134
Werner, H., 357
Wertheimer, M., 60
West, R. F., 231
West, W. C., 235
Westendorf, T., 54
Westmacott, R., 167
Wetherick, N. E., 279
Whaley, C. P., 18
Wharton, C. M., 323, 324, 325
Whelan, S. M., 304
White, M. J., 365
White, S. J., 254
Whitney, P., 227, 232, 294
Whorf, B. L., 219
Wichert, A., 302
Wickelgren, W. A., 312, 320, 330
Wickens, C. D., 66
Wickens, D. D., 109
Wiebe, D., 319, 320
Wiener, E. L., 82
Wiener, H. J., 285
Wiggs, C. L., 207
Wigstrom, H., 29
Wiley, C., 37
Wiley, J., 127, 257, 330
Wilhelm, O., 129
Wilkening, F., 303
Wilkes-Gibbs, D., 267
Willander, J., 154
Willems, J. L., 341
Williams, R. W., 26
Williams, S. C. R., 341
Williamson, V. J., 119
Wilson, B., 115
Wilson, M. P., 68, 106, 231
Wilson, R. S., 378
Wilson, T. D., 352
Windschitl, P. D., 287
Winer, G. A., 51
Wingfield, A., 144, 242, 381, 385
Winkielman, P., 348
Wippich, W., 133
Wiscott, R., 143
Witt, J. K., 68, 88
Witzki, A. H., 116

Wixted, J. T., 189
Wolitzky-Taylor, K. B., 341
Wolters, G., 82
Wondoloski, T. L., 361
Wong, K. Y., 285
Woo, H. K., 321
Wood, N. L., 93, 105
Wood, P. K., 105
Woodworth, R. S., 279
Worthy, D. A., 387
Wraga, M., 119
Wright, A. A., 108
Wroe, S., 217
Wurm, L. H., 145
Wynn, V., 388

X

Xie, X., 359
Xu, F., 372, 373
Xu, J., 140

Y

Y'dewalle, G., 120
Yamauchi, T., 183
Yang, D., 134
Yang, S-J., 68
Yantis, S., 57, 84
Yap, M. J., 14
Yaro, C., 54
Yates, M., 166
Yaure, R. G., 390
Yaxley, R. H., 262
Ye, H. H., 359
Yechiam, E., 284
Yeo, R. A., 39
Yonelinas, A. P., 133, 142
Yopchick, J. E., 299
Young, L. T., 97
Young, S. E., 116
Yu, D., 117
Yuille, J. C., 146
Yukalov, V. I., 302

Z

Zacks, J. L., 191, 380
Zacks, J. M., 380
Zacks, R. T., 96, 191, 192, 290, 380, 381, 385
Zanesco, A. P., 82
Zaragoza, M. S., 199, 202
Zavagnin, M., 380
Zbrodoff, N. J., 101
Zeelenberg, R., 337, 338, 348
Zelinski, E. M., 381
Zhang, W., 35
Zhou, K., 84
Zilles, K., 348
Zimny, S., 151, 152
Zola-Morgan, S., 153, 208
Zucco, G., 125
Zwaan, R. A., 150, 249, 250, 262, 263, 264, 384, 385

Subject Index

A

Absent-mindedness, 188, 203
Abstract–congruent sentence, 234
Accuracy, 14–15
Acetylcholine, 28, 78
Action potential, 27
    propagation of, 27
Action slips, 104
Adaptive thinking, 295–298
Additive factors method, 111
Ad hoc categories, 184–185
Advantage of clause recency, 260
Advantage of first mention, 259
Age categories, 355
Age-related neurological changes, 377
Agnosia, 66–71
Agraphia, 242
"Aha!" reaction, 319
Alertness, 78, 81–82
Algorithm, 284–286
All-or-none principle, 27
Alzheimer's disease, 28, 192
Ambiguous sentences, 227, 228
American Sign Language (ASL), 106, 119, 363
Amnesia, 153, 206–210
AMPA receptors, 29
Amygdala, 30–31, 336
Analogy, 319–325
    neurocognition in, 323–325
    problems, 321–323
Anaphoric reference, 258
Animal communication, 215–217
Animal learning in laboratory, 8
Anomia, 25, 169–170, 241–242
Anomic aphasia. *See* Anomia
Antecedent, 258, 259–260
Anterograde amnesia, 206, 207–208, 332
Anticipations of psychology, 6
Aphasia, 240–244
Apperceptive agnosia, 71
Arbitrariness of language, 214–215
Arousal, 81–82
Arithmetic, 368
Articles, 260
Articulation
    manner of, 220
    place of, 220
Articulatory loop, 117
Articulatory suppression effect, 117
Association neuron, 26
Associative agnosia, 71
Associative interference, 190–191
Atmosphere heuristic, 278–279
Attention, 3, 6, 11
    in children, 356–359
    controlled, 88, 91–97
    emotional guidance of, 337–339
    Engle's controlled model, 121–123
    forms of input, 81–90
    as a limited mental resource, 80
    as mental process, 79–80
    as mental resource, 97–103
    multiple meanings of, 78–80
    in older adults, 379–380
    processes, 356–358
    selection models, 93–97
    selective, 93
    video games and, 88
Attentional blink, 97, 340
Attention capture, 83–84
Attention-consuming counting task, 110
Attention-directing information, 85
Audition, 71, 75, 356
Auditory cortex, 73, 241
Auditory pattern recognition, 74–76
Auditory perception, 71–76
Auditory persistence, 75
Auditory sensation, 71–76
Auditory sensory memory, 72–74
Authorized inference, 263
Autobiographical memory, 152–155
    infantile amnesia, 153
    involuntary memory, 154
    psychologists as subjects, 152–153
    reminiscence bump, 153–154
Automaticity, 97
    disadvantages of, 101–103
    role of practice in, 101
Automatic priming, 100
Automatic processing, 97–100, 230–231
Availability heuristic, 289–290
Axons, 50
    of the ganglion cells, 50–51

B

Babbling stage, 363
Backward inferences, 264
Backward masking, 174
Bartlett's research on memory, 176
Basic number sense, 356
Bayes's theorem, 289
Beethoven's musical fame and deafness, 3
Behaviorism, 8
Belief bias, 276, 277
Benefit, 293
    of connectionist model, 169–170
    of semantic priming, 171–172
Beta movement, 60, 61
Biases, 188, 286–287
    hindsight, 292–293
    in the representativeness heuristic, 288
    typical, 291–292
Bilingualism, 238–239
Bipolar cells, 50
Blocking, 52, 127, 188, 203
Bottom-up, data-driven processing system, 61
Boundary extension, 120
Bound morphemes, 232, 241
Brain
    and connectionism, 42–43
    cortical structures, 31–32
    damage, 24–25, 243–244
    related disruptions of language and cognition, 241
    subcortical structures, 30–31
Brain–cognition relationships, 21, 24–25
Bridging inference, 250, 262
Broca's aphasia, 240–241
Broca's area, 116, 241, 242
Brodmann's areas, 32
Brown–Peterson task, 107, 108, 109, 128, 138, 142
Bruner, Jerome, 371–372

C

Case grammar, 235, 236
Case roles, 236
Categorical perception, 221, 223
Categorization
    characteristics of human categories, 142
    classic view of, 181–182
    explanation-based theories, 184–185
    probabilistic theories, 183–184
Categorical syllogism, 276–278
Category-specific deficit, 169
Central executive, 116, 117
Central tendency, 182
Cerebral cortex, 30, 31, 43, 71
Cerebral lateralization, 33, 166
Channel capacity, 12
Characteristic features, 163
Child-directed speech, 366
Childhood
    attention, 356–359
    decision-making ability, 369–373
    language processing, 362–366
    learning of new words, 365–367
    linguistic performance and linguistic competence, 364–365
    memory, 359–362
    neurological changes, 355–356
    perception, 356–359
    problem-solving ability, 369–373
    stages of language acquisition, 363–364
Chomsky's transformational grammar
    limitations, 228–230
    phrase structure grammar, 227–228
    transformational rules, 228
Classic view of categorization, 181–182
Clause recency, 259, 260
Coarticulation, 223, 225
Cognition, 37
    cross-fertilization for, 25
    defined, 4
Cognitive aging, 375–389
    and decision making, 386–387
    and emotion, 388–389
    and language processing changes, 383–386
    and memory, 380–383
    and neurological and cognitive changes in, 375–378
    perception and attention, 378–380

Cognitive aging (*continued*)
    and problem solving, 388
    and reasoning, 386
    studies, 376
    successful, 378
Cognitive demons, 59–60
Cognitive operations, 309
Cognitive psychology, 1–2, 20
    history of, 5–11
    and information processing, 11–12
    relevance of computing, 11
    verbal learning, influence on, 9–10
Cognitive Psychology, 4
Cognitive revolution, 1
    1950s, 1, 8
Cognitive science, 1, 2, 20
Competence, 218, 364–365
Composite memory, 199
Computational demons, 59
Computer analogy, 18
Computer-based technique for modeling complex systems. *See* Neural network models
Conceptual knowledge, 159
Conceptually driven pattern recognition, 61–62
Concrete–congruent sentence, 234
Concrete operations stage, 369–370
Condition–action pair, 328
Conditional reasoning, 275, 279–283
Conduction aphasia, 241
Cones, 50
Cone synapses, 50
Confirmation bias, 281
Conjugate reinforcement, 360
Conjunction fallacy, 297–298
Conjunction search, 86, 87, 356
Connectionism, 42–43, 168–170
Connectionist modeling, 24, 62–66
    of four-letter words, 65
    hidden units, 63
    input units, 65
    output units, 63
    of semantic memory loss, 169
    terminology, 64
Connectionist models, 42,
    top-down and bottom-up influences in, 63–66
    of word recognition, 65
Conscious/controlled processing, 97–98
Consequent of conditional clause, 279
Consolidation, 29, 135–136
Content processing, 147
Context, 146–148
Context effects, 18–20, 223–224
    event-related potentials (ERPs) and, 234
Contralaterality, 32–33, 50, 51, 71, 90
Contra lateral visual field, 50
Contrasting input, 88
Controlled attention, 88
Controlled priming, 173–174
Control processes, 16, 339
Convergence of synapses, 28
Conversation
    cognitive factors, 266–269
    cooperative principle, 266, 268
    empirical effects in, 270–271
    indirect replies, 270–271

indirect requests, 270
online theories, 268–269
rules or maxims, 266–268
structure of, 266
topic maintenance, 268
Convertible, concept of, 259
Cooperative principle, 266, 268
Corpus callosum, 30
Correlated attributes, 182
Cortex
    specialization in, 35–37
    visual pathways in, 36
Counterfactual reasoning, 291–293. *See also* Heuristics
Cross-sectional study, 376
Cryptomnesia, 147
Culture, influence on cognitive neurological processing, 35

D

Data-driven processing *vs* conceptually driven processing, 21
Decay, 54, 108, 188
Decision demon, 59, 60
Decision making, 9, 20, 183
    and problem solving, 369–373
Decisions, 283–286
    algorithms and heuristics, 284–286
    fast and slow, 285
    risky, 293–295
Declarative memory, 132–133, 156, 208, 359, 360
Deep structure representation, 227
Defining feature, 163
De Memoria, 5
Dendrites, 26, 27
Depth of processing, 140
    challenges to, 141–143
Digital computer, 11
Direct theory, 268, 269
Displacement, 215
Dissociation, 24
    double, 25
    of episodic and semantic memory, 206–207
    lack of, 25
    simple, 25
Distance effect/discriminability effect, 47
Distributed practice, 139–140
Divergence of synapses, 28
Domain knowledge, 302–305
Donders's subtractive method, 111
Dorsal pathway, 37
Dorsolateral prefrontal cortex (DLPFC), 116, 129
Double dissociation, 25
Downhill change, 292
Dual coding hypothesis, 166
Dual task/dual message procedure, 93
Dynamic icons, 55
Dysexecutive syndrome, 117
Dysfluencies, 218–219
Dyslexia, 41, 242

E

Early selection theory of attention, 94, 95
Easterbrook hypothesis, 346

Ebbinghaus's forgetting curve, 7, 361
Echoic memory, 72–73
Ecological validity, 4
Educational psychology, 6
Effector cells, 26
Einstellung, 318
Elaborative rehearsal, 141, 145
Electrical neuroimaging measures, 38–40
Electroencephalogram (EEG) recordings, 39
Elimination by aspects, 290–291
Embodied cognition, 37, 150, 263
Embodied perception, 67–68
Embodied semantics, 166
Embodiment in speech perception, 225
Emotion
    and ageing, 388
    decision making and, 349–352
    defined, 335
    information about words and situations, 348–349
    language and, 347–349
    memory and, 341–347
    neurological underpinnings, 335–337
    self-control and, 341
    types of, 335
Emotional Stroop task, 340–341
Empiricism, 6
Enactive representation, 371
Enactment effect, 144
Encoding, 16, 17, 18
Encoding specificity, 146–147
Engle's controlled attention model, 121–123
Enhancement process, 251
Epilepsy, 34, 207
Episodic buffer, 116, 121, 122, 133
Episodic long-term memory, 147
Episodic memory, 133
    boosting, 143–145
    dissociation of, 206–207
    preliminary issues of, 133–137
    retrieving, 188–194
    storing information in, 138–143
Event-related potentials (ERPs), 39, 165, 166, 167, 174–175, 234, 239, 240
    in simple comprehension task, 235
    studies of syntactic and semantic processing, 237, 239
Excitatory neurotransmitters, 28
Executive attention, 78–79
Explanation-based theories, 184–185
Explicit memory, 83, 132–133, 208–209
Explicit processing, 82
Extinction, 10
Eye–mind assumption, 252
Eye tracker, 55, 251–252, 253, 310

F

Facilitation, 171
    of semantic priming, 171–172, 173
Fallacies, 286–287
False memory, 195–196
Fan effect, 190, 191, 192
Fast and frugal heuristics, 295–298
Feature-based pattern recognition, 59–61, 71
Feature lists, 163
Feature search, 86, 356
Feedback, 119, 304
Feeling of knowing judgment, 137

Figure-ground principle, 57
Filtering, 93
Fixations, 52
  early parts of, 55
  in reading, 252
Fixation–saccade cycle, 52
Flashbulb memories, 345
Flexibility, 215
Focal attention, 56
Forgetting, 54, 107–110
Formal operations stage, 370
Form errors, 281
Forward inferences, 264
Fovea, 50
Free morphemes, 222
Free recall, 9, 110, 143, 144, 147, 195
Frontal lobes, 31, 32, 116, 117, 122, 184, 320, 337
Functional fixedness, 316–317
Functionalism, 7
Functional magnetic resonance imaging (fMRI), 84, 184, 320, 343
Fusiform face area, 32

G

Gammaaminobutyric acid (GABA), 29
Garden path sentences, 245
Gaze duration and reading, 251–253
General problem solver (GPS), 328–329
  limitations, 328–329
General world knowledge, 133, 159, 275
Generation effect, 143
Geons, 66, 71
Gernsbacher's structure-building framework, 248–249
Gestalt grouping principles, 57–58
  closure and other principles, 57–58
  figure-ground perceptual, 57
Gestures, 272
Given-new strategy, 231
Glutamate, 29
  short- and long-term effects of, 29
Goal directedness, 311
Good continuation, principle of, 58
Go-with-the-default heuristic, 295
Grammar, 218
Grammatical transformations, 227–230
Graphs, interpretation of, 13
Grice's conversational maxims, with two additional rules, 268
Guiding analogies, 12

H

Habituation, 84
Hemineglect, 88–91
Hemispheric encoding/retrieval asymmetry (HERA) model, 207
Hemispheric specialization, 33–34
Heuristics, 284–286
  algorithms and, 283–286
  atmosphere, 278–279
  availability, 289–290
  classic, 286–293
  counterfactual reasoning and, 291–293
  defined, 278
  elimination by aspects, 290–291
  fast and frugal, 295–296
  go-with-the-default, 295
  prototypicality, 305
  recognition, 296

representativeness, 287–289
satisficing, 296, 297
simulation, 290
"take the best," 296
undoing, 291–293
Hindsight bias, 292–293
Hippocampus, 30, 136, 153, 208
Human categories, characteristics of, 182
Human eye, 50
Human memory, Ebbinghaus understanding of, 135
Hypothesis testing, 224–225

I

Icon, 55
Iconic memory, 54, 55
Iconic representation, 57, 371
Iconic system, 215
Idea groupings, 198
Idioms, 271
If–then pair, 328
If–then statement, 279
Ill-defined problems, 314
Imagery, 144, 146, 332
Immediacy assumption, 252
Impetus theory, 303, 304
Implanted memories, 201
Implication, 208
Implicit memory, 82, 132–133
  amnesia and, 206–210
  vs explicit memory, 83
Implicit processing, 82
Inattention blindness, 52, 92
Incubation, 320
Independent stages of processing, 17
Individual differences, 125
  reference and inference processes, 264
Infantile amnesia, 153, 157, 359
Inference, 262–263
  authorized, 262
  bridging, 262
  unauthorized, 262
Inferred processes, 10
Information processing
  and cognitive psychology, 11–12
  measuring of, 12–15
Inhibition, 94–97
  of return, 87
  of semantic priming, 171–172
Inhibitory neurotransmitters, 29
Input attention, forms of, 81–91
  alertness, 81–82
  arousal, 81–82
  attention capture, 83–84
  contrasting input, 88
  controlled attention, 88
  hemineglect, 88–91
  orienting reflex, 83–84
  spotlight attention, 86
  visual search, 86–87
Insight for problem solving, 319–321
  neurocognition in, 323–325
Integration, 99–100, 196–199
Intensity of the emotion, 335
  memory and, 346
Interference, 54–55, 56, 73–74, 75, 107–108
  associative, 190–191
  proactive, 108–109
  retroactive, 108–109

Inter-neuron, 26
Introspection, 6
Intrusions, 15
Intuitive cognitive analysis, 2
Invariances, 363

J

James, William, 7–8
Just noticeable difference (JND), 46

L

Language. See also Phonology; Syntax
  arbitrariness of, 214–215
  brain and, 239–244
  defined, 213
  emotion and, 347–349
  learning of, 240
  levels of analysis, 218–219
  lexical factors, 232–235
  Miller's levels of analysis, 217
  semantic, 235–239
  universals, 213–215
Language comprehension
  comprehension research, 247
  conceptual and rule knowledge, 247–250
  levels of, 249–250
  metacomprehension abilities, 248, 250
  online comprehension task, 247–248
  as structure-building framework, 248–249
Language processing
  children, 362–366
  older adults, 383–386
Lateralization in humans, 33–34
  split-brain research, 34–35
Late selection theory of attention, 94
Learning, 29
Letter span task, 125
Levels of processing, 140
Levels of representation, 151
Lexical ambiguity, 18, 234
Lexical decision task, 17–18, 163, 166, 172–173, 232–233, 244, 264, 283
  and word frequency, 17–18
Lexical knowledge, 233
Lexical memory, 169, 233
Lexical representation, 232–233
Lifespan perspective, 354–355
Limited-capacity channels, 18
Limited domain knowledge, 302–305
Linguistic competence, 364–365
Linguistic performance, 364
Linguistic relativity hypothesis. See Sapir-Whorf linguistic relativity hypothesis
Linguistics, 10–11, 212
Longitudinal study, 354
Long-term memory (LTM), 16, 17, 19, 82, 84, 99, 110, 116, 119, 120, 121, 132, 138, 139, 140, 142, 144, 145, 159, 192, 208, 209
  automatic priming of, 100
  influence of context on cognition, 19
  working memory and, 127–128
Long-term potentiation (LTP), 29

M

Magnetic resonance imaging (MRI), 25, 217. See also Functional magnetic resonance imaging (fMRI)
Maintenance rehearsal, 141, 142, 143

Manner of articulation, 219–220
Massed practice, 106
Math skills, 366–369
    counting, 367–368
    numerical magnitude, 367
Means–end analysis
    basics, 262–263
    general problem solver, 264–265
    Tower of Hanoi, 263–264
Memory, 359–362, 380–383
    associations, 9
    defined, 3–4
    development, 359–362
    distortion effects, 201
    emotion and, 341–347
    flashbulb, 345
    intensity and, 346
    irony of, 203–205
    long-term. *See* Long-term memory (LTM)
    mood-congruent, 344
    perceptual, 356
    processes, 51
    role of practice and, 100
    seven sins of, 187–188
    short-term. *See* Short-term memory (STM)
Memory impairment, 169
Memory-related errors, 282–283
Mental addition task, 124
Mental imagery, 126, 144
Mental lexicon, 169, 173, 232
Mental models, 278, 279, 282–283, 302–303
Mental processes, 1, 2, 8, 9, 10, 11, 14, 17–18, 20, 25, 31, 37, 42, 45, 56, 67, 93, 105, 111, 113, 119, 156, 215, 219, 248, 253
    attention as, 79–80
Mental representations of
    three-dimensional objects, 66
Mental rotation, 119
Metabolic neuroimaging measures, 40–41
Metacognition, 137
Metacomprehension abilities, 248, 250
Metaphors, 271
Metamemory, 133, 136–137, 156–157
Method of loci, 134, 143
Mind wandering, 79–80
Mirror neurons, 36, 68, 225
Misattribution, 188, 200–201
Misinformation acceptance, 200–202
Misleading information effect, 199–200
Missionary–cannibals problem, 328
Mnemonics
    classic, 133
    defined, 133
    effectiveness of, 133–135
    power of, 134
    techniques of, 134
Modality effect, 77
Monitoring pressure, 351
Mood-congruent memories, 344
Morphemes, 222, 232
Motor theory of speech perception, 225
Multiconstraint theory, 323, 324
Multiword stage, 363
Multiple-choice recognition test, 140, 150
Myelin sheath, 26

N

Naive physics, 303–304
Naming, 215
N4 effect for semantic anomalies, 39
Negative priming, 94–97
Negative set, 317–318
Neobehaviorism, 8
Neocortex, 30, 31
    division of labor in, 34
    lobes of, 32
    principles of functioning in, 32–34
Network, 161–162
Neural communication, 26–29
Neural network models, 189
Neurogenesis, 29–30
Neuroimaging measures
    electrical, 37, 38–40
    of lesioning of the brain, 41
    metabolic, 40–41
    structural, 37, 38
Neurological changes during childhood, 355–356
Neurology, basic, 25–30
Neurons, 26
    depolarization of, 27
    estimate, in humans, 26
    and learning, 29–30
    mirror, 36
    in retina, 50
    structures of, 26
    working of, 37
Neurotransmitter, 28, 29
NMDA receptors, 29
Node of semantic network, 161–162
Nodes of Ranvier, 26
Nondeclarative memory, 133
Nonoverlapping stages of processing, 18
Nonreinforced response, 10
Norepinephrine, 28
Numerical magnitude, 367, 368

O

Object files, 57
Object permanence, 361
Object recognition
    and agnosia, 69–70
    by components, 66–68
    context and embodied perception, 68
    implications for cognitive science, 71
Occipital lobe, 31, 36
Older adults,
    attention, 379–380
    cognitive changes in, 376
    decision-making ability, 386–387
    and emotions, 388–389
    language processing, 383–386
    memory, 380–383
    neurological changes, W-20–W-21
    perception, 378–379
    problem-solving skill, 388
    reasoning in, 386
Online reading, 250–253
    basic effects, 253–255
    benefits of, 255–256
    factors affecting, 256–257
Operators in problem solving, 313
Orienting, 78
Orienting reflex, 83–84
Overextension, 360
Overconfidence in memory, 201
Overlearning, 7, 101, 104, 230
Overload procedure, 93

P

Paired-associate learning, 9, 146, 189–190
Pandemonium, 59–60
Parallel distributed processing (PDP) models, 42,
Parallel processing, 18, 169
Parallel search, 114
Parietal lobe, 31, 32, 35, 37, 73, 78, 84, 87, 90, 116, 119
Parse, 66, 75, 227, 228
Part-set cuing, 193–195
Pathways, 36–37, 50, 51, 71, 84, 87, 149, 162, 165, 190, 208, 241, 258, 304, 312, 313, 331
Pattern recognition, 57–61
    conceptually driven, 61–62
    feature analysis/feature detection, 59–61
    Gestalt grouping principles, 57–58
    template matching process, 58–59
Peg word mnemonic, 134
Perception, 3, 6, 8, 337
    in children, 356
    in older adults, 378–379
Perceptual memory, 356
Perceptual processes, 2, 51, 61, 121
    embodied aspect of, 68
Perceptual symbols, 168
Performance, 218
Persistence, 161, 188
Phi phenomenon, 60, 61
Phoneme, 219–222
Phonemic differences, 221
Phonological loop, 116, 117–119
Phonological similarity effect, 117–119
Phonological store, 117
Phonology, 219–226. *See also* Language
    combination of phonemes, 222
    English consonants and vowels, 219
    phonemes, 219–222
Phrase structure grammar, 227–228
Place of articulation, 220
Plans and the Structure of Behavior, 12
Plausible deniability, 271
Polysemy, 233–234
    priming and, 234
Pop-out effect, 86, 356, 357, 358
P6 or P600 effect, 39
Positivity effect, 388
Positron emission tomography (PET), 25, 40, 116, 207, 324
Posner's spatial cuing task, 85
Prefrontal cortex, 79, 116, 129, 159, 324, 336, 337, 348
Preoperational stage, 369
Presbycusis, 379
Primacy effect, 110, 139
Prime, 171
Priming, 18, 169
    defined, 171
    polysemy and, 234
    semantic. *See* Semantic priming
Principles of Physiological Psychology, 6
Proactive interference (PI), 108–109, 128, 189, 190
Probabilistic theories, 183–184
Problem of invariance, 223
Problem solving, 8, 12, 20, 369
    analogy and, 321–322
    automatic processing, 330
    characteristics, 310–311

contradictions and, 331
decision making and, 369–373
difficulties in, 316–318
domain knowledge and, 329–330
example, 314–315
Gestalt psychology and, 315–318
goal and, 313–314
inferences and, 330
insight for, 319
means–end analysis method, 325–329
operators, 313
practice, role of, 332–333
problem representation and, 332
problem space, 311–312
searching relations among problems, 331–332
studying, 310
subgoal heuristic for, 330
suggestions for improving, 329–333
systematic plan, role of, 330
vocabulary of, 311–315
working backward for, 331
Problem space, 311–312
Processing fluency, 201
Processing resources, 305–306
Process model, 17–18
Production system model, 328
Productivity principle, 215
Proposition
defined, 148–149
elaborated, 149
nature of, 148–150
nodes of, 149
remembering, 149–150
Propositional textbase, 249
Prosody, 231–232, 347–348
Prosopagnosia, 32, 69
Prospective memory, 155–156
Prototype, 183, 185
Prototypicality heuristic, 305
Proximity, 58
Pseudo-events, 202
Psycholinguistics, 212, 216, 221, 227, 230, 233
Psychological refractory period, 97
Psychophysics, 45–49
detection and absolute thresholds, 45–46
physical and mental differences, 47–48
signal detection theory, 48–49
Pure word deafness, 241, 242

R

Reading, 250–251
gaze duration and, 251–253
online, 253–256
variables affecting, 256–257
Reasoning, 3, 185
conditional. *See* Conditional reasoning
counterfactual, 291–293
formal logic and, 275–283
lack-of-knowledge, 3
limitations in, 302–305
moral, 128
syllogistic. *See* Syllogistic reasoning
types of, 305
working memory and, 128–129
Reasoning errors, 281, 305
Recall task, 110, 142, 144
Recency effect, 110, 139

Receptor cells, 26
Recognition by components (RBC) theory, 66–68
evidence for, 67
shortcomings, 67–68
Recognition heuristic, 296
Recognition task, 110–111
Reconsolidation, 343, 344
Reconstructive memory, 150, 176
Recovered memory, 202–203
Reductionism, 4
Reductionistic approach, 310
Reference, 258–260
simple, 259
Rehearsal, 9, 107, 110, 116, 117, 138–140
frequency and memory benefit, 138–139
maintenance, 141
serial position and, 139
Reinforced response, 10
Reinforcement, 10
Relearning task, 141, 156
Reminiscence bump, 153–154
REM (dreaming) sleep, 29
Repeated name penalty, 259
Repetition priming, 133, 209, 210
Representational momentum, 120
Representativeness heuristic, 287–289
biases in, 288
laws of large and small numbers, 287–288
stereotypes, 288–289
Repressed memory, 202–203
Resource theories, 97
Response time (RT), 6, 13, 14, 111, 187, 324, 356, 357
Retina, 50
Retrieval cues, 134, 192–193, 194
Retrieval failure, 188, 193
Retroactive interference (RI), 108–109
Retrograde amnesia, 206
Ribot's Law, 206
Right-hemisphere damage, 243–244
Right hemisphere language, 243
Rods, 50

S

Saccades, 52
Saccadic eye movement, 56, 57
Sapir-Whorf linguistic relativity hypothesis, 218, 219
Satisficing heuristic, 330
Savings score, 7
Schemata, 176–177
Schizophrenia, 96
Scripts, 177–180
Search errors, 281
Second-order theory, 269
Selbst-Beobachtung method, 6
Selecting, 93
Selective attention, 91, 93
Self-reference effect, 143
Selfridge's model of Pandemonium, 59–60
Semantic anomaly, 39
Semantic congruity effect, 47–48
Semantic features, 94, 163, 237
Semantic grammar theory, 238–239
Semanticity, 214, 216
Semantic, 235–239
Semantic memory, 24, 133, 148, 159, 360
amount of knowledge, 166–167

dissociation of, 206–207
embodied semantics, 166
feature comparison models, 162–163
loss, 169–170
recognition tasks in, 163–165
schemata and scripts, 175–180
semantic networks, 161–162
test of, 163–165
Semantic priming
across trials, 171
automatic, 173–174
concepts related to, 171–172
controlled, 173–174
empirical demonstrations of, 172–173
facilitation/benefit of, 171–172
implicitness of, 174–175
inhibition of, 172
lag between prime and target, 172
lexical decision task, 172–173
principles, 171
Semantic relatedness, 165–167
Semantic roles, 236
Semantic search, 162
Sensation, 51
Sense of hearing. *See* Auditory perception
Sensitivity of hearing, 71–72
Sensorimotor stage, 369
Sensory cortex, 35
Sensory memory, 16, 19
Sentence verification task, 163
Sequence of operations, 316
Sequential stages of processing, 18
Serial exhaustive search, 114
Serial learning, 9
Serial position curve, 14, 110, 111, 139
Serial position effect, 110
Serial recall, 110
Serial self-terminating search, 114
Shadowing task, 94, 99
Short-term memory (STM), 16–17, 361
capacity, 105–110
defined, 105
forgetting from, 107–110
retrieval, 110–114
scanning, 110–114
Sternberg task, 111–113
working memory. *See* Working memory
Sign languages, 106, 119, 231, 363
Signal detection theory, 48–49
Silent vocalization condition, 75
Similarity, 58
Simple memory span, 112
Simulation heuristic, 290, 291, 292
Single cell recording, 38
Situation model, 150–152, 191–192, 260–265
Situation-model-level processing, 249
Situation model representation, 249
Skinner, B. F., 8, 10
Sleep, 29
Social psychology, 6, 8
Socioemotional Selectivity Theory, 388
Soma, 26
Source memory, 201
Source misattribution, 200–201
Source monitoring, 147–148
Span of apprehension, 106
Specialization, 33–34
in cortex, 35–37
Speech act, 263–264

Speech errors, 232
Speech perception, 222–226
    effect of context, 223–224
    embodiment in, 225
    problem of invariance, 223
    top-down and bottom-up processes, 224–225
Split-brain research, 34–35
Spotlight attention, 85, 86
Spreading activation, 161, 162
Standard memory tasks, 142
Standard theory of memory, 16–17
State-dependent learning, 147
Stereotypes, 288–289
Stereotype threat, 350–352
Sternberg task, 111–114
    limitations to conclusions, 114
    results, 113
    sample, 112
Stimulus onset asynchrony (SOA), 171
Stimulus–response (S–R) behaviorism, 9
Stress and performance, 352
Stroop effect, 98, 127
Stroop task, 98
Structuralism, 6–7
Structural neuroimaging measures, 38
Subjective organization, 144
Suffix effect, 74
Suggestibility, 188, 345
Suppression of concepts, 251
Surface form of representation, 152, 249
Surface structure of sentences, 227
Sustained attention, 82
Syllogism, 128, 276–278
Syllogistic reasoning, 275
    biases of, 276–278
    theories of, 278–279
Symbolic comparisons, 47
Symbolic distance effect, 47, 48
Symbolic representation, 371, 372
Synapses, 27, 28, 169, 336
    cone, 50
Syntactic processing, 240
Syntax, 226–232
    cognitive psychology of, 230–231
    defined, 226
    interaction with semantics, 236–237
    phrase order, 226
    prosody, 231–232

T

Tabula rasa, 6
"Take the best" heuristic, 296
Target, 171
Task effects, 141–143
Template matching process, 59
Templates, 59
Temporal lobe, 30, 31, 37, 69, 71, 73, 241, 242, 343
Testing effect, 140
Thalamus, 30
Themes of cognition, 21
Theory of mind, 268
Thinking, 2–3
"Three-eared man" procedure, 74
Time-based prospective memory, 155
Tip-of-the-tongue (TOT) phenomenon, 157, 170, 242
Titchener, Edward, 6–7
Top-down, conceptually driven processing, 19, 61–66, 75, 77, 88, 100, 224–225
Topic maintenance, 268
Tower of Hanoi, 326–328
    four-disk version, 327–328
    three-disk version, 326–327
Transcranial magnetic stimulation (TMS), 39
Transformational grammar. *See* Chomsky's transformational grammar
Transformational rules, 228, 229
Transience, 188, 203
Trans-saccadic memory, 56–57
Typicality effects, 185

U

Unauthorized inference, 262
Updating processes, 265

V

Valence of an emotion, 335, 343
Valid arguments, 280
Ventral pathway, 36–37, 84
Verbal behavior, 10
Verbal learning, 9–10
Verbal protocol, 20, 310, 315, 333
Verbatim memory, 150
Verbatim mental representation, 249
Verbatim repetition, 149
Video games, and attention, 88
Vigilance phenomena, 82
Visual attention, 55–56, 84, 85, 86, 99
Visual field, 50, 85, 90, 325
Visual imagery, 125, 144, 332
Visual memory span task, 124
Visual persistence, 54
Visual search, 85–88, 339–340, 356
Visual sensation and perception, 50–57
    fixation–saccade cycle, 52
    retina, 50
    synesthesia, 53–54
    trans-saccadic memory, 56–57
    visual attention, 55–56
    visual information, gathering of, 51–52
    visual sensory memory, 54–55
Visual sensory memory/iconic memory, 54–55
    amount and duration of storage, 54
    interference, 54–55
Visuo-spatial sketch pad, 116, 119–121
Vividness, 60
Vocal–auditory channel, 214
Voluntary attentive processes, 84
Von Ebbinghaus, Hermann, 7
Von Restorff effect, 344–345
Vygotsky, Lev, 370–371

W

Weapon focus effect, 353
Well-defined problems, 314
Wernicke's aphasia, 241
"What" pathway, 71, 84
"What" system, 37
"Where" pathway, 84
Word frequency effect, 18
Word stem completion task, 82
Working memory (WM), 114–115
    assessment of, 123–127
    attention and, 127
    capacity, 129
    components of, 116–123
    defined, 115
    long-term memory and, 127–128
    overview, 129
    reasoning and, 128–129
    span of, 125–126
Working memory span, 123, 125–126, 128, 129
World War II and psychology, 8–9
Written language, 5, 58, 66, 213, 231, 241, 248, 321
Wundt, Wilhelm, 6

Y

Yerkes-Dodson law, 346

Z

Zone of proximal development, 371

Brodmann's Areas on the Lateral Surface of the Brain

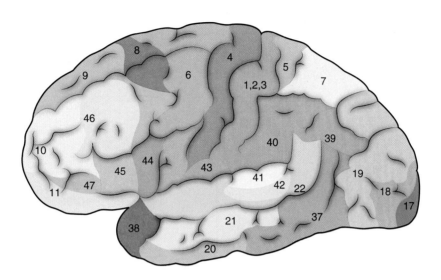

Brodmann's Areas on the Medial Surface of the Brain

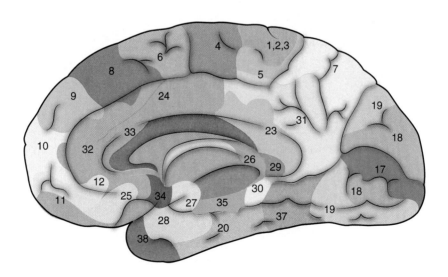

BLUE	ORANGE
RED	GREEN
PURPLE	ORANGE
GREEN	YELLOW
YELLOW	PURPLE
ORANGE	BLUE
RED	YELLOW
YELLOW	RED
BLUE	ORANGE
PURPLE	GREEN
ORANGE	ORANGE
GREEN	GREEN
ORANGE	PURPLE
PURPLE	RED
BLUE	BLUE
YELLOW	RED
GREEN	PURPLE
RED	YELLOW

Stroop color word lists.

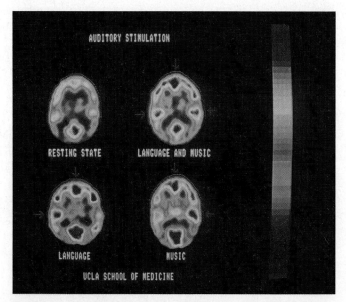

Positron emission tomography (PET) images of the brain when a person is engaged in listening to four different kinds of stimuli. The red areas reveal areas of more brain activity (relative to a resting baseline).